## Continuum Thermodynamics and Material Modelling

This concise and rigorous textbook introduces students to the subject of continuum thermodynamics, providing a complete treatment of the subject with practical applications to material modelling.

- Presents mathematical prerequisites and the foundations of continuum mechanics, taking the student step-by-step through the subject to allow full understanding of the theory.
- Introduces more advanced topics such as theories for the investigation of material models, showing how they relate to real-world practical applications.
- Numerous examples and illustrations, alongside end-of-chapter problems with helpful hints, help describe complex concepts and mathematical derivations.

This is the ideal, accessible introduction to continuum thermodynamics for senior undergraduate and graduate students in mechanical, aeronautical and civil engineering.

**Kari Santaoja** is Senior University Lecturer emeritus of Mechanical Engineering, Aalto University, Finland. He is a member of the Finnish Association for Structural Mechanics.

**J. N. Reddy** is Distinguished Professor, Regents Professor, and holder of the O'Donnell Foundation Chair IV of Mechanical Engineering at Texas A&M University. He is internationally recognized for his research and education in applied and computational mechanics and is the author of several hundred journal articles and twenty-four textbooks, including *Finite Element and Finite Volume Methods for Heat Transfer and Fluid Dynamics* (Cambridge University Press: 2022), *Introduction to Continuum Mechanics* (Cambridge University Press: 2013), and *Principles of Continuum Mechanics* (Cambridge University Press: 2009). He has been awarded the Leonardo da Vinci Award, the IACM (Gauss–Newton) Medal, the ASME Timoshenko Medal, the ASCE Theodore von Karman Medal, and the USACM John von Neumann Medal, among other awards. He is a member of the US National Academy of Engineering and a foreign member of five other academies.

"An indispensable resource for engineers and researchers alike, *Continuum Thermodynamics and Material Modelling* by Kari Santaoja and J. N. Reddy navigates the complexities of material behavior with unparalleled clarity. A masterful blend of theory and application, this book is set to redefine the landscape of engineering education. This book is a must-read for those eager to elevate their understanding of engineering principles."

George Z. Voyiadjis, *Louisiana State University*

"This book accomplishes what few textbooks at its level even attempt: systematically incorporating thermodynamics into continuum mechanics. The authors take special care to connect continuum thermodynamics with thermostatics, which most readers will have encountered through previous study. Detailed analyses of classical material models, such as heat conduction, creep, and fluid mechanics, highlight the utility of the approach."

Myron B. Allen, *University of Wyoming*

"This is the only book covering continuum mechanics and thermodynamics, or 'continuum thermodynamics,' from the fundamental principles that first-year graduate students can understand. This is essential for students and scientists researching fatigue and damage modelling. Chapters 6, 7, and 8 cover continuum thermodynamics fundamentals. The application of continuum thermodynamics is discussed in Chapters 9, 10, and 11. Hence, I strongly recommend this book for the graduate mechanics curriculum as an essential book."

Sameer B. Mulani, *University of Alabama*

"This book provides a bridge from standard continuum mechanics to the state of the art in material modeling theory. Without compromising on pedagogy, the authors skillfully introduce the concepts and historical motivations of nonequilibrium thermodynamics in the context of material modelling. The detailed derivations and broad scope of applications – ranging from damage mechanics to fluids – make this text a valuable resource for the novice as well as the experienced practitioner."

Brandon Runnels, *Iowa State University*

# Continuum Thermodynamics and Material Modelling

KARI SANTAOJA
*Aalto University, Finland*

J. N. REDDY
*Texas A&M University*

CAMBRIDGE
UNIVERSITY PRESS

CAMBRIDGE
UNIVERSITY PRESS

Shaftesbury Road, Cambridge CB2 8EA, United Kingdom

One Liberty Plaza, 20th Floor, New York, NY 10006, USA

477 Williamstown Road, Port Melbourne, VIC 3207, Australia

314–321, 3rd Floor, Plot 3, Splendor Forum, Jasola District Centre, New Delhi – 110025, India

103 Penang Road, #05–06/07, Visioncrest Commercial, Singapore 238467

Cambridge University Press is part of Cambridge University Press & Assessment, a department of the University of Cambridge.

We share the University's mission to contribute to society through the pursuit of education, learning and research at the highest international levels of excellence.

www.cambridge.org
Information on this title: www.cambridge.org/highereducation/ISBN/9781316517352

DOI: 10.1017/9781009043052



First published 2024

*A catalogue record for this publication is available from the British Library*

*A Cataloging-in-Publication data record for this book is available from the Library of Congress*

ISBN 978-1-316-51735-2 Hardback

Additional resources for this publication at www.cambridge.org/santaoja-reddy

# Contents

# Symbols

| Symbol | Meaning |
|---|---|
| $\vec{b}$ | body force vector |
| $c$ | specific heat capacity |
| $\mathbf{C}$ | constitutive tensor for a Hookean solid (fourth-order tensor) |
| $\underline{\mathbf{d}}$ | rate of deformation tensor |
| $D$ | damage |
| $\mathrm{d}A$ | differential area |
| $\mathrm{d}s$ | differential line element |
| $\mathrm{d}V$ | differential volume |
| $E$ | Young's modulus |
| $\mathbf{E}$ | Green–Lagrange strain tensor |
| $\mathcal{E}$ | state |
| $F$ | yield function |
| $\vec{F}$ | force vector |
| $g$ | specific Gibbs free energy |
| $g^{\mathrm{de}}$ | specific damage-elastic Gibbs free energy |
| $h(\vec{X})$ | function showing the inhomogeneity of the solid material |
| $(\vec{i}_1, \vec{i}_2, \vec{i}_3)$ | basis of the coordinate systems $(x_1, x_2, x_3)$ and $(X_1, X_2, X_3)$ |
| $\mathbf{I}$ | fourth-order identity tensor |
| $\mathbf{I}^{\mathrm{s}}$ | fourth-order symmetric identity tensor |
| $J(\vec{X}, t)$ | Jacobian determinant between coordinates $\underline{x}_i(t)$ and $X_i$ |
| $K$ | kinetic energy |
| $m$ | mass; mass of the subsystem |
| $\mathrm{m}^{\mathrm{b}}$ | mass of the system |
| $\vec{n}$ | outward unit normal in the current configuration, $\underline{v}^{\mathrm{b}}(t)$ |
| $\vec{N}$ | outward unit normal in the IG-CtP configuration $\mathrm{V}_0^{\mathrm{b}}$ |
| $p$ | thermodynamic pressure |
| $p_{\mathrm{mech}}$ | mechanical pressure |
| $P^{\mathrm{ext}}$ | power input |
| $\vec{q}$ | heat flux vector |
| $Q$ | heat input rate |
| $Q^r$ | number of microcracks in a microcrack group "$r$" |

| Symbol | Meaning |
|---|---|
| $r$ | heat source per unit mass |
| $s$ | specific entropy (entropy per unit mass) |
| $\dot{s}$ | specific entropy rate |
| $S$ | entropy |
| $\mathbf{S}$ | compliance tensor for a Hookean deformation |
| $\mathbf{S}^{\mathrm{d}}$ | compliance tensor due to damage |
| $\tilde{\mathbf{S}}$ | effective compliance tensor (for damaged material) |
| $t$ | time |
| $\underline{\vec{t}}(\underline{\vec{x}}(t), t, \underline{\vec{n}})$ | traction vector (in the current configuration $\underline{v}^{\mathrm{b}}(t)$) |
| $T$ | absolute temperature |
| $\vec{T}(\vec{X}, t, \vec{N})$ | traction vector (in the IG-CtP configuration $\mathrm{V}_0^{\mathrm{b}}$) |
| $u$ | specific internal energy (internal energy per unit mass) |
| $\vec{u}$ | displacement vector of a material point |
| $U$ | internal energy |
| $\underline{v}(t)$ | volume of an (arbitrary) subsystem $m$ in the current configuration $\underline{v}^{\mathrm{b}}(t)$ |
| $\underline{\vec{v}}(\underline{\vec{x}}(t), t)$ | velocity of a material point, velocity vector |
| $\vec{v}(\vec{x}, t)$ | fluid velocity (vector) field |
| $\|\vec{v}\|$ | absolute value of the vector $\vec{v}$ |
| $\underline{v}^{\mathrm{b}}(t)$ | current configuration; volume of the system $\mathrm{m}^{\mathrm{b}}$ in the current configuration $\underline{v}^{\mathrm{b}}(t)$ |
| $\mathrm{V}^{\mathrm{cv}}$ | control volume |
| $V_0$ | volume of an (arbitrary) subsystem $m$ in the IG-CtP configuration $\mathrm{V}_0^{\mathrm{b}}$ |
| $\mathrm{V}_0^{\mathrm{b}}$ | IG-CtP configuration; volume of the system (body) $\mathrm{m}^{\mathrm{b}}$ in the IG-CtP configuration $\mathrm{V}_0^{\mathrm{b}}$ |
| $\underline{\mathbf{W}}$ | tensor or vorticity tensor |
| $(x_1, x_2, x_3)$ | coordinates of a fixed point in space |
| $(\underline{x}_1(t), \underline{x}_2(t), \underline{x}_3(t))$ | current material coordinates of point $P(t)$ moving with deformation |
| $(X_1, X_2, X_3)$ | initial material coordinate system |
| $\langle x \rangle$ | Macaulay bracket: $\langle x \rangle = 0$, when $x < 0$; and $\langle x \rangle = x$, when $x \geq 0$ |
| $\mathbf{1}$ | second-order identity tensor |

| Symbol | Meaning |
|---|---|
| $\hat{\alpha}$ | linear coefficient of thermal expansion |
| $\boldsymbol{\alpha}$ | represents the other internal variables (a second-order tensor) |
| $\hat{\boldsymbol{\alpha}}$ | rate of the internal variable $\boldsymbol{\alpha}$ |
| $\boldsymbol{\beta}$ | internal force (a second-order tensor); conjugate to the internal variable $\boldsymbol{\alpha}$ |
| $\delta_{ij}$ | Kronecker delta |
| $\boldsymbol{\varepsilon}$ | strain tensor |
| $\boldsymbol{\varepsilon}^{\mathrm{d}}$ | damage strain tensor |
| $\boldsymbol{\varepsilon}^{\mathrm{de}}$ | damage-elastic strain tensor |
| $\boldsymbol{\varepsilon}^{\mathrm{e}}$ | elastic strain tensor |
| $\boldsymbol{\varepsilon}^{\mathrm{i}}$ | inelastic strain tensor (irreversible strain tensor); internal variable |
| $\boldsymbol{\varepsilon}^{\mathrm{p}}$ | plastic strain tensor |
| $\boldsymbol{\varepsilon}^{\mathrm{v}}$ | creep strain tensor, viscous strain tensor |
| $\overset{\circ}{\boldsymbol{\varepsilon}}$ | fluid strain rate tensor |
| $\theta$ | arbitrary function; can be temperature $T$ or stress tensor $\boldsymbol{\sigma}$ |
| $\theta(\vec{x}, t)$ | arbitrary function described by the spatial coordinates $x_i$ |
| $\underline{\theta}(\underline{\vec{x}}(t), t)$ | arbitrary function described by the current material coordinates $\underline{x}_i(t)$ |
| $\theta(\vec{X}, t)$ | arbitrary function described by the initial material coordinates $X_i$ |
| $\overset{\circ}{\lambda}$ | plasticity multiplier |
| $\nu$ | Poisson's ratio |
| $\rho$ | density of the material during deformation |
| $\rho_0$ | density of the material before deformation |
| $\boldsymbol{\sigma}$ | stress tensor (small deformation stress tensor) |
| $\underline{\boldsymbol{\sigma}}^{\mathrm{c}}(\underline{\vec{x}}(t), t)$ | Cauchy stress tensor |
| $\underline{\boldsymbol{\sigma}}^{\mathrm{cd}}(\underline{\vec{x}}(t), t)$ | Cauchy dissipative stress tensor |
| $\boldsymbol{\sigma}^{\mathrm{d}}(\vec{X}, t)$ | dissipative stress tensor (for solids) |
| $\boldsymbol{\sigma}^{\mathrm{d}}(\vec{x}, t)$ | viscous stress tensor (for fluids) |
| $\boldsymbol{\sigma}^{\mathrm{f}}(\vec{x}, t)$ | fluid stress tensor |
| $\boldsymbol{\sigma}^{\square}(\vec{X}, t)$ | first Piola-Kirchhoff stress tensor |
| $\Phi$ | dissipation |
| $\Phi^{\mathrm{fluid}}$ | fluid dissipation function |
| $\Phi_{\mathrm{mech}}$ | mechanical dissipation |
| $\Phi_{\mathrm{ther}}$ | thermal dissipation |

| Symbol | Meaning |
|---|---|
| $\varphi$ | specific dissipation function (dissipation potential) |
| $\varphi^{\mathrm{c}}$ | specific complementary dissipation function |
| $\varphi^{\mathrm{c}}_{\mathrm{mech}}$ | specific complementary dissipation function (mechanical part) |
| $\varphi_{\mathrm{mech}}$ | specific dissipation function (mechanical part) |
| $\varphi_{\mathrm{ther}}$ | specific dissipation function (thermal part) |
| $\psi$ | specific Helmholtz free energy |
| $\boldsymbol{\omega}$ | rotation tensor |
| $\mathring{\boldsymbol{\omega}}$ | fluid vorticity tensor |

## Special Notation

| Symbol | Meaning |
|---|---|
| $\mathrm{D}/\mathrm{D}t$ | material (time) derivative operator |
| $H(x)$ | $H(x) = 0$, when $x < 0$; and $H(x) = 1$, when $x \geq 0$ |
| $J_{\mathrm{vM}}(\cdot)$ | von Mises operator |
| $\mathrm{sgn}(x)$ | $\mathrm{sgn}(x) = -1$, when $x < 0$; and $\mathrm{sgn}(x) = 1$, when $x \geq 0$ |
| $\vec{\nabla}$ | vector operator del |
| $\overleftarrow{\nabla}$ | vector operator del acting on the preceding quantity $\theta$ (which can be a scalar, vector, or tensor) |
| $\vec{\theta}$ | enclosed quantity $\theta$ is a vector |
| $\underline{\theta}$ | underlined quantity ($\theta$) is described in the current configuration $\underline{v}^{\mathrm{b}}(t)$; i.e., $\underline{\theta} = \underline{\theta}(\underline{\vec{x}}(t), t)$ |
| $\dot{\theta}$ | material (time derivative) $\mathrm{D}\theta/\mathrm{D}t$ of a function |
| $\partial/\partial t$ | partial time derivative operator |
| $\partial(\ )$ | boundary of the domain ( ) |
| $:=$ | definition |
| $(\cdot)^{\mathrm{T}}$ | tensor transpose of the enclosed quantity |
| **AX** $n$ | $n$th axiom |
| **Def** | definition |
| **BL** $n$ | $n$th basic law |
| IG-CtP | initial geometry at current time and position |

# Preface

Engineering, a branch of science, is a problem-solving discipline, and it is concerned with studies of fundamental understanding of a variety of phenomena and systems in nature and with the design, building, and manufacturing of devices, systems, and processes for human safety, convenience, and entertainment. Engineers construct mathematical models of physical phenomena and develop analytical and numerical approaches to determine their responses. All mathematical models are approximate and obey the laws of physics and/or axioms that govern the phenomena. The most difficult part of engineering analysis, design, and manufacturing is to come up with a suitable mathematical model that accounts for the salient features (which affect the design) of the phenomena being studied. It is in this context that a course on continuum thermodynamics and materials modelling equips engineers and applied scientists with the background needed to formulate a suitable mathematical model and evaluate it in light of experimental evidence.

Books on continuum mechanics and elasticity fall into three categories: two extreme cases are those that are highly mathematical for beginners and those that are very short on explaining concepts and that lack mathematical rigor. The books in the middle are those that do a good job of explaining the concepts and illustrating with examples. However, they tend to cover few standard topics and leave much of the useful material, especially topics on thermodynamics and constitutive theories, to the more advanced books, which are not accessible to most readers because there is no continuity between what they know and what they see in these advanced books. The present textbook provides a more complete treatment of continuum thermodynamics with applications to material modelling, in a manner easily understandable to beginning graduate students.

*Continuum Thermodynamics and Material Modelling* has the objective of covering topics from continuum mechanics and thermodynamics, and it facilitates an easy and thorough understanding of the topics covered. The book offers a concise yet rigorous treatment of the subject of continuum thermodynamics at the introductory level. To facilitate student learning, mathematical prerequisites are set early in the book. First, the foundations of continuum mechanics are reviewed. The close relationship between solid mechanics and fluid mechanics is highlighted. The discussion continues with an introduction to the basics of thermostatics, which is then extended to cover thermodynamics as well. Theories for the investigation of material models are introduced. Material models covering Fourier's law of heat conduction and porous materials are then examined.

A brief discussion of the continuum concept, first and second laws of thermodynamics, and reversible and irreversible macro processes are presented in Chapter 1. Chapter 2 covers some of the mathematical preliminaries. These include the notion of vectors and tensors, calculus of vectors and tensors, and the Voigt summation convention. The kinematics of a continuum, including descriptions of motion, introduction to the Green–Lagrange strain tensor, the deformation gradient, and the Reynolds transport theorem, are presented in Chapter 3.

Chapter 4 introduces fundamental and derived quantities and fundamental units. One axiom and two basic laws for continuum mechanics are also introduced. Local forms of the axiom of conservation of mass are derived for the initial material description, current material description, and spatial description. Different stress measures are also discussed. The laws of balance of linear and angular momenta are introduced. Local forms of the laws are formulated for the material and spatial descriptions. The Extended Betti's theorem is also derived. Thermostatics and thermodynamics, continuum thermodynamics with internal variables, the first and second laws of thermodynamics and the resulting equations (e.g., heat equation and Clausius–Duhem inequality), the principle of maximum dissipation, material models, and gradient theory are covered in Chapter 5.

Chapter 6 introduces the axioms of thermostatics and thermodynamics in the forms that apply to continuum thermodynamics. These include the first law of thermodynamics, the second law of thermodynamics, the axiom of caloric equation of state, the axiom of local accompanying state, and the principle of maximum dissipation. The local forms of the first and second laws of thermodynamics are derived for solids and for fluids. The heat equation and the Clausius–Duhem inequality are derived for solids and fluids. The principle of maximum dissipation is used in formatting the normality rule for different cases. Special cases of continuum thermodynamics and step-by-step instructions for validating a material model are presented in Chapter 7. The chapter also gives rigorous instructions for the material model validation procedure. Chapter 8 contains formulations for material models of heat conduction and time-independent deformation.

Chapter 9 contains a study of material models for viscoelastic and viscoplastic deformation. They are time-dependent processes and therefore are usually modelled by a set of differential equations. In Chapter 10, the concept of the representative volume element RVE is introduced, the response of a Hookean matrix deformation with penny-shaped microcracks and spherical microvoids. The chapter also provides core information on the continuum damage mechanics. Brief descriptions of the Gurson–Tvergaard material model and a nonlocal material model for creep and damage are also included.

Finally, Chapter 11 studies the foundations of fluid mechanics and derives some preliminary results. Continuum thermodynamic derivations of the Navier–Poisson law of a Newtonian fluid and the Navier–Stokes equations are presented. Also, the ideal gas law is presented separately for continuum thermodynamics and for thermostatics.

Overall, the book covers material that is not generally found in most books on continuum mechanics or thermodynamics. We hope that this book fills the void in the literature and provides a comprehensive treatment of continuum thermodynamics and materials modelling.

The book may be used as a textbook for a single course on continuum mechanics or for a two-course sequence on continuum mechanics and thermodynamics. The book is fairly self-contained, requiring an engineering background consistent with a first-year graduate student of aerospace, civil, and mechanical engineering.

This book has its origins in the first author's (Kari Santaoja) course material for Continuum Mechanics and Material Modelling. Over the last twenty years, feedback and questions from students have helped him enhance its coverage and quality. The second author's (J. N. Reddy) book, *An Introduction to Continuum Mechanics* (Cambridge University Press, 2013) also provides a detailed background in continuum mechanics. An integration of concepts from these two books has culminated in a more balanced treatment of continuum thermodynamics and materials modeling in the present book.

The first author extends his deepest thanks to all his students, whose thoughtful comments and keen interest made this project so worthwhile. Thanks are also due to Dr. Ville Lilja for his many valuable comments on the manuscript of this book. The first author is also deeply indebted to Professor Emeritus Eero-Matti Salonen, who for the past thirty years has provided immense support through numerous discussions and his insightful comments on the text. The authors are pleased to thank Holly Monteith for her help in checking the language of the manuscript and Mrs. Soili Pallasaho for help in proofreading the manuscript in its final stages. Of course, the authors realize that a book of this mathematical complexity is bound to have errors that they may have missed. The authors sincerely request readers to bring any errors they find to the authors' attention (e.g., by sending an email to kari.santaoja@alumni.aalto.fi). In addition to mathematical and linguistic flaws, we are seeking to identify pedagogically weak parts of the book. These can be badly written equations, undefined variables, requirement of prerequisites that students do not have, or unclear texts or figures the message of which is difficult to understand. We will try to reformulate these cases and include them in future copies of the book. Thank you for your cooperation. Support material for lectures is planned to be uploaded to the webpage www.continuum-thermodynamics.com.

# 1 Introduction

## 1.1 General Remarks

This book deals with continuum thermodynamics, a subject that studies thermodynamics of processes in continuous systems. The major topic here is the evaluation of continuum mechanics and especially material models for solid materials. The definitions of stress and strain are vital parts of continuum mechanics. Detailed studies of the relationship between these two quantities (i.e., constitutive relations), however, do not belong to most continuum mechanics books or courses, but constitutive equations enter continuum mechanics studies as if they were given by an external agent. Thermodynamics is the ingredient that together with continuum mechanics forms a complete theory which ties up the definitions of stress and strain with material models.

Continuum mechanics is a branch of physical science concerned with the deformations and motions of continuous material media (called bodies) under the influence of external effects (or stimuli). In this book, our study focuses on cases in which the external effects are forces having a mechanical and/or thermal origin.

The adjective *continuous* refers to the simplifying concept underlying the analysis: we disregard the molecular structure of matter and picture it as being without gaps or empty spaces. We further suppose that all the mathematical functions entering the theory are continuous functions, except possibly at a finite number of interior surfaces separating regions of continuity. This statement implies that the derivatives of the functions are continuous too, if they enter the theory, since all functions entering the theory are assumed to be continuous. This hypothetical continuous material is called a continuous medium or simply a *continuum*. See Malvern [57] and Reddy [84].

In continuum damage mechanics, for example, which studies the influence of microcracks and cavities on the deformation of the body, the microcracks are not described as a jump in the values of the quantities, but the influence on the microcracks is smoothed over a finite volume of the material. This leads to a mathematical theory with continuous functions, although it describes the response of a discontinuous material. Clearly, this kind of theory cannot describe the micro-scale behavior of the material, but it can be accurate enough at the macroscale and provide an acceptable material model for engineering purposes.

The science of thermodynamics is a branch of physics. It describes natural processes in which changes in temperature play an important part. Such processes involve the transformation of energy from one form to another. Consequently, thermodynamics deals with the laws that govern such transformations of energy. See Kestin [37].

This does not mean that the merits of thermodynamics are derived solely from constitutive equations that model thermal effects. Continuum thermodynamics is an excellent platform for plasticity, creep, continuum damage mechanics, and so on.

True thermodynamics, as distinct from equilibrium or reversible thermodynamics (also called non-equilibrium or irreversible thermodynamics), is one of the newest disciplines of physics. The theoretical principles were published by Lars Onsager in *The Physical Review* in 1931, at time when quantum mechanics, for example, was already a well-defined formal discipline. Non-equilibrium thermodynamics is the universal theory of macroscopic processes, and as these processes are irreversible, it is hardly necessary to stress its importance, both theoretical and practical. See Szabó [108].

Equilibrium thermodynamics or reversible thermodynamics is usually called *thermostatics*, which practice is also followed here.

Continuum mechanics deals with deformable bodies. In its early stages, it was confined to a few special materials and to particular situations, namely, to ideal liquids or to elastic solids under isothermal or adiabatic conditions. In these special cases it is possible to solve the basic problem, that is, to determine the flow and pressure distributions or the deformation and stress fields in purely mechanical terms. This is due to the fact that the solution can be developed from a set of differential equations that do not contain the energy balance. Anyone working on continuum mechanics knows that sooner or later, they become involved in thermodynamics. The reason for this is that, in general, a complete set of differential equations must contain the energy balance. Since part of the energy exchange takes place as heat flow, the appropriate form of the energy balance is the first fundamental law of thermodynamics, and it becomes clear, therefore, that it is generally impossible to separate the mechanical aspect of a problem from the thermodynamic processes accompanying the motion. To obtain a solution, the fundamental laws of both mechanics and thermodynamics must be applied. See Ziegler ([121], p. v.)

Continuum mechanics has always been a field theory, even in its rudimentary forms like hydraulics or strength of materials. To treat even such a simple problem as the bending of a beam, one must recognize that the states of strain and stress depend on position and possibly on time. The object of thermodynamics, on the other hand, has always been a finite volume, for example, a mole, and the state within the body has been tactile, assumed to be the same throughout the entire volume. It is surprising that this philosophy has been maintained even in the age of statistical and quantum mechanics, although it is clearly inconsistent with the first fundamental law in this common form: at least a part of the heat supply appearing in this law is due to heat flow through the surface of the body. As long as this process goes on, the temperature of the material points near the surface differs from the material points further inside the body; the state of the body is therefore not homogeneous. See Ziegler ([121], p. vi). Ziegler continues that there are the following two ways out of this dilemma.

The *historical way,* still dominating vast areas of teaching in thermodynamics, consists in the restriction of infinitely slow processes. In place of actual processes, one considers sequences of (homogeneous) equilibrium states. Except for a few special cases, such idealized processes are practically reversible, and this explains why in classical thermodynamics (or rather thermostatics), the limiting case of reversibility plays such a dominant role. However, an engineer engaged in the construction of thermomechanical machinery cannot themselves to infinitely slow processes and hence has never taken this restriction seriously. The situation strongly resembles the one in pre-Newtonian mechanics with its attempts to develop dynamics from purely static concepts. See Ziegler ([121], p. vi).

The *modern way* out of the dilemma is different but surprisingly simple: instead of infinitely slow processes, one considers infinitesimal elements of the body in which a process takes place, admitting that the state variables differ from element to element. In other words, one conceives thermodynamics as a field theory in much the same way as continuum mechanics has been treated for more than 200 years. In such a field theory, reasonably fast processes can be treated with the same ease as slow ones, and restriction to reversible processes becomes unnecessary. Finally, this field theory is the proper form in which thermodynamics and continuum mechanics are easily amalgamated. See Ziegler ([121], pp. vi and vii).

The present book studies continuum thermodynamics from the material modelling point of view. When material models are used in engineering work, other branches of science beyond thermomechanical evaluation of the material model are needed. These are experimental work, investigation of micromechanisms of deformation, determination of the values for material parameters, and mathematical simulation as outlined in Figure 1.1.

In order to obtain information on the response of a material, it is necessary to carry out experiments, whose significance cannot be overemphasized. Firstly, it is the only way to get information about an event in order to produce the necessary assumptions required in the derivation of its material model. Secondly, the comparison of a mathematically simulated response with the observed one shows the validity of the constitutive model formulated. Finally, it is usually the only way to determine the values of the parameters in the material model. However, experimental work alone cannot be used as a predictive tool. This is because the experimental data found under certain conditions are not valid under other conditions, and new experiments have to be carried out (e.g., the results of a creep test at 500 °C cannot be used to predict creep at 800 °C).

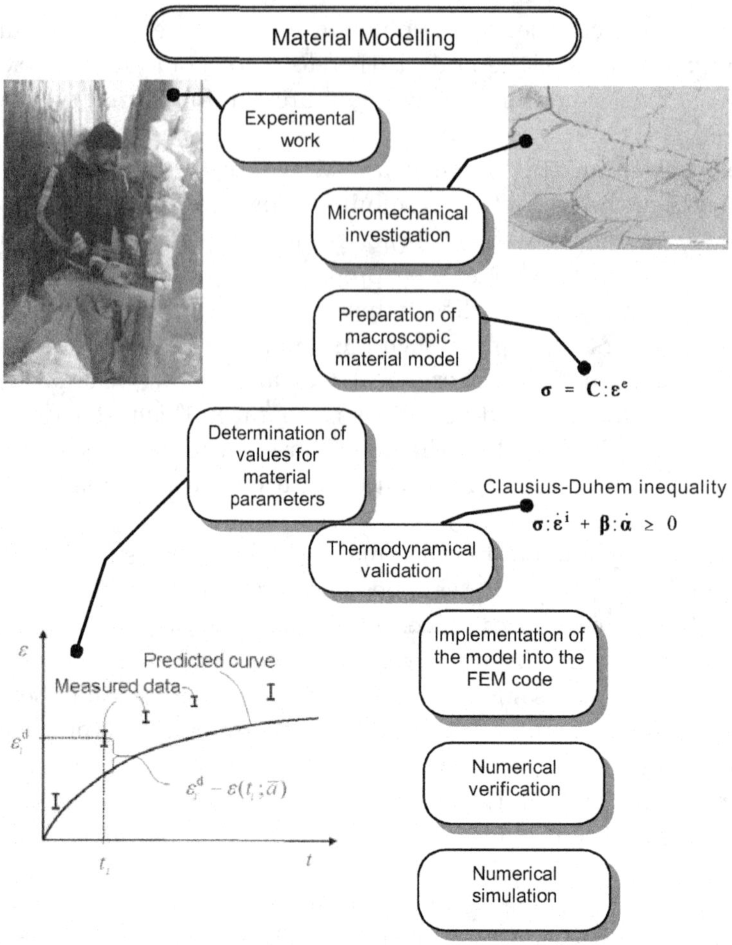


**Figure 1.1** Different branches of science related to material modelling. (top right) J. Rantala, VTT Technical Research Centre of Finland.

This difficulty can be avoided with a valid mathematical simulation which easily allows repetition of the calculation with new parameter values.

In order for the material model to be used as a predictive tool, the values of the material parameters have to be determined. Norton's law, for example,

$$\dot{\varepsilon}^{\mathrm{v}} = \overset{\circ}{\varepsilon}_{\mathrm{re}} \left( \frac{\sigma}{\sigma_{\mathrm{re}}} \right)^n , \tag{1.1}$$

is useless if the values of the material parameters $\overset{\circ}{\varepsilon}_{\mathrm{re}}$, $\sigma_{\mathrm{re}}$, and $n$ are not known. Determination of these material parameters requires experimental data, mathematical material model, and a nonlinear curve-fitting computer code. Determination of the material parameters is not an easy task since modern constitutive equations consist of a set of differential equations and therefore have many material parameters.

Investigation of the micromechanisms of deformation is the foundation of a material model. The macroscopic constitutive equation must be linked to the microscopic deformation of the material. Every variable of the model has to describe a certain microscopic process. The form of the model has to follow the mapping from micromechanisms to the macroscopic response of the material. A micromechanical model is a simplification of reality. The idea is to model only the dominant deformation mechanisms relevant for the objective of the study. An investigation of micromechanics also provides the variables for the constitutive equation.

The previous paragraph can be extended to many scientific theories in general. Before a new scientific theory can be accepted, at least the following two criteria must be met: the ability (a) to predict measured results and the ability (b) to provide reasoning. The latter is especially important since most of the theories were ditched before their ability to predict measurements was evaluated.

A mathematical simulation can be carried out either analytically or numerically. When an analytical solution is sought, usually plenty of simplified assumptions have to be made. Generally, such assumptions are very restrictive to the quality of the results. In other words, they often lead to a solution that cannot describe the natural event well enough. Thus, in many cases, the value of an analytical study is very limited.

Numerical simulations, however, give the possibility to include in the simulation many important details that have to be omitted when analytical methods are applied. This is the overwhelming power of numerical simulations. The main problems with a numerical simulation have been computing capacity and a lack of a suitable mathematical model describing the natural event (here, the constitutive equation). The first obstacle has been largely overcome by the rapid advances of computer architecture. The evaluation of constitutive equations is the topic of this book.

Only mathematical models provide a tool for predicting events. The major drawback of mathematical investigation is that it is a very difficult and long-term task. Only a small number of scientists seem capable of doing it successfully. Therefore, it has become popular to avoid theoretical work and concentrate on doing "practical work that can be utilized" (computation even in cases in which the theoretical basis of the applied equations is extremely weak or weak experimental work) instead of deriving "boring and useless theory."

Usually, the final goal of material modelling is to determine the response of the considered system under new conditions. This predictive work can be done by mathematical simulation, which nowadays is mainly computation. However, successful simulations are based on a reliable constitutive equation, and the predictions are never better than the model behind them. The most arduous work is done by the human brain, not by microprocessors or laboratory facilities.

The material model has to be thermodynamically correct, that is, it has to fulfill certain basic laws and axioms of nature. Those are the balance laws of continuum mechanics and the first and second laws of thermodynamics. It is generally accepted that a body has to be in equilibrium. This means that the balance law of momentum is followed. The first and second laws of thermodynamics also have to be fulfilled. If the model does not fulfill the requirements set by the basic laws and axioms of nature, the model must be abandoned, as the response it describes cannot exist in nature. One has to remember that although a material model is thermomechanically correct, it may not useful from a practical point of view. Thermodynamics shows only whether a constitutive equation is possible. Although thermodynamics is exploited, a material model does not need to show any temperature effects, but usually the validation is carried out to ensure that the energy dissipation given by the material model does not violate the second law of thermodynamics.

The present book provides an introduction to "continuum thermodynamics with internal variables" – the most promising dialect of continuum thermodynamics for validation of material models. The first part of the book concentrates on general theory, and the latter part studies different material models. In this book, displacements, rotations, and deformations are assumed to be small. These assumptions are not the restrictions set by continuum thermodynamics. They are made only in this work in order to prepare a simpler and more understandable formulation of the theory.

## 1.2 Form of the Second Law of Thermodynamics

This section studies how the form of the second law of thermodynamics differs from the form of the other basic laws and axioms and outlines why the second law of thermodynamics is referred to as an "arrow of time." This is because the second law of thermodynamics gives the direction for a process, whereas the other basic laws and axioms do not take a stand regarding the direction of time, that is, the direction of a process. This is evaluated through an example.

Using Newton's (second) law of motion, the motion of a particle P along the $x$-axis under the action of a force $N$ is investigated. The mass of the particle is denoted by $m_0$. For this uniaxial case, Newton's second law of motion can be expressed as follows:

$$\vec{F} = \frac{\mathrm{d}}{\mathrm{d}t}(m_0\,\vec{v}) \quad \Rightarrow \quad \vec{F} = m_0\,\vec{a} \quad \Rightarrow \quad N = m_0\,\ddot{x} \quad \Rightarrow \quad \ddot{x} = \frac{N}{m_0}, \tag{1.2}$$

where $\vec{F}$ is the force vector, $\mathrm{d}/\mathrm{d}t$ is the time derivative operator, $\vec{v}$ is the velocity vector, $\vec{a}$ is the acceleration vector, and the double-dot $\ddot{\theta}$ represents the second derivative of a quantity $\theta$ with respect to time $t$. Integration of Expression $(1.2)_4$ with respect to time gives

$$\dot{x} = \frac{N}{m_0}\,t + C_1 \qquad \text{and} \qquad x = \frac{1}{2}\frac{N}{m_0}t^2 + C_1\,t + C_2\,. \tag{1.3}$$

The motion of the particle is studied by two observers. The observers are Mike and Judith. The difference between Mike and Judith lies in their experience of the flow of time $t$. The main difference is that they experience the course of time in opposite directions and therefore see the movement of the particle in different ways. Next, Mike's and Judith's observations are studied in more detail. Mike's observations are used here as the reference observations.

**Mike**

According to Mike, the position of particle P at time $t = 0$ is $x = 0$, and it has no velocity. Mathematically, these initial conditions and their consequences can be written as

$$x = 0 \quad \text{at} \quad t = 0\,, \qquad \text{which yields} \qquad C_2 = 0 \tag{1.4}$$

and

$$\dot{x} = 0 \quad \text{at} \quad t = 0\,, \qquad \text{which yields} \qquad C_1 = 0\,. \tag{1.5}$$

Substitution of integration constants $C_2$ and $C_1$ from Equations $(1.4)_2$ and $(1.5)_2$ into Expressions(1.3) gives

$$\dot{x} = \frac{N}{m_0}\,t \qquad \text{and} \qquad x = \frac{1}{2}\frac{N}{m_0}\,t^2\,. \tag{1.6}$$

The position and the motion of the particle P at time $t = \Delta t$ are of interest. The application time of the applied force $N$ is $t = 0$. Based on Equations (1.6), the following is achieved:

$$v_{\Delta t} = \frac{N}{m_0}\,\Delta t \qquad \text{and} \qquad x_{\Delta t} = \frac{1}{2}\frac{N}{m_0}\,(\Delta t)^2, \tag{1.7}$$

where $v_{\Delta t}$ and $x_{\Delta t}$ stand for the velocity and the position of the particle P at time $t = \Delta t$, respectively.

**Judith**

Judith experiences the course of time $t$ in an opposite "direction." Her time is denoted by $\tilde{t}$. There is a simple linear relationship between the two time measures. This is given as

$$\tilde{t} = -t + \Delta t\,. \tag{1.8}$$

Definition (1.8) gives

$$\tilde{t} = 0 \quad \text{at} \quad t = \Delta t \qquad \text{and} \qquad \tilde{t} = \Delta t \quad \text{at} \quad t = 0\,. \tag{1.9}$$

Equation (1.9) shows that the time measures $t$ and $\tilde{t}$ have opposite "directions" and different "origins." Opposite directions means that if the time $t$ measures a certain time interval with a positive number, then the time $\tilde{t}$ measures the same time interval with a negative number (in this case, their absolute values are equal).

According to Judith, the position of particle P at time $\tilde{t} = 0$ is $x_{\Delta t}$, and it has a velocity $v_{\Delta t}$. Thus the initial conditions for Judith equal the conditions of Mike at time $t = \Delta t$. Due to different time "directions," Mike living in time space $t$ experiences the motion of particle P differently than Judith living in time space $\tilde{t}$. They experience the velocity direction of particle P oppositely. Mike experiences the motion toward the positive $x$-axis, whereas Judith experiences the motion toward the negative $x$-axis. This originates from the fact that the time measures $t$ and $\tilde{t}$ have opposite "directions" [see Definition (1.8)] and therefore the first derivatives with respect to time measures $t$ and $\tilde{t}$ have opposite signs. The second derivatives have the same sign, and therefore Mike and Judith experience the acceleration of particle P as equal.

The preceding discussion on velocity and acceleration can be written mathematically as follows:

$$\frac{\mathrm{d}x}{\mathrm{d}t} = \frac{\mathrm{d}x}{\mathrm{d}\tilde{t}}\frac{\mathrm{d}\tilde{t}}{\mathrm{d}t} = \frac{\mathrm{d}x}{\mathrm{d}\tilde{t}}(-1) = -\frac{\mathrm{d}x}{\mathrm{d}\tilde{t}}, \qquad \text{which is} \qquad v = -\tilde{v} \tag{1.10}$$

and

$$\begin{aligned} \frac{\mathrm{d}^2x}{\mathrm{d}t^2} &= \frac{\mathrm{d}}{\mathrm{d}t}\frac{\mathrm{d}x}{\mathrm{d}t} = \frac{\mathrm{d}}{\mathrm{d}t}\left(-\frac{\mathrm{d}x}{\mathrm{d}\tilde{t}}\right) = \frac{\mathrm{d}}{\mathrm{d}\tilde{t}}\left(-\frac{\mathrm{d}x}{\mathrm{d}\tilde{t}}\right)\frac{\mathrm{d}\tilde{t}}{\mathrm{d}t} \\ &= \frac{\mathrm{d}}{\mathrm{d}\tilde{t}}\left(-\frac{\mathrm{d}x}{\mathrm{d}\tilde{t}}\right)(-1) = \frac{\mathrm{d}^2x}{\mathrm{d}\tilde{t}^2}, \qquad \text{which is} \qquad a = \tilde{a}\,. \end{aligned} \tag{1.11}$$

Equations (1.3) also holds for Judith. The first and second time derivatives denoted by $\dot{}$ and $\ddot{}$, respectively, have to be replaced by the time derivatives $^\circ$ and $^{\circ\circ}$, which represent time derivatives with respect to time $\tilde{t}$. Thus, for Judith, Expressions (1.3) are replaced by

$$\mathring{x} = \frac{N}{m_0}\tilde{t} + D_1 \qquad \text{and} \qquad x = \frac{1}{2}\frac{N}{m_0}\tilde{t}^2 + D_1\,\tilde{t} + D_2\,. \tag{1.12}$$

Mathematically, the preceding initial conditions for Judith and their consequences can be written as

$$x = x_{\Delta t} \quad \text{at} \quad \tilde{t} = 0\,, \qquad \text{which yields} \qquad D_2 = x_{\Delta t} \tag{1.13}$$

and

$$\mathring{x} = v_{\Delta t} \quad \text{at} \quad \tilde{t} = 0\,, \qquad \text{which yields} \qquad D_1 = v_{\Delta t}\,. \tag{1.14}$$

Substitution of integration constants $D_2$ and $D_1$ from Equations $(1.13)_2$ and $(1.14)_2$ into Equations (1.12) yields

$$\tilde{v} = \mathring{x} = \frac{N}{m_0}\tilde{t} + v_{\Delta t} \qquad \text{and} \qquad x = \frac{1}{2}\frac{N}{m_0}\tilde{t}^2 + v_{\Delta t}\,\tilde{t} + x_{\Delta t}\,. \tag{1.15}$$

The state experienced by Judith is evaluated at time $\tilde{t} = \Delta\tilde{t}$. Since $\Delta\tilde{t} = -\Delta t$ [see Definition (1.8)], Equations (1.15) give

$$\tilde{v}_{\Delta\tilde{t}} = \frac{N}{m_0}(-\Delta t) + v_{\Delta t} \quad \text{and} \quad x_{\Delta\tilde{t}} = \frac{1}{2}\frac{N}{m_0}(-\Delta t)^2 + v_{\Delta t}(-\Delta t) + x_{\Delta t}\,. \tag{1.16}$$

Substitution of the velocity $v_{\Delta t}$ and the position $x_{\Delta t}$ from Equations (1.7) into Equations (1.16) gives

$$\tilde{v}_{\Delta\tilde{t}} = \frac{N}{m_0}(-\Delta t) + v_{\Delta t} = \frac{N}{m_0}(-\Delta t) + \frac{N}{m_0}\Delta t = 0 \tag{1.17}$$

and

$$\begin{aligned} x_{\Delta\tilde{t}} &= \frac{1}{2}\frac{N}{m_0}(-\Delta t)^2 + v_{\Delta t}(-\Delta t) + x_{\Delta t} \\ &= \frac{1}{2}\frac{N}{m_0}(-\Delta t)^2 + \frac{N}{m_0}\Delta t(-\Delta t) + \frac{1}{2}\frac{N}{m_0}(\Delta t)^2 = 0\,. \end{aligned} \tag{1.18}$$

Initial conditions (1.4) and (1.5) show that Mike experiences particle P to be at rest at time $t = 0$ at the point $x = 0$. Result (1.7) shows that Mike experiences particle P to move during the time interval $\Delta t$ to the position $x_{\Delta t}$ and to have the velocity of $v_{\Delta t}$ in the direction of the positive $x$-axis. Initial Conditions (1.13) and (1.14) show that Judith experiences particle P at time $\tilde{t} = 0$ to be at the position $x_{\Delta t}$ and to move towards the origin, that is, to move in the direction of the negative $x$-axis. According to Results (1.17) and (1.18), after the time interval $\Delta\tilde{t}$, Judith experiences particle P to be at the origin at rest. Thus, Mike and Judith experience realistic processes, although their time measures are opposite, as shown in Figure 1.2.

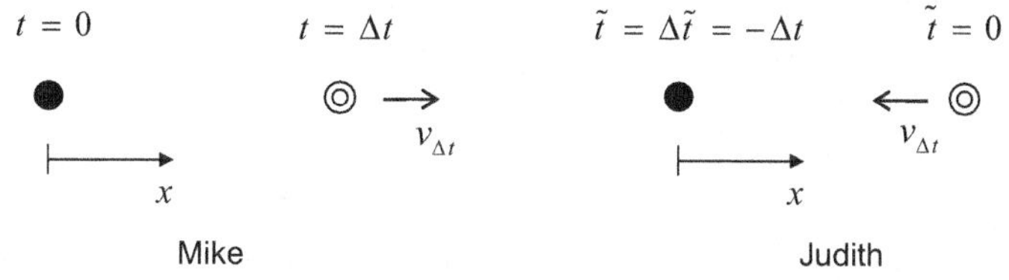


**Figure 1.2** A process experienced by Mike and a process experienced by Judith.

Unfortunately, the second law of thermodynamics does not allow us to create such a simple example as was given earlier. This is because the manipulation of the second law of thermodynamics into a form needed to show the "arrow of time" requires too heavy a mathematical derivation to be illustrative. Therefore, the following brief introduction is given: in order to demonstrate how the second law of thermodynamics provides the arrow of time, the Clausius–Duhem inequality is studied. The Clausius–Duhem inequality is a consequence of the second law of thermodynamics containing information from the other basic laws and axioms of nature. When a pure thermal process (i.e., no deformation occurs) is studied, the Clausius–Duhem inequality takes the following appearance:

$$-\frac{1}{T}\vec{q}\cdot\vec{\nabla}T \geq 0, \tag{1.19}$$

where $T$ is the absolute temperature, $\vec{q}$ is the heat flux vector, and $\vec{\nabla}$ is the vector differential operator, del. When only the $x$-direction is investigated, Clausius–Duhem Inequality (1.19) reduces to

$$-\frac{1}{T}q_x\frac{\partial T}{\partial x} \geq 0. \tag{1.20}$$

Figure 1.3 illustrates the message of Inequality (1.20). Since the absolute temperature $T$ is positive, the quantities $q_x$ and $\partial T/\partial x$ have to take opposite signs in order to satisfy Inequality (1.20). This means that the second law of thermodynamics sets a restriction on the direction of the process. Time $t$ is not an independent variable, but it is defined using the rate of a process. At the dawn of civilization, time $t$ was measured by following the movements of the sun, moon, and stellar system. In thermodynamical language, these movements are processes. Thus the rate of a process defines the variable time $t$. The following definition of a second supports this point of view.

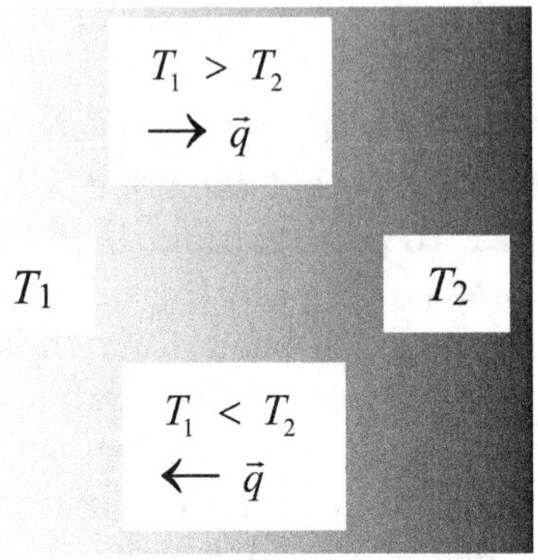


**Figure 1.3** Clausius–Duhem inequality determines uniquely the direction of a process and thus the direction of time $t$.

A second is defined as the duration of 9,192,631,770 periods of radiation corresponding to the transition between the two hyperfine levels of the ground state of the caesium-133 atom. A ground state means a level of zero magnetic flux [26].

The foregoing discussion can be summarized as follows: since the second law of thermodynamics determines the direction of a process, it also determines the direction of time, and therefore the second law of thermodynamics is called the "arrow of time."

## 1.3 Reversible Process Gives an Irreversible Process

This section evaluates two subvolumes filled with gas at different temperatures. After removal of the wall between the two subvolumes, the gases mix, leading to a gas at homogeneous temperature. This simple example is used to demonstrate how a reversible process on a microscale can lead to irreversible process on a macroscale. A reversible process is a process that reverses in such a way that no permanent changes will occur. An irreversible process causes permanent changes.

Comparing the preceding example, the two processes over opposite courses of time with the Clausius–Duhem inequality may lead to the notion that they are from two wholly different ways of modelling nature. This is not the case, as can be seen from the following example, which shows how a microscale reversible process obeying Newtonian mechanics can lead to a macroscale irreversible process with increasing disorder.

Gas inside a volume is considered, the gas being modelled by a group of atoms. Since the temperature of the gas is above absolute zero, the atoms move within the volume. Every atom has its own speed, but the average speed of the atoms is defined as the temperature of the gas. Thus the higher the average speed of the atoms is, the higher is the temperature of the gas.

The atoms are modelled as spherical elastic objects. They move within the gas and collide with the other atoms or with the walls of the volume. The collision is assumed always to be elastic. This means that a collision between two atoms having different speeds changes their speeds and directions of movement, but no energy dissipation occurs. Thus the collision of two atoms is a pure reversible process. Based on Newtonian mechanics, the collision of two atoms could occur in the opposite direction. The same holds true for the collision of atoms with the walls of the volume. This means that the microscale process is a reversible one. On the macroscale, however, the situation is different, as is demonstrated next.

The gas described earlier is assumed to fill the volume shown in Figure 1.4(a). The volume is separated into two subvolumes by a wall. The gas on both sides of the wall are the same, but their temperatures are different. On the right side of the wall, the temperature of the gas is higher than it is on the left side. This means that the average velocity of the atoms on the right side is higher than it ia on the left.

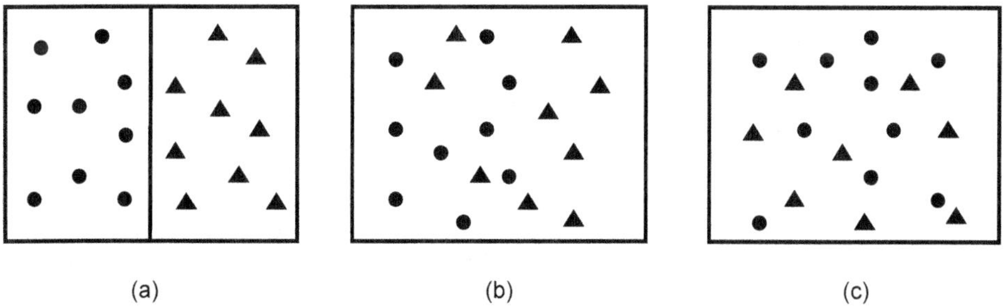


**Figure 1.4** Increasing disorder (entropy) of gas.

Next, the wall between the two subvolumes is taken away. The removal is assumed to be ideal in such a way that it does not affect the atoms. Now the atoms moving from the right side of the volume do not collide with the wall between the two subvolumes (which no longer exists) but are able to enter the left side until they collide with the atoms on that side. The same holds for the atoms on the left side of the volume. This means that the right-side atoms and the left-side atoms start to mix, as shown in Figure 1.4(b). The hotter gas mixes with the cooler gas. However, this process is not only the mixing of atoms. When two atoms having different speeds collide, the speed of the lower-speed atom increases, whereas the speed of the higher-speed atom decreases, cancelling out the speed difference. It becomes increasingly difficult to distinguish which atoms were from the hotter side of the wall and which were from the cooler side before the removal of the wall.

Finally, all the differences between these two gases are eliminated, as shown in Figure 1.4(c). Although the individual collisions of atoms are purely elastic and therefore reversible, the macroscopic process is irreversible. The atoms will never reorganize so that the right side consists of a hotter gas and the left side of a cooler gas. Actually, this does happen, but it takes such a long time that the universe is too young for such a reorganization to occur in practice. Thus, in practice, *the macroscopic process is irreversible, although the microscopic process is reversible.* In thermodynamics, this increase of disorder of gas is called the increase of entropy.

## 1.4 Structure of Continuum Thermodynamics

The structure of continuum thermodynamics is given here. Continuum thermodynamics is based on the following basic laws [**BL**] and axioms [**AX**].

As shown in Figure 1.5, continuum thermodynamics consists of three parts: continuum mechanics, thermostatics, and thermodynamics extension to thermostatics. The coming chapters give a detailed discussion of the basic laws and axioms with detailed mathematical derivations.

## 1.5 Summary

This chapter briefly outlined the content of continuum mechanics and thermodynamics. It pointed out that although continuum mechanics gives definitions for stress and strain, continuum thermodynamics provides a tool with which to write relations between these two notions. These relations are known as the material models, also called constitutive equations. The role of continuum thermodynamics is in the validation of the material models. Although continuum thermodynamics is related to temperature, the present approach is also for material models without thermal effects. The chapter also stressed the importance of experiments in material modelling.

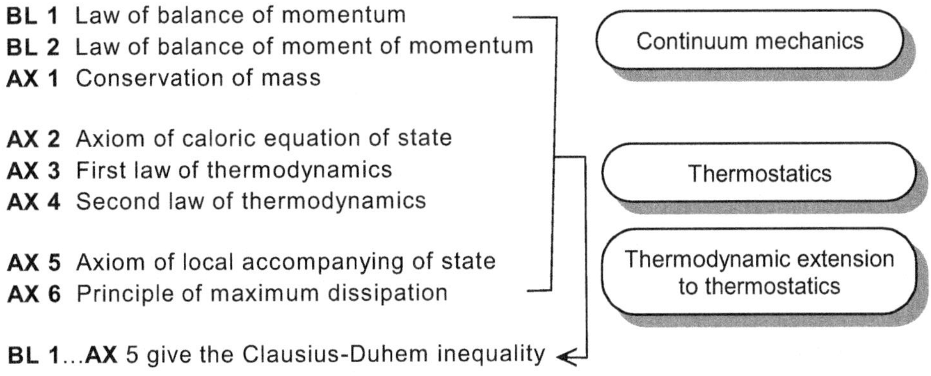


**Figure 1.5** Basic laws and axioms of continuum thermodynamics and the Clausius–Duhem inequality.

The high performance of modern computers allows the introduction of reliable material models, and therefore numerical simulation is today the main tool in structural analysis. Newton's second law of motion is used to demonstrate that usually the basic laws and axioms are symmetric in time; that is, the direction of time can be reversed, and the basic laws and axioms give realistic predictions. The only exception is the second law of thermodynamics, according to which the heat conduction in a body is from a hotter part to a colder part; that is, it gives the direction of the processes. Finally, the structure of continuum thermodynamics is given.

# 2
# Mathematical Prerequisites

## 2.1 Preliminary Comments

This chapter outlines some mathematical concepts useful in the rest of this book. Since the content of the present chapter covers only those mathematical expressions that are needed, the content of this chapter is limited and therefore does not form any complete presentation of the mathematical tools for continuum mechanics and continuum thermodynamics. A broader view of mathematics for continuum mechanics and for continuum thermodynamics is given in the books by Reddy [84] and Surana [105].

Since the book is written using tensor algebra, the first coordinate system, which is applicable with tensor notations, is defined. To keep the expressions short, Einstein summation convention is introduced. Several examples of this are given to demonstrate the Einstein summation convention and its benefit in shortening expressions. Special attention is paid to the "dot product" and "double-dot product" since they are present in almost all expressions in this book. Many special tensors are defined. Differentiation of tensors is expressed using three different techniques: first, the directional derivative with the Gibbs notation of tensors is utilized; second, the index representation of the tensors is used for formulation of the derivatives; and finally, some examples are given to demonstrate cases in which the derivative can be derived of tensors with the Gibbs notation (i.e., use of boldface or vector notation), but without applying the directional derivative. At the end of the chapter, the Voigt notation is introduced. The Voigt notation (i.e., index notation) is important to know, because instead of tensorial form, the computer code is based on the Voigt notation.

## 2.2 Rectangular Cartesian Coordinate System and Some General Notations

This section introduces the rectangular Cartesian coordinate system and gives some fundamental mathematical concepts and tools useful in the subsequent chapters. We begin with the notation used.

Scalar-valued variables are denoted in italics, such as $v$ and $\theta$. An arrow above an italicized quantity denotes the variable as a vector, such as $\vec{v}$. Second-order and higher-order tensors are denoted in bold regular font, such as $\mathbf{s}$ and $\boldsymbol{\varepsilon}$.

Gibbs notation, that is, boldface notation, is preferred here, because it provides the same appearance of expressions in any coordinate system. However, index notation is useful in some mathematical operations and computations.

If a superscript or a subscript is a variable, it is italicized; if it is not a variable, it is not italicized. Thus the indices $s$ and $t$ in the scalar component $\sigma_{st}$ are italicized, since they take values 1, 2, or 3 for a three-dimensional case and 1 or 2 for a two-dimensional case, but the superscript e in the notation $\boldsymbol{\varepsilon}^{\mathrm{e}}$ is written in roman font, since it refers to "elastic" and does not have any value. The same logic holds for the notation for transpose, which takes the appearance $(\cdot)^{\mathrm{T}}$.

Since boldface letters are difficult to write by hand, the authors propose the following notations for handwriting. A double-sided arrow above a quantity denotes the variable to be a second-order tensor. The stress tensor $\overleftrightarrow{\sigma}$ and the strain tensor $\overleftrightarrow{\varepsilon}$ are second-order tensors. The constitutive tensor for Hookean deformation is a fourth-order tensor, and it can be handwritten as $\overset{\Leftrightarrow}{\mathrm{C}}$ . Someone may prefer to write ${}^{2}\varepsilon$ and ${}^{4}\mathrm{C}$.

The Macaulay brackets $\langle \cdot \rangle$ have the following property:

$$\langle x \rangle := \begin{cases} 0 & \text{when } x < 0 \\ x & \text{when } x \geq 0\,. \end{cases} \tag{2.1}$$

The Heaviside function (unit-step function) $H(\cdot)$ is defined by

$$H(x) := \begin{cases} 0 & \text{when } x < 0 \\ 1 & \text{when } x \geq 0. \end{cases} \tag{2.2}$$

The signum function sgn $(\cdot)$ is defined by

$$\mathrm{sgn}(x) := \begin{cases} -1 & \text{when } x < 0 \\ 1 & \text{when } x \geq 0. \end{cases} \tag{2.3}$$

Kronecker delta $\delta_{ij}$ is defined to be zero when $i$ and $j$ have different values and to be unity when $i = j$ (a fixed value):

$$\delta_{ij} := \begin{cases} 0 & \text{when the values for } i \text{ and } j \text{ differ; e.g., } \delta_{12} = 0\,, \\ 1\,, & \text{when } i \text{ and } j \text{ take the same values; e.g., } \delta_{11} = 1\,. \end{cases} \tag{2.4}$$

In this book, vectors and tensors are represented in rectangular Cartesian coordinate systems. This means that the coordinate axes are perpendicular to each other, that they are rectilinear, and that the basis of the coordinate system is composed of unit vectors. However, when the Gibbs notation is used, the quantities and expressions are valid for all coordinate systems. The Gibbs notation uses boldface letters and arrows above vector quantities. When Gibbs notation is used, the velocity vector is given by $\vec{v}$, whereas the state of stress for the small deformation case is expressed by the stress tensor $\boldsymbol{\sigma}$. Thus the obtained results are also valid in the cylindrical coordinate system, for example.

Traditionally, the rectangular Cartesian coordinate axes are denoted by $(x, y, z)$ and the corresponding base vectors by $(\vec{\imath}, \vec{\jmath}, \vec{k})$. Dot products between the base vectors $\vec{\imath}$, $\vec{\jmath}$, and $\vec{k}$ are

$$\begin{aligned} \vec{\imath} \cdot \vec{\imath} &= \vec{\jmath} \cdot \vec{\jmath} = \vec{k} \cdot \vec{k} = 1 \,, \\ \vec{\imath} \cdot \vec{\jmath} &= \vec{\jmath} \cdot \vec{k} = \vec{k} \cdot \vec{\imath} = 0 \,, \\ \vec{\jmath} \cdot \vec{\imath} &= \vec{k} \cdot \vec{\jmath} = \vec{\imath} \cdot \vec{k} = 0 \,. \end{aligned} \tag{2.5}$$

The notations $(x, y, z)$ and $(\vec{\imath}, \vec{\jmath}, \vec{k})$ are not convenient in tensor algebra. Therefore the coordinate system is denoted by $(x_1, x_2, x_3)$, and its basis is denoted by $(\vec{\imath}_1, \vec{\imath}_2, \vec{\imath}_3)$, as shown in Figure 2.1. The basis $(\vec{\imath}_1, \vec{\imath}_2, \vec{\imath}_3)$ consists of mutually orthogonal unit vectors $\vec{\imath}_1, \vec{\imath}_2,$ and $\vec{\imath}_3$, and it obeys Result (2.5), viz.

$$\vec{\imath}_i \cdot \vec{\imath}_j = \delta_{ij} \qquad i, j = 1, 2, 3 \,. \tag{2.6}$$

The values 1, 2, 3 for the indices $i$ and $j$ represent a three-dimensional case. For a two-dimensional case (e.g., two-dimensional elasticity) the indices of the base vectors may take only values 1 and 2. The dot product is studied in more detail in Section 2.3.

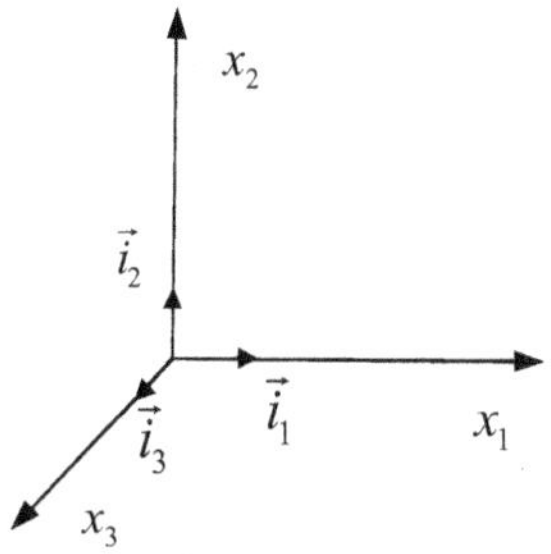


**Figure 2.1** Rectangular Cartesian frame.

When the coordinate system $(x, y, z)$ is used, a vector $\vec{b}$ is defined as follows:

$$\vec{b} := b_{\mathrm{x}} \vec{\imath} + b_{\mathrm{y}} \vec{\jmath} + b_{\mathrm{z}} \vec{k} \,, \tag{2.7}$$

where $b_{\mathrm{x}}$ is the scalar component of the vector $\vec{b}$ in the direction of the $x$-axis. Components $b_{\mathrm{y}}$ and $b_{\mathrm{z}}$ are defined similarly.

When the earlier introduced coordinate frame $(x_1, x_2, x_3)$ is used, the representation of vector $\vec{b}$ reads

$$\vec{b} := b_1 \vec{\imath}_1 + b_2 \vec{\imath}_2 + b_3 \vec{\imath}_3 \,. \tag{2.8}$$

In Definition (2.8), the notation $b_1$ refers to the scalar component of vector $\vec{b}$ in the direction of the $x_1$-axis. Components $b_2$ and $b_3$ are defined similarly.

The Einstein summation convention is explained as follows: whenever the same italicized letter appears twice (and not more) as an index in a term, it means a sum over the values of the index. If the value of the index is not specified, the value is assumed to be 3. Thus, for two scalar components of vectors $\vec{u}$ and $\vec{v}$, we have

$$u_i \, v_i = u_m \, v_m := u_1 \, v_1 + u_2 \, v_2 + u_3 \, v_3 \,. \tag{2.9}$$

When the Einstein summation convention is applied, Definition (2.8) can be written as

$$\vec{b} = b_i \vec{\imath}_i \qquad \text{in place of} \qquad \vec{b} = \sum_{i=1}^{3} b_i \vec{\imath}_i \,. \tag{2.10}$$

As Expression (2.9) shows, the sum is independent of the indices $i$ and $m$. Therefore summation indices $i$ and $m$ in Definition (2.9) are called the *dummy indices.* The same quantity can have more than one index; for example, in such a case, the Einstein summation convention gives

$$\sigma_{ii} = \sigma_{11} + \sigma_{22} + \sigma_{33} \,. \tag{2.11}$$

**Example 2.1:** What is the value of $\delta_{ss}$?

**Solution:** Based on the Einstein summation convention [see Expression (2.9) and Definition (2.4)] the following is achieved:

$$\delta_{ss} = \delta_{11} + \delta_{22} + \delta_{33} = 1 + 1 + 1 = 3 \,. \quad \blacksquare \tag{2.12}$$

Equation (2.4) and the result of **Example 2.1** give

$$\delta_{11} = \delta_{22} = \delta_{33} = 1 \qquad \text{but} \qquad \delta_{ii} = 3 \,. \tag{2.13}$$

The following equation is examined for its meaning:

$$\beta_j = \frac{\partial a_{ij}}{\partial x_i} \qquad \Rightarrow \qquad \beta_j = \frac{\partial a_{1j}}{\partial x_1} + \frac{\partial a_{2j}}{\partial x_2} + \frac{\partial a_{3j}}{\partial x_3} \,. \tag{2.14}$$

In Expression (2.14), there is an index ($j$) that did not take part in the summation. These kind of indices are called *free indices.* Free indices have to be present in all the terms on both sides of an equation. As for dummy indices, free indices take the values $1, 2, 3$, representing the three dimensions of the space, and the values $1, 2$ are for a two-dimensional case. This implies that in the three-dimensional case, Equations (2.14) give three equations. They are obtained for the values $j = 1$ , $j = 2$, and $j = 3$:

$$\begin{aligned}
\beta_1 &= \frac{\partial a_{11}}{\partial x_1} + \frac{\partial a_{21}}{\partial x_2} + \frac{\partial a_{31}}{\partial x_3} \qquad \text{for} \quad j = 1 \,, \\
\beta_2 &= \frac{\partial a_{12}}{\partial x_1} + \frac{\partial a_{22}}{\partial x_2} + \frac{\partial a_{32}}{\partial x_3} \qquad \text{for} \quad j = 2 \,, \\
\beta_3 &= \frac{\partial a_{13}}{\partial x_1} + \frac{\partial a_{23}}{\partial x_2} + \frac{\partial a_{33}}{\partial x_3} \qquad \text{for} \quad j = 3 \,.
\end{aligned} \tag{2.15}$$

The "open product" of vectors $\vec{b}$ and $\vec{c}$ is a second-order tensor denoted by $\vec{b}\,\vec{c}$. If $\mathbf{g} = \vec{b}\,\vec{c}$ and the basis $(\vec{\imath}_1, \vec{\imath}_2, \vec{\imath}_3)$ is used, the following definition can be written:

$$\textbf{Def} \qquad \mathbf{g} = \vec{b}\,\vec{c} = b_i\,\vec{i}_i\,c_s\,\vec{i}_s = b_i\,c_s\,\vec{i}_i\,\vec{i}_s = g_{is}\,\vec{i}_i\,\vec{i}_s\,, \tag{2.16}$$

with the notation $g_{is} = b_i c_s$. Equation (2.16) shows that the scalar components $b_i$ and $c_s$ have to have different indices, because the notation $b_i\,\vec{i}_i\,c_i\,\vec{i}_i$ is meaningless. Thus

$$\cancel{b_i\,\vec{i}_i\,c_i\,\vec{i}_i} \quad \text{is incorrect, but} \quad b_i\,\vec{i}_i\,c_s\,\vec{i}_s + a_i\,\vec{i}_i\,v_s\,\vec{i}_s \quad \text{is correct.} \tag{2.17}$$

The message is that no more than two of the same indices are allowed in one term, but the other terms in an expression can have the same pair of indices, as in $b_i\,\vec{i}_i\,c_s\,\vec{i}_s + a_i\,\vec{i}_i\,v_s\,\vec{i}_s$. The separator between two adjacent terms can be a plus "+" operator or minus "−" operator. The open product of two vectors is called a *dyad*, which is a second-order tensor.

Expression (2.16) shows that a scalar component can be moved to the other side of a base vector. In Expression (2.16) the scalar component $c_j$ is moved from the right side of the base vector $\vec{i}_i$ to the left side. Thus the notation on the far right of Expression (2.16), that is, $g_{ij}\,\vec{i}_i\,\vec{i}_j$, could be written $\vec{i}_i\,\vec{i}_j\,g_{ij}$, although it looks awkward. *The order of the base vectors is not allowed to be interchanged.*

The aim of the following study is to show that the order of the scalar components in a tensor can be changed without changing the meaning of the tensor. The second-order tensor $\mathbf{g}$ introduced in Equation (2.16) is used as an example. The corresponding second-order tensor with interchanged order of the scalar components is written for comparison. A two-dimensional case is studied here to reduce the number of terms. Based on the Einstein summation convention, the following relations are obtained:

$$\begin{aligned} \mathbf{g} = \vec{b}\,\vec{c} = b_i\,c_j\,\vec{i}_i\,\vec{i}_j &= b_1\,c_1\,\vec{i}_1\,\vec{i}_1 + b_1\,c_2\,\vec{i}_1\,\vec{i}_2 + b_2\,c_1\,\vec{i}_2\,\vec{i}_1 + b_2\,c_2\,\vec{i}_2\,\vec{i}_2\,, \\ c_j\,b_i\,\vec{i}_i\,\vec{i}_j &= c_1\,b_1\,\vec{i}_1\,\vec{i}_1 + c_2\,b_1\,\vec{i}_1\,\vec{i}_2 + c_1\,b_2\,\vec{i}_2\,\vec{i}_1 + c_2\,b_2\,\vec{i}_2\,\vec{i}_2\,, \\ &= b_i\,\vec{i}_i\,c_j\,\vec{i}_j = \left(b_1\,\vec{i}_1 + b_2\,\vec{i}_2\right)\left(c_1\,\vec{i}_1 + c_2\,\vec{i}_2\right). \end{aligned} \tag{2.18}$$

The equality $b_1\,c_1 = c_1\,b_1$ holds, since the quantities $b_i$ and $c_j$ are scalars and the multiplication of scalars is commutative, for example, $7 \times 8 = 8 \times 7$. The same result covers the other scalar components $b_i\,c_j$ and $c_j\,b_i$. Thus the notations $b_i\,c_j\,\vec{i}_i\,\vec{i}_j$ and $c_j\,b_i\,\vec{i}_i\,\vec{i}_j$ are equal, which means that the order of the scalar components can be changed. It is important to note that the scalar components keep their indices even when the order is changed. Usually, the scalar components of tensors cannot be separated into several parts, as was done in Equation (2.18). The scalar components of the stress tensor $\boldsymbol{\sigma}$ and the strain tensor $\boldsymbol{\epsilon}$, for example, are such quantities.

The role of first-order tensors (i.e., vectors) is easy to clarify. The velocity of a particle, for example, is a vector. The magnitude $v$ of the velocity vector

$\vec{v}$ gives the speed of the particle, whereas the propagation direction is given by the arrow of the velocity vector $\vec{v}$. Similar physical explanation holds for the force vector $\vec{F}$. Stress tensor $\boldsymbol{\sigma}$, for example, is a second-order tensor. This means that it has two base vectors. For a stress, it is important to know the plane on which it is acting. The plane area can be viewed as a vector, with the magnitude of the vector being the area and its direction given by the vector normal to the plane. This gives the first directional information for the stress tensor $\boldsymbol{\sigma}$ and therefore the first index. The second index shows the direction of the force acting on the plane. Thus, in order to define the stress tensor $\boldsymbol{\sigma}$, two vectors are needed, and therefore the stress tensor $\boldsymbol{\sigma}$ is a second-order tensor. Some writers define the meanings of the indices of the stress tensor $\boldsymbol{\sigma}$ opposite (transpose) to what is stated here. Both notations are correct, of course, but for a reader, one definition, as adopted here, is desirable.

Let $\vec{v}$ be an arbitrary vector, $\mathbf{g}$ an arbitrary second-order tensor, and $\mathbf{C}$ an arbitrary fourth-order tensor. These quantities can be written in terms of their components as

$$\vec{v} = v_s \, \vec{i}_s \,, \qquad \mathbf{g} = g_{tu} \, \vec{i}_t \, \vec{i}_u \,, \qquad \text{and} \qquad \mathbf{C} = C_{ijkl} \, \vec{i}_i \, \vec{i}_j \, \vec{i}_k \, \vec{i}_l \,, \tag{2.19}$$

where the dummy indices $s, t, u, i, \ldots$ can be replaced by other indices. The number of base vectors in the notations of Notations (2.19) indicates the order of a tensor. Thus a scalar is a zeroth-order tensor and a vector is a first-order tensor.

The component form of the second-order tensor $\mathbf{g}$ given by Expression $(2.19)_2$ is

$$\begin{aligned} \mathbf{g} &= g_{tu} \, \vec{i}_t \, \vec{i}_u \\ &= g_{11} \, \vec{i}_1 \, \vec{i}_1 + g_{12} \, \vec{i}_1 \, \vec{i}_2 + g_{13} \, \vec{i}_1 \, \vec{i}_3 + g_{21} \, \vec{i}_2 \, \vec{i}_1 + g_{22} \, \vec{i}_2 \, \vec{i}_2 + g_{23} \, \vec{i}_2 \, \vec{i}_3 \\ &\quad + g_{31} \, \vec{i}_3 \, \vec{i}_1 + g_{32} \, \vec{i}_3 \, \vec{i}_2 + g_{33} \, \vec{i}_3 \, \vec{i}_3 \,. \end{aligned} \tag{2.20}$$

Component Form (2.20) shows the effectiveness of the notations $\mathbf{g}$ and $g_{tu} \, \vec{i}_t \, \vec{i}_u$. They compress long expressions drastically. The second and the third lines of Expression (2.20) show that the second-order tensor $\mathbf{g}$ is expressed in terms of its components $g_{ij}$ in the three-dimensional space. In the case of a two-dimensional space, Expression (2.20) is replaced by

$$\mathbf{g} = g_{\alpha\beta} \, \vec{i}_\alpha \, \vec{i}_\beta = g_{11} \, \vec{i}_1 \, \vec{i}_1 + g_{12} \, \vec{i}_1 \, \vec{i}_2 + g_{21} \, \vec{i}_2 \, \vec{i}_1 + g_{22} \, \vec{i}_2 \, \vec{i}_2 \,. \tag{2.21}$$

The roles of the base vectors in a second-order tensor, such as a stress tensor $\boldsymbol{\sigma}$, were discussed earlier. A question may arise on the role of a fourth-order tensor in mechanics. Actually, the fourth-order tensor $\mathbf{C}$ selected in Expression (2.19) is to show that there are fourth-order tensors as well. The uniaxial form of Hooke's law takes the form $\sigma = E \, \varepsilon^{\mathrm{e}}$, where $E$ is Young's modulus and $\varepsilon^{\mathrm{e}}$ is the elastic strain, whereas when tensor notations are applied, the three-dimensional form of Hooke's law reads

$$\boldsymbol{\sigma} = \mathbf{C} : \boldsymbol{\varepsilon}^{\mathrm{e}} \,, \tag{2.22}$$

where $\mathbf{C}$ is the constitutive tensor for a Hookean deformation and $\boldsymbol{\varepsilon}^{\mathrm{e}}$ is the elastic strain tensor. The tensor $\mathbf{C}$ is a fourth-order tensor, and the strain tensor $\boldsymbol{\varepsilon}^{\mathrm{e}}$ is a second-order tensor. The notation : stands for the "double-dot product" defined later in this chapter.

Let $\mathbf{c}$ be an arbitrary second-order tensor and $\vec{b}$ an arbitrary vector. The Kronecker delta $\delta_{ij}$ has the following properties:

$$\delta_{ij}\, c_{si} = c_{sj}\,, \quad \delta_{ij}\, c_{is} = c_{js}\,, \quad \delta_{ij}\, b_j = b_i\,, \quad \text{and} \quad \delta_{ij}\, \delta_{is} = \delta_{sj} = \delta_{js}\,. \tag{2.23}$$

The idea in Equations (2.23) is that if the scalar component has the same index as the Kronecker delta $\delta_{ij}$, the index of the scalar component is replaced with the other index of the Kronecker delta $\delta_{ij}$. Results (2.23) are also valid for higher-order tensors and the Kronecker delta $\delta_{ij}$ itself. The mutual "location" of the indices in the Kronecker delta $\delta_{ij}$ does not play any role.

Relations (2.23) hold for more complicated cases, as the following examples show:

$$\delta_{st}\, a_{si}\, c_{ki} = a_{ti}\, c_{ki} \qquad \text{and} \qquad \delta_{st}\, a_{ni}\, c_{ks} = a_{ni}\, c_{kt}\,. \tag{2.24}$$

Using Definition (2.4) and the summation convention as in Equation (2.9), Results (2.23) can be proved by a simple way, as the following example shows.

**Example 2.2:** By using Definition (2.4), show that Result $(2.23)_2$ holds.

**Solution:** Based on Definition (2.4) and the Einstein summation convention, the following is achieved (the tensor $\mathbf{c}$ is replaced with the tensor $\mathbf{g}$):

$$\delta_{ij}\, g_{is} = \delta_{1j}\, g_{1s} + \delta_{2j}\, g_{2s} + \delta_{3j}\, g_{3s}\,. \tag{2.25}$$

In Equation (2.25) the free index $j$ takes the values 1, 2, and 3. The same holds for the free index $s$ as well, but it is not under consideration. Based on the different values for the free index $j$, the following three equations are obtained:

$$\begin{aligned} \delta_{ij}\, g_{is} &= \delta_{11}\, g_{1s} + \delta_{21}\, g_{2s} + \delta_{31}\, g_{3s} = g_{1s}\,, \quad && j = 1\,, \\ \delta_{ij}\, g_{is} &= \delta_{12}\, g_{1s} + \delta_{22}\, g_{2s} + \delta_{32}\, g_{3s} = g_{2s}\,, \quad && j = 2\,, \\ \delta_{ij}\, g_{is} &= \delta_{13}\, g_{1s} + \delta_{23}\, g_{2s} + \delta_{33}\, g_{3s} = g_{3s}\,, \quad && j = 3\,. \end{aligned} \tag{2.26}$$

The right side of Expression $(2.23)_2$ gives

$$\begin{aligned} &g_{js} \;\Rightarrow\; g_{1s}\,,\ \text{when } j = 1; \quad g_{js} \;\Rightarrow\; g_{2s}\,,\ \text{when } j = 2; \\ \text{and}\quad &g_{js} \;\Rightarrow\; g_{3s}\,,\ \text{when } j = 3\,. \end{aligned} \tag{2.27}$$

The right sides of Expressions (2.26) and (2.27) are equal for every $j = 1, 2,$ and 3. Therefore the left sides of Expressions (2.26) and (2.27) are equal for every $j = 1, 2,$ and 3. This gives

$$\delta_{ij}\, g_{is} = g_{js} \qquad j = 1, 2, 3\,. \tag{2.28}$$

Therefore Expression (2.28) reduces to

$$\delta_{ij}\, g_{is} = g_{js}\,. \tag{2.29}$$

Equation (2.29) is that aimed at in Expression $(2.23)_2$. ■

## 2.3 Dot Product and Double-Dot Product

This section introduces the dot product and the double-dot product and gives some expressions related to these two operations.

The dot product $\cdot$ of two tensors of any order (excluding scalars) is defined by

$$\textbf{Def} \qquad \circ\circ\circ\; \vec{i}_s \cdot (Scal\,Comp)\, \vec{i}_t \;\circ\circ\circ = \circ\circ\circ\, \vec{i}_s \cdot \vec{i}_t\, (Scal\,Comp) \;\circ\circ\circ$$
$$= \circ\circ\circ\, \delta_{st}\, (Scal\,Comp) \;\circ\circ\circ\,. \tag{2.30}$$

The notation $\circ\circ\circ$ represents any set of scalar components (*ScalComp*) and/or base vectors. In Expression (2.30) the (*ScalComp*) is first moved to the right, and then Property (2.6) is applied between the two adjacent base vectors. The order of the base vectors of a tensor cannot be changed. The order of a scalar and a base vector is interchangeable. The key point of the dot operator is that it is an operation between two adjacent base vectors between which the dot operator is located.

Based on Expressions (2.6) and $(2.23)_3$ and Definition (2.30), the dot product of the vector $\vec{v}$ with itself reads

$$\vec{v}\cdot\vec{v} = v_i\,\vec{i}_i \cdot v_s\vec{i}_s = v_i\,v_s\,\delta_{is} = v_s\,v_s = v_1\,v_1 + v_2\,v_2 + v_3\,v_3\,. \tag{2.31}$$

The following example further demonstrates the calculation of dot products $\cdot$.

**Example 2.3:** Let **A** and **B** be two arbitrary third-order tensors. Calculate the dot product $\mathbf{A}\cdot\mathbf{B}$.

**Solution:** Based on Definition (2.30) and Property (2.6), the dot product $\mathbf{A}\cdot\mathbf{B}$ takes the form

$$\mathbf{A}\cdot\mathbf{B} = A_{ijk}\,\vec{i}_i\,\vec{i}_j\,\underset{\uparrow}{\vec{i}_k} \cdot B_{stu}\,\underset{\uparrow}{\vec{i}_s}\,\vec{i}_t\,\vec{i}_u = A_{ijk}\,\vec{i}_i\,\vec{i}_j\,\delta_{ks}\,B_{stu}\,\vec{i}_t\,\vec{i}_u\,, \tag{2.32}$$

where the "double arrow" shows the operating base vectors. The double arrow does not belong to the standard notation, but it is added here for educational purposes. Manipulation used in Manipulation (2.32) utilized the definition in Definition (2.30). The indices in Dot Product (2.32) are according to the authors. There are two possible ways to continue the right side of Dot Product (2.32): the index $k$ in $A_{ijk}$ can be replaced with index $s$, or the index $s$ in $B_{stu}$ can be replaced by index $k$. Both ways give the same result since the dummy indices of the tensors do not have any fixed values and can be replaced by other indices. Of course, the indices on both sides of the equals sign have to be mutually compatible. The first way leads to

$$\begin{aligned}\mathbf{A}\cdot\mathbf{B} &= A_{ijk}\,\vec{i}_i\,\vec{i}_j\,\vec{i}_k\cdot B_{stu}\,\vec{i}_s\,\vec{i}_t\,\vec{i}_u = A_{ijk}\vec{i}_i\,\vec{i}_j\,\underset{\uparrow\quad\uparrow}{\delta_{ks}\,B_{stu}}\,\vec{i}_t\,\vec{i}_u\\ &= A_{ijs}\,B_{stu}\,\vec{i}_i\,\vec{i}_j\,\vec{i}_t\,\vec{i}_u\,,\end{aligned} \tag{2.33}$$

where the double arrow shows the operating indices. In Expression (2.33) the quantity $B_{stu}$ is moved to the front of the base vectors, leading to the "standard" look of a tensor. The order of the base vectors $\vec{i}_i\,\vec{i}_j\,\vec{i}_t\,\vec{i}_u$ cannot be changed. ■

**Example 2.4:** What are other ways than Expression (2.33) to continue from Expression (2.32)?

**Solution:** The quantity $\delta_{ks}$ can operate with the quantity $B_{stu}$ as follows:

$$\mathbf{A}\cdot\mathbf{B} = A_{ijk}\,\vec{i}_i\,\vec{i}_j\,\vec{i}_k\cdot B_{stu}\,\vec{i}_s\,\vec{i}_t\,\vec{i}_u = A_{ijk}\,\vec{i}_i\,\vec{i}_j\,\underset{\uparrow\;\uparrow}{\delta_{ks}\,B_{stu}}\,\vec{i}_t\,\vec{i}_u = A_{ijk}\,B_{ktu}\,\vec{i}_i\,\vec{i}_j\,\vec{i}_t\,\vec{i}_u \tag{2.34}$$

The right sides of Expressions (2.33) and (2.34) take equal values, since the indices $s$ and $k$ are dummy indices. ■

The expression $\vec{n}\cdot\boldsymbol{\sigma}^{\mathrm{c}} = \vec{t}$ is evaluated to demonstrate what is meant by the concept "every term of an equation has to have the same set of base vectors and those base vectors have to have the same order." The preceding expression takes the form

$$\vec{n}\cdot\boldsymbol{\sigma}^{\mathrm{c}} = \vec{t} \qquad \Leftrightarrow \qquad n_s\,\vec{i}_s\cdot\sigma^{\mathrm{c}}_{tn}\,\vec{i}_t\,\vec{i}_n = n_s\,\delta_{st}\,\sigma^{\mathrm{c}}_{tn}\,\vec{i}_n = n_s\,\sigma^{\mathrm{c}}_{sn}\,\vec{i}_n = \vec{t}\,. \tag{2.35}$$

In order to guarantee that the sides of Equation $(2.35)_2$ are mutually compatible, the vector $\vec{t}$ has to be written in terms of its components as $\vec{t} = t_n\,\vec{i}_n$. This ensures that the base vectors (in this case, it is $\vec{i}_n$ ) are the same and that the quantities on both sides of the expression are tensors of the same order. Equations (2.35) can now be written in the form

$$\vec{n}\cdot\boldsymbol{\sigma}^{\mathrm{c}} = \vec{t} \qquad \Leftrightarrow \qquad n_s\,\sigma^{\mathrm{c}}_{sn}\,\vec{i}_n = t_n\,\vec{i}_n\,. \tag{2.36}$$

There are two sums in Expression $(2.36)_2$: over $n$ and over $s$. Therefore, the following is achieved:

$$\begin{aligned}&(n_1\,\sigma^{\mathrm{c}}_{11}+n_2\,\sigma^{\mathrm{c}}_{21}+n_3\,\sigma^{\mathrm{c}}_{31})\,\vec{\imath}_1\\&+(n_1\,\sigma^{\mathrm{c}}_{12}+n_2\,\sigma^{\mathrm{c}}_{22}+n_3\,\sigma^{\mathrm{c}}_{32})\,\vec{\imath}_2\\&+(n_1\,\sigma^{\mathrm{c}}_{13}+n_2\,\sigma^{\mathrm{c}}_{23}+n_3\,\sigma^{\mathrm{c}}_{33})\,\vec{\imath}_3=t_1\,\vec{\imath}_1+t_2\,\vec{\imath}_2+t_3\,\vec{\imath}_3\,.\end{aligned}\tag{2.37}$$

Comparison of Equation $(2.36)_1$ with Expression (2.37) displays the power of tensorial notation. The information of Expression (2.37) is squeezed into tiny Expression $(2.36)_1$. In Expression (2.33), for example, there is a drastic compression due to tensorial notations.

Since the scalar components related to the base vectors $\vec{\imath}_1, \vec{\imath}_2$, and $\vec{\imath}_3$ take identical values on both sides of Expression (2.37), we arrive at the following:

$$\begin{aligned}n_1\,\sigma^{\mathrm{c}}_{11}+n_2\,\sigma^{\mathrm{c}}_{21}+n_3\,\sigma^{\mathrm{c}}_{31}&=t_1,\\n_1\,\sigma^{\mathrm{c}}_{12}+n_2\,\sigma^{\mathrm{c}}_{22}+n_3\,\sigma^{\mathrm{c}}_{32}&=t_2,\\n_1\,\sigma^{\mathrm{c}}_{13}+n_2\,\sigma^{\mathrm{c}}_{23}+n_3\,\sigma^{\mathrm{c}}_{33}&=t_3\,.\end{aligned}\tag{2.38}$$

The preceding evaluation can be extended for general second-order tensor equations as follows:

$$(Scal\,Comp①)\,\vec{\imath}_s\,\vec{\imath}_t=(Scal\,Comp②)\,\vec{\imath}_s\,\vec{\imath}_t+(Scal\,Comp③)\,\vec{\imath}_s\,\vec{\imath}_t\,.\tag{2.39}$$

As expressed, on both sides of the equals sign, there has to be an equal number of base vectors in the same order.

Sometimes, after calculation, the following form may be obtained:

$$n_s\,\sigma^{\mathrm{c}}_{sn}\,\vec{\imath}_n=t_k\,\vec{\imath}_k\,.\tag{2.40}$$

The summation indices of base vectors in Equation (2.40) are not equal. In order to fulfill the concept of Expression (2.39), the summation index $k$ is replaced with the summation index $n$, which leads to

$$n_s\,\sigma^{\mathrm{c}}_{sn}\,\vec{\imath}_n=t_n\,\vec{\imath}_n\,.\tag{2.41}$$

Equation (2.41) has the same form as Equation $(2.36)_2$ and can therefore be manipulated in the same way. The dummy index $n$ on the left side of Expression (2.40) can be replaced with $k$. This is another way to get the same base vector component on both sides of Expression (2.40).

Equation (2.41) can be manipulated as follows:

$$n_s\,\sigma^{\mathrm{c}}_{sn}\,\vec{\imath}_n=t_n\,\vec{\imath}_n\qquad\Rightarrow\qquad(n_s\,\sigma^{\mathrm{c}}_{sn}-t_n)\,\vec{\imath}_n=\vec{0}=0_n\,\vec{\imath}_n\,,\tag{2.42}$$

where the zero vector $\vec{0}$ is defined to have zero scalar components, as shown in Equation $(2.42)_2$. The scalar components on the left and right sides of Expression $(2.42)_2$ have to be equal. This is

$$(n_s\,\sigma^{\mathrm{c}}_{sn}-t_n)=0_n\qquad\Rightarrow\qquad n_s\,\sigma^{\mathrm{c}}_{sn}=t_n\,.\tag{2.43}$$

The preceding can be argued as follows: when the indices for the base vectors $\vec{i}_n$, $\vec{i}_s$, and so on in every term on both sides of an expression are the same and have the same order, the scalar components have to be equal. For Equations (2.39) and (2.41). this means the following:

$$Scal\,Comp① = Scal\,Comp② + Scal\,Comp③ \qquad \text{and} \qquad n_s\,\sigma^c_{sn} = t_n\,. \quad (2.44)$$

In Equation (2.39), there is one term on the left side of Equation (2.39) and two on the right. Equation $(2.44)_2$ gives Result (2.38).

If the indices of base vectors differ, an expression similar to Equation (2.37) has to be written in order to get Scalar Expressions (2.38). Equation $(2.43)_2$ is just a shorter way to obtain Equation (2.38).

Sometimes it is important to show the order of the dot products by parentheses, whereas sometimes parentheses are not needed. Let the quantities $\vec{a}$, $\vec{c}$, and $\vec{e}$ be arbitrary vectors and $\mathbf{r}$ be an arbitrary second-order tensor. The following two quantities and one relation are given:

$$\theta_1 := \vec{c}\cdot\vec{a}\,, \qquad \text{and} \qquad \theta_2 := \vec{a}\cdot\vec{c}\,, \quad \text{where} \quad \vec{a} = \mathbf{r}\cdot\vec{e}\,. \qquad (2.45)$$

Substitution of the vector $\vec{a}$ from Equation $(2.45)_3$ into Equations $(2.45)_1$ and $(2.45)_2$ gives

$$\theta_1 = \vec{c}\cdot\mathbf{r}\cdot\vec{e} \qquad \text{and} \qquad \theta_2 = (\mathbf{r}\cdot\vec{e})\cdot\,\vec{c}\,. \qquad (2.46)$$

Equation $(2.46)_1$ does not require parentheses to show the order of the dot product operations, since the vector $\vec{c}$ operates with the first base vector of the second-order tensor $\mathbf{r}$ and the vector $\vec{e}$ operates with the second base vector of the second-order tensor $\mathbf{r}$. Quantity $\theta_2$ requires a more careful evaluation. Due to the parentheses, the dot product $\mathbf{r}\cdot\vec{e}$ has to be calculated first. That is, the vector $\vec{e}$ operates with the second base vector of the second-order tensor $\mathbf{r}$. After that the vector $\vec{c}$ operates with the first base vector of the second-order tensor $\mathbf{r}$. This makes it seem as though the quantities $\theta_1$ and $\theta_2$ are equal, as they are. Furthermore, the notation $\theta_3 := \mathbf{r}\cdot(\vec{e}\cdot\vec{c})$ is meaningless. These two statements are proved in the following example.

**Example 2.5:** Study the values of the quantities $\theta_1$, $\theta_2$, and $\theta_3$ by applying the component forms of the vectors $\vec{a}$, $\vec{c}$, $\vec{e}$ and the second-order tensor $\mathbf{r}$.

**Solution:** Equation $(2.46)_1$ takes the appearance

$$\theta_1 = \vec{c}\cdot\mathbf{r}\cdot\vec{e} = \underset{\uparrow}{c_m\,\vec{i}_m}\cdot \underset{\uparrow\;\Uparrow}{r_{st}\,\vec{i}_s\,\vec{i}_t}\cdot \underset{\Uparrow}{e_v\,\vec{i}_v} = c_m\,\delta_{ms}\,r_{st}\,\delta_{tv}\,e_v = c_s\,r_{st}\,e_t\,. \qquad (2.47)$$

In Expression (2.47) the the arrows and the double arrows show the operating pairs of the base vectors. Expression (2.47) shows that there is no doubt on the operating couples of the base vectors. The right side of Expression (2.47) shows that the quantity $\theta_1$ is a scalar.

Equation $(2.46)_2$ takes the appearance

$$\theta_2 = (\mathbf{r}\cdot\vec{e})\cdot\vec{c} = (r_{st}\,\vec{i}_s\,\vec{i}_t\cdot e_v\,\vec{i}_v)\cdot c_m\,\vec{i}_m. \tag{2.48}$$

The dot product between the parentheses has to be performed first. This yields

$$\begin{aligned}\theta_2 = (\mathbf{r}\cdot\vec{e})\cdot\vec{c} &= (r_{st}\,\vec{i}_s\,\underset{\uparrow}{\vec{i}_t}\cdot e_v\,\underset{\uparrow}{\vec{i}_v})\cdot c_m\,\vec{i}_m = (r_{st}\,\vec{i}_s\,\delta_{tv}\,e_v)\cdot c_m\,\vec{i}_m \\ &= (r_{st}\,\vec{i}_s\,e_t)\cdot c_m\,\vec{i}_m\,.\end{aligned} \tag{2.49}$$

Now the calculation can be continued as follows:

$$\theta_2 = (\mathbf{r}\cdot\vec{e})\cdot\vec{c} = (r_{st}\,\underset{\uparrow}{\vec{i}_s}\,e_t)\cdot c_m\,\underset{\uparrow}{\vec{i}_m} = r_{st}\,\delta_{sm}\,e_t\,c_m = r_{st}\,e_t\,c_s\,. \tag{2.50}$$

Comparison of the right sides of Expressions (2.47) and (2.50) shows that they are equal, and therefore the left sides are equal, $\theta_1 = \theta_2$. This can be seen, for example, by moving in Expression (2.50) the quantity $c_s$ in front of the quantity $r_{st}$. This can be done since the right sides of Expressions (2.47) and (2.50) are multiplications of three scalars, as the example $3 \times 7 \times 8 = 8 \times 3 \times 7$ shows.

The order of the scalar components $r_{st}$, $e_t$, and $c_s$ in the expression $r_{st}\,e_t\,c_s$ of Equation (2.50) can be interchanged without a change in its meaning or value. This is an acceptable act, since the expression $r_{st}\,e_t\,c_s$ is a scalar. In other words, we have

$$c_s\,r_{st}\,e_t = e_t\,r_{st}\,c_s = e_t\,c_s\,r_{st} = r_{st}\,e_t\,c_s.$$

Thus, when $\mathbf{r}$ is a second-order tensor and $\vec{c}$ and $\vec{e}$ are vectors, we have

$$\vec{c}\cdot\mathbf{r}\cdot\vec{e} = (\vec{c}\cdot\mathbf{r})\cdot\vec{e} = \vec{e}\cdot(\vec{c}\cdot\mathbf{r}) = \vec{c}\cdot(\mathbf{r}\cdot\vec{e}) = (\mathbf{r}\cdot\vec{e})\cdot\vec{c}.$$

The fact that $\theta_3 := \mathbf{r}\cdot(\vec{e}\cdot\vec{c})$ is meaningless can be shown as follows:

$$\theta_3 := \mathbf{r}\cdot(\vec{e}\cdot\vec{c}) = \mathbf{r}\cdot(e_v\,\vec{i}_v\cdot c_m\,\vec{i}_m) = \mathbf{r}\cdot(e_v\,\delta_{vm}\,c_m) = \mathbf{r}\cdot(e_v\,c_v)\,. \tag{2.51}$$

According to Expression (2.51), there is no base vector in the term between the parentheses (i.e., $e_v\,c_v$) to operate with the second-order tensor $\mathbf{r}$. Thus the quantity $\theta_3$ does not exist, since its definition is meaningless. ■

The double-dot product : of two tensors of any order ($\geq 2$) is defined by

$$\textbf{Def} \quad \circ\circ\circ\, \vec{i}_m \vec{i}_n : (Scal\,Comp)\, \vec{i}_s \vec{i}_t \,\circ\circ\circ$$
$$= \circ\circ\circ\, \vec{i}_m\, \vec{i}_n : \vec{i}_s\, \vec{i}_t\, (Scal\,Comp)\, \circ\circ\circ$$
$$= \circ\circ\circ\, (\vec{i}_m \cdot \vec{i}_s)\, (\vec{i}_n \cdot \vec{i}_t)\, (Scal\,Comp)\, \circ\circ\circ$$
$$= \circ\circ\circ\, \delta_{ms}\, \delta_{nt}\, (Scal\,Comp)\, \circ\circ\circ\,. \tag{2.52}$$

The notation $\circ\circ\circ$ represents any set of scalar components (*ScalComp*) and/or base vectors. In Equation (2.52) the scalar component (*ScalComp*) is first moved to the right, and then the double-dot product operator : is allowed to operate. In the double-dot product, the two pairs of base vectors operate so that there is a dot product between the first base vectors of the pairs and another dot product between the second base vectors of the pairs. This can be seen in the second line of Equation (2.52).

The double-dot product $\mathbf{A}:\mathbf{B}$ of two third-order tensors takes the form

$$\mathbf{A}:\mathbf{B} = A_{ijk}\, \vec{i}_i\, \underset{\uparrow}{\overset{\downarrow}{\vec{i}_j}}\, \vec{i}_k : B_{stu}\, \vec{i}_s\, \underset{\uparrow}{\overset{\downarrow}{\vec{i}_t}}\, \vec{i}_u = A_{ijk}\, \vec{i}_i\, \delta_{js}\, \delta_{kt}\, B_{stu}\, \vec{i}_u\,, \tag{2.53}$$

where the double arrows show the operating base vectors. Manipulation (2.53) utilized Definition (2.52). As shown in Equation (2.53), there are different possibilities to operate with the Kronecker deltas $\delta_{js}$ and $\delta_{kt}$. One can write, for example,

$$\mathbf{A}:\mathbf{B} = A_{ijk}\, \vec{i}_i\, \vec{i}_j\, \vec{i}_k : B_{stu}\, \vec{i}_s\, \vec{i}_t\, \vec{i}_u = A_{ijk}\, \vec{i}_i\, \delta_{js}\, \delta_{kt}\, B_{stu}\, \vec{i}_u$$
$$= A_{isk}\, B_{sku}\, \vec{i}_i\, \vec{i}_u\,. \tag{2.54}$$

**Example 2.6:** When expressed by the Gibbs notation, the strain-energy density $w(\boldsymbol{\varepsilon}^{\mathrm{e}})$ for a Hookean deformation takes the following appearance:

$$w(\boldsymbol{\varepsilon}^{\mathrm{e}}) := \frac{1}{2}\, \boldsymbol{\varepsilon}^{\mathrm{e}} : \mathbf{C} : \boldsymbol{\varepsilon}^{\mathrm{e}}\,, \tag{2.55}$$

where the elastic strain tensor $\boldsymbol{\varepsilon}^{\mathrm{e}}$ is a second-order tensor and the constitutive tensor for a Hookean deformation $\mathbf{C}$ is a fourth-order tensor.

(a) Write the strain-energy density $w(\boldsymbol{\varepsilon}^{\mathrm{e}})$ in the index form for a material response showing Hooke's law.

(b) Substitute Hooke's law,

$$\boldsymbol{\sigma} = \mathbf{C} : \boldsymbol{\varepsilon}^{\mathrm{e}}\,, \tag{2.56}$$

into Equation (2.55) and write the obtained expression in the index form. The stress tensor $\boldsymbol{\sigma}$ is a second-order tensor.

**Solution:** The following forms for the quantities present in Definition (2.55) are written:

$$\boldsymbol{\varepsilon}^{\mathrm{e}} = \varepsilon^{\mathrm{e}}_{km}\,\vec{i}_k\,\vec{i}_m\,, \qquad \mathbf{C} = C_{stuw}\,\vec{i}_s\,\vec{i}_t\,\vec{i}_u\,\vec{i}_w\,, \qquad \text{and} \qquad \boldsymbol{\varepsilon}^{\mathrm{e}} = \varepsilon^{\mathrm{e}}_{pq}\,\vec{i}_p\,\vec{i}_q\,. \tag{2.57}$$

The reason to have different indices in the two expressions for $\boldsymbol{\varepsilon}^{\mathrm{e}}$ stems from the fact that only two of the same indices are allowed in one term in an expression.

(a) Substitution of the quantities given in Expression (2.57) into Definition (2.55) yields

$$\begin{aligned} w(\boldsymbol{\varepsilon}^{\mathrm{e}}) :&= \tfrac{1}{2}\,\boldsymbol{\varepsilon}^{\mathrm{e}} : \mathbf{C} : \boldsymbol{\varepsilon}^{\mathrm{e}} = \tfrac{1}{2}\,\varepsilon^{\mathrm{e}}_{km}\,\vec{i}_k\,\vec{i}_m : C_{stuw}\,\vec{i}_s\,\vec{i}_t\,\vec{i}_u\,\vec{i}_w : \varepsilon^{\mathrm{e}}_{pq}\,\vec{i}_p\,\vec{i}_q \\ &= \tfrac{1}{2}\,\varepsilon^{\mathrm{e}}_{km}\,\delta_{ks}\,\delta_{mt}\,C_{stuw}\,\delta_{up}\,\delta_{wq}\,\varepsilon^{\mathrm{e}}_{pq} = \tfrac{1}{2}\,\varepsilon^{\mathrm{e}}_{st}\,C_{stpq}\,\varepsilon^{\mathrm{e}}_{pq}\,. \end{aligned} \tag{2.58}$$

The aim of this problem was to show the appearance of the strain-energy density $w(\boldsymbol{\varepsilon}^{\mathrm{e}})$ when index notation is used, since many books and papers utilize index notation. However, note that the strain-energy expression is invariant (i.e., all of the indices are dummies).

(b) If Hooke's law $\boldsymbol{\sigma} = \mathbf{C} : \boldsymbol{\varepsilon}^{\mathrm{e}}$, Model (2.56), is substituted into Definition (2.55), the following result is obtained:

$$\hat{w} := \tfrac{1}{2}\,\boldsymbol{\varepsilon}^{\mathrm{e}} : \boldsymbol{\sigma} = \tfrac{1}{2}\,\varepsilon^{\mathrm{e}}_{km}\,\vec{i}_k\,\vec{i}_m : \sigma_{pq}\,\vec{i}_p\,\vec{i}_q = \tfrac{1}{2}\,\varepsilon^{\mathrm{e}}_{km}\,\delta_{kp}\,\delta_{mq}\,\sigma_{pq} = \tfrac{1}{2}\,\varepsilon^{\mathrm{e}}_{km}\,\sigma_{km}\,. \tag{2.59}$$

The notation $w(\boldsymbol{\varepsilon}^{\mathrm{e}})$ shows that the strain-energy density $w(\boldsymbol{\varepsilon}^{\mathrm{e}})$ depends on the elastic strain tensor $\boldsymbol{\varepsilon}^{\mathrm{e}}$ but not on the stress tensor $\boldsymbol{\sigma}$. Thus the variable $\hat{w}$ is not the strain-energy density $w(\boldsymbol{\varepsilon}^{\mathrm{e}})$, although $w(\boldsymbol{\varepsilon}^{\mathrm{e}})$ and $\hat{w}$ take the same value. Unfortunately, some authors express that the Form (2.59) stands for the strain-energy density. Figure 2.2 shows the uniaxial linear relation between the stress $\sigma$ and elastic strain $\varepsilon^{\mathrm{e}}$. The shadowed area shows the amount of the strain-energy density $w(\varepsilon^{\mathrm{e}})$ when the uniaxial stress $\sigma$ is applied.

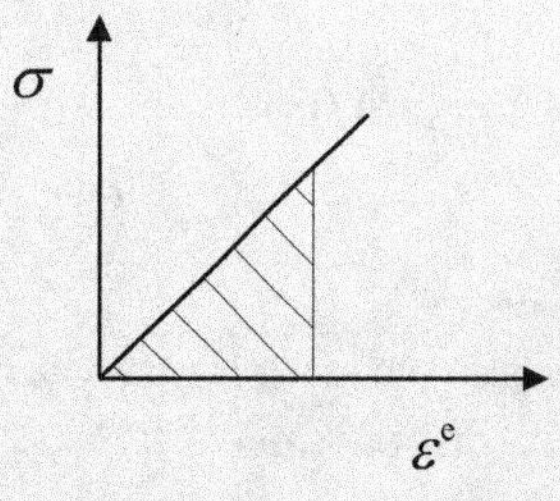


**Figure 2.2** Response of an elastic material obeying Hooke's law.

Manipulation carried out in Expression (2.59) can be repeated for the term $\frac{1}{2}\,\boldsymbol{\sigma} : \boldsymbol{\varepsilon}^{\mathrm{e}}$. This yields

$$\tfrac{1}{2}\,\boldsymbol{\sigma} : \boldsymbol{\varepsilon}^{\mathrm{e}} = \tfrac{1}{2}\,\sigma_{km}\,\varepsilon^{\mathrm{e}}_{km} \qquad \Rightarrow \qquad \tfrac{1}{2}\,\boldsymbol{\sigma} : \boldsymbol{\varepsilon}^{\mathrm{e}} = \tfrac{1}{2}\,\sigma_{km}\,\varepsilon^{\mathrm{e}}_{km} = \tfrac{1}{2}\,\varepsilon^{\mathrm{e}}_{km}\,\sigma_{km}\,. \tag{2.60}$$

The latter equality in Expressions $(2.60)_2$ utilizes the fact that the order of the scalar components $\varepsilon^{\mathrm{e}}_{km}$ and $\sigma_{km}$ of the tensors $\boldsymbol{\varepsilon}^{\mathrm{e}}$ and $\boldsymbol{\sigma}$ is possible to interchange, since $7 \times 8 = 8 \times 7$.

Based on Expressions (2.59) and (2.60), the following holds:

$$\hat{w} := \tfrac{1}{2}\,\boldsymbol{\varepsilon}^{\mathrm{e}} : \boldsymbol{\sigma} = \tfrac{1}{2}\,\boldsymbol{\sigma} : \boldsymbol{\varepsilon}^{\mathrm{e}}\,. \quad \blacksquare \tag{2.61}$$

Example 2.6 leads to the introduction of two expressions. First is the form for the complementary strain-energy density for Hookean deformation, denoted by $w^{c}(\boldsymbol{\sigma})$. Second, the double-dot product of two second-order tensors is commutative, as shown in Equation (2.61), and is given for arbitrary second-order tensors $\mathbf{c}$ and $\mathbf{r}$. Thus,

$$w^{c}(\boldsymbol{\sigma}) := \tfrac{1}{2}\,\boldsymbol{\sigma} : \mathbf{S} : \boldsymbol{\sigma} \qquad \text{and} \qquad \mathbf{c} : \mathbf{r} = \mathbf{r} : \mathbf{c}\,, \tag{2.62}$$

where $\mathbf{S}$ is the compliance tensor for a Hookean deformation. As Definition $(2.62)_1$ shows, complementary strain-energy density $w^{c}(\boldsymbol{\sigma})$ depends only on the stress tensor $\boldsymbol{\sigma}$.

## 2.4 Special Tensors and Their Relations

In this section we introduce several special tensors, including the second- and fourth-order identity tensors, and give relations that are needed in the forthcoming sections of this book.

Tensor $\mathbf{1}$ is called the second-order identity tensor $\mathbf{1}$. It is defined as follows: the dot product between the second-order identity tensor $\mathbf{1}$ and any vector $\vec{u}$ yields that vector itself. That is,

**Def**
$$\mathbf{1}\cdot\vec{u} = \vec{u}\cdot\mathbf{1} = \vec{u}\,. \tag{2.63}$$

The second-order identity tensor $\mathbf{1}$ has the following appearance when represented in terms of its components:

$$\mathbf{1} = \delta_{ij}\,\vec{\imath}_i\,\vec{\imath}_j. \tag{2.64}$$

For the second-order identity tensor $\mathbf{1}$ the following is achieved:

$$\mathbf{1} : \mathbf{c} = \mathbf{c} : \mathbf{1} = c_{ii} = c_{11} + c_{22} + c_{33} \qquad \text{and} \qquad \mathbf{1} : \mathbf{1} = \delta_{ii} = 3\,, \tag{2.65}$$

and finally,

$$\mathbf{1}\cdot\mathbf{c} = \mathbf{c}\cdot\mathbf{1} = \mathbf{c}\,. \tag{2.66}$$

**Example 2.7:** Show that the first of Relations (2.66) holds.

**Solution:** The following index forms are written:

$$\mathbf{1} = \delta_{ij}\,\vec{\imath}_i\,\vec{\imath}_j \qquad \text{and} \qquad \mathbf{c} = c_{st}\,\vec{\imath}_s\,\vec{\imath}_t\,. \tag{2.67}$$

Substitution of Expressions (2.67) into the first of Relations (2.66) yields

$$\mathbf{1}\cdot\mathbf{c} = \delta_{ij}\,\vec{\imath}_i\,\vec{\imath}_j\cdot c_{st}\,\vec{\imath}_s\,\vec{\imath}_t = \delta_{ij}\,\vec{\imath}_i\,\delta_{js}\,c_{st}\,\vec{\imath}_t = \delta_{is}\,c_{st}\,\vec{\imath}_i\,\vec{\imath}_t = c_{it}\,\vec{\imath}_i\,\vec{\imath}_t = \mathbf{c}\,. \quad \blacksquare \tag{2.68}$$

The tensor $\mathbf{I}$ is called the fourth-order identity tensor $\mathbf{I}$. It is defined as follows: the double-dot product between the fourth-order identity tensor $\mathbf{I}$ and any second-order tensor $\mathbf{c}$ yields that second-order tensor itself. This is

**Def** $$\mathbf{I}:\mathbf{c} = \mathbf{c}:\mathbf{I} = \mathbf{c}\,. \tag{2.69}$$

The fourth-order identity tensor $\mathbf{I}$ is given in terms of its components as follows:

$$\mathbf{I} = \delta_{ik}\,\delta_{jl}\,\vec{i}_i\,\vec{i}_j\,\vec{i}_k\,\vec{i}_l \quad \text{or} \quad \mathbf{I} = I_{ijkl}\,\vec{i}_i\,\vec{i}_j\,\vec{i}_k\,\vec{i}_l\,, \quad \text{where} \quad I_{ijkl} = \delta_{ik}\,\delta_{jl}\,. \tag{2.70}$$

It is important to notice the order of the indices on the right side of Expression $(2.70)_1$.

Let $\mathbf{A}$ be an arbitrary fourth-order tensor. The following equalities hold:

$$\mathbf{I}:\mathbf{I} = \mathbf{I} \quad \text{and} \quad \mathbf{I}:\mathbf{A} = \mathbf{A}:\mathbf{I} = \mathbf{A}\,. \tag{2.71}$$

The inverse $\mathbf{c}^{-1}$ of the second-order tensor $\mathbf{c}$ and the inverse $\mathbf{A}^{-1}$ of the fourth-order tensor $\mathbf{A}$ are defined by (if the inverses exist)

**Def** $$\mathbf{c}^{-1}\cdot\mathbf{c} = \mathbf{c}\cdot\mathbf{c}^{-1} = \mathbf{1} \quad \text{and} \quad \mathbf{A}^{-1}:\mathbf{A} = \mathbf{A}:\mathbf{A}^{-1} = \mathbf{I}\,. \tag{2.72}$$

The tensor transpose of an arbitrary second-order tensor $\mathbf{c}$ is denoted by $\mathbf{c}^{\mathrm{T}}$ and defined by

**Def** $$\vec{a}\cdot\mathbf{c}^{\mathrm{T}}\cdot\vec{b} = \vec{b}\cdot\mathbf{c}\cdot\vec{a}\,, \tag{2.73}$$

where $\vec{a}$ and $\vec{b}$ are arbitrary vectors. If the second-order tensor $\mathbf{c}$ is expressed as

$$\mathbf{c} = c_{in}\,\vec{i}_i\,\vec{i}_n\,, \qquad \text{then} \qquad \mathbf{c}^{\mathrm{T}} := c_{ni}\,\vec{i}_i\,\vec{i}_n\,. \tag{2.74}$$

Note that in $\mathbf{c}^{\mathrm{T}}$ the indices $i$ and $n$ change their positions only in the indices in $c_{ni}$, whereas the order of the base vectors $\vec{i}_i$ and $\vec{i}_n$ remains the same. If also the order of the base vectors $\vec{i}_i$ and $\vec{i}_n$ is changed, no transpose is taken, and therefore the tensor equals the original one.

Transpose of the quantity $a_{is}\,b_{st}\,\vec{i}_i\,\vec{i}_t$ reads

$$(a_{is}\,b_{st}\,\vec{i}_i\,\vec{i}_t)^{\mathrm{T}} = a_{ts}\,b_{si}\,\vec{i}_i\,\vec{i}_t\,. \tag{2.75}$$

A second-order tensor $\mathbf{c}$ is symmetric if

$$\mathbf{c}^{\mathrm{T}} = \mathbf{c}\,, \qquad \text{that is,} \qquad c_{in} = c_{ni}\,. \tag{2.76}$$

A second-order tensor $\mathbf{c}$ is anti-symmetric or skew-symmetric if

$$\mathbf{c}^{\mathrm{T}} = -\,\mathbf{c}\,, \qquad \text{that is,} \qquad c_{in} = -\,c_{ni}\,. \tag{2.77}$$

Since the transpose of a vector is the vector itself, the following holds:

$$\text{if} \quad \vec{a} = \mathbf{c}\cdot\vec{b}\,, \quad \text{then} \quad \vec{a}^{\mathrm{T}} = (\mathbf{c}\cdot\vec{b})^{\mathrm{T}} = \vec{b}^{\mathrm{T}}\cdot\mathbf{c}^{\mathrm{T}} \quad \Rightarrow \quad \vec{a} = \vec{b}\cdot\mathbf{c}^{\mathrm{T}}\,, \tag{2.78}$$

where $\vec{a}$ and $\vec{b}$ are arbitrary vectors and $\mathbf{c}$ is an arbitrary second-order tensor.

**Example 2.8:** Show that Expression (2.78) holds.

**Solution:** Let the second-order tensor $\mathbf{c}$ and the vector $\vec{b}$ take the following index forms:

$$\mathbf{c} := c_{st}\,\vec{i}_s\,\vec{i}_t \qquad \text{and} \qquad \vec{b} := b_n\,\vec{i}_n\,. \tag{2.79}$$

Expressions (2.74) and (2.79) give

$$\mathbf{c}\cdot\vec{b} = c_{st}\,\vec{i}_s\,\vec{i}_t\cdot b_n\,\vec{i}_n = c_{st}\,b_t\,\vec{i}_s \tag{2.80}$$

and

$$\vec{b}\cdot\mathbf{c}^{\mathrm{T}} = b_n\,\vec{i}_n\cdot c_{ts}\,\vec{i}_s\,\vec{i}_t = b_s\,c_{ts}\,\vec{i}_t\,. \tag{2.81}$$

When the indices $s$ and $t$ are interchanged in Expression (2.81), the right sides of Equations (2.80) and (2.81) are equal, and therefore the left sides are equal as well. Thus Expressions (2.78) hold. ■

**Example 2.9:** Let $\boldsymbol{\sigma}$ and $\boldsymbol{\varepsilon}$ be symmetric second-order tensors and $\boldsymbol{\omega}$ be a skew-symmetric second-order tensor. Calculate the double-dot products $\boldsymbol{\sigma}:\boldsymbol{\varepsilon}$ and $\boldsymbol{\sigma}:\boldsymbol{\omega}$. In what order are the obtained products? Do the double-dot products have any special features?

**Solution:** The component forms of the tensors $\boldsymbol{\sigma}$ and $\boldsymbol{\varepsilon}$ read

$$\boldsymbol{\sigma} := \sigma_{st}\,\vec{i}_s\,\vec{i}_t \qquad \text{and} \qquad \boldsymbol{\varepsilon} := \varepsilon_{ru}\,\vec{i}_r\,\vec{i}_u\,. \tag{2.82}$$

Notations (2.82) give

$$\boldsymbol{\sigma}:\boldsymbol{\varepsilon} = \sigma_{st}\,\vec{i}_s\,\vec{i}_t : \varepsilon_{ru}\,\vec{i}_r\,\vec{i}_u = \sigma_{st}\,\delta_{sr}\,\delta_{tu}\,\varepsilon_{ru} = \sigma_{st}\,\varepsilon_{st}\,. \tag{2.83}$$

Expression (2.83) shows that the product $\boldsymbol{\sigma}:\boldsymbol{\varepsilon}$ is a scalar, a zeroth-order tensor, since the number of the base vectors indicates the order of the tensor. Also, the product $\boldsymbol{\sigma}:\boldsymbol{\omega}$ is a zeroth-order tensor (i.e., a scalar) as well.

The double-dot product of two symmetric second-order tensors does not need to have any general properties (beyond that it is a scalar). The double-dot product of a symmetric second-order tensor and a skew-symmetric second-order tensor is worth evaluating.

Expressions $(2.76)_2$ and $(2.77)_2$ give

$$\sigma_{in} = \sigma_{ni} \qquad \text{and} \qquad \omega_{in} = -\omega_{ni}\,. \tag{2.84}$$

The formulation of the product $\boldsymbol{\sigma}:\boldsymbol{\omega}$ follows the same steps taken when the product $\boldsymbol{\sigma}:\boldsymbol{\varepsilon}$ was written. Based on Expression (2.83), the following reads:

$$\begin{aligned}\boldsymbol{\sigma}:\boldsymbol{\omega} = \sigma_{st}\,\omega_{st} &= \sigma_{11}\,\omega_{11} + \sigma_{12}\,\omega_{12} + \sigma_{13}\,\omega_{13}\\ &+ \sigma_{21}\,\omega_{21} + \sigma_{22}\,\omega_{22} + \sigma_{23}\,\omega_{23}\\ &+ \sigma_{31}\,\omega_{31} + \sigma_{32}\,\omega_{32} + \sigma_{33}\,\omega_{33}\,.\end{aligned} \tag{2.85}$$

By taking into consideration Property $(2.84)_1$, the terms of Expression (2.85) can be rearranged as follows:

$$\boldsymbol{\sigma} : \boldsymbol{\omega} = \sigma_{st}\,\omega_{st} = \sigma_{11}\,\omega_{11} + \sigma_{22}\,\omega_{22} + \sigma_{33}\,\omega_{33} + \sigma_{12}\,(\omega_{12} + \omega_{21}) + \sigma_{13}\,(\omega_{13} + \omega_{31}) + \sigma_{23}\,(\omega_{23} + \omega_{32})\,. \tag{2.86}$$

Property $(2.84)_2$ implies that the terms between parentheses in Equation (2.86) vanish. Therefore Equation (2.86) reduces to

$$\boldsymbol{\sigma} : \boldsymbol{\omega} = \sigma_{11}\,\omega_{11} + \sigma_{22}\,\omega_{22} + \sigma_{33}\,\omega_{33}\,. \tag{2.87}$$

Property $(2.84)_2$ gives

$$\omega_{11} = -\omega_{11} \equiv 0\,, \qquad \omega_{22} = -\omega_{22} \equiv 0\,, \qquad \text{and} \qquad \omega_{33} = -\omega_{33} \equiv 0\,. \tag{2.88}$$

Based on Equations (2.88), the right side of Equation (2.87) vanishes. This leads to

$$\boldsymbol{\sigma} : \boldsymbol{\omega} = 0\,. \tag{2.89}$$

Equation (2.89) means that the double-dot product between a symmetric second-order tensor and a skew-symmetric second-order tensor vanishes. ■

If

**Def** $$\vec{a}\,\vec{b} : \mathbf{A} = \vec{b}\,\vec{a} : \mathbf{A} \qquad \text{and/or} \qquad \mathbf{B} : \vec{a}\,\vec{b} = \mathbf{B} : \vec{b}\,\vec{a} \tag{2.90}$$

hold for arbitrary vectors $\vec{a}$ and $\vec{b}$, then the fourth-order tensor $\mathbf{A}$ is said to be minor symmetric in the first pair of indices and the fourth-order tensor $\mathbf{B}$ minor symmetric in the second pair of indices. These mean that

$$A_{klst} = A_{lkst} \qquad \text{and/or} \qquad B_{klst} = B_{klts}\,. \tag{2.91}$$

If

**Def** $$\mathbf{c} : \mathbf{A} : \mathbf{g} = \mathbf{g} : \mathbf{A} : \mathbf{c} \tag{2.92}$$

holds for arbitrary second-order tensors $\mathbf{c}$ and $\mathbf{g}$, then the fourth-order tensor $\mathbf{A}$ is said to be major symmetric. This means that

$$A_{klst} = A_{stkl}. \tag{2.93}$$

The tensor transpose of an arbitrary fourth-order tensor $\mathbf{A}$ is denoted by $\mathbf{A}^{\mathrm{T}}$ and is defined by

**Def** $$\mathbf{c} : \mathbf{A}^{\mathrm{T}} : \mathbf{g} = \mathbf{g} : \mathbf{A} : \mathbf{c}\,, \tag{2.94}$$

where **c** and **g** are arbitrary second-order tensors. If the fourth-order tensor **A** is expressed as

$$\mathbf{A} = A_{inkl}\,\vec{i}_i\,\vec{i}_n\,\vec{i}_k\,\vec{i}_l\,, \qquad \text{then} \qquad \mathbf{A}^{\mathrm{T}} = A_{klin}\,\vec{i}_i\,\vec{i}_n\,\vec{i}_k\,\vec{i}_l\,. \tag{2.95}$$

A fourth-order major-symmetric tensor **A** has a property

$$\mathbf{A} = \mathbf{A}^{\mathrm{T}}\,. \tag{2.96}$$

If **A** and **B** are arbitrary fourth-order tensors, the following holds:

$$(\mathbf{A}:\mathbf{B})^{\mathrm{T}} = \mathbf{B}^{\mathrm{T}}:\mathbf{A}^{\mathrm{T}}. \tag{2.97}$$

**Example 2.10:** Show that Property (2.97) holds.

**Solution:** Let the tensors **A** and **B** have the following index forms:

$$\mathbf{A} = A_{opqr}\,\vec{i}_o\,\vec{i}_p\,\vec{i}_q\,\vec{i}_r \qquad \text{and} \qquad \mathbf{B} = B_{stuw}\,\vec{i}_s\,\vec{i}_t\,\vec{i}_u\,\vec{i}_w\,. \tag{2.98}$$

Equation (2.98) gives

$$\mathbf{A}:\mathbf{B} = A_{opqr}\,\vec{i}_o\,\vec{i}_p\,\vec{i}_q\,\vec{i}_r : B_{stuw}\,\vec{i}_s\,\vec{i}_t\,\vec{i}_u\,\vec{i}_w = A_{opst}\,B_{stuw}\,\vec{i}_o\,\vec{i}_p\,\vec{i}_u\,\vec{i}_w\,. \tag{2.99}$$

Based on Definition (2.95), Equation (2.99) gives

$$(\mathbf{A}:\mathbf{B})^{\mathrm{T}} = A_{uwst}\,B_{stop}\,\vec{i}_o\,\vec{i}_p\,\vec{i}_u\,\vec{i}_w = B_{stop}\,A_{uwst}\,\vec{i}_o\,\vec{i}_p\,\vec{i}_u\,\vec{i}_w\,. \tag{2.100}$$

Equations (2.98) and (2.95) give

$$\begin{aligned}\mathbf{B}^{\mathrm{T}}:\mathbf{A}^{\mathrm{T}} &= B_{uwst}\,\vec{i}_s\,\vec{i}_t\,\vec{i}_u\,\vec{i}_w : A_{qrop}\,\vec{i}_o\,\vec{i}_p\,\vec{i}_q\,\vec{i}_r = B_{uwst}\,\vec{i}_s\,\vec{i}_t\,A_{qrop}\,\delta_{uo}\,\delta_{wp}\,\vec{i}_q\,\vec{i}_r \\ &= B_{uwst}\,A_{qruw}\,\vec{i}_s\,\vec{i}_t\,\vec{i}_q\,\vec{i}_r\,. \end{aligned} \tag{2.101}$$

The base vectors on the right sides of Expressions (2.100) and (2.101) are not equal. In order to compare the right sides of these expressions, they should be the same and have the same order. Thus the first step is to make the following index changes to Expression (2.101):

$$s \to o\,, \qquad t \to p\,, \qquad u \to m\,, \qquad \text{and} \qquad w \to k\,. \tag{2.102}$$

Based on Changes (2.102), Expression (2.101) takes the following change of appearance:

$$\mathbf{B}^{\mathrm{T}}:\mathbf{A}^{\mathrm{T}} = B_{uwst}\,A_{qruw}\,\vec{i}_s\,\vec{i}_t\,\vec{i}_q\,\vec{i}_r \quad \Rightarrow \quad \mathbf{B}^{\mathrm{T}}:\mathbf{A}^{\mathrm{T}} = A_{qrmk}\,B_{mkop}\,\vec{i}_o\,\vec{i}_p\,\vec{i}_q\,\vec{i}_r\,. \tag{2.103}$$

The order of the scalar components $B_{mkop}$ and $A_{qrmk}$ is interchanged in Expression $(2.103)_2$. The second change of indices reads

$$q \to u \qquad \text{and} \qquad r \to w\,. \tag{2.104}$$

Now Expression $(2.103)_2$ takes the following appearance:

$$\mathbf{B}^{\mathrm{T}} : \mathbf{A}^{\mathrm{T}} = A_{qrmk}\, B_{mkop}\, \vec{i}_o\, \vec{i}_p\, \vec{i}_q\, \vec{i}_r \quad \Rightarrow \quad \mathbf{B}^{\mathrm{T}} : \mathbf{A}^{\mathrm{T}} = A_{uwmk}\, B_{mkop}\, \vec{i}_o\, \vec{i}_p\, \vec{i}_u\, \vec{i}_w \,. \tag{2.105}$$

The right sides of Expressions (2.100) and $(2.103)_2$ are equal except for the dummy indices, which are not connected to the base vectors. Since these indices can be arbitrarily chosen, the left sides are equal as well, and Expression (2.97) is proved to hold. ■

It may be difficult to give indices for the tensors in such a way that after mathematical derivation the indices are the ones aimed at. This is the case with Equation (2.101). Therefore a good practice in the preceding example might be to take totally new indices for Expression (2.101), then, after getting the result [Expression (2.101) in this case], to change the indices as was done in Expressions (2.102) and (2.104) in such a way that the final indices are the same.

The tensor $\mathbf{I}^{\mathrm{s}}$ is called the fourth-order symmetric identity tensor. It is defined as follows: the double-dot product between the fourth-order symmetric identity tensor $\mathbf{I}^{\mathrm{s}}$ and any second-order tensor $\mathbf{c}$ gives the expression

**Def** $$\mathbf{I}^{\mathrm{s}} : \mathbf{c} = \mathbf{c} : \mathbf{I}^{\mathrm{s}} = \tfrac{1}{2}\left(\mathbf{c} + \mathbf{c}^{\mathrm{T}}\right) . \tag{2.106}$$

Definition (2.106) implies that the double-dot product $\mathbf{I}^{\mathrm{s}} : \mathbf{c}$ is symmetric for a nonsymmetric tensor $\mathbf{c}$ as well. The fourth-order symmetric identity tensor $\mathbf{I}^{\mathrm{s}}$ is given in terms of its components as follows:

$$\mathbf{I}^{\mathrm{s}} = \tfrac{1}{2}\left(\delta_{ik}\, \delta_{jl} + \delta_{il}\, \delta_{jk}\right) \vec{i}_i\, \vec{i}_j\, \vec{i}_k\, \vec{i}_l \,. \tag{2.107}$$

The deviatoric stress tensor $\mathbf{s}$ is defined by

$$\mathbf{s} := \mathbf{K} : \boldsymbol{\sigma} , \qquad \text{where the fourth-order tensor } \mathbf{K} := \mathbf{I} - \tfrac{1}{3}\mathbf{1}\,\mathbf{1} . \tag{2.108}$$

By following the concept given by Expressions (2.16) and (2.64), the notation $\mathbf{1}\,\mathbf{1}$ reads

$$\mathbf{1}\,\mathbf{1} = \delta_{ij}\, \vec{i}_i\, \vec{i}_j\, \delta_{kl}\, \vec{i}_k\, \vec{i}_l = \delta_{ij}\, \delta_{kl}\, \vec{i}_i\, \vec{i}_j\, \vec{i}_k\, \vec{i}_l \,. \tag{2.109}$$

**Example 2.11:** Write the components of the deviatoric stress tensor $\mathbf{s}$ in the form $s_{ij} = \cdots$, and so on, and furthermore in the form $s_{11} = \cdots$ and $s_{12} = \cdots$.

**Solution:** Definition (2.69) and Equation $(2.65)_1$ give

$$\mathbf{I} : \boldsymbol{\sigma} = \boldsymbol{\sigma} \qquad \text{and} \qquad \mathbf{1} : \boldsymbol{\sigma} = \boldsymbol{\sigma} : \mathbf{1} = \sigma_{tt} \,. \tag{2.110}$$

The double-dot product given by Definition $(2.108)_1$ is formulated. Based on Definitions (2.108), the following result is obtained:

$$\mathbf{s} := \mathbf{K} : \boldsymbol{\sigma} = \left(\mathbf{I} - \tfrac{1}{3}\mathbf{1}\,\mathbf{1}\right) : \boldsymbol{\sigma} = \mathbf{I} : \boldsymbol{\sigma} - \tfrac{1}{3}\mathbf{1}\,\mathbf{1} : \boldsymbol{\sigma}\,. \tag{2.111}$$

By taking Expressions (2.110) into consideration, Equation (2.111) can be cast in the form

$$\mathbf{s} = \mathbf{I} : \boldsymbol{\sigma} - \tfrac{1}{3}\mathbf{1}\,\mathbf{1} : \boldsymbol{\sigma} = \boldsymbol{\sigma} - \tfrac{1}{3}\mathbf{1}\,\sigma_{tt} \quad \left(= \boldsymbol{\sigma} - \tfrac{1}{3}\sigma_{tt}\,\mathbf{1}\right). \tag{2.112}$$

Within the parentheses in Expression (2.112) the order of the scalar $\sigma_{tt}$ and the tensor $\mathbf{1}$ is interchanged. This follows the idea given in Section 2.2. Equation (2.112) is written in terms of its components. When Equation (2.64) is applied, the following is achieved:

$$s_{ij}\,\vec{i}_i\,\vec{i}_j = \sigma_{ij}\,\vec{i}_i\,\vec{i}_j - \tfrac{1}{3}\delta_{ij}\,\vec{i}_i\,\vec{i}_j\,\sigma_{tt} = \sigma_{ij}\,\vec{i}_i\,\vec{i}_j - \tfrac{1}{3}\sigma_{tt}\,\delta_{ij}\,\vec{i}_i\,\vec{i}_j\,. \tag{2.113}$$

In Equation (2.112) the scalar $\sigma_{tt}$ is moved over the base vectors $\vec{i}_i$ and $\vec{i}_j$ from the right to the left. Based on the concept given in Section 2.2, this is an acceptable act. According to this concept, the order of the base vectors cannot be interchanged. However, the interchange of the terms $\sigma_{tt}$ and $\delta_{ij}$ is possible, since the term $\sigma_{tt}$ is a scalar. The interchange is already possible to carry out in manipulation of Equation (2.112) (as was done within the parentheses).

The scalar components on the left side of Expression (2.113) equal those on the right side of Expression (2.113). This is

$$s_{ij} = \sigma_{ij} - \tfrac{1}{3}\sigma_{tt}\,\delta_{ij}\,. \tag{2.114}$$

Equation (2.114) yields

$$\begin{aligned} s_{11} &= \sigma_{11} - \tfrac{1}{3}\sigma_{tt}\,\delta_{11} = \sigma_{11} - \tfrac{1}{3}(\sigma_{11} + \sigma_{22} + \sigma_{33})\,,\\ s_{12} &= \sigma_{12} - \tfrac{1}{3}\sigma_{tt}\,\delta_{12} = \sigma_{12}\,, \end{aligned} \tag{2.115}$$

and so on. ■

By using Expressions $(2.65)_1$, $(2.65)_2$, and (2.112), it can be shown that

$$\mathbf{1} : \mathbf{s} = \mathbf{s} : \mathbf{1} \equiv 0 \qquad \text{and, in general,} \qquad \mathbf{1} : \mathbf{b} = \mathbf{b} : \mathbf{1} \equiv 0\,, \tag{2.116}$$

where $\mathbf{b}$ is the deviatoric part of the second-order tensor $\boldsymbol{\beta}$. Equation $(2.116)_2$ holds for all deviatoric second-order tensors.

The mechanical pressure $p_{mech}$ is defined by

$$p_{mech} := -\tfrac{1}{3}\mathbf{1} : \boldsymbol{\sigma}\,. \tag{2.117}$$

The state of stress can be expressed as a sum of the deviatoric stress tensor $\mathbf{s}$ and of the mechanical pressure $p_{\text{mech}}$ as follows:

$$\boldsymbol{\sigma} = \mathbf{s} - p_{\text{mech}}\,\mathbf{1}\,. \tag{2.118}$$

The term "mechanical" in front of the term "pressure" is crucial since the notion of thermodynamical pressure denoted by $p$ is introduced later in this book.

Separation (2.118) shows that the stress tensor $\boldsymbol{\sigma}$ can be separated into deviatoric part $\mathbf{s}$, which causes elastic distortion or plastic yield, and hydrostatic part $p_{\text{mech}}$, which causes elastic dilatation (volume change).

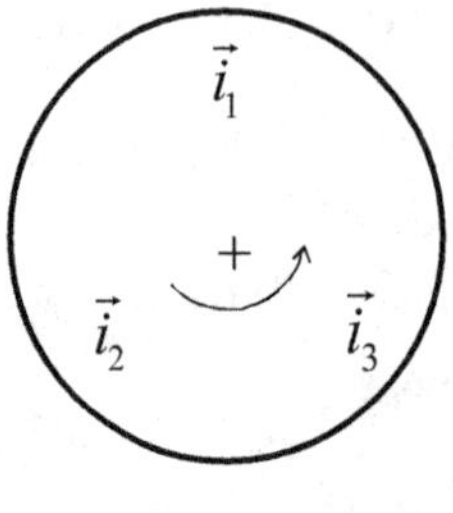

**Figure 2.3** Sequence of the base vectors $(\vec{\imath}_1, \vec{\imath}_2, \vec{\imath}_3)$ to give a positive value for the cross product.

The preceding definitions allow the following to be written: an arbitrary second-order tensor $\mathbf{a}$ can be separated into the spherical part $\mathbf{a}^{\text{sp}}$ and the deviatoric part $\mathbf{a}^{\text{dev}}$ as follows:

$$\mathbf{a} := \mathbf{a}^{\text{sp}} + \mathbf{a}^{\text{dev}}\,. \tag{2.119}$$

The deviatoric part of the second-order tensor $\mathbf{a}^{\text{dev}}$ is defined by

$$\mathbf{a}^{\text{dev}} := \mathbf{K} : \mathbf{a}\,. \tag{2.120}$$

Definitions (2.119) and (2.120) give the spherical part of the tensor $\mathbf{a}$,

$$\mathbf{a}^{\text{sp}} = \mathbf{a} - \mathbf{a}^{\text{dev}} = \mathbf{I}:\mathbf{a} - (\mathbf{I} - \tfrac{1}{3}\,\mathbf{1}\,\mathbf{1}):\mathbf{a} = \tfrac{1}{3}\,\mathbf{1}\,\mathbf{1}:\mathbf{a}\,. \tag{2.121}$$

Manipulation (2.121) made use of Definition (2.69). Expression (2.119) is the definition of the spherical tensor $\mathbf{a}^{\text{sp}}$.

Malvern [[57], Eq. (2.3.22)] gives the following expression for the cross product of vectors $\vec{c}$ and $\vec{b}$:

$$\vec{c} \times \vec{b} = e_{mnr}\, c_n\, b_r\, \vec{\imath}_m\,, \tag{2.122}$$

where the permutation symbol $e_{mnr}$ (or Levi–Civita tensor) is defined as having the value $0, +1, -1$ as follows:

$$e_{mnr} := \begin{cases} 0, & \text{when any two indices are equal}\,, \\ 1, & \text{when } m,\ n,\ r \text{ are } 1,\ 2,\ 3 \text{ or an even} \\ & \text{permutation of } 1,\ 2,\ 3 \text{ (that is, } 2,\ 3,\ 1 \text{ or } 3,\ 1,\ 2)\,, \\ -1, & \text{when } m,\ n,\ r \text{ are an odd permutation} \\ & \text{of } 1,\ 2,\ 3\,\text{(that is, } 3,\ 2,\ 1, \text{ or } 2,\ 1,\ 3 \text{ or } 1,\ 3,\ 2)\,. \end{cases} \tag{2.123}$$

Figure 2.3 provides an easy way to remember the message of Expression (2.123).

The von Mises operator $J_{\text{vM}}(\cdot)$ is introduced. When acting on a second-order tensor, $J_{\text{vM}}(\cdot)$ may be defined by [cf. Lin [52], p. 15 and Eq. (4.2.8)]

$$J_{\text{vM}}(\boldsymbol{\gamma}) := \sqrt{\tfrac{3}{2}\,\mathbf{g}:\mathbf{g}}\,, \tag{2.124}$$

where $\mathbf{g}$ denotes the deviatoric part of any second-order tensor $\boldsymbol{\gamma}$.

The magnitude (or absolute value) of a vector $\vec{v}$ is denoted by $|\vec{v}|$, and it is always positive or zero. The notation $|\vec{v}|$ is defined by

$$|\vec{v}| := \sqrt{\vec{v}\cdot\vec{v}}\,. \tag{2.125}$$

A vector of unit length is called a unit vector. Since the vector $\vec{n}$ is a unit vector, the following holds:

$$|\vec{n}| := \sqrt{\vec{n}\cdot\vec{n}} = 1\,. \tag{2.126}$$

## 2.5 Differentiation of Tensors

This section studies the differentiation of tensors and gives the relations needed in this book.

The following Lemma A is introduced. Let $f(x,\cdot) \in \mathrm{C}^1$ be a function of the independent variable $x$ and some other independent variables denoted by $\cdot$. By extending Marsden and Hughes ([58], p. 185), the following can be expressed: The partial directional derivative of $f(x,\cdot)$ in the direction of $u$ is denoted by $\partial_{\mathrm{X}} f(x,\cdot) * u$, and it obeys the following expression:

$$\frac{\partial f(x,\cdot)}{\partial x} * u = \partial_{\mathrm{X}} f(x,\cdot) * u = \frac{\partial}{\partial h} f(x + h\,u,\cdot)\Big|_{h=0}\,, \tag{2.127}$$

where the variable $u$ is a tensor of the same order as the variable $x$ and $h$ is a scalar. The notation $*$ stands for multiplication, a dot product, a double-dot product, and so on, depending on the form of the quantities $x$ and $u$. For scalars $x$ and $u$, the notation $*$ stands for multiplication; for vectors $x$ and $u$, the notation $*$ stands for a dot product; for second-order tensors $x$ and $u$, the notation $*$ stands for a double-dot product, and so on.

Derivative (2.127) provides a very convenient way to calculate the derivatives of tensor-valued functions $f(x,\cdot)$ with respect to tensorial variables $x$ of any order. Sometimes it may be challenging to formulate the result obtained on the right side of Derivative (2.127) in a form where the variable $u$ is on the right side of the expression, as given on the left side of Derivative (2.127). The following examples show how to apply Derivative (2.127) for calculating derivatives. Leigh ([49], p. 12) gives an expression similar to that given in Derivative (2.127).

**Example 2.12:** Let $\mathbf{c}$ be an arbitrary second-order tensor and $\mathbf{A}$ an arbitrary fourth-order tensor. Calculate the derivative $\partial[\,\mathbf{c}:\mathbf{A}:\mathbf{c}\,]\,/\,\partial\mathbf{c}$ using Directional Derivative (2.127).

**Solution:** Based on Derivative (2.127), the following is written:

$$\begin{aligned}\frac{\partial}{\partial \mathbf{c}}\big[\mathbf{c}:\mathbf{A}:\mathbf{c}\big]:\mathbf{u} &= \frac{\partial}{\partial h}\Big[(\mathbf{c}+h\,\mathbf{u}):\mathbf{A}:(\mathbf{c}+h\,\mathbf{u})\Big]_{h=0}\\ &= \Big[\mathbf{u}:\mathbf{A}:(\mathbf{c}+h\,\mathbf{u})+(\mathbf{c}+h\,\mathbf{u}):\mathbf{A}:\mathbf{u}\Big]_{h=0}\\ &= \mathbf{u}:\mathbf{A}:\mathbf{c}+\mathbf{c}:\mathbf{A}:\mathbf{u} = \mathbf{c}:\mathbf{A}^{\mathrm{T}}:\mathbf{u}+\mathbf{c}:\mathbf{A}:\mathbf{u}\\ &= \mathbf{c}:(\mathbf{A}+\mathbf{A}^{\mathrm{T}}):\mathbf{u}\,. \end{aligned} \tag{2.128}$$

Manipulation (2.128) utilized the equality

$$\mathbf{c}:\mathbf{A}^{\mathrm{T}}:\mathbf{u} = \mathbf{u}:\mathbf{A}:\mathbf{c}\,, \tag{2.129}$$

which follows Definition (2.94) for the tensor transpose of an arbitrary fourth-order tensor. Based on the manipulation of Manipulation (2.128), the following result is obtained:

$$\frac{\partial}{\partial \mathbf{c}}\big[\mathbf{c}:\mathbf{A}:\mathbf{c}\big] = \mathbf{c}:(\mathbf{A}+\mathbf{A}^{\mathrm{T}})\,. \tag{2.130}$$

If the fourth-order tensor $\mathbf{A}$ is a major symmetric tensor, that is, $\mathbf{A}^{\mathrm{T}} = \mathbf{A}$, Equation (2.130) reduces to

$$\text{If } A_{stkl} = A_{klst}\,, \quad \text{then} \quad \frac{\partial}{\partial \mathbf{c}}\big[\mathbf{c}:\mathbf{A}:\mathbf{c}\big] = 2\,\mathbf{c}:\mathbf{A} = 2\,\mathbf{A}:\mathbf{c}\,. \quad \blacksquare \tag{2.131}$$

**Example 2.13:** Let $\mathbf{c}$ be an arbitrary second-order tensor and $\mathbf{A}$ to be an arbitrary fourth-order tensor. Calculate the derivative $\partial[\mathbf{A}:\mathbf{c}]\,/\,\partial\,\mathbf{c}$ using Directional Derivative (2.127).

**Solution:** Based on Derivative (2.127), the following is written:

$$\frac{\partial}{\partial \mathbf{c}}\big[\mathbf{A}:\mathbf{c}\big]:\mathbf{u} = \frac{\partial}{\partial h}\Big[\mathbf{A}:(\mathbf{c}+h\,\mathbf{u})\Big]_{h=0} = \big[\mathbf{A}:\mathbf{u}\big]_{h=0} = \mathbf{A}:\mathbf{u}\,. \tag{2.132}$$

Based on Manipulation (2.132), the following is achieved:

$$\frac{\partial}{\partial \mathbf{c}}\big[\mathbf{A}:\mathbf{c}\big] = \mathbf{A}\,. \quad \blacksquare \tag{2.133}$$

**Example 2.14:** Let $\mathbf{c}$ be an arbitrary second-order tensor and $\mathbf{1}$ a second-order identity tensor. Calculate the derivative $\partial[\,\mathbf{1}:\mathbf{c}\,]^2/\,\partial\,\mathbf{c}$ using Directional

Derivative (2.127).

**Solution:** Based on Derivative (2.127), the following is written:

$$\begin{aligned}\frac{\partial}{\partial \mathbf{c}}[\mathbf{1}:\mathbf{c}]^2:\mathbf{u} &= \frac{\partial}{\partial \mathbf{c}}\Big\{[\mathbf{1}:\mathbf{c}][\mathbf{1}:\mathbf{c}]\Big\}:\mathbf{u}\\ &= \frac{\partial}{\partial h}\Big\{[\mathbf{1}:(\mathbf{c}+h\,\mathbf{u})][\mathbf{1}:(\mathbf{c}+h\,\mathbf{u})]\Big\}_{h=0}\\ &= \Big\{[\mathbf{1}:\mathbf{u}][\mathbf{1}:(\mathbf{c}+h\,\mathbf{u})]+[\mathbf{1}:(\mathbf{c}+h\,\mathbf{u})][\mathbf{1}:\mathbf{u}]\Big\}_{h=0}\\ &= [\mathbf{1}:\mathbf{u}][\mathbf{1}:\mathbf{c}]+[\mathbf{1}:\mathbf{c}][\mathbf{1}:\mathbf{u}] = 2[\mathbf{1}:\mathbf{c}]\mathbf{1}:\mathbf{u}.\end{aligned} \tag{2.134}$$

Based on Manipulation (2.134), the following is achieved:

$$\frac{\partial}{\partial \mathbf{c}}[\mathbf{1}:\mathbf{c}]^2 = 2[\mathbf{1}:\mathbf{c}]\mathbf{1}. \quad \blacksquare \tag{2.135}$$

**Example 2.15:** Let $\mathbf{s}$ be a second-order deviatoric stress tensor defined by Definition $(2.108)_1$, $\mathbf{s} \coloneqq \mathbf{K}:\boldsymbol{\sigma}$. Calculate the derivative $\partial[\mathbf{s}:\mathbf{s}]/\partial\boldsymbol{\sigma}$ using Directional Derivative (2.127).

**Solution:** Based on Definition $(2.108)_1$ and Derivative (2.127), the following is written:

$$\begin{aligned}\frac{\partial}{\partial \boldsymbol{\sigma}}\{\mathbf{s}:\mathbf{s}\}:\mathbf{u} &= \frac{\partial}{\partial \boldsymbol{\sigma}}\{[\mathbf{K}:\boldsymbol{\sigma}]:[\mathbf{K}:\boldsymbol{\sigma}]\}:\mathbf{u}\\ &= \frac{\partial}{\partial h}\{[\mathbf{K}:(\boldsymbol{\sigma}+h\,\mathbf{u})]:[\mathbf{K}:(\boldsymbol{\sigma}+h\,\mathbf{u})]\}_{h=0}.\end{aligned} \tag{2.136}$$

Equation (2.136) yields

$$\begin{aligned}\frac{\partial}{\partial \boldsymbol{\sigma}}\{\mathbf{s}:\mathbf{s}\}:\mathbf{u} &= \frac{\partial}{\partial h}\Big\{[\mathbf{K}:(\boldsymbol{\sigma}+h\,\mathbf{u})]:[\mathbf{K}:(\boldsymbol{\sigma}+h\,\mathbf{u})]\Big\}_{h=0}\\ &= \Big\{[\mathbf{K}:\mathbf{u}]:[\mathbf{K}:(\boldsymbol{\sigma}+h\,\mathbf{u})]+[\mathbf{K}:(\boldsymbol{\sigma}+h\,\mathbf{u})]:[\mathbf{K}:\mathbf{u}]\Big\}_{h=0}\\ &= [\mathbf{K}:\mathbf{u}]:[\mathbf{K}:\boldsymbol{\sigma}]+[\mathbf{K}:\boldsymbol{\sigma}]:[\mathbf{K}:\mathbf{u}]\\ &= [\mathbf{K}:\mathbf{u}]:\mathbf{s}+\mathbf{s}:[\mathbf{K}:\mathbf{u}].\end{aligned} \tag{2.137}$$

In the last two terms, $[\mathbf{K}:\mathbf{u}]:\mathbf{s}$ and $\mathbf{s}:[\mathbf{K}:\mathbf{u}]$, the deviatoric stress tensor $\mathbf{s}$ operates with the two first indices of the fourth-order tensor $\mathbf{K}$. This means that these terms are equal and Manipulation (2.137) can be written in the form

$$\frac{\partial}{\partial \boldsymbol{\sigma}}\{\mathbf{s}:\mathbf{s}\}:\mathbf{u} = [\mathbf{K}:\mathbf{u}]:\mathbf{s}+\mathbf{s}:[\mathbf{K}:\mathbf{u}] = 2\,\mathbf{s}:[\mathbf{K}:\mathbf{u}] = 2\,\mathbf{s}:\mathbf{K}:\mathbf{u}. \tag{2.138}$$

Expressions $(2.108)_2$, (2.69), and (2.116) read

$$\mathbf{K} := \mathbf{I} - \tfrac{1}{3}\,\mathbf{1}\,\mathbf{1}\,, \qquad \mathbf{I}:\mathbf{c} = \mathbf{c}:\mathbf{I} = \mathbf{c}\,, \qquad \text{and} \qquad \mathbf{1}:\mathbf{s} = \mathbf{s}:\mathbf{1} \equiv 0\,. \tag{2.139}$$

Based on Expressions (2.139),

$$\mathbf{s}:\mathbf{K} = \mathbf{s}:(\mathbf{I} - \tfrac{1}{3}\,\mathbf{1}\,\mathbf{1}) = \mathbf{s}:\mathbf{I} - \mathbf{s}:\tfrac{1}{3}\,\mathbf{1}\,\mathbf{1} = \mathbf{s} - 0\,\mathbf{1} = \mathbf{s}\,, \tag{2.140}$$

and Derivative (2.138) reduces to

$$\frac{\partial}{\partial \boldsymbol{\sigma}}\{\mathbf{s}:\mathbf{s}\}:\mathbf{u} = [\,\mathbf{K}:\mathbf{u}\,]:\mathbf{s} + \mathbf{s}:[\,\mathbf{K}:\mathbf{u}\,] = 2\,\mathbf{s}:[\,\mathbf{K}:\mathbf{u}\,] = 2\mathbf{s}:\mathbf{K}:\mathbf{u} = \mathbf{2}\,\mathbf{s}:\mathbf{u}\,. \tag{2.141}$$

Based on Manipulation (2.141), the following is achieved:

$$\frac{\partial}{\partial \boldsymbol{\sigma}}\{\mathbf{s}:\mathbf{s}\} = 2\,\mathbf{s}\,. \qquad \blacksquare \tag{2.142}$$

**Example 2.16:** Let $\mathbf{c}$ be an arbitrary second-order tensor and $\vec{a}$ and $\vec{b}$ arbitrary vectors. Calculate the derivative $\partial(\vec{a}\cdot\mathbf{c}\cdot\mathbf{c}\cdot\vec{b})\,/\,\partial\mathbf{c}$ using Directional Derivative (2.127).

**Solution:** Based on Derivative (2.127), the following is written:

$$\begin{aligned} \frac{\partial(\vec{a}\cdot\mathbf{c}\cdot\mathbf{c}\cdot\vec{b})}{\partial\mathbf{c}}:\mathbf{u} &= \frac{\partial}{\partial h}\Big[\,\vec{a}\cdot(\mathbf{c}+h\,\mathbf{u})\cdot(\mathbf{c}+h\,\mathbf{u})\cdot\vec{b}\,\Big]_{h=0} \\ &= \Big[\,\vec{a}\cdot\mathbf{u}\cdot(\mathbf{c}+h\,\mathbf{u})\cdot\vec{b} + \vec{a}\cdot(\mathbf{c}+h\,\mathbf{u})\cdot\mathbf{u}\cdot\vec{b}\,\Big]_{h=0} \\ &= \vec{a}\cdot\mathbf{u}\cdot\mathbf{c}\cdot\vec{b} + \vec{a}\cdot\mathbf{c}\cdot\mathbf{u}\cdot\vec{b}\,. \end{aligned} \tag{2.143}$$

The quantities on the last line of Expression (2.143) can be written in a new format as follows:

$$\begin{aligned} \vec{a}\,(\mathbf{c}\cdot\vec{b}):\mathbf{u} &= a_i\,\vec{i}_i\,(c_{sm}\,\vec{i}_s\,\vec{i}_m\cdot b_q\,\vec{i}_q) : u_{kt}\,\vec{i}_k\,\vec{i}_t = a_i\,\vec{i}_i\,(c_{sm}\,b_m\,\vec{i}_s) : u_{kt}\,\vec{i}_k\,\vec{i}_t \\ &= a_i\,(c_{sm}\,b_m)\,\delta_{ik}\,\delta_{st}\,u_{kt} = a_i\,c_{sm}\,b_m\,u_{is} = a_i\,u_{is}\,c_{sm}\,b_m \\ &= \vec{a}\cdot\mathbf{u}\cdot\mathbf{c}\cdot\vec{b} \end{aligned} \tag{2.144}$$

and

$$\begin{aligned} (\vec{a}\cdot\mathbf{c})\,\vec{b}:\mathbf{u} &= (a_q\,\vec{i}_q\cdot c_{si}\,\vec{i}_s\,\vec{i}_i)\,b_k\,\vec{i}_k : u_{mt}\,\vec{i}_m\,\vec{i}_t = (a_s\,c_{si}\,\vec{i}_i)\,b_k\,\vec{i}_k : u_{mt}\,\vec{i}_m\,\vec{i}_t \\ &= (a_s c_{s\,i})\,b_k\,\delta_{im}\,\delta_{kt}\,u_{mt} = a_s\,c_{si}\,b_k\,u_{ik} = a_s\,c_{si}\,u_{ik}\,b_k \\ &= \vec{a}\cdot\mathbf{c}\cdot\mathbf{u}\cdot\vec{b}\,. \end{aligned} \tag{2.145}$$

In Manipulations (2.144) and (2.145) the order of the scalar components is changed. Based on Results (2.144) and (2.145), Derivative (2.143) takes the form

$$\frac{\partial(\vec{a}\cdot\mathbf{c}\cdot\mathbf{c}\cdot\vec{b})}{\partial\mathbf{c}}:\mathbf{u}=\vec{a}\cdot\mathbf{u}\cdot\mathbf{c}\cdot\vec{b}+\vec{a}\cdot\mathbf{c}\cdot\mathbf{u}\cdot\vec{b}=\vec{a}\,(\mathbf{c}\cdot\vec{b}):\mathbf{u}+(\vec{a}\cdot\mathbf{c})\,\vec{b}:\mathbf{u}$$
$$=[\vec{a}\,(\mathbf{c}\cdot\vec{b})+(\vec{a}\cdot\mathbf{c})\,\vec{b}]:\mathbf{u}\,. \tag{2.146}$$

Manipulation (2.146) gives

$$\frac{\partial(\vec{a}\cdot\mathbf{c}\cdot\mathbf{c}\cdot\vec{b})}{\partial\mathbf{c}}=\vec{a}\,(\mathbf{c}\cdot\vec{b})+(\vec{a}\cdot\mathbf{c})\,\vec{b}\,. \qquad \blacksquare \tag{2.147}$$

**Example 2.17:** Let $\mathbf{c}$ be an arbitrary second-order tensor. Calculate the derivative $\partial\mathbf{c}/\partial\mathbf{c}$ using Directional Derivative (2.127).

**Solution:** Based on Derivative (2.127) and Definition(2.69), $\mathbf{I}:\mathbf{c}=\mathbf{c}:\mathbf{I}=\mathbf{c}$, the following result is obtained:

$$\frac{\partial\mathbf{c}}{\partial\mathbf{c}}:\mathbf{u}=\frac{\partial}{\partial h}\Big[\mathbf{c}+h\,\mathbf{u}\Big]_{h=0}=[\,\mathbf{u}\,]_{h=0}=\mathbf{u}=\mathbf{I}:\mathbf{u}\,. \tag{2.148}$$

Equation (2.148) gives

$$\frac{\partial\mathbf{c}}{\partial\mathbf{c}}=\mathbf{I}\,. \qquad \blacksquare \tag{2.149}$$

**Example 2.18:** Let $\vec{a}$ be an arbitrary vector. Calculate the derivative $\partial\vec{a}/\partial\vec{a}$ using Directional Derivative (2.127).

**Solution:** Based on Derivative (2.127) and Definition (2.63), $\mathbf{1}\cdot\vec{u}=\vec{u}\cdot\mathbf{1}=\vec{u}$, the following is written:

$$\frac{\partial\vec{a}}{\partial\vec{a}}\cdot\vec{u}=\frac{\partial}{\partial h}\left[\vec{a}+h\,\vec{u}\right]_{h=0}=[\,\vec{u}\,]_{h=0}=\vec{u}=\mathbf{1}\cdot\vec{u}\,. \tag{2.150}$$

Equation (2.150) gives

$$\frac{\partial\vec{a}}{\partial\vec{a}}=\mathbf{1}. \qquad \blacksquare \tag{2.151}$$

Theorem 1 of **Appendix A** [see Derivatives (A.2) and (A.3)] reads

$$\frac{\partial J_{\mathrm{vM}}(\boldsymbol{\sigma}-\boldsymbol{\beta}^1)}{\partial\boldsymbol{\sigma}}=\frac{3}{2}\,\frac{\mathbf{s}-\mathbf{b}^1}{J_{\mathrm{vM}}(\boldsymbol{\sigma}-\boldsymbol{\beta}^1)} \tag{2.152}$$

and

$$\frac{\partial J_{\mathrm{vM}}(\boldsymbol{\sigma}-\boldsymbol{\beta}^1)}{\partial \boldsymbol{\beta}^1} = -\frac{3}{2}\frac{\mathbf{s}-\mathbf{b}^1}{J_{\mathrm{vM}}(\boldsymbol{\sigma}-\boldsymbol{\beta}^1)}, \tag{2.153}$$

where $\mathbf{b}^1$ is the deviatoric part of the second-order tensor $\boldsymbol{\beta}^1$.

According to Theorem 1 of **Appendix B** [see Equation (B.1)], the derivative of a second-order tensor $\mathbf{a}(\mathbf{e})$ with respect to a second-order tensor $\mathbf{g}$, where $\mathbf{e} = f(\mathbf{g})$, is obtained by the following chain rule:

$$\frac{\partial \mathbf{a}(\mathbf{e})}{\partial \mathbf{g}} = \frac{\partial \mathbf{e}}{\partial \mathbf{g}} : \frac{\partial \mathbf{a}(\mathbf{e})}{\partial \mathbf{e}}. \tag{2.154}$$

The partial derivative of a scalar-valued function $f(x(t))$ with respect to the variable $t$ is studied. Thus the following “standard form” is achieved:

$$\frac{\partial f(x(t))}{\partial t} = \frac{\partial f(x(t))}{\partial x}\frac{\partial x(t)}{\partial t}. \tag{2.155}$$

There is a certain difference on the right sides of Derivatives (2.154) and (2.155). The “mutual” terms have a different order. If Derivative (2.155) had the same form as Derivative (2.154), the former would have the following appearance:

$$\frac{\partial f(x(t))}{\partial t} = \frac{\partial x(t)}{\partial t}\frac{\partial f(x(t))}{\partial x}. \tag{2.156}$$

Derivative (2.156) is correct, since the order of the scalar can be interchanged, but a similar interchange in Derivative (2.154) is not possible. The same holds for the following derivatives as well.

Theorems 2.1, 2.2, 3.1, and 3.2 of **Appendix B** [see Equations (B.6) and (B.7)] give

$$\frac{\partial \vec{a}(\mathbf{e})}{\partial \mathbf{c}} = \frac{\partial \mathbf{e}}{\partial \mathbf{c}} : \frac{\partial \vec{a}(\mathbf{e})}{\partial \mathbf{e}} \quad \text{and} \quad \frac{\partial a(\mathbf{e})}{\partial \mathbf{c}} = \frac{\partial \mathbf{e}}{\partial \mathbf{c}} : \frac{\partial a(\mathbf{e})}{\partial \mathbf{e}} \tag{2.157}$$

and furthermore

$$\frac{\partial \vec{a}(\vec{e})}{\partial \vec{b}} = \frac{\partial \vec{e}}{\partial \vec{b}} \cdot \frac{\partial \vec{a}(\vec{e})}{\partial \vec{e}} \quad \text{and} \quad \frac{\partial a(\vec{e})}{\partial \vec{b}} = \frac{\partial \vec{e}}{\partial \vec{b}} \cdot \frac{\partial a(\vec{e})}{\partial \vec{e}}, \tag{2.158}$$

where $\vec{a}$ is an arbitrary vector, $a$ is an arbitrary scalar, $\vec{e}$ and $\vec{b}$ are arbitrary vectors, and $\mathbf{e}$ and $\mathbf{c}$ are arbitrary second-order tensors.

In this book the forms of the variables are given in two ways. If the value of the variable $\theta$, for example, temperature, changes with time $t$, the value of the variable $\theta$ can be expressed by the position of the observed point and by time $t$. If the position is given with the coordinates $(x_1, x_2, x_3)$, we write it as $\theta(x_1, x_2, x_3, t)$. In order to shorten the notation, the notation $\theta(\vec{x}, t)$ is used:

$$\theta = \theta(x_1, x_2, x_3, t) \quad \Leftrightarrow \quad \theta = \theta(\vec{x}, t), \tag{2.159}$$

where $\theta$ can be a scalar, vector, or tensor of any order. Form $(2.159)_2$ is a shorter notation for Expression $(2.159)_1$.

The foregoing notation $\theta = \theta(\vec{x}, t)$ may have two problems. The first relates to the symbol $\theta$. On the left side of Expression $(2.159)_2$, it is the value of the function $\theta$ at the point $(x_1,\, x_2,\, x_3)$ at time $t$, and on the right side, it is a function. A careful writer would prefer to write Expression $(2.159)_2$ in the form $\theta = \hat{\theta}(\vec{x}, t)$, for example. However, it is common practice to write it as in Expression $(2.159)_2$. This practice is followed in this text.

The second problem is related to the replacement of the notation $x_1, x_2, x_3$ with $\vec{x}$. The former notation refers to a point having the coordinates $x_1, x_2, x_3$, and the latter notation is the position vector of the same point. The base vectors in the vector $\vec{x}$ create a problem. The set $x_1, x_2, x_3$ does not include base vectors, whereas the vector $\vec{x}$ does. One might use the notation $\bar{x}$ to represent the set $x_1, x_2, x_3$. The benefit of the notation $\vec{x}$ is that it clearly refers to a point, but its meaning may be unclear to the reader. Usually, the notation $\theta(\vec{x}, t)$ does not cause any confusion. Section 3.6 shows how to avoid mathematical difficulties created by the notation $\theta(\vec{x}, t)$. The notation $\vec{x}$ is an exception. The other vectors and higher-order tensors, being independent variables, are exactly as written. A good example is Equation (2.55), which is

$$w(\boldsymbol{\varepsilon}^{\mathrm{e}}) := \tfrac{1}{2}\, \boldsymbol{\varepsilon}^{\mathrm{e}} : \mathbf{C} : \boldsymbol{\varepsilon}^{\mathrm{e}} \,. \tag{2.160}$$

On the other hand, sometimes this rule is not clear. According to Definition (2.124), the von Mises value of the second-order tensor $\boldsymbol{\gamma}$ has the following appearance:

$$J_{\mathrm{vM}}(\boldsymbol{\gamma}) := \sqrt{\tfrac{3}{2}\, \mathbf{g} : \mathbf{g}}\;, \tag{2.161}$$

where $\mathbf{g}$ denotes the deviatoric part of any second-order tensor $\boldsymbol{\gamma}$. Based on Definition (2.120), $\mathbf{a}^{\mathrm{dev}} := \mathbf{K} : \mathbf{a}$, Definition (2.161) takes the form

$$J_{\mathrm{vM}}(\boldsymbol{\gamma}) = \sqrt{\tfrac{3}{2}\,(\mathbf{K} : \boldsymbol{\gamma}) : (\mathbf{K} : \boldsymbol{\gamma})}\;, \tag{2.162}$$

which is the form at which we are aiming.

Sometimes it is more convenient to express functions in terms of independent variables giving the values for the functions. For example, the specific Helmholtz free energy $\psi$ (not yet introduced in this book) can be written as either $\psi = \psi(\vec{x}, t)$ or $\psi = \psi(\boldsymbol{\varepsilon} - \boldsymbol{\varepsilon}^{\mathrm{i}}, \vec{D}, T)$, where $\boldsymbol{\varepsilon}$ is the strain tensor, $\boldsymbol{\varepsilon}^{\mathrm{i}}$ is the inelastic strain tensor (internal variable), $\vec{D}$ is another internal variable, and $T$ is the absolute temperature. If the specific Helmholtz free energy $\psi$ is written into the form $\psi = \psi(\boldsymbol{\varepsilon} - \boldsymbol{\varepsilon}^{\mathrm{i}}, \vec{D}, T)$, according to Chain Rules $(2.157)_1$ and $(2.158)_1$, we arrive at the following:

$$\dot{\psi}(\boldsymbol{\varepsilon} - \boldsymbol{\varepsilon}^{\mathrm{i}}, \vec{D}, T) = \frac{\mathrm{D}(\boldsymbol{\varepsilon} - \boldsymbol{\varepsilon}^{\mathrm{i}})}{\mathrm{D}t} : \frac{\partial \psi(\cdots)}{\partial(\boldsymbol{\varepsilon} - \boldsymbol{\varepsilon}^{\mathrm{i}})} + \frac{\mathrm{D}\vec{D}}{\mathrm{D}t} \cdot \frac{\partial \psi(\cdots)}{\partial \vec{D}} + \frac{\mathrm{D}T}{\mathrm{D}t}\, \frac{\partial \psi(\cdots)}{\partial T} \,. \tag{2.163}$$

The preceding notation $\mathrm{D}/\mathrm{D}t$ stands for the material time derivative operator, which describes how the material experiences the rate of change of its state. The state can be the absolute temperature $T$, for example. If $\mathbf{a}$ and $\mathbf{c}$ are arbitrary second-order tensors, the following holds: $\mathbf{c}:\mathbf{a} = \mathbf{a}:\mathbf{c}$. On the right side of Derivative (2.163), in the first two terms, there is a double-dot product between two second-order tensors. This implies

$$\frac{\mathrm{D}(\boldsymbol{\varepsilon}-\boldsymbol{\varepsilon}^{\mathrm{i}})}{\mathrm{D}t} : \frac{\partial\psi(\cdots)}{\partial(\boldsymbol{\varepsilon}-\boldsymbol{\varepsilon}^{\mathrm{i}})} = \frac{\partial\psi(\cdots)}{\partial(\boldsymbol{\varepsilon}-\boldsymbol{\varepsilon}^{\mathrm{i}})} : \frac{\mathrm{D}(\boldsymbol{\varepsilon}-\boldsymbol{\varepsilon}^{\mathrm{i}})}{\mathrm{D}t} \tag{2.164}$$

and a similar expression for vectors,

$$\frac{\mathrm{D}\vec{D}}{\mathrm{D}t} \cdot \frac{\partial\psi(\cdots)}{\partial\vec{D}} = \frac{\partial\psi(\cdots)}{\partial\vec{D}} \cdot \frac{\mathrm{D}\vec{D}}{\mathrm{D}t}. \tag{2.165}$$

The last term on the right side of Derivative (2.163) consists of two scalar-valued quantities the order of which can be interchanged [since $7 \times 8 = 8 \times 7$]. Based on this fact and on Expressions (2.164) and (2.165), Derivative (2.163) can be written in the form

$$\dot{\psi}(\boldsymbol{\varepsilon}-\boldsymbol{\varepsilon}^{\mathrm{i}}, \vec{D}, T) = \frac{\partial\psi(\cdots)}{\partial(\boldsymbol{\varepsilon}-\boldsymbol{\varepsilon}^{\mathrm{i}})} : \frac{\mathrm{D}(\boldsymbol{\varepsilon}-\boldsymbol{\varepsilon}^{\mathrm{i}})}{\mathrm{D}t} + \frac{\partial\psi(\cdots)}{\partial\vec{D}} \cdot \frac{\mathrm{D}\vec{D}}{\mathrm{D}t} + \frac{\partial\psi(\cdots)}{\partial T}\frac{\mathrm{D}T}{\mathrm{D}t}, \tag{2.166}$$

which yields

$$\dot{\psi}(\boldsymbol{\varepsilon}-\boldsymbol{\varepsilon}^{\mathrm{i}}, \vec{D}, T) = \frac{\partial\psi(\cdots)}{\partial(\boldsymbol{\varepsilon}-\boldsymbol{\varepsilon}^{\mathrm{i}})} : (\dot{\boldsymbol{\varepsilon}}-\dot{\boldsymbol{\varepsilon}}^{\mathrm{i}}) + \frac{\partial\psi(\cdots)}{\partial\vec{D}} \cdot \dot{\vec{D}} + \frac{\partial\psi(\cdots)}{\partial T}\dot{T}. \tag{2.167}$$

Form (2.167) is applied in this book, since that form is commonly used in the literature. The new quantities present in Expressions (2.163)–(2.167) are studied later in greater detail.

The result of the differentiation of tensorial expressions is sometimes difficult to obtain without detailed derivation, and the obtained result can be a surprise, even for an experienced researcher. Derivative (2.147) is a good example. It reads

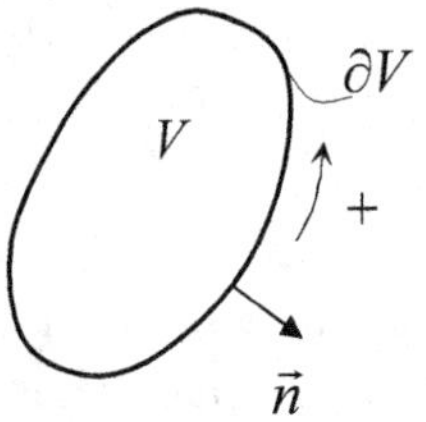


**Figure 2.4** Volume $V$, surface $\partial V$, unit outward normal vector $\vec{n}$, and positive integration path.

$$\frac{\partial(\vec{a}\cdot\mathbf{c}\cdot\mathbf{c}\cdot\vec{b})}{\partial\mathbf{c}} = \vec{a}\,(\mathbf{c}\cdot\vec{b}) + (\vec{a}\cdot\mathbf{c})\,\vec{b}, \tag{2.168}$$

where $\vec{a}$ and $\vec{b}$ are arbitrary vectors and $\mathbf{c}$ is an arbitrary second-order tensor. Thus detailed derivations are always needed.

The vector operator "del" denoted by $\nabla$ is defined in the (spatial) rectangular Cartesian coordinate system $(x_1, x_2, x_3)$ as follows:

$$\vec{\nabla}(\vec{x}) := \frac{\partial}{\partial\vec{x}} \qquad \text{or} \qquad \overleftarrow{\nabla}(\vec{x}) := \frac{\overleftarrow{\partial}}{\partial\vec{x}}. \tag{2.169}$$

The index forms of Definitions (2.169) are given in Definitions (2.185). A more detailed study of the vector operator "del" will be carried out in Chapter 3. Some illustrative examples of how to apply this operator will be given as well.

Modifying Coburn ([17], p. 74), the generalized Gauss's theorem is as follows: if $f$ is a continuous field (scalar, vector, etc.), and if $f$ possesses continuous partial derivatives in the closed subdomain $V$, then

$$\int_V \vec{\nabla} * f \, \mathrm{d}V = \oint_{\partial V} \vec{n} * f \, \mathrm{d}A \qquad \text{or} \qquad \int_V f * \overleftarrow{\nabla} \, \mathrm{d}V = \oint_{\partial V} f * \vec{n} \, \mathrm{d}A, \quad (2.170)$$

where $\partial V$ is the surface of the volume $V$, $\vec{n}$ is the outward unit normal to the volume $V$, and the "star product operator" $*$ in Theorem (2.170) represents a dot product operator $\cdot$, or a cross product operator $\times$, or it can be an empty space. Figure 2.4 shows the positive direction of the outward unit vector $\vec{n}$ and the positive integration path around the volume $V$. In the case of the function $f$ being a scalar, Theorem (2.170) takes the form

$$\int_V \vec{\nabla} f \, \mathrm{d}V = \oint_{\partial V} \vec{n} \, f \, \mathrm{d}A \qquad \text{and} \qquad \int_V f \overleftarrow{\nabla} \, \mathrm{d}V = \oint_{\partial V} f \, \vec{n} \, \mathrm{d}A. \quad (2.171)$$

Equations (2.170) and (2.171) (being in invariant form) are independent of a coordinate system.

## 2.6 Differentiation of Tensors Using Index Notation

Sometimes it is difficult to create derivatives of tensor-valued functions with respect to the tensorial variables. This can be seen in **Example 2.16**, which required a quite complicated derivation, carried out in Expressions (2.144) and (2.145). If such manipulations cannot be done, the index representation can be applied when derivatives are calculated. This section provides some tools for these actions.

In this book, partial derivatives of certain scalar-valued functions with respect to vectors or second-order tensors are considered. For example, the partial derivative of the specific Helmholtz free energy $\psi$ with respect to the strain tensor $\boldsymbol{\varepsilon}$ (a second-order tensor) is a second-order tensor, the components of which are the partial derivatives of the specific Helmholtz free energy $\psi$ with respect to the strain tensor components $\varepsilon_{in}$. A corresponding equation can be written for a partial derivative of a scalar with respect to a vector. Mathematically, these can be written in the forms

$$\frac{\partial \psi}{\partial \boldsymbol{\varepsilon}} = \vec{i}_i \, \vec{i}_n \, \frac{\partial \psi}{\partial \varepsilon_{in}} \qquad \text{and} \qquad \frac{\partial \psi}{\partial \vec{q}} = \vec{i}_n \, \frac{\partial \psi}{\partial q_n}. \quad (2.172)$$

Expressions (2.172) utilize the definition for a derivative of a quantity (tensor of any order) with respect to a second-order tensor and the definitions for a derivative of a quantity (tensor of any order) with respect to a vector. These definitions are as follows:

$$\textbf{Def} \qquad \frac{\partial}{\partial \mathbf{c}} := \vec{\imath}_i\,\vec{\imath}_n\,\frac{\partial}{\partial c_{in}} \qquad \text{and} \qquad \frac{\partial}{\partial \vec{v}} := \vec{\imath}_n\,\frac{\partial}{\partial v_n}\,, \tag{2.173}$$

where $\mathbf{c}$ is an arbitrary second-order tensor and $\vec{v}$ is an arbitrary vector.

Definition $(2.173)_1$ gives

$$\frac{\partial \mathbf{a}}{\partial \mathbf{c}} = \vec{\imath}_i\,\vec{\imath}_n\,\frac{\partial}{\partial c_{in}}\,a_{st}\,\vec{\imath}_s\,\vec{\imath}_t = \frac{\partial a_{st}}{\partial c_{in}}\,\vec{\imath}_i\,\vec{\imath}_n\,\vec{\imath}_s\,\vec{\imath}_t\,, \tag{2.174}$$

since the base vectors $\vec{\imath}_s$ and $\vec{\imath}_t$ are constants with respect to $c_{in}$.

Sometimes the derivative operator acts on the preceding quantity. In such a case, Definition (2.173) is replaced by the following one:

$$\textbf{Def} \qquad \frac{\overleftarrow{\partial}}{\partial \mathbf{c}} := \frac{\overleftarrow{\partial}}{\partial c_{in}}\,\vec{\imath}_i\,\vec{\imath}_n \qquad \text{and} \qquad \frac{\overleftarrow{\partial}}{\partial \vec{v}} := \frac{\overleftarrow{\partial}}{\partial v_n}\,\vec{\imath}_n\,. \tag{2.175}$$

The left arrow $\leftharpoonup$ on $\partial$ indicate, that the vector operator acts on the preceding quantity (i.e., the quantity on the left).

Definition $(2.175)_1$ gives

$$\begin{aligned} \mathbf{a}\,\frac{\overleftarrow{\partial}}{\partial \mathbf{c}} &= a_{st}\,\vec{\imath}_s\,\vec{\imath}_t\,\frac{\overleftarrow{\partial}}{\partial c_{in}}\,\vec{\imath}_i\,\vec{\imath}_n = a_{st}\,\frac{\overleftarrow{\partial}}{\partial c_{in}}\,\vec{\imath}_s\,\vec{\imath}_t\,\vec{\imath}_i\,\vec{\imath}_n = \frac{a_{st}\,\overleftarrow{\partial}}{\partial c_{in}}\,\vec{\imath}_s\,\vec{\imath}_t\,\vec{\imath}_i\,\vec{\imath}_n \\ &= \frac{\partial a_{st}}{\partial c_{in}}\,\vec{\imath}_s\,\vec{\imath}_t\,\vec{\imath}_i\,\vec{\imath}_n\,. \end{aligned} \tag{2.176}$$

Manipulation (2.176) exploited the following:

$$\frac{a_{st}\,\overleftarrow{\partial}}{\partial c_{in}} = \frac{\partial a_{st}}{\partial c_{in}}\,, \qquad \text{or, in general,} \qquad \frac{a_{st}\,\overleftarrow{\partial}}{\partial e} = \frac{\partial a_{st}}{\partial e}\,, \tag{2.177}$$

where $\mathbf{a}$ is an arbitrary tensor (scalar, vector, or higher-order tensor) and $e$ is a scalar or scalar component of a vector or a tensor.

**Example 2.19:** Derive the relation between the quantities on the left sides of Expressions (2.174) and (2.176).

**Solution:** According to Expression (2.95), the tensor transpose of an arbitrary fourth-order tensor $\mathbf{A}$ denoted by $\mathbf{A}^{\mathrm{T}}$ has the following property:

$$\mathbf{A} = A_{inkl}\,\vec{\imath}_i\,\vec{\imath}_n\,\vec{\imath}_k\,\vec{\imath}_l \qquad \text{by} \qquad \mathbf{A}^{\mathrm{T}} = A_{klin}\,\vec{\imath}_i\,\vec{\imath}_n\,\vec{\imath}_k\,\vec{\imath}_l\,. \tag{2.178}$$

Comparison of Derivatives (2.174) and (2.176) with Definition (2.178) yields

$$\mathbf{a}\,\frac{\overleftarrow{\partial}}{\partial \mathbf{c}} = \left(\frac{\partial \mathbf{a}}{\partial \mathbf{c}}\right)^{\mathrm{T}}. \qquad \blacksquare \tag{2.179}$$

**Example 2.20:** Derive the relation between the quantities

$$\vec{a}\,\frac{\overleftarrow{\partial}}{\partial \vec{v}} \qquad \text{and} \qquad \frac{\partial \vec{a}}{\partial \vec{v}}, \tag{2.180}$$

where $\vec{a}$ and $\vec{v}$ are arbitrary vectors.

**Solution:** According to Expression (2.74), if the second-order tensor $\mathbf{c}$ is expressed as

$$\mathbf{c} = c_{in}\,\vec{i}_i\,\vec{i}_n\,, \qquad \text{then} \qquad \mathbf{c}^{\mathrm{T}} := c_{ni}\,\vec{i}_i\,\vec{i}_n\,. \tag{2.181}$$

Definitions $(2.173)_2$ and $(2.175)_2$ lead to

$$\frac{\partial \vec{a}}{\partial \vec{v}} = \vec{i}_n\,\frac{\partial}{\partial v_n}\,a_s\,\vec{i}_s = \frac{\partial a_s}{\partial v_n}\,\vec{i}_n\,\vec{i}_s \tag{2.182}$$

and

$$\vec{a}\,\frac{\overleftarrow{\partial}}{\partial \vec{v}} = a_n\,\vec{i}_n\,\frac{\overleftarrow{\partial}}{\partial v_s}\,\vec{i}_s = \frac{a_n\,\overleftarrow{\partial}}{\partial v_s}\,\vec{i}_n\,\vec{i}_s = \frac{\partial a_n}{\partial v_s}\,\vec{i}_n\,\vec{i}_s\,, \tag{2.183}$$

where Expression (2.177) is exploited. Comparison of Derivatives (2.182) and (2.183) with Definition (2.181) leads to

$$\vec{a}\,\frac{\overleftarrow{\partial}}{\partial \vec{v}} = \left(\frac{\partial \vec{a}}{\partial \vec{v}}\right)^{\mathrm{T}}. \qquad \blacksquare \tag{2.184}$$

When index notation is applied, the vector operator "del" denoted by $\vec{\nabla}$ is defined in the (spatial) rectangular Cartesian coordinate system $(x_1, x_2, x_3)$ as follows:

$$\vec{\nabla}(\vec{x}) := \vec{i}_m\,\frac{\partial}{\partial x_m} \qquad \text{or} \qquad \overleftarrow{\nabla}(\vec{x}) := \frac{\overleftarrow{\partial}}{\partial x_m}\,\vec{i}_m\,. \tag{2.185}$$

Based on Result (2.184) and Notations (2.185), the following holds:

$$\vec{u}\,\overleftarrow{\nabla} = (\vec{\nabla}\,\vec{u})^{\mathrm{T}}. \tag{2.186}$$

The following results can also be written:

$$\frac{\partial c_{st}}{\partial c_{in}} = \delta_{is}\,\delta_{nt} \qquad \text{and} \qquad \frac{\partial a_i}{\partial a_j} = \delta_{ji}\,, \tag{2.187}$$

where $\mathbf{c}$ is an arbitrary second-order tensor and $\vec{a}$ is an arbitrary vector.

**Example 2.21:** Show that Derivative $(2.187)_1$ is correct.

**Solution:** Result (2.149) is

$$\frac{\partial \mathbf{c}}{\partial \mathbf{c}} = \mathbf{I}\,, \tag{2.188}$$

where $\mathbf{c}$ is an arbitrary second-order tensor. According to Expression $(2.70)_1$, the fourth-order identity tensor $\mathbf{I}$ has the following property:

$$\mathbf{I} = \delta_{ik}\,\delta_{jl}\,\vec{i}_i\,\vec{i}_j\,\vec{i}_k\,\vec{i}_l \qquad \Rightarrow \qquad \mathbf{I} = \delta_{is}\,\delta_{nt}\,\vec{i}_i\,\vec{i}_n\,\vec{i}_s\,\vec{i}_t\,, \tag{2.189}$$

where the indices $(i, j, k)$ are replaced with the indices $(s, n, t)$. Based on Derivative (2.174), we arrive at the following:

$$\frac{\partial \mathbf{c}}{\partial \mathbf{c}} = \frac{\partial c_{st}}{\partial c_{in}}\,\vec{i}_i\,\vec{i}_n\,\vec{i}_s\,\vec{i}_t\,. \tag{2.190}$$

Substituting Expression $(2.189)_2$ into Derivative (2.188) and comparing the obtained result with (2.190) gives

$$\frac{\partial c_{st}}{\partial c_{in}} = \delta_{is}\,\delta_{nt}\,. \tag{2.191}$$

The equality of Derivatives $(2.187)_1$ and (2.191) proves Derivative $(2.187)_1$. ■

**Example 2.22:** Show that Derivative $(2.187)_2$ is correct.

**Solution:** Result (2.151) reads

$$\frac{\partial \vec{a}}{\partial \vec{a}} = \mathbf{1}\,. \tag{2.192}$$

where $\vec{a}$ is an arbitrary vector. According to Expression (2.64), the second-order identity tensor $\mathbf{1}$ has the following property:

$$\mathbf{1} = \delta_{ij}\,\vec{i}_i\,\vec{i}_j \qquad \Rightarrow \qquad \mathbf{1} = \delta_{ns}\,\vec{i}_n\,\vec{i}_s\,, \tag{2.193}$$

Based on Definition $(2.173)_2$, the following is achieved:

$$\frac{\partial \vec{a}}{\partial \vec{a}} = \vec{i}_n\,\frac{\partial}{\partial a_n}\,a_s\,\vec{i}_s = \frac{\partial a_s}{\partial a_n}\,\vec{i}_n\,\vec{i}_s\,. \tag{2.194}$$

Comparison of Expressions (2.192), $(2.193)_2$, and (2.194) gives

$$\frac{\partial a_s}{\partial a_n} = \delta_{ns}\,. \tag{2.195}$$

Result (2.195) proves Derivative $(2.187)_2$. ■

To demonstrate the process to calculate derivatives of tensor-valued functions with respect to tensors, **Example 2.16** and **Example 2.17** are recalculated using the index form.

**Example 2.23:** Let $\mathbf{c}$ be an arbitrary second-order tensor and $\vec{a}$ and $\vec{b}$ arbitrary vectors. Calculate the derivative $\partial(\vec{a}\cdot\mathbf{c}\cdot\mathbf{c}\cdot\vec{b})/\partial\mathbf{c}$ using the index form.

**Solution:** The following tensors are introduced:

$$\vec{a} = a_i\,\vec{i}_i\,, \qquad \mathbf{c} = c_{mn}\,\vec{i}_m\,\vec{i}_n\,, \qquad \text{and} \qquad \vec{b} = b_u\,\vec{i}_u\,. \tag{2.196}$$

Substitution of Tensors (2.196) into the function $(\vec{a}\cdot\mathbf{c}\cdot\mathbf{c}\cdot\vec{b})$ yields (the second tensor $\mathbf{c}$ takes indices $s$ and $t$)

$$\begin{aligned} \vec{a}\cdot\mathbf{c}\cdot\mathbf{c}\cdot\vec{b} &= a_i\,\vec{i}_i\cdot c_{mn}\,\vec{i}_m\,\vec{i}_n\cdot c_{st}\,\vec{i}_s\,\vec{i}_t\cdot b_u\,\vec{i}_u = a_i\,\delta_{im}\,c_{mn}\,\delta_{ns}\,c_{st}\,\delta_{tu}\,b_u \\ &= a_m\,c_{ms}\,c_{st}\,b_t\,. \end{aligned} \tag{2.197}$$

Now, the derivative can be formulated by taking Definition $(2.173)_1$ and Expression $(2.187)_1$ into account. The following result is achieved:

$$\begin{aligned} \frac{\partial}{\partial\mathbf{c}}(\vec{a}\cdot\mathbf{c}\cdot\mathbf{c}\cdot\vec{b}) &= \frac{\partial}{\partial\mathbf{c}}(a_m\,c_{ms}\,c_{st}\,b_t) = \vec{i}_i\,\vec{i}_n\,\frac{\partial}{\partial c_{in}}(a_m\,c_{ms}\,c_{st}\,b_t) \\ &= \vec{i}_i\,\vec{i}_n\,a_m\,\frac{\partial}{\partial c_{in}}(c_{ms}\,c_{st})\,b_t = \vec{i}_i\,\vec{i}_n\,a_m\,(\delta_{im}\,\delta_{ns}\,c_{st} + c_{ms}\,\delta_{is}\,\delta_{nt})\,b_t \\ &= \vec{i}_i\,\vec{i}_n\,a_m\,(\delta_{im}\,c_{nt} + c_{mi}\,\delta_{nt})\,b_t = \vec{i}_i\,\vec{i}_n\,(a_i\,c_{nt}\,b_t + a_m\,c_{mi}\,b_n)\,. \end{aligned} \tag{2.198}$$

The result of Expression (2.198) can be written as follows:

$$\frac{\partial}{\partial\mathbf{c}}(\vec{a}\cdot\mathbf{c}\cdot\mathbf{c}\cdot\vec{b}) = \vec{i}_i\,\vec{i}_n\,(a_i\,c_{nt}\,b_t + a_m\,c_{mi}\,b_n) = \vec{a}\,(\mathbf{c}\cdot\vec{b}) + (\vec{a}\cdot\mathbf{c})\,\vec{b}\,. \tag{2.199}$$

Expression (2.199) equals that of Result (2.147). The parentheses in Expression (2.199) arc not necessary to add but may help in interpreting the result. ■

**Example 2.24:** Let $\mathbf{c}$ be an arbitrary second-order tensor. Calculate the derivative $\partial\mathbf{c}/\partial\mathbf{c}$ using index form.

**Solution:** The following tensor is introduced:

$$\mathbf{c} = c_{st}\,\vec{i}_s\,\vec{i}_t\,. \tag{2.200}$$

Index Form (2.200), Definition $(2.187)_1$, and Expression$(2.189)_2$ give

$$\frac{\partial\mathbf{c}}{\partial\mathbf{c}} = \vec{i}_i\,\vec{i}_n\,\frac{\partial c_{st}}{\partial c_{in}}\,\vec{i}_s\,\vec{i}_t = \delta_{is}\,\delta_{nt}\,\vec{i}_i\,\vec{i}_n\,\vec{i}_s\,\vec{i}_t = \mathbf{I}\,. \tag{2.201}$$

Result (2.201) coincides with Result (2.149). ■

## 2.7 Differentiation of Tensors Using Gibbs Notation

Sometimes derivatives of tensor-valued functions with respect to the tensorial variables can be determined by using the Gibbs notation without using the directional derivative introduced in Section 2.5.

**Example 2.25:** Let $\mathbf{s}$ be a second-order deviatoric stress tensor given by Definition $(2.108)_1$, $\mathbf{s} := \mathbf{K} : \boldsymbol{\sigma}$. Calculate the derivative $\partial[\mathbf{s} : \mathbf{s}]/\partial\boldsymbol{\sigma}$ using neither Directional Derivative (2.127) nor the index notation introduced in Section 2.6.

**Solution:** This problem is the same as in **Example 2.15**. Definitions (2.108) read

$$\mathbf{s} := \mathbf{K} : \boldsymbol{\sigma}\,, \qquad \text{where the fourth-order tensor} \qquad \mathbf{K} = \mathbf{I} - \tfrac{1}{3}\mathbf{1}\,\mathbf{1}\,. \tag{2.202}$$

Derivative (2.149) is

$$\frac{\partial \mathbf{c}}{\partial \mathbf{c}} = \mathbf{I}\,, \tag{2.203}$$

where $\mathbf{c}$ is an arbitrary second-order tensor. Based on Definition $(2.202)_1$ and Derivative (2.203), the following result is obtained:

$$\begin{aligned}\frac{\partial}{\partial\boldsymbol{\sigma}}\{\mathbf{s}:\mathbf{s}\} &= \frac{\partial}{\partial\boldsymbol{\sigma}}\Big\{(\mathbf{K}:\boldsymbol{\sigma}):(\mathbf{K}:\boldsymbol{\sigma})\Big\} = \left(\mathbf{K}:\frac{\partial\boldsymbol{\sigma}}{\partial\boldsymbol{\sigma}}\right):(\mathbf{K}:\boldsymbol{\sigma}) + (\mathbf{K}:\boldsymbol{\sigma}):\left(\mathbf{K}:\frac{\partial\boldsymbol{\sigma}}{\partial\boldsymbol{\sigma}}\right)\\ &= (\mathbf{K}:\mathbf{I}):(\mathbf{K}:\boldsymbol{\sigma}) + (\mathbf{K}:\boldsymbol{\sigma}):(\mathbf{K}:\mathbf{I}) = (\mathbf{K}:\mathbf{I}):\mathbf{s} + \mathbf{s}:(\mathbf{K}:\mathbf{I})\\ &= \mathbf{K}:(\mathbf{I}:\mathbf{s}) + (\mathbf{s}:\mathbf{K}):\mathbf{I}\,.\end{aligned} \tag{2.204}$$

The third line of Expression (2.204) was possible to write because within the second line after the second equals sign is a term $(\mathbf{K}:\mathbf{I}):\mathbf{s}$, where the double-dot product between $\mathbf{K}$ and $\mathbf{I}$ is performed with the first two base vectors of $\mathbf{I}$ and the double-dot product between $\mathbf{I}$ and $\mathbf{s}$ is performed with the latter two base vectors of $\mathbf{I}$. Thus there is no need to have the parentheses either at the end of the second line or on the third line. The same is true of the last terms on the second and third lines. The parentheses on the third line are only for clarification.

Definition (2.69) and Result (2.140) are

$$\textbf{Def} \qquad \mathbf{I}:\mathbf{c} = \mathbf{c}:\mathbf{I} = \mathbf{c} \qquad \text{and} \qquad \mathbf{s}:\mathbf{K} = \mathbf{s}\,. \tag{2.205}$$

Based on Expressions (2.205), Derivative (2.204) takes the form

$$\frac{\partial}{\partial\boldsymbol{\sigma}}\{\mathbf{s}:\mathbf{s}\} = \mathbf{K}:(\mathbf{I}:\mathbf{s}) + (\mathbf{s}:\mathbf{K}):\mathbf{I} = \mathbf{K}:\mathbf{s} + \mathbf{s}:\mathbf{I} = \mathbf{s} + \mathbf{s} = 2\,\mathbf{s}\,. \tag{2.206}$$

Manipulation (2.206) exploited the result $\mathbf{K} : \mathrm{s} = \mathrm{s}$, which is easy to see as being correct from Equation (2.140). ■

## 2.8 Miscellaneous Expressions

By extending the definition for homogeneous functions given, for example, by Widder ([117], pp. 19 and 20), the following is achieved: function $\theta(x, y, z, u, v)$ is homogeneous of degree $\omega$ in variables $x$, $y$, and $z$ in a region $R$ if and only if, for $x$, $y$, and $z$ in $R$ and for every positive value of $k$, the following holds:

$$\theta(k\,x, k\,y, k\,z, u, v) := k^{\omega}\,\theta(x, y, z, u, v)\,. \tag{2.207}$$

Sometimes the definition is assumed to hold for every real $k$, and if the values of $k$ are restricted to being positive, the function $\theta(x, y, z, u, v)$ is said to be a positive homogeneous function.

Euler's theorem on homogeneous functions (see the original form in Widder [117], p. 20) for the preceding extended definition reads

$$\omega\,\theta(x, y, z, u, v) = \frac{\partial\theta}{\partial x}\,x + \frac{\partial\theta}{\partial y}\,y + \frac{\partial\theta}{\partial z}\,z\,. \tag{2.208}$$

The quantities $x$, $y$, and $z$ are not coordinates, as this book uses different sets of coordinates.

## 2.9 Voigt Notation

This section introduces the Voigt notation.

So far, the Gibbs notation (boldface letters) and the index notation are used. They are not convenient in the implementation of material models in computer code. Although not yet introduced in this book, Hooke's law is studied as an example. When the Gibbs notation is used, Hooke's law takes the following appearance:

$$\boldsymbol{\sigma} = \mathbf{C} : \boldsymbol{\varepsilon}^{\mathrm{e}}\,. \tag{2.209}$$

As discussed before, the stress tensor $\boldsymbol{\sigma}$ and the elastic strain tensor $\boldsymbol{\varepsilon}^{\mathrm{e}}$ are second-order tensors, and the constitutive tensor for a Hookean material $\mathbf{C}$ is a fourth-order tensor. For a three-dimensional case, the components of the stress tensors can be collected into the following matrix:

$$\begin{bmatrix} \sigma_{11} & \sigma_{12} & \sigma_{13} \\ \sigma_{21} & \sigma_{22} & \sigma_{23} \\ \sigma_{31} & \sigma_{32} & \sigma_{33} \end{bmatrix}\,. \tag{2.210}$$

The corresponding quantity to Expression (2.210) for the constitutive tensor for a Hookean deformation $\mathbf{C}$ is a fourth-order tensor, and therefore it has eighty-one components. To handle Hooke's law is a remarkable task for a computer, and its programming requires remarkable coding effort. Therefore an easier and a computationally faster method is sought. The Voigt notation is the answer to these needs. From the preliminary solid mechanics course, it is known that the small deformation stress tensor $\boldsymbol{\sigma}$ is symmetric, $\tau_{xy} = \tau_{yx}$. By index notation, it means that $\sigma_{12} = \sigma_{21}$. The symmetry of the stress tensor $\boldsymbol{\sigma}$ can in general be written $\sigma_{ij} = \sigma_{ji}$. This means that Expression (2.210) has redundant information, since some of its components have equal values, although their definitions differ.

Instead of Expression (2.210), the following notation is used to reduce the number of stress components:

$$\{\sigma\} = \{\sigma_{11}, \sigma_{22}, \sigma_{33}, \sigma_{12}, \sigma_{13}, \sigma_{23}\}^{\mathrm{T}}, \tag{2.211}$$

where the column vector is expressed by the notation $\{\cdot\}$ and the row vector is expressed by the notation $\{\cdot\}$. Since the small deformation elastic strain tensor $\boldsymbol{\varepsilon}^{\mathrm{e}}$ is modelled as a symmetric tensor, we arrive at the following:

$$\{\varepsilon^{\mathrm{e}}\} = \{\varepsilon^{\mathrm{e}}_{11}, \varepsilon^{\mathrm{e}}_{22}, \varepsilon^{\mathrm{e}}_{33}, \varepsilon^{\mathrm{e}}_{12}, \varepsilon^{\mathrm{e}}_{13}, \varepsilon^{\mathrm{e}}_{23}\}^{\mathrm{T}}. \tag{2.212}$$

When Notations (2.211) and (2.212) are used, the constitutive tensor for a Hookean deformation $\mathbf{C}$ is replaced by a matrix. Based on the foregoing discussion, Hooke's law (for an isotropic material) takes the following appearance:

$$\begin{Bmatrix} \sigma_{11} \\ \sigma_{22} \\ \sigma_{33} \\ \sigma_{12} \\ \sigma_{13} \\ \sigma_{23} \end{Bmatrix} = \frac{E}{(1+\nu)(1-2\nu)} \begin{bmatrix} 1-\nu & \nu & \nu & 0 & 0 & 0 \\ \nu & 1-\nu & \nu & 0 & 0 & 0 \\ \nu & \nu & 1-\nu & 0 & 0 & 0 \\ 0 & 0 & 0 & 1-2\nu & 0 & 0 \\ 0 & 0 & 0 & 0 & 1-2\nu & 0 \\ 0 & 0 & 0 & 0 & 0 & 1-2\nu \end{bmatrix} \begin{Bmatrix} \varepsilon^{\mathrm{e}}_{11} \\ \varepsilon^{\mathrm{e}}_{22} \\ \varepsilon^{\mathrm{e}}_{33} \\ \varepsilon^{\mathrm{e}}_{12} \\ \varepsilon^{\mathrm{e}}_{13} \\ \varepsilon^{\mathrm{e}}_{23} \end{Bmatrix} \tag{2.213}$$

where $E$ is Young's modulus and $\nu$ is Poisson's ratio. Expression (2.213) is easy to program and does not require many lines of computer code. Expressions (2.211)–(2.213) are not quite the same as those which are usually called Voigt notation.

Before introducing the Voigt notation, a few words are in order on the method the Abaqus finite element method computer code uses to organize the stress and strain components for structural analysis. Small deformations are studied. Abaqus has two different codes: Abaqus standard and Abaqus explicit. Unfortunately, the cores of the codes are different, which has an effect on the programming of the material models.

For the plane stress condition and for the three-dimensional computation, the Abaqus UMAT subroutine (for the Abaqus standard) has to be programmed by using the following vectors. See Abaqus Analysis User's Guide ([1], pp. 22.2.1–5 and 22.2.1–2):

$$\begin{Bmatrix} \varepsilon_{11} \\ \varepsilon_{22} \\ \gamma_{12} \end{Bmatrix} \quad \text{and} \quad \begin{Bmatrix} \sigma_{11} \\ \sigma_{22} \\ \sigma_{12} \end{Bmatrix} \qquad \text{or} \qquad \begin{Bmatrix} \varepsilon_{11} \\ \varepsilon_{22} \\ \varepsilon_{33} \\ \gamma_{12} \\ \gamma_{13} \\ \gamma_{23} \end{Bmatrix} \quad \text{and} \quad \begin{Bmatrix} \sigma_{11} \\ \sigma_{22} \\ \sigma_{33} \\ \sigma_{12} \\ \sigma_{13} \\ \sigma_{23} \end{Bmatrix} . \tag{2.214}$$

On the other hand, the Abaqus VUMAT subroutine (for the Abaqus explicit) uses for the plane stress condition and for three-dimensional computation the following vectors. See Abaqus User Subroutines Reference Guide ([1], p. 1.2.22–2):

$$\begin{Bmatrix} \varepsilon_{11} \\ \varepsilon_{22} \\ \varepsilon_{33} \\ \varepsilon_{12} \end{Bmatrix} \quad \text{and} \quad \begin{Bmatrix} \sigma_{11} \\ \sigma_{22} \\ \sigma_{33} \\ \sigma_{12} \end{Bmatrix} \qquad \text{or} \qquad \begin{Bmatrix} \varepsilon_{11} \\ \varepsilon_{22} \\ \varepsilon_{33} \\ \varepsilon_{12} \\ \varepsilon_{23} \\ \varepsilon_{31} \end{Bmatrix} \quad \text{and} \quad \begin{Bmatrix} \sigma_{11} \\ \sigma_{22} \\ \sigma_{33} \\ \sigma_{12} \\ \sigma_{23} \\ \sigma_{31} \end{Bmatrix} . \tag{2.215}$$

It is important to note that not only does the number of stress and strain components differ, but also the shear strain representations are different. Abaqus standard uses engineering shear strain components, for example, $\gamma_{12}$, whereas the Abaqus explicit applies the tensorial shear strain components, for example, $\varepsilon_{12}$. The expressions between the engineering shear strain components, $\gamma_{12}$, and the tensorial shear components, $\varepsilon_{12}$, and so on read

$$\gamma_{12} = 2\,\varepsilon_{12} , \qquad \gamma_{13} = 2\,\varepsilon_{13} , \qquad \text{and} \qquad \gamma_{23} = 2\,\varepsilon_{23} . \tag{2.216}$$

When the Voigt notation is used, the two indices in the stress and strain components are replaced with one index. This is (see, e.g., Reddy [84], Section 6.2.2)

$$\begin{aligned} \sigma_1 &= \sigma_{11} , & \sigma_2 &= \sigma_{22} , & \sigma_3 &= \sigma_{33} , \\ \sigma_4 &= \sigma_{23} , & \sigma_5 &= \sigma_{13} , & \sigma_6 &= \sigma_{12} \end{aligned} \tag{2.217}$$

and

$$\begin{aligned} \varepsilon_1 &= \varepsilon_{11} , & \varepsilon_2 &= \varepsilon_{22} , & \varepsilon_3 &= \varepsilon_{33} , \\ \varepsilon_4 &= 2\,\varepsilon_{23} , & \varepsilon_5 &= 2\,\varepsilon_{13} , & \varepsilon_6 &= 2\,\varepsilon_{12} . \end{aligned} \tag{2.218}$$

Now, Hooke's Law (2.213) takes the form

$$\sigma_s = C_{st}\,\varepsilon_t^{\mathrm{e}} , \tag{2.219}$$

where the nonzero components $C_{st}$ are

$$C_{11} = C_{22} = C_{33} = \frac{(1-\nu)}{(1+\nu)\,(1-2\,\nu)}\,E , \tag{2.220}$$

$$C_{12} = C_{13} = C_{21} = C_{23} = C_{31} = C_{32} = \frac{\nu}{(1+\nu)\,(1-2\,\nu)}\,E\,, \tag{2.221}$$

and

$$C_{44} = C_{55} = C_{66} = \frac{(1-2\,\nu)}{2(1+\nu)\,(1-2\,\nu)}\,E\,. \tag{2.222}$$

When expressed in matrix form, the Voigt form of Hooke's law takes the following appearance:

$$\{\sigma\} = [\,C\,]\,\{\,\varepsilon^{\rm e}\,\}\,, \tag{2.223}$$

where

$$\{\sigma\} = \{\sigma_1,\, \sigma_2,\, \sigma_3,\, \sigma_4,\, \sigma_5,\, \sigma_6\}^{\rm T} \;\; \text{and} \;\; \{\varepsilon^{\rm e}\} = \{\varepsilon^{\rm e}_1,\, \varepsilon^{\rm e}_2,\, \varepsilon^{\rm e}_3,\, \varepsilon^{\rm e}_4,\, \varepsilon^{\rm e}_5,\, \varepsilon^{\rm e}_6\}^{\rm T}\,, \tag{2.224}$$

and finally,

$$[\,C\,] = \begin{bmatrix} C_{11} & C_{12} & C_{13} & 0 & 0 & 0 \\ C_{21} & C_{22} & C_{23} & 0 & 0 & 0 \\ C_{31} & C_{32} & C_{33} & 0 & 0 & 0 \\ 0 & 0 & 0 & C_{44} & 0 & 0 \\ 0 & 0 & 0 & 0 & C_{55} & 0 \\ 0 & 0 & 0 & 0 & 0 & C_{66} \end{bmatrix}. \tag{2.225}$$

## 2.10 Summary

In this chapter we introduced all the mathematical expressions that the reader will need to study successfully the entire contents of this book. Since tensor notations are used here, the main effort is on tensor algebra and its application.

Tensor notations may be new to a student reading this book. To be able to understand the message behind the equations written with tensor algebra, the student may need to practice by redoing the solved examples in this chapter (the reader may also consult the book by Reddy [84] for additional examples). This is especially important because this book is written with the Gibbs notation, which means that no indices of the tensors are usually given. The guidelines for performing the dot product and double-dot product help change the expressions given by the Gibbs notation into the index notation, which may be easier to realize for a newcomer to continuum mechanics and continuum thermodynamics. However, the expressions written with the Gibbs notations are clearer, since the missing indices allow the reader to concentrate on the message of the equations.

## Problems

**2.1** Prove the following expressions:

$$\text{(a)}\;\; \delta_{\rm ij}\,{\rm b_j} = {\rm b_i} \qquad \text{and} \qquad \text{(b)}\;\; \delta_{\rm ij}\,\delta_{\rm is} = \delta_{\rm js}\,. \tag{1}$$

**Hint.** Apply the Einstein summation convention.

**2.2** Let us assume that you have applied mathematical operations and obtained the following expression:

$$a_t\,\beta_{tu}\,\vec{e}_u = c_s\,\vec{e}_s\,, \tag{1}$$

where $\vec{e}_1$, $\vec{e}_2$, and $\vec{e}_3$ are the base vectors.

(a) Give the relations between the scalar components of the variables (not the base vectors) in the form $v_i \ldots = u_i \ldots$, where the dots $\ldots$ represent other potential variables, and then

(b) replace the indices with the values 1, 2, and 3.

**2.3** Verify the following identities:

(a) $e_{mjk}\,e_{njk} = 2\,\delta_{mn}$ (1)

(b) $e_{ijk}\,e_{ijk} = 6$ (1)

(c) $a_i\,a_j\,e_{ijk} = 0$ (1)

where $e_{ijk}$ is the permutation symbol [see Definition (2.123)] and $a_i$ are the components of a vector $\vec{a}$.

**Hint.** Please, notice the $e - \delta$ identity, that is, $e_{ijk}\,e_{mnk} = \delta_{im}\,\delta_{jn} - \delta_{in}\,\delta_{jm}$.

**2.4** Show that the following expression holds:

$$\underline{\vec{n}} \times \underline{\vec{a}} = \underline{\vec{n}} \cdot (\mathbf{1} \times \underline{\vec{a}})\,, \tag{1}$$

where $\mathbf{1}$ is the second-order identity tensor given by Definition (2.63) and $\underline{\vec{n}}$ and $\underline{\vec{a}}$ are arbitrary vectors.

**Hint.** The cross product $\times$ between a second-order tensor and a vector is defined to be an act between the second base vector of the second-order tensor and the base vector of the vector.

**2.5** Show that the following relation holds:

$$\mathbf{c} \cdot \mathbf{1} = \mathbf{c}\,, \tag{1}$$

where $\mathbf{c}$ is an arbitrary second-order tensor and $\mathbf{1}$ is a second-order identity tensor.

**2.6** Show that for arbitrary second-order tensors $\mathbf{a}$ and $\mathbf{c}$, the following holds:

$$(\mathbf{a} \cdot \mathbf{c})^{\mathrm{T}} = \mathbf{c}^{\mathrm{T}} \cdot \mathbf{a}^{\mathrm{T}}\,, \tag{1}$$

where the superscript T stands for the tensor transpose.

**2.7** Let $\mathbf{A}$ be a second-order tensor and $\mathbf{B}$ be a third-order tensor.

(a) Write product $\mathbf{A}\!:\!\mathbf{B}$ in all the four sets of indices.

(b) Why do these results express the same solution? Show the equality by comparing two results.

**Hint.** Apply the index form and the Einstein summation convention.

**2.8** Perform the following double-dot products and give their results:

$$\text{(a)}\ \mathbf{1}\!:\!\boldsymbol{\sigma} \qquad \text{and} \qquad \text{(b)}\ \mathbf{1}\!:\!\mathbf{1}\,, \tag{1}$$

where $\mathbf{1}$ is a second-order identity tensor and $\boldsymbol{\sigma}$ is an arbitrary second-order tensor.

**2.9** Is the fourth-order identity tensor $\mathbf{I}$ a symmetric tensor in the first pair of indices?

**2.10** Is the fourth-order identity tensor $\mathbf{I}$ a symmetric tensor in the second pair of indices?

**2.11** Show that the fourth-order identity tensor $\mathbf{I}^{\mathrm{s}}$ is a minor symmetric tensor in the first pair of indices.

**2.12** Show that Definition $(2.90)_1$ for the minor symmetry of the first pair of indices for an arbitrary fourth-order tensor $\mathbf{A}$ is correct. Definition $(2.90)_1$ reads

**Def** $$\vec{a}\vec{b} : \mathbf{A} = \vec{b}\vec{a} : \mathbf{A}\,, \tag{1}$$

where $\vec{a}$ and $\vec{b}$ are arbitrary vectors.

**2.13** Show that the following theorem holds:

$$\mathbf{r}\!:\!\mathbf{c} = (\mathbf{r}\!\cdot\!\mathbf{c}^{\mathrm{T}})\!:\!\mathbf{1}\,, \tag{1}$$

where $\mathbf{r}$ and $\mathbf{c}$ are arbitrary second-order tensors and $\mathbf{1}$ is the second-order identity tensor.

**2.14** Show that the following theorem holds:

$$\mathbf{r}\!:\!\mathbf{c} = (\mathbf{c}\!\cdot\!\mathbf{r}^{\mathrm{T}})\!:\!\mathbf{1}\,, \tag{1}$$

where $\mathbf{r}$ and $\mathbf{c}$ are arbitrary second-order tensors and $\mathbf{1}$ is the second-order identity tensor.

**2.15** The fourth-order tensor $\mathbf{A}$ is said to be $(\cdots)$ if the following holds:

**Def** $$\mathbf{c} : \mathbf{A} : \mathbf{g} = \mathbf{g} : \mathbf{A} : \mathbf{c}\,, \tag{1}$$

where $\mathbf{c}$ and $\mathbf{g}$ are arbitrary second-order tensors. Perform the double-dot products and study the scalar components of the tensor $\mathbf{A}$. What is the aforementioned property $(\cdots)$ of the tensor $\mathbf{A}$?

**2.16** Show that the following definition holds:

**Def** $$\mathbf{I}^{\mathrm{s}} : \mathbf{c} = \tfrac{1}{2}\,(\mathbf{c} + \mathbf{c}^{\mathrm{T}})\,, \tag{1}$$

if the fourth-order symmetric identity tensor $\mathbf{I}^{\mathrm{s}}$ takes the appearance

$$\mathbf{I}^{\mathrm{s}} = \tfrac{1}{2}\,(\delta_{ik}\,\delta_{jl} + \delta_{il}\,\delta_{jk})\,\vec{i}_i\,\vec{i}_j\,\vec{i}_k\,\vec{i}_l \tag{2}$$

and $\mathbf{c}$ is an arbitrary second-order tensor.

**2.17** Calculate the expression

$$\mathbf{I}^{\mathrm{s}} : \boldsymbol{\sigma} \tag{1}$$

using the index form for the tensors $\mathbf{I}^{\mathrm{s}}$ and $\boldsymbol{\sigma}$. The notation $\boldsymbol{\sigma}$ stands for the small deformation stress tensor, which is a symmetric second-order tensor.

**2.18** The quantities $\mathbf{a}$, $\mathbf{b}$, and $\mathbf{e}$ are assumed to be second-order tensors. Show that the following expression is correct by carrying out the double-dot products and by using the necessary derivatives. Use the index form for $\mathbf{a}$, $\mathbf{b}$, and $\mathbf{e}$.

$$\frac{\partial}{\partial \mathbf{e}}\,(\mathbf{a} : \mathbf{b}) = \frac{\partial \mathbf{a}}{\partial \mathbf{e}} : \mathbf{b} + \mathbf{a} : \frac{\mathbf{b}\overleftarrow{\partial}}{\partial \mathbf{e}}\,. \tag{1}$$

**2.19** Let $\mathbf{c}$ be an arbitrary second-order tensor and $\vec{a}$ be an arbitrary vector.

(a) Calculate the derivative

$$\frac{\partial(\vec{a}\cdot\mathbf{c}\cdot\vec{a})}{\partial \vec{a}}\,. \tag{1}$$

First perform the dot products, then make use of Definition $(2.173)_2$.

(b) What is the result if the tensor $\mathbf{c}$ is symmetric?

**2.20** Calculate the following:

$$\text{(a)}\;\; \vec{\nabla}(\vec{x})\,\vec{x} \qquad \text{and} \qquad \text{(b)}\;\; \vec{\nabla}(\vec{x})\cdot\vec{x}\,. \tag{1}$$

**Hint.** Express operator $\vec{\nabla}$ and variable $\vec{x}$ in the component form.

**2.21** Show that the following relation holds:

$$\mathbf{1} : \vec{\nabla}\vec{v} = \vec{\nabla}\cdot\vec{v}\,, \tag{1}$$

where $\mathbf{1}$ is the second-order identity tensor, $\vec{\nabla}$ is the vector operator del, and $\vec{v}$ is an arbitrary vector.

**2.22** Show that the following relation holds:

$$\vec{\nabla}(\vec{\nabla}\cdot\vec{v}) = \vec{\nabla}\cdot(\vec{v}\overleftarrow{\nabla})\,, \tag{1}$$

where $\vec{\nabla}$ is the vector operator del and $\vec{v}$ is an arbitrary vector.

**2.23** Show that the following equality holds:

$$\vec{\nabla}\cdot(\mathbf{h}\cdot\vec{e}) = (\vec{\nabla}\cdot\mathbf{h})\cdot\vec{e} + \mathbf{h} : \vec{\nabla}\vec{e}\,, \tag{1}$$

where $\vec{\nabla}$ is the vector operator del, $\mathbf{h}$ is a second-order tensor, and $\vec{e}$ is a vector.

**2.24** Calculate (simplify) the following relations ($x^2 = \vec{x}\cdot\vec{x}$, $\vec{a}$ is a vector). The notation $x_i$ stands for the axis of the coordinate system ($x_1$, $x_2$, $x_3$). **Hint.** Notice that the vector $\vec{a}$ is the acceleration vector, for example, expressed in the ($x_1$, $x_2$, $x_3$) frame. Use the index notation.

(a) $\vec{\nabla}(\vec{x})(x^n)$ (1)

(b) $\vec{\nabla}(\vec{x})\,(\vec{x}\cdot\vec{a})$ (1)

(c) $\vec{\nabla}(\vec{x})\cdot(\vec{x}\times\vec{a})$ (1)

**2.25** Let $\mathbf{c}$ be an arbitrary second-order tensor and $\vec{a}$ be an arbitrary vector.

(a) Calculate the derivative

$$\frac{\partial(\vec{a}\cdot\mathbf{c}\cdot\vec{a})}{\partial\vec{a}} \tag{1}$$

using the directional derivative defined in Section 2.5.

(b) What is the result if the tensor $\mathbf{c}$ is symmetric?

**2.26** Let $\mathbf{c}$ be an arbitrary second-order tensor and $\vec{a}$ and $\vec{b}$ be arbitrary vectors. Calculate the derivative

$$\frac{\partial(\vec{a}\cdot\mathbf{c}\cdot\vec{b})}{\partial\mathbf{c}} \tag{1}$$

using the directional derivative defined in Section 2.5.

**2.27** Calculate the following derivative using the directional derivative concept given in Section 2.5:

$$\frac{\partial(\vec{N}\cdot\boldsymbol{\sigma}\cdot\boldsymbol{\sigma}\cdot\vec{N})}{\partial\vec{N}}, \tag{1}$$

where $\boldsymbol{\sigma}$ is the small deformation stress tensor, being a symmetric second-order tensor.

**Hint.** After applying Lemma A [see Directional Derivative (2.127)], you get two terms. One of the terms is fine, but the other is not, and therefore you have to calculate the dot products. Try to do that without introducing the base vectors. This term can be studied without calculating the dot products, but it is a much more difficult approach.

**2.28** Lemma A [see Directional Derivative (2.127)] introduces the directional derivative for calculation of the derivatives of tensor-valued functions $f(x,\cdot)$ with respect to tensorial variables $x$ of any order. In **Example 2.18**, for example, the product rule for the directional derivative is applied. Derive the product rule for the case in which the function $f(x,\cdot)$ has the form

$$f(x,\cdot) = k(x,\cdot)\odot g(x,\cdot), \tag{1}$$

where the notation $\odot$ stands for an empty space, a dot product, a double-dot product, and so on.

# 3 Kinematics of a Continuum and Integral Rules

## 3.1 Preliminary Comments

The present chapter is dedicated to the study of the kinematics of a continuous material. This means that the motion and deformation of a system (body) are studied. There are two descriptions of motion to investigate the response of a material (see Reddy [84]). The description of motion can be applied to the matter (i.e., fixed material) or to the space (i.e., fixed space).

For solids, the investigation attached to the matter is a natural choice, since the deformation history of a material point affects the present response of that particular material point. Theories for description of plastic yield and creep, for example, usually need information on the deformation history of the material points. This requires that the material models be written for material points, not for points in space.

On the other hand, in fluid mechanics, the motion of material particles can be quite complex. The description of their displacements may be difficult and may not be of any real value either. In such situations, we generally take the approach of monitoring changes in the quantities of interest describing the state of the matter over time at a fixed position in space. Thus, in this approach, the observer is stationed at a fixed location in space and monitors the changes in the state of the matter over time at this fixed location. As the state of the matter changes over time, the fixed location of the observer is occupied by different material particles over time. In this approach, the observer never knows which particles are occupying this position but simply monitors the state of the particle that happens to be at this location at time $t$. See Reddy [84] and Surana [105].

The investigative approach relating to matter has three coordinate systems. When the initial coordinate system is used, the quantities related to a material point are expressed by the coordinates which that particular material point had before loading was applied. Since the investigation approach is attached to the material, the term *initial material coordinates* is adopted.

When the current coordinates are used, on the other hand, the quantities refer to the current position of the material points. This investigation approach is related to the material points, and, therefore, the term *current material coordinates* is used. In this approach, the spatial coordinates are used, which applies to the fluid mechanics. The difference between the current material coordinates and the reference coordinate system is discussed in Section 3.2.

When the convective coordinates are used, the coordinate system is attached to the material in such a way that the frame deforms with the deforming material. This implies that the coordinate system, which may initially be a rectangular Cartesian one, does not keep its character during deformation. The initially perpendicular coordinates do not stay perpendicular, and the rectilinear coordinate axes do not stay rectilinear during deformation but will have a curved appearance. The benefit of convective coordinates is that the coordinates of material points do not change during deformation; on the other hand, the mathematical treatment of equations is more complex than when the preceding two material coordinates are used. Although the use of the convective coordinates is related to the material and therefore suitable for the modelling of solids, it is not studied here because of its mathematical complexity. Figure 3.1 shows how the convective rectangular Cartesian coordinate system deforms with the deformation of the body $\mathrm{m^b}$. A two-dimensional sketch is given for the sake of clarity. It is not common practice to distinguish between the preceding three material coordinates by adding the prefix "initial," "current," or "convective"; however, for teaching purposes, this is done here.

When spatial coordinates are used, the point of observation is attached to the space. Traditionally, this approach is used in fluid mechanics. A system $\mathrm{m^b}$ is a collection of matter imagined to be bounded from its surroundings by a clearly defined boundary. If the system does not exchange matter with its surroundings, it is said to be closed, whereas an open system can exchange matter with its surroundings. The boundary of a system $\mathrm{m^b}$ can deform. An (arbitrary) part of the system $\mathrm{m^b}$ is called a subsystem and is denoted by $m$. The term *system* is used in thermodynamics. A more continuum mechanics–oriented equivalent term for a system $\mathrm{m^b}$ is *body*.

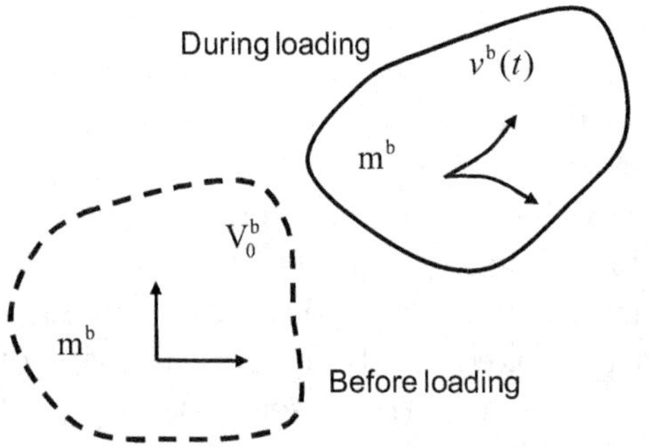


**Figure 3.1** Convective coordinates in a two-dimensional body.

The state $\mathcal{E}$ of a material point is known when the values of all the variables needed for characterization of that particular material point are known. These variables can be the stress tensor $\boldsymbol{\sigma}$ and the absolute temperature $T$, for example.

The concept of "configuration" is a vital part of solid mechanics. The definition of the concept of "configuration" follows that by Reddy ([84], p. 82) and reads, "The region occupied by the continuum at a given time $t$ is termed a configuration." Based on Reddy's definition, the term *configuration* defines the geometry and the position of a system $\mathrm{m^b}$. This means that configuration is a pure geometrical concept and therefore does not tell anything of the state $\mathcal{E}$ of the system $\mathrm{m^b}$. Before loading, the system $\mathrm{m^b}$ is said to be in the initial configuration, whereas during loading, the system $\mathrm{m^b}$ is in the current configuration. Due to the definition of the term *configuration* for a rigid body motion, the geometries of the initial configuration and current configuration are equal, whereas the positions differ; that is, the configurations are different. Later in this book, the third configuration is introduced.

## 3.2 Various Coordinates and Coordinate Systems

The present section introduces the initial material, the current material, and the spatial coordinates and provides some mathematical equations related to them. The convective material coordinates are not studied.

In Figure 3.2 the system (body) $\mathrm{m^b}$ is shown in the initial configuration $\mathrm{V_0^b}$ and in the current configuration $\underline{v}^{\mathrm{b}}(t)$. Before loading, that is, at time $t_0$, the system $\mathrm{m^b}$ is in the initial configuration $\mathrm{V_0^b}$, whereas during deformation, that is, at time $t$, the system $\mathrm{m^b}$ is in the current configuration $\underline{v}^{\mathrm{b}}(t)$. The quantities $\mathrm{V_0^b}$ and $\underline{v}^b(t)$ refer to the configurations and also give the volumes of the system $\mathrm{m^b}$ at $t_0$ and at $t$. A material point P is studied in the initial configuration $\mathrm{V_0^b}$ and in the current configuration $\underline{v}^{\mathrm{b}}(t)$. The notations $\partial\mathrm{V_0^b}$ and $\partial\underline{v}^b(t)$ stand for the boundaries of $\mathrm{V_0^b}$ and $\underline{v}^b(t)$, respectively. The investigation shown in Figure 3.2 is carried out in the rectangular Cartesian coordinate system, but the results are valid for other coordinate systems as well.

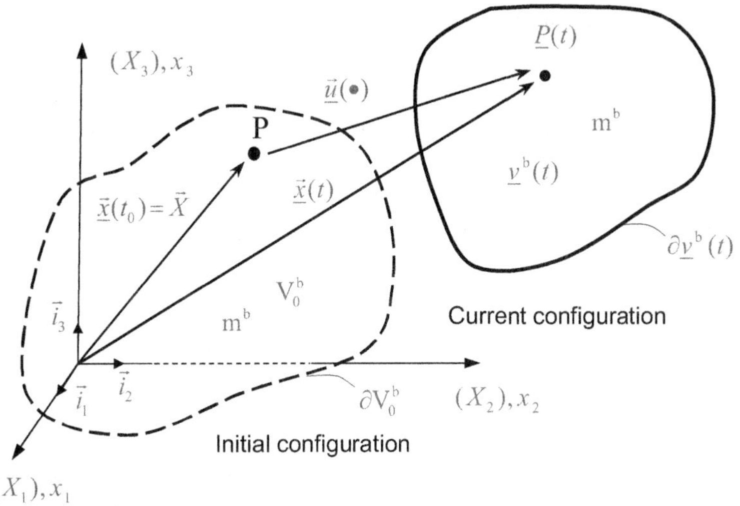


**Figure 3.2** System $\mathrm{m^b}$ in the initial configuration $\mathrm{V_0^b}$ and in the current configuration $\underline{v}^b(t)$.

According to Figure 3.2, the position of material point P at current time $t$

is denoted by $\underline{P}(t)$ and is expressed mainly by the coordinates $\underline{x}_i(t)$. Since the system $\mathrm{m}^{\mathrm{b}}$ is in the initial configuration $\mathrm{V}_0^{\mathrm{b}}$ at time $t_0$, the position of material point P in the initial configuration is $\vec{\underline{x}}(t_0)$, as shown in Figure 3.2. The basis of the coordinate system $(x_1, x_2, x_3)$ is $(\vec{i}_1, \vec{i}_2, \vec{i}_3)$. Thus the following holds:

$$\vec{\underline{x}}(t) = \underline{x}_i(t)\,\vec{i}_i\,. \tag{3.1}$$

Equation (3.1) shows that the position of material point $\underline{P}(t)$ varies with time $t$. Thus the current material coordinates $\underline{x}_i(t)$ vary with time, since they follow the moving material point $\underline{P}(t)$.

It is important to understand the difference between the coordinate system $(x_1, x_2, x_3)$ and the coordinates of point $\underline{P}(t)$. The coordinates of point $\underline{P}(t)$ are $(\underline{x}_1(t), \underline{x}_2(t), \underline{x}_3(t))$, since point $\underline{P}(t)$ moves in space. As already discussed, fluid flow is often described with respect to a fixed point in space, the coordinates of which do not depend on time $t$ and therefore read $(x_1, x_2, x_3)$.

As can be seen in Figure 3.2 and in the previous equations, the quantities related to the current configuration $\underline{v}^{\mathrm{b}}(t)$ are underlined. Thus temperature, for example, at the current configuration $\underline{v}^{\mathrm{b}}(t)$ takes the form $\underline{T}(\vec{\underline{x}}(t), t)$, whereas for fluids, it takes the appearance $T(\vec{x}, t)$. The "small" difference between these two notations points to many differences in the derivation of equations for a material response, as the coming pages will show.

The quantities $\underline{T}(\vec{\underline{x}}(t), t)$ and $T(\vec{x}, t)$ give information on the state $\mathcal{E}$. The temperature of the material point $\underline{P}(t)$ in the current configuration $\underline{v}^{\mathrm{b}}(t)$ is given by $\underline{T}(\vec{\underline{x}}(t), t)$, whereas $T(\vec{x}, t)$ gives the temperature at a fixed point in space. They do not give all information on the state $\mathcal{E}$, since the strain tensor $\underline{\boldsymbol{\varepsilon}}$, for example, belongs to the set of state variables as well, that is, the variables that define the state $\mathcal{E}$ of the system $\mathrm{m}^{\mathrm{b}}$.

The confusion between the coordinate systems (frames) and the coordinates of material (or spatial) points could be avoided by writing the position vector of the moving material point $\underline{P}(t)$ in the form $\vec{r}(t) = r_1(t)\,\vec{i}_1 + r_2(t)\,\vec{i}_2 + r_3(t)\,\vec{i}_3$. In that case, the notation $\underline{T}(\vec{\underline{x}}(t), t)$ would be replaced with $\underline{T}(\vec{\underline{r}}(t), t)$, for example, and the coordinates of point $\underline{P}(t)$ would be $(\underline{r}_1(t), \underline{r}_2(t), \underline{r}_3(t))$. This would make the notations much clearer, but it would create difficulties in the forthcoming derivations. To minimize the pain to the reader in the following pages, the notations introduced in the previous paragraphs, the foregoing equations, and Figure 3.2 have to be accepted. Furthermore, the notations are, if not exactly the same, somewhat in harmony with the generally accepted notations.

Based on Figure 3.2, we arrive at the following equation:

$$\vec{\underline{x}}(t) = \vec{\underline{x}}(t_0) + \vec{\underline{u}}(\underline{x}(t)) \qquad \Rightarrow \qquad \vec{\underline{u}}(\vec{\underline{x}}(t)) = \vec{\underline{x}}(t) - \vec{\underline{x}}(t_0). \tag{3.2}$$

Equation (3.2) is for the current material coordinates $\underline{x}_i(t)$. The term *current material coordinates* $\underline{x}_i(t)$ stems from the fact that the coordinates $\underline{x}_i(t)$ give the current position of the material point $\underline{P}(t)$ during deformation.

In continuum thermodynamics (in continuum mechanics), however, it is a common practice to replace the notation $\underline{\vec{x}}(t_0)$ with the notation $\vec{X}$, which reads

$$\vec{X} := \underline{\vec{x}}(t_0) \qquad \text{and} \qquad X_i := \underline{x}_i(t_0)\,, \qquad \text{where} \quad \vec{X} := X_i\,\vec{i}_i. \tag{3.3}$$

Substitution of Notations (3.3) into Equation $(3.2)_2$ leads to

$$\underline{\vec{u}}(\underline{\vec{x}}(t)) := \underline{\vec{x}}(t) - \vec{X}(\underline{\vec{x}}(t_0)) \qquad \Rightarrow \qquad \underline{\vec{x}}(t) = \vec{X}(\underline{\vec{x}}(t_0)) + \underline{\vec{u}}(\underline{\vec{x}}(t)). \tag{3.4}$$

Equations (3.4) are still for the current material coordinates, although the term $\vec{X}(\underline{\vec{x}}(t_0))$ may cause confusion. The vector $\vec{X}$ is a position vector of point P in the initial configuration $\mathrm{V}_0^{\mathrm{b}}$, but in Equations (3.4), it is written in terms of the coordinates for point $\underline{P}(t)$ in the current configuration $\underline{v}^{\mathrm{b}}(t)$. Equations (3.2) may help to understand Equations (3.4). Equation $(3.4)_1$ is the definition for the displacement vector $\underline{\vec{u}}$.

When the initial material coordinates $X_i$ are used, the displacement $\underline{\vec{u}}$ of the material point $\underline{P}(t)$ is defined as follows (see Figure 3.2):

$$\underline{\vec{u}}(\vec{X},t) := \underline{\vec{x}}(\vec{X},t) - \vec{X} \qquad \Rightarrow \qquad \underline{\vec{x}}(\vec{X},t) = \vec{X} + \underline{\vec{u}}(\vec{X},t)\,. \tag{3.5}$$

The corresponding frame for the initial material coordinates $X_i$ is the coordinate system $(X_1, X_2, X_3)$. This term stems from the property that also during deformation the material points keep their original (initial) material coordinates $X_i$. This makes it easier to prepare material models in cases in which the deformation history, for example, affects the current response of the matter.

As already discussed, since the position of point P within the system $\mathrm{m}^{\mathrm{b}}$ in the initial configuration $\mathrm{V}_0^{\mathrm{b}}$ does not depend on time $t$, the position vector $\vec{X}$ does not depend on time $t$.

Even though in Figure 3.2, and according to Expressions $(3.2)_1$ and $(3.4)_1$, the vector $\underline{\vec{u}}$ is a straight line, it does not mean that the path of the material point $\underline{P}(t)$ is a straight line. The path of the material point $\underline{P}(t)$, and therefore that of the moving tip of the displacement vector $\underline{\vec{u}}(\cdot)$, shows the nonlinear displacement history of the material point $\underline{P}(t)$. Thus the vector $\underline{\vec{u}}(\cdot)$ does not represent a real displacement, but is a vector from the initial position of point P to its current position $\underline{x}_i(t)$. The dot in the notation $\underline{\vec{u}}(\cdot)$ represents either $(X_i, t)$ when the initial material coordinates are used or $\underline{x}_i(t)$ when the current material coordinates are used.

The preceding equations show the difference between the initial material coordinates and the current material coordinates. Due to deformation, the volume of the system $\mathrm{m}^{\mathrm{b}}$ varies with time, which is expressed by variable $t$ in the quantity $\underline{v}^{\mathrm{b}}(t)$. In the initial configuration, however, the volume of the system $\mathrm{m}^{\mathrm{b}}$ does not vary, which is shown by the quantity $\mathrm{V}_0^{\mathrm{b}}$. The same holds for the position vector of a material point. When the material point $\underline{P}(t)$ is studied in the current configuration $\underline{v}^{\mathrm{b}}(t)$, its position vector reads $\underline{\vec{x}}(t)$, whereas the position vectors $\underline{\vec{x}}(t_0)$ and $\vec{X}$ give the position of the point P as it was in the initial configuration $\mathrm{V}_0^{\mathrm{b}}$.

It is common practice to draw illustrations similar to Figure 3.2 and denote the axes by $X_i$ and $x_i$. This may cause confusion, since the role of the coordinates $X_i$ is restricted. The coordinates $\underline{x}_i(t)$ refer to the material point $\underline{P}(t)$ in the current configuration $\underline{v}^{\mathrm{b}}(t)$ and, as a special case, when $t = t_0$ at the same point in the initial configuration $\mathrm{V}_0^{\mathrm{b}}$. Thus all the material points given in Figure 3.2 can be expressed by the coordinates $\underline{x}_i(t)$. The coordinates $X_i$, however, are restricted to an equation of the points of the system $\mathrm{m}^{\mathrm{b}}$ in the initial configuration $\mathrm{V}_0^{\mathrm{b}}$. Thus, when coordinates $X_i$ are applied in Figure 3.2, the current configuration $\underline{v}^{\mathrm{b}}(t)$ should be eliminated. This was the reason for the authors to put the coordinates $X_i$ in parentheses. As shown in Figure 3.2 and in the following figures, the frames $(X_1, X_2, X_3)$ and $(x_1, x_2, x_3)$ have the same origin, the same orientation, the same metric, and the same base vectors. This is due to the definition $\underline{\vec{x}}(t_0) = \vec{X}$ shown in Figure 3.2. Later in this book, the coordinates $X_i$ are used for other purposes as well.

The introduction of the absolute temperature $T$ and the strain tensor $\boldsymbol{\varepsilon}$ was an early act, since these quantities have not yet been defined in this book. However, they may help in understanding the message of this section, which, we hope, makes this small flaw acceptable.

## 3.3 IG-CtP Configuration and Material Descriptions

At the start of this chapter, the initial configuration $\mathrm{V}_0^{\mathrm{b}}$ and the current configuration $\underline{v}^{\mathrm{b}}(t)$ were introduced. The present section extends the viewpoint by introducing a configuration called "initial geometry at current time and position," IG-CtP configuration $\mathrm{V}_0^{\mathrm{b}}$. Material descriptions are also introduced.

Before loading, the system $\mathrm{m}^{\mathrm{b}}$ is in the initial configuration $\mathrm{V}_0^{\mathrm{b}}$. This means that the state variables, such as the absolute temperature $T$, take the values they had before loading. When the initial material coordinates $X_i$ are used, the value of the absolute temperature at point P is obtained from the equation for $T(\vec{X}, t_0)$. Thus the quantity $T(\vec{X}, t_0)$ gives the absolute temperature $T$ before loading (i.e., in the initial configuration $\mathrm{V}_0^{\mathrm{b}}$).

During loading, the material point is in the current configuration $\underline{v}^{\mathrm{b}}(t)$. Temperature $T$ may change, and therefore the absolute temperature is expressed by $\underline{T}(\underline{\vec{x}}(t), t)$. In the equation for $\underline{T}(\underline{\vec{x}}(t), t)$, the first variable $t$ (i.e., the variable $t$ in $\underline{\vec{x}}(t)$) tells us that the position of the point $\underline{P}(t)$ varies with time, and the second variable $t$ tells us that the value for the absolute temperature $T$ may vary with time $t$ independently of the varying position of the point $\underline{P}(t)$. If the position of point $\underline{P}(t)$ does not change when the system $\mathrm{m}^{\mathrm{b}}$ is heated by an external agent, the latter variable $t$ in the notation $\underline{T}(\underline{\vec{x}}(t), t)$ gives the temperature change.

The basic laws and axioms are written in forms that study the response of an arbitrarily selected but then nonchanging entity of matter in the current configuration $\underline{v}^{\mathrm{b}}(t)$. This means that the set of material points does not change once selected. Since, during deformation, the position of the matter changes in space, the current material coordinates $\underline{x}_i(t)$ create a vital frame for the study of basic laws and axioms. Thus basic laws and axioms are expressed by the current material coordinates $\underline{x}_i(t)$.

When solids are investigated, usually the initial material coordinates $X_i$ are used to define the state of the system $\mathrm{m^b}$. This means that in the description of the current value of the absolute temperature $T$, instead of the current material coordinates $\underline{x}_i(t)$, the initial material coordinates $X_i$ are used. This leads to the introduction of the notation $T(\vec{X},t)$. It stands for the current value of the absolute temperature $T$, but it is expressed in the initial material coordinates $X_i$. However, the initial material coordinates $X_i$ give the geometry of the system $\mathrm{m^b}$ (body) as it was before loading (i.e, in the initial geometry). In order to obtain the equation for $T(\vec{X},t)$ from the equation for $\underline{T}(\vec{\underline{x}}(t),t)$, a coordinate transformation is needed. The coordinate transformation does not change the position of the system $\mathrm{m^b}$ in space. For the preceding two reasons, in this book, the notation $T(\vec{X},t)$ is called the current value of the absolute temperature expressed in the "initial geometry in the current position." Thus the configuration "initial geometry at current time and position," IG-CtP configuration $\mathrm{V_0^b}$, is defined.

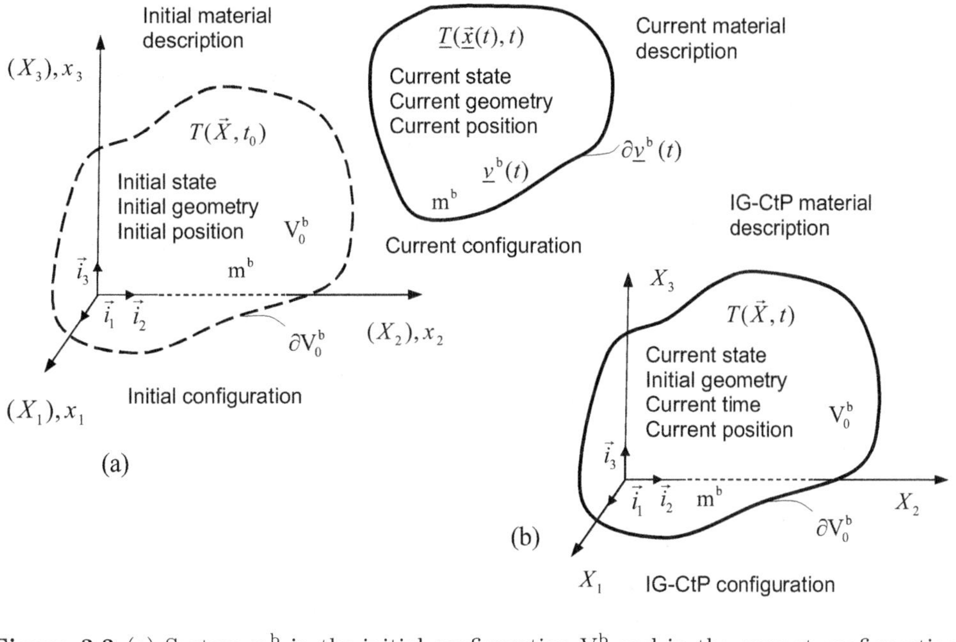


**Figure 3.3** (a) System $\mathrm{m^b}$ in the initial configuration $\mathrm{V_0^b}$ and in the current configuration $\underline{v}^b(t)$. (b) System $\mathrm{m^b}$ in IG-CtP configuration $\mathrm{V_0^b}$.

Figure 3.3 shows, what in this book is meant by the terms *configuration* and *description.* The term *configuration* gives the geometry and the position of the system $\mathrm{m^b}$. The term *description* adds the values of the variables (the state) and how they are expressed. The IG-CtP description gives the current value of the temperature $T$, but it is expressed by the initial material coordinates $X_i$, in the form $T(\vec{X},t)$, in the IG-CtP configuration $\mathrm{V}_0^{\mathrm{b}}$. This description is called "initial geometry at current time and position," as shown in Figure 3.3(b).

The mapping from $\underline{T}(\underline{\vec{x}}(t),t)$ to $T(\vec{X},t)$, for example, is a simple coordinate transformation, as will be shown in Section 3.10. Since the geometries of the initial configuration and the IG-CtP configuration coincide, $\mathrm{V}_0^{\mathrm{b}}$, they are expressed by the same notation $\mathrm{V}_0^{\mathrm{b}}$. This is acceptable, since the geometries, that is, the shapes, of these two appearances are the same. The differences lie in the positions of these two appearances and in the values of the variables. Figure 3.3(a) shows the initial configuration $\mathrm{V}_0^{\mathrm{b}}$ and the current configuration $\underline{v}^{\mathrm{b}}(t)$. It is an extended copy of Figure 3.3 with the notations needed for the present study. The IG-CtP configuration $\mathrm{V}_0^{\mathrm{b}}$ cannot be drawn in Figure 3.3(a), since it can be shown only in the initial material coordinate system $(X_1, X_2, X_3)$. Therefore Figure 3.3(b) is given. Comparison of Figures 3.3(a) and 3.3(b) shows that the initial configuration $\mathrm{V}_0^{\mathrm{b}}$ in Figure 3.3(a) has the same shape (geometry) as IG-CtP configuration $\mathrm{V}_0^{\mathrm{b}}$ given in Figure 3.3(b). In order to point out the different states and coordinates, the absolute temperature $T$ is expressed in Figure 3.3.

The coordinate transformation between the current configuration $\underline{v}^{\mathrm{b}}(t)$ and the IG-CtP configuration $\mathrm{V}_0^{\mathrm{b}}$ is equal to that between the initial configuration $\mathrm{V}_0^{\mathrm{b}}$ and the current configuration $\underline{v}^{\mathrm{b}}(t)$. The coordinate transformation is defined by the mapping from the initial configuration $\mathrm{V}_0^{\mathrm{b}}$ to the current configuration $\underline{v}^{\mathrm{b}}(t)$ and applied for determination of the IG-CtP configuration $\mathrm{V}_0^{\mathrm{b}}$ when the current configuration $\underline{v}^{\mathrm{b}}(t)$ is known.

## 3.4 Mathematical Prerequisites for the Material Descriptions

In this section we give some mathematical equations related to the material descriptions that are needed in the forthcoming sections of this book. When the IG-CtP material description (initial geometry at current time and position) is used, the mathematical derivation is carried out in the IG-CtP configuration $\mathrm{V}_0^{\mathrm{b}}$. However, the Green–Lagrange strain tensor, for example, is derived by comparing vectors between material points in the current configuration $\underline{v}^{\mathrm{b}}(t)$ with the vectors between the same material points in the initial configuration $\mathrm{V}_0^{\mathrm{b}}$.

The origin of the coordinate system $(X_1, X_2, X_3)$ associated with the initial configuration deviates from that of the coordinate system $(X_1^{①}, X_2^{①}, X_3^{①})$ associated with the IG-CtP configuration $\mathrm{V}_0^{\mathrm{b}}$. Therefore, in order to distinguish these two frames, the coordinate system associated with the IG-CtP configuration $\mathrm{V}_0^{\mathrm{b}}$ should be denoted by $(X_1^{①}, X_2^{①}, X_3^{①})$ and the displacement vector expressed in this coordinate system should be denoted by $\vec{u}^{①}$. The equations for the strain tensors for solids have terms such as gradients of the displacement vector with respect to the coordinate. In the frames $(X_1, X_2, X_3)$ and

$(X_1^{①}, X_2^{①}, X_3^{①})$ the terms have the same appearance. Thus the strain tensors defined in the coordinate system $(X_1, X_2, X_3)$ can be used in the derivations carried out in the frames $(X_1^{①}, X_2^{①}, X_3^{①})$. This implies that in the forthcoming pages of this book, the derivations for the IG-CtP configuration $\mathrm{V}_0^{\mathrm{b}}$ will be carried out using the frame $(X_1, X_2, X_3)$. This replacement was already done in Figure 3.3.

The vector operator "del" denoted by $\nabla$, which is given by Definitions (2.169), takes the following form in the rectangular Cartesian coordinates $(\underline{x}_1(t), \underline{x}_2(t), \underline{x}_3(t))$:

$$\vec{\underline{\nabla}}(\underline{\vec{x}}) := \frac{\partial}{\partial \underline{\vec{x}}(t)} = \vec{i}_m \frac{\partial}{\partial \underline{x}_m(t)} \qquad \text{or} \qquad \overleftarrow{\underline{\nabla}}(\underline{\vec{x}}) := \frac{\overleftarrow{\partial}}{\partial \underline{\vec{x}}(t)} = \frac{\overleftarrow{\partial}}{\partial \underline{x}_m(t)} \vec{i}_m \,. \tag{3.6}$$

A similar definition holds for $\nabla$ when the initial material coordinates $(X_1, X_2, X_3)$ are used:

$$\vec{\nabla}(\vec{X}) := \frac{\partial}{\partial \vec{X}} = \vec{i}_m \frac{\partial}{\partial X_m} \qquad \text{or} \qquad \overleftarrow{\nabla}(\vec{X}) := \frac{\overleftarrow{\partial}}{\partial \vec{X}} = \frac{\overleftarrow{\partial}}{\partial X_m} \vec{i}_m \,. \tag{3.7}$$

**Example 3.1:** What are the following derivatives in terms of the base vectors?

$$\vec{\underline{\nabla}}(\underline{\vec{x}})\, \underline{\vec{v}}(\underline{\vec{x}}(t), t) \qquad \text{and} \qquad \underline{\vec{v}}(\underline{\vec{x}}(t), t)\, \overleftarrow{\underline{\nabla}}(\underline{\vec{x}}) \,. \tag{3.8}$$

**Solution:** Substitution of $\underline{\vec{v}}(\underline{\vec{x}}(t), t) = \underline{v}_s(\underline{\vec{x}}(t), t)\, \vec{i}_s = \vec{i}_s\, \underline{v}_s(\underline{\vec{x}}(t), t)$ and Definition (3.6) into Terms (3.8) yields

$$\vec{\underline{\nabla}}(\underline{\vec{x}})\, \underline{\vec{v}}(\underline{\vec{x}}(t), t) = \vec{i}_m \frac{\partial}{\partial \underline{x}_m(t)} \underline{v}_s(\underline{\vec{x}}(t), t)\, \vec{i}_s = \frac{\partial \underline{v}_s(\underline{\vec{x}}(t), t)}{\partial \underline{x}_m(t)} \vec{i}_m \vec{i}_s \tag{3.9}$$

and

$$\underline{\vec{v}}(\underline{\vec{x}}(t), t)\, \overleftarrow{\underline{\nabla}}(\underline{\vec{x}}) = \vec{i}_s\, \underline{v}_s(\underline{\vec{x}}(t), t) \frac{\overleftarrow{\partial}}{\partial \underline{x}_m(t)} \vec{i}_m = \frac{\partial \underline{v}_s(\underline{\vec{x}}(t), t)\overleftarrow{\partial}}{\partial \underline{x}_m(t)} \vec{i}_s \vec{i}_m = \frac{\partial \underline{v}_s(\underline{\vec{x}}(t), t)}{\partial \underline{x}_m(t)} \vec{i}_s \vec{i}_m \tag{3.10}$$

■

The state of the material at a point $\underline{P}(t)$ is evaluated. The state is assumed to be expressed by a set of field variables. A scalar-valued field variable denoted by $\theta$ is examined. The quantity $\theta$ can be, for example, the absolute temperature $T$. When the initial material coordinates $(X_1, X_2, X_3)$ are used, the preceding quantity is expressed by $\theta(\vec{X}, t)$, whereas the use of the current material coordinates $(\underline{x}_1(t), \underline{x}_2(t), \underline{x}_3(t))$ leads to the appearance $\underline{\theta}(\underline{\vec{x}}(t), t)$. In these notations, $\vec{X}$ and $\underline{\vec{x}}(t)$ refer to the material point, and $t$ refers to the time dependence.

There are two types of time derivatives: the partial time derivative and the material (time) derivative. The material time derivative $\mathrm{D}/\mathrm{D}t$ describes how the material experiences the rate of change of its state. The state can be the absolute temperature $T$, for example. The material derivative of function $\underline{\theta} = \underline{\theta}(\underline{\vec{x}}(t), t)$ takes the form

$$\frac{\mathrm{D}\underline{\theta}(\underline{\vec{x}}(t),t)}{\mathrm{D}t} = \frac{\partial\underline{\theta}(\underline{\vec{x}}(t),t)}{\partial t} + \frac{\partial\underline{\vec{x}}(t)}{\partial t} \cdot \frac{\partial\underline{\theta}(\underline{\vec{x}}(t),t)}{\partial\underline{\vec{x}}(t)}\,. \tag{3.11}$$

**Example 3.2:** Show that Derivative (3.11) is correct.

**Solution:** Based on the chain rule for the scalar-valued quantities, Derivative (3.11) can be written in the following form:

$$\begin{aligned}\frac{\mathrm{D}\underline{\theta}(\underline{\vec{x}}(t),t)}{\mathrm{D}t} &= \frac{\mathrm{D}\underline{\theta}(\underline{x}_1(t),\underline{x}_2(t),\underline{x}_3(t),t)}{\mathrm{D}t} = \frac{\partial\underline{\theta}(\underline{x}_1(t),\underline{x}_2(t),\underline{x}_3(t),t)}{\partial t}\\ &+ \frac{\partial\underline{\theta}(\underline{x}_1(t),\underline{x}_2(t),\underline{x}_3(t),t)}{\partial\underline{x}_1(t)}\frac{\partial\underline{x}_1(t)}{\partial t} + \frac{\partial\underline{\theta}(\underline{x}_1(t),\underline{x}_2(t),\underline{x}_3(t),t)}{\partial\underline{x}_2(t)}\frac{\partial\underline{x}_2(t)}{\partial t}\\ &+ \frac{\partial\underline{\theta}(\underline{x}_1(t),\underline{x}_2(t),\underline{x}_3(t),t)}{\partial\underline{x}_3(t)}\frac{\partial\underline{x}_3(t)}{\partial t}\,.\end{aligned} \tag{3.12}$$

Next, the Einstein summation convention is applied to join the last three terms together. This gives

$$\frac{\mathrm{D}\underline{\theta}(\underline{\vec{x}}(t),t)}{\mathrm{D}t} = \frac{\partial\underline{\theta}(\underline{x}_1(t),\underline{x}_2(t),\underline{x}_3(t),t)}{\partial t} + \frac{\partial\underline{\theta}(\underline{x}_1(t),\underline{x}_2(t),\underline{x}_3(t),t)}{\partial\underline{x}_i(t)}\frac{\partial\underline{x}_i(t)}{\partial t}\,. \tag{3.13}$$

The order of the quantities in the latter term on the right side of Equation (3.13) is interchanged, and the following is achieved:

$$\frac{\mathrm{D}\underline{\theta}(\underline{\vec{x}}(t),t)}{\mathrm{D}t} = \frac{\partial\underline{\theta}(\underline{x}_1(t),\underline{x}_2t),\underline{x}_3(t),t)}{\partial t} + \frac{\partial\underline{x}_i(t)}{\partial t}\frac{\partial\underline{\theta}(\underline{x}_1(t),\underline{x}_2(t),\underline{x}_3(t),t)}{\partial\underline{x}_i(t)}\,. \tag{3.14}$$

Equation (3.14) can be written with the following appearance:

$$\frac{\mathrm{D}\underline{\theta}(\underline{\vec{x}}(t),t)}{\mathrm{D}t} = \frac{\partial\underline{\theta}(\underline{\vec{x}}(t),t)}{\partial t} + \frac{\partial\underline{x}_i(t)}{\partial t}\frac{\partial\underline{\theta}(\underline{\vec{x}}(t),t)}{\partial\underline{x}_i(t)}\,. \tag{3.15}$$

Definition $(2.173)_2$ gives

$$\frac{\partial}{\partial\underline{\vec{x}}(t)} := \vec{i}_n\,\frac{\partial}{\partial\underline{x}_n(t)}\,. \tag{3.16}$$

Based on Derivative (3.16), the following is achieved:

$$\frac{\partial\underline{\vec{x}}(t)}{\partial t}\cdot\frac{\partial\underline{\theta}(\underline{\vec{x}}(t),t)}{\partial\underline{\vec{x}}(t)} = \frac{\partial}{\partial t}\underline{x}_s(t)\,\vec{i}_s\cdot\vec{i}_n\frac{\partial\underline{\theta}(\underline{\vec{x}}(t),t)}{\partial\underline{x}_n(t)} = \frac{\partial\underline{x}_s(t)}{\partial t}\frac{\partial\underline{\theta}(\underline{\vec{x}}(t),t)}{\partial\underline{x}_s(t)}\,. \tag{3.17}$$

Substitution of Result (3.17) into Derivative (3.15) yields

$$\frac{\mathrm{D}\underline{\theta}(\underline{\vec{x}}(t),t)}{\mathrm{D}t} = \frac{\partial\underline{\theta}(\underline{\vec{x}}(t),t)}{\partial t} + \frac{\partial\underline{\vec{x}}(t)}{\partial t}\cdot\frac{\partial\underline{\theta}(\underline{\vec{x}}(t),t)}{\partial\underline{\vec{x}}(t)}\,. \tag{3.18}$$

Comparison of Derivatives (3.11) and (3.18) shows that they are equal. ■

Introduction of Operator $(3.6)_1$ allows Derivative (3.11) to take the form

$$\frac{\mathrm{D}\underline{\theta}(\underline{\vec{x}}(t),t)}{\mathrm{D}t} = \frac{\partial \underline{\theta}(\underline{\vec{x}}(t),t)}{\partial t} + \underline{\vec{v}}(\vec{x}(t),t)\cdot\underline{\vec{\nabla}}(\underline{\vec{x}})\,\underline{\theta}(\underline{\vec{x}}(t),t)\,, \tag{3.19}$$

where the velocity vector $\underline{\vec{v}}(\underline{\vec{x}}(t),t)$ is defined to be

$$\underline{\vec{v}}(\underline{\vec{x}}(t),t) := \frac{\partial \underline{\vec{x}}(t)}{\partial t} \quad \text{and the following holds:} \quad \vec{v}(\vec{X},t) = \underline{\vec{v}}(\underline{\vec{x}}(t),t)\,. \tag{3.20}$$

The velocity vector $\underline{\vec{v}}(\underline{\vec{x}}(t),t)$ is the velocity of the material point as it moves through a space with deformation. Equation (3.19) holds for vectors $\underline{\vec{\theta}}(\underline{\vec{x}}(t),t)$ and tensors $\underline{\boldsymbol{\theta}}(\underline{\vec{x}}(t),t)$ of any order.

**Example 3.3:** Verify that the result in Derivative (3.19) is correct.

**Solution:** Substitution of Definition $(3.6)_1$ and Velocity $(3.20)_1$ into Expression (3.19) yields the result

$$\frac{\mathrm{D}\underline{\theta}(\underline{\vec{x}}(t),t)}{\mathrm{D}t} = \frac{\partial \underline{\theta}(\underline{\vec{x}}(t),t)}{\partial t} + \frac{\partial \underline{\vec{x}}(t)}{\partial t}\cdot\frac{\partial \underline{\theta}(\underline{\vec{x}}(t),t)}{\partial \underline{\vec{x}}(t)}\,. \tag{3.21}$$

Derivatives (3.11) and (3.21) are equal. Thus Derivative (3.19) is correct. ■

The material time derivative of a function $\theta = \theta(\vec{X},t)$ is

$$\frac{\mathrm{D}\theta(\vec{X},t)}{\mathrm{D}t} = \frac{\partial \theta(\vec{X},t)}{\partial t} + \frac{\partial \vec{X}}{\partial t}\cdot\frac{\partial \theta(\vec{X},t)}{\partial \vec{X}} = \frac{\partial \theta(\vec{X},t)}{\partial t}\,. \tag{3.22}$$

The derivation in Expression (3.22) exploits the fact that the vector $\vec{X}$ is a constant with respect to time $t$, which gives $\partial\vec{X}/\partial t = 0$. Expression (3.22) holds for vectors $\vec{\theta}(\vec{X},t)$ and tensors $\boldsymbol{\theta}(\vec{X},t)$ of any order.

**Example 3.4:** Study a uniaxial case and assume that the relation between the current material coordinates $\underline{x}_i(t)$ and the initial material coordinates $X_i$ reads

$$\underline{x}(t) = X + a\,X\,t + b\,X\,t^2\,, \tag{3.23}$$

where $a$ and $b$ are constants giving correct units for the terms on the right side of Expression (3.23). Show that the velocity of a material point has the form $\underline{\vec{v}}(\underline{\vec{x}}(t),t)$.

**Solution:** Based on Definition $(3.20)_1$ and Expressions $(3.20)_2$ and (3.23), the velocity of a material point $\underline{\vec{v}}$ in the uniaxial case is

$$\frac{\partial \underline{x}(t)}{\partial t} = v(X,t) = a\,X + 2\,b\,X\,t = (a+2\,b\,t)\,X\,. \tag{3.24}$$

Relation (3.23) gives

$$X = \frac{\underline{x}(t)}{1 + a\,t + b\,t^2}\,. \tag{3.25}$$

Substitution of the initial material coordinate $X$ from Equation (3.25) into Velocity (3.24) yields

$$\underline{v} = \frac{(a + 2\,b\,t)\,\underline{x}(t)}{1 + a\,t + b\,t^2}\,. \tag{3.26}$$

On the right side of Expression (3.26) are the terms of the form $\underline{x}(t)$ and $t$. The same must hold for the left side as well, which means that

$$\underline{v} = \underline{v}(\underline{x}(t), t)\,. \tag{3.27}$$

Expression (3.27) is such that the velocity of the material point in the current configuration $\underline{v}^{\mathrm{b}}(t)$ has the form $\underline{v}(\underline{x}(t), t)$, which is the result for which we are aiming. ■

## 3.5 Green–Lagrange Strain Tensor E

In this section the Green–Lagrange strain tensor **E** is introduced. Since it is for finite deformation, the reader may skip this section and continue with Section 3.6.

The fundamental concept of the formulation of strain tensors is the comparison of the vectorial distance between material points in the initial configuration $\mathrm{V}_0^{\mathrm{b}}$ and in the current configuration $\underline{v}^b(t)$. This comparison is done in this section in the derivation of the Green–Lagrange strain tensor **E**, but the same concept holds for derivation of other strain tensors as well.

### 3.5.1 Taylor Series

This subsection introduces the Taylor series.

Let the function $g(x, y, z)$ and all its derivatives be continuous within the interval $\hat{h}$. The values of the function $g(x, y, z)$ and all its derivatives are assumed to be known at the point $(x_0, y_0, z_0)$ being within the interval $\hat{h}$. The value of the neighboring point $(x_0 + \Delta_{\mathrm{x}}, y_0 + \Delta_{\mathrm{y}}, z_0 + \Delta_{\mathrm{z}})$ to the point $(x_0, y_0, z_0)$ can be obtained by the following Taylor series:

$$\begin{aligned} g(x_0 + \Delta_{\mathrm{x}}, y_0 + \Delta_{\mathrm{y}}, z_0 + \Delta_{\mathrm{z}}) &= g(x_0, y_0, z_0) \\ &+ \frac{\partial g(x_0, y_0, z_0)}{\partial x}\,\Delta_{\mathrm{x}} + \frac{\partial g(x_0, y_0, z_0)}{\partial y}\,\Delta_{\mathrm{y}} + \frac{\partial g(x_0, y_0, z_0)}{\partial z}\,\Delta_{\mathrm{z}} \\ &+ \frac{1}{2!}\frac{\partial^2 g(x_0, y_0, z_0)}{\partial x^2}\,\Delta_{\mathrm{x}}^2 + \frac{1}{2!}\frac{\partial^2 g(x_0, y_0, z_0)}{\partial y^2}\,\Delta_{\mathrm{y}}^2 + \frac{1}{2!}\frac{\partial^2 g(x_0, y_0, z_0)}{\partial z^2}\,\Delta_{\mathrm{z}}^2 \end{aligned}$$

$$+\frac{1}{2!}\frac{\partial^2 g(x_0,y_0,z_0)}{\partial x\,\partial y}\,\Delta_\mathrm{x}\,\Delta_\mathrm{y}+\frac{1}{2!}\frac{\partial^2 g(x_0,y_0,z_0)}{\partial y\,\partial x}\,\Delta_\mathrm{y}\,\Delta_\mathrm{x}+\cdots. \tag{3.28}$$

Series (3.28) has an infinite number of terms, and the values for the derivatives are calculated at the point $(x_0, y_0, z_0)$.

The point $(x_0, y_0, z_0)$ is denoted by the symbol P and the point $(x_0+\Delta_\mathrm{x}, y_0+\Delta_\mathrm{y}, z_0+\Delta_\mathrm{z})$ by the symbol Q. At the same time the scalar-valued quantity $g$ is replaced by the vector $\vec{g}$ and the coordinates $(x, y, z)$ by the coordinates $X_i$. This means that P and Q are points in the initial configuration $\mathrm{V}_0^\mathrm{b}$. Based on the foregoing assumptions, Equation (3.28) reduces to

$$\begin{aligned}\vec{g}(\mathrm{Q}) = \vec{g}(\mathrm{P}) &+ \frac{\partial\vec{g}(\mathrm{P})}{\partial X_1}\,\Delta X_1+\frac{\partial\vec{g}(\mathrm{P})}{\partial X_2}\,\Delta X_2+\frac{\partial\vec{g}(\mathrm{P})}{\partial X_3}\,\Delta X_3\\ &+\frac{1}{2!}\frac{\partial^2\vec{g}(\mathrm{P})}{\partial X_1^2}\,(\Delta X_1)^2+\frac{1}{2!}\frac{\partial^2\vec{g}(\mathrm{P})}{\partial X_2^2}\,(\Delta X_2)^2+\frac{1}{2!}\frac{\partial^2\vec{g}(\mathrm{P})}{\partial X_3^2}\,(\Delta X_3)^2+\cdots,\end{aligned} \tag{3.29}$$

where the notations $\Delta X_1$, $\Delta X_2$, and $\Delta X_3$ are distances between points P and Q. The position of point P is $(X_1, X_2, X_3)$, and the distance between points P and Q is expressed by the differential vector $\mathrm{d}\vec{X}$, which reads

$$\mathrm{d}\vec{X} := \mathrm{d}X_1\,\vec{\imath}_1+\mathrm{d}X_2\,\vec{\imath}_2+\mathrm{d}X_3\,\vec{\imath}_3=\mathrm{d}X_s\,\vec{\imath}_s\,. \tag{3.30}$$

Figure 3.4 shows points P and Q and the coordinate system $(X_1, X_2, X_3)$. Since the distance between P and Q can be given by a differential vector $\mathrm{d}\vec{X}$, only the first row of Series (3.29) applies, and the following result is obtained:

$$\vec{g}(\mathrm{Q}) = \vec{g}(\mathrm{P})+\frac{\partial\vec{g}(\mathrm{P})}{\partial X_1}\,\mathrm{d}X_1+\frac{\partial\vec{g}(\mathrm{P})}{\partial X_2}\,\mathrm{d}X_2+\frac{\partial\vec{g}(\mathrm{P})}{\partial X_3}\,\mathrm{d}X_3\,, \tag{3.31}$$

or

$$\vec{g}(\mathrm{Q}) = \vec{g}(\mathrm{P})+\frac{\partial\vec{g}(\mathrm{P})}{\partial X_n}\,\mathrm{d}X_n \tag{3.32a}$$

$$= \vec{g}(\mathrm{P})+\mathrm{d}X_n\,\frac{\partial\,\vec{g}(\mathrm{P})}{\partial X_n}\,, \tag{3.32b}$$

where summation on repeated indices is implied. Equation (3.32a) exploits the fact that in the multiplication, the order of the scalars can be interchanged. Expression (3.32b) can be cast in the forms

$$\vec{g}(\mathrm{Q}) = \vec{g}(\mathrm{P})+\mathrm{d}\vec{X}\cdot\vec{\nabla}(\vec{X})\,\vec{g}(\mathrm{P}) \tag{3.33a}$$

and

$$\vec{g}(\mathrm{Q}) = \vec{g}(\mathrm{P})+\vec{g}(\mathrm{P})\overleftarrow{\nabla}(\vec{X})\cdot\mathrm{d}\vec{X}\,. \tag{3.33b}$$

The discussion presented may raise the question of potential limitations of the present approach, where the vector from point P to point Q is assumed to be a differential and therefore the higher-order terms in the Taylor series are neglected in Expressions (3.26), (3.27), and (3.31)–(3.33b). Does this approach lose some information? This question is studied as an example where the uniaxial Green–Lagrange strain component is derived by taking more terms from the Taylor series. Actually, from a physics point of view, such an approach is clearer and therefore more acceptable than the present one where differential vectors are applied.

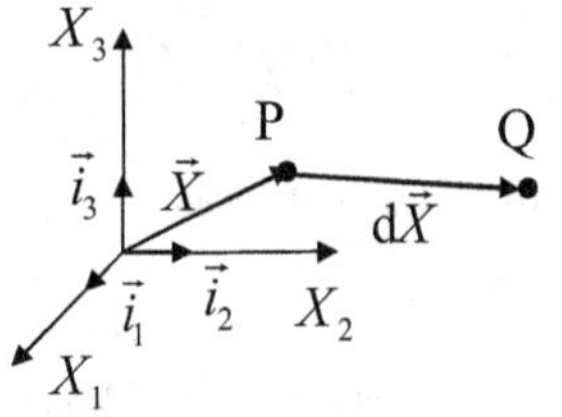


**Figure 3.4** Points P and Q in the material frame $(X_1, X_2, X_3)$.

**Example 3.5:** Show that Equations (3.33a) and (3.33b) hold and that Expressions (3.31), (3.32a), and (3.32b) are correct.

**Solution:** Definitions (3.7) are

$$\vec{\nabla}(\vec{X}) := \frac{\partial}{\partial \vec{X}} = \vec{i}_m \frac{\partial}{\partial X_m} \quad \text{or} \quad \overleftarrow{\nabla}(\vec{X}) := \frac{\overleftarrow{\partial}}{\partial \vec{X}} = \frac{\overleftarrow{\partial}}{\partial X_m} \vec{i}_m . \tag{3.34}$$

Equations (3.30) and $(3.34)_1$ give

$$\mathrm{d}\vec{X} \cdot \vec{\nabla}(\vec{X}) \vec{g}(\mathrm{P}) = \mathrm{d}X_s \vec{i}_s \cdot \vec{i}_m \frac{\partial}{\partial X_m} \vec{g}\,(\mathrm{P}) = \mathrm{d}X_m \frac{\partial \vec{g}(\mathrm{P})}{\partial X_m} = \frac{\partial\, \vec{g}(\mathrm{P})}{\partial X_m} \mathrm{d}X_m . \tag{3.35}$$

Equations (3.30) and $(3.34)_2$ give

$$\vec{g}(\mathrm{P}) \overleftarrow{\nabla}(\vec{X}) \cdot \mathrm{d}\vec{X} = \vec{g}(\mathrm{P}) \frac{\overleftarrow{\partial}}{\partial X_m} \vec{i}_m \cdot \mathrm{d}X_s \vec{i}_s = \frac{\partial \vec{g}\,(\mathrm{P})}{\partial X_m} \mathrm{d}X_m , \tag{3.36}$$

where Expression $(2.177)_2$ is exploited. Substitution of Expressions (3.35) and (3.36) into Equations (3.32a) and (3.32b) gives Results (3.33a) and (3.33b) ■.

Since the right sides of Expressions (3.35) and (3.36) are equal, the left sides are equal as well; that is,

$$\vec{g}(\mathrm{P}) \overleftarrow{\nabla}\,(\vec{X}) \cdot \mathrm{d}\vec{X} = \mathrm{d}\vec{X} \cdot \vec{\nabla}(\vec{X}) \vec{g}(\mathrm{P}) . \tag{3.37}$$

### 3.5.2 Displacement from Initial to Current Configuration

This subsection gives equations for the displacements of material points P, Q, and R in the initial configuration $\mathrm{V}_0^{\mathrm{b}}$ to the points $\underline{P}(t)$, $\underline{Q}(t)$, and $\underline{R}(t)$ of the current configuration $\underline{v}^{\mathrm{b}}(t)$. The equations for the differential line elements between points $\underline{P}(t)$ and $\underline{Q}(t)$ and between points $\underline{P}(t)$ and $\underline{R}(t)$ are also given.

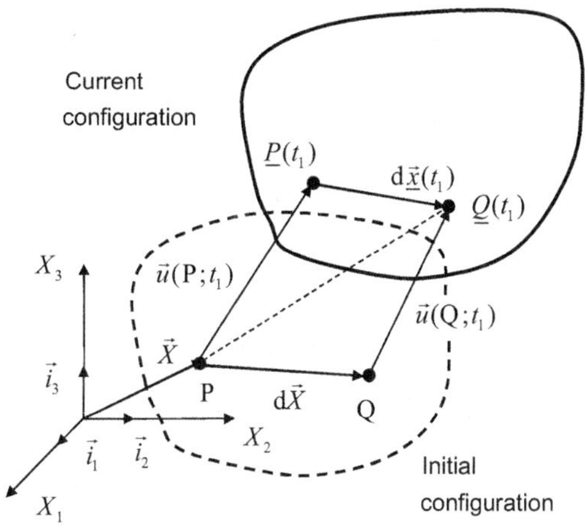


**Figure 3.5** Differential vectors between P and Q in the initial configuration $V_0^b$ and between $\underline{P}(t)$ and $\underline{Q}(t)$ in the current configuration $\underline{v}^b(t)$.

The system $m^b$ is deforming due to the application of a load. Deformation is studied at time $t_1$. Point P moves with deformation from position P to a new position so that it will be at time $t_1$ in position $\underline{P}(t_1)$ shown in Figure 3.5. Similarly, point Q moves from position Q to a new position $\underline{Q}(t_1)$. According to Figure 3.5, the differential vector from point P to point Q is $\mathrm{d}\vec{X}$. In Figure 3.5 the differential vector from point $\underline{P}(t_1)$ to point $\underline{Q}(t_1)$ is denoted by $\mathrm{d}\vec{\underline{x}}(t_1)$. Keep in mind that $\mathrm{d}\vec{\underline{x}}(t_1)$ is not an image of $\mathrm{d}\vec{X}$ in the current configuration. The image of $\mathrm{d}\vec{X}$ in the current configuration is a curved element. Furthermore, the differential vector $\mathrm{d}\vec{X}$ does not depend on time $t$, since the positions of points P and Q being in the initial configuration $V_0^b$ do not depend on time $t$. On the other hand, the positions of points $\underline{P}(t_1)$ and $\underline{Q}(t_1)$ depend on time $t$, which leads to the time dependency of $\mathrm{d}\vec{\underline{x}}(t_1)$.

Figure 3.5 also shows the displacement vectors $\vec{u}(\mathrm{P};t_1)$ and $\vec{u}(\mathrm{Q};t_1)$, where the semicolon indicates that the quantity after the semicolon, $t_1$, is a parameter. The displacement vector $\vec{u}(\mathrm{P};t_1)$ is a vector from point P to point $\underline{P}(t_1)$, whereas the displacement vector $\vec{u}(\mathrm{Q};t_1)$ is a vector from point Q to point $\underline{Q}(t_1)$. The displacement vectors $\vec{u}(\mathrm{P};t_1)$ and $\vec{u}(\mathrm{Q};t_1)$ do not show the paths of the material points during deformation but are vectors from points in the initial configuration $V_0^b$ to their locations in the current configuration $\underline{v}^b(t)$. The paths of the material points during deformation may be nonlinear.

Thus, based on Equation (3.33b), the displacement vector $\vec{u}(\mathrm{Q};t_1)$ reads

$$\vec{u}(\mathrm{Q};t_1) = \vec{u}(\mathrm{P};t_1) + \vec{u}\,(\mathrm{P};t_1)\overleftarrow{\nabla}(\vec{X})\cdot \mathrm{d}\vec{X}\,. \tag{3.38}$$

In Equation (3.38) the quantities are expressed by the initial material coordinates $X_i$.

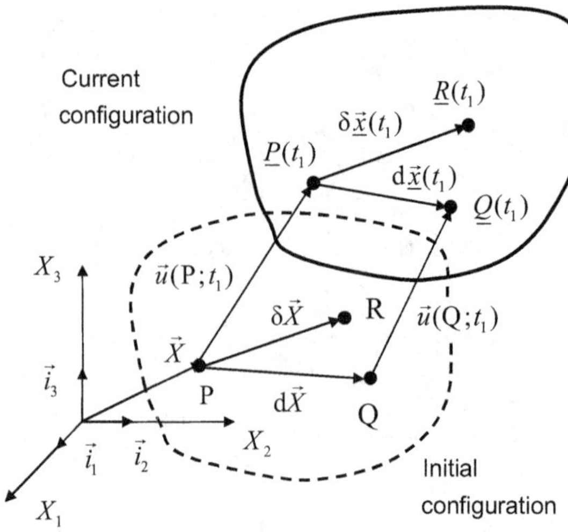


**Figure 3.6** Differential vectors between P, Q, and R in the initial configuration $\mathrm{V}_0^{\mathrm{b}}$ and between $\underline{P}(t_1)$, $\underline{Q}(t_1)$, and $\underline{R}(t_1)$ in the current configuration $\underline{v}^{\mathrm{b}}(t)$.

The vector $\mathrm{P}-\underline{Q}(t_1)$ shown by a dashed line in Figure 3.5 can be expressed as a sum of vectors $\mathrm{P}-\underline{P}(t_1)+\underline{P}(t_1)-\underline{Q}(t_1)$ or $\mathrm{P}-\mathrm{Q}+\mathrm{Q}-\underline{Q}(t_1)$. This gives the following relation:

$$\vec{u}(\mathrm{P};t_1)+\mathrm{d}\underline{\vec{x}}(t_1)=\mathrm{d}\vec{X}+\vec{u}(\mathrm{Q};t_1)\,. \tag{3.39}$$

Substitution of Expression (3.38) into Equation (3.39) gives

$$\vec{u}(\mathrm{P};t_1)+\mathrm{d}\underline{\vec{x}}(t_1)=\mathrm{d}\vec{X}+\vec{u}(\mathrm{P};t_1)+\vec{u}(\mathrm{P};t_1)\overleftarrow{\nabla}(\vec{X})\cdot\mathrm{d}\vec{X}\,, \tag{3.40}$$

which yields

$$\mathrm{d}\underline{\vec{x}}(t_1)=\mathrm{d}\vec{X}+\vec{u}(\mathrm{P};t_1)\overleftarrow{\nabla}(\vec{X})\cdot\mathrm{d}\vec{X}\,. \tag{3.41}$$

The displacement of point R in the initial configuration $\mathrm{V}_0^{\mathrm{b}}$ to point $\underline{R}(t_1)$ in the current configuration $\underline{v}^{\mathrm{b}}(t)$ is studied next. Points R and $\underline{R}(t_1)$ are shown in Figure 3.6. The study uses the same steps as taken earlier. In order to distinguish the differential vectors, the notation $\delta$ is introduced. Thus the symbol $\delta$ is not a variational symbol used in variational calculus. Based on Result (3.41), the following can be written:

$$\delta\underline{\vec{x}}(t_1)=\delta\vec{X}+\vec{u}(\mathrm{P};t_1)\overleftarrow{\nabla}(\vec{X})\cdot\delta\vec{X}\,. \tag{3.42}$$

### 3.5.3 Explicit Form for the Green–Lagrange Strain Tensor E

Using the results obtained in the previous subsections, the explicit form of the Green–Lagrange strain tensor **E** is derived here. In order to simplify the notations, the dependence on current time $t_1$ is omitted from the following derivation. However, once the Green–Lagrange strain tensor **E** has been derived, a stricter form of it will be given.

The Green–Lagrange strain tensor is defined to be

**Def** $$\mathrm{d}\underline{\vec{x}}(t_1)\cdot\delta\underline{\vec{x}}(t_1) - \mathrm{d}\vec{X}\cdot\delta\vec{X} = 2\,\mathrm{d}\vec{X}\cdot\mathbf{E}(\vec{X},t_1)\cdot\delta\vec{X}\,. \tag{3.43}$$

In order to make the following derivation clearer, the time $t_1$ is dropped in the following derivation. Substitution of $\mathrm{d}\underline{\vec{x}}$ and $\delta\underline{\vec{x}}$ from Expressions (3.41) and (3.42) into the left side of Definition (3.43) gives

$$\begin{aligned}
&\mathrm{d}\underline{\vec{x}}\cdot\delta\underline{\vec{x}} - \mathrm{d}\vec{X}\cdot\delta\vec{X}\\
&\quad= \Big[\mathrm{d}\vec{X} + \vec{u}(\mathrm{P})\overleftarrow{\nabla}(\vec{X})\cdot\mathrm{d}\vec{X}\Big]\cdot\Big[\delta\vec{X} + \vec{u}(\mathrm{P})\overleftarrow{\nabla}(\vec{X})\cdot\delta\vec{X}\Big] - \mathrm{d}\vec{X}\cdot\delta\vec{X}\\
&\quad= \mathrm{d}\vec{X}\cdot\delta\vec{X} + \mathrm{d}\vec{X}\cdot\Big[\vec{u}(\mathrm{P})\overleftarrow{\nabla}(\vec{X})\cdot\delta\vec{X}\Big] + \Big[\vec{u}(\mathrm{P})\overleftarrow{\nabla}(\vec{X})\cdot\mathrm{d}\vec{X}\Big]\cdot\delta\vec{X}\\
&\quad+ \Big[\vec{u}(\mathrm{P})\overleftarrow{\nabla}(\vec{X})\cdot\mathrm{d}\vec{X}\Big]\cdot\Big[\vec{u}(\mathrm{P})\overleftarrow{\nabla}(\vec{X})\cdot\delta\vec{X}\Big] - \mathrm{d}\vec{X}\cdot\delta\vec{X}\,.
\end{aligned} \tag{3.44}$$

The term $\mathrm{d}\vec{X}\cdot\delta\vec{X}$ cancels with the same on the left side of the equality. Thus Expression (3.44) reduces to

$$\begin{aligned}
\mathrm{d}\underline{\vec{x}}\cdot\delta\underline{\vec{x}} - \mathrm{d}\vec{X}\cdot\delta\vec{X} &= \mathrm{d}\vec{X}\cdot\Big[\vec{u}(\mathrm{P})\overleftarrow{\nabla}(\vec{X})\cdot\delta\vec{X}\Big] + \Big[\vec{u}(\mathrm{P})\overleftarrow{\nabla}(\vec{X})\cdot\mathrm{d}\vec{X}\Big]\cdot\delta\vec{X}\\
&\quad+ \Big[\vec{u}(\mathrm{P})\overleftarrow{\nabla}(\vec{X})\cdot\mathrm{d}\vec{X}\Big]\cdot\Big[\vec{u}(\mathrm{P})\overleftarrow{\nabla}(\vec{X})\cdot\delta\vec{X}\Big]\,.
\end{aligned} \tag{3.45}$$

The right side of Expression (3.45) has the form $\mathrm{d}\vec{X}\cdot(\text{something})\cdot\delta\tilde{\mathrm{X}}$. Since the differential vectors $\mathrm{d}\vec{X}$ and $\delta\vec{X}$ are arbitrary, comparison of both sides of Definition (3.43) requires the left side of Definition (3.43) to be written in the form $\mathrm{d}\vec{X}\cdot(\text{something})\cdot\delta\tilde{\mathrm{X}}$. This implies that the right side of Expression (3.45) must be rewritten in the form $\mathrm{d}\vec{X}\cdot(\text{something})\cdot\delta\tilde{\mathrm{X}}$. To make the equations easier to read, symbol P is dropped. This should not cause any confusion, as the values of all the quantities in Expression (3.45) are calculated at point P. This means that the value of the Green–Lagrange strain tensor is calculated at point P in the initial configuration $\mathrm{V}_0^{\mathrm{b}}$.

The first term on the right side of Expression (3.45) takes the following form:

$$\mathrm{d}\vec{X}\cdot\Big[\vec{u}\overleftarrow{\nabla}(\vec{X})\cdot\delta\vec{X}\Big] = \mathrm{d}\vec{X}\cdot\Big[\vec{u}\overleftarrow{\nabla}(\vec{X})\Big]\cdot\delta\vec{X}\,. \tag{3.46}$$

Equation (3.37) allows the second term on the right side of Expression (3.45) to be manipulated as follows:

$$\Big[\vec{u}\overleftarrow{\nabla}(\vec{X})\cdot\mathrm{d}\vec{X}\Big]\cdot\delta\vec{X} = \Big[\mathrm{d}\vec{X}\cdot\vec{\nabla}(\vec{X})\vec{u}\Big]\cdot\delta\vec{X} = \mathrm{d}\vec{X}\cdot\Big[\vec{\nabla}(\vec{X})\vec{u}\Big]\cdot\delta\vec{X}\,. \tag{3.47}$$

The first equality in Expression (3.47) requires a more detailed evaluation. On the very left of Equation (3.47), there is a dot product between the quantities $\vec{u}$ and $\delta\vec{X}$. The same holds for the term after the first equals sign.

Expression (3.37) is used for manipulation of the third term of Expression (3.45). The following is obtained:

$$\begin{aligned}\left[\vec{u}\overleftarrow{\nabla}(\vec{X})\cdot\mathrm{d}\vec{X}\right]\cdot\left[\vec{u}\overleftarrow{\nabla}(\vec{X})\cdot\delta\vec{X}\right] &= \left[\mathrm{d}\vec{X}\cdot\vec{\nabla}(\vec{X})\vec{u}\right]\cdot\left[\vec{u}\overleftarrow{\nabla}(\vec{X})\cdot\delta\vec{X}\right]\\ &= \mathrm{d}\vec{X}\cdot\left[\vec{\nabla}(\vec{X})\vec{u}\right]\cdot\left[\vec{u}\overleftarrow{\nabla}(\vec{X})\right]\cdot\delta\vec{X} = \mathrm{d}\vec{X}\cdot\left[\left(\vec{\nabla}(\vec{X})\vec{u}\right)\cdot\left(\vec{u}\overleftarrow{\nabla}(\vec{X})\right)\right]\cdot\delta\vec{X}\,.\end{aligned} \tag{3.48}$$

On the far left side of Expression (3.48), the dot product between the square brackets involves the dot product operation between the gradients of the displacement vector $\vec{u}$. By using the forward gradient operator (which is the transpose of the backward gradient operator), one can obtain what is given in the second line of Expression (3.48).

Substitution of Expressions (3.46)–(3.48) into Equation (3.45) yields

$$\begin{aligned}\mathrm{d}\underline{\vec{x}}(t_1)\cdot\delta\underline{\vec{x}}(t_1) - \mathrm{d}\vec{X}\cdot\delta\vec{X} &= \mathrm{d}\vec{X}\cdot\left[\vec{u}\overleftarrow{\nabla}(\vec{X})\right]\cdot\delta\vec{X} + \mathrm{d}\vec{X}\cdot\left[\vec{\nabla}(\vec{X})\vec{u}\right]\cdot\delta\vec{X}\\ &\quad + \mathrm{d}\vec{X}\cdot\left[\left(\vec{\nabla}(\vec{X})\vec{u}\right)\cdot\left(\vec{u}\overleftarrow{\nabla}(\vec{X})\right)\right]\cdot\delta\vec{X}\,.\end{aligned} \tag{3.49}$$

Expression (3.49) gives

$$\mathrm{d}\underline{\vec{x}}(t_1)\cdot\delta\underline{\vec{x}}(t_1) - \mathrm{d}\vec{X}\cdot\delta\vec{X} = \mathrm{d}\vec{X}\cdot\left[\vec{u}\overleftarrow{\nabla}(\vec{X}) + \vec{\nabla}(\vec{X})\vec{u} + \left(\vec{\nabla}(\vec{X})\vec{u}\right)\cdot\left(\vec{u}\overleftarrow{\nabla}(\vec{X})\right)\right]\cdot\delta\vec{X}\,. \tag{3.50}$$

Comparison of Definition (3.43) with Expression (3.50) gives

$$\mathbf{E} = \frac{1}{2}\left[\vec{u}\overleftarrow{\nabla}(\vec{X}) + \vec{\nabla}(\vec{X})\vec{u} + \left(\vec{\nabla}(\vec{X})\vec{u}\right)\cdot\left(\vec{u}\overleftarrow{\nabla}(\vec{X})\right)\right]. \tag{3.51}$$

Equation (3.51) is expressed by the initial material coordinates $X_i$ and is derived for point P at time $t_1$. This gives

$$\begin{aligned}\mathbf{E}(\mathrm{P};t_1) = \frac{1}{2}\Big[&\vec{u}(\mathrm{P};t_1)\overleftarrow{\nabla}(\vec{X}) + \vec{\nabla}(\vec{X})\vec{u}(\mathrm{P};t_1)\\ &+ \left(\vec{\nabla}(\vec{X})\vec{u}(\mathrm{P};t_1)\right)\cdot\left(\vec{u}(\mathrm{P};t_1)\overleftarrow{\nabla}(\vec{X})\right)\Big].\end{aligned} \tag{3.52}$$

The preceding study is repeated at times $t_2$, $t_3$, and so on, so that finally, all the moments are evaluated. Furthermore, if the position of point P is given by $\vec{X}$, Expression (3.52) yields

$$\mathbf{E}(\vec{X},t) = \frac{1}{2}\left[\vec{u}(\vec{X},t)\overleftarrow{\nabla}(\vec{X}) + \vec{\nabla}(\vec{X})\vec{u}(\vec{X},t) + \left(\vec{\nabla}(\vec{X})\vec{u}(\vec{X},t)\right)\cdot\left(\vec{u}(\vec{X},t)\overleftarrow{\nabla}(\vec{X})\right)\right]. \tag{3.53}$$

The appearance of Expression (3.53) is heavy; thus Form (3.51) or the following even simpler equation is used in this book:

$$\mathbf{E} = \frac{1}{2}\left[\vec{u}\overleftarrow{\nabla} + \vec{\nabla}\vec{u} + (\vec{\nabla}\vec{u})\cdot(\vec{u}\overleftarrow{\nabla})\right]. \tag{3.54}$$

The use of Expression (3.54) requires considerable expertise, since the user has to be aware of the coordinates. The Green–Lagrange strain tensor **E** is defined in the initial configuration $V_0^b$ but applied in the initial geometry at the current time and position configuration, that is, in the IG-CtP configuration $V_0^b$.

**Example 3.6:** What are the scalar components of the Green–Lagrange strain tensor **E**?

**Solution:** Definitions for the vector operator "del" from Definitions (3.7) are recalled:

$$\vec{\nabla}(\vec{X}) := \frac{\partial}{\partial \vec{X}} = \vec{i}_m \frac{\partial}{\partial X_m} \quad \text{or} \quad \overleftarrow{\nabla}(\vec{X}) := \frac{\overleftarrow{\partial}}{\partial \vec{X}} = \frac{\overleftarrow{\partial}}{\partial X_m} \vec{i}_m \,. \tag{3.55}$$

The displacement vector $\vec{u}$ can be written as follows:

$$\vec{u} = u_s \, \vec{i}_s = \vec{i}_r \, u_r \,. \tag{3.56}$$

Based on Expressions (3.55) and (3.56), we arrive at the following:

$$\vec{\nabla}(\vec{X})\vec{u} = \vec{i}_m \frac{\partial}{\partial X_m} (u_s \, \vec{i}_s) = \frac{\partial u_s}{\partial X_m} \vec{i}_m \, \vec{i}_s \tag{3.57}$$

and

$$\vec{u}\overleftarrow{\nabla}(\vec{X}) = (\vec{i}_r u_r) \frac{\overleftarrow{\partial}}{\partial X_n} \vec{i}_n = \frac{\partial u_r}{\partial X_n} \vec{i}_r \, \vec{i}_n \,. \tag{3.58}$$

The dummy indices in Equations (3.57) and (3.58) are changed. In Equation (3.57) the replacements $m \to i$ and $s \to j$ are made, whereas in Equation (3.58) the replacements are as follows: $r \to i$ and $n \to j$. This leads to the following:

$$\vec{\nabla}(\vec{X})\vec{u} = \frac{\partial u_j}{\partial X_i} \vec{i}_i \, \vec{i}_j \quad \text{and} \quad \vec{u}\overleftarrow{\nabla}(\vec{X}) = \frac{\partial u_i}{\partial X_j} \vec{i}_i \, \vec{i}_j \,. \tag{3.59}$$

Equations (3.57) and (3.58) yield

$$\left(\vec{\nabla}(\vec{X})\,\vec{u}\right) \cdot \left(\vec{u}\,\overleftarrow{\nabla}(\vec{X})\right) = \left(\frac{\partial u_s}{\partial X_m} \vec{i}_m \, \vec{i}_s\right) \cdot \left(\frac{\partial u_r}{\partial X_n} \vec{i}_r \, \vec{i}_n\right) = \frac{\partial u_s}{\partial X_m} \frac{\partial u_s}{\partial X_n} \vec{i}_m \, \vec{i}_n \,. \tag{3.60}$$

The dummy index in Equation (3.60) is changed as follows: $m \to i$ and $n \to j$. Thus Equation (3.60) takes the form

$$\left(\vec{\nabla}(\vec{X})\,\vec{u}\right) \cdot \left(\vec{u}\,\overleftarrow{\nabla}(\vec{X})\right) = \frac{\partial u_s}{\partial X_i} \frac{\partial u_s}{\partial X_j} \vec{i}_i \, \vec{i}_j \,. \tag{3.61}$$

Terms (3.59) and (3.61) are substituted into Expression (3.53) to obtain

$$\mathbf{E} = \frac{1}{2} \left[ \frac{\partial u_i}{\partial X_j} + \frac{\partial u_j}{\partial X_i} + \frac{\partial 1 u_s}{\partial X_i} \frac{\partial \, u_s}{\partial X_j} \right] \vec{i}_i \, \vec{i}_j \,. \tag{3.62}$$

Equation (3.62) results in the components

$$E_{ij} = \frac{1}{2}\left[\frac{\partial u_i}{\partial X_j} + \frac{\partial u_j}{\partial X_i} + \frac{\partial u_s}{\partial X_i}\frac{\partial u_s}{\partial X_j}\right]. \tag{3.63}$$

For example, Equation (3.63) gives (for $i = 1$ and $j = 1$)

$$\begin{aligned} E_{11} &= \frac{1}{2}\left[\frac{\partial u_1}{\partial X_1} + \frac{\partial u_1}{\partial X_1} + \frac{\partial u_s}{\partial X_1}\frac{\partial u_s}{\partial X_1}\right] \\ &= \frac{\partial u_1}{\partial X_1} + \frac{1}{2}\left[\left(\frac{\partial u_1}{\partial X_1}\right)^2 + \left(\frac{\partial u_2}{\partial X_1}\right)^2 + \left(\frac{\partial u_3}{\partial X_1}\right)^2\right] \end{aligned} \tag{3.64}$$

and (for $i = 1$ and $j = 2$)

$$\begin{aligned} E_{12} &= \frac{1}{2}\left[\frac{\partial u_1}{\partial X_2} + \frac{\partial u_2}{\partial X_1} + \frac{\partial u_s}{\partial X_1}\frac{\partial u_s}{\partial X_2}\right] \\ &= \frac{1}{2}\left[\frac{\partial u_1}{\partial X_2} + \frac{\partial u_2}{\partial X_1} + \frac{\partial u_1}{\partial X_1}\frac{\partial u_1}{\partial X_2} + \frac{\partial u_2}{\partial X_1}\frac{\partial u_2}{\partial X_2} + \frac{\partial u_3}{\partial X_1}\frac{\partial u_3}{\partial X_2}\right]. \ \blacksquare \end{aligned} \tag{3.65}$$

**Example 3.7:** Derive the scalar component $E_{11}$ for a case in which, under uniaxial loading, the distance between two points grows from $L_0$ to $L$. Figure 3.7 shows a case in which a uniaxial elongation is applied in direction $X_1$. In order to make Figure 3.7 clearer, a rigid body motion is assumed in the $X_2$ direction. This does not affect the strain components.

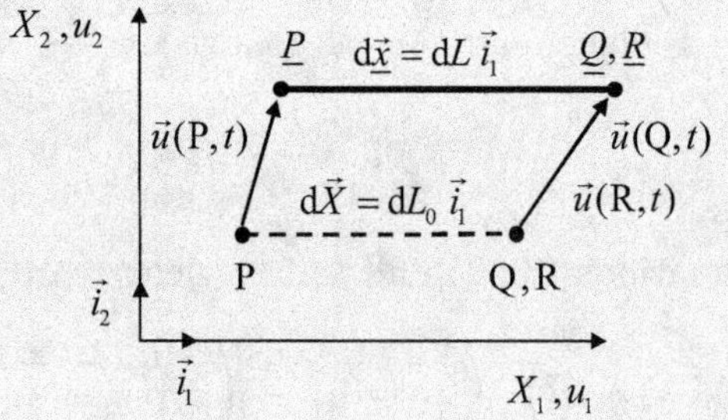


**Figure 3.7** Deformation due to uniaxial tension.

**Solution:** Based on Figure 3.7, the following result is obtained:

$$d\vec{X} = \delta\vec{X} = dL_0\,\vec{i}_1 \quad \text{and} \quad d\underline{\vec{x}} = \delta\underline{\vec{x}} = dL\,\vec{i}_1\,. \tag{3.66}$$

The notation d in the lengths $dL_0$ and $dL(t)$ shows that these quantities are infinitesimal (whatever the meaning of infinitesimal is in physics). The definition for the Green–Lagrange strain tensor, Definition (3.43), is written for the current time $t$ as

**Def**
$$d\underline{\vec{x}}(t)\cdot\delta\underline{\vec{x}}(t) - d\vec{X}\cdot\delta\vec{X} = 2\,d\vec{X}\cdot\mathbf{E}(\vec{X},t)\cdot\delta\vec{X}\,. \tag{3.67}$$

Substitution of Quantities (3.66) into Definition (3.67) gives

$$\mathrm{d}L\,\vec{i}_1 \cdot \mathrm{d}L\,\vec{i}_1 - \mathrm{d}L_0\,\vec{i}_1 \cdot \mathrm{d}L_0\,\vec{i}_1 = 2\,\mathrm{d}L_0\,\vec{i}_1 \cdot E_{st}\,\vec{i}_s\,\vec{i}_t \cdot \mathrm{d}L_0\,\vec{i}_1\,, \tag{3.68}$$

which leads to

$$(\,\mathrm{d}L\,)^2 - (\,\mathrm{d}L_0\,)^2 = 2\,\mathrm{d}L_0\,E_{11}\,\mathrm{d}L_0 \qquad \Rightarrow \qquad E_{11} = \frac{(\,\mathrm{d}L\,)^2 - (\,\mathrm{d}L_0\,)^2}{2\,(\,\mathrm{d}L_0\,)^2}\,. \quad \blacksquare \tag{3.69}$$

**Example 3.8:** Derive the scalar component $E_{12}$ by studying a case in which the response of a material point is in the $(X_1, X_2)$ plane. Figure 3.8 shows how, during pure shear deformation, the vectors between points P and Q and P and R in the initial configuration $\mathrm{V}_0^\mathrm{b}$ are replaced by the vectors between points $\underline{P}(t)$ and $\underline{Q}(t)$ and $\underline{P}(t)$ and $\underline{R}(t)$ in the current configuration $\underline{v}^b(t)$. Since the idea is to study pure shear deformation, points P and $\underline{P}(t)$ coincide, as shown in Figure 3.8.

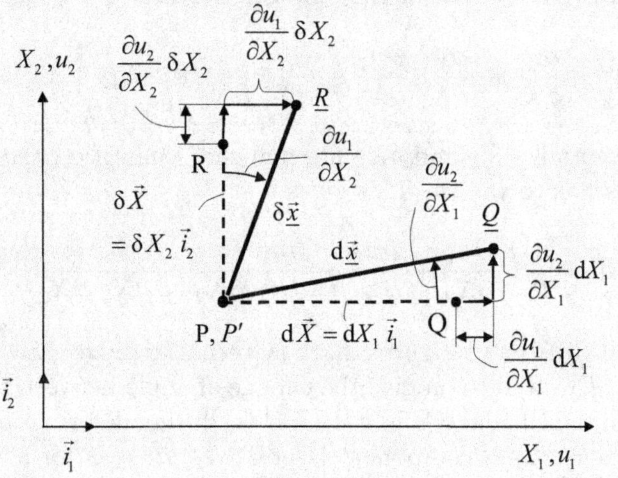


**Figure 3.8** Deformation in the $(X_1, X_2)$ plane.

**Solution:** Based on Figure 3.8, we arrive at:

$$\mathrm{d}\underline{\vec{x}} = \mathrm{d}X_1\,\vec{i}_1 + \frac{\partial u_1}{\partial X_1}\,\mathrm{d}X_1\,\vec{i}_1 + \frac{\partial u_2}{\partial X_1}\,\mathrm{d}X_1\,\vec{i}_2 \tag{3.70}$$

and

$$\delta\underline{\vec{x}} = \delta X_2\,\vec{i}_2 + \frac{\partial u_2}{\partial X_2}\,\delta X_2\,\vec{i}_2 + \frac{\partial u_1}{\partial X_2}\,\delta X_2\,\vec{i}_1\,. \tag{3.71}$$

The definition for the Green–Lagrange strain tensor **E**, Definition (3.43), is written for the current time $t$ as

$$\textbf{Def} \qquad \mathrm{d}\underline{\vec{x}}(t)\cdot\delta\underline{\vec{x}}(t) - \mathrm{d}\vec{X}\cdot\delta\vec{X} = 2\,\mathrm{d}\vec{X}\cdot\mathbf{E}(\vec{X},t)\cdot\delta\vec{X}\,. \tag{3.72}$$

Substitution of Quantities (3.70) and (3.71) into Definition (3.72) gives

$$\left[\mathrm{d}X_1\,\vec{i}_1 + \frac{\partial u_1}{\partial X_1}\,\mathrm{d}X_1\,\vec{i}_1 + \frac{\partial u_2}{\partial X_1}\,\mathrm{d}X_1\vec{i}_2\right]\cdot\left[\delta X_2\,\vec{i}_2 + \frac{\partial u_2}{\partial X_2}\,\delta X_2\,\vec{i}_2 + \frac{\partial u_1}{\partial X_2}\,\delta X_2\,\vec{i}_1\right]$$
$$- \mathrm{d}X_1\,\vec{i}_1\cdot\delta X_2\,\vec{i}_2 = 2\,\mathrm{d}X_1\,\vec{i}_1\cdot E_{st}\,\vec{i}_s\,\vec{i}_t\cdot\delta X_2\,\vec{i}_2 \tag{3.73}$$

which yields

$$\mathrm{d}X_1\,\frac{\partial u_1}{\partial X_2}\,\delta X_2 + \mathrm{d}X_1\,\frac{\partial u_1}{\partial X_1}\,\frac{\partial u_1}{\partial X_2}\,\delta X_2 + \frac{\partial u_2}{\partial X_1}\,\mathrm{d}X_1\,\delta X_2 + \mathrm{d}X_1\,\frac{\partial u_2}{\partial X_1}\,\frac{\partial u_2}{\partial X_2}\,\delta X_2$$
$$= 2\,\mathrm{d}X_1\,E_{12}\,\delta X_2\,. \tag{3.74}$$

Expression (3.74) can be cast in the following form:

$$\mathrm{d}X_1\left(\frac{\partial u_1}{\partial X_2} + \frac{\partial u_2}{\partial X_1} + \frac{\partial u_1}{\partial X_1}\,\frac{\partial u_1}{\partial X_2} + \frac{\partial u_2}{\partial X_1}\,\frac{\partial u_2}{\partial X_2} - 2\,E_{12}\right)\delta X_2 = 0\,. \tag{3.75}$$

Since the differentials $\mathrm{d}X_1$ and $\delta X_2$ are non-zero quantities, the term between the parentheses has to vanish. This gives

$$E_{12} = \frac{1}{2}\left(\frac{\partial u_1}{\partial X_2} + \frac{\partial u_2}{\partial X_1} + \frac{\partial u_1}{\partial X_1}\,\frac{\partial u_1}{\partial X_2} + \frac{\partial u_2}{\partial X_1}\,\frac{\partial u_2}{\partial X_2}\right)\,. \tag{3.76}$$

Based on Result (3.76) and Figure 3.8, the Green–Lagrange shear strain components $E_{12}$, $E_{23}$, $E_{31}$, and so on give the change of angle between two orthogonal line segments in the current configuration $\underline{v}^{\mathrm{b}}(t)$ during deformation. If the angle is acute, the shear strain component is positive, whereas for an obtuse angle, the shear strain component is negative. ■

Actually, the foregoing derivation did not study the change of line elements but rather the change of position of the material points Q and R to the positions $\underline{Q}(t)$ and $\underline{R}(t)$. The line elements were then drawn from point P to point $\underline{Q}(t)$ and from point P to point $\underline{R}(t)$. Finally, these line elements were compared with those drawn between the points P and Q and P and R.

**Example 3.7** showed the reason for introducing the three points P, Q, and R in the initial configuration $\mathrm{V}_0^{\mathrm{b}}$ and for studying their change of position during deformation. It is possible to study only points P and Q and define the Green–Lagrange strain tensor to be $\mathrm{d}\underline{\vec{x}}(t)\cdot\mathrm{d}\underline{\vec{x}}(t) - \mathrm{d}\vec{X}\cdot\mathrm{d}\vec{X} = 2\,\mathrm{d}\vec{X}\cdot\mathrm{E}(\vec{X},t)\cdot\mathrm{d}\vec{X}$. This gives a mathematically correct result but Definition (3.38) provides a physically stronger foundation.

Derivation of the explicit form of the Green–Lagrange strain tensor **E** involved expressing vectors from point P to point Q and from point P to point R by the differential vectors $\mathrm{d}\vec{X}$ and $\delta\vec{X}$ [cf. Equations (3.29) and (3.31)]. This means that the higher-order terms present in the Taylor series were neglected. It may lead to the question, Does the introduction of the differentials $\mathrm{d}\vec{X}$ and $\delta\vec{X}$ lead to a reduced accuracy of the strain tensor, or does it create some other limitations? The question may derive from the unclear physical interpretation of the concept "differential." Since the differentials $\mathrm{d}\vec{X}$, $\delta\vec{X}$, $\mathrm{d}\underline{\vec{x}}(t)$ and $\delta\underline{\vec{x}}(t)$ are infinitesimal quantities Definition (3.43) for the Green–Lagrange strain tensor $\mathbf{E}(\vec{X},t)$ contains the idea that $\mathbf{E}(\vec{X},t)$ is obtained as a limit when the lengths of $\mathrm{d}\vec{X}$ and $\delta\vec{X}$ approach zero. Result $(3.69)_2$ is utilized.

The following simplified uniaxial example demonstrates how to derive the explicit form for the Green–Lagrange strain tensor **E** without introducing differentials.

**Example 3.9:** Derive the component $E_{11}$ of the Green–Lagrange strain tensor **E** for a bar under uniaxial tensile loading (see Figure 3.9). In order to simplify notations and make the study clearer, the coordinates are denoted by $X$ and $\underline{x}$, instead of by $X_1$ and $\underline{x}_1$. The displacement in the horizontal direction is denoted by $u$. For this particular simplified uniaxial case, component $E_{11}$ of the Green–Lagrange strain tensor **E** at point P is obtained from the following relation:

$$E_{11}(\mathrm{P}) = \lim_{\Delta X \to 0} \frac{\Delta \underline{x}^2 - \Delta X^2}{2\,\Delta X^2}\,. \tag{3.77}$$

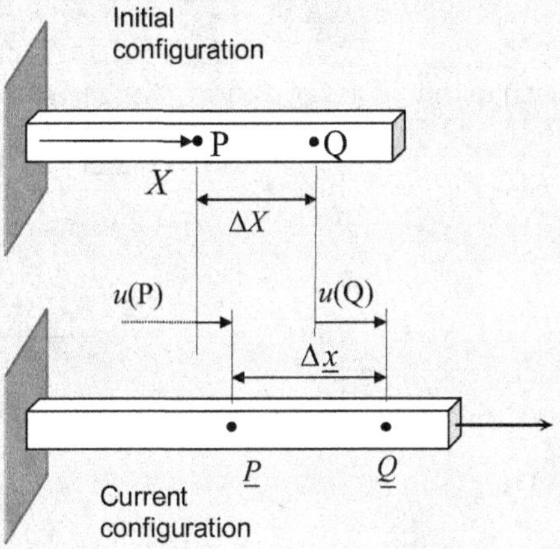


**Figure 3.9** Rod under uniaxial tensile loading.

**Solution:** Figure 3.9 shows a uniaxial tensile bar in the initial configuration $V_0^b$ and in the current configuration $\underline{v}^b(t)$. Comparison of Definition (3.43) with Expression (3.77) shows that they have similar appearances.

Figure 3.9 gives

$$\Delta\underline{x} = \Delta X + u(\mathrm{Q}) - u(\mathrm{P})\,. \tag{3.78}$$

The displacement at point Q is approximated by the Taylor series as follows: let the function $f(x)$ and all its derivatives be continuous within the interval $\hat{h}$. The values of function $f(x)$ and all its derivatives are assumed to be known at point $x_0$ being within the interval $\hat{h}$. The value of the neighbouring point $(x_0 + \Delta)$ to point $(x_0)$ can be obtained by the following Taylor series:

$$f(x_0+\Delta) = f(x_0) + \frac{\partial f(x_0)}{\partial x}\Delta + \frac{1}{2!}\frac{\partial^2 f(x_0)}{\partial x^2}\Delta^2 + \frac{1}{3!}\frac{\partial^3 f(x_0)}{\partial x^3}\Delta^3 + \cdots . \tag{3.79}$$

Based on Figure 3.9, we arrive at the following:

$$u(\mathrm{P}) = u(X). \tag{3.80}$$

Assuming that $f(x_0) = u(\mathrm{P}) = u(X)$, $f(x_0 + \Delta X) = u(\mathrm{Q})$, $\Delta = \Delta X$, and by taking the first three terms on the right side of Taylor Series (3.79), the following is obtained:

$$u(\mathrm{Q}) \approx u(X) + \frac{\partial u(X)}{\partial X}\Delta X + \frac{1}{2!}\frac{\partial^2 u(X)}{\partial X^2}\Delta X^2 . \tag{3.81}$$

Substitution of Terms (3.80) and (3.81) into Expression (3.78) gives

$$\begin{aligned}\Delta \underline{x} &= \Delta X + u(\mathrm{Q}) - u(\mathrm{P}) \\ &= \Delta X + u(X) + \frac{\partial u(X)}{\partial X}\Delta X + \frac{1}{2!}\frac{\partial^2 u(X)}{\partial X^2}\Delta X^2 - u(X) \\ &= \Delta X + \frac{\partial u(X)}{\partial X}\Delta X + \frac{1}{2!}\frac{\partial^2 u(X)}{\partial X^2}\Delta X^2 .\end{aligned} \tag{3.82}$$

Substitution of Expression (3.82) into Expression (3.77) yields

$$\begin{aligned}E_{11}(\mathrm{P}) :&= \lim_{\Delta X \to 0} \frac{\Delta \underline{x}^2 - \Delta X^2}{2\,\Delta X^2} \\ &= \lim_{\Delta X \to 0} \frac{\left[\Delta X + \frac{\partial u(X)}{\partial X}\Delta X + \frac{1}{2!}\frac{\partial^2 u(X)}{\partial X^2}\Delta X^2\right]^2 - \Delta X^2}{2\,\Delta X^2} .\end{aligned} \tag{3.83}$$

Equation (3.83) gives

$$E_{11}(\mathrm{P}) = \lim_{\Delta X \to 0} \frac{2\frac{\partial u(X)}{\partial X}\Delta X^2 + \frac{2}{2!}\frac{\partial^2 u(X)}{\partial X^2}\Delta X^3 + \left[\frac{\partial u(X)}{\partial X}\Delta X + \frac{1}{2!}\frac{\partial^2 u(X)}{\partial X^2}\Delta X^2\right]^2}{2\,\Delta X^2} . \tag{3.84}$$

Equation (3.84) further gives

$$E_{11}(\mathrm{P}) = \lim_{\Delta X \to 0}\left\{\frac{\partial u(X)}{\partial X} + \frac{1}{2!}\frac{\partial^2 u(X)}{\partial X^2}\Delta X + \frac{1}{2}\left[\frac{\partial u(X)}{\partial X} + \frac{1}{2!}\frac{\partial^2 u(X)}{\partial X^2}\Delta X\right]^2\right\}, \tag{3.85}$$

which yields

$$E_{11}(\mathrm{P}) = \lim_{\Delta X \to 0} \left\{ \frac{\partial u(X)}{\partial X} + \frac{1}{2!}\frac{\partial^2 u(X)}{\partial X^2}\Delta X + \frac{1}{2}\left(\frac{\partial u(X)}{\partial X}\right)^2 \right.$$
$$\left. + \frac{1}{2!}\frac{\partial u(X)}{\partial X}\frac{\partial^2 u(X)}{\partial X^2}\Delta X + \frac{1}{2}\left(\frac{1}{2}\right)^2\left(\frac{\partial^2 u(X)}{\partial X^2}\right)^2 \Delta X^2 \right\}. \tag{3.86}$$

Expression (3.86) gives the final result, which reads

$$E_{11}(\mathrm{P}) = E_{11}(X) = \frac{\partial u(X)}{\partial X} + \frac{1}{2}\left(\frac{\partial u(X)}{\partial X}\right)^2. \tag{3.87}$$

Result (3.87) equals the uniaxial case of $E_{11}$ given in Equation (3.64). It can easily be shown that introducing even higher-order terms than those in Equation (3.81) does not give any additional terms beyond Equation (3.87). The example showed that the Green–Lagrange strain tensor can be derived without introducing differentials. ■

### 3.5.4 Green–Lagrange Strain Tensor E in Terms of the Deformation Gradient F

The present subsection introduces the deformation gradient **F** and derives the expression for the Green–Lagrange strain tensor **E** in terms of the deformation gradient **F**.

By extending Expression (3.41) to cover all the moments of time, and taking into account that the position of point P is $\vec{X}$, Expression (3.41) gives

$$\mathrm{d}\vec{\underline{x}}(t) = \mathrm{d}\vec{X} + \vec{u}(\vec{X},t)\overleftarrow{\nabla}(\vec{X})\cdot\mathrm{d}\vec{X}\,. \tag{3.88}$$

Displacement gradient **H** and deformation gradient **F** are defined to be

$$\mathbf{H} := \vec{u}(\vec{X},t)\overleftarrow{\nabla}(\vec{X}) \tag{3.89}$$

and

$$\mathbf{F} := \mathbf{1} + \mathbf{H} = \mathbf{1} + \vec{u}(\vec{X},t)\overleftarrow{\nabla}(\vec{X})\,. \tag{3.90}$$

Definition (2.63) allows the following to be written:

$$\mathrm{d}\vec{X} = \mathbf{1}\cdot\mathrm{d}\vec{X}\,, \qquad \mathrm{d}\vec{X} = \mathrm{d}\vec{X}\cdot\mathbf{1}\,, \qquad \text{and} \qquad \delta\vec{X} = \mathbf{1}\cdot\delta\vec{X}\,. \tag{3.91}$$

Substitution of Expression $(3.91)_1$ into Equation (3.88) yields

$$\mathrm{d}\vec{\underline{x}}(t) = \mathbf{1}\cdot\mathrm{d}\vec{X} + \vec{u}(\vec{X},t)\overleftarrow{\nabla}(\vec{X})\cdot\mathrm{d}\vec{X}\,, \tag{3.92}$$

and furthermore,

$$\mathrm{d}\underline{\vec{x}}(t) = \left[\mathbf{1} + \vec{u}(\vec{X},t)\overleftarrow{\nabla}(\vec{X})\right] \cdot \mathrm{d}\vec{X} \,. \tag{3.93}$$

Substitution of Definitions (3.89) and (3.90) into Expression (3.93) gives

$$\mathrm{d}\underline{\vec{x}}(t) = [\,\mathbf{1} + \mathbf{H}\,] \cdot \mathrm{d}\vec{X} \qquad \text{and} \qquad \mathrm{d}\underline{\vec{x}}(t) = \mathbf{F} \cdot \mathrm{d}\vec{X} \,. \tag{3.94}$$

Based on Property (2.78), Equation $(3.94)_2$ gives

$$\mathrm{d}\underline{\vec{x}}(t) = \mathrm{d}\vec{X} \cdot \mathrm{F}^{\mathrm{T}} \,. \tag{3.95}$$

**Example 3.10:** Derive the Green–Lagrange strain tensor **E** in terms of the deformation gradient **F**.

**Solution:** Definition (3.43) of the Green–Lagrange strain tensor **E** is applied to the current time $t$:

**Def** $$\mathrm{d}\underline{\vec{x}}(t) \cdot \delta\underline{\vec{x}}(t) - \mathrm{d}\vec{X} \cdot \delta\vec{X} = 2\,\mathrm{d}\vec{X} \cdot \mathbf{E}(\vec{X},t) \cdot \delta\vec{X} \,. \tag{3.96}$$

Based on Equation $(3.94)_2$, we arrive at the following:

$$\delta\underline{\vec{x}}(t) = \mathbf{F} \cdot \delta\vec{X} \,. \tag{3.97}$$

Substitution of Equations (3.95) and (3.97) into Definition (3.96) yields

$$\mathrm{d}\vec{X} \cdot \mathbf{F}^{\mathrm{T}} \cdot \mathbf{F} \cdot \delta\vec{X} - \mathrm{d}\vec{X} \cdot \delta\vec{X} = 2\,\mathrm{d}\vec{X} \cdot \mathbf{E} \cdot \delta\vec{X} \,. \tag{3.98}$$

Substitution of Terms $(3.91)_2$ and $(3.91)_3$ into Equation (3.98) yields

$$\mathrm{d}\vec{X} \cdot (\mathbf{F}^{\mathrm{T}} \cdot \mathbf{F} - \mathbf{1} \cdot \mathbf{1}) \cdot \delta\vec{X} = 2\,\mathrm{d}\vec{X} \cdot \mathbf{E} \cdot \delta\vec{X} \,, \tag{3.99}$$

which leads to

$$\mathrm{d}\vec{X} \cdot \left(\mathbf{F}^{\mathrm{T}} \cdot \mathbf{F} - \mathbf{1} \cdot \mathbf{1} - 2\,\mathbf{E}\right) \cdot \delta\vec{X} = 0 \,. \tag{3.100}$$

Since $\mathrm{d}\vec{X}$ and $\delta\vec{X}$ are arbitrary nonzero vectors, the term in parentheses in Equation (3.100) has to be a zero second-order tensor. This gives

$$\mathbf{E} = \frac{1}{2}\left(\mathbf{F}^{\mathrm{T}} \cdot \mathbf{F} - \mathbf{1}\right) , \tag{3.101}$$

where the equation $\mathbf{1} \cdot \mathbf{1} = \mathbf{1}$ is exploited [cf. Definition (2.66)]. ■

Since the IG-CtP configuration $\mathrm{V}_0^{\mathrm{b}}$ is defined by obtaining it from the current configuration $\underline{v}^b(t)$ with the same coordinate transformation, which is between the initial configuration $\mathrm{V}_0^{\mathrm{b}}$ and the current configuration $\underline{v}^{\mathrm{b}}(t)$, the study carried out in this subsection can be interpreted as being performed in the initial configuration $V_0^b$ but the results are applied in the IG-CtP configuration $V_0^b$. This means that the Green–Lagrange strain tensor $\mathbf{E}$ can be interpreted as now defined in the IG-CtP configuration $V_0^b$.

## 3.6 Spatial Description

The present section introduces the spatial description and spatial coordinates $x_i$, and provides some mathematical results related to it. The mathematical expressions for solids and for fluids have different appearances. Solids are described using a material description, which means that the quantities are expressed in terms of the material coordinates $X_i$ or $\underline{x}_i(t)$. This practice is followed in this book in such a way that the response of the solids is described by the initial material coordinates $X_i$. Although the basic laws and axioms are the same for solids and fluids, the variables in the local forms for the basic laws and axioms differ. For example, the current density $\rho$ is an important quantity when fluids are studied, whereas the solid mechanics prefers to know the value for the stain $\varepsilon$. The present book follows the standard procedure, where, the equations for fluids are given by spatial description.

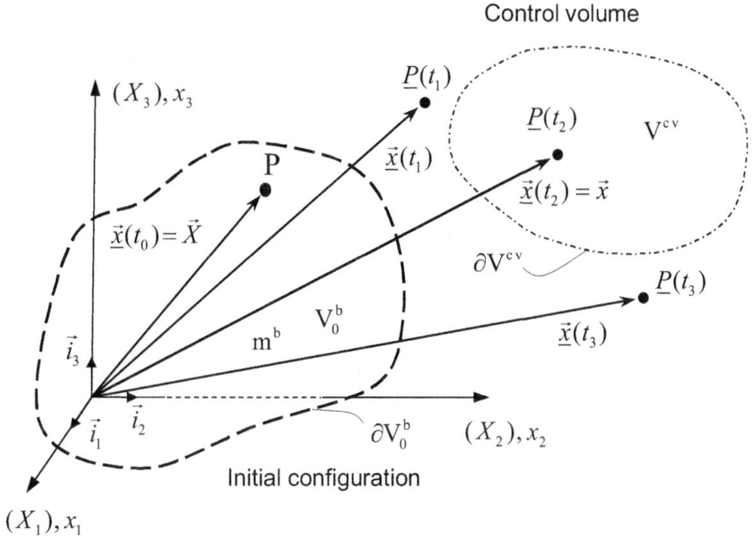


**Figure 3.10** Material point $P(t)$ moving through the control volume $V^{cv}$.

In Figure 3.10, a material point of the system (body) $m^b$ is shown to move in space with deformation of the system $m^b$. To simplify the figure's appearance, the current configurations $\underline{v}^b(t)$ of the system $m^b$ are not shown; only the control volume $V^{cv}$ is. The control volume $V^{cv}$ is a vital tool when equations for fluid mechanics are derived. In contrast to the current configuration $\underline{v}^b(t)$ showing the present position and the present geometry of the system $m^b$, control volume

$\mathrm{V^{cv}}$ has a certain fixed position and geometry in space. This means that during the deformation, there is a flow of matter (more precisely fluid) through the control volume $\mathrm{V^{cv}}$. As shown later in this book, mathematical operations with the basic laws and axioms for solids and fluids take different approaches. This stems from the difference between the current configuration $\underline{v}^{\mathrm{b}}(t)$ and the control volume $\mathrm{V^{cv}}$.

As Figure 3.10 shows, the current material coordinates $\underline{x}_i(t)$ show the position of the moving material point $\underline{P}(t)$, but the spatial coordinates $x_i$ point to a fixed position in space. In the particular case shown in Figure 3.10, at time $t_2$, the material point $P$ is in the position of the pointed by the spatial coordinates $x_i$. This implies that $\underline{\vec{x}}(t_2) = \vec{x}$, as sketched in Figure 3.10.

When expressed in the spatial coordinates $x_i$, the derivation of the material time derivative of the function $\theta$ takes a longer and a more complicated path compared to that when the material coordinates $\underline{x}_i(t)$ or $X_i$ are utilized. This will be shown next.

As already discussed, the material point $\underline{P}(t)$ is assumed to be at time $t_2$ at the point $(x_1, x_2, x_3)$. The values of the function $\theta$ and all its derivatives with respect to its independent variables $(x_1, x_2, x_3, t)$ are assumed to be known at time $t_2$ at this point. Since the material particle is a part of fluid flow, its position changes with time in such a way that after a short period of time $\Delta t$, its position reads $(x_1 + \Delta x_1, x_2 + \Delta x_2, x_3 + \Delta x_3)$. The value of the function $\theta$ associated to the material particle $\underline{P}(t_2)$ is obtained by the Taylor series. Thus, we arrive at the following:

$$\begin{aligned}\theta(x_1 + \Delta x_1, x_2 + \Delta x_2, x_3 + \Delta x_3, t_2 + \Delta t) &= \theta(x_1, x_2, x_3, t_2)\\ &+ \frac{\partial \theta(x_1, x_2, x_3, t_2)}{\partial x_1}\Delta x_1 + \frac{\partial \theta(\cdot)}{\partial x_2}\Delta x_2 + \frac{\partial \theta(\cdot)}{\partial x_3}\Delta x_3 + \frac{\partial \theta(\cdot)}{\partial t}\Delta t \\ &+ \frac{1}{2!}\left[\frac{\partial^2 \theta(\cdot)}{\partial x_1^2}\Delta x_1^2 + \frac{\partial^2 \theta(\cdot)}{\partial x_1\, \partial x_2}\Delta x_1\, \Delta x_2 + \cdots + \frac{\partial \theta(\cdot)}{\partial t^2}\Delta t^2\right] + \cdots .\end{aligned} \tag{3.102}$$

The key point in Equation (3.102) is that the values of the derivatives are expressed in the fixed point $(x_1, x_2, x_3)$ in space.

The following notation is introduced:

$$\Delta\theta(x_1, x_2, x_3, t_2) := \theta(x_1 + \Delta x_1, x_2 + \Delta x_2, x_3 + \Delta x_3, t_2 + \Delta t) - \theta(x_1, x_2, x_3, t_2). \tag{3.103}$$

The following shorter notations are introduced:

$$\theta(\vec{x}, t_2) := \theta(x_1, x_2, x_3, t_2) \quad \text{and} \quad \Delta\theta(\vec{x}, t_2) := \Delta\theta(x_1, x_2, x_3, t_2)\,. \tag{3.104}$$

Substitution of Notations (3.104) into Taylor Series (3.102) yields

$$\Delta\theta(\vec{x},t_2) = \frac{\partial\theta(\vec{x},t_2)}{\partial x_1}\,\Delta x_1 + \frac{\partial\theta(\cdot)}{\partial x_2}\,\Delta x_2 + \frac{\partial\theta(\cdot)}{\partial x_3}\,\Delta x_3 + \frac{\partial\theta(\cdot)}{\partial t}\,\Delta t$$
$$+ \frac{1}{2!}\left[\frac{\partial^2\theta(\cdot)}{\partial x_1^2}\,\Delta x_1^2 + \frac{\partial^2\theta(\cdot)}{\partial x_1\,\partial x_2}\,\Delta x_1\,\Delta x_2 + \cdots + \frac{\partial\theta(\cdot)}{\partial t^2}\,\Delta t^2\right] + \cdots . \tag{3.105}$$

Equation (3.105) can be rewritten in the following form:

$$\Delta\theta(\vec{x},t_2) = \Delta x_i \frac{\partial\theta(\vec{x},t_2)}{\partial x_i} + \frac{\partial\theta(\vec{x},t_2)}{\partial t}\,\Delta t$$
$$+ \frac{1}{2!}\left[\frac{\partial^2\theta(\cdot)}{\partial x_1^2}\,\Delta x_1^2 + \frac{\partial^2\theta(\cdot)}{\partial x_1\,\partial x_2}\,\Delta x_1\,\Delta x_2 + \cdots + \frac{\partial\theta(\cdot)}{\partial t^2}\,\Delta t^2\right] + \cdots , \tag{3.106}$$

where the Einstein summation convention is utilized. Expression (3.106) is divided by the time increment $\Delta t$, and the Einstein summation convention is replaced by the dot product. Thus the following equation is achieved:

$$\frac{\Delta\theta(\vec{x},t_2)}{\Delta t} = \frac{\Delta\vec{x}}{\Delta t}\cdot\frac{\partial\theta(\vec{x},t_2)}{\partial\vec{x}} + \frac{\partial\theta(\vec{x},t_2)}{\partial t}$$
$$+ \frac{1}{2!}\left[\frac{\partial^2\theta(\cdot)}{\partial x_1^2}\,\frac{\Delta x_1}{\Delta t}\,\Delta x_1 + \frac{\partial^2\theta(\cdot)}{\partial x_1\,\partial x_2}\,\frac{\Delta x_1}{\Delta t}\,\Delta x_2 + \cdots + \frac{\partial\theta(\cdot)}{\partial t^2}\,\Delta t\right] + \cdots . \tag{3.107}$$

The material time derivative for the function $\theta$ at the point $\vec{x}$ at time $t_2$ reads

$$\frac{\mathrm{D}\theta(\vec{x},t_2)}{\mathrm{D}t} := \lim_{\Delta t\to 0}\frac{\theta(\cdots,t_2+\Delta t)-\theta(\cdots,t_2)}{\Delta t} = \lim_{\Delta t\,\to\,0}\frac{\Delta\theta(\vec{x},t_2)}{\Delta t}\,. \tag{3.108}$$

In Definition $(3.108)_1$ the three dots $\cdots$ represent terms depending on the spatial coordinates $x_i$ and the increments $\Delta x_i$. The increments $\Delta\vec{x}$, $\Delta x_1$, $\Delta x_2$, $\Delta x_3$, and $\Delta t$ are connected together in such a way that during the period of time $\Delta\,t$, the fluid particle takes the increments $\Delta\vec{x}$, $\Delta x_1$, $\Delta x_2$, $\Delta x_3$. So when $\Delta t$ approaches zero, the displacement increments $\Delta\vec{x}$, $\Delta x_1$, $\Delta x_2$, $\Delta x_3$ approach zero as well. Thus the velocity of the fluid particle passing the point $(x_1,x_2,x_3)$ is defined as follows:

$$\vec{v}(\vec{x},t_2) := \lim_{\Delta t\,\to\,0}\frac{\Delta\vec{x}}{\Delta t}\,. \tag{3.109}$$

Expression $(3.108)_2$, Definition (3.109), and Series (3.107) yield

$$\frac{\mathrm{D}\theta(\vec{x},t_2)}{\mathrm{D}t} = \lim_{\Delta t\,\to\,0}\frac{\Delta\theta(\vec{x},t_2)}{\Delta t} = \vec{v}(\vec{x},t_2)\cdot\frac{\partial\theta(\vec{x},t_2)}{\partial\vec{x}} + \frac{\partial\theta(\vec{x},t_2)}{\partial t}\,. \tag{3.110}$$

The terms on the right side of Derivative (3.10) are reordered, which gives

$$\frac{\mathrm{D}\theta(\vec{x},t_2)}{\mathrm{D}t} = \frac{\partial\theta(\vec{x},t_2)}{\partial t} + \vec{v}(\vec{x},t_2)\cdot\frac{\partial\theta(\vec{x},t_2)}{\partial\vec{x}}\,. \tag{3.111}$$

The preceding evaluation can be repeated for all the points of the control volume $V^{cv}$ at all the moments of time $t$. Thus Expression (3.111) yields

$$\frac{D\theta(\vec{x},t)}{Dt} = \frac{\partial\theta(\vec{x},t)}{\partial t} + \vec{v}(\vec{x},t) \cdot \frac{\partial\theta(\vec{x},t)}{\partial\vec{x}}, \tag{3.112}$$

where the notation $\vec{v}(\vec{x},t)$ stands for the velocities of the fluid particles when they pass a certain point $\vec{x}$ in space at time $t$. The quantity $\vec{v}(\vec{x},t)$ is called the fluid velocity field.

The vector operator "del" given in Definitions (2.185) is recalled. It has the following appearance:

$$\vec{\nabla}(\vec{x}) := \vec{i}_m \frac{\partial}{\partial x_m} \qquad \text{or} \qquad \overleftarrow{\nabla}(\vec{x}) := \frac{\overleftarrow{\partial}}{\partial x_m}\vec{i}_m . \tag{3.113}$$

Based on Definition $(3.113)_1$, Derivative (3.112) takes the form

$$\frac{D\theta(\vec{x},t)}{Dt} = \frac{\partial\theta(\vec{x},t)}{\partial t} + \vec{v}(\vec{x},t)\cdot\vec{\nabla}(\vec{x})\theta(\vec{x},t) . \tag{3.114}$$

The preceding derivation has a minor flaw. Actually, the investigation was carried out by studying a certain point of flow at a given time $t_2$. A careful writer would have to refer to a certain point $\vec{x}^2$, for example. For the sake of clearer notation this was not done, but was replaced by the foregoing "for all the points."

## 3.7 On Material and Spatial Descriptions and Time Derivatives

This section compares material and spatial descriptions by preparing a simple example and discusses the time derivatives used with these descriptions.

In solids, the response of a material point is the key factor, and it usually depends on the history experienced by the matter. Since the material (time) derivative $D\theta/Dt$ expresses how the material experiences the rate of change of its state, in solid mechanics, the material derivatives $D\theta/Dt$ play an important role. Material models, for instance, are written in terms of material derivatives. Fluid mechanics, on the other hand, studies fluid flow with respect to a fixed position in space. Since the partial time derivative $\partial\theta/\partial t$ expresses how an external observer experiences the rate of change of a state at a spatial point, the spatial time derivatives $\partial\theta/\partial t$ are for fluid mechanics. As Siikonen [102] discussed, this does not apply to theoretical derivations in fluid mechanics where material time derivatives are used. The numerical implementations, of course, use partial time derivatives [102].

The following material time derivative appearances were derived in this book. The material time derivative of the function $\theta$ when initial material coordinates $X_i$ are used, Expression (3.22), reads

$$\frac{\mathrm{D}\theta(\vec{X},t)}{\mathrm{D}t} = \frac{\partial\theta(\vec{X},t)}{\partial t} + \frac{\partial\vec{X}}{\partial t} \cdot \frac{\partial\theta(\vec{X},t)}{\partial\vec{X}} = \frac{\partial\theta(\vec{X},t)}{\partial t}, \tag{3.115}$$

where the fact that the vector $\vec{X}$ is a constant with respect to time $t$, which gives $\partial\vec{X}/\partial t = 0$, is exploited.

The material time derivative of the function $\theta$ when the current material description is used, Expression (3.19), reads

$$\frac{\mathrm{D}\underline{\theta}(\underline{\vec{x}}(t),t)}{\mathrm{D}t} = \frac{\partial\underline{\theta}(\underline{\vec{x}}(t),t)}{\partial t} + \underline{\vec{v}}(\underline{\vec{x}}(t),t)\cdot\underline{\vec{\nabla}}(\underline{\vec{x}})\underline{\theta}(\underline{\vec{x}}(t),t)\,. \tag{3.116}$$

When the material time derivative of the function $\theta$ for the spatial description was derived, the following Expression (3.114) was obtained:

$$\frac{\mathrm{D}\theta(\vec{x},t)}{\mathrm{D}t} = \frac{\partial\theta(\vec{x},t)}{\partial t} + \vec{v}(\vec{x},t)\cdot\frac{\partial\theta(\vec{x},t)}{\partial\vec{x}}\,. \tag{3.117}$$

A careful reader may notice the "mismatch" between Derivatives (3.115) and (3.117). In Expression (3.115) the second term in the middle of Derivative (3.115) is shown to vanish, since the initial material coordinates $X_i$ do not depend on time $t$. This is due to the property of the initial material coordinates $X_i$. Although the material point $\underline{P}(t)$ moves with deformation, the initial coordinates $X_i$ of the material point $\underline{P}(t)$ do not change. The same argumentation does not apply to the spatial description, as can be seen in Equation (3.117). The difference between these two descriptions is that when the IG-CtP material description is used, the current position of the material point $\underline{P}(t)$ is always expressed by the initial material coordinates $X_i$, which do not change with time $t$. Therefore, the derivative $\partial\vec{X}/\partial t$ vanishes. The "corresponding" term in spatial description, is the fluid velocity field $\vec{v}(\vec{x},t)$. In spatial description there is fluid flow through the point $\vec{x}$. So, in spatial description, there is real fluid flow (matter has a velocity), and therefore the velocity term $\vec{v}(\vec{x},t)$ does not vanish. This may be an annoying but small difference between physics and mathematics. As already mentioned, the notation $\vec{v}(\vec{x},t)$ stands for the velocities of the fluid particles when they pass a certain point $\vec{x}$ in space.

The right sides of Material Derivatives (3.116) and (3.117) have two terms. The partial derivative of the function $\theta$ with respect to time $t$, $\partial\theta/\partial t$, gives the local rate of change of value of the function $\theta$, and the latter term gives the convective rate of change for the value of the function $\theta$.

In order to clarify the investigation approaches of the current material description and the spatial description, an example is prepared. To make the investigation easier to follow, a moving rigid rod is evaluated as an example. Furthermore, for the sake of the spatial description, the position of the observer is assumed to be fixed at a certain point. The state under consideration is the temperature $T$. As Figure 3.11 shows, in the initial configuration $V_0^b$, the temperature at the top of the rod is $T_1$ and at the bottom $T_2$ such that $T_1 \neq T_2$. Furthermore, the rod is moving downward, and the temperature of the rod $T$ is changing in such a way that the end temperatures in the current configuration $\underline{v}^b(t)$ are $\underline{T}_1 + \Delta\underline{T}_1$ and $\underline{T}_2 + \Delta\underline{T}_2$.

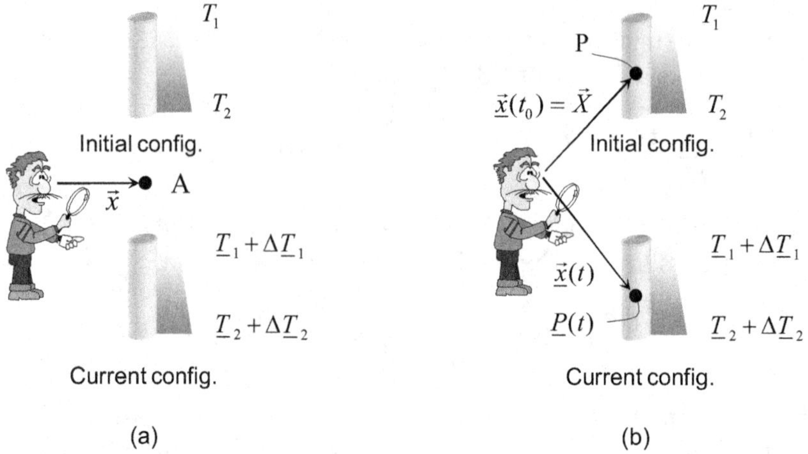


**Figure 3.11** The moving rigid rod has different and changing end temperatures. (a) Spatial description (b) current material description.

Figure 3.11(a) demonstrates the spatial description. The observer is concentrating to discover the temperature $T$ change at the point A having a fixed position in space $\vec{x}$ and therefore a fixed position with respect to the observer. Therefore the temperature change that the observer notices has two sources. The first part stems from the temperature change of the rod. The second part of the noticed temperature change is due to the movement of the rod, which implies that at different moments, different parts of the rod are observed. Since the different parts of the rod have different temperatures $T$, the movement of the rod is seen as a changing temperature $T$. The partial time derivative $\partial T(\vec{x}, t)/\partial t$ expresses the temperature change observed by spatial description. As already said, the partial time derivative $\partial T(\vec{x}, t)/\partial t$ shows the change of temperature of the rod plus the change of temperature due to a different part of the rod being evaluated (since the rod is moving with respect to point A).

If the material time derivative of the temperature $T$ is under consideration, the effect of the movement of the rod has to be neglected. This is the role of the second term (convective term) on the right sides of Derivative (3.117).

When the investigation approach is the current material description, as shown in Figure 3.11(b), the observation is focused on a certain material point $P$. Before loading (i.e., in the initial configuration) $V_0^b$, attention is placed on the point P expressed by the position vector $\underline{\vec{x}}(t_0) = \vec{X}$ and the eyes of the observer follow the moving rod in such a way that at time $t$, they observe the current position of the point $\underline{P}(t)$, the position vector of which is $\underline{\vec{x}}(t)$. Thus the observer notices how the material point P experiences the change of its state $\mathcal{E}$, in this case, the change of the temperature $T$. Therefore, for the current material description, the material time derivative $\mathrm{D}T/\mathrm{D}t$ is a correct choice, since in this case it expresses the change of temperature noticed by an observer following with their eyes the movement (deformation in the general case) attached to the material point P and therefore moving with the rod.

## 3.8 On Strain Measures and Strain Rate Measures

This section introduces the strain measures and strain rate measures needed in this book. The Green–Lagrange strain tensor $\mathbf{E}$ was derived in Section 3.5. It was given in Expressions (3.53) and (3.54),

$$\mathbf{E}(\vec{X},t) = \frac{1}{2}\left[\vec{u}(\vec{X},t)\overleftarrow{\nabla}(\vec{X}) + \vec{\nabla}(\vec{X})\vec{u}(\vec{X},t) + \left(\vec{\nabla}(\vec{X})\vec{u}(\vec{X},t)\right)\cdot\left(\vec{u}(\vec{X},t)\overleftarrow{\nabla}(\vec{X})\right)\right] \tag{3.118}$$

or shortly

$$\mathbf{E} = \frac{1}{2}\left[\vec{u}\overleftarrow{\nabla} + \vec{\nabla}\vec{u} + (\vec{\nabla}\vec{u})\cdot(\vec{u}\overleftarrow{\nabla})\right]. \tag{3.119}$$

When solids are studied, the present book assumes small deformations and rotations. In the case of small deformations and rotations, the higher-order terms are neglected as small quantities, and therefore Expression (3.119) reduces to a small deformation strain tensor $\boldsymbol{\varepsilon}$, viz.

$$\boldsymbol{\varepsilon}(\vec{X},t) := \frac{1}{2}\left[\vec{u}(\vec{X},t)\overleftarrow{\nabla}(\vec{X}) + \vec{\nabla}(\vec{X})\vec{u}(\vec{X},t)\right] \quad \text{or} \quad \boldsymbol{\varepsilon} := \frac{1}{2}\left(\vec{u}\overleftarrow{\nabla} + \vec{\nabla}\vec{u}\right). \tag{3.120}$$

The corresponding rotation tensor $\boldsymbol{\omega}$ is defined to be

$$\boldsymbol{\omega}(\vec{X},t) := \frac{1}{2}\left[\vec{u}(\vec{X},t)\overleftarrow{\nabla}(\vec{X}) - \vec{\nabla}(\vec{X})\vec{u}(\vec{X},t)\right] \quad \text{or} \quad \boldsymbol{\omega} := \frac{1}{2}\left(\vec{u}\overleftarrow{\nabla} - \vec{\nabla}\vec{u}\right). \tag{3.121}$$

Notations $(3.120)_1$ and $(3.121)_1$ are given to show that the mathematical description of solids is carried out by the initial material coordinates $X_i$, whereas Notations $(3.120)_2$ and $(3.121)_2$ are simpler forms of the definitions.

Equations (3.120) give

$$\frac{\mathrm{D}\boldsymbol{\varepsilon}(\vec{X},t)}{\mathrm{D}t} = \dot{\boldsymbol{\varepsilon}} = \frac{1}{2}\left[\dot{\vec{u}}(\vec{X},t)\overleftarrow{\nabla}(\vec{X}) + \vec{\nabla}(\vec{X})\dot{\vec{u}}(\vec{X},t)\right], \quad \text{or} \quad \dot{\boldsymbol{\varepsilon}} := \frac{1}{2}\left(\dot{\vec{u}}\overleftarrow{\nabla} + \vec{\nabla}\dot{\vec{u}}\right). \tag{3.122}$$

Strain tensor $\boldsymbol{\varepsilon}$ and the rotation tensor $\boldsymbol{\omega}$ are important players in solid mechanics. They are based on the displacement vector $\vec{u}$, as Definitions (3.120) and (3.121) show. Instead of the displacement vector $\vec{u}$, the scope of the fluid mechanics is on the fluid velocity field $\vec{v}$. Therefore the fluid strain rate tensor

$\overset{\circ}{\boldsymbol{\varepsilon}}$ and the fluid vorticity tensor $\overset{\circ}{\boldsymbol{\omega}}$ are introduced. They are defined by the equations

$$\overset{\circ}{\boldsymbol{\varepsilon}}(\vec{x},t) := \frac{1}{2}\left[\vec{v}(\vec{x},t)\overleftarrow{\nabla}(\vec{x}) + \vec{\nabla}(\vec{x})\,\vec{v}(\vec{x},t)\right] \quad \text{or} \quad \overset{\circ}{\boldsymbol{\varepsilon}} := \frac{1}{2}\left(\vec{v}\overleftarrow{\nabla} + \vec{\nabla}\vec{v}\right) \tag{3.123}$$

and

$$\overset{\circ}{\boldsymbol{\omega}}(\vec{x},t) := \frac{1}{2}\left[\vec{v}(\vec{x},t)\overleftarrow{\nabla}(\vec{x}) - \vec{\nabla}(\vec{x})\,\vec{v}(\vec{x},t)\right] \quad \text{or} \quad \overset{\circ}{\boldsymbol{\omega}} := \frac{1}{2}\left(\vec{v}\overleftarrow{\nabla} - \vec{\nabla}\vec{v}\right). \tag{3.124}$$

When the finite deformation of the solids is investigated, the rate of deformation tensor $\underline{\mathbf{d}}$ and the spin tensor or vorticity tensor $\underline{\mathbf{W}}$ are important quantities. They are defined as

$$\underline{\mathbf{d}}(\underline{\vec{x}}(t),t) := \frac{1}{2}\left[\underline{\vec{v}}(\underline{\vec{x}}(t),t)\underline{\overleftarrow{\nabla}}(\underline{\vec{x}}) + \underline{\vec{\nabla}}(\underline{\vec{x}})\,\underline{\vec{v}}(\underline{\vec{x}}(t),t)\right] \quad \text{or} \quad \underline{\mathbf{d}} := \frac{1}{2}\left(\underline{\vec{v}}\underline{\overleftarrow{\nabla}} + \underline{\vec{\nabla}}\underline{\vec{v}}\right) \tag{3.125}$$

and

$$\underline{\mathbf{W}}(\underline{\vec{x}}(t),t) := \frac{1}{2}\left[\underline{\vec{v}}(\underline{\vec{x}}(t),t)\overleftarrow{\nabla}(\underline{\vec{x}}) - \vec{\nabla}(\underline{\vec{x}})\,\underline{\vec{v}}(\underline{\vec{x}}(t),t)\right] \quad \text{or} \quad \underline{\mathbf{W}} := \frac{1}{2}\left(\underline{\vec{v}}\underline{\overleftarrow{\nabla}} - \underline{\vec{\nabla}}\underline{\vec{v}}\right). \tag{3.126}$$

The line beneath Definitions $(3.125)_2$ and $(3.126)_2$ shows that they are expressed by the current material coordinates $\underline{x}_i(t)$. Some writers [see, e.g., Reddy [84], Eq. (3.4.11)] do not use notations like $\vec{u}\overleftarrow{\nabla}$ and $\vec{v}\overleftarrow{\nabla}$ but write instead $(\vec{\nabla}\vec{u})^{\mathrm{T}}$ and $(\vec{\nabla}\vec{v})^{\mathrm{T}}$. These notations are equal, as shown in Expression (2.186).

## 3.9 Leibnitz Integral Rule for the Current Material Description

The present section studies the material derivative of a volume integral for the current material description and therefore derives the Leibnitz integration rule. Since the scope of this section is purely mathematical, except that the flow of matter is investigated, the derived results are called the Leibnitz integration rule.

The concept of the Leibnitz integral rule is to change the order of the material derivative operator $\mathrm{D}/\mathrm{D}t$ and the integration $\int$. This is needed when the integral forms of the basic laws and axioms are transformed into the forms of differential equations; in other words, global forms of the basic laws and axioms are transformed into local forms.

Later in this book, the equations obtained here will be recast in another form by applying the continuity equation, which is the local form of the axiom of conservation of mass. This latter form is called the Reynolds transport theorem. However, the present derivation assumes that there is no mass creation or destruction. Thus one of the axioms of the present book, the axiom of conservation of mass, is utilized. It is unfair to use the axiom already here, since it has not yet been introduced. The reason to adopt the axiom and to derive the results of the present and the next section was to make Chapter 4 more readable, as the derivations in these sections are quite extensive and provide only two simple tools for Chapter 4. Furthermore, the introduction of the mass creation/destroy terms would have made the following derivations unnecessarily complicated.

It is not a common practice to distinguish between the Leibnitz integration rule and the Reynolds transport theorem, but for educational purposes, it is done here. Based on Boas ([11], pp. 262 and 263) and Holzapfel ([29], p. 74), the following equation is obtained:

$$\int_{\underline{v}(t)} \underline{\theta}(\underline{\vec{x}}(t), t)\, \mathrm{d}\underline{v} = \int_{V_0} \theta(\vec{X}, t)\, J(\vec{X}, t)\, \mathrm{d}V\,, \tag{3.127}$$

where the volumes $\underline{v}(t)$ and $V_0$ are occupied by the same material points. Function $\underline{\theta}(\underline{\vec{x}}(t), t)$ can be a scalar, vector, or tensor of any order.

In Expression (3.127), the notation $J(\vec{X}, t)$ stands for the Jacobian determinant between the current material coordinates $\underline{x}_i(t)$ and the initial material coordinates $X_i$. It reads

$$J(\vec{X}, t) := \begin{vmatrix} \dfrac{\partial \underline{x}_1(t)}{\partial X_1} & \dfrac{\partial \underline{x}_1(t)}{\partial X_2} & \dfrac{\partial \underline{x}_1(t)}{\partial X_3} \\ \dfrac{\partial \underline{x}_2(t)}{\partial X_1} & \dfrac{\partial \underline{x}_2(t)}{\partial X_2} & \dfrac{\partial \underline{x}_2(t)}{\partial X_3} \\ \dfrac{\partial \underline{x}_3(t)}{\partial X_1} & \dfrac{\partial \underline{x}_3(t)}{\partial X_2} & \dfrac{\partial \underline{x}_3(t)}{\partial X_3} \end{vmatrix}. \tag{3.128}$$

Expression (3.128) shows that the coordinate transformation varies with time $t$. In case mass $m$ is destroyed or created, the relation between the current material coordinates $\underline{x}_i(t)$ and the initial material coordinates $X_i$ is not unique, and therefore Expressions (3.127) and (3.128) are not valid.

Next, the material time derivative of the volume integral $I(t)$ is considered:

$$\frac{\mathrm{D}}{\mathrm{D}t} I(t) = \frac{\mathrm{D}}{\mathrm{D}t} \int_{\underline{v}(t)} \underline{\theta}(\underline{\vec{x}}(t), t)\, \mathrm{d}\underline{v}\,. \tag{3.129}$$

Based on Expressions (3.127) and (3.129), the following is obtained:

$$\begin{aligned} \frac{\mathrm{D}}{\mathrm{D}t} I(t) &= \frac{\mathrm{D}}{\mathrm{D}t} \int_{\underline{v}(t)} \underline{\theta}(\underline{\vec{x}}(t), t)\, \mathrm{d}\underline{v} = \frac{\mathrm{D}}{\mathrm{D}t} \int_{V_0} \theta(\vec{X}, t)\, J(\vec{X}, t)\, \mathrm{d}V \\ &= \int_{V_0} \frac{\mathrm{D}}{\mathrm{D}t} \Big( \theta(\vec{X}, t)\, J(\vec{X}, t) \Big)\, \mathrm{d}V \\ &= \int_{V_0} \left( \frac{\mathrm{D}\theta(\vec{X}, t)}{\mathrm{D}t}\, J(\vec{X}, t) + \theta(\vec{X}, t)\, \frac{\mathrm{D}J(\vec{X}, t)}{\mathrm{D}t} \right) \mathrm{d}V\,. \end{aligned} \tag{3.130}$$

The steps used in Manipulation (3.129) exploit the fact that the integration domain $V_0$ does not depend on time $t$ and therefore the order of differentiation and integration can be interchanged. Based on Expression (3.127), the first term on the third line of Expression (3.130) takes the form

$$\int_{V_0} \frac{\mathrm{D}\theta(\vec{X},t)}{\mathrm{D}t}\, J(\vec{X},t)\,\mathrm{d}V = \int_{\underline{v}(t)} \frac{\mathrm{D}\underline{\theta}(\underline{\vec{x}}(t),t)}{\mathrm{D}t}\,\mathrm{d}\underline{v}\,. \tag{3.131}$$

The second term on the third line of Expression (3.130) is simplified as follows:

$$\begin{aligned}\int_{V_0} \theta(\vec{X},t)\,\frac{\mathrm{D}J(\vec{X},t)}{\mathrm{D}t}\,\mathrm{d}V &= \int_{V_0} \theta(\vec{X},t)\,\dot{J}(\vec{X},t)\,\mathrm{d}V \\ &= \int_{V_0} \theta(\vec{X},t)\,\frac{\dot{J}(\vec{X},t)}{J(\vec{X},t)}\,J(\vec{X},t)\,\mathrm{d}V\,. \end{aligned} \tag{3.132}$$

Equation (C.12) of **Appendix C** reads

$$\frac{\dot{J}(\vec{X},t)}{J(\vec{X},t)} = \underline{\vec{\nabla}}(\underline{\vec{x}})\cdot\underline{\vec{v}}(\underline{\vec{x}}(t),t)\,. \tag{3.133}$$

The message of Equation (3.133) is that the left-side quantity expressed in the coordinates $X_i$ takes the appearance given on the right side when expressed in the coordinates $\underline{x}_i(t)$ . With the aim of Equations (3.127) and (3.133), Evaluation (3.132) can be continued as follows:

$$\begin{aligned}\int\limits_{V_0} \theta(\vec{X},t)\,\frac{\mathrm{D}J(\vec{X},t)}{\mathrm{D}t}\,\mathrm{d}V &= \int\limits_{V_0} \theta(\vec{X},t)\,\frac{\dot{J}(\vec{X},t)}{J(\vec{X},t)}\,J(\vec{X},t)\,\mathrm{d}V \\ &= \int\limits_{\underline{v}(t)} \underline{\theta}(\underline{\vec{x}}(t),t)\,\underline{\vec{\nabla}}(\underline{\vec{x}})\cdot\underline{\vec{v}}(\underline{\vec{x}}(t),t)\,\mathrm{d}\underline{v}\,. \end{aligned} \tag{3.134}$$

Substitution of Terms (3.131) and (3.134) into Expression (3.130) gives

$$\begin{aligned}\frac{\mathrm{D}}{\mathrm{D}t}I(t) = \frac{\mathrm{D}}{\mathrm{D}t}\int\limits_{\underline{v}(t)} \underline{\theta}(\underline{\vec{x}}(t),t)\,\mathrm{d}\underline{v} &= \int\limits_{V_0}\left(\frac{\mathrm{D}\theta(\vec{X},t)}{\mathrm{D}t}J(\vec{X},t) + \theta(\vec{X},t)\,\frac{\mathrm{D}J(\vec{X},t)}{\mathrm{D}t}\right)\mathrm{d}V \\ &= \int\limits_{\underline{v}(t)}\left(\frac{\mathrm{D}\underline{\theta}(\underline{\vec{x}}(t),t)}{\mathrm{D}t} + \underline{\theta}(\underline{\vec{x}}(t),t)\,\underline{\vec{\nabla}}(\underline{\vec{x}})\cdot\underline{\vec{v}}(\underline{\vec{x}}(t),t)\right)\mathrm{d}\underline{v}\,. \end{aligned} \tag{3.135}$$

In Expression (3.135), the order of the material derivative operator $\mathrm{D}/\mathrm{D}t$ and the integration $\int$ is interchanged, which was the aim of this section. Expression (3.135) is the Leibnitz integral rule, also called the Reynolds transport theorem (see, e.g., Reddy [84], p. 186).

## 3.10 Reynolds Transport Theorem for the Spatial Description

This section studies the material derivative of a volume integral for the spatial description, and therefore derives the Reynolds transport theorem for the spatial description. Since the derivation is quite complex, the reader may just study Results (3.156) and (3.158) and revert later to this section, if necessary.

The concept of the Reynolds transport theorem for the spatial description has two tasks. First, the order of the material derivative operator $\mathrm{D}/\mathrm{D}t$ and the integration $\int$ is interchanged. Second, the quantities expressed by the current material coordinates $\underline{x}_i(t)$ are replaced with the same quantities but expressed by the spatial coordinates $\underline{x}_i$. This is needed when the integral forms of the basic laws and axioms are transformed into the forms of differential equations for fluids; in other words, global forms of the basic laws and axioms are transformed into local forms.

Later in this book, the expressions obtained here will be recast in another form by applying the continuity equation, which is the local form of the axiom of conservation of mass. This latter expression is called the Reynolds transport theorem as well. However, also the present derivation assumes that there is no mass creation or destruction. Thus the axiom of conservation of mass is utilized.

Based on Boas ([11], pp. 262 and 263) and Holzapfel ([29], p. 74), the following result is obtained:

$$\int_{\underline{v}(t)} \underline{\theta}(\vec{\underline{x}}(t), t)\, \mathrm{d}\underline{v} = \int_{\mathrm{V^{cv}}} \theta(\vec{x}, t)\, J(\vec{x}, t)\, \mathrm{d}V\,, \tag{3.136}$$

where the volumes $\underline{v}(t)$ and $\mathrm{V^{cv}}$ are occupied by the same material particles. Function $\underline{\theta}(\vec{\underline{x}}(t), t)$ can be a scalar, vector, or tensor of any order. In Expression (3.136), the notation $J(\vec{x}, t)$ stands for the Jacobian determinant between the current material coordinates $\underline{x}_i(t)$ and the spatial coordinates $x_i$. The form of the Jacobian determinant $J(\vec{x}, t)$ (although the coordinates are different) is given in Expression (3.128) and therefore does not need to be given here.

Since the quantity $\theta$ takes equal value, at a point independently of the expressing coordinates, the following holds:

$$\underline{\theta}(\vec{\underline{x}}(t), t) = \theta(\vec{x}, t)\,. \tag{3.137}$$

The following volume integral is considered:

$$I(t) := \int_{\underline{v}(t)} \underline{\theta}(\vec{\underline{x}}(t), t)\, \mathrm{d}\underline{v}\,. \tag{3.138}$$

The material time derivative of the volume integral $I(t)$ in Expression (3.138) is

$$\frac{\mathrm{D}}{\mathrm{D}t} I(t) = \frac{\mathrm{D}}{\mathrm{D}t} \int_{\underline{v}(t)} \underline{\theta}(\vec{\underline{x}}(t), t)\, \mathrm{d}\underline{v}\,. \tag{3.139}$$

Equation (3.139) describes the time rate of change of $I(t)$ for a certain entity of matter $m$ occupying the volume $\underline{v}(t)$ at time $t$. In contrast to the previous section, here the Reynolds transport theorem is written in terms of the spatial coordinates $x_i$. Therefore a different path for the derivation has to be taken.

The value for the integral $I(t)$ depends only on $t$. Therefore the following result can be obtained:

$$\frac{\mathrm{d}}{\mathrm{d}t} I(t) = \frac{\mathrm{D}}{\mathrm{D}t} I(t) . \tag{3.140}$$

However, sometimes there are differences between the quantities in Expression (3.138). Equation (3.135) expresses how the material within the volume $\mathrm{V}^{\mathrm{cv}}$ experiences the rate of change of the quantity $\theta$, whereas the left side of Equation (3.140) may not be associated with a certain material entity. Therefore the present text uses the notation $\mathrm{D}/\mathrm{D}t$ to stress that the material time derivative is sought.

The definition of the time derivative, where $\Delta t$ is the time increment, provides the following equation (see, e.g., Fulks [23], p. 56):

$$\frac{\mathrm{d}}{\mathrm{d}t} I(t) := \lim_{\Delta t \to 0} \frac{I(t+\Delta t) - I(t)}{\Delta t} . \tag{3.141}$$

Case A: First, the case in which matter $m_1$ occupies the volume $\mathrm{V}^{\mathrm{cv}}$ at time $t = t_1$ is evaluated. This means that the control volume $\mathrm{V}^{\mathrm{cv}}$ equals the volume $\underline{v}(t_1)$ at time $t_1$. Within the volume $\mathrm{V}^{\mathrm{cv}} = \underline{v}(t_1)$ the coordinates $\underline{x}_i(t_1)$ and $x_i$ of a material point are equal. This means that the Jacobian determinant $J(\vec{x}, t_1)$ takes the value 1 and Expressions (3.136) and (3.138) give

$$I(t_1) = \int_{\underline{v}(t_1)} \underline{\theta}(\underline{\vec{x}}(t_1), t_1)\, \mathrm{d}\underline{v} = \int_{\mathrm{V}^{\mathrm{cv}}} \theta(\vec{x}, t_1)\, \mathrm{d}V . \tag{3.142}$$

Equation (3.142) gives

$$\frac{\mathrm{D}}{\mathrm{D}t} I(t_1) = \frac{\mathrm{D}}{\mathrm{D}t} \int_{\underline{v}(t)} \underline{\theta}(\underline{\vec{x}}\,(t_1), t_1)\, \mathrm{d}\underline{v} = \frac{\mathrm{D}}{\mathrm{D}t} \int_{\mathrm{V}^{\mathrm{cv}}} \theta(\vec{x}, t_1)\, \mathrm{d}V . \tag{3.143}$$

Figure 3.12(a) illustrates the position of the material entity $m_1$ at time $t = t_1$ (shown by the solid boundary) and at time $t = t_1 + \Delta t$ (shown by the dashed boundary). The volume the surface of which is shown by the solid line is called the control volume and is denoted by $\mathrm{V}^{\mathrm{cv}}$; its boundary is denoted by $\partial \mathrm{V}^{\mathrm{cv}}$. The control volume $\mathrm{V}^{\mathrm{cv}}$ is a fixed volume in space. The *control volume* stems from fluid mechanics.

Figure 3.12(a) shows that at time $t = t_1$, the material collection $m_1$ occupies the control volume $\mathrm{V}^{\mathrm{cv}}$, whereas at time $t = t_1 + \Delta t$, part of the matter entity $m_1$ is outside the control volume $\mathrm{V}^{\mathrm{cv}}$ and part of the control volume $\mathrm{V}^{\mathrm{cv}}$ is filled by matter that does not belong to $m_1$. Thus, based on Expressions (3.137) and

(3.138), we arrive at the following:

$$\begin{aligned} I(t_1+\Delta t) &= \int_{\underline{v}(t_1+\Delta t)} \underline{\theta}(\underline{\vec{x}}(t_1+\Delta t), t_1+\Delta t)\ \mathrm{d}\underline{v} = \int_{\underline{v}(t_1+\Delta t)} \theta(\vec{x}, t_1+\Delta t)\ \mathrm{d}V \\ &= \int_{\underline{v}(t_1)} \theta(\vec{x}, t_1+\Delta t)\ \mathrm{d}V + \oint_{\partial\underline{v}(t_1)} \theta(\vec{x}, t_1+\Delta t)\, v^{\mathrm{n}}(\vec{x}, t_1)\Delta t\ \mathrm{d}A\,. \end{aligned} \tag{3.144}$$

The second line of Expression (3.144) can be interpreted as follows. The first term represents the value of Integral $(3.142)_2$ at time $t = t_1 + \Delta t$ but for the mass at time $t = t_1$. The second term adds the influence of transport of the mass across the boundary $\partial\mathrm{V}^{\mathrm{cv}} = \partial\underline{v}(t_1)$. Figure 3.12(b) shows a detailed study for the second term on the latter line of Expression (3.144). The notation $v^{\mathrm{n}}$ stands for the outward normal speed of material particles on the non-changing surface $\partial\mathrm{V}^{\mathrm{cv}} = \partial\underline{v}(t_1)$. Therefore, during the time increment $\Delta t$, the surface particles have moved by the distance $v^{\mathrm{n}}\,\Delta t$. The differential surface element on the surface $\partial\mathrm{V}^{\mathrm{cv}} = \partial\underline{v}(t_1)$ is denoted by $\mathrm{d}A$. Therefore, the differential volume of the material particles moved out and in through the surface $\partial\mathrm{V}^{\mathrm{cv}} = \partial\underline{v}(t_1)$ is $v^{\mathrm{n}}\,\Delta t\,\mathrm{d}A$, as given in the second term on the latter line of Expression (3.144).

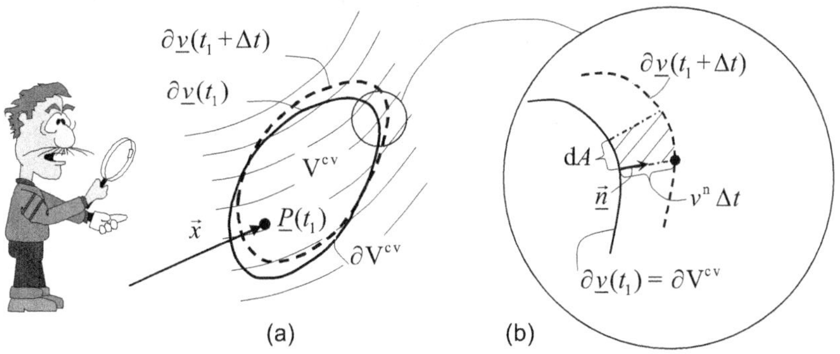


**Figure 3.12** (a) Material entity $m_1$ at time $t = t_1$ and at time $t = t_1 + \Delta t$. (b) Material particles moving out from the volume $\underline{v}(t_1) = \mathrm{V}^{\mathrm{cv}}$. Inspector: Corel images.

Since the volumes $\underline{v}(t_1)$ and $\mathrm{V}^{\mathrm{cv}}$ are equal, Expression (3.144) yields

$$I(t_1+\Delta t) = \int_{\mathrm{V}^{\mathrm{cv}}} \theta(\vec{x}, t_1+\Delta t)\ \mathrm{d}V + \oint_{\partial\mathrm{V}^{\mathrm{cv}}} \theta(\vec{x}, t_1+\Delta t)\, v^{\mathrm{n}}(\vec{x}, t_1)\,\Delta t\ \mathrm{d}A\,. \tag{3.145}$$

The outward unit normal vector to the boundary $\partial\mathrm{V}^{\mathrm{cv}} = \partial\underline{v}(t_1)$ is denoted by $\vec{n}$. Thus we arrive at the following:

$$v^{\mathrm{n}}(\vec{x}, t_1) = \vec{v}(\vec{x}, t_1)\cdot\vec{n}(\vec{x})\,, \tag{3.146}$$

where $\vec{v}(\vec{x}, t_1)$ is the velocity of material particles. Figures 3.13(a) and 3.13(b) demonstrate that Expression (3.146) gives the correct sign for the outward normal speed of material particles $v^{\mathrm{n}}$.

In Figure 3.13(a), the velocity of material particles $\vec{v}(\vec{x}, t_1)$ is assumed to be outwards from the volume $V^{cv} = \underline{v}(t_1)$ and therefore the angle between the velocity field $\vec{v}(\vec{x}, t_1)$ and the outward unit vector $\vec{n}$ is acute, giving a positive value for $v^{\mathrm{n}}$. Since the role of the material time operator $\mathrm{D}/\mathrm{D}t$ is to study how the matter experiences the rate of change of the quantity $\theta$, the term corresponding to the one sketched in Figure 3.12(b) has to be added to the integral. Thus, since the variables $v^{\mathrm{n}}$ and $\Delta t$ are positive, the sign in front of the second term on the latter line has to be positive.

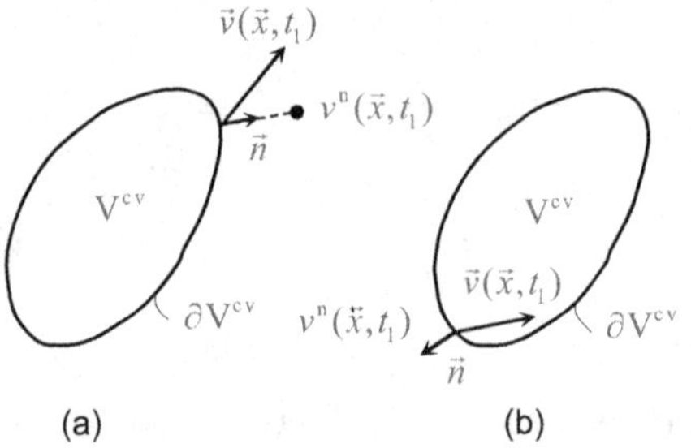


**Figure 3.13** (a) Material flows out from the volume $\mathrm{V^{cv}} = v(t_1)$. (b) Material flows into the volume $\mathrm{V^{cv}} = v(t_1)$.

In Figure 3.13(b) the angle between the velocity field $\vec{v}(\vec{x}, t_1)$ and the outward unit vector $\vec{n}$ is blunt, and therefore the variable $v^{\mathrm{n}}$ takes a negative value. In Figure 3.13(b) a piece of material, which does not belong to $m_1$, is entering the volume $\mathrm{V^{cv}} = \underline{v}(t_1)$, and therefore its influence has to be subtracted from the integral. This is the case, since the variable $v^{\mathrm{n}}$ takes a negative value for the case shown in Figure 3.13(b).

Substitution of Expression (3.146) into Equation (3.145) gives

$$I(t_1{+}\Delta t) = \int\limits_{\mathrm{V^{cv}}} \theta(\vec{x}, t_1{+}\Delta t)\, \mathrm{d}V + \oint\limits_{\partial\mathrm{V^{cv}}} \theta(\vec{x}, t_1{+}\Delta t)\, \vec{v}(\vec{x}, t)\cdot\vec{n}(\vec{x})\, \Delta t\, \mathrm{d}A\,. \quad (3.147)$$

Based on the mean value theorem (see, e.g., Apostol [6], p. 93), the following result is obtained:

$$\theta(\vec{x}, t_1 + \Delta t) = \theta(\vec{x}, t_1) + \frac{\partial\theta(\vec{x}, t_1 + \kappa\,\Delta t)}{\partial t}\,\Delta t\,, \quad (3.148)$$

where the coefficient $\kappa \in [0, 1]$ is a positive real number. The sufficient condition for the satisfaction of Expression (3.148) is that $\theta(\vec{x}, t)$ is a continuous function possessing a continuous partial derivative, $\theta(\vec{x}, t) \in \mathrm{C}^1$, in the closed interval $t \in [t_1, t_1 + \Delta t]$. The value for the coefficient $\kappa$ is not known, but there exists at least one value for $\kappa$ that satisfies Equation (3.148).

Substitution of Expression (3.148) into Expression (3.147) yields

$$I(t_1 + \Delta t) = \int_{\mathrm{V^{cv}}} \theta(\vec{x}, t_1 + \Delta t)\; \mathrm{d}V + \oint_{\partial\mathrm{V^{cv}}} \theta(\vec{x}, t_1 + \Delta t)\, \vec{v}(\vec{x}, t_1)\cdot\vec{n}(\vec{x})\, \Delta t\; \mathrm{d}A$$

$$= \int_{V^{cv}} \theta(\vec{x}, t_1)\, dV + \int_{V^{cv}} \frac{\partial \theta(\vec{x}, t_1 + \kappa\, \Delta t)}{\partial t} \Delta t\, dV$$
$$+ \oint_{\partial V^{cv}} \theta(\vec{x}, t_1)\, \vec{v}(\vec{x}, t_1) \cdot \vec{n}(\vec{x})\, \Delta t\, dA$$
$$+ \oint_{\partial V^{cv}} \frac{\partial \theta(\vec{x}, t_1 + \kappa\, \Delta t)}{\partial t} \vec{v}(\vec{x}, t_1) \cdot \vec{n}(\vec{x})\, (\Delta t)^2\, dA\,. \tag{3.149}$$

Expression (3.149), Notation $(3.142)_2$, and Definition (3.141) allow the following to be written:

$$\frac{d}{dt} I(t_1) = \lim_{\Delta t \to 0} \frac{1}{\Delta t} \left[ \int_{V^{cv}} \theta(\vec{x}, t_1)\, dV + \int_{V^{cv}} \frac{\partial \theta(\vec{x}, t_1 + \kappa\, \Delta t)}{\partial t} \Delta t\, dV \right.$$
$$+ \oint_{\partial V^{cv}} \theta(\vec{x}, t_1)\, \vec{v}(\vec{x}, t_1) \cdot \vec{n}(\vec{x})\, \Delta t\, dA$$
$$+ \oint_{\partial V^{cv}} \frac{\partial \theta(\vec{x}, t_1 + \kappa \Delta t)}{\partial t} \vec{v}(\vec{x}, t_1) \cdot \vec{n}(\vec{x})\, (\Delta t)^2\, dA$$
$$\left. - \int_{V^{cv}} \theta(\vec{x}, t_1)\, dV \right]. \tag{3.150}$$

Expression (3.150) reduces to

$$\frac{d}{dt} I(t_1) = \lim_{\Delta t \to 0} \left[ \int_{V^{cv}} \frac{\partial \theta(\vec{x}, t_1 + \kappa\, \Delta t)}{\partial t} dV + \oint_{\partial V^{cv}} \theta(\vec{x}, t_1)\, \vec{v}(\vec{x}, t_1)\, \vec{n}(\vec{x})\, dA \right.$$
$$\left. + \oint_{\partial V^{cv}} \frac{\partial \theta(\vec{x}, t_1 + \kappa\, \Delta t)}{\partial t} \vec{v}(\vec{x}, t_1) \cdot \vec{n}(\vec{x})\, \Delta t\, dA \right]. \tag{3.151}$$

The value of $\Delta t$ is let to approach zero. The integrand of the first term changes as follows:

$$\frac{\partial \theta(\vec{x}, t_1 + \kappa\, \Delta t)}{\partial t} \quad \to \quad \frac{\partial \theta(\vec{x}, t_1)}{\partial t}\,. \tag{3.152}$$

The third term on the right side of Expression (3.151) is studied next. Since the other terms besides the term $\Delta t$ in the integrand of the third term are bounded, the third term vanishes when $\Delta t \to 0$. Thus Derivative (3.151) yields

$$\frac{d}{dt} I(t_1) = \int_{V^{cv}} \frac{\partial \theta(\vec{x}, t_1)}{\partial t}\, dV + \oint_{\partial V^{cv}} \theta(\vec{x}, t_1)\, \vec{v}(\vec{x}, t_1) \cdot \vec{n}(\vec{x})\, dA\,. \tag{3.153}$$

This case studied the situation in which matter $m_1$ occupies the control volume $V^{cv}$ at time $t = t_1$.

Case B: Next, the case in which the matter $m_2$ occupies the volume $V^{cv}$ at time $t = t_2$ is evaluated. The evaluation of this case takes the same steps as were taken in Case A. Therefore Result (3.153) can be modified for the present case, and we arrive at the following:

$$\frac{d}{dt} I(t_2) = \int_{V^{cv}} \frac{\partial \theta(\vec{x}, t_2)}{\partial t}\, dV + \oint_{\partial V^{cv}} \theta(\vec{x}, t_2)\, \vec{v}(\vec{x}, t_2) \cdot \vec{n}(\vec{x})\, dA\,. \tag{3.154}$$

General Case: Based on Cases A and B, the following general form can be written:

$$\frac{\mathrm{d}}{\mathrm{d}t} I(t) = \int_{\mathrm{V^{cv}}} \frac{\partial \theta(\vec{x},t)}{\partial t}\, \mathrm{d}V + \oint_{\partial \mathrm{V^{cv}}} \theta(\vec{x},t)\, \vec{v}(\vec{x},t) \cdot \vec{n}(\vec{x})\; \mathrm{d}A\,. \tag{3.155}$$

Finally, by applying Relation (3.140) to Expression (3.155) and by taking Expression (3.143) into account, the following result is obtained:

$$\frac{\mathrm{D}}{\mathrm{D}t} \int_{\underline{v}(t)} \theta(\underline{\vec{x}}(t),t)\, \mathrm{d}\underline{v} = \int_{\mathrm{V^{cv}}} \frac{\partial \theta(\vec{x},t)}{\partial t}\, \mathrm{d}V + \oint_{\partial \mathrm{V^{cv}}} \theta(\vec{x},t)\, \vec{v}(\vec{x},t) \cdot \vec{n}(\vec{x})\, \mathrm{d}A\,. \tag{3.156}$$

Expression (3.156) is the Reynolds transport theorem for the spatial description.

Generalized Gauss's Theorem $(2.170)_1$ reads

$$\int_V \vec{\nabla} * f \,\mathrm{d}V = \oint_{\partial V} \vec{n} * f \,\mathrm{d}A\,. \tag{3.157}$$

Replacing the star product $*$ with the dot product $\cdot$ and substituting $f = \theta(\vec{x},t)\, \vec{v}(\vec{x},t)$ and $V = \mathrm{V^{cv}}$ in Theorem (3.157) allows Equation (3.156) to take the form

$$\frac{\mathrm{D}}{\mathrm{D}t} \int_{\underline{v}(t)} \theta(\underline{\vec{x}}(t),t)\, \mathrm{d}\underline{v} = \int_{\mathrm{V^{cv}}} \left[ \frac{\partial \theta(\vec{x},t)}{\partial t} + \vec{\nabla}(\vec{x}) \cdot [\theta(\vec{x},t)\, \vec{v}(\vec{x},t)] \right] \mathrm{d}V\,. \tag{3.158}$$

Equation (3.158) utilized the fact that the dot product between two vectors is commutative, $\vec{f} \cdot \vec{n} = \vec{n} \cdot \vec{f}$. Equation (3.158) is another form of the Reynolds transport theorem for the spatial description.

Equations (3.156) and (3.158) have the appearance for which we aimed at the start of the present section. The order of the material derivative operator $\mathrm{D}/\mathrm{D}t$ and the integration $\int$ is interchanged, and the quantities expressed by the current material coordinates $\underline{x}_i(t)$ are replaced with the same quantities but expressed by the spatial coordinates $x_i$.

## 3.11 Summary

The beginning of this chapter introduced the coordinate systems $(x_1, x_2, x_3)$ and $(X_1, X_2, X_3)$, since solid and fluid mechanics apply different coordinates. Before loading, (i.e., at time $t_0$), the system (body) $\mathrm{m^b}$ is in the initial configuration $\mathrm{V_0^b}$, whereas during deformation; that is, at time $t$, the system $\mathrm{m^b}$ is in the current configuration $\underline{v}^{\mathrm{b}}(t)$. Due to deformation, the position of the material point $\underline{P}(t)$ being in the current configuration $\underline{v}^{\mathrm{b}}(t)$ varies with time $t$. Therefore its position is written in the form $(\underline{x}_1(t), \underline{x}_2(t), \underline{x}_3(t))$. For clarity, sometimes the quantities expressed in the current configuration $\underline{v}^{\mathrm{b}}(t)$ are given without time $t$ and the time dependency is shown only by underlining the quantity.

When loading is applied, the system $\mathrm{m^b}$ deforms and the quantities, such as temperature $T$ and stress tensor $\boldsymbol{\sigma}$, take their present values $\underline{T}$ and $\underline{\boldsymbol{\sigma}}$. The mass of the system $\mathrm{m^b}$ does not change during deformation, but its geometry and volume $\underline{v}^{\mathrm{b}}(t)$ do.

Often, solids are studied in the geometry the system had before loading took place. In order to do that, a coordinate transformation from coordinates $(\underline{x}_1(t), \underline{x}_2(t), \underline{x}_3(t))$ to coordinates $(X_1, X_2, X_3)$ is carried out. This new configuration is the IG-CtP configuration $\mathrm{V}_0^{\mathrm{b}}$, that is, initial geometry at current time and position. Although the geometry is the one that existed when loading took place, the quantities have their current values, and the position of the system $\mathrm{m^b}$ does not change. When the quantities are expressed in the current configuration $\underline{v}^{\mathrm{b}}(t)$, they are written in the form $\underline{T}(\underline{\vec{x}}(t), t)$ and $\underline{\boldsymbol{\sigma}}(\underline{\vec{x}}(t), t)$, where the notation $\underline{\vec{x}}(t)$ refers to the current coordinates $(\underline{x}_1(t), \underline{x}_2(t), \underline{x}_3(t))$ of the material point $\underline{P}(t)$. In IG-CtP configuration $\mathrm{V}_0^{\mathrm{b}}$, the values of the quantities are the current ones, that is, the same as in the current configuration $\underline{v}^{\mathrm{b}}(t)$, but expressed as $T(\vec{X}, t)$ and $\boldsymbol{\sigma}(\vec{X}, t)$, where the notation $\vec{X}$ refers to the coordinates $(X_1, X_2, X_3)$.

The preceding configurations are related to the material description, which is applied for solids. Section 3.4 derives equations for the material (time) derivatives of the quantities expressed in the forms $\underline{T}(\underline{\vec{x}}(t), t)$ and $T(\vec{X}, t)$. The material (time) derivative expresses how the material point experiences the change of its state (absolute temperature $T$, for example) with the times $t$.

Green–Lagrange strain tensor $\mathbf{E}$ is derived in two different ways: Sections 3.5.1–3.5.3 utilize the Taylor series with strict geometrical evaluation, whereas Section 3.5.4 applies the deformation gradient $\mathbf{F}$. The latter approach is a gateway to the finite deformation theory, since it displays the practice of the mathematical derivation in the finite deformation theory. However, the physics of the Green–Lagrange strain tensor $\mathbf{E}$ is hidden behind the mathematical treatment. Therefore, the first approach may give a more detailed understanding of the physical background of strain tensors; it is therefore strongly suggested that this be studied at least once during a person's career in continuum mechanics.

Section 3.7 studies the spatial description, which is for fluids. In fluid mechanics the control volume $\mathrm{V^{cv}}$ is a vital part of the theory. Control volume $\mathrm{V^{cv}}$ is a fixed volume in space. In the spatial description the idea is that the fluid flow is studied through a fixed point in space. Therefore the preceding quantities now take the following appearances: $T(\vec{x}, t)$ and $\boldsymbol{\sigma}(\vec{x}, t)$. The spatial description has its own form for the material (time) derivative given in Section 3.6.

Sections 3.7 and 3.8 compare the material and spatial descriptions, time derivatives, strain measures, and strain rate measures.

The basic laws and axioms of continuum thermodynamics contain terms in which the material derivative operator acts on quantities written in the integral forms. In order to write the local forms of these basic laws and axioms, the material derivative operator has to be "moved" on the right side of the integral sign. This was the topic for the rest of this chapter.

## Problems

**3.1** Show that the following expression holds:

$$\mathbf{1} : \overset{\circ}{\boldsymbol{\varepsilon}}(\vec{x},t) = \vec{\nabla}(\vec{x}) \cdot \vec{v}(\vec{x},t) \qquad \Leftrightarrow \qquad \mathbf{1} : \overset{\circ}{\boldsymbol{\varepsilon}} = \vec{\nabla} \cdot \vec{v} \,. \tag{1}$$

**3.2** According to Malvern [[57], Eq. (4.6.1a)], the deformation gradient **F** can be written in the following two forms:

$$\mathbf{F} = \mathbf{R} \cdot \mathbf{U} = \mathbf{V} \cdot \mathbf{R} \,, \tag{1}$$

where **R** denotes the orthogonal rotation tensor, **U** is the right stretch tensor, and **V** is the left stretch tensor. The rotation tensor **R** obeys [see Malvern [57], Eq. (4.6.1c)]

$$\mathbf{R} \cdot \mathbf{R}^{\mathrm{T}} = \mathbf{R}^{\mathrm{T}} \cdot \mathbf{R} = \mathbf{1} \,, \tag{2}$$

where **1** is the second-order identity tensor. It can be shown that the following holds:

$$\mathbf{U}^{\mathrm{T}} = \mathbf{U} \,. \tag{3}$$

According to Expression (3.101), the Green–Lagrange strain tensor **E** takes the following appearance:

$$\mathbf{E} := \frac{1}{2} \left[ \mathbf{F}^{\mathrm{T}} \cdot \mathbf{F} - \mathbf{1} \right] . \tag{4}$$

In case of pure rotation, the following holds:

$$\mathbf{U} = \mathbf{1} \,. \tag{5}$$

What is the value of the Green–Lagrange strain tensor **E** in case of pure rotation?

**Hint.** Don't worry; the solution is very short.

**3.3** According to Malvern [[57], Eq. (4.6.1a)], the deformation gradient **F** can be written in the following two forms:

$$\mathbf{F} = \mathbf{R} \cdot \mathbf{U} = \mathbf{V} \cdot \mathbf{R} \,, \tag{1}$$

where **R** denotes the orthogonal rotation tensor, **U** is the right stretch tensor, and **V** is the left stretch tensor. The rotation tensor **R** obeys [see Malvern [57], Eq. (4.6.1c)]

$$\mathbf{R} \cdot \mathbf{R}^{\mathrm{T}} = \mathbf{R}^{\mathrm{T}} \cdot \mathbf{R} = \mathbf{1} \,, \tag{2}$$

where $\mathbf{1}$ is the second-order identity tensor. It can be shown that the following holds:

$$\mathbf{U}^{\mathrm{T}} = \mathbf{U}\,. \tag{3}$$

According to Expression (3.101), the Green–Lagrange strain tensor $\mathbf{E}$ takes the following appearance:

$$\mathbf{E} := \frac{1}{2}\left[\mathbf{F}^{\mathrm{T}}\cdot\mathbf{F} - \mathbf{1}\right]. \tag{4}$$

Show that the value of the Green–Lagrange strain tensor $\mathbf{E}$ is independent of rotation $\mathbf{R}$.

**Hint.** Don't worry; the solution is very short. You may need to apply the following: for the second-order tensors $\mathbf{c}$ and $\mathbf{g}$ the following holds:

$$(\mathbf{c}\cdot\mathbf{g})^{\mathrm{T}} = \mathbf{g}^{\mathrm{T}}\cdot\mathbf{c}^{\mathrm{T}}\,. \tag{3}$$

**3.4** Establish the relation

$$\dot{\mathbf{E}} = \mathbf{F}^{\mathrm{T}}\cdot\mathbf{d}\cdot\mathbf{F}\,, \tag{1}$$

where $\mathbf{E}$ is the Green–Lagrange strain tensor, given in Definition (3.118), and $\mathbf{d}$ is the rate of deformation tensor given in Definition (3.125).

**3.5** The displacement vector $\vec{u}(\vec{X})$ of a point $\underline{x}$ in a body in equilibrium has the form $\vec{u}(\vec{X}) = u_1\,\vec{i}_1 + u_2\,\vec{i}_2 + u_3\,\vec{i}_3$, with the components

$$\begin{aligned} u_1 &= X_1\left[X_1^2\,X_2 + c_1\left(2\,c_2^3 + 3\,c_2^2\,X_2 - X_2^3\right)\right], \\ u_2 &= -X_2\left(2\,c_2^3 + \frac{3}{2}\,c_2^2\,X_2 - \frac{1}{4}\,X_2^3 + \frac{3}{2}\,c_1\,X_1^2\,X_2\right), \\ u_3 &= 0\,, \end{aligned} \tag{1}$$

where $c_1$ and $c_2$ are constants. Find the the components of

(a) the small deformation strain tensor $\varepsilon$ and

(b) the Green–Lagrange strain tensor $\mathbf{E}$.

**3.6** The displacement vector $\vec{u}(\vec{X},t)$ of a material point $\underline{x}$ in a body at time $t$ has the form

$$\vec{u}(\vec{X},t) = k(t)\,X_1\,X_2\,\vec{i}_1 + k(t)\,X_1\,X_2\,\vec{i}_2 + 2\,k(t)\,(X_1 + X_2)\,X_3\,\vec{i}_3\,, \tag{1}$$

where $k(t)$ is a scalar-valued quantity taking such small values that the small deformation theory is applicable. Derive the components of the small deformation strain tensor.

**3.7** The two-dimensional displacement vector $\vec{u}(\vec{X},t)$ has the following form:

$$\vec{u}(\vec{X},t) = k_1(t)\,X_1\left(X_1^2\,X_2 - k_2(t)\,X_2^3\right)\vec{\imath}_1 - k_1(t)\,X_2\left(X_2^3 - X_1^2\,X_2\right)\vec{\imath}_2\,, \qquad (1)$$

where $k_1(t)$ and $k_2(t)$ are scalar-valued quantities.

(a) Derive the small deformation strain tensor components.

(b) Derive the Green–Lagrange strain tensor components.

**3.8** During deformation, the position vector of the material point $\underline{P}(t)$ has the following components:

$$\begin{aligned} \underline{x}_1(t) &= \frac{1}{2}\,(X_1 + X_2)\,e^{t/t_0} + \frac{1}{2}\,(X_1 - X_2)\,e^{-t/t_0}\,, \\ \underline{x}_2(t) &= \frac{1}{2}\,(X_1 + X_2)\,e^{t/t_0} - \frac{1}{2}\,(X_1 - X_2)\,e^{-t/t_0}\,, \qquad (1) \\ \underline{x}_3(t) &= X_3\,, \end{aligned}$$

where the role of the constant $t_0$ is to make the exponent dimensionless.

(a) Derive the components of the velocity vector $\vec{v}(\vec{X},t)$ in terms of the initial material coordinates $X_i$ and time $t$.

(b) Derive the components of the velocity vector $\underline{\vec{v}}(\underline{\vec{x}}(t),t)$ in terms of the current material coordinates $\underline{x}_i(t)$ and time $t$.

(c) Derive the components of the rate of deformation tensor $\mathbf{d}$.

(d) Calculate the values of the components of the rate of deformation tensor $\mathbf{d}$ at time $t = 0$ s and $t = 0.05$ s. Assume that $t_0 = 1$ s.

(e) Derive the components of the displacement vector $\vec{u}(\vec{X},t)$ in terms of the initial material coordinates $X_i$ and time $t$.

(f) Assume that the small deformation theory holds. Derive the components of the small deformation strain tensor $\boldsymbol{\varepsilon}$ in terms of the initial material coordinates $X_i$ and time $t$.

(g) Give the components obtained in Step (e) in terms of the hyperbolic functions.

(h) Calculate the values of the small deformation strain tensor $\boldsymbol{\varepsilon}$ at time $t = 0$ s and $t = 0.05$ s. Assume that $t_0 = 1$ s.

(i) Derive the components of the strain rate tensor $\dot{\boldsymbol{\varepsilon}}$ associated with the small deformation theory.

(j) Calculate the values of the strain rate tensor $\dot{\boldsymbol{\varepsilon}}$ at time $t = 0$ s and $t = 0.05$ s. Assume that $t_0 = 1$ s.

(k) Compare the values of the components of the rate of deformation tensor $\mathbf{d}$ and the strain rate tensor $\dot{\boldsymbol{\varepsilon}}$ at time $t = 0$ s and $t = 0.05$ s. Assume that $t_0 = 1$ s. The values are given in Steps (d) and (j).

**3.9** Show that

$$\vec{v}\cdot\vec{\nabla}\vec{v} = \vec{\nabla}\big(\tfrac{1}{2}\,\vec{v}\cdot\vec{v}\big) - \vec{v}\times\vec{\nabla}\times\vec{v}\,, \tag{1}$$

where $\vec{v}$ is the velocity vector. Expression (1) is for fluid mechanics.

**Hint:** You may start with the right side and arrive at the left side of the equality.

**3.10** Show that

$$\begin{aligned}\dot{\underline{\vec{v}}}(\underline{\vec{x}}(t),t) &= \frac{\partial\vec{v}(\vec{x},t)}{\partial t} + \frac{1}{2}\,\vec{\nabla}(\vec{x})\big[\vec{v}(\vec{x},t)\cdot\vec{v}(\vec{x},t)\big] + 2\,\overset{\circ}{\boldsymbol{\omega}}(\vec{x},t)\cdot\vec{v}(\vec{x},t)\\ &= \frac{\partial\vec{v}(\vec{x},t)}{\partial t} + \frac{1}{2}\,\vec{\nabla}(\vec{x})\big[\vec{v}(\vec{x},t)\cdot\vec{v}(\vec{x},t)\big] + 2\,\vec{\omega}(\vec{x},t)\times\vec{v}(\vec{x},t)\,, \end{aligned} \tag{1}$$

where $\underline{\vec{v}}(\underline{\vec{x}}(t),t)$ is the velocity vector in the current material description and $\vec{v}(\vec{x},t)$ is the velocity vector in the spatial description. The material time derivative of $\underline{\vec{v}}(\underline{\vec{x}}(t),t)$ is obtained from $\dot{\underline{\vec{v}}}(\underline{\vec{x}}(t),t) = \mathrm{D}\underline{\vec{v}}(\underline{\vec{x}}(t),t)/\mathrm{D}t$. The fluid vorticity tensor denoted by $\overset{\circ}{\boldsymbol{\omega}}(\vec{x},t)$ is given in Definition (3.124) and the fluid vorticity vector $\vec{\omega}(\vec{x},t)$ is defined $\vec{\omega} = \frac{1}{2}\vec{\nabla}\times\vec{v}$. The components of the fluid worticity tensor $\overset{\circ}{\boldsymbol{\omega}}(\vec{x},t)$ and the fluid vorticity vector $\vec{\omega}(\vec{x},t)$ are related by $\overset{\circ}{\omega}_{ij} = -\,e_{ijk}\,\omega_k\,.$

# 4 Continuum Mechanics

## 4.1 Introduction

Although this book focuses primarily on the evaluation of the material models for solid materials, for the sake of completeness, the theoretical foundations and some models of fluid mechanics are also discussed, which significantly broadens the scope of this book. In this regard, continuum mechanics plays an important role.

First, in this chapter, fundamental and derived quantities and fundamental units are introduced. One axiom and two basic laws for continuum mechanics are also introduced. Local forms of the axiom of conservation of mass are derived for the initial material description, current material description, and spatial description. Different stress measures are discussed. The law of balance of linear momentum and law of balance of moment of momentum (i.e., angular momentum) are introduced. Local forms of the laws of momenta are formulated for the IG-CtP material description, for current material coordinates $\underline{x_i}(t)$, and for fluids (i.e., in the spatial description). Local forms for the law of moment of momentum are discussed briefly. Equilibrium equations for beams are derived from those of continuum mechanics. The Cauchy tetrahedron is used in formulating the relation between the Cauchy stress tensor $\underline{\boldsymbol{\sigma}}^{\mathrm{c}}$ and the traction vector $\underline{\vec{t}}$. Also, the extended Betti's theorem is derived.

## 4.2 Roles of Mathematics and Physics

In this book we study continuum thermodynamics, which is a branch of physics. Physics considers natural events and often tries to model them. Although physics utilizes mathematics, there is a fundamental difference between physics and mathematics. Physics is based on basic laws and axioms that are assumed to hold for natural events, but in mathematics, the axioms are pure theoretical constructions without any necessary connection to nature. This book adopts the results of mathematics as they are given. However, the Riemann integral, for example, needs the introduction of length $\ell$. Thus, when using the results of mathematics, one must make sure that all the details in these operations are defined. When vectors are defined, the area and volume can be constructed by a mathematical approach (see, e.g., Widder [117], pp. 215 and 216). Thus, after the introduction of length and a vector, the Riemann integration, volume, and area are assumed to be known.

When we say that physics is an exact science, we mean that its laws are expressed in the form of mathematical equations that describe and predict the results of accurate measurements. The advantage of a quantitative physical theory is a practical one of giving us the power to accurately predict and control natural phenomena. By comparing the results of accurate measurements with the numerical predictions of the theory, we can gain considerable confidence that the theory or model is valid, and we can determine in what respects it needs to be modified. It is often possible to explain a given phenomenon in several rough qualitative ways, and if we are content with that, it may be impossible to decide which theory is correct. But if a theory can predict correctly the results of measurements to four or five digits (or even two or three), self-evident figures, the theory can hardly be very far wrong. Rough agreement might be a coincidence, but close agreement is unlikely to be. Furthermore, there have been many cases in the history of science when small but significant discrepancies between theory and accurate measurements have led to the development of new and more far-reaching theories. Such slight discrepancies would not even have been detected if we had been content with a merely qualitative explanation of the phenomena. See Symon ([107], p. 1).

A good example of how a very small discrepancy between theory and measurement led to the construction of a more sophisticated theory was the introduction of the general theory of relativity. Despite the many triumphs of Newtonian mechanics in dynamic astronomy, a few phenomena remain that are in apparent disagreement with it; the best known concerns the orbit of the planet Mercury. This difficulty was overcome when Einstein created the general theory of relativity.

The following simple example shows how the theory of relativity enhanced the accuracy of some results in Newtonian mechanics. Kinetic energy of a body is considered.

The equivalence between mass $m$ and energy $E$ by Einstein is

$$E = m\,c^2\,, \tag{4.1}$$

where $m$ is the mass of the body and $c$ is the speed of the light. The mass of a body $m$ can be divided into two parts: Rest mass $m_0$ and mass $m - m_0$, which is associated with the kinetic energy of the body. The energy of the body $E$ is divided into the kinetic energy $E^{\mathrm{K}}$ and the rest energy $E^0$, as follows:

$$E = E^{\mathrm{K}} + E^0\,, \qquad \text{where} \qquad E^{\mathrm{K}} := (m - m_0)\,c^2 \quad \text{and} \quad E^0 := m_0\,c^2\,. \tag{4.2}$$

According to special relativity, the mass $m$ changes with the velocity $v$, as follows:

$$m = \frac{m_0}{\sqrt{1 - \left(\frac{v}{c}\right)^2}} = m_0 \left[1 - \left(\frac{v}{c}\right)^2\right]^{-1/2}\,. \tag{4.3}$$

The Maclaurin series of the function $(1+x)^\alpha$ is defined to be (Abramowitz and Stegun [2], p. 15)

$$(1+x)^\alpha = 1 + \alpha\,x + \frac{\alpha\,(\alpha - 1)}{2!}\,x^2 + \cdots\,, \qquad \text{where } -1 < x < 1\,. \tag{4.4}$$

Comparing the left side of Expression (4.4) with the expression for mass, Equation (4.3), the following can be written: $\alpha = -1/2$ and $x = -(v/c)^2$. Thus

Equation (4.3) yields

$$m = m_0 \left\{ 1 - \tfrac{1}{2} \left[ - \left( \frac{v}{c} \right)^2 \right] + \frac{-\tfrac{1}{2} \left( -\tfrac{1}{2} - 1 \right)}{2!} \left[ - \left( \frac{v}{c} \right)^2 \right]^2 + \cdots \right\} . \tag{4.5}$$

Substitution of Maclaurin Series (4.5) into Equation $(4.2)_2$ gives

$$\begin{aligned} E^{\mathrm{K}} &= m\,c^2 - m_0\,c^2 \\ &= \left[ m_0\,c^2 + \tfrac{1}{2}\,m_0 \left( \frac{v}{c} \right)^2 c^2 + \tfrac{3}{8}\,m_0 \left( \frac{v}{c} \right)^4 c^2 + \cdots \right] - m_0\,c^2 \\ &= \tfrac{1}{2}\,m_0\,v^2 + \tfrac{3}{8}\,m_0\,v^2 \left( \frac{v}{c} \right)^2 + \cdots = \tfrac{1}{2}\,m_0\,v^2 \left[ 1 + \tfrac{3}{4} \left( \frac{v}{c} \right)^2 + \cdots \right] . \end{aligned} \tag{4.6}$$

Based on Newtonian mechanics, the kinetic energy of a body $E^{\mathrm{KN}}$ is

$$E^{\mathrm{KN}} = \tfrac{1}{2}\,m_0\,v^2 . \tag{4.7}$$

Comparison of Results (4.6) and (4.7) shows that Newtonian mechanics provides only an approximation for the kinetic energy $E^{\mathrm{K}}$. However, for speeds used by a human being, the result from Newtonian mechanics has great accuracy and hence is adequate for most practical purposes.

## 4.3 Fundamental and Derived Quantities and Fundamental Units

This section introduces fundamental quantities, which form the foundation for continuum mechanics. The way to define the other quantities called "derived quantities" is discussed. Fundamental units for continuum mechanics are introduced as well.

The symbols that are to appear in the equations that express the laws of a science must represent quantities that can be expressed in numerical terms. Hence the concepts in terms of which an exact science is to be developed must be given precise numerical meanings. If a definition of a quantity (mass $m$, for example) is to be given, the definition must be such as to specify precisely how the value of the quantity is to be determined in any given case. A qualitative remark about its meaning may be helpful, but it is not sufficient as a definition. Sometimes a new concept can be defined in terms of others whose meanings are known, in which case there is no problem. For example,

$$Momentum := Mass \times Velocity \tag{4.8}$$

gives a perfectly precise definition of "momentum" provided "mass" and "velocity" are assumed to be precisely defined already. But this kind of definition will not do for all terms in a theory, since we must start somewhere with a set of basic concepts or "primitive concepts" terms whose meanings are assumed to be known. The first concept to be introduced in a theory cannot be defined in the above way, since we do not have anything on the right side of the equation. See Symon ([107], pp. 1–2).

Thus, before a mathematical derivation can be started, some quantities for the right side of the equation have to be chosen. According to Symon ([107], p. 11), those quantities are called "fundamental quantities." Derived quantities are defined by equations similar to Equation (4.8) having on the right side fundamental quantities and/or derived quantities already derived. It is customary to choose mass $m$, length $\ell$, and time $t$ as the fundamental quantities in mechanics, although there is nothing sacred in this choice. For these fundamental quantities, fundamental units have to be defined (Symon [107], p. 11). Symon ([107], p. 11) continued: "We could equally well choose some other three quantities, or even more or fewer than three quantities, as fundamental." In this book the set kilogram-meter-second, that is, MKS or [kg]-[m]-[s] is used. The units for the derived quantities are then obtained from equations where the right side units are already defined. For example, the component of the velocity in the $\ell$-direction $v_\ell$,

$$v_\ell := \frac{\mathrm{d}\ell}{\mathrm{d}t}, \tag{4.9}$$

is defined as a distance $\ell$ (length) divided by time $t$. Hence, in this case, the unit of velocity $v_\ell$ is m/s.

Although in mechanics it is popular to speak about three fundamental quantities, they are not sufficient for the construction of mechanics. Something else is needed. Therefore Symon ([107], p. 8) proposed to take force $\vec{F}$ as a fundamental quantity. One could express it such that the force vector $\vec{F}$ is defined by

$$\vec{F} := \frac{\mathrm{D}}{\mathrm{D}t}(m\,\vec{v}) \qquad \Rightarrow \qquad \vec{a} = \frac{\vec{F}}{m} \quad \text{for constant } m\,, \tag{4.10}$$

where $\vec{a}$ is the acceleration. The problem with Definition $(4.10)_1$ is that it is Newton's second law of motion and therefore a basic law. If Definition $(4.10)_1$ is used as a definition for a force vector $\vec{F}$ in particle mechanics, Definition $(4.10)_1$ should be used first to define the force vector $\vec{F}$ and later to operate as a basic law. The same contradiction would be in continuum mechanics where Newton's second law of motion is replaced by the law of balance of momentum. Symon [107] solved this problem by accepting force vector $\vec{F}$ into the set of fundamental quantities. Its unit, the newton [ N ], is not a fundamental unit but is defined as follows:

$$[\mathrm{N}] := \left[\frac{\mathrm{kg\,m}}{\mathrm{s}^2}\right]. \tag{4.11}$$

Symon [107] studied particle mechanics, and therefore mass $m$ and force vector $\vec{F}$ are natural choices for the set of fundamental quantities. In continuum mechanics and in continuum thermodynamics, however, mass $m$ and (point) force vector $\vec{F}$ are derived quantities, but length $\ell$ and time $t$ are fundamental quantities.

Instead of mass $m$, density of the material $\underline{\rho}(\vec{\underline{x}}(t), t)$ is defined here to be the fundamental quantity for continuum mechanics and for continuum thermodynamics. The mass of a certain entity of matter $m$ within the volume $\underline{v}(t)$ is obtained as an integral of the density $\underline{\rho}(\vec{\underline{x}}(t), t)$ over the volume $\underline{v}(t)$, as follows:

$$m := \int_{\underline{v}(t)} \underline{\rho}(\vec{\underline{x}}(t), t) \, \mathrm{d}\underline{v} \, . \tag{4.12}$$

Expression (4.12) is the definition of the mass $m$. Since the length $\underline{\ell}$ is a fundamental quantity, the integration on the right side of Definition (4.12) can be constructed, and therefore the right side of Definition (4.12) exists. Since the fundamental unit for the density $\underline{\rho}(\vec{\underline{x}}(t), t)$ is $\mathrm{kg/m^3}$, according to Definition (4.12) the unit for the mass $m$ is kg. There is no particular unit for density $\underline{\rho}(\vec{\underline{x}}(t), t)$. Density $\underline{\rho}(\vec{\underline{x}}(t), t)$ has the unit $\mathrm{kg/m^3}$, which is a combination of two fundamental units.

Some writers introduce mass $m$ instead of density $\rho$ into the set of fundamental quantities. This is not a good choice, because, as mentioned by Mikkola [63], "The density $\rho$ is the quantity for continuous matter whereas the mass $m$ is associated with certain entity of matter." This is expressed in Definition (4.12).

In continuum mechanics and in continuum thermodynamics the force vector $\vec{F}$ is a resultant force obtained by integration of a surface force field over an area or by integration of a body force vector field over a volume. Thus the (point) force vector $\vec{F}$ cannot be a fundamental quantity for continuum mechanics or for continuum thermodynamics. Instead of the force vector $\vec{F}$, the traction vector $\vec{\underline{t}}(\vec{\underline{x}}(t), t, \vec{\underline{n}})$ (or surface traction vector, or tractions for short) and the body force vector $\vec{\underline{b}}(\vec{\underline{x}}(t), t)$ are adopted here into the set of fundamental quantities. Since their units are $\mathrm{N/m^2}$ and N/kg, no additional fundamental units are introduced.

Figure 4.1 gives the fundamental quantities and the units associated to them for continuum mechanics. Later the set is extended to cover continuum thermodynamics as well. The quantities in Figure 4.1 are underlined to show that they are defined in current configuration $\underline{v}^{\mathrm{b}}(t)$. Thus, for example, the notations $\vec{\underline{b}}(\vec{\underline{x}}(t), t)$ and $\vec{\underline{b}}$ have the same meaning.

| | | | | |
|---|---|---|---|---|
| Density $\underline{\rho}$<br>kilogram /$\mathrm{m^3}$<br>[kg/meter$^3$] | Length $\underline{\ell}$<br>meter [m] | Time $t$<br>second [s] | Traction vector $\vec{\underline{t}}$<br>[N/$\mathrm{m^2}$] | Body force vector $\vec{\underline{b}}$<br>[N/kg] |

**Figure 4.1** Fundamental quantities and the units associated to them for continuum mechanics.

The introduction of vector fields also requires the introduction of a coordinate system. The coordinate system is a mathematical concept, although the fundamental quantity length $\underline{\ell}$ is needed for scaling the axes.

A vector (field) is a quantity having magnitude, direction, and sense. The magnitude of a vector (field) is expressed by its length, as Figure 4.2 shows.

Mathematically, the absolute value $|\vec{a}|$ gives the magnitude for the vector $\vec{a}$ [see Definition (2.125)]. Its direction is shown by its inclination with respect to the coordinate system. The sense of the vector (field) is given by an expression similar to Equation $(4.10)_2$. If the force vector $\vec{F}$ and the mass $m$ were defined, Equation $(4.10)_2$ would give the magnitude and direction for the vector $\vec{a}$.

Matter has a molecular structure. Continuum mechanics disregards this fact and models matter as continuous. Functions that describe the response of matter are continuous, except possibly a finite number of interior surfaces separating regions of continuity. This means that also all pores and holes are smoothed out, and their influence on the response of the material is described by continuous functions. This hypothetical continuous material is called a continuous medium or continuum.

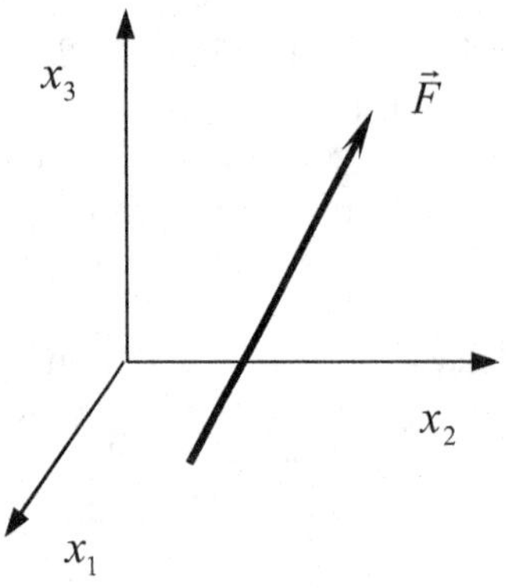


**Figure 4.2** A vector has its length, direction, and sense.

Physical bodies or systems, denoted here by $m^{\mathrm{b}}$, appear in the form of a collection of material points interconnected by some internal forces. A solid, a fluid, and so on are bodies. A particle is a mathematical model for a point mass. It has no size, but it does have mass, and its location may be specified.

Quantities are terms the values of which can be measured. This means that not all the terms in physics are quantities, but they are concepts. All the quantities are concepts, but not vice versa. Concepts that cannot be derived but have to be taken somewhere outside physical theories are called primitive concepts. The electron, for example, is a concept of physics, but it is not a quantity. The mass of an electron, however, is a quantity. See Symon ([107], p. 2) and Salonen ([86], p. 11).

The preceding implies that density $\underline{\rho}$, length $\underline{\ell}$, time $t$, the traction vector $\underline{\vec{t}}$, and the body force vector $\underline{\vec{b}}$ are primitive concepts but the force vector, for example, is a derived concept. These are also quantities. However, in the few preceding pages, the defined continuous material, system $\mathrm{m}^{\mathrm{b}}$, and the particle are (primitive) concepts but not quantities. The coordinate system is also a primitive concept.

## 4.4 Basic Laws and Axioms for Newtonian and Continuum Mechanics

This section introduces the basic laws and axiom for Newtonian mechanics and derives Euler's first and second laws. Euler's first law is the law of balance of momentum, and Euler's second law is the law of balance of moment of momentum. Some other names exist in the literature.

In specifying physical quantities, we assume that there exists a frame of reference (reference frame, reference coordinate system) with respect to which measurement can be made. The so-called inertial reference indicates the coordinate system the origin of which is attached to a fixed star in space. The directions of the coordinate axes are attached to other fixed stars. Any other system that moves uniformly and without rotation relative to the fixed star may be used as inertial reference. In most of our engineering work, however, the coordinate system is attached to the surface of the earth, since the error incurred due to rotation of the earth and the variation of this motion around the sun is very slight. This paragraph is modified from Huang ([30], p. 6).

In the present work the term *basic law* is used for relations that are generally accepted and can be experimentally verified to hold for a wide range of natural events. It should be kept in mind that the following basic laws are not "facts" but form a picture of nature that has been observed to be a good approximation of "reality." This uncertainty originates from the fact that the "theory of everything" is unclear for mankind and therefore it is difficult to know what relations are consequences for the theory of everything (i.e., which are the basic laws) and which are not. *Axiom*, on the other hand, refers to an idea that may be widely accepted/verified to hold but is known to have some validity limitations. Axioms form with the basic laws the foundation of a theory. Conservation of mass, for example, is a fundamental equation in Newtonian mechanics, but in the theory of relativity derived by Einstein, it is replaced by the balance of mass and energy $E = m\,c^2$, where $c$ is the speed of light in a vacuum. Since the processes which will be modelled here obey the conservation of mass, this assumption is acceptable in the present context. Therefore here the conservation of mass falls into the category of axioms. The notations **AX** and **BL** refer to an axiom and to a basic law, respectively.

According to Synge and Griffith ([106], pp. 27 and 28), the three fundamental laws, on which Newtonian particle mechanics is based are the following:

(1) *Law of motion.* Relative to a basic frame of reference, a particle of mass $m$, subject to a force $\vec{F}$, moves in accordance with the equation

$$\vec{F} = \frac{\mathrm{D}}{\mathrm{D}t}\,(m\vec{v})\,. \tag{4.13}$$

(2) *Law of action and reaction.* When two particles exert forces on each other, the forces are equal in magnitude and act along the line joining the particles.

(3) *Law of the parallelogram of forces.* When two forces $\vec{P}$ and $\vec{Q}$ act on the particle, they are together equivalent to a single force $\vec{F} = \vec{P} + \vec{Q}$, the vector being defined by the parallelogram construction.

Figure 4.3 illustrates the fundamental laws of Newtonian particle mechanics. Since particle mechanics does not introduce the concepts of initial configuration, current configuration, and so on, the preceding variables are not underlined.

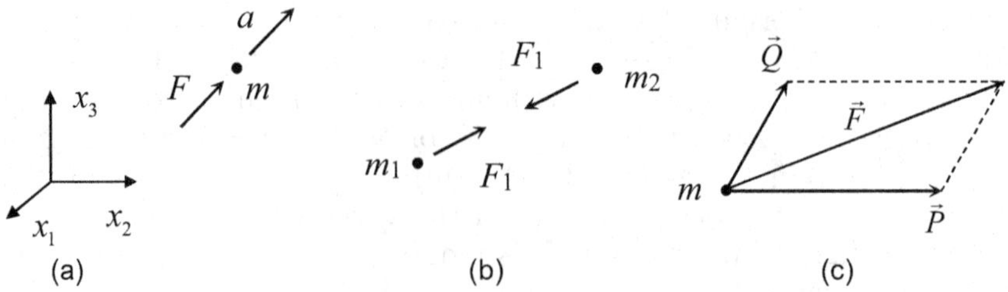


**Figure 4.3** (a) Law of motion. (b) Law of action and reaction. (c) Law of parallelogram of forces.

As the vocabulary of the text by Synge and Griffith [106] shows, the given laws are expressed for particle mechanics. For continuum mechanics, which is the topic of this work, the first two laws are replaced by the laws given by Euler (1750 and 1755). These laws are adopted here. They are studied next.

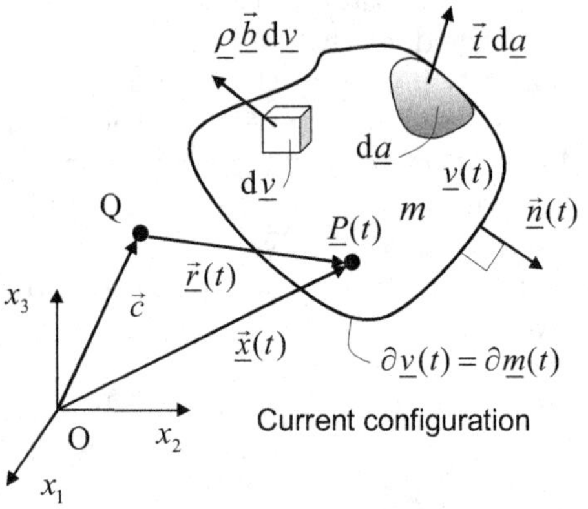


**Figure 4.4** Forces acting on an arbitrary subsystem $m$ in the current configuration $\underline{v}^{\mathrm{b}}(t)$.

An arbitrarily selected but then non-changing subsystem (part) $m$ of the system (body) $\mathrm{m}^{\mathrm{b}}$ in the current (deformed) configuration $\underline{v}^{b}(t)$ is investigated in Figure 4.4. The volume of the subsystem in the current configuration is $\underline{v}(t)$. The subsystem $m$ is loaded by the traction vector $\underline{\vec{t}}$ and by the body force vector $\underline{\vec{b}}$. The resultant force due to $\underline{\vec{t}}$ and $\underline{\vec{b}}$ is denoted by $\underline{\vec{F}}_E$, and it has the following appearance:

$$\underline{\vec{F}}_{\mathrm{E}} := \oint_{\partial\underline{m}(t)} \underline{\vec{t}}\,\mathrm{d}(\partial m) + \int_m \underline{\vec{b}}\,\mathrm{d}m \quad\Rightarrow\quad \underline{\vec{F}}_{\mathrm{E}} = \oint_{\partial\underline{v}(t)} \underline{\vec{t}}\,\mathrm{d}\underline{a} + \int_{\underline{v}(t)} \rho\underline{\vec{b}}\,\mathrm{d}\underline{v}\,. \quad (4.14)$$

The time dependence in the notation $\partial\underline{m}(t)$ stems from the fact that the mass $m$ remains the same as its boundary deforms. The odd notation $\mathrm{d}(\partial m)$ in the first term on the right side of Definition $(4.14)_1$ denotes the differential surface area of the mass. The first term in Form $(4.14)_2$ exploits the equality

$\partial \underline{m}(t) = \partial \underline{v}(t)$ (see Figure 4.4) and that the differential surface area of the mass $\mathrm{d}(\partial m)$ is replaced by the differential area $\mathrm{d}\underline{a}$. The second term exploits the fact that $\mathrm{d}m = \underline{\rho}(\underline{\vec{x}}(t))\,\mathrm{d}\underline{v}(t)$. This, possibly a strange notation, shows that during deformation, the density $\underline{\rho}$ can take different values in different parts of the matter $m$ and can vary with time $t$ as well. Furthermore, the volume occupied by the matter $m$ and its surface vary with time $t$, but the mass $m$ itself remains the same. The quantities $\underline{\vec{t}}$, $\underline{\vec{b}}$, and $\underline{\rho}$ have the forms $\underline{\vec{t}} = \underline{\vec{t}}(\underline{\vec{x}}(t), t, \underline{\vec{n}})$, $\underline{\vec{b}} = \underline{\vec{b}}(\underline{\vec{x}}(t), t)$, and $\underline{\rho} = \underline{\rho}(\underline{\vec{x}}(t), t)$. It is important to note that the integration in Definition $(4.14)_1$ is carried out around and over a certain amount of mass $m$ moving and deforming with the deformation of the system $\mathrm{m^b}$. In thermodynamic language, a closed subsystem is considered.

The momentum (or linear momentum) $\underline{\vec{G}}$ by subsystem having a mass $m$ and a velocity field $\underline{\vec{v}}(\underline{\vec{x}}(t), t)$ is defined to be

$$\underline{\vec{G}} := \int_m \underline{\vec{v}}\,\mathrm{d}m \qquad \Rightarrow \qquad \underline{\vec{G}} = \int_{\underline{v}(t)} \underline{\rho}\,\underline{\vec{v}}\,\mathrm{d}\underline{v}\,, \tag{4.15}$$

where the velocity vector $\underline{\vec{v}}(\underline{\vec{x}}(t), t)$ is given by Definition $(3.20)_1$, that is, $\underline{\vec{v}}(\underline{\vec{x}}(t), t) := \partial \underline{\vec{x}}(t)/\partial t$.

**BL 1** *Law of balance of momentum.* Using a natural language, this law can be expressed by modifying Narasimhan's statement ([66], p. 320) as "the time rate of change of the momentum of a body as it undergoes a deformation or flow is in balance with the sum of the surface forces on the bounding surface of the body and the body force acting on the body." It can be written in the following mathematical form:

**BL 1**
$$\underline{\vec{F}}_{\mathrm{E}} = \frac{\mathrm{D}}{\mathrm{D}t}\underline{\vec{G}}\,. \tag{4.16}$$

Substitution of Definitions $(4.14)_1$ and $(4.15)_1$ or Expressions $(4.14)_2$ and $(4.15)_2$ into Basic Law (4.16) [**BL 1**] yields

**BL 1**
$$\oint_{\partial \underline{m}(t)} \underline{\vec{t}}\,\mathrm{d}(\partial m) + \int_m \underline{\vec{b}}\,\mathrm{d}m = \frac{\mathrm{D}}{\mathrm{D}t}\int_m \underline{\vec{v}}\,\mathrm{d}m \tag{4.17}$$

or

**BL 1**
$$\oint_{\partial \underline{v}(t)} \underline{\vec{t}}\,\mathrm{d}\underline{a} + \int_{\underline{v}(t)} \underline{\rho}\,\underline{\vec{b}}\,\mathrm{d}\underline{v} = \frac{\mathrm{D}}{\mathrm{D}t}\int_{\underline{v}(t)} \underline{\rho}\,\underline{\vec{v}}\,\mathrm{d}\underline{v}\,. \tag{4.18}$$

There are three forms for the law of balance of momentum, that is, Basic Laws(4.16)–(4.18). This is because Form (4.18) is usually referred to as the law of balance of momentum. This practice is also followed in this book.

A body $\mathrm{m^b}$, the volume of which is $\underline{v}^{\mathrm{b}}(t)$, is in pure translation. This implies that all the material points of the body have equal velocity $\underline{\vec{v}}(\underline{\vec{x}}(t), t)$ and that therefore the velocity field $\underline{\vec{v}}(\underline{\vec{x}}(t), t)$ can be moved to the front of the integral sign on the right sides of Expressions (4.17) and (4.18). Now the integration can be carried out and it gives $m$. Therefore the term on the right sides of Expressions (4.17) and (4.18) reduces to $\frac{\mathrm{D}}{\mathrm{D}t}(\underline{\vec{v}}\,m)$. Thus the law of balance of momentum, that is, Basic Law (4.18) [or Expressions (4.17) and (4.16)], is an extension of Newton's Law of Motion (4.13).

The resultant moment of the loading given by Figure 4.4 about an arbitrary point Q is denoted by $\underline{\vec{M}}_{\mathrm{E}}$ and has the following appearance:

$$\underline{\vec{M}}_{\mathrm{E}} := \oint_{\partial \underline{m}(t)} (\underline{\vec{r}} \times \underline{\vec{t}}\,)\, \mathrm{d}(\partial m) + \int_{m} (\underline{\vec{r}} \times \underline{\vec{b}}\,)\, \mathrm{d}m\,, \tag{4.19}$$

which yields

$$\underline{\vec{M}}_{\mathrm{E}} = \oint_{\partial \underline{v}(t)} (\underline{\vec{r}} \times \underline{\vec{t}}\,)\, \mathrm{d}\underline{a} + \int_{\underline{v}(t)} (\underline{\vec{r}} \times \rho\, \underline{\vec{b}}\,)\, \mathrm{d}\underline{v}\,. \tag{4.20}$$

In Definition (4.19) the vector $\underline{\vec{r}}(t)$ is the vector from an arbitrary fixed point Q to a material point $\underline{P}(t)$. Since the position of the material point $\underline{P}(t)$ varies with deformation, the value of the vector $\underline{\vec{r}}(t)$ varies with time $t$.

The moment of momentum $\underline{\vec{H}}$ by subsystem having a mass $m$ and a velocity field $\underline{\vec{v}}(\underline{\vec{x}}(t), t)$ is defined to be

$$\underline{\vec{H}} := \int_{m} (\underline{\vec{r}} \times \underline{\vec{v}}\,)\, \mathrm{d}m\,, \qquad \text{which yields} \quad \underline{\vec{H}} = \int_{\underline{v}(t)} (\underline{\vec{r}} \times \rho\, \underline{\vec{v}}\,)\, \mathrm{d}\underline{v}\,. \tag{4.21}$$

**BL 2** *Law of balance of moment of momentum.* Using natural language, this law can be expressed by modifying Narasimhan's statement ([66], p. 320) as "the time rate of change of the moment of momentum about an arbitrary point of a body undergoing a deformation or flow is in balance with the flux of the moment of momentum of the traction vector $\underline{\vec{t}}$ and the moment of momentum of the body force vector $\underline{\vec{b}}$." It can be written in the following mathematical form:

**BL2**

$$\underline{\vec{M}}_{\mathrm{E}} = \frac{\mathrm{D}}{\mathrm{D}t}\, \underline{\vec{H}}\,. \tag{4.22}$$

Substitution of Definition (4.20) and Expression $(4.21)_2$ into Basic Law (4.22) [**BL 2**] yields

**BL 2**

$$\oint_{\partial \underline{v}(t)} (\underline{\vec{r}} \times \underline{\vec{t}}\,)\, \mathrm{d}\underline{a} + \int_{\underline{v}(t)} (\underline{\vec{r}} \times \rho\, \underline{\vec{b}}\,)\, \mathrm{d}\underline{v} = \frac{\mathrm{D}}{\mathrm{D}t} \int_{\underline{v}(t)} (\underline{\vec{r}} \times \rho\, \underline{\vec{v}}\,)\, \mathrm{d}\underline{v}\,. \tag{4.23}$$

The definitions of the quantities are already discussed with Basic Law (4.18) [**BL 1**]. The third form, being similar to Form (4.17), is not given, since it is usually not applied.

From this point forward, Form (4.23) is referred to as the law of balance of moment of momentum (or law of balance of angular momentum).

According to McGill and King ([62], pp. 586–588), the *law of action and reaction* is obtained from Basic Laws (4.18) [**BL 1**] and (4.23) [**BL 2**]. Furthermore, as shown, the *law of motion* is obtained from Basic Law (4.18) [**BL 1**].

The *law of the parallelogram of forces* is replaced by the assumption that the force is a vector. Thus the three fundamental laws of particle mechanics are replaced in continuum mechanics by another set containing three terms.

The given set of Newton's laws are not the only axioms for particle mechanics. By using Hamilton's principle, which leads to Lagrange's equations, one can get an alternative set of fundamental axioms for particle mechanics (see, e.g., Landau and Lifshitz [43], Chapters II and III). According to Symon ([107], p. 3), the equations of Lagrange and of Hamilton are not new physical theories, for they may be derived from Newton's laws, but they are different ways of expressing the same physical theory. He continues: They use more advanced mathematical concepts, they are in some respects more elegant than Newton's formulation, and they are in some cases more powerful in that they allow solutions of some problems, based directly on Newton's laws, that would be very difficult.

A vital part of Newtonian mechanics is the assumption that mass can be neither created nor destroyed. As already mentioned, in his general theory of relativity, Einstein showed the relation between mass $m$ and energy $E$ by writing the famous equation $E = m\,c^2$. This Newtonian axiom reads as follows.

**AX 1** *Conservation of mass.* Using a natural language, this law can be stated as: "the mass of a subsystem $m$ before deformation or flow is equal to the mass of the subsystem $m$ after deformation or flow." It can be expressed in the following mathematical form:

**AX 1**
$$\frac{\mathrm{D}m}{\mathrm{D}t} = 0\,. \tag{4.24}$$

## 4.5 Global to Local Form for Solids: IG-C$t$P Material Description

As Expressions (4.17), (4.19), and $(4.21)_1$ show, basic laws and axioms are written in integral form for an arbitrarily selected but then nonchanging entity of matter of the subsystem $m$ which is in the current configuration $\underline{v}^{\mathrm{b}}(t)$. In other words, once the material points are selected to form subsystem $m$, the same material points remain within subsystem $m$ during the deformation. In thermodynamic language, closed subsystems are considered. These integral expressions are called "global forms of the basic laws and axioms." Furthermore, the quantities are expressed by the current material coordinates $\underline{\vec{x}}(t)$. Global forms are difficult to apply in mathematical derivations, but local forms are needed. Local forms are consequences of global forms but expressed as partial differential equations that apply for every point of the system $\mathrm{m}^{\mathrm{b}}$. Such expressions are called field equations, being capable of giving values for quantities that have different values at different points in the system $\mathrm{m}^{\mathrm{b}}$. A differential equation for beam bending is a good example of a field equation.

For the IG-CtP material description, the quantities have to be written in terms of initial material coordinates $X_i$, which means that a coordinate transformation is needed. This section gives a general view of how to transform expressions from global form to local form but does not give a detailed derivation. The law of balance of momentum is studied, but investigation of the other laws and axioms takes the same steps. These steps will be numbered by I, II, and so on. A similar procedure applies for spatial description, that is, for fluids.

When the independent variables are given, the law of balance of momentum, Expression (4.18), takes the following appearance:

**BL 1**
$$\oint_{\partial \underline{v}(t)} \underline{\vec{t}}(\underline{\vec{x}}(t), t, \underline{\vec{n}})\, \mathrm{d}\underline{a} + \int_{\underline{v}(t)} \underline{\rho}(\underline{\vec{x}}(t), t)\, \underline{\vec{b}}(\underline{\vec{x}}(t), t)\, \mathrm{d}\underline{v} = \frac{\mathrm{D}}{\mathrm{D}t} \int_{\underline{v}(t)} \underline{\rho}(\underline{\vec{x}}(t), t)\, \underline{\vec{v}}(\underline{\vec{x}}(t))\, \mathrm{d}\underline{v}\,. \tag{4.25}$$

Step I. Write Basic Law (4.25) [**BL 1**] in terms of the material coordinates $X_i$. The following is achieved:

$$\oint_{\partial V_0} \vec{T}(\vec{X}, t, \vec{N})\, \mathrm{d}A + \int_{V_0} \rho_0(\vec{X})\, \vec{b}(\vec{X}, t)\, \mathrm{d}V = \frac{\mathrm{D}}{\mathrm{D}t} \int_{V_0} \rho_0(\vec{X})\, \vec{v}(\vec{X}, t)\, \mathrm{d}V\,. \tag{4.26}$$

The key point between Expressions (4.25) and (4.26) is that the integration domain has been changed.

The challenge in Equation (4.26) is that the definition of the traction vector $\vec{T}(\vec{X}, t, \vec{N})$ is given in such a way that the first terms in Expressions (4.25) and (4.26) are equal. By applying the axiom of conservation of mass, Section 4.11 derives an expression for converting the latter two integrals in Expression (4.25) to the forms given in Expression (4.26).

Step II. Change the first integral in Expression (4.26) into the volume integral and move the material time operator $\frac{\mathrm{D}}{\mathrm{D}t}$ to the right side of the integral sign $\int$.

We arrive at the following:

$$\int_{V_0} \vec{\nabla}(\vec{X}) \cdot \boldsymbol{\sigma}^{\square}(\vec{X}, t)\, \mathrm{d}V + \int_{V_0} \rho_0(\vec{X})\, \vec{b}(\vec{X}, t)\, \mathrm{d}V = \int_{V_0} \rho_0(\vec{X})\, \frac{\mathrm{D}\vec{v}(\vec{X}, t)}{\mathrm{D}t}\, \mathrm{d}V\,. \tag{4.27}$$

This movement is called the Reynolds transport theorem, and it is derived in Section 4.13.

Step III. Collect all the terms of Expression (4.27) under one integral sign on the left side of the equals sign.

The following is achieved for every $V_0$ within $\mathrm{V}_0^{\mathrm{b}}$:

$$\int_{V_0} \left[ \vec{\nabla}(\vec{X}) \cdot \boldsymbol{\sigma}^{\square}(\vec{X}, t) + \rho_0(\vec{X})\, \vec{b}(\vec{X}, t) - \rho_0(\vec{X})\, \frac{\mathrm{D}\vec{v}(\vec{X}, t)}{\mathrm{D}t} \right] \mathrm{d}V = \vec{0}\,. \tag{4.28}$$

The phrase "for every $V_0$ within $\mathrm{V}_0^{\mathrm{b}}$" means that the volume $V_0$ (i.e., volume of the subsystem $m$) is an arbitrary subvolume of the volume $\mathrm{V}_0^{\mathrm{b}}$ (i.e., volume of the system $\mathrm{m}^{\mathrm{b}}$).

Step IV. Show that Integrand (4.28) has to vanish at every point of the system $\mathrm{m^b}$.

We arrive at the following:

$$\vec{\nabla}(\vec{X})\cdot\boldsymbol{\sigma}^{\square}(\vec{X},t)+\rho_0(\vec{X})\,\vec{b}(\vec{X},t)=\rho_0(\vec{X})\,\frac{\mathrm{D}\vec{v}(\vec{X},t)}{\mathrm{D}t};\quad \text{Everywhere within } \mathrm{V_0^b}. \tag{4.29}$$

This is studied in Section 4.7.

Step V. Assume small deformations. The following is obtained:

$$\vec{\nabla}(\vec{X})\cdot\boldsymbol{\sigma}(\vec{X},t)+\rho_0(\vec{X})\,\vec{b}(\vec{X},t)=\rho_0(\vec{X})\,\frac{\mathrm{D}\vec{v}(\vec{X},t)}{\mathrm{D}t},\quad \text{Everywhere within } \mathrm{V_0^b}. \tag{4.30}$$

It is a common practice to leave out the phrase "everywhere within $\mathrm{V_0^b}$" in Equation (4.30), but it is given for clarifying the study that will be carried out in the next section. The forthcoming sections give detailed investigations of the given steps.

## 4.6 Axiom of Conservation of Mass for Solids: Initial Material Description

The present section derives the consequence of the axiom of conservation of mass for solids written in terms of the initial material coordinates $X_i$.

The axiom of conservation of mass, Axiom (4.24) [**AX 1**], reads

**AX 1**
$$\frac{\mathrm{D}m}{\mathrm{D}t}=0\,. \tag{4.31}$$

Equation (4.31) does not take a stand on the description and is thus valid for both solids and fluids. The mass of a subsystem $m$ is obtained from the equation

$$m=\int_m \mathrm{d}m\,. \tag{4.32}$$

Figure 4.5 sketches a subsystem $m$ the volume of which in the current configuration is denoted by $\underline{v}(t)$. When the current material coordinates $\underline{x}_i(t)$ are used, Expression (4.12) gives the following definition for the mass of the subsystem $m$:

$$m:=\int_{\underline{v}(t)}\underline{\rho}(\underline{\vec{x}}(t),t)\,\mathrm{d}\underline{v}\,, \tag{4.33}$$

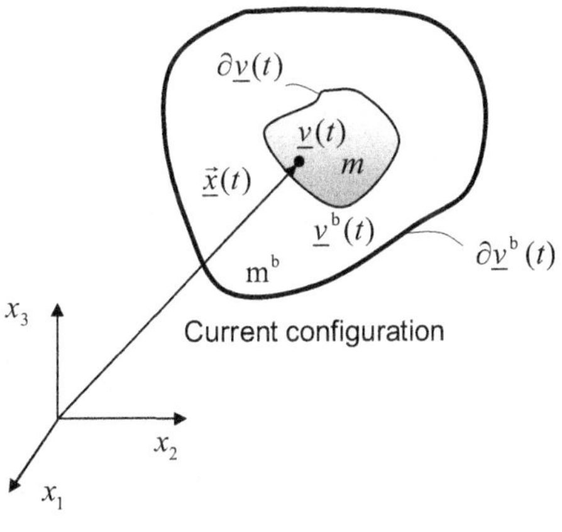


**Figure 4.5** Arbitrary subsystem $m$ of the system $\mathrm{m^b}$.

where the integration is carried out in the current configuration $\underline{v}^{\mathrm{b}}(t)$ and where the notation $\rho(\vec{\underline{x}}(t), t)$ refers to the density in the current configuration $\underline{v}^{\mathrm{b}}(t)$. As shown in Figure 4.5, the position vector $\vec{\underline{x}}(t)$ points to the material points within the volume of the current configuration $\underline{v}^{\mathrm{b}}(t)$. The crucial thing here is that the mass $m$ is an arbitrarily selected but then nonchanging, fixed part of the total mass of the system $\mathrm{m}^{\mathrm{b}}$, and therefore the subsystem $m$ is an arbitrary part of the system $\mathrm{m}^{\mathrm{b}}$.

Axiom (4.31) [**AX 1**] and Definition (4.33) yield

$$\frac{\mathrm{D}}{\mathrm{D}t} \int_{\underline{v}(t)} \rho(\vec{\underline{x}}(t), t)\, \mathrm{d}\underline{v} = 0\,. \tag{4.34}$$

The standard procedure to continue from Equation (4.34) is to apply the Reynolds transport theorem, which allows the derivative operator $\frac{\mathrm{D}}{\mathrm{D}t}$ to be moved to the right side of the integral operator. This approach will be taken later with other basic laws and axioms, but for Expression (4.34), it is of little practical value, and the following steps will be taken here instead.

When studied in the initial configuration $\mathrm{V}_0^{\mathrm{b}}$, the mass of the subsystem $m$ is obtained from

$$m = \int_m \mathrm{d}m = \int_{V_0} \rho_0(\vec{X}, t_0)\, \mathrm{d}V\,, \tag{4.35}$$

where $V_0$ and $\rho_0(\vec{X}, t_0)$ are the volume and density of the subsystem in the initial configuration $\mathrm{V}_0^{\mathrm{b}}$, that is, before deformation. The quantity $t_0$ is added to stress that the value of $\rho_0$ is determined before deformation.

The axiom of conservation of mass, Axiom (4.31) [**AX 1**], was given as follows: the mass of a subsystem $m$ before deformation or flow is equal to the mass of the subsystem $m$ during deformation or flow. The mass of the subsystem $m$ is given by Definition (4.33) and Equation (4.35). To make these equations comparable, they have to be expressed within the same coordinates. Since it is common practice to evaluate solids with the initial material coordinates $X_i$, a coordinate transformation in Equation (4.33) is done next. Based on Boas ([11], pp. 262 and 263) and Holzapfel ([29], p. 74), the following equation is obtained:

$$m = \int_{\underline{v}(t)} \rho(\vec{\underline{x}}(t), t)\, \mathrm{d}\underline{v} = \int_{V_0} \rho(\vec{X}, t)\, J(\vec{X}, t)\, \mathrm{d}V\,. \tag{4.36}$$

In Expression (4.36), the notation $J(\vec{X}, t)$ stands for the Jacobian determinant between the current material coordinates $\underline{x}_i(t)$ and the initial material coordinates $X_i$. It reads

$$J(\vec{X}, t) := \begin{vmatrix} \dfrac{\partial \underline{x}_1(t)}{\partial X_1} & \dfrac{\partial \underline{x}_1(t)}{\partial X_2} & \dfrac{\partial \underline{x}_1(t)}{\partial X_3} \\ \dfrac{\partial \underline{x}_2(t)}{\partial X_1} & \dfrac{\partial \underline{x}_2(t)}{\partial X_2} & \dfrac{\partial \underline{x}_2(t)}{\partial X_3} \\ \dfrac{\partial \underline{x}_3(t)}{\partial X_1} & \dfrac{\partial \underline{x}_3(t)}{\partial X_2} & \dfrac{\partial \underline{x}_3(t)}{\partial X_3} \end{vmatrix}\,. \tag{4.37}$$

The axiom of conservation of mass tells that the right sides of Expressions (4.35) and (4.36) are equal. This is written as follows:

$$\int_{V_0} \rho(\vec{X},t)\, J(\vec{X},t)\, \mathrm{d}V = \int_{V_0} \rho_0(\vec{X},t_0)\, \mathrm{d}V\,. \tag{4.38}$$

Equation (4.38) yields

$$\int_{V_0} \left[\rho(\vec{X},t)\, J(\vec{X},t) - \rho_0(\vec{X},t_0)\right] \mathrm{d}V = 0\,. \tag{4.39}$$

Since the subsystem $m$ is an arbitrary part of the system $\mathrm{m^b}$, Integrand (4.39) must vanish for every $V_0$ within the volume $\mathrm{V_0^b}$. Thus the term in the square brackets vanishes at every point of the system $\mathrm{m^b}$, and the following local form is achieved:

$$\rho(\vec{X},t)\, J(\vec{X},t) - \rho_0(\vec{X},t_0) = 0\,. \tag{4.40}$$

Equation (4.40) is a result of the axiom of conservation of mass, Axiom (4.31) [**AX 1**], for solids. Equation (4.40) is used when the equations for the basic laws and axioms are transformed from the current material coordinates $\underline{x}_i(t)$ to the initial material coordinates $X_i$. Although Equation (4.40) was derived by comparing the initial configuration $\mathrm{V_0^b}$ with the current configuration $\underline{v}^{\mathrm{b}}(t)$, it can be applied when the investigation is transferred from the current configuration $\underline{v}^{\mathrm{b}}(t)$ to the IG-CtP configuration $\mathrm{V_0^b}$.

## 4.7 From Volume Integral to Field Equation

This section studies step IV of Section 4.5, that is, how to show that the integrand of a volume integral over an arbitrary subsystem vanishes. The result of this section was already utilized when Equation (4.40) was obtained from Equation (4.39). It is also used in the forthcoming pages when deriving the local forms of the basic laws and axioms. To keep the study simple, a uniaxial case is investigated.

The reason why Integrand (4.40) must vanish is illustrated by the following uniaxial problem. Integral (4.39) is replaced by its uniaxial counterpart,

$$\int_{x_a}^{x_b} f(x)\, \mathrm{d}x = 0\,, \quad \text{for every subinterval } [x_a, x_b] \text{ within interval } [a,b]\,, \tag{4.41}$$

where $f(x)$ is a continuous function within the interval $[a,b]$ and the subinterval $[x_\mathrm{a}, x_\mathrm{b}]$ is an arbitrary part of the interval $[a,b]$. The interval $[a,b]$ is the uniaxial counterpart for the system (body) $\mathrm{V_0^b}$, whereas the subinterval $[x_\mathrm{a}, x_\mathrm{b}]$ is the counterpart of the subsystem $V_0$, the latter being an arbitrary part of the system $\mathrm{V_0^b}$. Figure 4.5 illustrates Integration (4.41).

In Figures 4.6(a) and 4.6(b), the value of Integral (4.41) equals the shadowed areas. Figure 4.6(a) shows a special case in which Condition (4.41) is satisfied, whereas in the case of Figure 4.6(b), Condition (4.41) is not satisfied. Since the subinterval $[x_\mathrm{a}, x_\mathrm{b}]$ is an arbitrary part of the interval $[a,b]$, the only way to satisfy Condition (4.41) is to require that the function $f(x)$ vanish everywhere within the interval $[a, b]$. This is

$$f(x) = 0\,. \qquad \text{Everywhere within the interval } [a, b]\,. \quad (4.42)$$

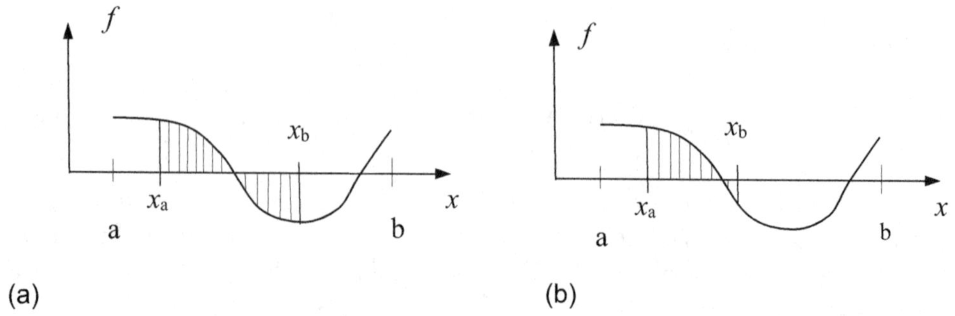


**Figure 4.6** The value of the integral of $f(x)$ equals the shadowed area.

The procedure for obtaining Equation (4.42) from Expression (4.41) is the same as when Equation (4.40) was obtained from Expression (4.39).

The studied system $\mathrm{V}_0^{\mathrm{b}}$ and its arbitrary subsystem $V_0$ are consequences obtained by coordinate transformation from the current material coordinates $\underline{x}_i(t)$ to the initial material coordinates $X_i$. The basic laws and axioms are written in terms of the current material coordinates $\underline{x}_i(t)$. Therefore, instead of $\mathrm{V}_0^{\mathrm{b}}$ and $V_0$, the original integration volumes are $\underline{v}^{\mathrm{b}}(t)$ and $\underline{v}(t)$. The coordinate transformation was already discussed in step I of Section 4.5 and is derived in Section 4.11.

The preceding evaluation demonstrated clearly that the basic laws and axioms are important to write for the subsystem $m$ [for its volume $\underline{v}(t)$] instead of the whole system (body) $\mathrm{m}^{\mathrm{b}}$ [for its volume $\underline{v}^{\mathrm{b}}(t)$]. Some writers are negligent in their derivation and do not make a difference between subsystem $m$ and system $\mathrm{m}^{\mathrm{b}}$. Since Results (4.29) and (4.42) hold for every subsystem $m$ [volume $V_0$ or originally $\underline{v}(t)$] (or subinterval $[x_{\mathrm{a}}, x_{\mathrm{b}}]$), they of course hold for the system $\mathrm{m}^{\mathrm{b}}$ [volume $\mathrm{V}_0^{\mathrm{b}}$ or originally volume $\underline{v}^{\mathrm{b}}(t)$] (or interval $[a, b]$).

The investigation showed how the global form, which is written for a finite amount of matter $m$, is changed to the local equations, which are written for every single point of the system (body) $\mathrm{m}^{\mathrm{b}}$. The obtained equations are called "field equations." The same method is applied many times in the forthcoming derivations.

The previous study assumed that the field quantities are continuous. This is not the case in general, but discontinuity surfaces are possible. The interface of two materials, for example, usually introduces a discontinuity surface for the value of density $\rho$. Since the material is assumed to contain a finite number of discontinuities, the arbitrary volumes $V$ in the previous investigations can be assumed to be free from discontinuities. When the point under consideration approaches a discontinuity surface, the derivatives present in the local forms of the basic laws and axioms have to be interpreted to be directional derivatives.

## 4.8 Axiom of Conservation of Mass: Current Material Description

The present section utilizes the axiom of conservation of mass and therefore derives its consequence, which is the continuity equation written in terms of the current material coordinates $\underline{x}_i(t)$.

The axiom of conservation of mass, Axiom (4.24) [**AX 1**], reads

**AX 1**
$$\frac{\mathrm{D}m}{\mathrm{D}t} = 0\,. \tag{4.43}$$

Equation (4.43) does not take a stand on the description. Thus Form (4.43) is valid for material and spatial descriptions.

When expressed by the current material coordinates $\underline{x}_i(t)$, the mass of the subsystem $m$ is obtained from Definition (4.12),

$$m := \int_{\underline{v}(t)} \underline{\rho}(\vec{\underline{x}}(t), t)\, \mathrm{d}\underline{v}\,, \tag{4.44}$$

where $\underline{v}(t)$ is the volume of the subsystem $m$, where the integration is carried out in the current configuration $\underline{v}^{\mathrm{b}}(t)$ and where the notation $\underline{\rho}(\vec{\underline{x}}(t), t)$ refers to the density in the current configuration $\underline{v}^{\mathrm{b}}(t)$.

The Leibnitz integral rule, Theorem (3.135), reads

$$\frac{\mathrm{D}}{\mathrm{D}t}\int_{\underline{v}(t)} \underline{\theta}(\vec{\underline{x}}(t), t)\, \mathrm{d}\underline{v} = \int_{\underline{v}(t)} \left[\frac{\mathrm{D}\underline{\theta}(\vec{\underline{x}}(t), t)}{\mathrm{D}t} + \underline{\theta}(\vec{\underline{x}}(t), t)\, \vec{\underline{\nabla}}(\vec{\underline{x}}) \cdot \vec{\underline{v}}(\vec{\underline{x}}(t), t)\right] \mathrm{d}\underline{v}\,. \tag{4.45}$$

Equations (4.43)–(4.45) give

$$\int_{\underline{v}(t)} \left[\frac{\mathrm{D}\underline{\rho}(\vec{\underline{x}}(t), t)}{\mathrm{D}t} + \underline{\rho}(\vec{\underline{x}}(t), t)\, \vec{\underline{\nabla}}(\vec{\underline{x}}) \cdot \vec{\underline{v}}(\vec{\underline{x}}(t), t)\right] \mathrm{d}\underline{v} = 0\,. \tag{4.46}$$

Since the volume $\underline{v}(t)$ of the subsystem $m$ is an arbitrary part of the volume $\underline{v}^{\mathrm{b}}(t)$ of the system $\mathrm{m}^{\mathrm{b}}$, Integrand (4.46) must vanish for every $\underline{v}(t)$ within the volume $\underline{v}^{\mathrm{b}}(t)$ (the details are given in Section 4.7). Thus the term in parentheses vanishes at every point of the material, and the following local form is achieved:

$$\frac{\mathrm{D}\underline{\rho}(\vec{\underline{x}}(t), t)}{\mathrm{D}t} + \underline{\rho}(\vec{\underline{x}}(t), t)\, \vec{\underline{\nabla}}(\vec{\underline{x}}) \cdot \vec{\underline{v}}(\vec{\underline{x}}(t), t) = 0\,, \quad \text{i.e.,} \quad \frac{\mathrm{D}\underline{\rho}}{\mathrm{D}t} + \underline{\rho}\, \vec{\underline{\nabla}} \cdot \vec{\underline{v}} = 0\,. \tag{4.47}$$

Equation (4.47) is the continuity equation. Another form for the continuity equation is derived next. Material Time Derivative (3.19) reads

$$\frac{\mathrm{D}\theta(\vec{\underline{x}}(t), t)}{\mathrm{D}t} = \frac{\partial\theta(\vec{\underline{x}}(t), t)}{\partial t} + \vec{\underline{v}}(\vec{\underline{x}}(t), t) \cdot \vec{\underline{\nabla}}(\vec{\underline{x}})\theta(\vec{\underline{x}}(t), t)\,. \tag{4.48}$$

Substituting $\theta = \rho$ into Derivative (4.48) yields

$$\frac{\mathrm{D}\underline{\rho}}{\mathrm{D}t} = \frac{\partial\underline{\rho}}{\partial t} + \vec{\underline{v}} \cdot \vec{\underline{\nabla}}\underline{\rho}\,. \tag{4.49}$$

Substitution of Derivative $\mathrm{D}\rho/\mathrm{D}t$ obtained from Expression (4.49) into Equation (4.47) yields

$$\frac{\partial \rho}{\partial t} + \underline{\vec{v}} \cdot \underline{\vec{\nabla}} \rho + \rho \underline{\vec{\nabla}} \cdot \underline{\vec{v}} = 0 \quad \Rightarrow \quad \frac{\partial \rho}{\partial t} + \underline{\vec{\nabla}} \cdot (\rho \underline{\vec{v}}) = 0. \tag{4.50}$$

Equation $(4.50)_2$ is another form for the continuity equation.

## 4.9 Axiom of Conservation of Mass for Fluids: Spatial Description

The present section utilizes the axiom of conservation of mass and therefore derives its consequence, which is the continuity equation written in terms of the spatial coordinates $x_i$.

**(a)** First, an approach present in many preliminary textbooks is taken. The derivation is carried out by studying the flow of fluid through a control volume $\mathrm{V}^{\mathrm{cv}}$ that is fixed in space; that is, a spatial description is applied. Instead of studying fluid flow through a parallelepiped, which is usually the case in preliminary textbooks, here the control volume $\mathrm{V}^{\mathrm{cv}}$ has an arbitrary shape.

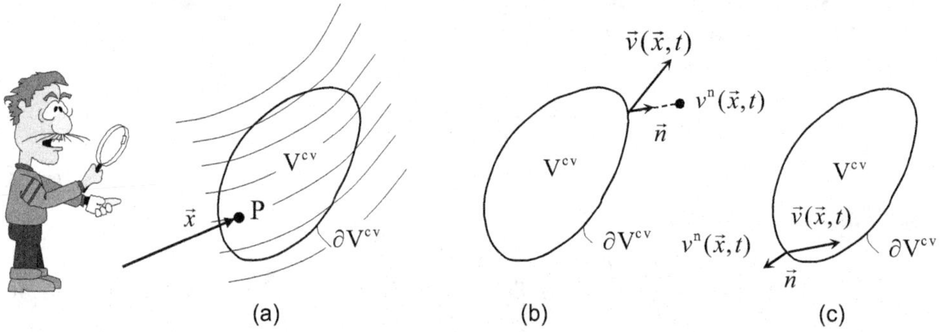


**Figure 4.7** (a) Control volume $\mathrm{V}^{\mathrm{cv}}$ fixed in space within the fluid flow. The normal component of the velocity $v^{\mathrm{n}}$ is (b) positive for outgoing flow and (c) negative for incoming flow. Inspector: Corel images.

Figure 4.7(a) shows an external observer monitoring the flow of fluid through a control volume $\mathrm{V}^{\mathrm{cv}}$ that is fixed in space. The boundary of the control volume is denoted by $\partial\mathrm{V}^{\mathrm{cv}}$. The streamlines show the movement of the fluid flow. The vectors in Figures 4.7(b) and 4.7(c) characterize the velocity of the fluid flow at two points of the boundary $\partial\mathrm{V}^{\mathrm{cv}}$. Figure 4.7(b) illustrates a velocity vector of the fluid $\vec{v}(\vec{x},t)$ on the right side of the control volume $\mathrm{V}^{\mathrm{cv}}$ where the flow of the fluid is directed out from the control volume $\mathrm{V}^{\mathrm{cv}}$. At the point of the boundary $\partial\mathrm{V}^{\mathrm{cv}}$ sketched in Figure 4.7(c), the flow of the fluid is directed into the control volume $\mathrm{V}^{\mathrm{cv}}$. The normal component of the velocity $\vec{v}(\vec{x},t)$ to the boundary $\partial\mathrm{V}^{\mathrm{cv}}$ is denoted by $v^{\mathrm{n}}(\vec{x},t)$. It is obtained as

$$v^{\mathrm{n}}(\vec{x},t) = \vec{v}(\vec{x},t) \cdot \vec{n}\,. \tag{4.51}$$

If the angle between two vectors is acute, the dot product of the vectors is positive, whereas if the angle is blunt, the dot product takes a negative value. Thus the dot product in Expression (4.51) and Figures 4.7(b) and 4.7(c) shows that the normal component of the velocity vector $v^{\mathrm{n}}(\vec{x}, t)$ is positive for outgoing fluid flow and negative for incoming fluid flow.

Based on the preceding evaluation, the net mass flow of the fluid out from the control volume $\mathrm{V}^{\mathrm{cv}}$ is

$$\begin{aligned} \text{Net mass flow out from } \mathrm{V}^{\mathrm{cv}} &= \oint_{\partial \mathrm{V}^{\mathrm{cv}}} \rho(\vec{x}, t)\, v^{\mathrm{n}}(\vec{x}, t)\, \mathrm{d}A \\ &= \oint_{\partial \mathrm{V}^{\mathrm{cv}}} \rho(\vec{x}, t)\, [\vec{v}(\vec{x}, t) \cdot \vec{n}]\, \mathrm{d}A \\ &= \oint_{\partial \mathrm{V}^{\mathrm{cv}}} \vec{n} \cdot [\rho(\vec{x}, t)\, \vec{v}(\vec{x}, t)]\, \mathrm{d}A\,. \end{aligned} \tag{4.52}$$

Simplification of Equation (4.52) made use of Equation (4.51) and the property that the dot product $\cdot$ is commutative. Equation (4.52) expresses that the density of the fluid $\rho(\vec{x}, t)$ and the velocities of the fluid particles $\vec{v}(\vec{x}, t)$ are dependent on position $\vec{x}$ and time $t$. The control volume $\mathrm{V}^{\mathrm{cv}}$, however, is fixed in space and time $t$, and therefore $\mathrm{V}^{\mathrm{cv}}$ is not dependent on the variables $\vec{x}$ and $t$.

Generalized Gauss's Theorem $(2.170)_1$ is

$$\int_V \vec{\nabla} * f\, \mathrm{d}V = \oint_{\partial V} \vec{n} * f\, \mathrm{d}A\,, \tag{4.53}$$

which yields

$$\int_{\mathrm{V}^{\mathrm{cv}}} \vec{\nabla} \cdot (\rho\, \vec{v})\, \mathrm{d}V = \oint_{\partial \mathrm{V}^{\mathrm{cv}}} \vec{n} \cdot (\rho\, \vec{v})\, \mathrm{d}A\,. \tag{4.54}$$

Based on Expression (4.54), Equation (4.52) gives

$$\text{Net mass flow out from } \mathrm{V}^{\mathrm{cv}} = \int_{\mathrm{V}^{\mathrm{cv}}} \vec{\nabla}(\vec{x}) \cdot [\, \rho(\vec{x}, t)\, \vec{v}(\vec{x}, t)\,]\, \mathrm{d}V\,. \tag{4.55}$$

The total mass $m(t)$ of the matter inside the control volume $\mathrm{V}^{\mathrm{cv}}$ is

$$m(t) = \int_{\mathrm{V}^{\mathrm{cv}}} \rho(\vec{x}, t)\, \mathrm{d}V\,. \tag{4.56}$$

Since the density of the fluid $\rho(\vec{x}, t)$ changes with time $t$, also the total mass $m(t)$ of the matter inside the control volume $\mathrm{V}^{\mathrm{cv}}$ varies with time $t$.

The rate (of increase) of the total mass in the control volume $\mathrm{V}^{\mathrm{cv}}$ is

$$\frac{\partial m(t)}{\partial t} = \frac{\partial}{\partial t} \int_{\mathrm{V}^{\mathrm{cv}}} \rho(\vec{x}, t)\, \mathrm{d}V = \int_{\mathrm{V}^{\mathrm{cv}}} \frac{\partial \rho(\vec{x}, t)}{\partial t}\, \mathrm{d}V\,. \tag{4.57}$$

The net mass flow out from the control volume $\mathrm{V^{cv}}$ in Expression (4.55) equals the negative value of the rate of the total mass in Expression (4.57). This is

$$\text{Net mass flow out from } \mathrm{V^{cv}} = -\,\frac{\partial m(t)}{\partial t}\,. \tag{4.58}$$

Substitution of Expressions (4.55) and (4.57) into Equation (4.58) gives

$$\int_{\mathrm{V^{cv}}} \vec{\nabla}(\vec{x})\cdot[\,\rho(\vec{x},t)\,\vec{v}(\vec{x},t)\,]\,\mathrm{d}V = -\int_{\mathrm{V^{cv}}} \frac{\partial \rho(\vec{x},t)}{\partial t}\,\mathrm{d}V\,. \tag{4.59}$$

Equation (4.59) yields

$$\frac{\partial \rho(\vec{x},t)}{\partial t} + \vec{\nabla}(\vec{x})\cdot[\,\rho(\vec{x},t)\,\vec{v}(\vec{x},t)\,] = 0\,. \tag{4.60}$$

Equation (4.60) is the continuity equation. The approach (b) discussed next will provide another form for the continuity equation.

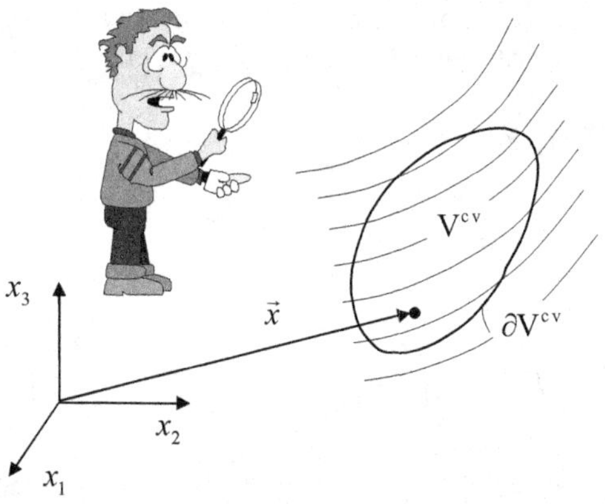


**Figure 4.8** System $\mathrm{m^b}$ loaded by the forces $F_1$ and $F_2$. Inspector: Corel images.

**(b)** Next, the foregoing continuity equation is derived by applying the Reynolds transport theorem.

The axiom of conservation of mass, Axiom (4.24) [**AX 1**], reads

$$\textbf{AX 1} \qquad \frac{\mathrm{D}m}{\mathrm{D}t} = 0\,. \tag{4.61}$$

Form (4.61) does not take a stand on the description. Thus Form (4.31) is valid for solids and for fluids.

When expressed by the current material coordinates $\underline{x}_i(t)$, the mass of the collection of fluid particles is obtained by using the same equation as for calculation of the mass of the entity of a solid material. Thus Definition (4.12) is valid for fluids as well. It reads

$$m := \int_{\underline{v}(t)} \rho(\vec{\underline{x}}(t),t)\,\mathrm{d}\underline{v}\,, \tag{4.62}$$

where $\underline{v}(t)$ is the current volume of the subsystem $m$, where the integration is carried out in the current configuration $\underline{v}^{\mathrm{b}}(t)$ and where the notation $\underline{\rho}(\vec{\underline{x}}(t),t)$ refers to the density in the current configuration $\underline{v}^{\mathrm{b}}(t)$.

Reynolds Transport Theorem (3.158) reads

$$\frac{\mathrm{D}}{\mathrm{D}t}\int_{\underline{v}(t)} \underline{\theta}(\vec{\underline{x}}(t),t)\,\mathrm{d}\underline{v} = \int_{\mathrm{V^{cv}}} \left\{ \frac{\partial \theta(\vec{x},t)}{\partial t} + \vec{\nabla}(\vec{x})\cdot[\,\theta(\vec{x},t)\,\vec{v}(\vec{x},t)\,] \right\} \mathrm{d}V\,. \tag{4.63}$$

Reynolds Transport Theorem (4.63) changes the view from the current material coordinates $\underline{\vec{x}}(t)$ to the in space–fixed spatial coordinates $\vec{x}$. As shown in Figure 4.8, the position vector $\vec{x}$ points to a fixed point in space. Expressions (4.61)–(4.63) give $[\theta(\vec{x},t) = \rho(\vec{x},t)]$

$$\int_{\mathrm{V^{cv}}} \left\{ \frac{\partial \rho(\vec{x},t)}{\partial t} + \vec{\nabla}(\vec{x}) \cdot [\, \rho(\vec{x},t)\, \vec{v}(\vec{x},t)\,] \right\} \mathrm{d}V = 0\,. \tag{4.64}$$

Since the control volume $\mathrm{V^{cv}}$ is an arbitrary part of the volume $\underline{v}^{\mathrm{b}}(t)$ of the system $\mathrm{m^b}$, Integrand (4.64) must vanish for every $\mathrm{V^{cv}}$ within the volume $\underline{v}^{\mathrm{b}}(t)$. Thus the term in square brackets vanishes at every point of the system $\mathrm{m^b}$, and the following local form is achieved (see Section 4.7 for details):

$$\frac{\partial \rho(\vec{x},t)}{\partial t} + \vec{\nabla}(\vec{x}) \cdot [\, \rho(\vec{x},t)\, \vec{v}(\vec{x},t)\,] = 0\,. \quad \text{Everywhere within } \underline{v}^{\mathrm{b}}(t)\,. \tag{4.65}$$

Equation (4.65) is the continuity equation, and it plays an important role in fluid mechanics. In the forthcoming equations, the phrase "everywhere within $\mathrm{m^b}$" is dropped. The phrase "everywhere within $\underline{v}^{\mathrm{b}}(t)$" will henceforth not be repeated.

The equation for Material Time Derivative (3.114) allows another form for the continuity equation to be derived. Derivative (3.114) reads

$$\frac{\mathrm{D}\theta(\vec{x},t)}{\mathrm{D}t} = \frac{\partial \theta(\vec{x},t)}{\partial t} + \vec{v}(\vec{x},t) \cdot \vec{\nabla}(\vec{x})\, \theta(\vec{x},t)\,. \tag{4.66}$$

For $\theta = \rho$, Derivative (4.66) gives

$$\frac{\mathrm{D}\rho(\vec{x},t)}{\mathrm{D}t} = \frac{\partial \rho(\vec{x},t)}{\partial t} + \vec{v}(\vec{x},t) \cdot \vec{\nabla}(\vec{x})\, \rho(\vec{x},t)\,. \tag{4.67}$$

Continuity Equation (4.65) yields

$$\frac{\partial \rho(\vec{x},t)}{\partial t} + \rho(\vec{x},t) \overleftarrow{\nabla}(\vec{x}) \cdot \vec{v}(\vec{x},t) + \rho(\vec{x},t)\, \vec{\nabla}(\vec{x}) \cdot \vec{v}(\vec{x},t) = 0\,. \tag{4.68}$$

Combination of Equations (4.67) and (4.68) gives another form of the continuity equation,

$$\frac{\mathrm{D}\rho(\vec{x},t)}{\mathrm{D}t} + \rho(\vec{x},t)\, \vec{\nabla}(\vec{x}) \cdot \vec{v}(\vec{x},t) = 0\,. \tag{4.69}$$

In a special case in which the material is assumed to be incompressible so that the density $\rho(\vec{x},t)$ remains constant, the continuity equation reduces to

$$\vec{\nabla}(\vec{x}) \cdot \vec{v}(\vec{x},t) = 0\,. \tag{4.70}$$

Result (4.70) holds for a stationary fluid field as well.

## 4.10 Domain Attached Either to Matter or to Space

This section looks at the differences between descriptions for solids and for fluids. Usually, the initial material description is for solids and the spatial description is for fluids.

The basic laws and axioms of continuum thermodynamics and continuum mechanics are expressed in the current configuration $\underline{v}^{\mathrm{b}}(t)$; that is, they have the form of the current material description. Solids are often studied by applying the IG-CtP (initial geometry at current time and position) material description, whereas fluids prefer the spatial description, as shown in Figure 4.9.

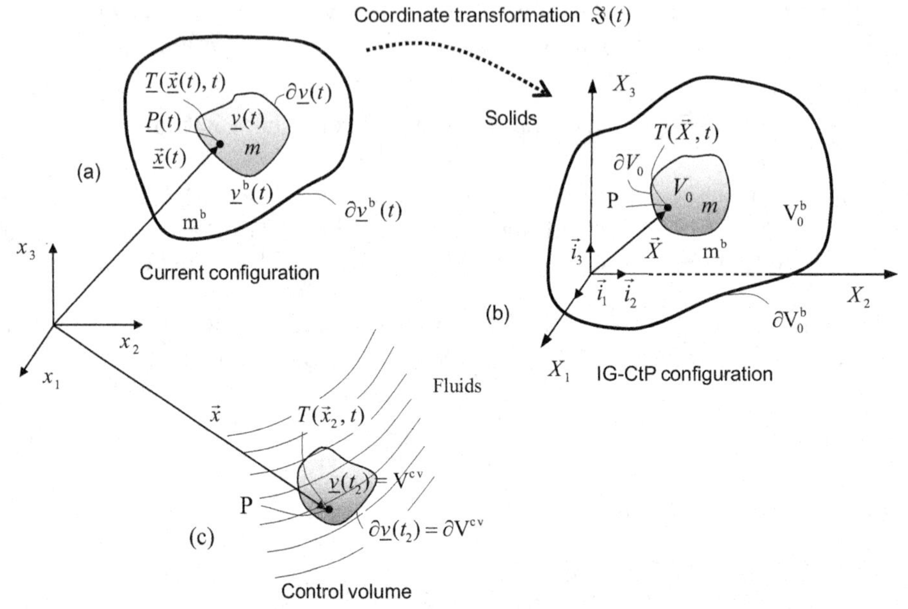


**Figure 4.9** (a) Basic laws and axioms are for current configuration $\underline{v}^{\mathrm{b}}(t)$. (b) Solids are described in the IG-CtP configuration $\mathrm{V}_0^{\mathrm{b}}$ and (c) fluids are described by the control volume $\mathrm{V}^{\mathrm{cv}}$.

Figure 4.9 shows that the absolute temperature $T$, for example, is expressed differently in the current configuration $\underline{v}^{\mathrm{b}}(t)$, the IG-CtP configuration $\mathrm{V}_0^{\mathrm{b}}$ and the control volume $\mathrm{V}^{\mathrm{cv}}$. These quantities are $\underline{T}(\underline{\vec{x}}(t), t)$, $T(\vec{X}, t)$, and $T(\vec{x}, t)$. According to Figure 4.9, the coordinate transformation from the current material coordinates $\underline{x}_i(t)$ to the initial material coordinates $X_i$ is possible at any time, but the subsystem of the current configuration $\underline{v}(t)$ equals the control volume $\mathrm{V}^{\mathrm{cv}}$ only at a certain moment, denoted here by $t_2$. This means that the point P is always within the volume $V_0$ but only temporarily within the control volume $\mathrm{V}^{\mathrm{cv}}$. Section 4.12 gives a detailed study of the spatial description and the control volume $\mathrm{V}^{\mathrm{cv}}$. The coordinate transformation between the coordinates $\underline{x}_i(t)$ and $X_i$ is denoted by $\Im(t)$. As Figure 4.9 and the notation $\Im(t)$ show, the coordinate transformation is time dependent; that is, every current configuration $\underline{v}^{\mathrm{b}}(t)$ has its own coordinate transformation $\Im(t)$.

## 4.11 Transformation of Current to IG-CtP Material Description

The basic laws and axioms have the form of the current material description, but it is common practice to write the field equations for solids in terms of the initial material coordinates $X_i$. In order to do that, a transformation equation from the current material description to the IG-CtP material description is needed. This section derives this expression, which is needed in step I of Section 4.5.

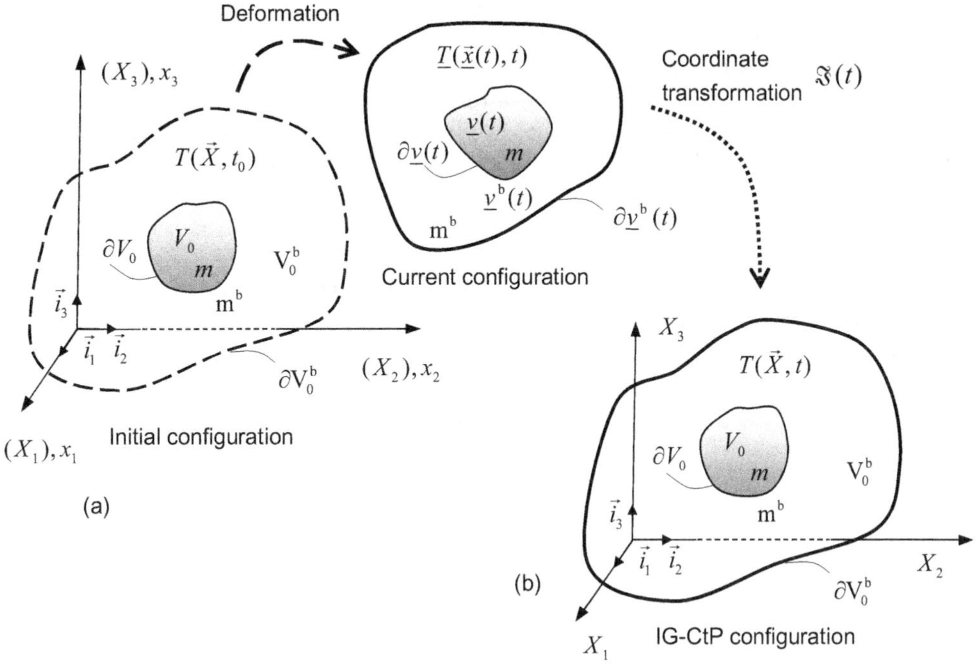


**Figure 4.10** (a) Subsystem $m$ in the initial configuration $\mathrm{V}_0^{\mathrm{b}}$ and in the current configuration $\underline{v}^{\mathrm{b}}(t)$. (b) Subsystem $m$ in the IG-CtP configuration $\mathrm{V}_0^{\mathrm{b}}$.

Figure 4.10 shows the system (body) $\mathrm{m}^{\mathrm{b}}$ (a) in the initial configuration $\mathrm{V}_0^{\mathrm{b}}$ and in the current configuration $\underline{v}^{\mathrm{b}}(t)$ and (b) in the IG-CtP configuration $\mathrm{V}_0^{\mathrm{b}}$. An arbitrary part of the system $\mathrm{m}^{\mathrm{b}}$ is called the subsystem and is denoted by $m$. The volume of the subsystem $m$ in the current configuration is $\underline{v}(t)$, and in the IG-CtP configuration $\mathrm{V}_0^{\mathrm{b}}$, it is $V_0$, as shown in Figures 4.10(a) and 4.10(b). Thus the volume of the subsystem $m$ varies with time $t$, whereas the mass $m$ remains constant.

Basic laws and axioms are the fundamental equations for continuum thermodynamics. They are written for the subsystem $m$ in integral form in the current configuration $\underline{v}^{\mathrm{b}}(t)$; that is, they are expressed as the current material coordinates $\underline{x}_i(t)$. Integration is performed over the mass of the subsystem $m$ or around the subsystem $m$. Integrals over the mass $m$ are then changed to volume integrals by applying the expression $\mathrm{d}m = \underline{\rho}(\cdot, t)\mathrm{d}\underline{v}(t)$. Time $t$ is added here to the term $\mathrm{d}\underline{v}(t)$ to show that, although the mass $m$ does not change, its volume $\underline{v}(t)$ changes with time $t$. Often, basic laws and axioms are written directly as volume integrals. This will cause confusion since the basic laws and axioms are for the nonchanging entity of matter and not for volume.

Thus the basic laws and axioms have a form in which the integrands are expressed in terms of the current material coordinates $\underline{x}_i(t)$ and in which the integration is changed to the volume or area integral either over the volume $\underline{v}(t)$ or around it. Since the current material coordinates $\underline{x}_i(t)$ are used, and since the domain of investigation is attached to the matter, basic laws and axioms are written for the current material description. In order to obtain equations for the IG-CtP material description, the integrands have to be expressed in terms of the material coordinates $X_i$. This leads to the volume and area integrals over the volume $V_0$, which is the volume of the subsystem $m$ in the initial geometry $\mathrm{V}_0^{\mathrm{b}}$. Since solid mechanics applies the IG-CtP material description, the concept for the following investigation involves a transformation from the current material coordinates $\underline{x}_i(t)$ to the initial material coordinates $X_i$. This will lead to equations where the current values of the quantities are studied in the IG-CtP configuration $\mathrm{V}_0^{\mathrm{b}}$.

After changing the integrals over mass $m$ into the volume integrals, some of the terms in the expressions for the basic laws and axioms of continuum thermodynamics have the following form:

$$\int_{\underline{v}(t)} \underline{\rho}(\vec{\underline{x}}(t),t)\, \underline{\theta}(\vec{\underline{x}}(t),t)\, \mathrm{d}\underline{v}\,. \tag{4.71}$$

When the coordinates $\underline{x}_i(t)$ are replaced by the coordinates $X_i$, the following is achieved (see, e.g., Boas [11], pp. 262 and 263):

$$\int_{\underline{v}(t)} \underline{\rho}(\vec{\underline{x}}(t),t)\, \underline{\theta}(\vec{\underline{x}}(t),t)\, \mathrm{d}\underline{v} = \int_{V_0} \rho(\vec{X},t)\, \theta(\vec{X},t)\, |\, J(\vec{X},t)\, |\, \mathrm{d}V\,. \tag{4.72}$$

In Expression (4.72), the notation $|J(\vec{X},t)|$ stands for the absolute value of the Jacobian determinant between the initial material coordinates $X_i$ and the current material coordinates $\underline{x}_i(t)$. According to Holzapfel ([29], p. 74), the Jacobian determinant is positive, and therefore Equation (4.72) takes the form

$$\int_{\underline{v}(t)} \underline{\rho}(\vec{\underline{x}}(t),t)\, \underline{\theta}(\vec{\underline{x}}(t),t)\, \mathrm{d}\underline{v} = \int_{V_0} \rho(\vec{X},t)\, \theta(\vec{X},t)\, J(\vec{X},t)\, \mathrm{d}V\,. \tag{4.73}$$

Equation (4.37) gives the Jacobian determinant $J(\vec{X},t)$.

Based on the axiom of conservation of mass, the following equation was derived in Expression (4.40):

$$\rho(\vec{X},t)\, J(\vec{X},t) - \rho_0(\vec{X},t_0) = 0\,. \tag{4.74}$$

Substituting Equation (4.74) into Integrand (4.73) gives

$$\int_{\underline{v}(t)} \underline{\rho}(\vec{\underline{x}}(t),t)\, \underline{\theta}(\vec{\underline{x}}(t),t)\, \mathrm{d}\underline{v} = \int_{V_0} \rho_0(\vec{X})\, \theta(\vec{X},t)\, \mathrm{d}V\,. \tag{4.75}$$

The variable $t_0$ in the latter term of Expression (4.74) is given to emphasize that the quantity $\rho_0(\vec{X}, t_0)$ is the density of the material in the initial configuration $\mathrm{V}_0^{\mathrm{b}}$. Thus, on the right side of Expression (4.75), the quantity $\rho_0(\vec{X})$ is the density of the material before loading was applied, whereas the quantity $\theta(\vec{X}, t)$ is the current value, but it is expressed in the initial material coordinates $X_i$. This means that the quantity $\theta(\vec{X}, t)$ represents the value of $\theta$ at current time $t$ (state) in the initial geometry. The density $\rho_0(\vec{X})$ is independent of time $t$, whereas the quantity $\theta(\vec{X}, t)$ depends on time $t$. The quantity $\theta(\vec{X}, t)$ can, for example, be the absolute temperature $T(\vec{X}, t)$. The preceding means that on the right side of Expression (4.75), the integration is carried out over the volume of the initial geometry.

It is important to note that the IG-CtP material description does not assume that during deformation the value of the density of material $\rho$ does not vary with time, that is, that rigid body motion is studied. Furthermore, small deformations are not assumed, but Expression (4.75) is valid for finite deformation theory. The presence of the initial density $\rho_0$ in Integrand (4.75) originates from a mathematical derivation. If the density of material $\rho$ is of interest, it can be calculated from the value of the strain tensor $\boldsymbol{\varepsilon}$.

## 4.12 Transformation between Current Material and Spatial Descriptions

The basic laws and axioms have the form of the current material description, but it is common practice to write the field equations for fluids in terms of the spatial coordinates $x_i$. In order to do that, a transformation equation from the current material description to the spatial description is needed. This section derives this expression.

The value of a quantity $\theta$ is independent of the coordinates. For the present coordinates, it means that

$$\underline{\theta}(\underline{\vec{x}}(t), t) = \theta(\vec{x}, t) \,. \tag{4.76}$$

Figure 4.11 shows the difference between the current material description and the spatial description. The position vector of the current material description $\underline{\vec{x}}(t)$ shows the current position of point $\underline{P}(t)$, whereas the position vector of the spatial description $\vec{x}$ points to a fixed point in space. It is assumed that at time $t_2$, the fluid particle P is at the point $\vec{x}$. Furthermore, the volume of the subsystem $\underline{v}(t)$ equals that of the control volume $\mathrm{V}^{\mathrm{cv}}$, that is, $\underline{v}(t_2) = \mathrm{V}^{\mathrm{cv}}$, as shown in Figure 4.11. If the control volume $\mathrm{V}^{\mathrm{cv}}$ is not any specific volume in space, it is possible to extend the previous concept and to say that for every volume of the subsystem $\underline{v}(t)$ at every moment $t$, there exists an equal control volume $\mathrm{V}^{\mathrm{cv}}$. This is $\underline{v}(t) = \mathrm{V}^{\mathrm{cv}}$. The inverse holds as well. If there is a particular control volume $\mathrm{V}^{\mathrm{cv}}$, it is possible to find a subsystem $\underline{v}(t)$ that equals the control volume $\mathrm{V}^{\mathrm{cv}}$ at time $t_3$, that is, $\underline{v}(t_3) = \mathrm{V}^{\mathrm{cv}}$. This can be repeated for any moment of time $t$, and therefore, in general, the following holds: $\underline{v}(t) = \mathrm{V}^{\mathrm{cv}}$. Thus, Equality (4.76) leads to

$$\int_{\underline{v}(t)} \underline{\theta}(\underline{\vec{x}}(t), t)\, \mathrm{d}\underline{v} = \int_{\mathrm{V}^{\mathrm{cv}}} \theta(\vec{x}, t)\, \mathrm{d}V\,. \tag{4.77}$$

Therefore the basic laws and axioms, which have the current material character, i.e., the form on the left side of Expression (4.77), can be changed to the spatial description, i.e., to the form on the right side of Expression (4.77), simply by replacing the volume integration over $\underline{v}(t)$ with the volume integration over $\mathrm{V}^{\mathrm{cv}}$ and replacing the coordinates $\underline{x}_i(t)$ with the coordinates $x_i$. This is an important concept for fluid mechanics. It is worth noting that the role of the notation $\theta$ in Expression (4.77) is to show that the integrands describe the same physical reality, although the coordinates differ. Thus the functional appearances of the quantities $\theta$ on the left and right sides of Expression (4.77) differ.

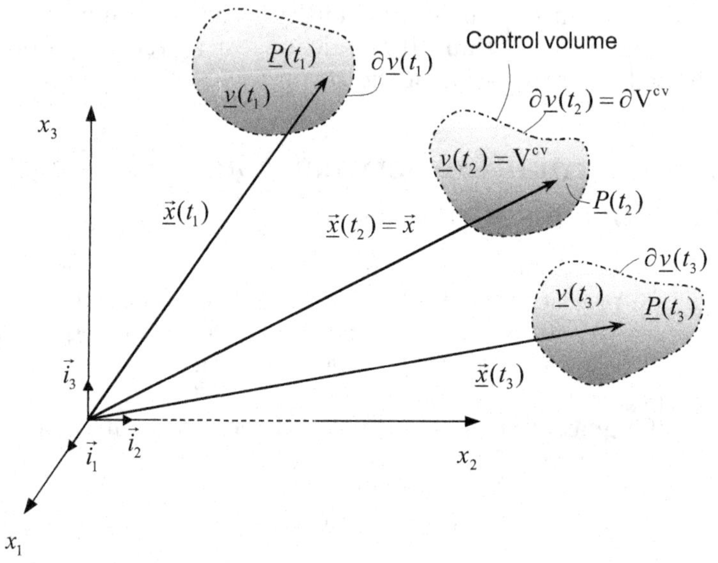


**Figure 4.11** Current material description and the spatial description.

However, not all the terms in the basic laws and axioms have the form of the left side of Expression (4.77), but some terms have a material time operator $\mathrm{D}/\mathrm{D}t$ in front of the integral sign. Thus these terms have the form

$$\frac{\mathrm{D}}{\mathrm{D}t}\int_{\underline{v}(t)} \underline{\theta}(\underline{\vec{x}}(t), t)\, \mathrm{d}\underline{v} \quad \text{and} \quad \frac{\mathrm{D}}{\mathrm{D}t}\int_{\underline{v}(t)} \underline{\rho}(\underline{\vec{x}}(t), t)\, \underline{\theta}(\underline{\vec{x}}(t), t)\, \mathrm{d}\underline{v}\,. \tag{4.78}$$

The main difference between the two expressions in Expression (4.78) lies in $\underline{\rho}(\underline{\vec{x}}(t), t)$ being a part of the integrand in the second expression while it is not in the first expression. The act of moving the operator $\mathrm{D}/\mathrm{D}t$ inside (i.e., the right side of) the integral operator is called the Reynolds transport theorem. The reason for introducing the material time derivative operator $\mathrm{D}/\mathrm{D}t$ is to study how a material experiences the change in its state. As shown in Figure 4.11, the fluid particles move through the control volume $V^{\mathrm{cv}}$ present on the right side of Equation (4.77), whereas the volume $\underline{v}(t)$ present in Term $(4.78)_1$ includes all the time the same material points. Therefore the derivation of the Reynolds transport theorem for the spatial description has its own special character compared to that for the current material description. Unfortunately, in both cases, the Reynolds transport theorem takes the same appearance, especially when the coordinates are not shown. Sometimes this has led to an incorrect understanding of the physics behind the mathematical derivation. The role of the following section is to derive the Reynolds transport theorem both for material descriptions and for the spatial description.

## 4.13 Material Time Derivative of a Volume Integral

The present section studies the material derivative of volume integrals and derives the Reynolds transport theorem for the terms present in the basic laws and axioms. The IG-CtP material description, the current material description and the spatial description, are considered. The focus of this section is on step II of Section 4.5.

When deriving local forms for the basic laws and axioms of continuum thermodynamics, the following material time derivatives of volume integrals are studied:

$$\frac{\mathrm{D}}{\mathrm{D}t}\int_{V_0}\rho_0(\vec{X})\,\theta(\vec{X},t)\,\mathrm{d}V \qquad \frac{\mathrm{D}}{\mathrm{D}t}\int_{\underline{v}(t)}\rho(\underline{\vec{x}}(t),t)\,\theta(\underline{\vec{x}}(t),t)\,\mathrm{d}\underline{v}\,. \tag{4.79}$$

Expressions (4.79) are worth investigating in more detail.

In continuum thermodynamics, when field equations are derived, the material time derivative operator $\mathrm{D}/\mathrm{D}t$ present in Expressions (4.79) has to be moved to the right side of the integral sign $\int$ to act on the integrand. When the Eulerian description is used, this "movement" is called the Reynolds transport theorem. For simplification, the same word is used here with the material descriptions.

The described movement is easier for the material description; therefore it is studied first. Since the integration domain $V_0$ does not depend on time $t$, the derivative operator $\mathrm{D}/\mathrm{D}t$ and the integral operator $\int$ are commutative; that is, the order of differentiation and integration can be interchanged without changing the meaning, and the following can be obtained:

$$\frac{\mathrm{D}}{\mathrm{D}t}\int_{V_0}\rho_0(\vec{X})\,\theta(\vec{X},t)\,\mathrm{d}V = \int_{V_0}\rho_0(\vec{X})\,\frac{\mathrm{D}\theta(\vec{X},t)}{\mathrm{D}t}\,\mathrm{d}V\,. \tag{4.80}$$

Since $\underline{\rho}\,\mathrm{d}\underline{v} = \mathrm{d}m$, where $\mathrm{d}m$ is the differential mass element, the following holds for the current material description:

$$\frac{\mathrm{D}}{\mathrm{D}t}\int_{\underline{v}(t)} \underline{\rho}(\underline{\vec{x}}(t),t)\,\underline{\theta}(\underline{\vec{x}}(t),t)\,\mathrm{d}\underline{v} = \frac{\mathrm{D}}{\mathrm{D}t}\int_{m} \underline{\theta}(\underline{\vec{x}}(t),t)\,\mathrm{d}m = \int_{m} \frac{\mathrm{D}\underline{\theta}(\underline{\vec{x}}(t),t)}{\mathrm{D}t}\,\mathrm{d}m$$

$$= \int_{\underline{v}(t)} \underline{\rho}(\underline{\vec{x}}(t),t)\,\frac{\mathrm{D}\underline{\theta}(\underline{\vec{x}}(t),t)}{\mathrm{D}t}\,\mathrm{d}\underline{v}\,. \tag{4.81}$$

Manipulation (4.81) exploits the fact that the derivative operator $\mathrm{D}/\mathrm{D}t$ and the integral $\int$ are commutative, since the mass of the subsystem $m$ does not depend on time $t$. Result (4.81) is called the Reynolds transport theorem.

Reynolds Transport Theorem (3.158) reads

$$\frac{\mathrm{D}}{\mathrm{D}t}\int_{\underline{v}(t)} \underline{\theta}(\underline{\vec{x}}(t),t)\,\mathrm{d}\underline{v} = \int_{\mathrm{V_{cv}}} \left\{ \frac{\partial\theta(\vec{x},t)}{\partial t} + \vec{\nabla}(\vec{x})\cdot[\,\theta(\vec{x},t)\,\vec{v}(\vec{x},t)\,] \right\} \mathrm{d}V\,. \tag{4.82}$$

Theorem (4.82) was already applied when the continuity equation in Section 4.9 was derived. Expression (4.82) or similar to it is often (see, e.g., Malvern [57], p. 211, or White [116], p. 90) called the Reynolds transport theorem.

The function $\theta$ in Expression (4.82) is replaced by the term $\rho\,\theta$, and we arrive at the following:

$$\frac{\mathrm{D}}{\mathrm{D}t}\int_{\underline{v}(t)} \underline{\rho}\,\underline{\theta}\,\mathrm{d}\underline{v} = \int_{\mathrm{V_{cv}}} \left[ \frac{\partial[\,\rho(\vec{x},t)\,\theta(\vec{x},t)\,]}{\partial t} + \vec{\nabla}(\vec{x})\cdot[\,\rho(\vec{x},t)\,\theta(\vec{x},t)\,\vec{v}(\vec{x},t)\,] \right] \mathrm{d}V\,, \tag{4.83}$$

where $\underline{\rho} = \rho\,(\underline{\vec{x}}(t),t)$ and $\underline{\theta} = \theta\,(\underline{\vec{x}}(t),t)$.

The terms of the integrand on the right side of Expression (4.83) can be expressed as follows:

$$\frac{\partial[\,\rho(\vec{x},t)\,\theta(\vec{x},t)\,]}{\partial t} = \frac{\partial[\,\rho\,\theta\,]}{\partial t} = \frac{\partial\rho}{\partial t}\,\theta + \rho\,\frac{\partial\theta}{\partial t} \tag{4.84}$$

and

$$\vec{\nabla}(\vec{x})\cdot[\,\rho(\vec{x},t)\,\theta(\vec{x},t)\,\vec{v}(\vec{x},t)\,] = \vec{\nabla}\cdot[\,\theta\,(\rho\,\vec{v})\,] = \rho\,\vec{v}\cdot\vec{\nabla}\,\theta + \theta\,\vec{\nabla}\cdot(\rho\,\vec{v})\,. \tag{4.85}$$

Substitution of Terms (4.84) and (4.85) into the integrands on the right side of Expression (4.83) yields

$$\frac{\mathrm{D}}{\mathrm{D}t}\int_{\underline{v}(t)} \underline{\rho}\,\underline{\theta}\,\mathrm{d}\underline{v} = \int_{\mathrm{V_{cv}}} \left\{ \frac{\partial[\,\rho(\vec{x},t)\,\theta(\vec{x},t)\,]}{\partial t} + \vec{\nabla}(\vec{x})\cdot[\,\rho(\vec{x},t)\,\theta(\vec{x},t)\,\vec{v}(\vec{x},t)\,] \right\} \mathrm{d}V$$

$$= \int_{\mathrm{V_{cv}}} \left\{ \rho\left[\frac{\partial\theta}{\partial t} + \vec{v}\cdot\vec{\nabla}\theta\right] + \theta\left[\frac{\partial\rho}{\partial t} + \vec{\nabla}\cdot(\rho\,\vec{v})\right] \right\} \mathrm{d}V\,. \tag{4.86}$$

Continuity Equation (4.65) and the equation for the material derivative of the function $\theta$, Expression (3.114), are recalled. They are

$$\frac{\partial\rho(\vec{x},t)}{\partial t} + \vec{\nabla}(\vec{x})\cdot[\,\rho(\vec{x},t)\,\vec{v}(\vec{x},t)\,] = 0 \tag{4.87}$$

and

$$\frac{\mathrm{D}\theta(\vec{x},t)}{\mathrm{D}t} = \frac{\partial\theta(\vec{x},t)}{\partial t} + \vec{v}(\vec{x},t)\cdot\vec{\nabla}(\vec{x})\,\theta(\vec{x},t)\,. \tag{4.88}$$

According to Continuity Equation (4.87), the second square-bracket term on the second line of Expression (4.86) vanishes. According to Expression (4.88), the first square-bracket term on the second line of Expression (4.86) is the material derivative of the function $\theta$. Thus Expression (4.86) reduces to

$$\frac{\mathrm{D}}{\mathrm{D}t}\int_{\underline{v}(t)} \underline{\rho}(\underline{\vec{x}}(t),t)\,\underline{\theta}(\underline{\vec{x}}(t),t)\,\mathrm{d}\underline{v} = \int_{\mathrm{V^{cv}}} \rho(\vec{x},t)\,\frac{\mathrm{D}\theta(\vec{x},t)}{\mathrm{D}t}\,\mathrm{d}V\,. \tag{4.89}$$

Expression (4.89) is the Reynolds transport theorem for fluid mechanics.

It is worth noting that Reynolds Transport Theorem (3.158) and Reynolds Transport Theorems (4.81) and (4.89) assume the conservation of mass. The time-independent constant mass $m$ in Expression (4.81) and the missing "mass generation term" in the derivation of the Reynolds transport theorem (see Section 3.10) indicate that.

## 4.14 Various Stress Measures

This section introduces stress measures for the current material description, for the IG-CtP material description and for the spatial description. Stress measures are needed in step I of Section 4.5.

The Cauchy stress theorem defines the Cauchy stress tensor $\underline{\boldsymbol{\sigma}}^{\mathrm{c}}$ as follows:

$$\underline{\vec{n}}(\underline{\vec{x}}(t))\cdot\underline{\boldsymbol{\sigma}}^{\mathrm{c}}(\underline{\vec{x}}(t),t) := \underline{\vec{t}}(\underline{\vec{x}}(t),t,\underline{\vec{n}}) \qquad \text{or} \qquad \underline{\vec{n}}\cdot\underline{\boldsymbol{\sigma}}^{\mathrm{c}} := \underline{\vec{t}}\,, \tag{4.90}$$

where $\underline{\vec{n}}(\underline{\vec{x}}(t))$ is the outward unit vector of the subsystem $m$ in the current configuration $\underline{v}^{\mathrm{b}}(t)$. As Figure 4.12 shows, the Cauchy stress tensor $\underline{\boldsymbol{\sigma}}^{\mathrm{c}}$ is defined in the current configuration $\underline{v}^{\mathrm{b}}(t)$.

Besides the Cauchy stress tensor $\underline{\boldsymbol{\sigma}}^{\mathrm{c}}$, Figure 4.12 provides elements for the definition of the first Piola–Kirchhoff stress tensor $\boldsymbol{\sigma}^{\square}(\vec{X},t)$. It is defined in the IG-CtP configuration $\mathrm{V}_0^{\mathrm{b}}$ as follows:

$$\vec{N}(\vec{X})\cdot\boldsymbol{\sigma}^{\square}(\vec{X},t) := \vec{T}(\vec{X},t,\vec{N})\,, \tag{4.91}$$

where $\vec{N}(\vec{X})$ is the outward unit vector of the subsystem $m$ and $\vec{T}(\vec{X},t,\vec{N})$ is the traction vector in the IG-CtP configuration $\mathrm{V}_0^{\mathrm{b}}$. There are two important messages related to Figure 4.12, as explained next.

First, the differential resultant forces $\mathrm{d}\underline{\vec{F}}$ and $\mathrm{d}\vec{F}$ of the traction vectors $\underline{\vec{t}}$ and $\vec{T}$ are equal, since the differential force field $\mathrm{d}\vec{F}$ (see Figure 4.13) is independent of the coordinates. This is

$$\underline{\vec{t}}(\underline{\vec{x}}(t),t,\underline{\vec{n}})\,\mathrm{d}\underline{a} = \mathrm{d}\underline{\vec{F}} = \mathrm{d}\vec{F} = \vec{T}(\vec{X},t,\vec{N})\,\mathrm{d}A\,. \tag{4.92}$$

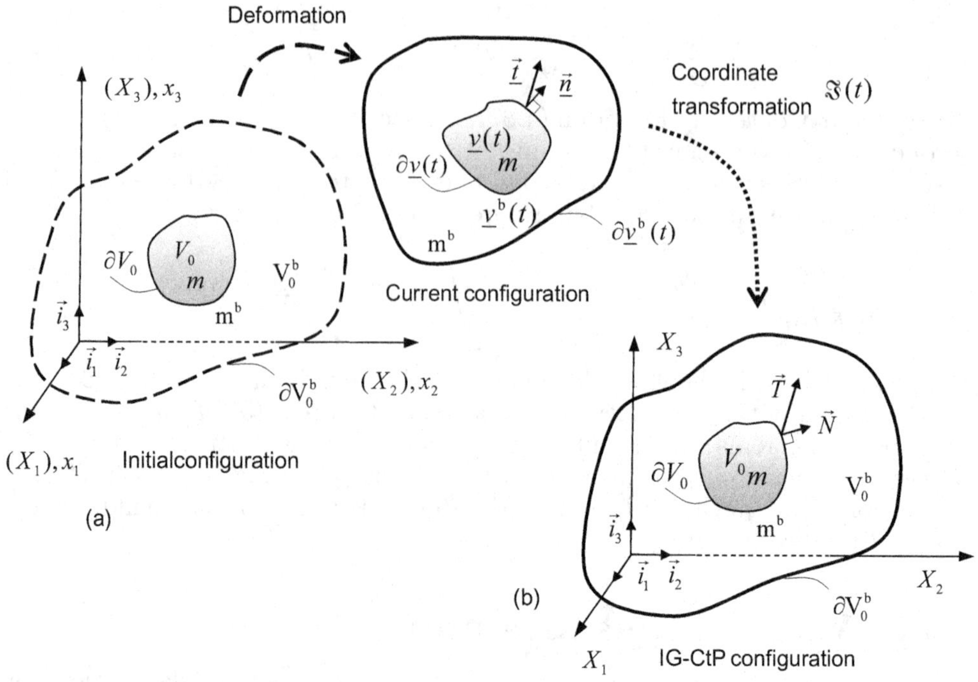


**Figure 4.12** Definitions for the Cauchy stress tensor $\underline{\boldsymbol{\sigma}}^{\mathrm{c}}$ and for the first Piola–Kirchhoff stress tensor $\boldsymbol{\sigma}^{\square}(\vec{X}, t)$.

As Expression (4.92) shows, the traction vectors $\underline{\vec{t}}$ and $\vec{T}$ are parallel but different in length, as shown also in Figure 4.12.

Second, the outward unit vectors to the subsystem $m$ [i.e., $\underline{\vec{n}}(\underline{\vec{x}}(t))$ and $\vec{N}(\vec{X})$] are different in the current configuration $\underline{v}^{\mathrm{b}}(t)$ and in the initial geometry at current time and position (IG-CtP configuration) $\mathrm{V}_0^{\mathrm{b}}$. Therefore, the Cauchy stress tensor $\underline{\boldsymbol{\sigma}}^{\mathrm{c}}$ and the first Piola–Kirchhoff stress tensor $\boldsymbol{\sigma}^{\square}(\vec{X}, t)$ are different, as Definitions (4.90) and (4.91) indicate.

As already discussed, the term *surface* in the (surface) traction vector $\underline{\vec{t}}$ does not necessarily mean that the vector $\underline{\vec{t}}$ acts on the surface of the system (body) $\partial\underline{v}^{\mathrm{b}}(t)$, but it usually acts on the surface of the subsystem $\partial\underline{v}(t)$. Since the subsystem $m$ is arbitrarily situated in the system $\mathrm{m}^{\mathrm{b}}$, the surfaces of the subsystems can lie anywhere within the system $\mathrm{m}^{\mathrm{b}}$, and therefore the Cauchy stress tensor $\underline{\boldsymbol{\sigma}}^{\mathrm{c}}(\underline{\vec{x}}(t), t)$ and the first Piola–Kirchhoff stress tensor $\boldsymbol{\sigma}^{\square}(\vec{X}, t)$ are defined everywhere in the system $\mathrm{m}^{\mathrm{b}}$ by Definitions (4.90) and (4.91). Thus, the present book proposes the term *traction vector* $\underline{\vec{t}}$.

Definitions (4.90) and (4.91) are necessary conditions in continuum mechanics, since they make it possible to convert, for example, global forms of the law of balance of momentum and the law of balance of moment of momentum to their local forms. This is shown, for example, in Section 4.15.

In Figure 4.13, there is a subsystem $m$ at time $t$ expressed (a) in the current material coordinates $\underline{x}_i(t)$ and (b) in the initial material coordinates $X_i$. Thus the piece of material in both figures is the same, and the moment of time $t$ is equal, but the coordinates differ. Due to the different coordinates, the shapes of the subsystem $m$ differ in Figures 4.13(a) and 4.13(b). The shape of the subsystem $m$ in Figure 4.13(a) changes with deformation, whereas in Figure 4.13(b), it takes its original form. Therefore the differential surface element $\mathrm{d}\underline{a}(\underline{\vec{x}}(t))$ and the outward unit normal $\underline{\vec{n}}(\underline{\vec{x}}(t))$ depend on time $t$. However, in the IG-CtP material description, the differential surface element $\mathrm{d}A(\vec{X})$ and the outward unit normal $\vec{N}(\vec{X})$ are time-independent.

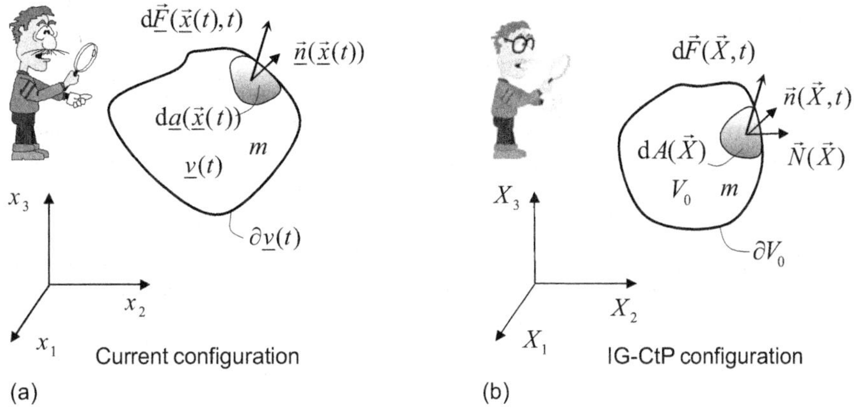


**Figure 4.13** Subdomain m expressed (a) in the current material coordinates $\underline{x}_i(t)$ and (b) in the initial material coordinates $X_i$. Inspector: Corel images. Special glasses added by Joni Hautala.

In Figure 4.13(a), the differential force vector $\mathrm{d}\vec{F}$ is a differential resultant force of the traction vector $\underline{\vec{t}}(\underline{\vec{x}}(t), t, \underline{\vec{n}})$ acting on the differential surface element $\mathrm{d}\underline{a}(\underline{\vec{x}}(t))$ when expressed in the current material coordinates $\underline{x}_i(t)$. As already stated, since the direction and the value of a force do not change in coordinate transformation, the differential resultant force vector $\mathrm{d}\vec{F}$ takes the same value and direction when expressed by the initial material coordinates $X_i$, as Figure 4.13 shows. When expressed by the initial material coordinates $X_i$, the differential force vector $\mathrm{d}\vec{F}$ is a differential resultant force of the traction vector $\vec{T}(\vec{X}, t, \vec{N})$ acting on the differential surface element $\mathrm{d}A$. The resultant force vector $\vec{F}$ is obtained by integration over the area, where the traction vectors $\underline{\vec{t}}(\underline{\vec{x}}(t), t, \underline{\vec{n}})$ and $\vec{T}(\vec{X}, t, \vec{N})$ are acting. This reads

$$\oint_{\partial\underline{v}(t)} \underline{\vec{t}}(\underline{\vec{x}}(t), t, \underline{\vec{n}})\,\mathrm{d}\underline{a} = \underline{\vec{F}} = \vec{F} = \oint_{\partial V_0} \vec{T}(\vec{X}, t, \vec{N})\,\mathrm{d}A. \tag{4.93}$$

Substitution of Definitions (4.90) and (4.91) into Expression (4.90) yields

$$\oint_{\partial\underline{v}^{\mathrm{b}}(t)} \underline{\vec{n}}(\underline{\vec{x}}(t)) \cdot \underline{\boldsymbol{\sigma}}^{\mathrm{c}}(\underline{\vec{x}}(t), t)\,\mathrm{d}\underline{a} = \underline{\vec{F}} = \vec{F} = \oint_{\partial V_0} \vec{N}(\vec{X}) \cdot \boldsymbol{\sigma}^{\square}(\vec{X}, t)\,\mathrm{d}A. \tag{4.94}$$

Based on Definitions (4.90) and (4.91) and Figures 4.13, the Cauchy stress tensor $\underline{\boldsymbol{\sigma}}^{\mathrm{c}}(\underline{\vec{x}}(t), t)$ is defined as a stress measure in which the true force $\underline{\vec{F}}$ acts on the true area $\underline{a}$, whereas the first Piola–Kirchhoff stress tensor $\boldsymbol{\sigma}^{\square}(\vec{X}, t)$ is defined as a stress measure in which the true force $\vec{F}$ acts on the nominal area $A$. Therefore the Cauchy stress tensor $\underline{\boldsymbol{\sigma}}^{\mathrm{c}}(\underline{\vec{x}}(t), t)$ is a true stress and the first Piola–Kirchhoff stress tensor $\boldsymbol{\sigma}^{\square}(\vec{X}, t)$ is a nominal stress. The preceding stress measures belong to the finite deformation theory. When small deformations are studied, the difference between these two stress measures vanishes and can be replaced by the stress tensor $\boldsymbol{\sigma}$, as is done in this book. More information on the stress measures for finite deformation theory is given, for example, by Reddy ([84], Chapter 4).

The inspector with the special glasses in Figure 4.13 sees the geometry of the system in the current configuration $\underline{v}^{\mathrm{b}}(t)$ as it was in the initial configuration $\mathrm{V}_0^{\mathrm{b}}$. This is only an illusion, since the real position of the system is at the current position in the current configuration $\underline{v}^{\mathrm{b}}(t)$. To make this clear, the phrase "initial geometry at the current time and position," the IG-CtP configuration, was introduced.

The foregoing stress measures are for solids. For fluids, no initial configuration or finite deformation theory exists. Thus, the definition for the stress measure is carried out in the control volume $\mathrm{V}^{\mathrm{cv}}$. The fluid stress tensor $\boldsymbol{\sigma}^{\mathrm{f}}$ is defined by

$$\vec{n}(\vec{x}, t) \cdot \boldsymbol{\sigma}^{\mathrm{f}}(\vec{x}, t) := \vec{t}(\vec{x}, t, \vec{n}), \qquad \text{or} \qquad \vec{n} \cdot \boldsymbol{\sigma}^{\mathrm{f}} := \vec{t}. \tag{4.95}$$

## 4.15 Law of Balance of Momentum for Solids: IG-CtP Material Description

This section derives equations of motion, which are local forms of the law of balance of momentum. The initial material coordinates $X_i$ are used. The next derivation follows the steps given in Section 4.5.

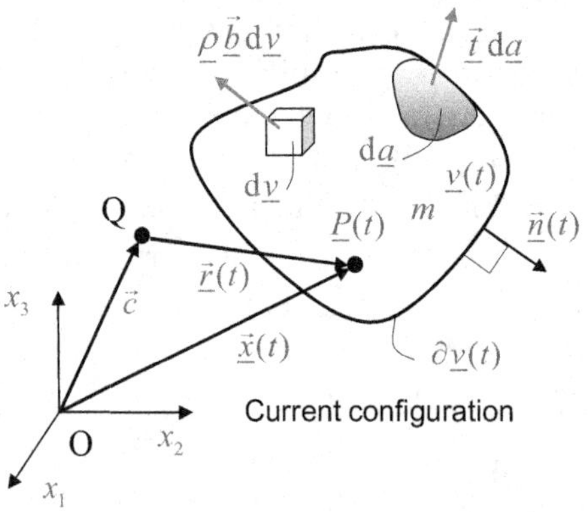


**Figure 4.14** Forces acting on an arbitrary subsystem $m$.

Figure 4.14 shows the differential forces $\underline{\vec{t}}\,\mathrm{d}\underline{a}$ and $\underline{\rho}\,\underline{\vec{b}}\,\mathrm{d}\underline{v}$ acting on an arbitrary subsystem $m$ in the current configuration $\underline{v}^{\mathrm{b}}(t)$. The notation $\mathrm{d}\underline{v}$ refers to a differential volume, and $\mathrm{d}\underline{a}$ is a differential area. According to Basic Law (4.18) [or (4.25)] [**BL 1**], the law of balance of momentum for a subsystem $m$ in the current configuration $\underline{v}^{\mathrm{b}}(t)$ reads

$$\textbf{BL 1} \qquad \oint_{\partial\underline{v}(t)} \underline{\vec{t}}(\underline{\vec{x}}(t), t, \underline{\vec{n}})\,\mathrm{d}\underline{a} + \int_{\underline{v}(t)} \underline{\rho}(\underline{\vec{x}}(t), t)\,\underline{\vec{b}}(\underline{\vec{x}}(t), t)\,\mathrm{d}\underline{v} = \frac{\mathrm{D}}{\mathrm{D}t}\int_{\underline{v}(t)} \underline{\rho}(\underline{\vec{x}}(t), t)\,\underline{\vec{v}}(\underline{\vec{x}}(t))\,\mathrm{d}\underline{v}\,. \qquad (4.96)$$

Step I. Write Basic Law (4.96) [**BL 1**] in terms of the initial material coordinates $X_i$.

Figure 4.13(a) shows that the terms of Basic Law (4.96) [**BL 1**] expressed in the current material coordinates $\underline{x}_i(t)$ have to be reformulated so that they are expressed in the initial material coordinates $X_i$ shown in Figure 4.13(b).

Equation (4.93) is recalled. It has the following appearance:

$$\oint_{\partial\underline{v}(t)} \underline{\vec{t}}(\underline{x}(t), t, \underline{\vec{n}})\,\mathrm{d}\underline{a} = \oint_{\partial V_0} \vec{T}(\vec{X}, t, \vec{N})\,\mathrm{d}A\,. \qquad (4.97)$$

Equation (4.75) is recalled:

$$\int_{\underline{v}(t)} \underline{\rho}(\underline{\vec{x}}(t), t)\,\underline{\theta}(\underline{\vec{x}}(t), t)\,\mathrm{d}\underline{v} = \int_{V_0} \rho_0(\vec{X})\,\theta(\vec{X}, t)\,\mathrm{d}V\,. \qquad (4.98)$$

Based on Equation (4.98), the second term on the left side and the term on the right side of Basic Law (4.96) [**BL 1**] take the forms

$$\int_{\underline{v}(t)} \underline{\rho}(\underline{\vec{x}}(t), t)\,\underline{\vec{b}}(\underline{\vec{x}}(t), t)\,\mathrm{d}\underline{v} = \int_{V_0} \rho_0(\vec{X})\,\vec{b}(\vec{X}, t)\,\mathrm{d}V \qquad (4.99)$$

and

$$\int_{\underline{v}(t)} \underline{\rho}(\underline{\vec{x}}(t), t)\,\underline{\vec{v}}(\underline{\vec{x}}(t))\,\mathrm{d}\underline{v} = \int_{V_0} \rho_0(\vec{X})\,\vec{v}(\vec{X}, t)\,\mathrm{d}V\,. \qquad (4.100)$$

The left side of Expression (4.100) exploits the fact that the velocity field $\underline{\vec{v}}(\underline{\vec{x}}(t))$ is the velocity of the material point $\underline{P}(t)$ and therefore it cannot be dependent on time $t$ independently of the position vector $\underline{\vec{x}}(t)$.

Substitution of Terms (4.97), (4.99), and (4.100) into Basic Law (4.96) [**BL 1**] gives:

$$\oint_{\partial V_0} \vec{T}(\vec{X}, t)\,\mathrm{d}A + \int_{V_0} \rho_0(\vec{X})\,\vec{b}(\vec{X}, t)\,\mathrm{d}V = \frac{\mathrm{D}}{\mathrm{D}t}\int_{V_0} \rho_0(\vec{X})\,\vec{v}(\vec{X}, t)\,\mathrm{d}V. \qquad (4.101)$$

Step II. Change the first integral in Expression (4.101) into a volume integral and move the material time operator to the right side of the integral sign.

Definition of the first Piola–Kirchhoff stress tensor $\boldsymbol{\sigma}^{\square}(\vec{X},t)$, Definition (4.91), is

$$\vec{N}(\vec{X}) \cdot \boldsymbol{\sigma}^{\square}(\vec{X},t) := \vec{T}(\vec{X},t,\vec{N})\,. \tag{4.102}$$

Substitution of Definition (4.102) into the first term of Expression (4.101) gives

$$\oint_{\partial V_0} \vec{T}(\vec{X},t,\vec{N})\,\mathrm{d}A = \oint_{\partial V_0} \vec{N}(\vec{X})\cdot\boldsymbol{\sigma}^{\square}(\vec{X},t)\,\mathrm{d}A\,. \tag{4.103}$$

Generalized Gauss's Theorem $(2.170)_1$ takes the following appearance for the initial material coordinates $X_i$:

$$\int_{V_0} \vec{\nabla}(\vec{X}) * f(\vec{X},t)\,\mathrm{d}V = \oint_{\partial V_0} \vec{N}(\vec{X}) * f(\vec{X},t)\,\mathrm{d}A\,. \tag{4.104}$$

Substituting $f(\vec{X},t) = \boldsymbol{\sigma}^{\square}(\vec{X},t)$ into Expression (4.104), replacing the star product $*$ by the dot product $\cdot$, and substituting the obtained result further into Expression (4.103) yields

$$\oint_{\partial V_0} \vec{T}(\vec{X},t,\vec{N})\,\mathrm{d}A = \oint_{\partial V_0} \vec{N}(\vec{X})\cdot\boldsymbol{\sigma}^{\square}(\vec{X},t)\,\mathrm{d}A = \int_{V_0} \vec{\nabla}(\vec{X})\cdot\boldsymbol{\sigma}^{\square}(\vec{X},t)\,\mathrm{d}V\,. \tag{4.105}$$

Reynolds Transport Theorem (4.80) is

$$\frac{\mathrm{D}}{\mathrm{D}t}\int_{V_0} \rho_0(\vec{X})\,\theta(\vec{X},t)\,\mathrm{d}V = \int_{V_0} \rho_0(\vec{X})\,\frac{\mathrm{D}\theta(\vec{X},t)}{\mathrm{D}t}\,\mathrm{d}V\,. \tag{4.106}$$

Based on Equation (4.106), the right side of Expression (4.101) takes the form

$$\frac{\mathrm{D}}{\mathrm{D}t}\int_{V_0} \rho_0(\vec{X})\,\vec{v}(\vec{X},t)\,\mathrm{d}V = \int_{V_0} \rho_0(\vec{X})\,\frac{\mathrm{D}\vec{v}(\vec{X},t)}{\mathrm{D}t}\,\mathrm{d}V\,. \tag{4.107}$$

Substitution of Terms (4.105) and (4.107) into Expression (4.101) yields

$$\int_{V_0} \vec{\nabla}(\vec{X})\cdot\boldsymbol{\sigma}^{\square}(\vec{X},t)\,\mathrm{d}V + \int_{V_0} \rho_0(\vec{X})\,\vec{b}(\vec{X},t)\,\mathrm{d}V = \int_{V_0} \rho_0(\vec{X})\,\frac{\mathrm{D}\vec{v}(\vec{X},t)}{\mathrm{D}t}\,\mathrm{d}V\,. \tag{4.108}$$

Step III. Collect all the terms of Expression (4.108) under one integral sign on the left side of the equals sign.

Equation (4.108) gives

$$\int_{V_0} \Big[\vec{\nabla}(\vec{X})\cdot\boldsymbol{\sigma}^{\square}(\vec{X},t) + \rho_0(\vec{X})\,\vec{b}(\vec{X},t) - \rho_0(\vec{X})\,\frac{\mathrm{D}\vec{v}(\vec{X},t)}{\mathrm{D}t}\Big]\mathrm{d}V = \vec{0}\,, \tag{4.109}$$

for every $V_0$ within $\mathrm{V}^{\mathrm{b}}$.

In Expression (4.109), the notation $\vec{0}$ stands for a zero vector that is defined to be a vector the components of which vanish. The phrase "for every $V_0$ within $\mathrm{V^b}$" is added to Expression (4.109) to point out that the volume $V_0$ (i.e., volume of the subsystem $m$) is an arbitrary subvolume of the volume $\mathrm{V^b}$ (i.e., volume of the system $\mathrm{m^b}$).

Step IV. Show that the integrand of Integral (4.109) has to vanish at every point of the volume $\mathrm{m^b}$.

The detailed study was given in Section 4.7. The proof is based on the fact that the volume $V_0$ is an arbitrary part of the system $\mathrm{V^b}$. We arrive at the following:

$$\vec{\nabla}(\vec{X})\cdot\boldsymbol{\sigma}^{\square}(\vec{X},t)+\rho_0(\vec{X})\,\vec{b}(\vec{X},t)=\rho_0(\vec{X})\,\frac{\mathrm{D}\vec{v}(\vec{X},t)}{\mathrm{D}t}\,. \tag{4.110}$$

Everywhere within $\mathrm{V^b}$.

Equation (4.110) is called the equations of motion. Equation (4.110) contains three scalar equations.

Step V. Assume small deformations.

When small displacements, deformations, and rotations are studied, the first Piola–Kirchhoff stress tensor $\boldsymbol{\sigma}^{\square}(\vec{X},t)$ reduces to the small deformation stress tensor $\boldsymbol{\sigma}(\vec{X},t)$. In the present book, the small deformation stress tensor $\boldsymbol{\sigma}(\vec{X},t)$ is called the stress tensor, for short. Thus, for small deformations and rotations, Equation (4.110) reduces to

$$\vec{\nabla}(\vec{X})\cdot\boldsymbol{\sigma}(\vec{X},t)+\rho_0(\vec{X})\,\vec{b}(\vec{X},t)=\rho_0(\vec{X})\,\frac{\mathrm{D}\vec{v}(\vec{X},t)}{\mathrm{D}t}\,. \tag{4.111}$$

Everywhere within $\mathrm{V^b}$.

It is common practice to leave out the phrase "Everywhere within $\mathrm{V^b}$" out in Equation (4.111).

In order to give a clearer picture of the obtained field equation, Expression (4.111) is written without independent variables as follows:

$$\vec{\nabla}\cdot\boldsymbol{\sigma}+\rho_0\,\vec{b}=\rho_0\,\frac{\mathrm{D}\vec{v}}{\mathrm{D}t}\qquad\text{or}\qquad\frac{\partial\sigma_{sk}}{\partial X_s}+\rho_0\,b_k=\rho_0\,\frac{\mathrm{D}v_k}{\mathrm{D}t}\,. \tag{4.112}$$

Equation $(4.112)_1$ contains three equations of motion. Equation $(4.112)_2$ gives the same set of equations, but in the index form. If the right sides of Equations $(4.112)_1$ and $(4.112)_2$ can be neglected as small quantities, Equations (4.112) will reduce to the form

$$\vec{\nabla}\cdot\boldsymbol{\sigma}+\rho_0\,\vec{b}=\vec{0}\qquad\text{or}\qquad\frac{\partial\sigma_{sk}}{\partial X_s}+\rho_0\,b_k=0\,. \tag{4.113}$$

Equations (4.113) are referred to as the equilibrium equations.

## 4.16 Remarks on Section 4.15

This section provides more details on the derivation carried out in Section 4.15. The outward unit normal to the differential surface element $\mathrm{d}\underline{a}(\underline{\vec{x}}(t))$ is denoted by $\underline{\vec{n}}(\underline{\vec{x}}(t))$. It is important to notice that the outward unit normal $\vec{n}$ expressed in the material coordinates $X_i$, that is, $\vec{n}(\vec{X},t)$, is not normal to the surface element $\mathrm{d}A(\vec{X})$. Therefore $\vec{n}(\vec{X},t)$ cannot be used in the generalized Gauss's theorem. This led to the introduction of the first Piola–Kirchhoff stress tensor $\boldsymbol{\sigma}^{\square}(\vec{X},t)$ and to the outward unit normal vector $\vec{N}(\vec{X})$.

Equation (4.94) gives

$$\oint_{\partial\underline{v}(t)} \underline{\vec{n}}(\underline{\vec{x}}(t),t)\cdot\underline{\boldsymbol{\sigma}}^{\mathrm{c}}(\underline{\vec{x}}(t),t)\,\mathrm{d}\underline{a} = \oint_{\partial V_0} \vec{N}(\vec{X})\cdot\boldsymbol{\sigma}^{\square}(\vec{X},t)\,\mathrm{d}A\,. \tag{4.114}$$

As Figure 4.13(b) and Expression (4.114) show, the first Piola–Kirchhoff stress tensor $\boldsymbol{\sigma}^{\square}(\vec{X},t)$ is an odd quantity. It is related to the fictitious surface area $\partial V_0$, which equals the area of the subsystem $m$ before loading took place. The Cauchy stress tensor $\underline{\boldsymbol{\sigma}}^{\mathrm{c}}(\underline{\vec{x}}(t),t)$, however, is a true stress tensor, since it is related to the actual (current) surface area $\partial\underline{v}(t)$ as shown in Figure 4.13(b) and Expression (4.114).

In addition to what was mentioned earlier, the first Piola–Kirchhoff stress tensor $\boldsymbol{\sigma}^{\square}(\vec{X},t)$ is not a symmetric tensor, and therefore, in the finite deformation theory it is replaced by the second Piola–Kirchhoff stress tensor. Neither of these pseudo stress tensors is for material modelling. When material models are prepared, the Cauchy stress tensor $\underline{\boldsymbol{\sigma}}^{\mathrm{c}}$ is an evident measure for stress. Many different stress, strain and strain rate measures give a wide variety of approaches for finite deformation theory.

When the basic laws and axioms are reformulated in terms of the material coordinates $X_i$, Expression (4.75) is used. It reads

$$\int_{\underline{v}(t)} \underline{\rho}(\underline{\vec{x}}(t),t)\,\underline{\theta}(\underline{\vec{x}}(t),t)\,\mathrm{d}\underline{v} = \int_{V_0} \rho_0(\vec{X})\,\theta(\vec{X},t)\,\mathrm{d}V\,. \tag{4.115}$$

A consequence of Expression (4.115) is that the density in the initial configuration $\rho_0(\vec{X})$ (i.e., before loading took place) is present in the local forms of the basic laws and axioms. This covers the finite deformation theory as well, as can be seen in Expression (4.110). Thus the density $\rho_0(\vec{X})$ in equations does not indicate the small deformation theory, but it is a consequence of the mathematical derivation. This means that the value for the density $\rho$ has to be determined from the strain tensor $\boldsymbol{\varepsilon}$.

Equation (4.115) gives (incorrectly) the idea that the appearances of the functions $\theta(\vec{X},t)$ and $\underline{\theta}(\underline{\vec{x}}(t),t)$ are the same. This is not true, as the following example shows. Let the function $\underline{\theta}(\underline{\vec{x}}(t),t)$ and the relationship between the coordinates $\underline{x}_i(t)$ and $X_i$ have the following uniaxial forms:

$$\underline{\theta}(\underline{x}_1(t),t) = \underline{x}_1(t)\,t \qquad \text{and} \qquad \underline{x}_1(t) = X_1(1+t^2)\,. \tag{4.116}$$

Substituting Relation $(4.116)_2$ into Equation $(4.116)_1$ gives

$$\theta(X_1,t) = X_1(1+t^2)\,t\,. \tag{4.117}$$

Equations $(4.116)_1$ and (4.117) show that the forms of the functions $\theta(\vec{X},t)$ and $\underline{\theta}(\vec{\underline{x}}(t),t)$ are different. Therefore a careful writer would use, for example, the notations $\tilde{\theta}(\vec{X},t)$ and $\underline{\theta}(\vec{\underline{x}}(t),t)$. In order to reduce the number of different notations, this practice is not followed here. Although the equations of the functions $\theta(\vec{X},t)$ and $\underline{\theta}(\vec{\underline{x}}(t),t)$ are different, their values are the same when the same material point is studied. It is worth noting that beyond the described differences between the functional forms, the stress measures also have different definitions, as discussed.

## 4.17 Balance of Momentum in the Current Material Coordinates

This section derives Cauchy equations of motions, which are the local forms of the law of balance of momentum. The current material coordinates $\underline{x}_i(t)$ are used.

Figure 4.14 shows the differential forces $\vec{\underline{t}}\,\mathrm{d}\underline{a}$ and $\underline{\rho}\,\vec{\underline{b}}\,\mathrm{d}\underline{v}$ acting on an arbitrary subsystem $m$ in the current configuration $\underline{v}^{\mathrm{b}}(t)$. The notation $\mathrm{d}\underline{v}$ refers to a differential volume, and $\mathrm{d}\underline{a}$ is a differential area. According to Basic Law (4.18) [or (4.25)] [**BL 1**], the law of balance of momentum for a subsystem $m$ in the current configuration $\underline{v}^{\mathrm{b}}(t)$ reads

**BL 1**
$$\oint_{\partial\underline{v}(t)} \vec{\underline{t}}(\vec{\underline{x}}(t),t,\vec{\underline{n}})\,\mathrm{d}\underline{a} + \int_{\underline{v}(t)} \underline{\rho}(\vec{\underline{x}}(t),t)\,\vec{\underline{b}}(\vec{\underline{x}}(t),t)\,\mathrm{d}\underline{v} = \frac{\mathrm{D}}{\mathrm{D}t}\int_{\underline{v}(t)} \underline{\rho}(\vec{\underline{x}}(t),t)\,\vec{\underline{v}}(\vec{\underline{x}}(t),t)\,\mathrm{d}\underline{v}\,. \tag{4.118}$$

It may be more convenient to make the derivation without the independent variables and then finally to write the obtained results in terms of the independent variables. Thus, Expression (4.118) is replaced by

**BL 1**
$$\oint_{\partial\underline{v}(t)} \vec{\underline{t}}\,\mathrm{d}\underline{a} + \int_{\underline{v}(t)} \underline{\rho}\,\vec{\underline{b}}\,\mathrm{d}\underline{v} = \frac{\mathrm{D}}{\mathrm{D}t}\int_{\underline{v}(t)} \underline{\rho}\,\vec{\underline{v}}\,\mathrm{d}\underline{v}\,. \tag{4.119}$$

By applying the generalized Gauss's theorem [see Theorem $(2.170)_1$], and the definition of the Cauchy stress tensor $\underline{\boldsymbol{\sigma}}^{\mathrm{c}}$ [Definition (4.90)], that is, $\vec{\underline{n}}\cdot\underline{\boldsymbol{\sigma}}^{\mathrm{c}} = \vec{\underline{t}}$, the first term of Basic Law (4.119) [**BL 1**] can be manipulated as follows:

$$\oint_{\partial\underline{v}(t)} \vec{\underline{t}}\,\mathrm{d}\underline{a} = \oint_{\partial\underline{v}(t)} \vec{\underline{n}}\cdot\underline{\boldsymbol{\sigma}}^{\mathrm{c}}\,\mathrm{d}\underline{a} = \int_{\underline{v}(t)} \vec{\underline{\nabla}}\cdot\underline{\boldsymbol{\sigma}}^{\mathrm{c}}\,\mathrm{d}\underline{v}\,. \tag{4.120}$$

Reynolds Transport Theorem (4.81) provides the following manipulation for the right side of Basic Law (4.119) [**BL 1**]:

$$\frac{\mathrm{D}}{\mathrm{D}t}\int_{\underline{v}(t)} \underline{\rho}\,\vec{\underline{v}}\,\mathrm{d}\underline{v} = \int_{\underline{v}(t)} \underline{\rho}\,\frac{\mathrm{D}\vec{\underline{v}}}{\mathrm{D}t}\,\mathrm{d}\underline{v}\,. \tag{4.121}$$

Substitution of Equations (4.120) and (4.121) into Basic Law (4.119) [**BL 1**] gives the following:

$$\int_{\underline{v}(t)} \left( \underline{\vec{\nabla}} \cdot \underline{\boldsymbol{\sigma}}^{\mathrm{c}} + \underline{\rho}\,\underline{\vec{b}} - \underline{\rho}\,\frac{\mathrm{D}\underline{\vec{v}}}{\mathrm{D}t} \right) \mathrm{d}\underline{v} = \vec{0}\,. \tag{4.122}$$

Since the volume $\underline{v}(t)$ of the subsystem $m$ is an arbitrary part of the volume $\underline{v}^{\mathrm{b}}(t)$ of the system $\mathrm{m^b}$, Integrand (4.122) must vanish for every $\underline{v}(t)$ within the volume $\underline{v}^{\mathrm{b}}(t)$ (the details are given in Section 4.7). Thus the term in parentheses vanishes at every point of the material, and the following local form is achieved:

$$\underline{\vec{\nabla}} \cdot \underline{\boldsymbol{\sigma}}^{\mathrm{c}} + \underline{\rho}\,\underline{\vec{b}} = \underline{\rho}\,\frac{\mathrm{D}\underline{\vec{v}}}{\mathrm{D}t} \qquad \text{or} \qquad \frac{\partial \underline{\sigma}^{\mathrm{c}}_{sk}}{\partial \underline{x}_s} + \underline{\rho}\,\underline{b}_k = \underline{\rho}\,\frac{\mathrm{D}\underline{v}_k}{\mathrm{D}t}\,. \tag{4.123}$$

Equations $(4.123)_1$ are the Cauchy equations of motion. Equations $(4.123)_2$ gives the three equations of motion in the index form. The term "Cauchy" stems from the Cauchy stress tensor $\underline{\boldsymbol{\sigma}}^{\mathrm{c}}$ being present in Expressions (4.123). When the independent variables are shown, Equation $(4.123)_1$ takes the following appearance:

$$\underline{\vec{\nabla}}(\underline{\vec{x}}) \cdot \underline{\boldsymbol{\sigma}}^{\mathrm{c}}(\underline{\vec{x}}(t),t) + \underline{\rho}(\underline{\vec{x}}(t),t)\,\underline{\vec{b}}(\underline{\vec{x}}(t),t) = \underline{\rho}(\underline{\vec{x}}(t),t)\,\frac{\mathrm{D}\underline{\vec{v}}(\underline{\vec{x}}(t),t)}{\mathrm{D}t}\,. \tag{4.124}$$

If the right sides of Equations $(4.123)_1$ and $(4.123)_2$ can be neglected as small quantities, Equations (4.123) reduce to the forms

$$\underline{\vec{\nabla}} \cdot \underline{\boldsymbol{\sigma}}^{\mathrm{c}} + \underline{\rho}\,\underline{\vec{b}} = \vec{0} \qquad \text{or} \qquad \frac{\partial \underline{\sigma}^{\mathrm{c}}_{sk}}{\partial \underline{x}_s} + \underline{\rho}\,\underline{b}_k = 0\,. \tag{4.125}$$

Equations (4.125) are called the equilibrium equations.

When studying small deformations and rotations, the coordinates $\underline{x}_i(t)$ and $X_i$ can be approximated to coincide. Furthermore, in such a case, the Cauchy stress tensor $\underline{\boldsymbol{\sigma}}^{\mathrm{c}}$ approaches the small deformation stress tensor $\boldsymbol{\sigma}$, that is, $\underline{\boldsymbol{\sigma}}^{\mathrm{c}} \rightarrow \boldsymbol{\sigma}$, and the current density $\underline{\rho}$ approaches the initial density $\rho_0$, that is, $\underline{\rho} \rightarrow \rho_0$. This might give an impression that for small deformation theory, Equation (4.124) reduces to

$$\vec{\nabla}(\vec{X}) \cdot \boldsymbol{\sigma}(\vec{X},t) + \rho_0(\vec{X})\,\vec{b}(\vec{X},t) = \rho_0(\vec{X})\,\frac{\mathrm{D}\vec{v}(\vec{X},t)}{\mathrm{D}t}\,, \tag{4.126}$$

which are the same as Equations of Motion (4.111). This is not the case, since in Equation (4.111), the density is exactly $\rho_0$, not approximately. The initial density $\rho_0$ enters into Equation (4.111) through mathematical derivation, as Section 4.15 shows, and is present in equations for the finite deformation theory as well.

## 4.18 Law of Balance of Momentum for Fluids: Spatial Description

This section derives equations of motion which are the local forms of the law of balance of momentum. The spatial coordinates $x_i$ are used. Figure 4.14 shows the differential forces $\underline{\vec{t}}\,\mathrm{d}\underline{a}$ and $\underline{\rho}\,\underline{\vec{b}}\,\mathrm{d}\underline{v}$ acting on an arbitrary subsystem $m$ in the current configuration $\underline{v}^{\mathrm{b}}(t)$. The notation $\mathrm{d}v$ refers to a differential volume, and $\mathrm{d}a$ is a differential area.

According to Basic Law (4.18) [or (4.25)] [**BL 1**], the law of balance of momentum for a subsystem $m$ in the current configuration $\underline{v}^{\mathrm{b}}(t)$ reads

**BL 1**
$$\oint_{\partial\underline{v}(t)} \underline{\vec{t}}(\underline{\vec{x}}(t),t,\underline{\vec{n}})\,\mathrm{d}\underline{a} + \int_{\underline{v}(t)} \underline{\rho}(\underline{\vec{x}}(t),t)\,\underline{\vec{b}}(\underline{\vec{x}}(t),t)\,\mathrm{d}\underline{v} = \frac{\mathrm{D}}{\mathrm{D}t}\int_{\underline{v}(t)} \underline{\rho}(\underline{\vec{x}}(t),t)\,\underline{\vec{v}}(\underline{\vec{x}}(t))\,\mathrm{d}\underline{v}\,. \tag{4.127}$$

The control volume $\mathrm{V}^{\mathrm{cv}}$ is introduced. It is assumed that the volume $\underline{v}(t_2)$ of the subsystem $m$ occupies exactly the control volume $\mathrm{V}^{\mathrm{cv}}$ at time $t_2$, as shown in Figure 4.15. It is possible to select another control volume $\mathrm{V}^{\mathrm{cv}}$ in such a way that the volume $\underline{v}(t_3)$ of the subsystem $m$ occupies exactly the control volume $\mathrm{V}^{\mathrm{cv}}$ at time $t_3$. Thus it is possible to introduce a control volume $\mathrm{V}^{\mathrm{cv}}$ for any volume $\underline{v}(t)$ of the subsystem $m$ at any time $t$.

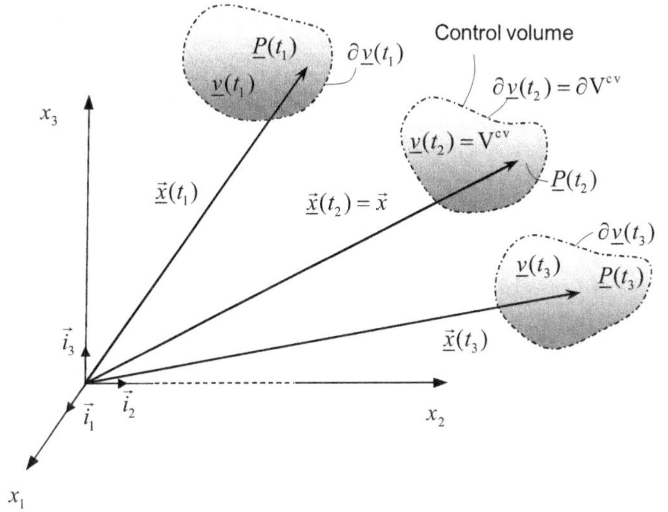


**Figure 4.15** Volume $\underline{v}(t)$ of the subsystem $m$ occupies the control volume $\mathrm{V}^{\mathrm{cv}}$ at time $t_2$.

The value of a quantity $\theta$ is independent of the coordinate system. For the present case, it means that

$$\underline{\theta}(\underline{\vec{x}}(t),t) = \theta(\vec{x},t)\,. \tag{4.128}$$

The preceding discussion allows the terms on the left side of Basic Law (4.127) [**BL 1**] to take the forms

$$\oint_{\partial \underline{v}(t)} \underline{\vec{t}}(\underline{\vec{x}}(t), t, \underline{\vec{n}})\, \mathrm{d}\underline{a} = \oint_{\partial \mathrm{V}^{\mathrm{cv}}} \vec{t}(\vec{x}, t, \vec{n})\, \mathrm{d}A = \oint_{\partial \mathrm{V}^{\mathrm{cv}}} \vec{t}\, \mathrm{d}A \tag{4.129}$$

and

$$\int_{\underline{v}(t)} \underline{\rho}(\underline{\vec{x}}(t), t)\, \underline{\vec{b}}(\underline{\vec{x}}(t), t)\, \mathrm{d}\underline{v} = \int_{\mathrm{V}^{\mathrm{cv}}} \rho(\vec{x}, t)\, \vec{b}(\vec{x}, t)\, \mathrm{d}V = \int_{\mathrm{V}^{\mathrm{cv}}} \rho\, \vec{b}\, \mathrm{d}V\,. \tag{4.130}$$

By applying the generalized Gauss's theorem [Theorem $(2.170)_1$], and the definition of the fluid stress tensor $\boldsymbol{\sigma}^{\mathrm{f}}$ [Definition (4.95)], that is, $\vec{n} \cdot \boldsymbol{\sigma}^{\mathrm{f}} := \vec{t}$, Expression (4.129) can be manipulated as follows:

$$\oint_{\partial \mathrm{V}^{\mathrm{cv}}} \vec{t}\, \mathrm{d}A = \oint_{\partial \mathrm{V}^{\mathrm{cv}}} \vec{n} \cdot \boldsymbol{\sigma}^{\mathrm{f}}\, \mathrm{d}A = \int_{\mathrm{V}^{\mathrm{cv}}} \vec{\nabla} \cdot \boldsymbol{\sigma}^{\mathrm{f}}\, \mathrm{d}V\,. \tag{4.131}$$

Reynolds Transport Theorem (4.89) is recalled. It reads

$$\frac{\mathrm{D}}{\mathrm{D}t} \int_{\underline{v}(t)} \underline{\rho}(\underline{\vec{x}}(t), t)\, \underline{\theta}(\underline{\vec{x}}(t), t)\, \mathrm{d}\underline{v} = \int_{\mathrm{V}^{\mathrm{cv}}} \rho(\vec{x}, t)\, \frac{\mathrm{D}\theta(\vec{x}, t)}{\mathrm{D}t}\, \mathrm{d}V = \int_{\mathrm{V}^{\mathrm{cv}}} \rho\, \frac{\mathrm{D}\theta}{\mathrm{D}t}\, \mathrm{d}V\,. \tag{4.132}$$

Substitution of $\theta = \vec{v}$ into Reynolds Transport Theorem (4.132) yields

$$\frac{\mathrm{D}}{\mathrm{D}t} \int_{\underline{v}(t)} \underline{\rho}\,(\underline{\vec{x}}(t), t)\, \underline{\vec{v}}(\underline{\vec{x}}(t), t)\, \mathrm{d}\underline{v} = \int_{\mathrm{V}^{\mathrm{cv}}} \rho(\vec{x}, t)\, \frac{\mathrm{D}\vec{v}(\vec{x}, t)}{\mathrm{D}t}\, \mathrm{d}V = \int_{\mathrm{V}^{\mathrm{cv}}} \rho\, \frac{\mathrm{D}\vec{v}}{\mathrm{D}t}\, \mathrm{d}V\,. \tag{4.133}$$

Substitution of Terms (4.130), (4.131), and (4.133) into Basic Law (4.127) [**BL 1**] yields

$$\int_{\mathrm{V}^{\mathrm{cv}}} \left( \vec{\nabla} \cdot \boldsymbol{\sigma}^{\mathrm{f}} + \rho\, \vec{b} - \rho\, \frac{\mathrm{D}\vec{v}}{\mathrm{D}t} \right) \mathrm{d}V = \vec{0}\,. \tag{4.134}$$

Since the control volume $\mathrm{V}^{\mathrm{cv}}$ is an arbitrary part of the volume $\underline{v}^{\mathrm{b}}(t)$, Integrand (4.134) must vanish for every $\mathrm{V}^{\mathrm{cv}}$ within the volume $\underline{v}^{\mathrm{b}}(t)$ (Section 4.7 gives a detailed study). Thus the term in parentheses vanishes at every point of the material, and the following local form is achieved:

$$\vec{\nabla} \cdot \boldsymbol{\sigma}^{\mathrm{f}} + \rho\, \vec{b} = \rho\, \frac{\mathrm{D}\vec{v}}{\mathrm{D}t} \qquad \text{or} \qquad \frac{\partial \sigma^{\mathrm{f}}_{sk}}{\partial x_s} + \rho\, b_k = \rho\, \frac{\mathrm{D}v_k}{\mathrm{D}t}\,. \tag{4.135}$$

In fluid mechanics, it is common practice to separate the fluid stress tensor $\boldsymbol{\sigma}^{\mathrm{f}}$ into two parts, as follows:

$$\boldsymbol{\sigma}^{\mathrm{f}} = -p\, \mathbf{1} + \boldsymbol{\sigma}^{\mathrm{d}}\,, \tag{4.136}$$

where $p$ is the thermodynamic pressure and $\boldsymbol{\sigma}^{\mathrm{d}}$ is the viscous stress tensor or dissipative stress tensor. It is important to note that the thermodynamical pressure $p$ is not the same as the mechanical pressure $p_{\mathrm{mech}}$, defined by Definition (2.117) as follows:

$$p_{\mathrm{mech}} := -\frac{1}{3}\, \mathbf{1} : \boldsymbol{\sigma} \qquad \Rightarrow \qquad p_{\mathrm{mech}} = -\frac{1}{3}\, \mathbf{1} : \boldsymbol{\sigma}^{\mathrm{f}}\,. \tag{4.137}$$

Substitution of Expression (4.136) into the first term of Equation (4.135) yields

$$\vec{\nabla}\cdot\boldsymbol{\sigma}^{\mathrm{f}} = \vec{\nabla}\cdot(-p\,\mathbf{1} + \boldsymbol{\sigma}^{\mathrm{d}}) = -\vec{\nabla}p + \vec{\nabla}\cdot\boldsymbol{\sigma}^{\mathrm{d}}\,, \tag{4.138}$$

where Definition (2.63), that is, $\vec{u}\cdot\mathbf{1} = \vec{u}$, is exploited. Substitution of Equation (4.138) into Expressions (4.135) gives

$$\rho\,\frac{\mathrm{D}\vec{v}}{\mathrm{D}t} = \rho\,\vec{b} + \vec{\nabla}\cdot\boldsymbol{\sigma}^{\mathrm{d}} - \vec{\nabla}p \quad \text{or} \quad \rho\,\frac{\mathrm{D}v_k}{\mathrm{D}t} = \rho\,b_k + \frac{\partial\sigma^{\mathrm{d}}_{sk}}{\partial x_s} - \frac{\partial p}{\partial x_k}\,. \tag{4.139}$$

Equations $(4.139)_1$ are the equations of motion. Equations $(4.139)_2$ give the same equations in the index form.

When the independent variables are shown, Expressions $(4.139)_1$ take the following appearance:

$$\rho(\vec{x},t)\,\frac{\mathrm{D}\vec{v}(\vec{x},t)}{\mathrm{D}t} = \rho(\vec{x},t)\,\vec{b}(\vec{x},t) + \vec{\nabla}(\vec{x})\cdot\boldsymbol{\sigma}^{\mathrm{d}}\,(\vec{x},t) - \vec{\nabla}(\vec{x})\,p\,. \tag{4.140}$$

If the left sides of Equations $(4.139)_1$ and $(4.139)_2$ can be neglected as small quantities, Equations (4.139) reduce to the forms

$$\vec{0} = \rho\,\vec{b} + \vec{\nabla}\cdot\boldsymbol{\sigma}^{\mathrm{d}} - \vec{\nabla}p \qquad \text{or} \qquad 0 = \rho\,b_k + \frac{\partial\sigma^{\mathrm{d}}_{sk}}{\partial x_s} - \frac{\partial p}{\partial x_k}\,. \tag{4.141}$$

The left sides of Equations of Motion (4.139) and (4.140) can be neglected as a small quantity for cases in which the Reynolds number is small, that is, when $Re << 1$. This is the case, for example, during settling of sediment particles near the ocean bottom. More examples can be found in Kundu and Cohen ([41], Chapter 9, Section 12).

## 4.19 Law of Balance of Moment of Momentum

This section derives the consequence for the law of balance of moment of momentum for the current material coordinates $\underline{x}_i(t)$. As a consequence, the result obtained, that is, that the Cauchy stress tensor $\underline{\boldsymbol{\sigma}}^{\mathrm{c}}$ is symmetric, is then extended to small deformations of solids by noting that also the small deformation stress tensor $\boldsymbol{\sigma}$ is symmetric. The same result holds for fluid stress tensor $\boldsymbol{\sigma}^{\mathrm{f}}$.

According to Expression (4.23), the law of balance of moment of momentum for subsystem $\underline{v}(t)$ takes the following appearance:

$$\textbf{BL2} \qquad \oint_{\partial\underline{v}(t)} (\underline{\vec{r}}\times\underline{\vec{t}}\,)\,\mathrm{d}\underline{a} + \int_{\underline{v}(t)} (\underline{\vec{r}}\times\underline{\rho}\,\underline{\vec{b}}\,)\,\mathrm{d}\underline{v} = \frac{\mathrm{D}}{\mathrm{D}t}\int_{\underline{v}(t)} (\underline{\vec{r}}\times\underline{\rho}\,\underline{\vec{v}}\,)\,\mathrm{d}\underline{v}\,, \tag{4.142}$$

where, according to Figure 4.16, $\underline{\vec{r}}(t)$ is a vector from an arbitrary fixed point Q to the material point $\underline{P}(t)$. This means that in Basic Law (4.142) [**BL 2**], the moment of the forces $\underline{\vec{t}}\,\mathrm{d}\underline{a}$ and $\underline{\rho}\,\underline{\vec{b}}\,\mathrm{d}\underline{v}$ is calculated about the point Q.

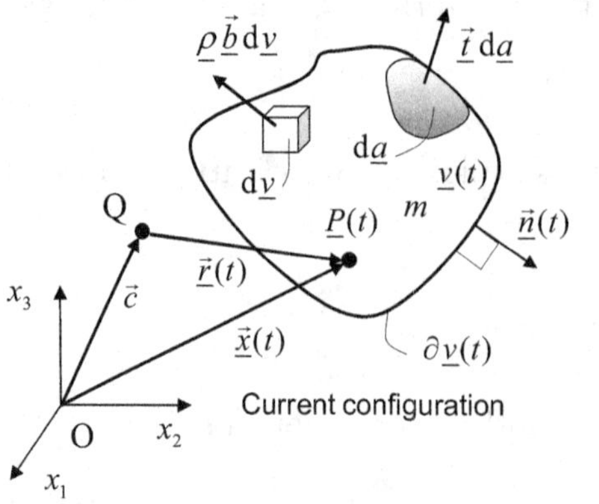


**Figure 4.16** Forces acting on an arbitrary subdomain $m$.

Vector $\vec{\underline{r}}(t)$ can be expressed as the sum of two vectors, namely, vectors $\vec{c}$ and $\vec{\underline{x}}(t)$, where $\vec{c}$ is the vector from the origin of the coordinate system $(x_1, x_2, x_3)$ to the point Q and $\vec{\underline{x}}(t)$ is the position vector of material point $\underline{P}(t)$ (see Figure 4.16). This is

$$\vec{\underline{r}}(t) = -\vec{c} + \vec{\underline{x}}(t)\,. \tag{4.143}$$

For clarity, the independent variables are dropped in the following derivation. Equation (4.143) provides the following form for Basic Law (4.142) [**BL 2**]:

$$\begin{aligned} -\vec{c}\times\left[\oint_{\partial\underline{v}(t)} \vec{\underline{t}}\,\mathrm{d}\underline{a} + \int_{\underline{v}(t)} \underline{\rho}\,\vec{\underline{b}}\,\mathrm{d}\underline{v} - \frac{\mathrm{D}}{\mathrm{D}t}\int_{\underline{v}(t)} \underline{\rho}\,\vec{\underline{v}}\,\mathrm{d}\underline{v}\right] \\ + \oint_{\partial\underline{v}(t)} (\,\vec{\underline{x}}\times\vec{\underline{t}}\,)\,\mathrm{d}\underline{a} + \int_{\underline{v}(t)} (\,\vec{\underline{x}}\times\underline{\rho}\,\vec{\underline{b}}\,)\,\mathrm{d}\underline{v} = \frac{\mathrm{D}}{\mathrm{D}t}\int_{\underline{v}(t)} (\,\vec{\underline{x}}\times\underline{\rho}\,\vec{\underline{v}}\,)\,\mathrm{d}\underline{v}\,. \end{aligned} \tag{4.144}$$

Equation (4.144) exploits the fact that the vector $\vec{c}$ is constant, as shown in Equation (4.143). This means that it can be moved outside the integral sign $\int$. Due to the law of balance of momentum {Basic Law (4.18) [**BL 1**]} the first row of Expression (4.144) vanishes. Thus, using the expression for the cross product $\times$ [Expression (2.122)], Basic Law (4.142) [**BL 2**] can be written as

$$\oint_{\partial\underline{v}(t)} (\,e_{ijk}\,\underline{x}_j\,\underline{t}_k\,)\,\mathrm{d}\underline{a}\,\vec{i}_i + \int_{\underline{v}(t)} (\,e_{ijk}\,\underline{x}_j\,\underline{\rho}\,\underline{b}_k\,)\,\mathrm{d}\underline{v}\;\vec{i}_i = \frac{\mathrm{D}}{\mathrm{D}t}\int_{\underline{v}(t)} (\,e_{ijk}\,\underline{\rho}\,\underline{v}_k\,)\,\mathrm{d}\underline{v}\,\vec{i}_i\,. \tag{4.145}$$

By utilizing the definition for the Cauchy stress tensor $\boldsymbol{\underline{\sigma}}^{\mathrm{c}}$ [Definition (4.90)] in the indicial form $\underline{n}_s\,\underline{\sigma}^{\mathrm{c}}_{sk} = \underline{t}_k$, and Generalized Gauss's Theorem $(2.170)_1$], the first integral of Expression (4.145) can be changed to a volume integral to give

$$\begin{aligned} \oint_{\partial\underline{v}(t)} (\,e_{ijk}\,\underline{x}_j\,\underline{t}_k\,)\,\mathrm{d}\underline{a} &= \oint_{\partial\underline{v}(t)} (\,e_{ijk}\,\underline{x}_j\,\underline{n}_s\,\underline{\sigma}^{\mathrm{c}}_{sk}\,)\,\mathrm{d}\underline{a} = e_{ijk}\oint_{\partial\underline{v}(t)} \underline{n}_s\,(\,\underline{x}_j\,\underline{\sigma}^{\mathrm{c}}_{sk}\,)\,\mathrm{d}\underline{a} \\ &= e_{ijk}\int_{\underline{v}(t)} \frac{\partial}{\partial\underline{x}_s}(\,\underline{x}_j\,\underline{\sigma}^{\mathrm{c}}_{sk}\,)\,\mathrm{d}\underline{v} \\ &= \int_{\underline{v}(t)} e_{ijk}\left(\frac{\partial\underline{x}_j}{\partial\underline{x}_s}\,\underline{\sigma}^{\mathrm{c}}_{sk} + \underline{x}_j\,\frac{\partial\underline{\sigma}^{\mathrm{c}}_{sk}}{\partial\underline{x}_s}\right)\mathrm{d}\underline{v}\,. \end{aligned} \tag{4.146}$$

Applying Results $(2.187)_2$ and $(2.23)_2$ gives

$$\frac{\partial\underline{x}_j}{\partial\underline{x}_s}\,\underline{\sigma}^{\mathrm{c}}_{sk} = \delta_{sj}\,\underline{\sigma}^{\mathrm{c}}_{sk} = \underline{\sigma}^{\mathrm{c}}_{jk}\,. \tag{4.147}$$

Based on Result (4.147), Manipulation (4.146) reduces to

$$\oint_{\partial \underline{v}(t)} (\, e_{ijk}\, \underline{x}_j\, \underline{t}_k \,)\, \mathrm{d}\underline{a} = \int_{\underline{v}(t)} e_{ijk} \left( \underline{\sigma}^{\mathrm{c}}_{jk} + \underline{x}_j \frac{\partial \underline{\sigma}^{\mathrm{c}}_{sk}}{\partial \underline{x}_s} \right) \mathrm{d}\underline{v}\,. \tag{4.148}$$

Reynolds Transport Theorem (4.81) provides the following simplification for the right side of Expression (4.145):

$$\begin{aligned} \frac{\mathrm{D}}{\mathrm{D}t} \int_{\underline{v}(t)} (\, e_{ijk}\, \underline{x}_j\, \underline{\rho}\, \underline{v}_k \,)\, \mathrm{d}\underline{v} &= \int_{\underline{v}(t)} e_{ijk}\, \underline{\rho}\, \frac{\mathrm{D}}{\mathrm{D}t} (\, \underline{x}_j\, \underline{v}_k \,)\, \mathrm{d}\underline{v} \\ &= \int_{\underline{v}(t)} e_{ijk}\, \underline{\rho} \left[ \underline{v}_j\, \underline{v}_k + \underline{x}_j \frac{\mathrm{D}\underline{v}_k}{\mathrm{D}t} \right] \mathrm{d}\underline{v}\,, \end{aligned} \tag{4.149}$$

where also the fact that $\mathrm{D}\underline{x}_j / \mathrm{D}t = \underline{v}_j$ is exploited.

Equation (D.1) of **Appendix D** shows that the double-dot product of a third-order tensor $\mathbf{c}$ that is skew-symmetric in the last two indices (i.e., $c_{ijk} = -\, c_{ikj}$) and a symmetric second-order tensor $\mathbf{h}$ ($h_{ij} = h_{ji}$) vanishes. Since the permutation symbol $e_{ijk}$ is skew-symmetric in the indices $jk$ [see Definition (2.123)] and the tensor $\underline{v}_j\, \underline{x}_k$ is symmetric,

$$e_{ijk}\, \underline{v}_j\, \underline{x}_k = 0 \tag{4.150}$$

holds, and Term (4.149) collapses to

$$\frac{\mathrm{D}}{\mathrm{D}t} \int_{\underline{v}(t)} (\, e_{ijk}\, \underline{x}_j\, \underline{\rho}\, \underline{v}_k \,)\, \mathrm{d}\underline{v} = \int_{\underline{v}(t)} (\, e_{ijk}\, \underline{\rho}\, \underline{x}_j \frac{\mathrm{D}\underline{v}_k}{\mathrm{D}t} )\, \mathrm{d}\underline{v}\,. \tag{4.151}$$

Substitution of Expressions (4.148) and (4.151) into Expression (4.145) yields

$$\int_{\underline{v}(t)} e_{ijk}\, \underline{\sigma}^{\mathrm{c}}_{jk}\, \mathrm{d}\underline{v}\, \vec{i}_i + \int_{\underline{v}(t)} \left[ e_{ijk}\, \underline{x}_j \left( \frac{\partial \underline{\sigma}^{\mathrm{c}}_{sk}}{\partial \underline{x}_s} + \underline{\rho}\, \underline{b}_k - \underline{\rho} \frac{\mathrm{D}\underline{v}_k}{\mathrm{D}t} \right) \right] \mathrm{d}\underline{v}\, \vec{i}_i = \vec{0}\,. \tag{4.152}$$

According to Cauchy equations of motion [Equation $(4.123)_2$], the term in parentheses in the second integral of Expression (4.152) vanishes, and we arrive at the following:

$$\int_{\underline{v}(t)} e_{ijk}\, \underline{\sigma}^{\mathrm{c}}_{jk}\, \vec{i}_i\, \mathrm{d}\underline{v} = \vec{0}\,, \tag{4.153}$$

where the base vector $\vec{i}_i$ is moved to the left to be part of the integrand. This can be done since the base vector $\vec{i}_i$ has a constant value with respect to the integration.

Since the volume $\underline{v}(t)$ of the subsystem $m$ is an arbitrary part of the volume $\underline{v}^{\mathrm{b}}(t)$ of the system $\mathrm{m}^{\mathrm{b}}$, the integrand in Expression (4.153) has to vanish everywhere within the volume $\underline{v}^{\mathrm{b}}(t)$. Thus Expression (4.153) gives

$$e_{ijk}\, \underline{\sigma}^{\mathrm{c}}_{jk} = 0\,. \tag{4.154}$$

The procedure to obtain Local Form (4.154) from Integral Form (4.153) (global form) was investigated in Section 4.7.

Definition of the permutation symbol $e_{ijk}$ [Definition (2.123)] allows Condition (4.154) to be written as

$$\begin{aligned} i = 1: &\qquad \underline{\sigma}^{\mathrm{c}}_{23} - \underline{\sigma}^{\mathrm{c}}_{32} = 0\,, \\ i = 2: &\qquad \underline{\sigma}^{\mathrm{c}}_{31} - \underline{\sigma}^{\mathrm{c}}_{13} = 0\,, \\ i = 3: &\qquad \underline{\sigma}^{\mathrm{c}}_{12} - \underline{\sigma}^{\mathrm{c}}_{21} = 0\,, \end{aligned} \tag{4.155}$$

which can be written as

$$\underline{\sigma}^{\mathrm{c}}_{ij} = \underline{\sigma}^{\mathrm{c}}_{ji} \qquad \text{or} \qquad \underline{\boldsymbol{\sigma}}^{\mathrm{c}} = (\underline{\boldsymbol{\sigma}}^{\mathrm{c}})^{\mathrm{T}}\,, \tag{4.156}$$

where the notation "$(\ )^{\mathrm{T}}$" displays a tensor transpose defined by Definition (2.73). Result (4.156) means that the Cauchy stress tensor $\underline{\boldsymbol{\sigma}}^{\mathrm{c}}$ is symmetric. The Cauchy stress tensor $\underline{\boldsymbol{\sigma}}^{\mathrm{c}}$ is defined in the current material coordinates $\underline{x}_i(t)$, that is,

$$\underline{\boldsymbol{\sigma}}^{\mathrm{c}} = \underline{\boldsymbol{\sigma}}^{\mathrm{c}}\left(\underline{\vec{x}}(t), t\right). \tag{4.157}$$

When small deformations and rotations are studied, the Cauchy stress tensor $\underline{\boldsymbol{\sigma}}^{\mathrm{c}}(\underline{\vec{x}}(t), t)$ and the first Piola–Kirchhoff stress tensor $\boldsymbol{\sigma}^{\square}(\vec{X}, t)$ reduce to the small deformation stress tensor $\boldsymbol{\sigma}(\vec{X}, t)$, or stress tensor for short. Thus, based on Result (4.157) and Definition (4.91), we arrive at the following:

$$\sigma_{ij} = \sigma_{ji} \qquad \text{or} \qquad \boldsymbol{\sigma} = \boldsymbol{\sigma}^{\mathrm{T}} \qquad \text{and} \qquad \vec{N} \cdot \boldsymbol{\sigma} = \vec{T}\,. \tag{4.158}$$

The same result can be derived for the fluid stress tensor $\boldsymbol{\sigma}^{\mathrm{f}}$ , that is,

$$\boldsymbol{\sigma}^{\mathrm{f}} = (\boldsymbol{\sigma}^{\mathrm{f}})^{\mathrm{T}}\,, \qquad \text{where} \qquad \boldsymbol{\sigma}^{\mathrm{f}} = \boldsymbol{\sigma}^{\mathrm{f}}(\vec{x}, t)\,. \tag{4.159}$$

**Example 4.1:** A cantilever beam subjected to a uniform constant load $\omega$ is studied in Figure 4.17. The weight of the beam can be neglected as a small quantity. The cross-sectional dimensions of the beam are the height $h$ and width $b$. The length of the beam is $\ell$. According to the beam theory, the expressions for the normal stress $\sigma_{11}(\vec{X}, t)$ $(= \sigma_{\mathrm{x}})$ and for the shear stress $\sigma_{12}(\vec{X}, t)$ $(= \tau_{\mathrm{xy}})$ take the following appearances:

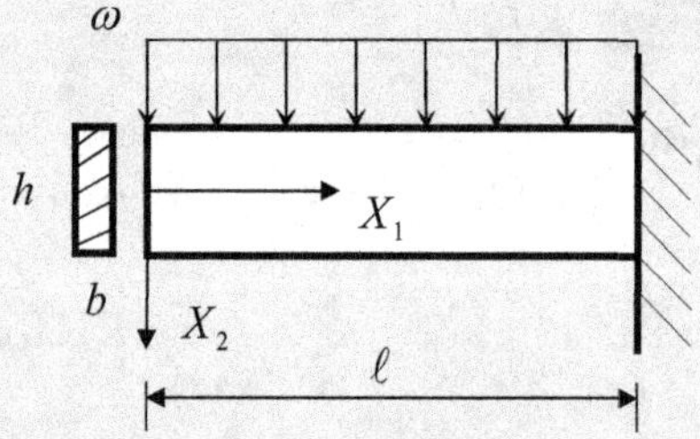


**Figure 4.17** Cantilever beam.

$$\begin{aligned} \sigma_{11}(\vec{X}, t) &= -\frac{6\,\omega\,X_1^2\,X_2}{b\,h^3} \\ \sigma_{12}(\vec{X}, t) &= \frac{3\,\omega\,X_1}{2\,bh^3}\left(1 - 4\frac{X_2^2}{h^2}\right). \end{aligned} \tag{4.160}$$

Does Solution (4.160) satisfy Equilibrium Equations (4.113)?

**Solution:** Equilibrium Equations (4.113) read

$$\vec{\nabla}\cdot\boldsymbol{\sigma}+\rho_0\,\vec{b}=\vec{0} \qquad \text{or} \qquad \frac{\partial\sigma_{sk}}{\partial X_s}+\rho_0\,b_k=0\,. \tag{4.161}$$

The component forms of Equilibrium Equations $(4.161)_2$ are

$$\begin{aligned}
k=1:&\qquad \frac{\partial\sigma_{11}}{\partial X_1}+\frac{\partial\sigma_{21}}{\partial X_2}+\frac{\partial\sigma_{31}}{\partial X_3}+\rho_0\,b_1=0\,,\\
k=2:&\qquad \frac{\partial\sigma_{12}}{\partial X_1}+\frac{\partial\sigma_{22}}{\partial X_2}+\frac{\partial\sigma_{32}}{\partial X_3}+\rho_0\,b_2=0\,,\\
k=3:&\qquad \frac{\partial\sigma_{13}}{\partial X_1}+\frac{\partial\sigma_{23}}{\partial X_2}+\frac{\partial\sigma_{33}}{\partial X_3}+\rho_0\,b_3=0\,.
\end{aligned} \tag{4.162}$$

According to the beam theory, the stress components having index 3 [in the present orientation of the coordinate system $(X_1, X_2, X_3)$] are assumed to be so small that they can be neglected as small quantities. The same holds for the normal stress $\sigma_{22}$. Thus, for the present coordinates $X_i$, the following holds:

$$\sigma_{13}=\sigma_{31}\approx\sigma_{23}=\sigma_{32}\approx\sigma_{33}=\sigma_{22}\approx 0\,. \tag{4.163}$$

As mentioned, the weight of the beam can be neglected as a small quantity. This reads

$$b_1\approx b_2\approx b_3\approx 0. \tag{4.164}$$

Substituting Values (4.163) and (4.164) into Equilibrium Equations (4.162), we obtain

$$\begin{aligned}
&\frac{\partial\sigma_{11}}{\partial X_1}+\frac{\partial\sigma_{21}}{\partial X_2}=0\\
&\frac{\partial\sigma_{12}}{\partial X_1}+\frac{\partial\sigma_{22}}{\partial X_2}=0\,.
\end{aligned} \tag{4.165}$$

Based on Expression (4.160) for the stress state, the first two derivative terms in Equilibrium Equations (4.165) take the forms

$$\frac{\partial\sigma_{11}}{\partial X_1}=-\frac{12\,\omega\,X_1\,X_2}{b\,h^3} \qquad \text{and} \qquad \frac{\partial\sigma_{21}}{\partial X_2}=\frac{\partial\sigma_{12}}{\partial X_2}=-\frac{12\,\omega\,X_1\,X_2}{b\,h^3}\,, \tag{4.166}$$

where in the latter equation the symmetry of the stress tensor $\boldsymbol{\sigma}$, that is, $\sigma_{21}=\sigma_{12}$, is exploited. The latter two derivatives take the forms

$$\frac{\partial\sigma_{12}}{\partial X_1}=\frac{3\,\omega}{2\,b\,h^3}\left(1-4\,\frac{X_2^2}{h^2}\right) \qquad \text{and} \qquad \frac{\partial\sigma_{22}}{\partial X_2}=0\,. \tag{4.167}$$

The latter equation of Expressions (4.167) results from the assumption that $\sigma_{22}\approx 0$. Substitution of Terms (4.166) and (4.167) into Set (4.165) gives

$$-\frac{12\,\omega\,X_1\,X_2}{b\,h^3}-\frac{12\,\omega X_1 X_2}{bh^3}\stackrel{?}{=}0 \qquad \text{and} \qquad \frac{3\,\omega}{2\,b\,h^3}\left(1-4\,\frac{X_2^2}{h^2}\right)\stackrel{?}{=}0. \tag{4.168}$$

In order to satisfy the equilibrium equations, the left side of Expressions (4.167) should be zero within the volume of the beam, that is, when $X_1 \in [0, \ell]$ and $X_2 \in [-h/2, h/2]$. From Equation (4.168) it is clear that this condition does not hold and therefore the beam in Figure 4.17 is not in equilibrium. This means that the state of stress given by Equations (4.160) is incorrect for the body. However, the equilibrium equations are satisfied at points $(X_1, X_2) = (0, \pm h/2)$, which are the upper and lower left corners of the beam. The equilibrium equations must be satisfied at all points of a body. Otherwise, the body is not in equilibrium. ∎

## 4.20 Equilibrium Equations and Stress Resultants for Beams

This section deals with equilibrium equations. Since small deformations and rotations are assumed, Equilibrium Equations (4.113) can be applied. It is shown how to derive relations between stress resultants for beams from Equilibrium Equations (4.113). Because a solid is studied, the investigation is carried out in the initial geometry at current time and position $\mathrm{V}_0^{\mathrm{b}}$, that is, in the IG-CtP configuration $\mathrm{V}_0^{\mathrm{b}}$.

Mathematical derivation is simplified by assuming the body force vector $\vec{b}$ can be neglected as a small quantity compared with any loads that are applied on the body. Therefore, Equilibrium Equations (4.113) reduce to

$$\vec{\nabla} \cdot \boldsymbol{\sigma} = \vec{0} \qquad \text{or} \qquad \frac{\partial \sigma_{sk}}{\partial X_s} = 0\,. \tag{4.169}$$

(Local) Equilibrium Equations $(4.169)_1$ are integrated over the differential length element of the beam $\mathrm{d}V$. This gives

$$\int_{\mathrm{d}V} \vec{\nabla} \cdot \boldsymbol{\sigma} \, \mathrm{d}V = \vec{0}\,. \tag{4.170}$$

Applying Generalized Gauss's Theorem $(2.170)_1$

$$\int_V \vec{\nabla} * f \, \mathrm{d}V = \oint_{\partial V} \vec{n} * f \, \mathrm{d}A \tag{4.171}$$

by replacing the star product operator $*$ by the dot product operator $\cdot$, the outward unit normal $\vec{n}$ by the outward unit normal $\vec{N}$, and the function $f$ by the stress tensor $\boldsymbol{\sigma}$ Integral (4.170) can be written as

$$\oint_{\partial \mathrm{d}V} \vec{N} \cdot \boldsymbol{\sigma} \, \mathrm{d}A = \vec{0}\,. \tag{4.172}$$

Figure 4.18 shows the differential length element of the beam $\mathrm{d}V$ under consideration. It is assumed that all the loads are in the $(X_1, X_2)$-plane. Therefore, the problem is two-dimensional.

Figure 4.18 introduces the $\eta$-coordinate. It is for making mathematical derivation clearer, and it equals the $X_1$-axis. According to Figure 4.18(b), the following holds:

$$\begin{aligned} &\text{Face } A^1: \quad \vec{N} = -\vec{\imath}_1 \qquad \text{and} \qquad \text{Face } A^2: \quad \vec{N} = \vec{\imath}_2 \\ &\text{Face } A^3: \quad \vec{N} = \vec{\imath}_1 \qquad \text{and} \qquad \text{Face } A^4: \quad \vec{N} = -\vec{\imath}_2\,. \end{aligned} \tag{4.173}$$

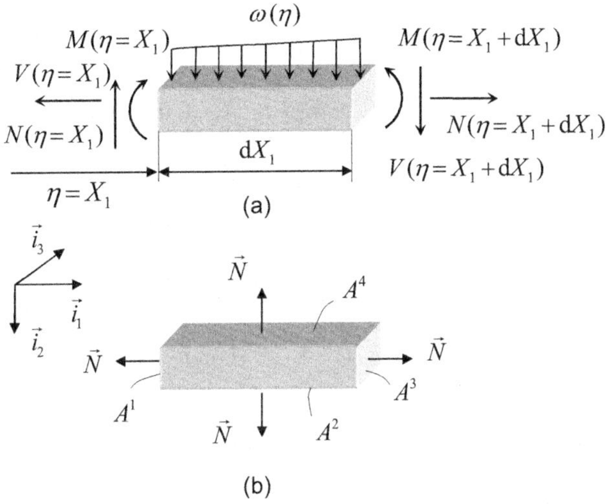


**Figure 4.18** (a) Applied loads for bending of a beam. (b) Positive outward unit vectors $\vec{n}$ for the faces $A^1$–$A^4$ of the beam.

Substitution of Normals (4.173) into Expression (4.172) gives

$$-\int_{A^1} \vec{\imath}_1 \cdot \boldsymbol{\sigma}\, \mathrm{d}A + \int_{A^2} \vec{\imath}_2 \cdot \boldsymbol{\sigma}\, \mathrm{d}A + \int_{A^3} \vec{\imath}_1 \cdot \boldsymbol{\sigma}\, \mathrm{d}A - \int_{A^4} \vec{\imath}_2 \cdot \boldsymbol{\sigma}\, \mathrm{d}A = \vec{0}\,. \tag{4.174}$$

Equation (4.174) exploits the facts that the surfaces $\vec{\imath}_3 = constant$ are free from stresses and the positive integration path over the surface $\partial \mathrm{d}V$ is in the anticlockwise direction. The dot product between the base vectors $\vec{\imath}_1$ and $\vec{\imath}_2$ and the stress tensor $\boldsymbol{\sigma}$ can be expressed as

$$\boldsymbol{\sigma} = \sigma_{ij}\, \vec{\imath}_i\, \vec{\imath}_j\,, \qquad \text{which gives} \qquad \begin{aligned} \vec{\imath}_1 \cdot \boldsymbol{\sigma} &= \vec{\imath}_1 \cdot \sigma_{ij}\, \vec{\imath}_i\, \vec{\imath}_j = \sigma_{1j}\, \vec{\imath}_j \\ \vec{\imath}_2 \cdot \boldsymbol{\sigma} &= \vec{\imath}_2 \cdot \sigma_{ij}\, \vec{\imath}_i\, \vec{\imath}_j = \sigma_{2j}\, \vec{\imath}_j\,. \end{aligned} \tag{4.175}$$

Substitution of Results $(4.175)_2$ into Equation (4.174) leads to

$$\left[ -\int_{A^1} \sigma_{1j}\, \mathrm{d}A + \int_{A^2} \sigma_{2j}\, \mathrm{d}A + \int_{A^3} \sigma_{1j}\, \mathrm{d}A - \int_{A^4} \sigma_{2j}\, \mathrm{d}A \right] \vec{\imath}_j = \vec{0}\,. \tag{4.176}$$

**(a)** The first component of Vector (4.176) is (i.e., when $j = 1$)

$$-\int_{A^1} \sigma_{11}\,\mathrm{d}A + \int_{A^2} \sigma_{21}\,\mathrm{d}A + \int_{A^3} \sigma_{11}\,\mathrm{d}A - \int_{A^4} \sigma_{21}\,\mathrm{d}A = 0\,. \tag{4.177}$$

It is assumed that on the top and bottom faces of the beam, that is, on $A^2$ and $A^4$, no shear stresses $\sigma_{21}$ exist. Therefore Expression (4.177) reduces to

$$-\int_{A^1} \sigma_{11}\,\mathrm{d}A + \int_{A^3} \sigma_{11}\,\mathrm{d}A = 0\,. \tag{4.178}$$

Resultant normal forces $N(\cdot)$ at the ends of the beam are defined by

$$N(\eta = X_1) := \int_{A^1} \sigma_{11}\,\mathrm{d}A \quad \text{and} \quad N(\eta = X_1 + \mathrm{d}X_1) := \int_{A^3} \sigma_{11}\,\mathrm{d}A\,. \tag{4.179}$$

Substitution of Definitions (4.179) into Expression (4.178) yields

$$-N(\eta = X_1) + N(\eta = X_1 + \mathrm{d}X_1) = 0\,. \tag{4.180}$$

Result (4.180) can be written in a more convenient form as follows:

$$-N(X_1) + N(X_1 + \mathrm{d}X_1) = 0. \tag{4.181}$$

Result (4.181) is the global equilibrium equation in the $X_1$-direction ($x$-direction in the elementary beam theory). Here it is written for a differential length element of the beam $\mathrm{d}V$, but it also holds for the whole beam. It is worth noting that although the stress tensor component $\sigma_{11}$ is dependent on all the space coordinates, that is, $\sigma_{11} = \sigma_{11}(X_1, X_2, X_3)$, the resultant normal force $N$ depends only on $X_1$, that is, $N = N(X_1)$. This originates from the definition of the resultant normal force,

$$N(X_1) := \int_{\mathrm{A^{cs}}} \sigma_{11}(X_1, X_2, X_3)\,\mathrm{d}A\,, \tag{4.182}$$

where $\mathrm{A^{cs}}$ is the area of the beam cross section situated in the $(X_2, X_3)$-plane.

**(b)** The second component of Vector (4.176) is (i.e., when $j = 2$)

$$-\int_{A^1} \sigma_{12}\,\mathrm{d}A + \int_{A^2} \sigma_{22}\,\mathrm{d}A + \int_{A^3} \sigma_{12}\,\mathrm{d}A - \int_{A^4} \sigma_{22}\,\mathrm{d}A = 0\,. \tag{4.183}$$

Resultant shear forces $V(\cdot)$ at the ends of the differential length element of the beam $\mathrm{d}V$ and the distributed load $\omega(\cdot)$ [force/length] along the beam are defined by

$$V(\eta = X_1) := \int_{A^1} \sigma_{12}\,\mathrm{d}A \quad \text{and} \quad V(\eta = X_1 + \mathrm{d}X_1) := \int_{A^3} \sigma_{12}\,\mathrm{d}A \tag{4.184}$$

and finally

$$\omega(\eta) := -\int_{-\frac{b}{2}}^{\frac{b}{2}} \sigma_{22}\,\mathrm{d}X_3 \qquad (\text{on the plane } A^4)\,. \tag{4.185}$$

In Definition (4.185), the notation $b$ stands for the width of the beam. Assuming that the stress $\sigma_{22}$ vanishes on the surface $A^2$ (bottom of the beam) and substituting Definitions (4.184) and (4.185) into Expression (4.183) gives

$$-V(\eta = X_1) + V(\eta = X_1 + \mathrm{d}X_1) + \int_{X_1}^{X_1+\mathrm{d}X_1} \omega(\eta)\,\mathrm{d}\eta = 0\,. \tag{4.186}$$

If function $f(x)$ is a derivable function and $\partial f(x)/\partial x$ is integrable within a closed interval $[\,a, b\,]$, then

$$\int_a^b \frac{\partial f(x)}{\partial x}\,\mathrm{d}x = f(b) - f(a)\,. \tag{4.187}$$

Expression (4.187) allows the first two terms of Equation (4.186) to be written in the form

$$-V(\eta = X_1) + V(\eta = X_1 + \mathrm{d}X_1) = \int_{X_1}^{X_1+\mathrm{d}X_1} \frac{\partial V(\eta)}{\partial \eta}\,\mathrm{d}\eta. \tag{4.188}$$

Substitution of Difference (4.188) into Expression (4.186) yields

$$\int_{X_1}^{X_1+\mathrm{d}X_1} \frac{\partial V(\eta)}{\partial \eta}\,\mathrm{d}\eta + \int_{X_1}^{X_1+\mathrm{d}X_1} \omega(\eta)\,\mathrm{d}\eta = 0\,, \tag{4.189}$$

which gives

$$\int_{X_1}^{X_1+\mathrm{d}X_1} \left[ \frac{\partial V(\eta)}{\partial \eta} + \omega(\eta) \right] \mathrm{d}\eta = 0\,. \tag{4.190}$$

Since the interval $[\,X_1, X_1 + \mathrm{d}X_1\,]$ is an arbitrary part of the beam length, the value of Integrand (4.190) is zero everywhere in the beam [see Section 4.7]. This is

$$\frac{\partial V(\eta)}{\partial \eta} + \omega(\eta) = 0\,, \qquad \text{which gives} \qquad \frac{\partial V(X_1)}{\partial X_1} = -\,\omega(X_1)\,. \tag{4.191}$$

In Equation $(4.191)_2$, the $\eta$-coordinate is replaced by the $X_1$-coordinate. Equation $(4.191)_2$ is the global equilibrium equation in the $X_2$-direction ($y$-direction in the elementary beam theory).

**(c)** In order to obtain the moment equilibrium equation, Expression $(4.169)_2$ is multiplied by the $X_2$-coordinate and integrated over the differential length element of the beam $\mathrm{d}V$. This leads to

$$\int_{\mathrm{d}V} \frac{\partial \sigma_{sk}}{\partial X_s}\,X_2\,\mathrm{d}V = 0\,. \tag{4.192}$$

The derivative of the product of the quantities $\sigma_{sk}$ and $X_2$ is integrated over the volume $\mathrm{d}V$. This gives

$$\int_{\mathrm{d}V} \frac{\partial}{\partial X_s}(\sigma_{sk}\,X_2)\,\mathrm{d}V = \int_{\mathrm{d}V} \frac{\partial \sigma_{sk}}{\partial X_s}\,X_2\,\mathrm{d}V + \int_{\mathrm{d}V} \sigma_{sk}\,\frac{\partial X_2}{\partial X_s}\,\mathrm{d}V\,. \tag{4.193}$$

Since, according to Equation $(2.187)_2$, $\partial X_2/\partial X_s = \delta_{s2}$, the second term on the right side of Integral Equation (4.193) takes a different form. Furthermore, the left side of Equation (4.193) is recast into a different appearance by applying Generalized Gauss's Theorem $(2.171)_1$. Therefore we arrive at the following:

$$\oint_{\partial \mathrm{d}V} N_s\,(\sigma_{sk}\,X_2)\,\mathrm{d}A = \int_{\mathrm{d}V} \frac{\partial \sigma_{sk}}{\partial X_s}\,X_2\,\mathrm{d}V + \int_{\mathrm{d}V} \sigma_{2k}\,\mathrm{d}V\,. \tag{4.194}$$

Equation (4.192) allows Equation (4.194) to be written in the form

$$\oint_{\partial \mathrm{d}V} N_s\,(\sigma_{sk}\,X_2)\mathrm{d}A - \int_{\mathrm{d}V} \sigma_{2k}\,\mathrm{d}V = 0\,. \tag{4.195}$$

Based on Expressions (4.173), the following is achieved:

$$A^1: \begin{array}{ll} s=1 & N_s=-1 \\ s=2 & N_s=0 \end{array} \qquad \text{and} \qquad A^2: \begin{array}{ll} s=1 & N_s=0 \\ s=2 & N_s=1 \end{array} \tag{4.196}$$

and furthermore,

$$A^3: \begin{array}{ll} s=1 & N_s=1 \\ s=2 & N_s=0 \end{array} \qquad \text{and} \qquad A^4: \begin{array}{ll} s=1 & N_s=0 \\ s=2 & N_s=-1\,. \end{array} \tag{4.197}$$

The information given by Expressions (4.196) and (4.197) is substituted into Equation (4.195). This leads to

$$-\int\limits_{A^1} \sigma_{1k}\,X_2\,\mathrm{d}A + \int\limits_{A^2} \sigma_{2k}\,X_2\,\mathrm{d}A + \int\limits_{A^3} \sigma_{1k}\,X_2\,\mathrm{d}A - \int\limits_{A^4} \sigma_{2k}\,X_2\,\mathrm{d}A - \int\limits_{\mathrm{d}V} \sigma_{2k}\,\mathrm{d}V = 0\,. \tag{4.198}$$

Substitution of $k=1$ into Equation (4.198) yields

$$-\int\limits_{A^1} \sigma_{11}\,X_2\,\mathrm{d}A + \int\limits_{A^2} \sigma_{21}\,X_2\,\mathrm{d}A + \int\limits_{A^3} \sigma_{11}\,X_2\,\mathrm{d}A - \int\limits_{A^4} \sigma_{21}\,X_2\,\mathrm{d}A - \int_{\mathrm{d}V} \sigma_{21}\,\mathrm{d}V = 0\,. \tag{4.199}$$

Resultant bending moments $M(\cdot)$ at the ends of the beam and resultant shear force $V(\cdot)$ along the beam are defined by

$$M(\eta = X_1) := \int_{A^1} \sigma_{11}\,X_2\,\mathrm{d}A \quad \text{and} \quad M(\eta = X_1 + \mathrm{d}X_1) := \int_{A^3} \sigma_{11}\,X_2\,\mathrm{d}A\,, \tag{4.200}$$

and furthermore,

$$V(\eta) := \int_{\mathrm{A^{cs}}} \sigma_{12}\,\mathrm{d}A\,. \tag{4.201}$$

Assuming that the shear stress $\sigma_{21} = \sigma_{12}$ vanishes on the top and bottom surfaces of the beam, and taking into account Definitions (4.200), Expression (4.199) gives

$$-M(\eta = X_1) + M(\eta = X_1 + \mathrm{d}X_1) - \int_{\mathrm{d}V} \sigma_{21}\,\mathrm{d}V = 0\,. \tag{4.202}$$

The last term of Expression (4.202) can be expressed as follows:

$$\int_{\mathrm{d}V} \sigma_{21}\,\mathrm{d}V = \int_{X_1}^{X_1+\mathrm{d}X_1} \left[ \int_{\mathrm{A^{cs}}} \sigma_{12}\,\mathrm{d}A \right] \mathrm{d}\eta = \int_{X_1}^{X_1+\mathrm{d}X_1} V(\eta)\,\mathrm{d}\eta\,, \tag{4.203}$$

where Definition (4.201) was used. Substitution of Expression (4.203) into Equation (4.202) yields

$$-\,M(\eta = X_1) + M(\eta = X_1 + \mathrm{d}X_1) - \int_{X_1}^{X_1+\mathrm{d}X_1} V(\eta)\,\mathrm{d}\eta = 0\,. \tag{4.204}$$

Equation (4.187) allows the first two terms of Equation (4.204) to be written in the form

$$-\,M(\eta = X_1) + M(\eta = X_1 + \mathrm{d}X_1) = \int_{X_1}^{X_1+\mathrm{d}X_1} \frac{\partial M(\eta)}{\partial \eta}\,\mathrm{d}\eta\,. \tag{4.205}$$

Substitution of Difference (4.205) into Expression (4.204) yields

$$\int_{X_1}^{X_1+\mathrm{d}X_1} \frac{\partial M(\eta)}{\partial \eta}\,\mathrm{d}\eta - \int_{X_1}^{X_1+\mathrm{d}X_1} V(\eta)\,\mathrm{d}\eta = 0\,, \tag{4.206}$$

which gives

$$\int_{X_1}^{X_1+\mathrm{d}X_1} \left[ \frac{\partial M(\eta)}{\partial \eta} - V(\eta) \right] \mathrm{d}\eta = 0\,. \tag{4.207}$$

Since the interval $[\,X_1, X_1 + \mathrm{d}X_1\,]$ is an arbitrary part of the beam length, the value of Integrand (4.207) is zero everywhere in the beam [see Section 4.7]. This is

$$\frac{\partial M(\eta)}{\partial \eta} - V(\eta) = 0\,, \qquad \text{which gives} \qquad \frac{\partial M(X_1)}{\partial X_1} = V(X_1)\,. \tag{4.208}$$

In Equation $(4.208)_2$, the $\eta$-coordinate is replaced by the $X_1$-coordinate. Equation $(4.208)_2$ is the global moment equilibrium equation around the $X_3$-axis (around the $z$-axis in the elementary beam theory).

Equilibrium Equations (4.181), $(4.191)_2$, and $(4.208)_2$, kinematic equations (i.e., equations between the strain and the displacement), and the material model (i.e., Hooke's law) form the foundation for the beam theory.

The concept of this section is to show that equilibrium equations for the beam theory can be derived from the law of balance of momentum and law of balance of moment of momentum, that is, the basic laws of continuum mechanics are adequate for the beam theory. No "extra" equations such as force equilibrium equations and moment equilibrium equations, which are the common tools for beam theory in bachelor-level books, are needed. This book follows the idea of Occam's razor by keeping the number of axioms and basic laws as small as possible. Therefore it is important to show that the basic laws of continuum mechanics are sufficient for derivation of the equilibrium equations for beam theory.

## 4.21 Cauchy Tetrahedron and the Cauchy Stress Tensor $\underline{\sigma}^{\mathrm{c}}$

This section shows that Definition $(4.90)_2$ for the Cauchy stress tensor $\underline{\sigma}^{\mathrm{c}}$, that is, $\vec{\underline{n}} \cdot \underline{\sigma}^{\mathrm{c}} := \vec{\underline{t}}$, is compatible with the law of balance of momentum, that is, Basic Law (4.18) [**BL 1**]. It is worth noting that actually this section should be placed before Definition (4.90) of the Cauchy stress tensor $\underline{\sigma}^{\mathrm{c}}$. However, since the present study is quite extensive, it would make the derivation earlier in this chapter difficult to follow. Therefore it is placed here. The study is carried out in the current configuration $\underline{v}^{\mathrm{b}}(t)$ and therefore the current material coordinates $\underline{x}_i(t)$ are used.

The law of balance of momentum {Basic Law (4.18) [**BL 1**]} is

**BL 1**
$$\oint_{\partial \underline{v}(t)} \vec{\underline{t}}\, \mathrm{d}\underline{a} + \int_{\underline{v}(t)} \underline{\rho}\, \vec{\underline{b}}\, \mathrm{d}\underline{v} = \frac{\mathrm{D}}{\mathrm{D}t} \int_{\underline{v}(t)} \underline{\rho}\, \vec{\underline{v}}\, \mathrm{d}\underline{v}\,. \tag{4.209}$$

The stress state at a point O is considered. Therefore, an infinitesimal tetrahedron OABC shown in Figure 4.19 is introduced. The tetrahedron OABC is placed so that the point O is in one of its corners. Three faces of the tetrahedron OABC are parallel to the coordinate planes $(x_1, x_2)$, $(x_2, x_3)$, and $(x_1, x_3)$, and the origin of the coordinate system $(x_1, x_2, x_3)$ is at the point O.

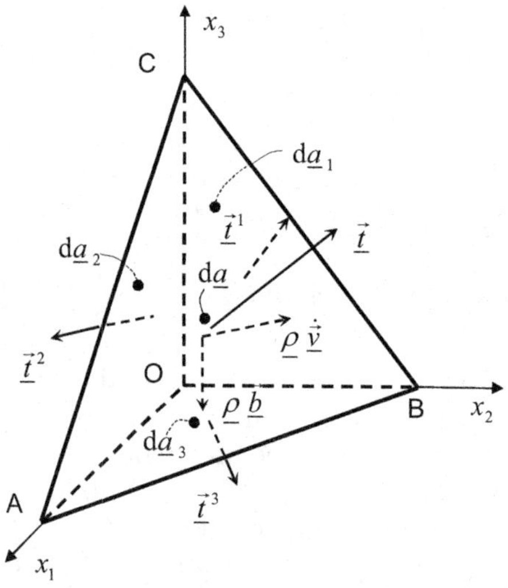


**Figure 4.19** Infinitesimal tetrahedron OABC for determination of the stress state at a point O.

The traction vector $\vec{\underline{t}}$ is a continuous function on the faces of the tetrahedron OABC. Figure 4.19 gives its values on the centroids of the triangular faces of the tetrahedron OABC. Also the volume quantities $\underline{\rho}\,\dot{\vec{\underline{v}}}$ and $\underline{\rho}\,\vec{\underline{b}}$ are continuous functions. Their values are given in Figure 4.19 on the centroid of the tetrahedron OABC. The areas of the faces of the tetrahedron are denoted by $\mathrm{d}\underline{a}$, $\mathrm{d}\underline{a}_1$,

$\mathrm{d}\underline{a}_2$, and $\mathrm{d}\underline{a}_3$. The volume of the tetrahedron OABC is $\mathrm{d}\underline{v} = 1/3h\mathrm{d}\underline{a}$, where $h$ is the height of the vertex O from the face $\mathrm{d}\underline{a}$.

The law of balance of momentum, that is, Basic Law (4.209) [**BL 1**], is applied for the Cauchy tetrahedron OABC. When integrating over the faces of the tetrahedron OABC, the values of the traction vectors $\vec{\underline{t}}$ are approximated by their values on the centroid of each face. Similarly, the volume forces $\underline{\rho}\dot{\vec{\underline{v}}}$ and $\underline{\rho}\vec{\underline{b}}$ are approximated by their values on the centroid of the tetrahedron OABC. The assumption that the traction vector $\vec{\underline{t}}$ remains unchanged on the faces of the tetrahedron does not cause any error. The differences between the applied values and the correct values are infinitesimal. Since at the end of this evaluation the size of the tetrahedron OABC is let to approach zero, the infinitesimal values vanish.

Equation (4.209) [**BL 1**] with the information given by Figure 4.19 yields

$$\vec{\underline{t}}\,\mathrm{d}\underline{a} + \vec{\underline{t}}^{\,1}\,\mathrm{d}\underline{a}_1 + \vec{\underline{t}}^{\,2}\,\mathrm{d}\underline{a}_2 + \vec{\underline{t}}^{\,3}\,\mathrm{d}\underline{a}_3 + \underline{\rho}\,\vec{\underline{b}}\,\tfrac{1}{3}\,h\,\mathrm{d}\underline{a} = \underline{\rho}\,\dot{\vec{\underline{v}}}\,\tfrac{1}{3}\,h\,\mathrm{d}\underline{a}\,. \tag{4.210}$$

The angle between the $x_1$-axis and the outward unit normal of the face ABC $\vec{\underline{n}}$ is denoted by $\alpha$. Similarly, the quantities $\beta$ and $\gamma$ are the angles between the $x_2$-axis, the $x_3$-axis, and the vector $\vec{\underline{n}}$. This gives

$$\begin{cases} \mathrm{d}\underline{a}_1 = \mathrm{d}\underline{a}\cos\alpha \\ \mathrm{d}\underline{a}_2 = \mathrm{d}\underline{a}\,\cos\beta \\ \mathrm{d}\underline{a}_3 = \mathrm{d}\underline{a}\,\cos\gamma \end{cases} \text{and} \quad \begin{cases} \underline{n}_{\mathrm{x}1} = \cos\alpha \\ \underline{n}_{\mathrm{x}2} = \cos\beta \\ \underline{n}_{\mathrm{x}3} = \cos\gamma \end{cases} \Rightarrow \quad \begin{cases} \mathrm{d}\underline{a}_1 = \underline{n}_{\mathrm{x}1}\,\mathrm{d}\underline{a} \\ \mathrm{d}\underline{a}_2 = \underline{n}_{\mathrm{x}2}\,\mathrm{d}\underline{a} \\ \mathrm{d}\underline{a}_3 = \underline{n}_{\mathrm{x}3}\,\mathrm{d}\underline{a}\,, \end{cases} \tag{4.211}$$

where the quantities $\underline{n}_{\mathrm{x}1}$, $\underline{n}_{\mathrm{x}2}$, and $\underline{n}_{\mathrm{x}3}$ are the scalar components of the outward unit normal vector $\vec{\underline{n}}$ along the $x_1$-, $x_2$-, and $x_3$-axes (see also Figure 4.20).

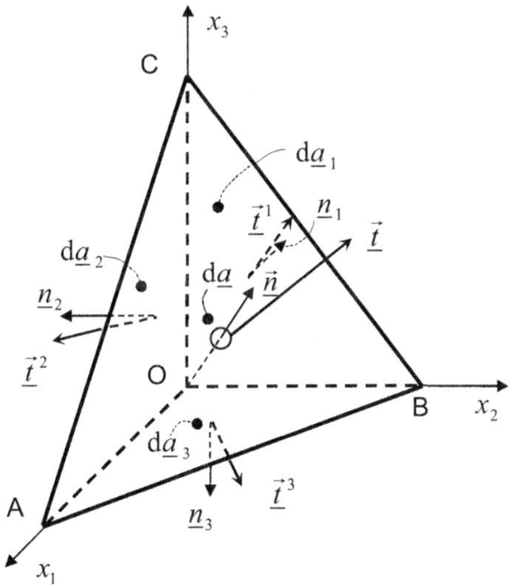


**Figure 4.20** Infinitesimal tetrahedron OABC for determination of the unit normals.

Substitution of Expression (4.211) into Equation (4.210) gives

$$\underline{\vec{t}}\,\mathrm{d}\underline{a} + \underline{\vec{t}}^{\,1}\underline{n}_{\mathrm{x}1}\,\mathrm{d}\underline{a} + \underline{\vec{t}}^{\,2}\underline{n}_{\mathrm{x}2}\,\mathrm{d}\underline{a} + \underline{\vec{t}}^{\,3}\underline{n}_{\mathrm{x}3}\,\mathrm{d}\underline{a} + \underline{\rho}\,\underline{\vec{b}}\,\tfrac{1}{3}\,h\,\mathrm{d}\underline{a} = \underline{\rho}\,\underline{\dot{\vec{v}}}\,\tfrac{1}{3}\,h\,\mathrm{d}\underline{a}\,. \tag{4.212}$$

Equation (4.212) is divided by the differential area $\mathrm{d}\underline{a}$, and we arrive at the following:

$$\underline{\vec{t}} + \underline{\vec{t}}^{\,1}\underline{n}_{\mathrm{x}1} + \underline{\vec{t}}^{\,2}\underline{n}_{\mathrm{x}2} + \underline{\vec{t}}^{\,3}\underline{n}_{\mathrm{x}3} + \underline{\rho}\,\underline{\vec{b}}\,\tfrac{1}{3}\,h = \underline{\rho}\,\underline{\dot{\vec{v}}}\,\tfrac{1}{3}\,h\,. \tag{4.213}$$

Now the size of the tetrahedron OABC is let to approach zero. This implies that the values of the functions $\underline{\vec{t}}$, $\underline{\vec{t}}^{\,1}$, $\underline{\vec{t}}^{\,2}$, and $\underline{\vec{t}}^{\,3}$ approach their values at the point O. Furthermore, the height $h$ approaches zero, making those terms containing it negligible compared to the other terms of Expression (4.213). Thus Expression (4.213) reduces to

$$\underline{\vec{t}} + \underline{\vec{t}}^{\,1}\underline{n}_{\mathrm{x}1} + \underline{\vec{t}}^{\,2}\underline{n}_{\mathrm{x}2} + \underline{\vec{t}}^{\,3}\underline{n}_{\mathrm{x}3} = \vec{0}\,. \tag{4.214}$$

The outward unit normals to the faces OAB, OAC, and OBC are toward the opposite directions of the components $\underline{n}_{\mathrm{x}1}$, $\underline{n}_{\mathrm{x}2}$, and $\underline{n}_{\mathrm{x}3}$ of the vector $\underline{\vec{n}}$, as shown in Figure 4.20. Thus is obtained

$$\underline{n}_{\mathrm{x}1} = -\,\underline{n}_1\,, \qquad \underline{n}_{\mathrm{x}2} = -\,\underline{n}_2\,, \qquad \text{and} \quad \underline{n}_{\mathrm{x}3} = -\,\underline{n}_3\,. \tag{4.215}$$

Substitution of Relations (4.215) into Expression (4.214) gives

$$\underline{\vec{t}} - \underline{\vec{t}}^{\,1}\underline{n}_1 - \underline{\vec{t}}^{\,2}\underline{n}_2 - \underline{\vec{t}}^{\,3}\underline{n}_3 = \vec{0}\,, \tag{4.216}$$

which yields

$$\underline{\vec{t}} = \underline{\vec{t}}^{\,1}\underline{n}_1 + \underline{\vec{t}}^{\,2}\underline{n}_2 + \underline{\vec{t}}^{\,3}\underline{n}_3\,. \tag{4.217}$$

Equation (4.217) contains three scalar equations, which are

$$\begin{aligned}
\underline{t}_1 &= \underline{t}_1^1\,\underline{n}_1 + \underline{t}_1^2\,\underline{n}_2 + \underline{t}_1^3\,\underline{n}_3\,,\\
\underline{t}_2 &= \underline{t}_2^1\,\underline{n}_1 + \underline{t}_2^2\,\underline{n}_2 + \underline{t}_2^3\,\underline{n}_3\,,\\
\underline{t}_3 &= \underline{t}_3^1\,\underline{n}_1 + \underline{t}_3^2\,\underline{n}_2 + \underline{t}_3^3\,\underline{n}_3\,.
\end{aligned} \tag{4.218}$$

The following notation is introduced:

$$\underline{\sigma}^{\mathrm{c}}_{ji} := \underline{t}^i_j\,. \tag{4.219}$$

Notation (4.219) allows Equations (4.218) to be written in the form

$$\begin{aligned}
\underline{t}_1 &:= \underline{\sigma}^{\mathrm{c}}_{11}\,\underline{n}_1 + \underline{\sigma}^{\mathrm{c}}_{12}\,\underline{n}_2 + \underline{\sigma}^{\mathrm{c}}_{13}\,\underline{n}_3\,,\\
\underline{t}_2 &:= \underline{\sigma}^{\mathrm{c}}_{21}\,\underline{n}_1 + \underline{\sigma}^{\mathrm{c}}_{22}\,\underline{n}_2 + \underline{\sigma}^{\mathrm{c}}_{23}\,\underline{n}_3\,,\\
\underline{t}_3 &:= \underline{\sigma}^{\mathrm{c}}_{31}\,\underline{n}_1 + \underline{\sigma}^{\mathrm{c}}_{32}\underline{n}_2 + \underline{\sigma}^{\mathrm{c}}_{33}\,\underline{n}_3\,.
\end{aligned} \tag{4.220}$$

Equation (4.220) can be written in terms of tensors as follows:

$$\underline{\vec{n}}\cdot\underline{\boldsymbol{\sigma}}^{\mathrm{c}} := \underline{\vec{t}}\,. \tag{4.221}$$

Equation (4.221) is the definition of the Cauchy stress tensor $\underline{\boldsymbol{\sigma}}^{\mathrm{c}}$ [see Definition $(4.90)_2$]. Equation (4.221) is called Cauchy's stress theorem or Cauchy's formula.

In continuum mechanics, the Cauchy tetrahedron is traditionally used for deriving the traction vector $\underline{\vec{t}}$ in cases in which the Cauchy stress tensor $\underline{\boldsymbol{\sigma}}^{\mathrm{c}}$ is known (see, e.g., Fung [24], pp. 63–65; Malvern [57], pp. 73–77; Reddy [84], pp. 152–155). The definitions given by Figure 4.12 and Definition (4.90) show that this should be vice versa, as noted by Reddy ([84], Section 4.2.3). Equation (4.221) is the definition for the Cauchy stress tensor $\underline{\boldsymbol{\sigma}}^{\mathrm{c}}$, since the traction vector $\underline{\vec{t}}$ is already defined.

There is a small flaw in the preceding derivation. The origin O of the coordinate system $(x_1, x_2, x_3)$ is assumed to be at the point where the Cauchy stress tensor $\underline{\boldsymbol{\sigma}}^{\mathrm{c}}$ is wanted to be defined. However, in Figure 4.9, for example, the origin of the coordinate system O is outside the system $\mathrm{m^b}$. This is not a real problem, since Figure 4.9 is just a sketch and the position of the frame $(x_1, x_2, x_3)$ is not defined in the mathematical derivations. Thus, in this section, its position can be adjusted freely. In order to get the definition for the Cauchy stress tensor $\underline{\boldsymbol{\sigma}}^{\mathrm{c}}$ at all the points within the system $\mathrm{m^b}$, the foregoing study must be carried out at every point of the system $\mathrm{m^b}$.

## 4.22 Basic Laws Set Requirements for Surface Integrals

The present section evaluates the requirements set by the surface integrals present in the basic laws. The law of balance of momentum with the current material coordinates $\underline{x}_i(t)$ is studied as an example.

The law of balance of momentum for subdomain $\underline{v}(t)$ {Basic Law (4.18) [**BL 1**]} is recalled. It is

**BL 1**
$$\oint_{\partial \underline{v}(t)} \underline{\vec{t}}\,\mathrm{d}\underline{a} + \int_{\underline{v}(t)} \underline{\rho}\,\underline{\vec{b}}\,\mathrm{d}\underline{v} = \frac{\mathrm{D}}{\mathrm{D}t}\int_{\underline{v}(t)} \underline{\rho}\,\underline{\vec{v}}\,\mathrm{d}\underline{v}\,. \tag{4.222}$$

In Basic Law (4.222) [**BL 1**], there are surface and volume integrals. All the integrands have to be written in a single volume integral, thus the surface integral has to be converted somehow into a volume integral. Generalized Gauss's Theorem $(2.170)_1$ is

$$\int_V \vec{\nabla} * f\,\mathrm{d}V = \oint_{\partial V} \vec{n} * f\,\mathrm{d}A\,, \tag{4.223}$$

where the "star product operator" $*$ in Theorem (4.223) represents a dot product operator $\cdot$, or a cross product operator $\times$, or it can be an empty space. Before writing Basic Law (4.222) [**BL 1**], all its quantities and mathematical operations are of course known. Therefore the traction vector $\underline{\vec{t}}$ is also known. In order to change the surface integral term

$$\oint_{\partial \underline{v}(t)} \underline{\vec{t}}\,\mathrm{d}\underline{a} \tag{4.224}$$

to a volume integral by the generalized Gauss's theorem, a new quantity denoted by $\underline{\phi}$ has to obey the property

$$\underline{\vec{n}} * \underline{\phi} := \underline{\vec{t}}. \tag{4.225}$$

If the star product operator $*$ was an empty space, the quantity $\underline{\phi}$ should be a scalar in order for the term $\underline{\vec{n}}*\underline{\phi}$ to be a vector. This means that the directions of the outward unit normal $\underline{\vec{n}}$ and of the traction vector $\underline{\vec{t}}$ coincide. This is not for solids, as Figure 4.12(a) shows. Figure 4.21 illustrates the important detail of Figure 4.12(a) for this study.

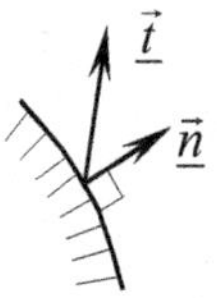


**Figure 4.21** Vectors $\underline{\vec{t}}$ and $\underline{\vec{n}}$ are usually directed differently in solids.

As the forthcoming evaluation will show, the idea of the quantity $\underline{\phi}$ being a scalar is for fluids since the direction of the surface traction vector $\underline{\vec{t}}$ associated to the thermodynamic pressure $p$ coincides with the direction of the outward unit normal $\underline{\vec{n}}$. Therefore, empty space is a candidate for the star product operator $*$.

If the star product operator $*$ was a cross product operator $\times$, the quantity $\underline{\phi}$ should be a vector in order for the term $\underline{\vec{n}}*\underline{\phi}$ to be a vector. A cross product of two vectors is a vector that is perpendicular to both vectors. Thus, at least in principle, this kind of a material behavior could exist. Thus the star product operator $*$ can be a cross product operator, $\times$.

If the star product $*$ was a dot product $\cdot$, the quantity $\underline{\phi}$ should be a second-order tensor in order for the term $\underline{\vec{n}} * \underline{\phi}$ to be a vector. This is the case in solid mechanics and therefore it is also a possible model.

Based on the preceding discussion and Definition (4.225), the following can be written:

$$\underline{\vec{n}}\,\underline{a} + \underline{\vec{n}} \times \underline{\vec{a}} + \underline{\vec{n}} \cdot \underline{\mathbf{a}} = \underline{\vec{t}}, \tag{4.226}$$

where the scalar $\underline{a}$, the vector $\underline{\vec{a}}$, and the second-order tensor $\underline{\mathbf{a}}$ are as yet unknown quantities. In order to simplify forthcoming derivations, the first two terms on the left side of Expression (4.226) are written as follows:

$$\underline{\vec{n}}\,\underline{a} = \underline{\vec{n}} \cdot \mathbf{1}\,\underline{a} \qquad \text{and} \qquad \underline{\vec{n}} \times \underline{\vec{a}} = \underline{\vec{n}} \times (\mathbf{1} \cdot \underline{\vec{a}}) = \underline{\vec{n}} \cdot (\mathbf{1} \times \underline{\vec{a}})\,. \tag{4.227}$$

Problem 2.4 is to show that Expression $(4.227)_2$ is correct. The cross product $\times$ between a second-order tensor and a vector is defined to be an act between the second base vector of the second-order tensor and the base vector of the vector.

Based on Expressions (4.227), the following form for Equation (4.226) is achieved:

$$\underline{\vec{n}} \cdot \mathbf{1}\,\underline{a} + \underline{\vec{n}} \cdot (\mathbf{1} \times \underline{\vec{a}}) + \underline{\vec{n}} \cdot \underline{\mathbf{a}} = \underline{\vec{t}}, \quad \text{which is} \quad \underline{\vec{n}} \cdot [\mathbf{1}\,\underline{a} + (\mathbf{1} \times \underline{\vec{a}}) + \underline{\mathbf{a}}] = \underline{\vec{t}}. \tag{4.228}$$

Equation $(4.228)_2$ can be written as

$$\underline{\vec{n}} \cdot \underline{\boldsymbol{\sigma}}^{\mathrm{c}} = \underline{\vec{t}}, \quad \text{where} \quad \underline{\boldsymbol{\sigma}}^{\mathrm{c}} := \mathbf{1}\,\underline{a} + (\mathbf{1} \times \underline{\vec{a}}) + \underline{\mathbf{a}}\,. \tag{4.229}$$

By using Definition $(4.229)_1$, the evaluation can be continued by following the steps taken in Section 4.17. Thus Equation of Motion $(4.123)_1$ holds. It is

$$\vec{\underline{\nabla}} \cdot \underline{\sigma}^{\mathrm{c}} + \underline{\rho} \, \vec{\underline{b}} = \underline{\rho} \, \frac{\mathrm{D}\vec{\underline{v}}}{\mathrm{D}t} \,. \tag{4.230}$$

The law of balance of momentum yields that the second-order tensor $\underline{\sigma}^{\mathrm{c}}$ is symmetric. Since the second-order identity tensor $\mathbf{1}$ is symmetric, based on Definition $(4.90)_2$, the first term of the tensor $\underline{\sigma}^{\mathrm{c}}$ is symmetric. It can be shown (Problem 2.4) that the term $\mathbf{1} \times \vec{\underline{a}}$ is symmetric only when the vector $\vec{\underline{a}}$ vanishes. This implies that the second term of Equation $(4.229)_2$ vanishes. When the tensor $\underline{\sigma}^{\mathrm{c}}$ is symmetric, the tensor $\underline{\mathbf{a}}$ has to be symmetric. Thus the preceding study gives

$$\underline{\sigma}^{\mathrm{c}} = \underline{a} \, \mathbf{1} + \underline{\mathbf{a}} \,, \tag{4.231}$$

where $\underline{\sigma}^{\mathrm{c}}$ is a second-order symmetric tensor.

## 4.23 Betti's Theorem

In this section, we derive Betti's theorem. In Section 4.23.1, Betti's theorem is derived for a case in which the system $\mathrm{m}^{\mathrm{b}}$ is loaded by the point forces. Section 4.23.2 studies a case in which the system $\mathrm{m}^{\mathrm{b}}$ is loaded by continuous loading. Both sections study Hookean deformation of the material. The assumption of small deformations and rotations is assumed to hold.

### 4.23.1 System $\mathrm{m}^{\mathrm{b}}$ Loaded by Point Forces

This section derives the Betti's theorem for a system $\mathrm{m}^{\mathrm{b}}$ loaded by point forces. For simplicity, the number of point forces is two, but the view can easily be extended for many point forces.

A system $\mathrm{m}^{\mathrm{b}}$ showing Hookean deformation is loaded by the point forces $F_1$ and $F_2$, as depicted in Figure 4.22. Two different loading cases are studied. In the loading case A, the force $F_1$ is applied first and kept constant while load $F_2$ is applied, as shown in Figures 4.23. The loading is assumed to be quasi-static; thus the inertia forces can be neglected as small quantities. On the other hand, loading case B describes a process in which the load $F_2$ is applied first and kept constant while $F_1$ is applied.

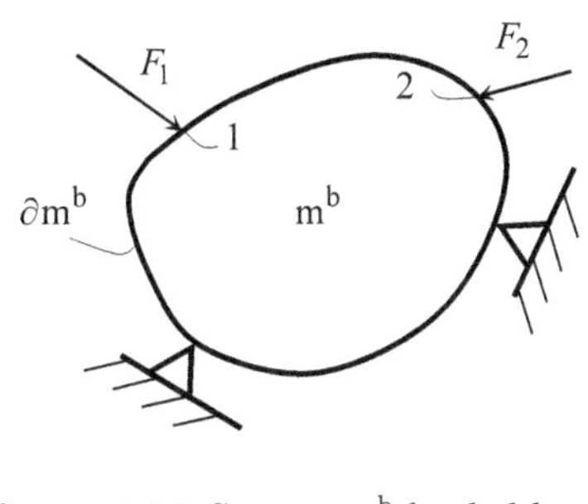


**Figure 4.22** System $\mathrm{m}^{\mathrm{b}}$ loaded by the forces $F_1$ and $F_2$ .

The displacement of the material point $s$ in the direction of force $F_s$ applied at the point $s$ due to force $F_t$ is denoted by $\Delta_{st}$.

Case A. Force $F_1$ is applied. It does the work $W_1$,

$$W_1 = \frac{1}{2} F_1 \, \Delta_{11} \,. \tag{4.232}$$

Force $F_2$ is applied while the value of force $F_1$ is kept constant. The total work done, denoted by $\bar{W}_1$, reads

$$\bar{W}_1 = \frac{1}{2} F_1 \Delta_{11} + \frac{1}{2} F_2 \Delta_{22} + F_1 \Delta_{12} . \tag{4.233}$$

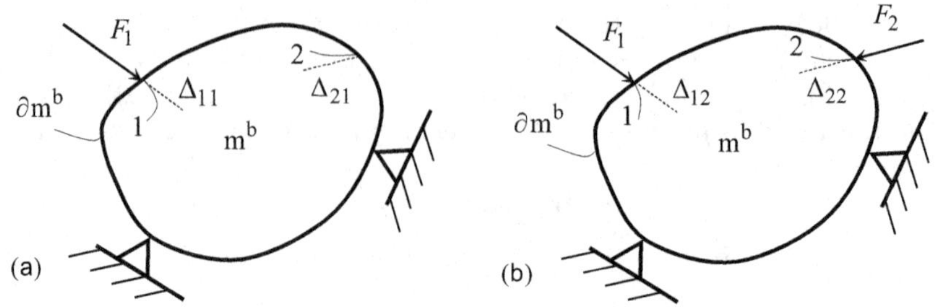


**Figure 4.23** (a) Growing force $F_1$ acting on the system $m^b$ and (b) growing force $F_2$ acting on the system $m^b$ whereas force $F_1$ keeps a constant value.

Case B. Force $F_2$ is applied. It does the work $W_2$,

$$W_2 = \frac{1}{2} F_2 \Delta_{22} . \tag{4.234}$$

Force $F_1$ is applied while the value of force $F_2$ is kept constant. The total work done, denoted by $\bar{W}_2$, reads

$$\bar{W}_2 = \frac{1}{2} F_2 \Delta_{22} + \frac{1}{2} F_1 \Delta_{11} + F_2 \Delta_{21} . \tag{4.235}$$

The work done during the loading cases A and B are equal. This is

$$\bar{W}_1 = \bar{W}_2 . \tag{4.236}$$

Substitution of Expressions (4.233) and (4.235) into Equation (4.236) leads to

$$\frac{1}{2} F_1 \Delta_{11} + \frac{1}{2} F_2 \Delta_{22} + F_1 \Delta_{12} = \frac{1}{2} F_2 \Delta_{22} + \frac{1}{2} F_1 \Delta_{11} + F_2 \Delta_{21} , \tag{4.237}$$

which yields

$$F_1 \Delta_{12} = F_2 \Delta_{21} . \tag{4.238}$$

Equation (4.238) is Betti's theorem. The physical meaning of the left side of Betti's theorem reads "the force $F_1$ times the displacement caused by force $F_2$ at the application point of the force $F_1$ and in the direction of force $F_1$." The right side has a similar meaning.

## 4.23.2 System $\mathrm{m^b}$ Loaded by the Surface Traction Vector $\vec{T}$ and Body Force Vector $\vec{b}$

This section derives Betti's theorem for a case in which the subsystem $V_0$ is loaded by the surface traction vector $\vec{T}$ and the body force vector $\vec{b}$. The response of the system $\mathrm{m^b}$ is assumed to obey Hooke's law.

Two loading cases, denoted by $\boldsymbol{\sigma}^{①}$ and $\boldsymbol{\sigma}^{②}$, are studied. The corresponding strain fields are denoted by $\boldsymbol{\varepsilon}^{①}$ and $\boldsymbol{\varepsilon}^{②}$, respectively. The following two quantities are studied:

$$\breve{w}^{①②} := \boldsymbol{\varepsilon}^{②} : \boldsymbol{\sigma}^{①} \qquad \text{and} \qquad \breve{w}^{②①} := \boldsymbol{\varepsilon}^{①} : \boldsymbol{\sigma}^{②} . \tag{4.239}$$

The first superscript in the term $\breve{w}^{①②}$ refers to loading case ① and the second one to loading case ②. The same rule holds for the quantity $\breve{w}^{②①}$. Hooke's law for these two loading cases takes the following appearances:

$$\boldsymbol{\sigma}^{①} = \mathbf{C} : \boldsymbol{\varepsilon}^{①} \qquad \text{and} \qquad \boldsymbol{\sigma}^{②} = \mathbf{C} : \boldsymbol{\varepsilon}^{②} . \tag{4.240}$$

Substitution of Constitutive Equations (4.240) into Definitions (4.239) gives

$$\breve{w}^{①②} = \boldsymbol{\varepsilon}^{②} : \mathbf{C} : \boldsymbol{\varepsilon}^{①} \qquad \text{and} \qquad \breve{w}^{②①} = \boldsymbol{\varepsilon}^{①} : \mathbf{C} : \boldsymbol{\varepsilon}^{②} . \tag{4.241}$$

Term $(4.241)_1$ can then take the form

$$\breve{w}^{①②} = (\boldsymbol{\varepsilon}^{②}) : (\mathbf{C} : \boldsymbol{\varepsilon}^{①}) . \tag{4.242}$$

The double-dot product between two second-order tensors is a scalar. Therefore, if $\mathbf{a}$ and $\mathbf{c}$ are arbitrary second-order tensors, the following holds:

$$\mathbf{a} : \mathbf{c} = a_{st}\, c_{st} = c_{st}\, a_{st} = \mathbf{c} : \mathbf{a} , \tag{4.243}$$

where the fact that the order of scalars in the multiplication can be interchanged is exploited. In Expression (4.242), the terms in the parentheses are second-order tensors. Therefore Expression (4.243) can be applied to Term (4.242), which now takes the following appearance:

$$\breve{w}^{①②} = (\boldsymbol{\varepsilon}^{②}) : (\mathbf{C} : \boldsymbol{\varepsilon}^{①}) = (\boldsymbol{\varepsilon}^{①} : \mathbf{C}^{\mathrm{T}}) : (\boldsymbol{\varepsilon}^{②}) . \tag{4.244}$$

The first term on the right side of Expression (4.244) requires a more detailed study. In the middle term of Expression (4.244), the double-dot product between $\boldsymbol{\varepsilon}^{②}$ and $\mathbf{C}$ is for the first two indices of the fourth-order constitutive tensor $\mathbf{C}$ for a Hookean deformation. The same holds for the rightmost term of Expression (4.244), since there is the transpose of the tensor $\mathbf{C}$. In the middle term of Expression (4.244), the double-dot product between $\boldsymbol{\varepsilon}^{①}$ and $\mathbf{C}$ is for the latter two indices of the tensor $\mathbf{C}$. The same holds for the rightmost term of Expression (4.244). Term (4.244) yields

$$\breve{w}^{①②} = (\boldsymbol{\varepsilon}^{②}) : (\mathbf{C} : \boldsymbol{\varepsilon}^{①}) = (\boldsymbol{\varepsilon}^{①} : \mathbf{C}^{\mathrm{T}}) : (\boldsymbol{\varepsilon}^{②}) = \boldsymbol{\varepsilon}^{①} : \mathbf{C}^{\mathrm{T}} : \boldsymbol{\varepsilon}^{②} = \boldsymbol{\varepsilon}^{①} : \mathbf{C} : \boldsymbol{\varepsilon}^{②} , \tag{4.245}$$

where the major symmetry of the tensor $\mathbf{C}$ is exploited.

Comparison of Expressions $(4.241)_2$ and (4.245) gives

$$\breve{w}^{①②} = \breve{w}^{②①}, \qquad \text{where} \qquad \breve{w}^{①②} = \boldsymbol{\sigma}^{①} : \boldsymbol{\varepsilon}^{②} \quad \text{and} \quad \breve{w}^{②①} = \boldsymbol{\sigma}^{②} : \boldsymbol{\varepsilon}^{①}, \tag{4.246}$$

where Definitions (4.239) and Property (4.243) are exploited. The order of the terms within the pairs $\boldsymbol{\sigma}^{①}$ and $\boldsymbol{\varepsilon}^{②}$ and $\boldsymbol{\sigma}^{②}$ and $\boldsymbol{\varepsilon}^{①}$ is interchanged to get more convenient forms for the expressions in the following derivation.

Equation (4.246) gives

$$\breve{W}^{①②} = \breve{W}^{②①}, \tag{4.247}$$

where

$$\breve{W}^{①②} := \int_{V_0} \breve{w}^{①②} \, \mathrm{d}V = \int_{V_0} \boldsymbol{\sigma}^{①} : \boldsymbol{\varepsilon}^{②} \, \mathrm{d}V \tag{4.248a}$$

and

$$\breve{W}^{②①} := \int_{V_0} \breve{w}^{②①} \, \mathrm{d}V = \int_{V_0} \boldsymbol{\sigma}^{②} : \boldsymbol{\varepsilon}^{①} \, \mathrm{d}V \,. \tag{4.248b}$$

Definitions (4.248a) and (4.248b) are obtained by integration over an arbitrary volume $V_0$ in the IG-CtP configuration.

According to Result (2.89), the double-dot product between the symmetric stress tensor $\boldsymbol{\sigma}^{①}$ and the skew-symmetric rotation tensor $\boldsymbol{\omega}^{②}$ vanishes. Thus the integrand of Expression (4.248a) can take the following appearance:

$$\boldsymbol{\sigma}^{①} : \boldsymbol{\varepsilon}^{②} = \boldsymbol{\sigma}^{①} : (\boldsymbol{\varepsilon}^{②} - \boldsymbol{\omega}^{②}) \,. \tag{4.249}$$

The strain tensor $\boldsymbol{\varepsilon}$ and the rotation tensor $\boldsymbol{\omega}$ are given in Definitions (3.120) and (3.121),

$$\boldsymbol{\varepsilon} := \frac{1}{2} (\vec{u}\overleftarrow{\nabla} + \vec{\nabla}\vec{u}) \qquad \text{and} \qquad \boldsymbol{\omega} := \frac{1}{2} (\vec{u}\overleftarrow{\nabla} - \vec{\nabla}\vec{u}) \,. \tag{4.250}$$

Definitions (4.250) gives

$$\boldsymbol{\varepsilon} - \boldsymbol{\omega} = \vec{\nabla}\vec{u} \,. \tag{4.251}$$

Substitution of the information given in Difference (4.251) into Equation (4.249) yields

$$\boldsymbol{\sigma}^{①} : \boldsymbol{\varepsilon}^{②} = \boldsymbol{\sigma}^{①} : (\boldsymbol{\varepsilon}^{②} - \boldsymbol{\omega}^{②}) = \boldsymbol{\sigma}^{①} : \vec{\nabla}\vec{u}^{②} \,. \tag{4.252}$$

Problem 2.23 is to show that the following holds:

$$\vec{\nabla} \cdot (\mathbf{h} \cdot \vec{e}) = (\vec{\nabla} \cdot \mathbf{h}) \cdot \vec{e} + \mathbf{h} : \vec{\nabla}\vec{e}, \tag{4.253}$$

where $\vec{\nabla}$ is the vector operator del, $\mathbf{h}$ is a second-order tensor, and $\vec{e}$ is a vector. By substituting $\mathbf{h} = \boldsymbol{\sigma}^{①}$ and $\vec{e} = \vec{u}^{②}$ into Derivative (4.253) and reordering the obtained terms, Equation (4.253) gives

$$\boldsymbol{\sigma}^{①} : \vec{\nabla}\vec{u}^{②} = \vec{\nabla} \cdot (\boldsymbol{\sigma}^{①} \cdot \vec{u}^{②}) - (\vec{\nabla} \cdot \boldsymbol{\sigma}^{①}) \cdot \vec{u}^{②} \,. \tag{4.254}$$

Substitution of $\boldsymbol{\sigma}^{①} : \vec{\nabla}\vec{u}^{②} = \boldsymbol{\sigma}^{①} : \boldsymbol{\varepsilon}^{②}$ obtained from Equation (4.252) into Equation (4.254) yields

$$\boldsymbol{\sigma}^{①} : \boldsymbol{\varepsilon}^{②} = \vec{\nabla} \cdot (\boldsymbol{\sigma}^{①} \cdot \vec{u}^{②}) - (\vec{\nabla} \cdot \boldsymbol{\sigma}^{①}) \cdot \vec{u}^{②} . \qquad (4.255)$$

Equations of Motion (4.112) read

$$\vec{\nabla} \cdot \boldsymbol{\sigma} + \rho_0 \vec{b} = \rho_0 \frac{\mathrm{D}\vec{v}}{\mathrm{D}t} \qquad \text{or} \qquad \vec{\nabla} \cdot \boldsymbol{\sigma} + \rho_0 \vec{b} = \rho_0 \dot{\vec{v}} . \qquad (4.256)$$

Substitution of Form $(4.256)_2$ into the last term of Equation (4.255) leads to

$$\boldsymbol{\sigma}^{①} : \boldsymbol{\varepsilon}^{②} = \vec{\nabla} \cdot (\boldsymbol{\sigma}^{①} \cdot \vec{u}^{②}) + \rho_0 \vec{b}^{①} \cdot \vec{u}^{②} - \rho_0 \dot{\vec{v}}^{①} \cdot \vec{u}^{②} . \qquad (4.257)$$

Substitution of Result (4.257) into Expression (4.248a) gives

$$\breve{W}^{①②} = \int_{V_0} \vec{\nabla} \cdot (\boldsymbol{\sigma}^{①} \cdot \vec{u}^{②}) \, \mathrm{d}V + \int_{V_0} \rho_0 \vec{b}^{①} \cdot \vec{u}^{②} \, \mathrm{d}V - \int_{V_0} \rho_0 \dot{\vec{v}}^{①} \cdot \vec{u}^{②} \, \mathrm{d}V . \quad (4.258)$$

Generalized Gauss's Theorem $(2.170)_1$ is

$$\int_V \vec{\nabla} * f \, \mathrm{d}V = \oint_{\partial V} \vec{n} * f \, \mathrm{d}A , \qquad (4.259)$$

where $f$ possesses continuous partial derivatives in the closed subdomain $V$, $\partial V$ is the surface of the volume $V$, $\vec{n}$ is the outward unit normal to the volume $V$, and the "star product operator" $*$ in Theorem $(2.170)_1$ represents a dot product operator $\cdot$, or a cross product operator $\times$, or it can be an empty space. The generalized Gauss's theorem is applied to the second term of Equation (4.258). This gives

$$\breve{W}^{①②} = \int_{\partial V_0} \vec{N} \cdot (\boldsymbol{\sigma}^{①} \cdot \vec{u}^{②}) \, \mathrm{d}A + \int_{V_0} \rho_0 \vec{b}^{①} \cdot \vec{u}^{②} \, \mathrm{d}V - \int_{V_0} \rho_0 \dot{\vec{v}}^{①} \cdot \vec{u}^{②} \, \mathrm{d}V . \qquad (4.260)$$

In the first term on the right side of Equation (4.260), the outward unit normal $\vec{N}$ operates with the first index of the stress tensor $\boldsymbol{\sigma}^{①}$ and the displacement vector $\vec{u}^{②}$ operates with the second index of the stress tensor $\boldsymbol{\sigma}^{①}$. Therefore the parentheses are not necessary, and Equation (4.260) can be therefore written in the form

$$\breve{W}^{①②} = \int_{\partial V_0} \vec{N} \cdot \boldsymbol{\sigma}^{①} \cdot \vec{u}^{②} \, \mathrm{d}A + \int_{V_0} \rho_0 \vec{b}^{①} \cdot \vec{u}^{②} \, \mathrm{d}V - \int_{V_0} \rho_0 \dot{\vec{v}}^{①} \cdot \vec{u}^{②} \, \mathrm{d}V . \quad (4.261)$$

When small deformations and rotations are studied, the first Piola–Kirchhoff stress tensor $\boldsymbol{\sigma}^{\square}$, reduces to the stress tensor $\boldsymbol{\sigma}$ and Definition (4.91) yields

$$\vec{N}(\vec{X}) \cdot \boldsymbol{\sigma}^{\square}(\vec{X}, t) := \vec{T}(\vec{X}, t, \vec{N}) \qquad \Rightarrow \qquad \vec{N} \cdot \boldsymbol{\sigma} = \vec{T} , \qquad (4.262)$$

where $\vec{T}$ is the traction vector. Substitution of Equation $(4.262)_2$ into the first term on the right side of Equation (4.261) yields

$$\breve{W}^{\textcircled{1}\textcircled{2}} = \int_{\partial V_0} \vec{T}^{\textcircled{1}} \cdot \vec{u}^{\textcircled{2}} \, dA + \int_{V_0} \rho_0 \vec{b}^{\textcircled{1}} \cdot \vec{u}^{\textcircled{2}} \, dV - \int_{V_0} \rho_0 \dot{\vec{v}}^{\textcircled{1}} \cdot \vec{u}^{\textcircled{2}} \, dV \, . \quad (4.263)$$

Result (4.263) is derived from Expression (4.248a), where the integrand is $\boldsymbol{\sigma}^{\textcircled{1}} : \boldsymbol{\varepsilon}^{\textcircled{2}}$. A corresponding derivation can be carried out by starting from Expression (4.248b) where the integrand reads $\boldsymbol{\sigma}^{\textcircled{2}} : \boldsymbol{\varepsilon}^{\textcircled{1}}$. This would lead to the following expression:

$$\breve{W}^{\textcircled{2}\textcircled{1}} = \int_{\partial V_0} \vec{T}^{\textcircled{2}} \cdot \vec{u}^{\textcircled{1}} \, dA + \int_{V_0} \rho_0 \vec{b}^{\textcircled{2}} \cdot \vec{u}^{\textcircled{1}} \, dV - \int_{V_0} \rho_0 \dot{\vec{v}}^{\textcircled{2}} \cdot \vec{u}^{\textcircled{1}} \, dV \, . \quad (4.264)$$

Based on Equation (4.247), the left sides of Equations (4.263) and (4.264) are equal. Therefore, the right sides of Equations (4.263) and (4.264) are equal as well. This is

$$\begin{aligned} &\int_{\partial V_0} \vec{T}^{\textcircled{1}} \cdot \vec{u}^{\textcircled{2}} dA + \int_{V_0} \rho_0 \vec{b}^{\textcircled{1}} \cdot \vec{u}^{\textcircled{2}} dV - \int_{V_0} \rho_0 \dot{\vec{v}}^{\textcircled{1}} \cdot \vec{u}^{\textcircled{2}} dV \\ &\qquad = \int_{\partial V_0} \vec{T}^{\textcircled{2}} \cdot \vec{u}^{\textcircled{1}} \, dA + \int_{V_0} \rho_0 \vec{b}^{\textcircled{2}} \cdot \vec{u}^{\textcircled{1}} \, dV - \int_{V_0} \rho_0 \dot{\vec{v}}^{\textcircled{2}} \cdot \vec{u}^{\textcircled{1}} \, dV \, . \end{aligned} \quad (4.265)$$

Equation (4.265) is the extended Betti's theorem.

The first term on the left side of Expression (4.265) is studied in more detail. Due to the integral, the displacement vector $\vec{u}^{\textcircled{2}}$ is calculated at the same point with the traction (force) vector $\vec{T}^{\textcircled{1}}$, and the dot product $\cdot$ between $\vec{T}^{\textcircled{1}}$ and $\vec{u}^{\textcircled{2}}$ gives the component of the displacement vector $\vec{u}^{\textcircled{2}}$ in the direction of the traction (force) vector $\vec{T}^{\textcircled{1}}$. The displacement vector $\vec{u}^{\textcircled{2}}$ is due to the traction (force) vector $\vec{T}^{\textcircled{2}}$. This is physically exactly the same as set out at the end of Section 4.23.1 for the term on the left side of Betti's Theorem (4.238) for the point forces.

## 4.24 Summary

Chapter 3 and Chapter 4 so far gave a detailed introduction to continuum mechanics. This extensive number of details could have obscured the reader's view of the broader picture, which this section hopefully provides. Thus only a small number of equations is given, and these are mainly copies of the results already given. Since the focus of the present section is to give a general view, many important details are dropped, and simplifications must be made. However, the positions of the missing derivations are given, which should help the reader deepen their knowledge, if that part of the text was not studied.

Before introducing a new branch of physics, some fundamental elements need presenting. They are fundamental quantities (variables), units, and primitive concepts. At the beginning of this chapter, the fundamental quantities and fundamental units of continuum mechanics were introduced and summarized in Figure 4.1. Figure 4.1 is copied to 4.24. The other variables are called derived variables, and they are derived from the fundamental variables. Besides the fundamental quantities and fundamental units, some primitive concepts were introduced, such as a coordinate system and continuous matter.

| | | | | |
|---|---|---|---|---|
| Density $\underline{\rho}$ kilogram /m³ [kg/meter³] | Length $\underline{\ell}$ meter [m] | Time $t$ second [s] | Traction vector $\underline{\vec{t}}$ [N/m²] | Body force vector $\underline{\vec{b}}$ [N/kg] |

**Figure 4.24** Fundamental quantities and the units associated to them for continuum mechanics.

Before loading, the system $\mathrm{m}^{\mathrm{b}}$ (body) is said to be in the initial configuration $\mathrm{V}_0^{\mathrm{b}}$. When loading is applied, usually the system moves and deforms. Since in general the amount of loading varies with time, the movement and the deformation of the system $\mathrm{m}^{\mathrm{b}}$ are time-dependent processes. Therefore the values of the quantities, temperature $T$, for example, vary with time $t$. During movement and deformation, the system $\mathrm{m}^{\mathrm{b}}$ is said to be in the current configuration $\underline{v}^{\mathrm{b}}(t)$. Figure 4.25 show, the system $\mathrm{m}^{\mathrm{b}}$ in the initial configuration $\mathrm{V}_0^{\mathrm{b}}$ and in the current configuration $\underline{v}^{\mathrm{b}}(t)$. Since these processes are time dependent, there are an infinite number of current configurations $\underline{v}^{\mathrm{b}}(t)$, one of those is shown in Figure 4.25(a). The quantities which refer to the current configuration $\underline{v}^{\mathrm{b}}(t)$ are underlined.

It is important to note that the basic laws and axioms are written for the current configuration $\underline{v}^{\mathrm{b}}(t)$ and for an arbitrarily selected but nonchanging entity of matter $m$. Since this entity of matter is an arbitrary part of the system $\mathrm{m}^{\mathrm{b}}$ (body), it is called subsystem $m$. When the mass points are once selected to belong to the subsystem $m$, they remain there during the deformation. Although the mass of the subsystem $m$ remains constant during deformation, its volume $\underline{v}(t)$ varies with time $t$ and its boundary $\partial\underline{v}(t)$ deforms, as shown in Figure 4.25.

In Figure 4.25, the position of point P and its position vector are shown in all three of the aforementioned configurations. Since the position of point P and temperature of the system $T$ vary with time $t$, they are expressed in the current configuration $\underline{v}^{\mathrm{b}}(t)$ in the forms $\underline{\vec{x}}(t)$ and $\underline{T}(\underline{\vec{x}}(t), t)$. The first time $t$ enters the equation for temperature $\underline{T}(\underline{\vec{x}}(t), t)$ to show that the position of point $\underline{\mathrm{P}}(t)$ can vary with time $t$. The second time $t$ is written to show that the temperature $\underline{T}(\underline{\vec{x}}(t), t)$ can vary, although the position of point $\underline{P}(t)$ does not change.

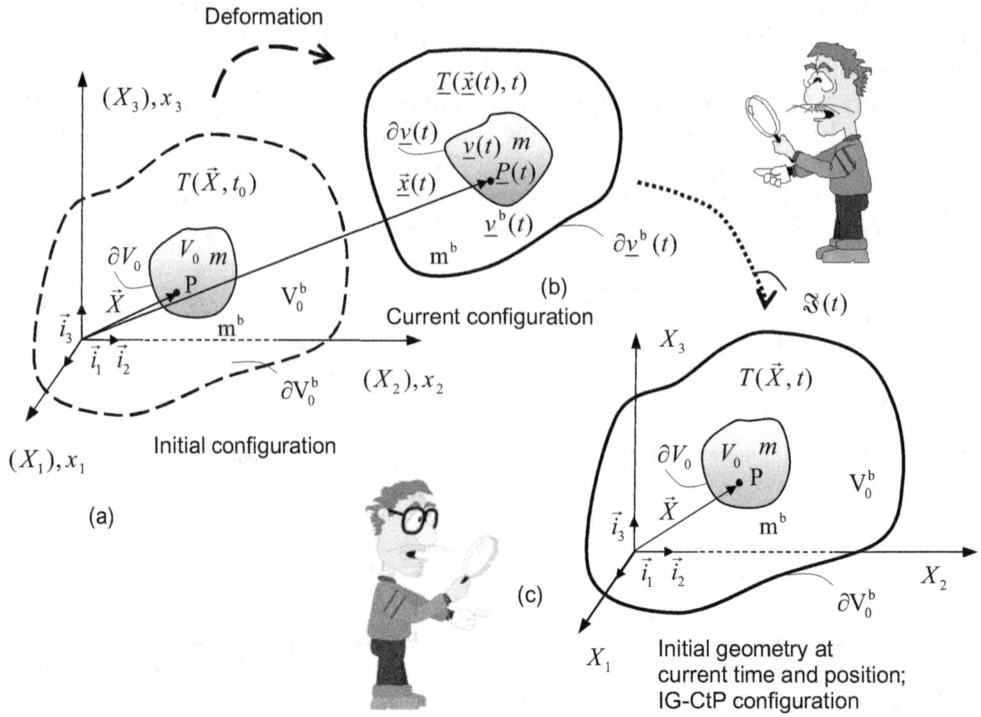


**Figure 4.25** System $\mathrm{m}^{\mathrm{b}}$ (body) in the initial configuration $\mathrm{V}_0^{\mathrm{b}}$, current configuration $\underline{v}^{\mathrm{b}}(t)$, and IG-CtP configuration $\mathrm{V}_0^{\mathrm{b}}$. Inspectors: Corel images. Special glasses added by Joni Hautala.

A vital part of continuum mechanics and continuum thermodynamics is the basic laws and axioms. There are two basic laws and one axiom for continuum mechanics.

The law of balance of momentum was given by Basic Law (4.18) [**BL 1**],

**BL 1**
$$\oint_{\partial \underline{v}(t)} \underline{\vec{t}}\, \mathrm{d}\underline{a} + \int_{\underline{v}(t)} \underline{\rho}\, \underline{\vec{b}}\, \mathrm{d}\underline{v} = \frac{\mathrm{D}}{\mathrm{D}t} \int_{\underline{v}(t)} \underline{\rho}\, \underline{\vec{v}}\, \mathrm{d}\underline{v}\,, \tag{4.266}$$

where $\underline{\vec{v}}$ is the current value of the velocity of material point $\underline{P}(t)$.

The law of balance of moment of momentum was given by Basic Law (4.23) [**BL 2**],

**BL 2**
$$\oint_{\partial \underline{v}(t)} (\underline{\vec{r}} \times \underline{\vec{t}}\,)\, \mathrm{d}\underline{a} + \int_{\underline{v}(t)} (\underline{\vec{r}} \times \underline{\rho}\, \underline{\vec{b}})\, \mathrm{d}\underline{v} = \frac{\mathrm{D}}{\mathrm{D}t} \int_{\underline{v}(t)} (\underline{\vec{r}} \times \underline{\rho}\, \underline{\vec{v}})\, \mathrm{d}\underline{v}\,, \tag{4.267}$$

where $\underline{\vec{r}}$ is the vector from an arbitrary point to point $\underline{P}(t)$.

The axiom of conservation of mass was given by Axiom (4.24) [**AX 1**] and reads

**AX 1**
$$\frac{\mathrm{D}m}{\mathrm{D}t} = 0\,. \tag{4.268}$$

The material points in the current configuration $\underline{v}^{\mathrm{b}}(t)$ are expressed by the frame $(x_1, x_2, x_3)$. It is important to distinguish between the coordinate system and the coordinates of a point. Since the position of point $\underline{P}(t)$ varies with time $t$, its coordinates in the current configuration $\underline{v}^{\mathrm{b}}(t)$ read $(\underline{x}_1(t), \underline{x}_2(t), \underline{x}_3(t))$, but the coordinate system is $(x_1, x_2, x_3)$. However, it is common practice in solid mechanics to express the values of the quantities in the geometry the system $\mathrm{m}^{\mathrm{b}}$ had before loading took place. It is for that purpose that the coordinate system $(X_1, X_2, X_3)$ was introduced. To be able to express the current values of the quantities by the coordinate system $(X_1, X_2, X_3)$, a coordinate transformation from the coordinates $(\underline{x}_1(t), \underline{x}_2(t), \underline{x}_3(t))$ to the coordinates $(X_1, X_2, X_3)$ is needed. A mathematician would use the term mapping instead of coordinate transformation. Since the coordinates $(\underline{x}_1(t), \underline{x}_2(t), \underline{x}_3(t))$ depend on time $t$, the transformation depends on time $t$ as well.

In Figure 4.25, the notation $\Im(t)$ denotes the coordinate transformation from the current material coordinates $\underline{x}_i(t)$ to the initial material coordinates $X_i$. The coordinate transformation $\Im(t)$ is time-dependent, since the current material coordinates $\underline{x}_i(t)$ of point $P(t)$ vary with time $t$.

In Figure 4.25(a), someone inspecting the system $\mathrm{m}^{\mathrm{b}}$ in the current configuration $\underline{v}^{\mathrm{b}}(t)$ would investigate it using the frame $(x_1, x_2, x_3)$. The person would see it moving and deforming in space, along with the underlying processes, and would be able to measure the changing values of the quantities, for example, absolute temperature $T$ at the current moment $t$.

Imagine the inspector then wearing special glasses allowing him/her to see the current geometry of the system $\mathrm{m}^{\mathrm{b}}$ having been replaced by its initial geometry, i.e., with the geometry the system $\mathrm{m}^{\mathrm{b}}$ had before loading took place. To do that, they use the coordinate system $(X_1, X_2, X_3)$. The inspector experiences the system $\mathrm{m}^{\mathrm{b}}$ as neither moving nor deforming, since there is a continuous sequence of the coordinate transformation from the coordinates $(\underline{x}_1(t), \underline{x}_2(t), \underline{x}_3(t))$ to the coordinates $(X_1, X_2, X_3)$. Thus the inspector needs an infinite number of glasses, since the coordinate transformations have different values for different moments $t$. However, they can see all the undergoing processes and measure the current values of the quantities, but the processes seem to undergo in the initial geometry of the system $\mathrm{m}^{\mathrm{b}}$, and the current values of the quantities are oddly associated to the geometry the system $\mathrm{m}^{\mathrm{b}}$ had before the application of loading.

Introduction of the coordinate system $(X_1, X_2, X_3)$ according to the preceding view may have created the concept that in solid mechanics, the investigation is carried out in the initial configuration $\mathrm{V}_0^{\mathrm{b}}$ . This is incorrect, since the initial configuration $\mathrm{V}_0^{\mathrm{b}}$ defines the position of the system $\mathrm{m}^{\mathrm{b}}$ to be where it was before the application of loading. Coordinate transformation does not move the system $\mathrm{m}^{\mathrm{b}}$ in space. Furthermore, in the initial configuration $\mathrm{V}_0^{\mathrm{b}}$, there were no loads. The question then arises of how to get loading for the initial configuration $\mathrm{V}_0^{\mathrm{b}}$ in order to perform analyses of solid materials.

To overcome these problems, this book introduces the "initial geometry at current time and position," the IG-CtP configuration $\mathrm{V}_0^{\mathrm{b}}$. The phrase "the IG-CtP description" is adopted when the values of the (state) variables are studied in the IG-CtP configuration $\mathrm{V}_0^{\mathrm{b}}$. With the introduction of the IG-CtP configuration $V_0^b$, it is clear that when the equations for the solids are derived, the values of the quantities are the current ones, although they are expressed in the initial material coordinates $(X_1, X_2, X_3)$. However, the geometrical quantities such as volume $V_0^b$ and density $\rho_0$ [density $\rho_0$ not shown in Figure 4.25(c)] have the values they had in the initial configuration $V_0^b$. This is visualized in Figure 4.25(c) by the inspector with the glasses. Thus the "drawback" of the representation with the frame $(X_1, X_2, X_3)$ is that it does not provide the current value of the density $\underline{\rho}(\vec{X}, t)$. Since the current value of the density $\underline{\rho}(\vec{X}, t)$ does not belong to the set of variables in the local form, it has to be calculated from the strain components. The introduction of a new concept is a small price to pay for making the foundations of solid mechanics clear and unique.

The preceding notations are clear but misleading: the quantities are not dependent on the vectors $\underline{\vec{x}}(t)$ or $\vec{X}$ but rather on the coordinates $(\underline{x}_1(t), \underline{x}_2(t), \underline{x}_3(t))$ or $(X_1, X_2, X_3)$. Thus, instead of $\underline{T}(\underline{\vec{x}}(t), t)$, one should really write $\underline{T}(\underline{x}_1(t), \underline{x}_2(t), \underline{x}_3(t), t)$, and instead of $T(\vec{X}, t)$ one should write $T(X_1, X_2, X_3, t)$, but $\underline{T}(\underline{\vec{x}}(t), t)$ and $T(\vec{X}, t)$ are used to make the notations shorter. Furthermore, the notations $\underline{\vec{x}}(t)$ and $\vec{X}$ are the position vectors of the points having the positions $(\underline{x}_1(t), \underline{x}_2(t), \underline{x}_3(t))$ and $(X_1, X_2, X_3)$. The same notation rule holds for equations for fluid mechanics as well.

Sometimes in fluid mechanics, the fluid flow is investigated using the equations written for the current configuration $\underline{v}^b(t)$, which means that the movement of individual fluid particles is studied. However, in the fluid mechanics literature, it is common practice to express the fluid flow as if there were an observer surveying the flow through a fixed point in space.

The second way to survey fluid flow requires expressing all quantities in terms of fixed coordinates $(x_1, x_2, x_3)$. Absolute temperature, for example, is now expressed in the form $T(\vec{x}, t)$ and the quantieties in general in the form $\theta(\vec{x}, t)$. As already discussed with the coordinates $(\underline{x}_1(t), \underline{x}_2(t), \underline{x}_3(t))$ or $(X_1, X_2, X_3)$, the absolute temperature of the fluid should be written in the form $T(x_1, x_2, x_3, t)$, but the form $T(\vec{x}, t)$ is interpreted to be a shorter notation for $T(x_1, x_2, x_3, t)$. When the quantities are expressed in the form $\underline{\theta}(\underline{\vec{x}}(t), t)$, this is called the current material description, since the notation $\underline{\vec{x}}(t)$ stands for the position of point $\underline{P}(t)$ in the current configuration $\underline{v}^b(t)$. When they are expressed in the form $\theta(\vec{x}, t)$ this is called the spatial description, since the notation $x_i$ stands for a spatial point fixed in space. Figure 4.26 studies the subsystem $m$ in current configuration(s) at time $t_1$, $t_2$, and $t_3$. The control volume $V^{cv}$ is fixed in space. At time $t_2$ the volume of the subsystem $\underline{v}(t)$ occupies exactly the control volume $V^{cv}$, i.e., $\underline{v}(t_2) = V^{cv}$, as shown in Figure 4.26. Similarly, the position vector $\underline{\vec{x}}(t)$ equals the position vector of the point fixed in space, i.e., $\underline{\vec{x}}(t_2) = \vec{x}$. This allows to move from the current material description to the spatial description, as shown later in this section.

The difference between the notations for the quantities expressed by the current material coordinates, that is, $\underline{\theta}(\underline{\vec{x}}(t), t)$, and those expressed by the spatial coordinates, that is, $\theta(\vec{x}, t)$, seems to be negligible, but this is not the case, as the derivation of the material time derivative of the quantity $\theta$ shows. The material time derivative expresses how the material experiences its state as changing. To make notations shorter, a uniaxial case is studied. The uniaxial counterparts for the above quantities are $\underline{\theta}(\underline{x}(t), t)$ and $\theta(x, t)$. More detailed studies were presented in Sections 3.4 and 3.6.

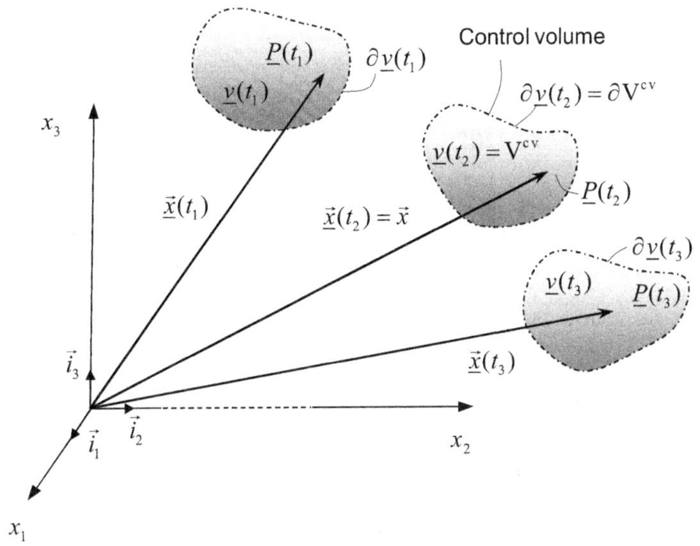


**Figure 4.26** Current material description and the spatial description.

For the current material description, the chain rule of derivation can be utilized as follows:

$$\frac{\mathrm{D}\underline{\theta}(\underline{x}(t), t)}{\mathrm{D}t} = \frac{\partial \underline{\theta}(\underline{x}(t), t)}{\partial x(t)} \frac{\partial x(t)}{\partial t} + \frac{\partial \underline{\theta}(\underline{x}(t), t)}{\partial t} = \frac{\partial \underline{\theta}(\underline{x}(t), t)}{\partial t} + \frac{\partial \underline{x}(t)}{\partial t} \frac{\partial \underline{\theta}(\underline{x}(t), t)}{\partial \underline{x}(t)}$$
$$= \frac{\partial \underline{\theta}(\underline{x}(t), t)}{\partial t} + \underline{v}(\underline{x}(t), t) \frac{\partial \underline{\theta}(\underline{x}(t), t)}{\partial \underline{x}(t)}, \tag{4.269}$$

where the velocity $\underline{v}(\underline{x}(t), t)$ is defined to be

$$\underline{v}(\underline{x}(t), t) := \frac{\partial \underline{x}(t)}{\partial t}. \tag{4.270}$$

For the quantity expressed by the spatial coordinate $x$, the chain rule cannot be applied, since the spatial coordinate $x$ is not dependent on time $t$. Based on Figure 4.26, the value for the material time derivative has to be calculated at time $t_2$ at point $\underline{P}(t_2)$ the spatial coordinate of which is $x$. The material time derivative of the function $\theta(x, t)$ at time $t_2$ is obtained from

$$\frac{\mathrm{D}\theta(x, t_2)}{\mathrm{D}t} := \lim_{\Delta t \to 0} \frac{\theta(x + \Delta x, t_2 + \Delta t) - \theta(x, t_2)}{\Delta t} = \lim_{\Delta t \to 0} \frac{\Delta \theta(x, t_2)}{\Delta t}, \tag{4.271}$$

since the fluid particle moves by the amount $\Delta x$ during the time increment $\Delta t$. Based on Expression (4.271), the derivation of the material derivative must have information on the fluid particle when it is at time $t_2 + \Delta t$ at the point $x + \Delta x$. The Taylor series provides a tool for expressing the value of the quantity $\theta(x + \Delta x, t_2 + \Delta t)$ in terms of $\theta(x, t_2)$ and its derivatives. For the present purpose, the Taylor series are

$$\begin{aligned}\theta(x+\Delta x, t_2+\Delta t) = \theta(x,t_2) &+ \frac{\partial\theta(x,t_2)}{\partial x}\Delta x + \frac{\partial\theta(x,t_2)}{\partial t}\Delta t \\ &+ \frac{1}{2!}\frac{\partial^2\theta(x,t_2)}{\partial x^2}(\Delta x)^2 + \frac{1}{2!}\frac{\partial^2\theta(x,t_2)}{\partial x\,\partial t}\Delta x\,\Delta t \\ &+ \frac{1}{2!}\frac{\partial^2\theta(x,t_2)}{\partial t\,\partial x}\Delta t\,\Delta x + \frac{1}{2!}\frac{\partial^2\theta(x,t_2)}{\partial t^2}\Delta t^2 + \cdots . \end{aligned} \tag{4.272}$$

The following notation is introduced:

$$\Delta\theta(x,t_2) := \theta(x+\Delta x, t_2+\Delta t) - \theta(x,t_2) . \tag{4.273}$$

When Expression (4.272) is divided by the time increment $\Delta t$ and Notations (4.271) and (4.273) are taken into account, the following is achieved:

$$\begin{aligned}\frac{\Delta\theta(x,t_2)}{\Delta t} = &\frac{\partial\theta(x,t_2)}{\partial x}\frac{\Delta x}{\Delta t} + \frac{\partial\theta(x,t_2)}{\partial t} \\ &+ \frac{1}{2!}\frac{\partial^2\theta(x,t_2)}{\partial x^2}\frac{\Delta x}{\Delta t}\Delta x + \frac{1}{2!}\frac{\partial^2\theta(x,t_2)}{\partial x\,\partial t}\Delta x \\ &+ \frac{1}{2!}\frac{\partial^2\theta(x,t_2)}{\partial t\,\partial x}\Delta x + \frac{1}{2!}\frac{\partial^2\theta(x,t_2)}{\partial t^2}\Delta t^2 + \cdots . \end{aligned} \tag{4.274}$$

When $\Delta t \to 0$, then $\Delta x \to 0$, and Expression (4.274) yields

$$\lim_{\Delta t\to 0}\frac{\Delta\theta(x,t_2)}{\Delta t} = \frac{\partial\theta(x,t_2)}{\partial x}\lim_{\Delta t\to 0}\frac{\Delta x}{\Delta t} + \frac{\partial\theta(x,t_2)}{\partial t} = \frac{\partial\theta(x,t_2)}{\partial t} + v(x,t_2)\frac{\partial\theta(x,t_2)}{\partial x}, \tag{4.275}$$

where

$$v(x,t_2) := \lim_{\Delta t\to 0}\frac{\Delta x}{\Delta t} . \tag{4.276}$$

Derivative (4.271) and Expression (4.275) yield

$$\frac{\mathrm{D}\theta(x,t_2)}{\mathrm{D}t} = \frac{\partial\theta(x,t_2)}{\partial t} + v(x,t_2)\frac{\partial\theta(x,t_2)}{\partial x} . \tag{4.277}$$

The foregoing evaluation can be repeated for all the points in the control volume $\mathrm{V}^{\mathrm{cv}}$ at all the moments of time $t$. Thus Derivative (4.277) yields

$$\frac{\mathrm{D}\theta(x,t)}{\mathrm{D}t} = \frac{\partial\theta(x,t)}{\partial t} + v(x,t)\frac{\partial\theta(x,t)}{\partial x} . \tag{4.278}$$

Material (time) derivative (4.278) at which we are aiming.

Equations (4.270) and (4.276) define two velocities. In Equation (4.276), an observer is observing fluid flow through a fixed point in space, the position vector of which is $\vec{x}$. At time $t_2$, the fluid velocity is $v(x, t_2)$.

Definition (4.270) gives the velocity of a fluid particle $\underline{\vec{v}}(\underline{x}(t), t)$. The fluid particle is moving in space, and its current position is given by the position vector $\underline{\vec{x}}(t)$. At time $t_2$ the fluid particle is in the position $\underline{\vec{x}}(t_2) = \vec{x}$ and the velocities are equal, that is, $\underline{\vec{v}}(\underline{x}(t_2), t_2) = v(x, t_2)$ .

The integral forms given by Expressions (4.266) and (4.267) are not suitable for analysis, but the corresponding differential equations were derived. The benefit of the differential equations is that they express the information of the basic laws and axioms at every point in the system $\mathrm{m^b}$. There are special techniques for solving differential equations analytically or numerically. Although Basic Laws (4.266) and (4.267) are valid for solids and fluids, the corresponding differential equations take different appearances. The difference is based on the different types of responses of solids and fluids, leading to the introduction of different sets of governing variables. Basic Law (4.266) [**BL 1**] is studied as an example. No exact derivation is performed, but some important differences between descriptions are pointed out.

In the procedure for creating equations in the local form (i.e., in the form of differential equations), all the terms of the basic laws and axioms must be written in the form of one integrand of the volume integral. Then, the value of the integrand is shown to be zero, or zero vector, depending on the form of the integrand.

In basic laws and axioms, the state $\mathcal{E}$ is expressed by the current material coordinates $\underline{x}_i(t)$ in the current configuration $\underline{v}^{\mathrm{b}}(t)$. Therefore the Cauchy stress tensor $\underline{\boldsymbol{\sigma}}^{\mathrm{c}}$ needs to be introduced. It is given by Definition (4.90),

$$\underline{\vec{n}}(\underline{\vec{x}}(t)) \cdot \underline{\boldsymbol{\sigma}}^{\mathrm{c}}(\underline{\vec{x}}(t), t) := \underline{\vec{t}}(\underline{\vec{x}}(t), t, \underline{\vec{n}}) \qquad \text{or} \qquad \underline{\vec{n}} \cdot \underline{\boldsymbol{\sigma}}^{\mathrm{c}} := \underline{\vec{t}}, \tag{4.279}$$

where $\underline{\vec{n}}$ is the outward unit normal to the subsystem $m$, if the quantities are expressed in the current material coordinates $\underline{x}_i(t)$. Now the first term of Basic Law (4.266) [**BL 1**] takes the form

$$\oint_{\partial \underline{v}(t)} \underline{\vec{t}} \, \mathrm{d}\underline{a} = \oint_{\partial \underline{v}(t)} \underline{\vec{n}} \cdot \underline{\boldsymbol{\sigma}}^{\mathrm{c}} \, \mathrm{d}\underline{a} \, . \tag{4.280}$$

The right side of Expression (4.280) can be transformed into the volume integral by applying Generalized Gauss's Theorem $(2.170)_1$. Thus Definition (4.279) is a crucial part of continuum mechanics, since without it, no local form can be derived.

When the initial material coordinates $X_i$ are used with the IG-CtP configuration $\mathrm{V_0^b}$, the first term of Basic Law (4.266) [**BL 1**] is manipulated, as given in Equation (4.93):

$$\oint_{\partial \underline{v}(t)} \underline{\vec{t}}(\underline{x}(t), t, \underline{\vec{n}}) \, \mathrm{d}\underline{a} = \underline{\vec{F}} = \vec{F} = \oint_{\partial \mathrm{V}_0} \vec{T}(\vec{X}, t, \vec{N}) \, \mathrm{d}A \, . \tag{4.281}$$

The concept of Expression (4.281) is that the resultant force $\vec{F}$ takes the same value in both coordinate systems, but the traction vectors $\underline{\vec{t}}$ and $\underline{\vec{t}}$ are different, since they have the form force over area and the surfaces on which they act are different. Definition (4.279) cannot now be applied, since the vector $\vec{n}(\vec{X},t)$ is not normal to the surface $\mathrm{d}A(\vec{X})$, as illustrated in Figure 4.27, which is a copy of Figure 4.13. Therefore, the first Piola–Kirchhoff stress tensor $\boldsymbol{\sigma}^{\square}(\vec{X},t)$ was introduced in Definition (4.91). It reads

$$\vec{N}(\vec{X})\cdot\boldsymbol{\sigma}^{\square}(\vec{X},t) := \vec{T}(\vec{X},t,\vec{N})\,, \tag{4.282}$$

where the vector $\vec{N}(\vec{X})$ is the unit normal vector to the surface $\mathrm{d}A(\vec{X})$. Definition (4.282) is substituted into the right side of Expression (4.281) and the generalized Gauss's theorem given in Expression $(2.170)_1$ is applied. This action gives a volume integral.

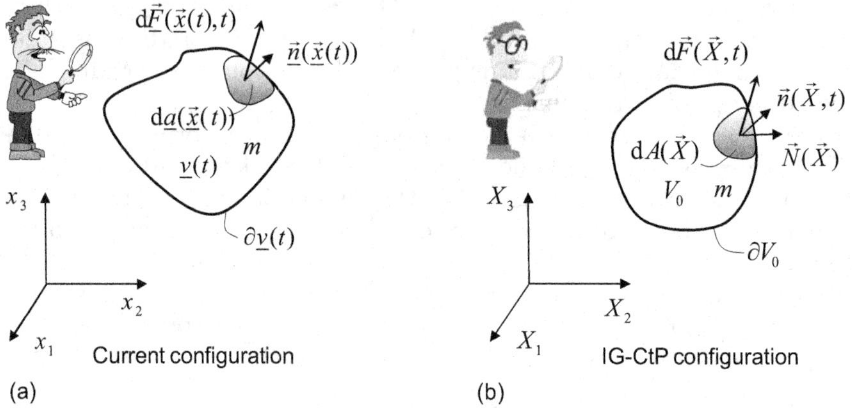


**Figure 4.27** Subdomain $m$ expressed (a) in the current material coordinates $\underline{x}_i(t)$ and (b) in the initial material coordinates $X_i$. Inspector: Corel images. Special glasses added by Joni Hautala.

The problem with the first Piola–Kirchhoff stress tensor $\boldsymbol{\sigma}^{\square}(\vec{X},t)$ is that it is not a symmetric tensor. The second problem is that it is a nominal stress tensor. Therefore it does not give a real value for the state of stress. It has a form of the engineering stress $\sigma^{\mathrm{eng}}$. When the notations of a preliminary book of mechanics of materials are used, the value for the engineering stress $\sigma^{\mathrm{eng}}$ can be obtained from

$$\sigma^{\mathrm{eng}} = \frac{N}{A_0}\,, \tag{4.283}$$

where $N$ is the applied force. In Expression (4.283), the value of the surface area is not the current one, which is $A$, but the one it had before the loading took place, that is, $A_0$. Thus, the engineering stress $\sigma^{\mathrm{eng}}$ and the first Piola–Kirchhoff stress tensor $\boldsymbol{\sigma}^{\square}(\vec{X},t)$ do not give the real values for the state of stress, and are therefore nominal stresses. The Cauchy stress tensor $\underline{\boldsymbol{\sigma}}^{\mathrm{c}}$, on the other hand, gives the real state of stress and therefore it is a true stress tensor. When small deformations and rotations are studied, as is done in this book, the first Piola–Kirchhoff stress tensor $\boldsymbol{\sigma}^{\square}(\vec{X},t)$ and the Cauchy stress tensor $\underline{\boldsymbol{\sigma}}^{\mathrm{c}}$ approach the engineering stress $\sigma^{\mathrm{eng}}$ (here it is only given in a uniaxial case) and the true area $A$ can be approximated by the initial area $A_0$, making the engineering stress $\sigma^{\mathrm{eng}}$ an acceptable tool for continuum mechanics and engineering.

The value of a quantity is independent of the coordinates. This yields

$$\underline{\theta}(\underline{\vec{x}}(t),t) = \theta(\vec{x},t)\,. \tag{4.284}$$

It is assumed that at time $t_2$, the volume of the subsystem $\underline{v}(t)$ equals that of the control volume $\mathrm{V}^{\mathrm{cv}}$, that is, $\underline{v}(t_2) = \mathrm{V}^{\mathrm{cv}}$, as shown in Figure 4.26. If the control volume $\mathrm{V}^{\mathrm{cv}}$ is not any specific volume in space, it is possible to extend the previous concept and to say that for every volume of the subsystem $\underline{v}(t)$ at every moment $t$ there exists an equal control volume $\mathrm{V}^{\mathrm{cv}}$. This is $\underline{v}(t) = \mathrm{V}^{\mathrm{cv}}$. The inverse holds as well. Thus Equality (4.280) leads to

$$\oint_{\partial\underline{v}(t)} \underline{\vec{t}}(\underline{\vec{x}}(t),t,\underline{\vec{n}})\,\mathrm{d}\underline{a} = \oint_{\partial\mathrm{V}^{\mathrm{cv}}} \vec{t}(\vec{x},t,\vec{n})\mathrm{d}A = \oint_{\partial\mathrm{V}^{\mathrm{cv}}} \vec{t}\,\mathrm{d}A\,. \tag{4.285}$$

In Expression (4.285), the domain of integration could be changed following the argumentation in the paragraph above it, and the change of the integrand was possible due to Equation (4.284). The fluid stress tensor $\boldsymbol{\sigma}^{\mathrm{f}}$ is given by Definition (4.95), which reads

$$\vec{n}(\vec{x},t)\cdot\boldsymbol{\sigma}^{\mathrm{f}}(\vec{x},t) := \vec{t}(\vec{x},t,\vec{n}), \qquad \text{or} \qquad \vec{n}\cdot\boldsymbol{\sigma}^{\mathrm{f}} := \vec{t}\,. \tag{4.286}$$

Definition(4.286) is substituted into Expression (4.285), and the derivation is continued by taking the same steps that were taken with the Cauchy stress tensor $\underline{\boldsymbol{\sigma}}^{\mathrm{c}}$. Similar to the Cauchy stress tensor $\underline{\boldsymbol{\sigma}}^{\mathrm{c}}$, the fluid stress tensor $\boldsymbol{\sigma}^{\mathrm{f}}$ gives the real state of stress. The only difference between the Cauchy stress tensor $\underline{\boldsymbol{\sigma}}^{\mathrm{c}}$ and the fluid stress tensor $\boldsymbol{\sigma}^{\mathrm{f}}$ is that they are expressed by different coordinates, as the comparison of Definitions $(4.279)_1$ and $(4.286)_1$ shows. The difference between these two stress measures is that the view of the Cauchy stress tensor $\underline{\boldsymbol{\sigma}}^{\mathrm{c}}$ is at point $\underline{P}(t)$, which moves with deformation, whereas the fluid stress tensor $\boldsymbol{\sigma}^{\mathrm{f}}$ gives the state of stress at a fixed point in space.

The second term of Law of Balance of Momentum (4.266) [**BL 1**] reads

$$\int_{\underline{v}(t)} \underline{\rho}\,\underline{\vec{b}}\,\mathrm{d}\underline{v}\,. \tag{4.287}$$

For the current material description, the second term, Term (4.287), has the form at which we are aiming.

For the IG-CtP description, Equation (4.75) was derived and has the following appearance:

$$\int_{\underline{v}(t)} \underline{\rho}(\underline{\vec{x}}(t),t)\,\underline{\theta}(\underline{\vec{x}}(t),t)\,\mathrm{d}\underline{v} = \int_{V_0} \rho_0(\vec{X})\,\theta(\vec{X},t)\,\mathrm{d}V\,, \tag{4.288}$$

where $\rho_0(\vec{X})$ stands for the density before loading. The derivation of Expression (4.288) exploits the axiom of conservation of mass [**AX 1**], that is, Axiom (4.268). Equation (4.288) is applied to Term (4.287).

For the spatial description, Expressions (4.77) is derived. It is

$$\int_{\underline{v}(t)} \underline{\theta}(\vec{\underline{x}}(t), t)\,\mathrm{d}\underline{v} = \int_{\mathrm{V^{cv}}} \theta(\vec{x}, t)\,\mathrm{d}V\,. \tag{4.289}$$

The argument given above Equation (4.285) applies to Expression (4.289) as well. Expression (4.289) can be applied for Term (4.287), and the form at which we are aiming is obtained.

The term on the right side of the law of balance of momentum obtained from Basic Law (4.266) [**BL 1**] is

$$\frac{\mathrm{D}}{\mathrm{D}t}\int_{\underline{v}(t)} \underline{\rho}(\vec{\underline{x}}(t), t)\,\underline{\theta}(\vec{\underline{x}}(t), t)\,\mathrm{d}\underline{v}\,. \tag{4.290}$$

The act to move the derivative operator $\mathrm{D}/\mathrm{D}t$ to the right side of the integral sign $\int$ is called the Reynolds transport theorem.

When the current material description is used, the right side of the law of balance of momentum [**BL 1**], Term (4.290), can be manipulated as shown by Reynolds Transport Theorem (4.81), as follows:

$$\begin{aligned}\frac{\mathrm{D}}{\mathrm{D}t}\int_{\underline{v}(t)} \underline{\rho}\,(\vec{\underline{x}}(t), t)\,\underline{\theta}(\vec{\underline{x}}(t), t)\,\mathrm{d}\underline{v} &= \frac{\mathrm{D}}{\mathrm{D}t}\int_{m} \underline{\theta}(\vec{\underline{x}}(t), t)\,\mathrm{d}m = \int_{m} \frac{\mathrm{D}\underline{\theta}(\vec{\underline{x}}(t), t)}{\mathrm{D}t}\,\mathrm{d}m \\ &= \int_{\underline{v}(t)} \underline{\rho}(\vec{\underline{x}}(t), t)\,\frac{\mathrm{D}\underline{\theta}(\vec{\underline{x}}(t), t)}{\mathrm{D}t}\,\mathrm{d}\underline{v}\,,\end{aligned} \tag{4.291}$$

where the fact that the mass of the subsystem $m$ does not change during deformation is exploited.

When the IG-CtP description is used, Equation (4.288) can be applied, which leads to the integration over a constant volume of $V_0$ having an arbitrarily selected but then nonchanging entity of matter. This makes it possible to move the derivative operator $\mathrm{D}/\mathrm{D}t$ into the other side of the integral sign $\int$, as was done in Manipulation (4.291).

When spatial description is used, Equation (4.289) can be applied to Term (4.290), and the following is achieved:

$$\frac{\mathrm{D}}{\mathrm{D}t}\int_{\underline{v}(t)} \underline{\rho}\,(\vec{\underline{x}}(t), t)\,\underline{\theta}(\vec{\underline{x}}(t), t)\,\mathrm{d}\underline{v} = \frac{\mathrm{D}}{\mathrm{D}t}\int_{\mathrm{V^{cv}}} \rho(\vec{x}, t)\,\theta(\vec{x}, t)\,\mathrm{d}V\,. \tag{4.292}$$

Since the control volume $\mathrm{V^{cv}}$ is a nonchanging volume (but in space), it is attractive just to assume that the derivative operator $\mathrm{D}/\mathrm{D}t$ can be moved to the other side of the integral sign $\int$. Unfortunately, this is not possible, since there is fluid flow through the control volume $\mathrm{V^{cv}}$, and the operator $\mathrm{D}/\mathrm{D}t$ expresses how the matter experiences the change of its state with time $t$. Figure 4.28 sketches the fluid flow through the control volume $\mathrm{V^{cv}}$. It is a copy of Figure 3.12, although in the original figure, the moment of observation is $t_1$.

The definition of the material time derivative is as follows:

$$\frac{\mathrm{D}}{\mathrm{D}t} I(t) := \lim_{\Delta t \to 0} \frac{I(t+\Delta t) - I(t)}{\Delta t}\,, \tag{4.293}$$

where in this case

$$I(t) = \int_{\mathrm{V^{cv}}} \rho(\vec{x},t)\,\theta(\vec{x},t)\mathrm{d}V\,. \tag{4.294}$$

As Figure 4.28 shows, the volumes $\underline{v}(t)$ and $\mathrm{V^{cv}}$ coincide at time $t = t_2$, that is, $\underline{v}(t_2) = \mathrm{V^{cv}}$. The problem is that the value for the quantity $I(t)$ has to be calculated at time $t = t_2 + \Delta t$; that is, the value for $I(t_2 + \Delta t)$ is needed. Unfortunately, at that time, part of the fluid is already outside the volume $\mathrm{V^{cv}}$ and new fluid has entered the volume $\mathrm{V^{cv}}$. This implies that a special technique to solve the right side of Equation (4.292) is needed, as Sections 3.10 and 4.13 show. After quite extensive derivation, the result given in Equation (4.89) is obtained. It is

$$\frac{\mathrm{D}}{\mathrm{D}t} \int_{\underline{v}(t)} \underline{\rho}(\underline{\vec{x}}(t),t)\,\underline{\theta}(\underline{\vec{x}}(t),t)\,\mathrm{d}\underline{v} = \int_{\mathrm{V^{cv}}} \rho(\vec{x},t)\,\frac{\mathrm{D}\theta(\vec{x},t)}{\mathrm{D}t}\,\mathrm{d}V\,. \tag{4.295}$$

Two different forms for the Reynolds transport theorem are given in Equations (4.291) and (4.295). Both move the material time operator $\mathrm{D}/\mathrm{D}t$ to the right side of the integral sign $\int$, and Expression (4.295) replaces the integration volume $\underline{v}(t)$ with the control volume $\mathrm{V^{cv}}$.

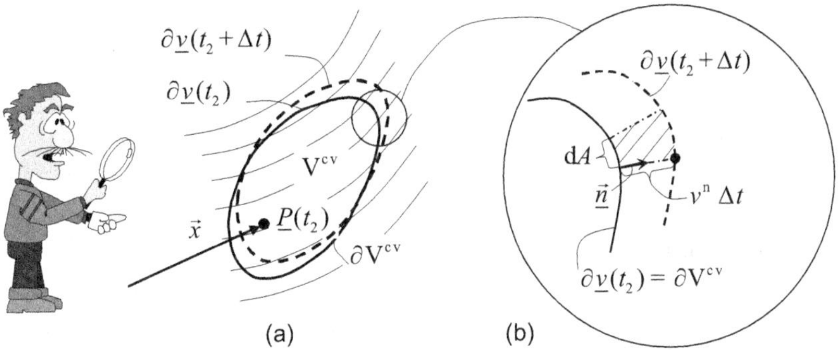


**Figure 4.28** (a) Material entity $m_1$ at time $t = t_2$ and at time $t = t_2 + \Delta t$. (b) Material particles moving out from the volume $\underline{v}(t_2) = \mathrm{V^{cv}}$.

Now all the terms have the correct form, and therefore all the terms separately for every description can be collected into one integrand of volume integrals. Then, according to Section 4.7, the integrand is zero (or zero vector in this case), and the local forms, that is, differential equations, are obtained.

When the current material coordinates $\underline{x}_i(t)$ are used, the local form of Law of Balance of Momentum (4.266) [**BL 1**] takes the form of Equation (4.123),

$$\underline{\vec{\nabla}} \cdot \underline{\boldsymbol{\sigma}}^{\mathrm{c}} + \underline{\rho}\,\underline{\vec{b}} = \underline{\rho}\,\frac{\mathrm{D}\underline{\vec{v}}}{\mathrm{D}t} \qquad \text{or} \qquad \frac{\partial \underline{\sigma}^{\mathrm{c}}_{sk}}{\partial \underline{x}_s} + \underline{\rho}\,\underline{b}_k = \underline{\rho}\,\frac{\mathrm{D}\underline{v}_k}{\mathrm{D}t}\,, \tag{4.296}$$

where the vector operator del follows the concept given in Definition $(2.169)_1$, which takes the following appearance for the current coordinates $\underline{x}_i(t)$:

$$\underline{\vec{\nabla}}(\underline{\vec{x}}) := \frac{\partial}{\partial \underline{\vec{x}}} \,. \tag{4.297}$$

When the response of solids is evaluated, the initial material coordinates $X_i$ are preferred. Therefore, the equations of motion, that is, the local forms (differential equations) corresponding to Basic Law (4.266) [**BL 1**], have the following appearance [see Expression (4.111)]:

$$\vec{\nabla}(\vec{X}) \cdot \boldsymbol{\sigma}(\vec{X}, t) + \rho_0(\vec{X})\, \vec{b}(\vec{X}, t) = \rho_0(\vec{X})\, \frac{\mathrm{D}\vec{v}(\vec{X}, t)}{\mathrm{D}t} \,, \tag{4.298}$$

where the stress tensor $\boldsymbol{\sigma}(\vec{X}, t)$ is the six-dimensional (three normal components and three shear components) counterpart for the engineering stress $\sigma^{\text{eng}}$ introduced in Equation (4.283).

In the derivation of Expressions (4.298), small deformations and rotations were assumed. The density of the material in the initial configuration $\rho_0(\vec{X})$ does not display the assumption of small deformations but is a consequence of the coordinate transformation. This means that when expressed in the initial coordinates $X_i$, the expressions for the finite deformation theory have the variable $\rho_0(\vec{X})$.

For fluids, the local forms of the basic laws and axioms are expressed by the spatial coordinates $x_i$. For Basic Law (4.266) [**BL 1**], the following differential form was achieved [see Expression (4.140)]:

$$\rho(\vec{x}, t)\, \frac{\mathrm{D}\vec{v}(\vec{x}, t)}{\mathrm{D}t} = \rho(\vec{x}, t)\, \vec{b}(\vec{x}, t) + \vec{\nabla}(\vec{x}) \cdot \boldsymbol{\sigma}^{\text{d}}(\vec{x}, t) - \vec{\nabla}(\vec{x})\, p \,, \tag{4.299}$$

where $\boldsymbol{\sigma}^{\text{d}}(\vec{x}, t)$ is the viscous stress tensor or dissipative stress tensor. It can be obtained from Equation (4.136),

$$\boldsymbol{\sigma}^{\text{f}} = -p\, \mathbf{1} + \boldsymbol{\sigma}^{\text{d}} \,, \tag{4.300}$$

where $p$ is the thermodynamic pressure. It is important to note that the thermodynamic pressure $p$ is not the same as the mechanical pressure $p_{\text{mech}}$, defined in Definition (2.117).

Application of Law of Balance of Moment of Momentum (4.267) [**BL 2**] gave the result that the Cauchy stress tensor $\underline{\boldsymbol{\sigma}}^{\text{c}}$ and the fluid stress tensor $\boldsymbol{\sigma}^{\text{f}}$ are symmetric. This result is given in Equations (4.156) and (4.159). Since, in case of small deformations and rotations, the Cauchy stress tensor $\underline{\boldsymbol{\sigma}}^{\text{c}}$ approaches the stress tensor $\boldsymbol{\sigma}$, the stress tensor $\boldsymbol{\sigma}$ is symmetric as well. Since the second-order identity tensor $\mathbf{1}$ [see Equation (2.64)] and the fluid stress tensor $\boldsymbol{\sigma}^{\text{f}}$ are symmetric, Expression (4.300) implies that the dissipative stress tensor $\boldsymbol{\sigma}^{\text{d}}$ is symmetric.

Axiom of Conservation of Mass (4.268) [**AX 1**] gives the continuity equation, which plays an important role in fluid mechanics. When the current material coordinates $\underline{x}_i(t)$ are used, Expressions (4.47) and $(4.50)_2$ are achieved. They are

$$\frac{\mathrm{D}\underline{\rho}(\underline{\vec{x}}(t),t)}{\mathrm{D}t}+\underline{\rho}(\underline{\vec{x}}(t),t)\,\underline{\vec{\nabla}}(\underline{\vec{x}})\cdot\underline{\vec{v}}(\underline{\vec{x}}(t),t)=0\,,\quad \text{i.e.,}\quad \frac{\mathrm{D}\underline{\rho}}{\mathrm{D}t}+\underline{\rho}\,\underline{\vec{\nabla}}\cdot\underline{\vec{v}}=0\,, \tag{4.301}$$

and

$$\frac{\partial\underline{\rho}}{\partial t}+\underline{\vec{\nabla}}\cdot(\underline{\rho}\,\underline{\vec{v}})=0\,. \tag{4.302}$$

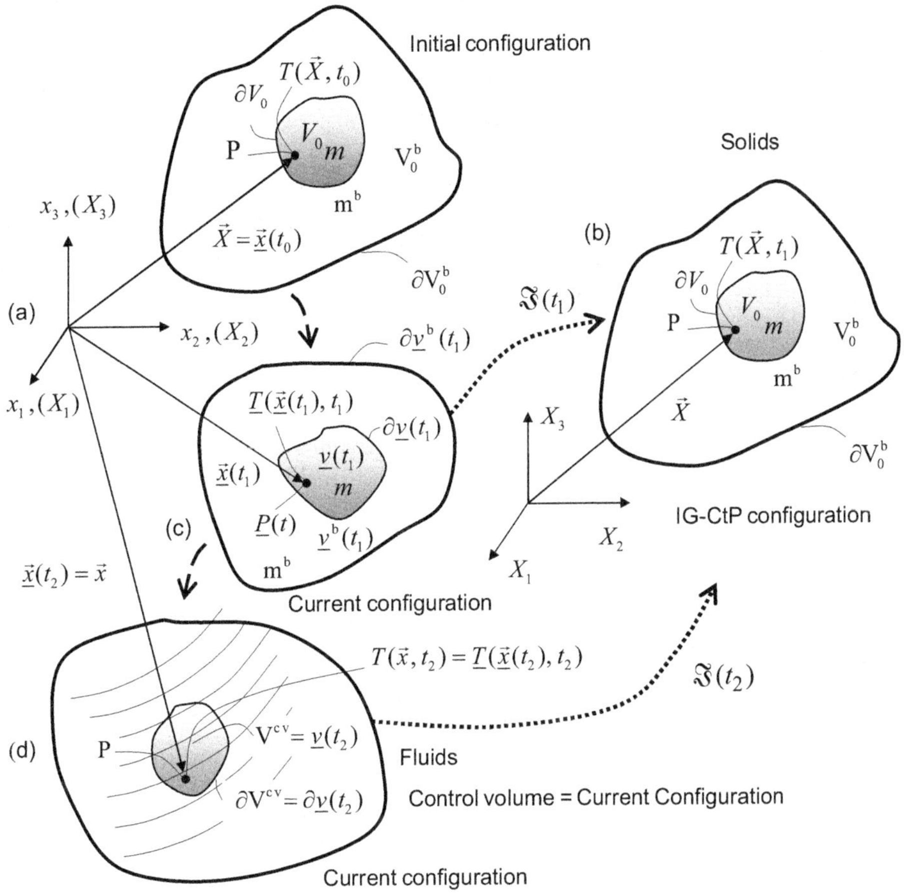


**Figure 4.29** Continuum mechanics in one figure.

When expressed in spatial coordinates $x_i$, the corresponding equations for Equations (4.301) and (4.302) are given in Equations (4.60) and (4.69). They are

$$\frac{\partial\rho(\vec{x},t)}{\partial t}+\vec{\nabla}(\vec{x})\cdot\left[\rho(\vec{x},t)\,\vec{v}(\vec{x},t)\right]=0 \tag{4.303}$$

and

$$\frac{\mathrm{D}\rho(\vec{x},t)}{\mathrm{D}t} + \rho(\vec{x},t)\,\vec{\nabla}(\vec{x})\cdot\vec{v}(\vec{x},t) = 0\,. \tag{4.304}$$

The derived local forms (i.e., differential forms) and definitions apply at every point of the system $\mathrm{m^b}$ (body).

The message of the present section is compressed into Figure 4.29, which shows the system $\mathrm{m^b}$ in the initial configuration $\mathrm{V_0^b}$ and at two different times, $t_1$ and $t_2$, in the current configuration $\underline{v}^{\mathrm{b}}(t)$. The counterpart IG-CtP configurations $\mathrm{V_0^b}$ for the current configurations $\underline{v}^{\mathrm{b}}(t)$ are given as well. At time $t_2$, the control volume $\mathrm{V^{cv}}$ equals the (volume of the) subsystem $\underline{v}(t)$, which makes it possible to switch the view from the current material coordinates $\underline{x}_i(t)$ to the spatial coordinates $x_i$.

Figure 4.29 is misleading. The system $\mathrm{m^b}$, being in the IG-CtP configuration $\mathrm{V_0^b}$, should not be visible, since it is present only in the $(X_1, X_2, X_3)$ coordinate system and the reader sees the processes and systems in the $(x_1, x_2, x_3)$ coordinate system. The system $\mathrm{m^b}$ in the initial configuration $\mathrm{V_0^b}$ is correctly shown in Figure 4.29, since for the initial configuration $\mathrm{V_0^b}$, the coordinates $X_i$ are only defined as $X_i = \underline{x}_i(t_0)$, as the position vector $\vec{X} = \underline{\vec{x}}(t_0)$ in the upper left corner of Figure 4.29 shows.

# Problems

**4.1** (a) Using the Einstein summation convention, write the equilibrium equations in the component form. Equilibrium Equations $(4.113)_2$ read

$$\frac{\partial \sigma_{sk}}{\partial X_s} + \rho_0\, b_k = 0\,. \tag{1}$$

**Hint.** There will be three equations; in the first one $k = 1$, in the second $k = 2$, and in the third $k = 3$. Note that $k$ is a free index and that $s$ is a dummy index.

(b) Substitute the following into the first ($k = 1$) equilibrium equation:

$$\sigma_{11} = \sigma_x\,, \qquad \sigma_{21} = \tau_{yx}\,, \qquad \sigma_{31} = \tau_{zx} \qquad \text{and} \qquad b_1 = b_x \tag{2}$$

and

$$X_1 = x\,, \qquad X_2 = y\,, \qquad \text{and} \qquad X_3 = z\,. \tag{3}$$

(c) Study the first term of the equilibrium equation obtained in (b). What is the physical meaning of the term?

**4.2** The strain tensor $\boldsymbol{\varepsilon}$ and the rotation tensor $\boldsymbol{\omega}$ are defined by Definitions $(3.120)_2$ and $(3.121)_2$. These definitions are

$$\boldsymbol{\varepsilon} := \frac{1}{2}\,(\vec{u}\overleftarrow{\nabla} + \vec{\nabla}\vec{u}) \qquad \text{and} \qquad \boldsymbol{\omega} := \frac{1}{2}\,(\vec{u}\overleftarrow{\nabla} - \vec{\nabla}\vec{u})\,. \tag{1}$$

Derive the terms $\varepsilon_{ij}$ and $\omega_{ij}$.

**4.3** Show that the following equality holds:

$$\underline{\boldsymbol{\sigma}}^{\mathrm{c}} : \underline{\vec{\nabla}\vec{v}} = \underline{\boldsymbol{\sigma}}^{\mathrm{c}} : \underline{\mathbf{d}} \,, \tag{1}$$

where $\underline{\boldsymbol{\sigma}}^{\mathrm{c}}$ is the Cauchy stress tensor, $\underline{\vec{\nabla}}$ is the vector operator del, $\underline{\vec{v}}$ is the velocity vector, and $\underline{\mathbf{d}}$ is the rate of deformation tensor.

**4.4** The following holds for scalar-valued quantities:

$$\frac{\partial}{\partial x}(\, f(x)\, g(x)\,) = g(x)\,\frac{\partial f(x)}{\partial x} + f(x)\,\frac{\partial g(x)}{\partial x} \,. \tag{1}$$

The key point in Equation (1) is that on the right side, the derivative operator $\partial/\partial x$ acts only on one function. Study the quantity

$$(\,\boldsymbol{\sigma} : \vec{u}\,)\overleftarrow{\nabla} \,. \tag{2}$$

Write a "similar equation" to Expression (1). Thus the solution has two terms, and the derivative operator acts only on one quantity in one term. **Hint.** Use the index notation, and lastly return to the Gibbs notation. The solution here has two lines, but you may need few more.

**4.5** Show that the following equality holds:

$$(\vec{e}\cdot\mathbf{h})\cdot\overleftarrow{\nabla} = \vec{e}\cdot(\,\mathbf{h}\cdot\overleftarrow{\nabla}\,) + \vec{e}\,\overleftarrow{\nabla} : \mathbf{h} \,, \tag{1}$$

where $\overleftarrow{\nabla}$ is the vector operator del acting on the preceding quantity, $\mathbf{h}$ is a second-order tensor, and $\vec{e}$ is a vector.

**Hint:** Use the index form for the above operator and quantities.

**4.6** The strain energy $W(\boldsymbol{\varepsilon}^{\mathrm{e}})$ of the subsystem $m$ in the IG-CtP configuration $\mathrm{V}_0^{\mathrm{b}}$ is obtained from

$$W(\boldsymbol{\varepsilon}^{\mathrm{e}}) := \int_{V_0} w(\boldsymbol{\varepsilon}^{\mathrm{e}})\mathrm{d}V \,, \tag{1}$$

where $w(\boldsymbol{\varepsilon}^{\mathrm{e}})$ is the strain-energy density and $\boldsymbol{\varepsilon}^{\mathrm{e}}$ is the elastic strain tensor. According to Equation (2.55), the strain-energy density $w(\boldsymbol{\varepsilon}^{\mathrm{e}})$ takes the form

$$w(\boldsymbol{\varepsilon}^{\mathrm{e}}) := \frac{1}{2}\,\boldsymbol{\varepsilon}^{\mathrm{e}} : \mathbf{C} : \boldsymbol{\varepsilon}^{\mathrm{e}}, \tag{2}$$

where $\mathbf{C}$ is the fourth-order constitutive tensor for Hookean deformation. Substitution of Definition (2) into Definition (1) yields

$$W(\boldsymbol{\varepsilon}^{\mathrm{e}}) := \int_{V_0} w(\boldsymbol{\varepsilon}^{\mathrm{e}})\,\mathrm{d}V = \int_{V_0} \frac{1}{2}\,\boldsymbol{\varepsilon}^{\mathrm{e}} : \mathbf{C} : \boldsymbol{\varepsilon}^{\mathrm{e}} \,. \tag{3}$$

Derive the Clapeyron's theorem taking the following steps:

(a) Write Hooke's Law (2.56) and substitute it into Equation (3). Change the order of the tensors in the expression you just obtained. Explain why this action is possible.

(b) Write the relation between the small deformation strain tensor $\boldsymbol{\varepsilon}$ and the displacement vector $\vec{u}$. Apply it to the elastic deformation.

(c) Copy Equation (1) of Problem 2.23. Replace the vector $\vec{e}$ with the displacement vector $\vec{u}$ and the second-order tensor $\mathbf{h}$ with the stress tensor $\boldsymbol{\sigma}$. Write the obtained expression in the form which allows you to get an expression present on the right side of the previous equation.

(d) Copy Equation (1) of Problem 4.5. Replace the vector $\vec{e}$ with the displacement vector $\vec{u}$ and the second-order tensor $\mathbf{h}$ with the stress tensor $\boldsymbol{\sigma}$. Identify the term that is almost the one in the integrand obtained in Step (b), and change it to be exactly the same. Explain why the change is possible.

(e) Substitute the terms obtained in Step (d) into the integrand obtained in Step (b). You will get four terms.

(f) The stress tensor $\boldsymbol{\sigma}$ is a symmetric second-order tensor. Use this property to show which of the four terms from steps (b)–(c) are equal.

(g) Change the four-term equation obtained in Step (e) to a two-term equation by applying the equalities obtained in Step (f). Select the terms in such a way that the rightmost term will be the displacement vector $\vec{u}$.

(h) Write the equations of motion for small deformation theory in such a way that the derivative term is on the left side and the other terms are on the right side of the equation.

(i) Select the term where the vector operator del operates only the stress tensor $\boldsymbol{\sigma}$ obtained in Step (g). Apply the term obtained in the previous step to the selected term.

(j) Apply the generalized Gauss's theorem.

(k) Apply the definition of the stress tensor $\boldsymbol{\sigma}$. Now you have derived the extended Clapeyron's theorem.

(l) Neglect the acceleration term as a small quantity to get the Clapeyron's theorem.

**4.7** The shear stress vector $\vec{\tau}$ is defined to have the following form:

$$\vec{\tau} := \vec{N} \cdot \boldsymbol{\sigma}^{\square} \cdot [\mathbf{1} - \vec{N}\,\vec{N}\,], \tag{1}$$

where $\vec{N}$ is the unit outward normal to the surface $\partial \mathrm{V}_0$, as shown in Figure P4.7. Notation $\boldsymbol{\sigma}^{\square}$ stands for the first Piola–Kirchhoff stress tensor. Expression (1) is written for the IG-CtP configuration $\mathrm{V}_0^{\mathrm{b}}$ (initial geometry

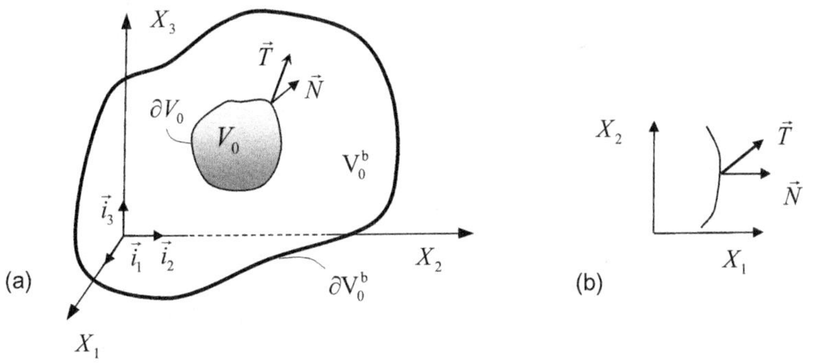


**Figure P4.7** (a) A system (body) $m^b$ and its subsystem $m$ in the IG-CtP configuration $V_0^b$. (b) Special orientation of the subsystem $m$.

at current time and position). Based on Expression (4.91), the first Piola–Kirchhoff stress tensor $\boldsymbol{\sigma}^{\square}$ is defined as

$$\vec{N} \cdot \boldsymbol{\sigma}^{\square} := \vec{T}\,, \tag{2}$$

where $\vec{T}$ is the traction vector in the IG-CtP configuration $V_0^b$. Substitution of Definition (2) into Definition (1) yields

$$\vec{\tau} = \vec{T} \cdot [\mathbf{1} - \vec{N}\vec{N}]. \tag{3}$$

Study a two-dimensional case, that is, where the indices take only the values 1 and 2. This is due to the shorter expressions.

(a) Write the right side of Expression (3) in the index form by performing the dot product.

(b) Apply the Einstein summation convention. Two-dimensional case.

(c) Assume that the outward unit normal $\vec{N}$ points towards the direction of the $X_1$ coordinate, as shown in Figure P4.7(b).

(d) Show that the shear stress vector $\vec{\tau}$ is perpendicular to the normal vector $\vec{N}$. Do not use index notation.

(e) What can you say about the orientations of the shear stress vector $\vec{\tau}$ and the normal stress vector $\vec{\sigma}$, the latter being defined by

$$\vec{\sigma} := \vec{N} \cdot \boldsymbol{\sigma} \cdot \vec{N}\vec{N}\,? \tag{1}$$

A verbal explanation is enough.

**4.8** A system (body) $m^b$, shown in Figure P4.8, is investigated in the current configuration $\underline{v}^b(t)$. The aim of this study is to investigate the application of the first term on the left side of the law of balance of momentum. In the first term of the law of balance of momentum, there is a surface integral around an arbitrary subsystem $m$, the integrand of which is the traction vector $\underline{\vec{t}}$. This term is given in Basic Law (4.18)] [**BL 1**]. The term reads

$$\oint_{\partial \underline{v}(t)} \underline{\vec{t}}\, \mathrm{d}\underline{a}\,, \tag{1}$$

where $\partial \underline{v}(t)$ is the surface of an arbitrary subsystem $m$.

Two potential arbitrary subsystems are studied. The surface of the subsystem $m^{①}$ equals the left lower corner of the surface of the system $\mathrm{m^b}$ and cuts the system $\mathrm{m^b}$ into two parts along the surface $\underline{S}^{①}$, as shown in Figure P4.8. The second candidate for an "arbitrary" sub-system is denoted by $m^{②}$. The surface subsystem $m^{②}$ equals that of the right lower corner of the surface of the system $\mathrm{m^b}$ and cuts the system $\mathrm{m^b}$ along the surface $\underline{S}^{②}$.

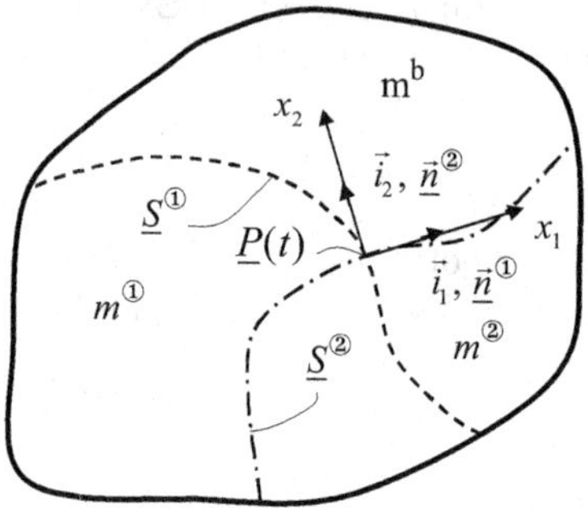


**Figure P4.8** A system (body) $\underline{v}^{\mathrm{b}}(t)$ is imagined to have two internal surfaces $\underline{S}^{①}$ and $\underline{S}^{②}$.

For clarity, the system $\mathrm{m^b}$ is assumed to be two-dimensional. Thus, instead of the surfaces $\underline{S}^{①}$ and $\underline{S}^{②}$, the focus is on the curves $\underline{S}^{①}$ and $\underline{S}^{②}$. This assumption does not restrict the generality of the obtained results but makes the problem easier to solve.

The aim of this study is to evaluate the values and directions of the surface traction vectors $\underline{\vec{t}}^{①}$ and $\underline{\vec{t}}^{②}$ in terms of the Cauchy stress tensor $\underline{\boldsymbol{\sigma}}^{\mathrm{c}}$ at point $\underline{P}(t)$ where the curves $\underline{S}^{①}$ and $\underline{S}^{②}$ intersect. To make the study more comprehensive, the values of the components of the Cauchy stress tensor $\underline{\boldsymbol{\sigma}}^{\mathrm{c}}$ are assumed to be known. Figure P4.8 introduces a coordinate system $(x_1, x_2)$. The basis of the frame $(x_1, x_2)$ is denoted by $(\vec{\imath}_1, \vec{\imath}_2)$. The origin of the coordinate system $(x_1, x_2)$ is at point $\underline{P}(t)$, and its orientation is set in such a way that the coordinates $(x_1, x_2)$ point in the directions of the normals of the curves of $\underline{S}^{①}$ and $\underline{S}^{②}$, as shown in Figure P4.8.

Take the following steps:

(a) Introduce the definition of the Cauchy stress tensor $\underline{\boldsymbol{\sigma}}^{\mathrm{c}}$. Derive a relation between the traction vector $\underline{\vec{t}}^{\,①}$ and the components of the Cauchy stress tensor $\underline{\boldsymbol{\sigma}}^{\mathrm{c}}$ in terms of the basis $(\vec{i}_1, \vec{i}_2)$. Write a similar equation for the traction vector $\underline{\vec{t}}^{\,②}$.

(b) Give the numerical values for the components of the Cauchy stress tensor $\underline{\boldsymbol{\sigma}}^{\mathrm{c}}$ at point $P(t)$. They are: $\underline{\sigma}_{11}^{\mathrm{c}} = 1\,\mathrm{MPa}$, $\underline{\sigma}_{22}^{\mathrm{c}} = 2\,\mathrm{MPa}$ and $\underline{\sigma}_{12}^{\mathrm{c}} = \underline{\sigma}_{21}^{\mathrm{c}} = 3\,\mathrm{MPa}$. Substitute these values in the expressions derived in Step (a).

(c) Draw the surface traction vectors $\underline{\vec{t}}^{\,①}$ and $\underline{\vec{t}}^{\,②}$ in a coordinate system where the $x_1$-axis is horizontal and the $x_2$-axis is vertical.

(d) Calculate the Cauchy principal stresses and the orientation of the principal planes with the following equations:

$$\underline{\sigma}_{1,2}^{\mathrm{c}} = \frac{\underline{\sigma}_{11}^{\mathrm{c}} + \underline{\sigma}_{22}^{\mathrm{c}}}{2} \pm \frac{1}{2}\sqrt{(\underline{\sigma}_{11}^{\mathrm{c}} - \underline{\sigma}_{22}^{\mathrm{c}})^2 + 4\,(\underline{\sigma}_{12}^{\mathrm{c}})^2} \tag{2}$$

and

$$\tan \underline{\theta}_1 = \frac{\underline{\sigma}_1^{\mathrm{c}} - \underline{\sigma}_{11}^{\mathrm{c}}}{\underline{\sigma}_{12}^{\mathrm{c}}} \qquad \text{and} \qquad \underline{\theta}_2 = \underline{\theta}_1 \pm \frac{\pi}{2}\,. \tag{3}$$

(e) Add the values and directions of the principal stresses $\underline{\sigma}_1^{\mathrm{c}}$ and $\underline{\sigma}_2^{\mathrm{c}}$ to the prepared drawing.

(f) Can you say something about the surface traction vectors $\underline{\vec{t}}^{\,①}$ and $\underline{\vec{t}}^{\,②}$? If not, do not worry. This task is for the professor.

**4.9** Derive the principle of virtual work (for a case where inertia forces can be neglected as small quantities) by following the step-by-step instructions given below. After the derivation you will be able to explain the background of the principle of virtual work (see Reddy [85]). It is worth noting that in the following derivation, the tensors are not expressed in terms of their components, that is, the index form is not used.

(a) Start your derivation from the physical background of the principle of virtual work. It consists of the following parts: local forms of the law of balance of momentum and law of balance of moment of momentum, the definition of the stress tensor $\boldsymbol{\sigma}$ and the definitions of the strain tensor $\boldsymbol{\varepsilon}$ and the rotation tensor $\boldsymbol{\omega}$. Thus write first equilibrium equations, Equations $(4.113)_1$ (since inertia forces are neglected); second, the symmetry of the stress tensor $\boldsymbol{\sigma}$, Equation $(4.158)_2$; third, the definition for the first Piola–Kirchhoff stress tensor $\boldsymbol{\sigma}^{\square}$, Definition (4.91); and fourth, the definitions for the strain tensor $\boldsymbol{\varepsilon}$ and for the rotation tensor $\boldsymbol{\omega}$ given by Definitions $(3.120)_2$ and $(3.121)_2$. Assume small deformations and rotations. Thus the first Piola–Kirchhoff stress tensor $\boldsymbol{\sigma}^{\square}$ approaches the stress tensor $\boldsymbol{\sigma}$.

This was the physics part of the derivation. The rest is pure mathematics.

(b) Multiply Equation $(4.113)_1$ from the right by the term $\cdot\vec{g}$, where $\vec{g}$ is a continuously differentiable function. Note that the quantity $\cdot\vec{g}$ contains two parts: the dot product operator $\cdot$ and function $\vec{g}$.

(c) Rewrite Equation (1) of Problem 2.23. Replace the quantities $\mathbf{h}$ (a second–order tensor) and $\vec{e}$ (a vector), found in Equation (1) of Problem 2.23, with variables that exist in the equation obtained in Step (b).

(d) Change the appearance of the equation obtained in Step (b) by using the equation in Step (c).

(e) Integrate the equation obtained in Step (d) over the volume of the entire body in the IG-CtP configuration $\mathrm{V}_0^{\mathrm{b}}$ expressed by the initial material coordinates $X_i$.

(f) Rewrite Generalized Gauss's Theorem $(2.170)_1$. Within the equation obtained in Step (e) there exists a term to which the generalised Gauss's theorem can be applied. Apply it.

(g) In Step (a) you wrote the definition for the first Piola–Kirchhoff stress tensor $\boldsymbol{\sigma}^{\square}$ and then assumed the small deformation theory, which allows the first Piola–Kirchhoff stress tensor $\boldsymbol{\sigma}^{\square}$ to be replaced with the stress tensor $\boldsymbol{\sigma}$. Apply the obtained expression.

(h) Replace an arbitrary continuously differentiable function $\vec{g}$ with a virtual displacement $\delta\vec{u}$ (whatever it is; see Reddy [84], pp. 314–316; and Reddy [85], pp. 135–147).

(i) Assume that equations similar to Equations (3.120) and (3.121) also hold for the variables $\delta\vec{u}$, $\delta\boldsymbol{\varepsilon}$, and $\delta\boldsymbol{\omega}$. Apply the information given below. This was the last step in obtaining the principle of virtual work (see Reddy [85], Chapter 4), which is

$$\delta W^{\mathrm{int}} = \delta W^{\mathrm{ext}} \,. \tag{1}$$

Write the terms $\delta W^{\mathrm{int}}$ and $\delta W^{\mathrm{ext}}$. Since the stress tensor $\boldsymbol{\sigma}$ is symmetric (i.e., $\boldsymbol{\sigma}^{\mathrm{T}} = \boldsymbol{\sigma}$ ) and the rotation tensor $\boldsymbol{\omega}$ is skew-symmetric (i.e., $\boldsymbol{\omega}^{\mathrm{T}} = -\boldsymbol{\omega}$), the following holds:

$$\boldsymbol{\sigma} : \delta\boldsymbol{\omega} = 0 \,. \tag{2}$$

# 5 Introduction to Continuum Thermodynamics

## 5.1 Preliminary Comments

Continuum thermodynamics is a science that combines continuum mechanics with thermodynamics. Thermodynamics, on the other hand, is based on thermostatics. This chapter, therefore, first looks at thermostatics, then extends its scope to thermodynamics; finally, Chapter 6 binds the obtained expressions with continuum mechanics to formulate continuum thermodynamics. Here, the first and second law of thermodynamics are written for homogeneous systems, that is, for thermostatics, whereas later, the focus is extended to non-homogeneous systems.

To remind the reader of the big picture, a summary of continuum thermodynamics is given:

$$\text{Continuum thermodynamics} = \text{Continuum mechanics} + \text{Thermodynamics.}$$

Continuum thermodynamics is based on the basic laws [**BL**] and axioms [**AX**] shown in Figure 5.1.

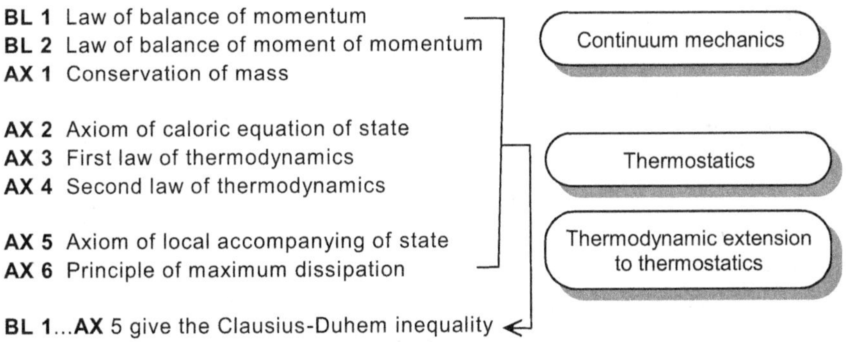


**Figure 5.1** Basic laws and axioms of continuum thermodynamics and the Clausius–Duhem inequality.

Thermostatics provides a path from continuum mechanics to continuum thermodynamics. Therefore, this chapter first explores thermostatics and then introduces its fundamental notions. The axiom of caloric equation of state, the first law of thermodynamics, and the second law of thermodynamics are introduced for thermostatics. From these fundamental notions, state equations are then derived. Thermostatics is formulated for equilibrium processes (or more exactly: it compares two states of equilibrium) within homogeneous systems (e.g., absolute temperature $T$ takes equal values everywhere within the system $\mathrm{m^b}$). Since these assumptions are too restrictive for material modelling, thermodynamics is needed to overcome the assumption of homogeneous systems and equilibrium. Thermodynamics allows the derivation of state equations for more general cases. However, in Chapter 6, they are extended to cover cases in which the material model is expressed with internal variables. At the end of this chapter, the study is extended to cover continuum thermodynamics. As shown in Figure 5.1, continuum thermodynamics covers the information obtained from continuum mechanics, thermostatics, and thermodynamics and, therefore, forms a tool for validation of material models. The Clausius–Duhem inequality is a validation tool to ensure that the material model does not violate the basic laws and axioms.

## 5.2 Background

A thermodynamic system $\mathrm{m^b}$, or system for short, is any collection of matter imagined to be isolated from the rest by a clearly defined boundary. It is not necessarily assumed that the boundary is rigid. On the contrary, in most cases, the boundary will deform when the system is subjected to some process that we have to understand and analyze (Kestin [37], Vol. 1: pp. 22 and 23).

A subsystem $m$ is an arbitrary part of the system $\mathrm{m^b}$. A subsystem $m$ is called closed when its boundary is not crossed by matter. In the contrary case, it is called open. The present study is restricted to closed subsystems. We study thermomechanical systems (i.e., thermomechanical response of materials), which means that the influence of, for example, electrical and chemical effects is neglected. A thermomechanical closed subsystem $m$ exchanges energy with its surroundings by exchange of heat and by work done by the volume and surface forces acting on the boundary $\partial m$.

Consider a system $\mathrm{m^b}$. When all the information required for a complete characterization of the system $\mathrm{m^b}$ for a purpose at hand is available, it is said that the state of the system is known. For example, for a certain homogeneous elastic body at rest, a complete description of its thermodynamic state $\mathcal{E}$ requires a specification of its material content, that is, the quantity of each chemical substance contained; its geometry in the natural or unstrained state $\mathcal{E}$; its deviation from the natural state $\mathcal{E}$, or strain field; its stress field; and, if some physical properties depend on whether the body is hot or cold, one extra independent quantity that fixes the degree of hotness or coldness. These quantities are called state variables. If a certain state variable can be expressed as a single-valued function of a set of other state variables, then the functional relationship is said to be an equation of state, and the variable so described is called a state function. The selection of a particular set of independent state variables is important in each problem, but the choice is to a certain extent arbitrary. (Fung [24], p. 341).

If the state $\mathcal{E}$ of a material point P within a subsystem $m$ is independent of the location of the point P, that is, the state $\mathcal{E}$ has the same value everywhere within the subsystem $m$, and if the material of the subsystem $m$ has the same properties everywhere, the subsystem $m$ is called a homogeneous subsystem. A thermodynamic subsystem $m$ in which there is no energy exchange with its surroundings is said to be isolated. A subsystem $m$ is said to be in thermodynamic equilibrium if this subsystem $m$ is isolated and it does not evolve with time (Maugin and Muschik [61], p. 219). If the state variables vary with time, the system is said to undergo a thermodynamic process or a process (Fung [24], p. 341). Maugin ([60], pp. 263 and 264) gives the following definition for reversible and irreversible processes: a thermodynamic process is said to be reversible if the inverse evolution of the system in time – that is, the succession of thermodynamic states $\mathcal{E}$ that the system has gone through – implies the reversal in time of the action of external stimuli. Otherwise, the thermodynamic process is said to be irreversible.

Thermostatics is the science that compares subsystems in thermodynamic equilibrium. For example, it describes the transition from a state $\mathcal{E}_1$ of equilibrium to another state $\mathcal{E}_2$ of equilibrium. Thermodynamics, in its main sense, is the study of phenomena outside a state of equilibrium, but actually not far outside this equilibrium. Everybody, of course, agrees that in the years 1890 to 1920 thermostatics was developed in an elegant mathematical form by Clausius, Gibbs, Duhem, and Carathéodory, in harmony with the experiments. Unfortunately, we cannot say the same about thermodynamics outside equilibrium; schools strongly disagree with each other on this subject. See Maugin ([60], pp. 263 and 264).

The preceding discussion gives rise to the question about the form of transition from a state $\mathcal{E}_1$ of equilibrium to another state $\mathcal{E}_2$ of equilibrium. According to Lavenda ([44], p. 4), the transition is a quasi-static process, which is a process that is carried out infinitely slowly so that the system can be considered as passing through a continuous series of equilibrium states. The processes described by Lavenda are not (yet) finished, since they are infinitely slow and since the age of the universe is finite. This may cast suspicion on the elegance of the formulation of thermostatics (cf. preceding discussion by Maugin [60]).

This study presents thermostatics as a theory for homogeneous subsystems, whereas the field theory lies within the framework of thermodynamics. Ziegler ([121], p. 56) argues that the lack of a thermodynamic field theory even at the end of the first half of the present century seems to be surprising. Maugin ([60], p. 263) states that already in 1920 thermostatics was developed in an elegant mathematical form. The comments by Ziegler and by Maugin support the concept that thermostatics is not a field theory. However, it has to be kept in mind that the assumption that thermostatics is a field theory would lead to the same description for continuum thermodynamics as is presented here. Only the path of the derivation would be different.

Unfortunately, *thermodynamics* is usually the term that covers thermostatics and thermodynamics. The basic laws of thermodynamics are good examples of this terminology; the reader must know whether the discussion is about thermostatics or thermodynamics. The problem is real, since – as discussed earlier by Maugin – thermostatics is a well-defined theory, whereas there are still some open questions in the theory of thermodynamics, which has led to the introduction of several different schools of thermodynamics.

Furthermore, even the term *thermodynamics* is misleading. Thermodynamics does not necessarily study the dynamics of systems, that is, the inertia effects are not included in the investigation. In general, the term *dynamics* refers to the effects of inertia, but in thermodynamics, it reflects the time dependency. The aim of thermodynamics is to study processes, that is, how the state of a system evolves with time.

Misunderstandings might occur, for example, when using thermostatics to study the Carnot and Otto processes. No processes are in fact modelled, but thermostatics is used to compare two states, and the path (i.e., process) between these two states is not included in the survey. Thus, time $t$ is not present in the expressions of thermostatics, making thermostatics similar to the elasticity theory. Therefore, thermostatics does not study real processes but compares equilibrium states. With the introduction of thermodynamics, time $t$ enters into the theory.

The introduction of entropy and internal energy are the main notions of thermodynamics beyond mechanics. The following discussion by Narasimhan ([66], pp. 345 and 346) clarifies the idea of entropy.

The concept of entropy occupies the core of thermodynamics. Before developing an analytical formulation of entropy, we present a motivation for the entropy concept. The first law of thermodynamics, namely, the law of conservation of energy, essentially states that the energy of a material system cannot be created or destroyed, but can only be transformed from one form into another. The first law does not, however, specify the manner in which this transformation may occur. For instance, there is no information furnished by the first law as to whether the energy transfer is reversible or irreversible. The latter question of reversibility of energy transfer becomes important in material systems in order to keep track of the amount of energy available for use. The information as regards the manner in which the transformation of energy occurs is furnished by the law of entropy, which is also referred to as the second law of thermodynamics.

Narasimhan [66] continues, consider the example of gasification of coal, in which the latter is heated to produce an energy source as coke oven gas. But a part of this available energy in the coal is transformed into hydrogen sulfide, ammonia, and other gases, which escape into external atmosphere. Although the total energy is not lost, the burned coal cannot be reheated to obtain a further amount of work. This means that the energy transfer occurs only in one direction, from an available into an unavailable form. Entropy may be interpreted as a measure of the loss in the amount of energy that is transformed irreversibly from a usable to an unusable form, in which it cannot be converted into work again. Similarly, a physical system with some initial order prevailing in its internal constituents tends to lose that order in an irreversible way upon heating, corresponding to the transformation of energy from available to unavailable form. A transformation from an ordered to a disordered state $\mathcal{E}$ is interpreted as an increase in entropy. Hence, for physical systems, the addition of energy which is drawn by them from their environment contributes to an increase in entropy. Production of entropy (the present book uses the term *dissipation*) is a physical system, therefore implying that the system has undergone irreversible changes, and conversely, irreversible changes in the system imply entropy production. A constant entropy implies only reversible changes in the system. The law of entropy is essentially the embodiment of the statement of increase of entropy in physical systems.

## 5.3 Thermostatics

This section sketches the foundations of thermostatics. Since thermostatics does not define the concepts of initial configuration, current configuration, and so on, these notions are not used here.

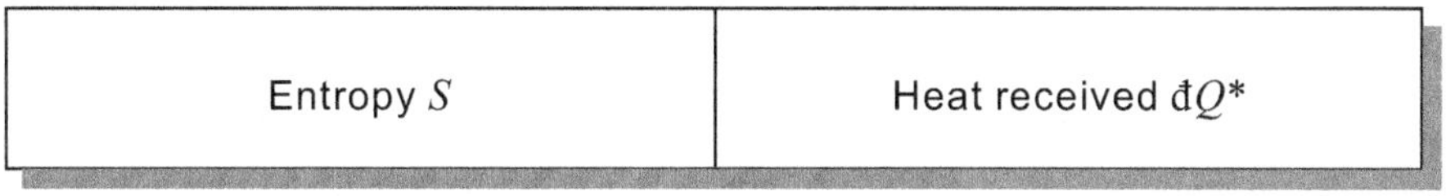


**Figure 5.2** Fundamental quantities of thermostatics beyond continuum mechanics.

Beyond expressions for mechanics, the foundations of thermostatics also looks at the following axioms of thermostatics:

**AX 2** *Axiom of caloric equation of state*,

**AX 3** *First law of thermodynamics*,

**AX 4** *Second law of thermodynamics*.

In thermostatics, there are one primitive concept and two fundamental quantities, beyond those in continuum mechanics. The primitive concept is the state

$\mathcal{E}$, and the fundamental quantities are entropy $S$ and the heat received $đQ^*$. The fundamental quantities are shown in Figure 5.2.

Since the unit for the heat received $đQ^*$ is Nm, introducing the heat received $đQ^*$ does not require introducing any new units beyond those already introduced for continuum mechanics. Entropy $S$, however, is a thermal variable, which therefore asks to introduce a thermal unit. However, the new thermal fundamental unit is not that associated to the new fundamental quantity entropy $S$, but it is related to a derived quantity, absolute temperature $T$. The kelvin is a natural choice for a fundamental unit, as it is well known to humans. Therefore, instead of entropy $S$, the absolute temperature $T$ is used in material models, as the following pages show. This work refers to $T$ as absolute temperature, or temperature for short.

The internal energy $U$ is a derived quantity in thermostatics. It is derived from the primitive concept state $\mathcal{E}$ and from the fundamental quantity entropy $S$ through the caloric equation of state.

Next, the axioms **AX 2**, **AX 3**, and **AX 4** are studied, one at a time, in more detail.

**AX 2** *Axiom of caloric equation of state (for thermostatics)*

Thermostatics is based on the assumption that the state $\mathcal{E}$ of a subsystem $m$ is described by a finite set of mutually independent variables $Y_\gamma$ where $\gamma = 1, 2, \ldots, n$ (which are mechanical, electrical, chemical, etc., depending on the modelled processes) and one thermal variable, which in the present work is the entropy $S$. Some scientists use thermodynamic temperature $T$ instead of entropy $S$ in the description of the state $\mathcal{E}$ (see, e.g., Malvern [57], p. 260). State $\mathcal{E}$ is expressed by the internal energy $U$ through the caloric equation of state

**AX 2**
$$U := U(Y_\gamma, S)\,. \tag{5.1}$$

**AX 3** *First law of thermodynamics*

According to Maugin ([60], pp. 266 and 267) the differential form of the first law of thermodynamics for a mechanical (i.e., electrical, chemical, and other processes are neglected) closed and homogeneous system reads

**AX 3**
$$\mathrm{d}U = đW + đQ^*\,, \tag{5.2}$$

where $đW$ is the work received and $đQ^*$ is the heat received. Since Axiom (5.2) [**AX 3**] is formulated for thermostatics, no kinetic energy term is presented. In Axiom (5.2) [**AX 3**], the notation $đ$ shows that $đW$ and $đQ^*$ are not perfect differentials. In physical terms, we can attribute a certain internal energy $U$ with a given thermodynamic state $\mathcal{E}$, but we cannot speak about the quantity of heat that the system possesses in that state (Lavenda [44], p. 5).

**AX 4** *Second law of thermodynamics*

Maugin ([60], p. 267) gives the second law of thermodynamics for closed systems and reversible processes. It has the following appearance:

**AX 4** $$đQ^* = T\,\mathrm{d}S\,. \tag{5.3}$$

One could express that the quantity $đQ^*$ is defined by Axiom (5.3). This problem is similar to that with force vector $\vec{F}$ where Newton's Second Law of Motion (4.13) [see also discussion around Equation (4.10)] was not used as a definition for the force vector $\vec{F}$. Thus, if Expression (5.3) is interpreted to be a definition for the heat received $đQ^*$, Equation (5.3) can no longer be used as an axiom of thermostatics, and vice versa. Therefore the heat received $đQ^*$ is assumed to be a fundamental quantity for thermostatics, and Expression (5.3) is introduced as an axiom.

Substitution of Second Law of Thermodynamics (5.3) [**AX 4**] into First Law of Thermodynamics (5.2) [**AX 3**] gives

$$\mathrm{d}U = đW + T\,\mathrm{d}S. \tag{5.4}$$

**Example 5.1:** Derive the state equations for thermostatics by studying the "gas inside a cylinder" case shown in Figure 5.3.

**Solution:** The volume and the pressure of the gas are denoted by $V$ and $p$. The investigated system of the $V$ in this example is defined as the volume bounded by the walls of the cylinder and the piston head.

For the present example, a pure mechanical study implies that the elementary work received $đW$ is

$$đW = -p\,\mathrm{d}V\,. \tag{5.5}$$

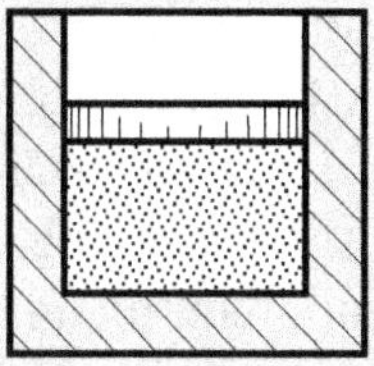

**Figure 5.3** Gas inside a cylinder.

Substitution of Equation (5.5) into Equation (5.4) yields

$$\mathrm{d}U = -p\,\mathrm{d}V + T\,\mathrm{d}S\,. \tag{5.6}$$

Equation (5.6) shows that for this particular example, the independent variables for the internal energy $U$ are the volume of the gas $V$ and the entropy $S$; that is,

$$U = U(V, S)\,. \tag{5.7}$$

Differentiation of Expression (5.7) gives

$$\mathrm{d}U = \frac{\partial U(V,S)}{\partial V}\,\mathrm{d}V + \frac{\partial U(V,S)}{\partial S}\,\mathrm{d}S\,. \tag{5.8}$$

Comparison of Forms (5.6) and (5.8) for the differential internal energy $\mathrm{d}U$ gives the following result:

$$p = -\frac{\partial U(V,S)}{\partial V} \quad \text{and} \quad T = \frac{\partial U(V,S)}{\partial S}\,, \tag{5.9}$$

which are the state equations. ■

## 5.4 Thermodynamics

This section gives the extension for thermodynamics beyond thermostatics. Since thermodynamics does not define the concepts of initial configuration, current configuration, and so on, these notions are not used here. However, in Chapter 6, where the content of the present section is combined with continuum mechanics to form continuum thermodynamics, the roles of different configurations become important, and therefore the notations related to the current configuration $\underline{v}^{\mathrm{b}}(t)$ have to be introduced. This may introduce a small but therefore acceptable flaw in the present study.

| Specific entropy | Heat flux vector | Heat source per unit mass |
|---|---|---|
| $s$ | $\vec{q}$ | $r$ |
| [Nm / K kg] | [Nm / s m²] | [Nm / s kg] |

**Figure 5.4** Fundamental quantities of thermodynamics beyond continuum mechanics and the units associated to them.

The entropy $S$, internal energy $U$, and heat received $đQ^*$ are quantities for thermostatics, since they are quantities for the whole system $\mathrm{m}^{\mathrm{b}}$. Since the goal of this book is to cover continuum thermodynamics, which is a field theory, rather than the preceding quantities, the corresponding field variables are introduced as fundamental quantities. The fundamental quantities of thermostatics are no longer used and have been replaced with the quantities given in Figure 5.4, which shows introduction of thermodynamics gives one new fundamental unit, which is the kelvin (K). Now Axiom (1) [**AX 2**] can be reformulated for thermodynamics as follows.

**AX 2** *Axiom of caloric equation of state (for thermodynamics)*

Thermodynamics is based on the assumption that the state $\mathcal{E}$ at the every point in a subsystem $m$ is described by a finite set of mutually independent field variables $\upsilon_\gamma$ where $\gamma = 1, 2, \ldots, n$ (which are mechanical, electrical, chemical, etc., depending on the modelled processes) and one thermal variable, which in the present work is the specific entropy $s$. State $\mathcal{E}$ is expressed by the specific internal energy $u$ through the caloric equation of state

**AX 2** $$u := u(\upsilon_\gamma, s)\,. \tag{5.10}$$

Since thermodynamics is an extension of thermostatics, it requires more fundamental information. The following concepts may not be the final ones but today seem to be the most popular. The two axioms for thermodynamics beyond thermostatics are as follows:

**AX 5** *Axiom of local accompanying state*

**AX 6** *Principle of maximum dissipation*

Axioms 5 and 6 are discussed briefly next.

**AX 5** *Axiom of local accompanying state*

By applying the axiom of local accompanying state, points of nonequilibrium state space are associated with points of equilibrium state space by means of a projection (Maugin and Muschik [61], p. 226). This projection is shown in Figure 5.5 (see Muschik [65], Figure 1.16).

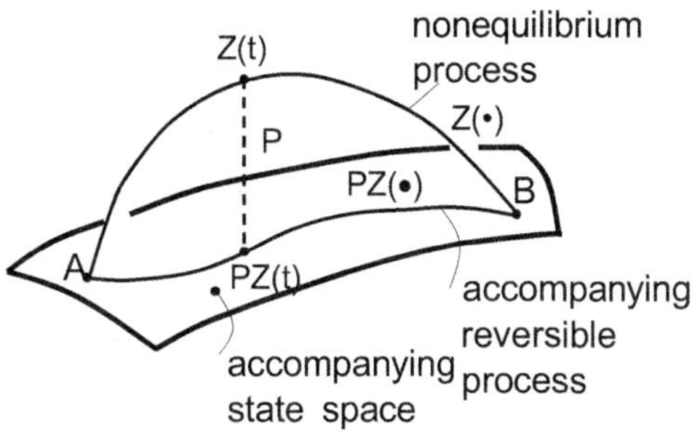


**Figure 5.5** Projection P maps the (non-equilibrium) process Z(·) point by point onto the equilibrium subspace, represented as a hypersurface in the state space. Consequently, the accompanying reversible process PZ(·) is parameterized by $t$. Modified from Muschik ([65], Figure 1.16.)

According to Maugin and Muschik ([61], p. 226), the axiom of local accompanying state is the most commonly accepted viewpoint, which consists in replacing the axiom of the local equilibrium state of classical irreversible thermodynamics with a somewhat straightforward generalization known under the name of the axiom of the local accompanying state.

The key concept of the axiom of local accompanying state is that it allows the introduction of state equations [cf. State Equations (5.9) of thermostatics] also outside the equilibrium. Thus state equations are derived for an accompanying process in the equilibrium state space by using thermostatics, and the obtained state equations are assumed to hold for the real nonequilibrium process.

The second extension of thermodynamics beyond thermostatics is discussed next.

**AX 6** *Principle of maximum dissipation*

With the introduction of the specific dissipation function $\varphi$, the principle of maximum dissipation provides the evolution equation for the mechanical variables and for the heat flux vector $\vec{q}$. There are several variations of this theme, but here the authors refer to the work by Ziegler ([120], p. 14).

The following two examples support Section 6.4, where controllable state variables are introduced. The idea of the following examples is that, although the processes under consideration are thermostatical processes, the state variables are for thermodynamics; that is, they are field quantities.

When processes within the framework of thermostatics were examined, the first law of thermodynamics and the second law of thermodynamics were combined into Expression (5.4):

$$\mathrm{d}U = đW + T\,\mathrm{d}S\,. \tag{5.11}$$

As discussed previously, thermodynamics utilizes field variables. Thus Equation (5.11) is divided by the mass of the matter $m$, and we arrive at the following:

$$\frac{\mathrm{d}U}{m} = \frac{đW}{m} + T\,\frac{\mathrm{d}S}{m}\,, \qquad \text{which yields} \qquad \mathrm{d}u = đw + T\,\mathrm{d}s\,. \tag{5.12}$$

Expression $(5.12)_2$ displays the fact that temperature $T$ divided by mass $m$ is not an acceptable variable for thermodynamics (if not at all an acceptable variable).

**Example 5.2:** Derive state equations for thermodynamics by studying a case "gas inside a cylinder," shown in Figure 5.6.

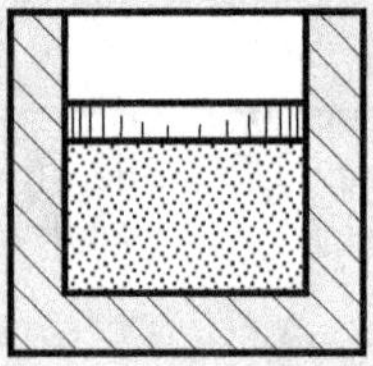

**Figure 5.6** Gas inside a cylinder.

**Solution:** For the case "some gas inside a cylinder" (see Figure 5.3), Expression (5.5) was achieved. It is

$$đW = -p\,\mathrm{d}V \quad \Rightarrow \quad đw = \frac{-p\,\mathrm{d}V}{m} = \frac{-p\,\mathrm{d}V}{\rho\,V}\,. \tag{5.13}$$

Since the mass (of the gas) $m$ does not change (**AX 1** *Axiom of conservation of mass*), the following relation holds:

$$\mathrm{d}m = \mathrm{d}\rho\,V + \rho\,\mathrm{d}V = 0 \quad \Rightarrow \quad \frac{\mathrm{d}V}{V} = -\frac{\mathrm{d}\rho}{\rho}\,. \tag{5.14}$$

Substitution of Expression $(5.14)_2$ into Expressions $(5.13)_2$ and of the obtained result further into Equation $(5.12)_2$ yields

$$\mathrm{d}u = \frac{1}{\rho^2}\,p\,\mathrm{d}\rho + T\,\mathrm{d}s\,. \tag{5.15}$$

According to Gibbs Equation (5.15), the specific internal energy $u$ is dependent on the density of the gas $\rho$ and on the specific entropy $s$, that is, $u = u(\rho, s)$. This gives

$$\mathrm{d}u = \frac{\partial u(\rho, s)}{\partial \rho}\,\mathrm{d}\rho + \frac{\partial u(\rho, s)}{\partial s}\,\mathrm{d}s\,. \tag{5.16}$$

Comparison of Gibbs Equation (5.15) with Equation (5.16) gives the following state equations:

$$p = \rho^2\,\frac{\partial u(\rho, s)}{\partial \rho} \qquad \text{and} \qquad T = \frac{\partial u(\rho, s)}{\partial s}\,. \quad \blacksquare \tag{5.17}$$

**Example 5.3:** Derive state equations for thermodynamics by studying a case "a bar under uniaxial tension or compression."

**Solution**: Based on Figure 5.7, the following can be written:

$$\text{đ}W = F\,\mathrm{d}\ell, \quad \text{which gives} \quad \text{đ}w = \frac{F\,\mathrm{d}\ell}{m} = \frac{F\,\mathrm{d}\ell}{\rho_0\,A_0\,\ell_0} = \frac{1}{\rho_0}\,\frac{F}{A_0}\,\frac{\mathrm{d}\ell}{\ell_0} = \frac{1}{\rho_0}\,\sigma\,\mathrm{d}\varepsilon\,, \tag{5.18}$$

where $F$ is the applied force, $\ell$ is the current length of the bar, $\ell_0$ is the initial length of the bar, and $A_0$ is the initial area of the cross section of the bar. Substitution of Expression $(5.18)_2$ into Equation $(5.12)_2$ gives the following Gibbs equation:

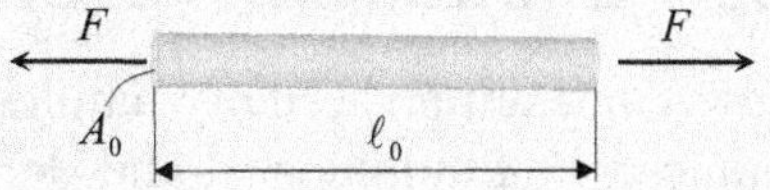


**Figure 5.7** Bar under uniaxial tension.

$$\mathrm{d}u = \frac{1}{\rho_0}\,\sigma\,\mathrm{d}\varepsilon + T\,\mathrm{d}s\,. \tag{5.19}$$

According to Gibbs Equation (5.19), the specific internal energy $u$ is dependent on the strain $\varepsilon$ and on the specific entropy $s$, that is, $u = u(\varepsilon, s)$. This gives

$$\mathrm{d}u = \frac{\partial u(\varepsilon, s)}{\partial \varepsilon}\,\mathrm{d}\varepsilon + \frac{\partial u(\varepsilon, s)}{\partial s}\,\mathrm{d}s\,. \tag{5.20}$$

Comparison of Gibbs Equation (5.19) with Equation (5.20) gives the following state equations:

$$\sigma = \rho_0\,\frac{\partial u(\varepsilon,\, s)}{\partial \varepsilon} \quad \text{and} \quad T = \frac{\partial u(\varepsilon,\, s)}{\partial s}\,. \;\blacksquare \tag{5.21}$$

Although the quantities $\sigma$ and $\varepsilon$ are field variables, they have the same values at all points of the bar. This means that the system is homogeneous. However, it is assumed that the results in Results (5.21) hold for nonhomogeneous systems. The study of a nonhomogeneous system would be too complicated to be educative. The same comment holds for the quantities present in **Example 5.2**.

Equations (5.17) and (5.21) show the difference between fluid mechanics and solid mechanics. In the equations for fluid mechanics, the density of the material is the current density $\rho(\vec{x}, t) = \rho(\underline{\vec{x}}(t_2), t_2)$, since the control volume $\mathrm{V}^{\mathrm{cv}}$ equals the subsystem $\underline{v}(t)$ at time $t_2$, that is, $\underline{v}(t_2) = \mathrm{V}^{\mathrm{cv}}$. The local forms of the expressions for solid mechanics apply the density of the material in the IG-CtP configuration $\mathrm{V}_0^{\mathrm{b}}$ (initial geometry at current time and position), which equals the value of the density in the initial configuration $\rho_0(\vec{X})$.

The origin of the density $\rho_0$ in State Equations (5.21) can be shown by a simple dimensional analysis. State Equation $(5.9)_1$ for thermostatics provides the following dimensional analysis to be carried out:

$$p = \frac{\partial U}{\partial V} = \left\{ \frac{\text{Energy}}{\text{Volume}} \right\} = \left[ \frac{\text{Nm}}{m^3} \right] = \left[ \frac{\text{N}}{m^2} \right] . \tag{5.22}$$

State Equation $(5.21)_1$ can be evaluated as follows:

$$\sigma = \rho_0 \frac{\partial u}{\partial \epsilon} = \left\{ \text{Density} \times \frac{\text{Energy per mass}}{\text{Pure number}} \right\} = \left[ \frac{\text{kg}}{m^3} \times \frac{\text{Nm/kg}}{1} \right] = \left[ \frac{\text{N}}{m^2} \right] . \tag{5.23}$$

## 5.5 Foundation of Continuum Thermodynamics

Since continuum thermodynamics is a combination of thermodynamics and mechanics, the fundamental and derived quantities and the fundamental units for continuum thermodynamics were already discussed in Sections 4.3, 4.24, and 5.4 (not exactly as shown on the next page). The aim of this section is to combine them. The basic laws and axioms for continuum mechanics and for continuum thermodynamics are listed, for example, in Section 5.3. They are also given in this section. The explicit forms for the basic laws and axiom of continuum mechanics were given in Section 4.4 and their local forms derived in the other sections of Chapter 4. The explicit forms for the axioms of thermodynamics and their local forms are introduced in the following chapter.

The reader may wonder at the structure of this book. Why was the introduction of continuum thermodynamics incremental? First, the foundations of continuum mechanics were discussed. Second, the ideas of thermostatics were evaluated. Third, thermostatics was extended to nonequilibrium processes, that is, thermodynamics was derived. Finally, all these theories are now put together, and the soup obtained is called continuum thermodynamics. Why did the authors not introduce straightforward continuum thermodynamics by giving its basics without dividing them into the different categories of continuum mechanics, thermostatics, and thermodynamics? The reason for categorizing the foundations of continuum thermodynamics is both educational and historical. Many people know continuum mechanics and may therefore find interesting the relationship between continuum mechanics and continuum thermodynamics.

The basic laws and axioms studied so far are not adequate to describe the continuum thermodynamic aspects of nature. Several fundamental quantities, fundamental units, and primitive concepts must first be introduced. First, the fundamental quantities, fundamental units, and primitive concepts were introduced for continuum mechanics. The fundamental quantities and their units were introduced in Figure 4.1. The quantities in Figure 4.1 are underlined to show that they are defined in the current configuration $\underline{v}^{\text{b}}(t)$. The derived quantities of continuum mechanics include the stress tensor $\boldsymbol{\sigma}$ and the strain tensor $\boldsymbol{\varepsilon}$. A primitive concept for continuum mechanics is that the matter is a continuum.

| Density $\underline{\rho}$ kilogram /m$^3$ [kg/meter$^3$] | Length $\underline{\ell}$ meter [m] | Time $t$ second [s] | Traction vector $\vec{\underline{t}}$ [N/m$^2$] | Body force vector $\vec{\underline{b}}$ [N/kg] |
|---|---|---|---|---|

**Figure 5.8** Fundamental quantities and the units associated to them for continuum mechanics.

When thermostatics was introduced, some new fundamental quantities were also introduced, but these were discarded with the introduction of thermodynamics. The role of thermostatics was to show the pathway for derivation of state equations, which was followed when state equations for continuum thermodynamics were introduced. The fundamental quantities for thermodynamics were given in Figure 5.4, and now takes the appearance shown in Figure 5.9.

| Specific entropy $s$ [Nm / K kg] | Heat flux vector $\vec{q}$ [Nm / s m$^2$] | Heat source per unit mass $r$ [Nm / s kg] |
|---|---|---|

**Figure 5.9** Fundamental quantities of thermodynamics beyond continuum mechanics and the units associated to them.

The role of the Cauchy heat flux vector $\vec{\underline{q}}^{\mathrm{c}}$ will be discussed in more detail in Section 6.2.1 on the first law of thermodynamics. Contrary to the forms given in Section 5.4, here the preceding quantities are underlined to show that they are defined in the current configuration $\underline{v}^{\mathrm{b}}(t)$.

The introduction of thermodynamics demanded the introduction of one thermal variable which, in this book, is the (absolute) temperature $\underline{T}$. Among others, the specific internal energy $\underline{u}$ is a derived quantity and the state $\mathcal{E}$ is a primitive concept.

The introduction of the specific entropy $\underline{s}$, specific internal energy $\underline{u}$ , Cauchy heat flux vector $\vec{\underline{q}}^{\mathrm{c}}$, and heat source per unit mass $\underline{r}$ allows the following definition of the corresponding quantities for an arbitrarily selected but then nonchanging entity of matter $m$: entropy

$$\underline{S}(t) := \int_m \underline{s}(\vec{\underline{x}}(t),t)\,\mathrm{d}m \quad \Rightarrow \quad \underline{S}(t) = \int_{\underline{v}(t)} \underline{\rho}(\vec{\underline{x}}(t),t)\,\underline{s}(\vec{\underline{x}}(t),t)\,\mathrm{d}\underline{v} \tag{5.24}$$

and internal energy

$$\underline{U}(t) := \int_m \underline{u}(\vec{\underline{x}}(t),t)\mathrm{d}m \quad \Rightarrow \quad \underline{U}(t) = \int_{\underline{v}(t)} \underline{\rho}(\vec{\underline{x}}(t),t)\,\underline{u}(\vec{\underline{x}}(t),t)\,\mathrm{d}\underline{v} \tag{5.25}$$

and, finally, heat input rate

$$\underline{Q}(t) := -\oint_{\partial m(t)} \vec{\underline{n}}(\vec{\underline{x}}(t),t)\cdot\vec{\underline{q}}^{\mathrm{c}}(\vec{\underline{x}}(t),t)\,\mathrm{d}(\partial m) + \int_m \underline{r}(\vec{\underline{x}}(t),t)\,\mathrm{d}m\,. \tag{5.26}$$

Definition (5.26) yields

$$\underline{Q}(t) = -\oint_{\partial\underline{v}(t)} \underline{\vec{n}}(\underline{\vec{x}}(t),t)\cdot\underline{\vec{q}}^{\,\mathrm{c}}(\underline{\vec{x}}(t),t)\,\mathrm{d}\underline{a} + \int_{\underline{v}(t)} \underline{\rho}(\underline{\vec{x}}(t),t)\,\underline{r}(\underline{\vec{x}}(t),t)\,\mathrm{d}\underline{v}\,. \quad (5.27)$$

Equations $(5.24)_2$ and $(5.25)_2$ and Expression (5.27) utilize the equalities $\partial\underline{m}(t) = \partial\underline{v}(t)$ and $\mathrm{d}m = \underline{\rho}(\cdot,t)\,\mathrm{d}\underline{v}(t)$. As Equations $(5.24)_1$, $(5.25)_1$, and (5.26) show, the quantities are defined in the current configuration $\underline{v}^{\mathrm{b}}(t)$, and the integration is carried out over or around the subsystem $m$ moving and deforming with deformation of the system $\mathrm{m}^{\mathrm{b}}$.

By applying Reynolds Transport Theorem (4.81), Expressionss $(5.24)_2$ and $(5.25)_2$ give

$$\underline{\dot{S}}(t) = \int_{\underline{v}(t)} \underline{\rho}(\underline{\vec{x}}(t),t)\,\underline{\dot{s}}(\underline{\vec{x}}(t),t)\,\mathrm{d}\underline{v}, \quad \underline{\dot{U}}(t) = \int_{\underline{v}(t)} \underline{\rho}(\underline{\vec{x}}(t),t)\,\underline{\dot{u}}(\underline{\vec{x}}(t),t)\,\mathrm{d}\underline{v}\,. \quad (5.28)$$

Equations (5.28) are slightly misleading. The quantities $\underline{S}(t)$, $\underline{U}(t)$, and $\underline{Q}(t)$ for thermodynamics are not the same as the quantities $S$, $U$, and $Q$ for thermostatics. The former are for an arbitrary nonhomogeneous subsystem $m$ and depend on time $t$, whereas the latter are for the whole homogeneous system $\mathrm{m}^{\mathrm{b}}$ and are time-independent quantities. Furthermore, in thermostatics, the notation state $\mathcal{E}$ stands for the state of whole system $\mathrm{m}^{\mathrm{b}}$, whereas in thermodynamics, which is a field theory, the state $\mathcal{E}$ refers to the state of a material point. Since this discrepancy may not cause any major confusion, the preceding notations can be settled on. The given basic laws [**BL N**] and axioms [**AX N**] of continuum thermodynamics are shown in Figure 5.10.

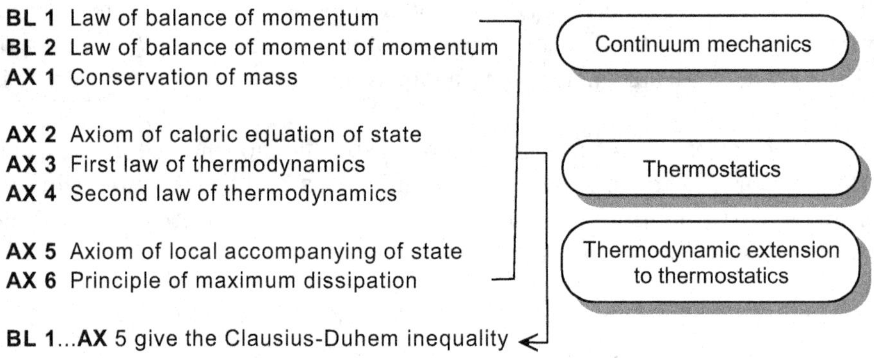


**Figure 5.10** Basic laws and axioms of continuum thermodynamics and the Clausius–Duhem inequality.

The Clausius–Duhem inequality is a validation tool to ensure that the material model does not violate the basic laws and axioms discussed in the preceding paragraphs. However, something is still missing. The following discussion is by

Astarita ([8], pp. 13–14). The authors have changed the text to fit the vocabulary of the present work. Comments by the authors are between curly braces { }.

Every branch of physical science is based on two sets of fundamental equations. The first set is that of basic laws and axioms of physics, which are postulated to hold valid for all bodies under all conceivable circumstances; {the axiom of} conservation of mass, the law of balance of momentum, and the law of balance of moment of momentum are typical examples. In thermodynamics, the axioms are the first and second laws. {The present work also expresses the axiom of caloric equation of state to belong to this set.} The large majority of basic laws and axioms of physics are principles of conservation of some quantity (mass, linear momentum, etc.); the first law of thermodynamics falls into this category, but the second law is an exception, since it is not a principle of conservation.

The second set of fundamental equations are the constitutive equations: these are relationships that are not supposed to hold for all bodies but are only to describe the behavior of some restricted class of bodies, or possibly of a larger class of bodies for a more restricted class of phenomena. A good example is the mechanics of rigid bodies; it is of course obvious that there are many bodies in nature that are not rigid (and perhaps one could argue that there are in actual fact no bodies that are truly rigid); however, the theory of rigid bodies is a useful abstraction that describes satisfactorily some phenomena as observed in nature. Constitutive equations are assumptions that may, or may not, adequately describe the behavior of real bodies.

To the foregoing discussion by Astarita, the authors add the following: the form of the constitutive equation is dependent on (a) the material, (b) the loading/environment, and (c) the required accuracy of the model. A detailed discussion is presented next.

- **(a) and (b)** The material and the loading/environment aspects have, for example, the following effects on material models: the response of structural steel, for example, can be simulated by Hooke's law, if the stress is below the yield strength of the material and the temperature is close to or below room temperature. For higher stress, however, the plastic yield must be described, and at elevated temperatures, creep may be the dominant deformation mechanism. On the other hand, at a stress level at which plastic flow is remarkable in structural steels, many ceramics obey Hooke's law.

- **(c)** The required accuracy of the constitutive equation is dependent on the application of the model. When simulating hysteresis loops caused by cyclic loading, the material model needs to be far more accurate than when evaluating the elastic-plastic bending of a beam.

It is worth noting that the present theory is an effective one. In an effective theory, the level of description is coarsened in such a way that many details are eliminated by describing processes with "averaged" variables. Thus atoms and molecules are replaced with continuous matter. When effective theories are used, some information will be lost, but far fewer computational resources are needed, as it is no longer necessary to model the response of every single molecule.

As obtained in State Equations $(5.17)_2$ and $(5.21)_2$, the quantity absolute temperature $T$ is defined as a derivative of the (specific) internal energy ($u$) $U$ with respect to the (specific) entropy ($s$) $S$. Furthermore, heat conduction is modelled as dependent on the temperature gradient. In "reality," temperature $T$ reflects the vibrations of molecules, whereas the heat conduction reflects the collision between them. In the hotter domain of matter, molecules vibrate more than in the cooler domain. Vibration extends to the cooler domain when the strongly vibrating molecules collide with those vibrating less, making the latter vibrate more. During this process, the vibration in hotter domains slows down, that is, dropping the temperature.

## 5.6 Major Dialects of Thermodynamics

This section discusses different theories describing processes outside the thermodynamic equilibrium by introducing different dialects of continuum thermodynamics. In modelling of materials, the major dialect is the theory called "continuum thermodynamics with internal variables," which is briefly discussed in the following section. To keep the discussion short, some concepts introduced here are formulated more rigorously later in this book.

Unfortunately, in the literature, the terminology has not yet been established; therefore the term *thermodynamics* also covers thermostatics and continuum thermodynamics. This can be seen in the present and the following chapter, were the terminology used in the original references has been left unchanged. Instead of the phrase "continuum thermodynamics," some writers use the word "thermomechanics."

The main problem of thermodynamics is the definition of the absolute temperature $T$ and entropy $S$ outside equilibrium. There are several variations of the theme to solve this problem and to formulate an elegant thermodynamic theory to model processes (far) outside equilibrium. The following discussion by Lebon et al. ([46], pp. 41 and 42) and Lebon ([45], p. 7) sketches the concepts of the main theories. Figure 5.11 and comments in curly braces { } are those of the authors.

Since the Second World War, two lines of thought have been developed in the field of nonequilibrium thermodynamics. The first is known as the classical thermodynamic theory of irreversible processes, in short, classical irreversible thermodynamics (CIT); the second one is referred to by its founders Truesdell, Coleman, and Noll ([111, 110]) as rational thermodynamics (RT). The foundations of CIT were laid down by Onsager in two celebrated papers [73, 74], but the theory owes much of its success to the Brussels school directed by Prigogine [79, 80]. It is worth recalling that both Onsager and Prigogine were awarded the Nobel Prize in Chemistry in 1968 and 1977, respectively. Rational and classical thermodynamics aim at the same objective: to derive constitutive equations of material systems driven out of equilibrium {thermodynamic equilibrium}.

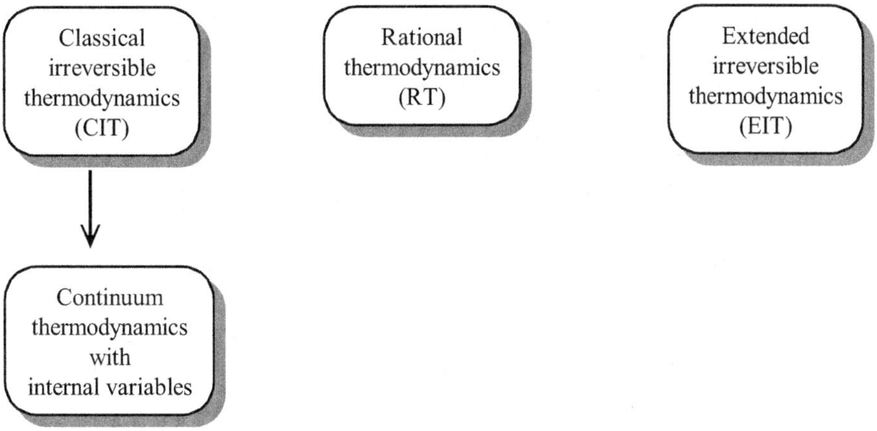


**Figure 5.11** Major theories for description of processes outside the thermodynamic equilibrium.

CIT borrows several results from classical thermodynamics {thermostatics} and is concerned with the class of materials and processes described by the "local equilibrium hypothesis." The latter states that the local and instantaneous relations between the thermal and mechanical properties of a material system are the same as for a uniform system at equilibrium. CIT is mainly applicable to situations "not too far" from equilibrium. More explicitly stated, it is supposed that the constitutive equations are expressed by means of linear relations between cause (also named force) and effect (the flux, according to the CIT terminology). An important pillar of the theory is provided by the Onsager symmetry relations between transport coefficients of coupling processes. Onsager's reciprocal relations were derived from the statistical theory of fluctuations and the hypothesis of microscopic reversibility. They receive a microscopic confirmation from the theory of gases and the theory of linear response; they were also confirmed by several experimental observations mainly in thermoelasticity and thermodiffusion.

It must be realized that CIT is unable to describe materials with memory and is not adequate for studying processes taking place far from equilibrium, in particular, high frequency and short wavelength phenomena. These restrictions are inherent to the local equilibrium hypothesis. According to CIT, the disturbances will propagate at an infinite velocity, in contradiction with the principle of causality, which demands that the cause precedes the effects: in CIT, cause and effect happen simultaneously. {The heat conduction equation evaluated in Section 8.2 is an excellent example of this phenomenon.}

The objective of RT is more ambitious than that of CIT as it seeks to describe a wider class of materials driven far from equilibrium. But this goal is achieved at the price of a greater mathematical complexity. In RT, absolute temperature and entropy are introduced as primitive concepts without a sound physical interpretation. The notion of state $\mathcal{E}$, expressing that any property evaluated at time $t$ can be written in terms of the state parameters given at the same time $t$, is given up and replaced by the notion of history or memory. Accordingly, not only are the properties of any system affected by the values of the variables at the present time but the values of the variables at the present time may also depend on their values in the past. As a consequence, the constitutive equations take the form of time-functionals. These functionals cannot, however, take any arbitrary possible form: there are restrictions placed by the second law of thermodynamics, such as the positiveness of the heat conductivity and the viscosity and the criterion of material frame-indifference, demanding that the constitutive equations are independent of the motion of any observer. The theory has met with impressive success among mathematicians and theoretical mechanicians because of its generality and its mathematical rigor.

The third approach is called extended irreversible thermodynamics (EIT) and it provides a mesoscopic and causal description of non-equilibrium processes: it was born out of the double necessity to go beyond the hypothesis of local equilibrium and to avoid the paradox of propagation of disturbances at infinite velocity. {At the present time, the theory may be considered as fully developed, and it is formulated in the book by Jou et al. [34]. These three major theories and the thermodynamics of internal variables are outlined in Figure 5.11.}

## 5.7 Continuum Thermodynamics with Internal Variables

The aim of this section is to sketch the foundations for continuum thermodynamics with internal variables. The present work utilizes a variation of CIT that can be called continuum thermodynamics with internal variables. Maugin and Muschik ([61], pp. 222 and 223) give the following description for this theory.

A fourth thermodynamics is the one that introduces internal variables as state variables. It stands somewhere in between the CIT and RT, but in fact it is the simplest generalization of CIT. The origin of the internal-variable thermodynamics may first be traced back in the kinetic description of physicochemical processes of evolution, but its spectacular development is related to rheological models and the elasto-viscoplasticity of deformable materials of the metallic type (alloys, polycrystals) (see, e.g., Lemaitre [50]; Chaboche [14]). As already mentioned, this approach is adopting a somewhat intermediary line between the two thermodynamics already sketched out. Essentially, it provides a new characterization of continuous media that, in order to define the thermodynamic state $\mathcal{E}$ of a subsystem $\underline{v}(t)$, introduces, in addition to the usual observable state variables ($\chi$; e.g., temperature and elastic strain), a certain number of internal variables collectively denoted by $\alpha$, which are supposed to describe the internal structure [hidden to the eye of the (untrained) external observer, who can see only a black box – hence also the alternate name of hidden variables, but this naming will be avoided as it sometimes creates confusion with variables so christened in certain re-interpretations of quantum mechanics]. It follows that the value, at moment $t$, of the dependent variables (e.g., the stress) becomes simultaneously a function both of the values of the independent observable variables and of the internal variables. This constitutive equation, say, $\sigma(\chi, \alpha)$ , where $\chi$ represents, as before, the controllable variables of state, must be complemented by an evolution equation that describes the temporal evolution of the variable $\alpha$. For instance, we can write the following:

$$\sigma = \sigma(\chi, \alpha) \qquad \text{state equation} \tag{5.29}$$

and

$$\dot{\alpha} = f(\chi, \alpha) + g(\chi, \alpha)\,\dot{\chi} \qquad \text{evolution equation}\,. \tag{5.30}$$

In fact, we may suppose that we have been able to select the $\alpha$s in such a way that $g(\chi, \alpha)$ might be identically zero and that an instantaneous variation of $\chi$ does not cause any instantaneous variation in the $\alpha$s [if $\chi$ is a strain, the hypothesis $g(\chi, \alpha) = 0$ corresponds to the fact that instantaneous strains are elastic or zero].

The authors' comments on the foregoing discussion by Maugin and Muschik [61] are as follows. First, the authors cannot agree with the comment on hidden variables. When mentioning internal variables that can be hidden to the eye of the (untrained) external observer, the writers may have been thinking of, for example, dislocation movement (plasticity or creep) in metals. Since metallic materials are not transparent, all internal deformation mechanisms are "hidden." On the other hand, damage associated with microcracking is often modelled by an internal variable called damage (see, e.g., Lemaitre [50]; Chaboche [14], pp. 58 and 346). However, microcracks in ice are visible to everyone, as Figure 5.12 clarifies. This implies that the division of variables into observable variables and internal variables is not acceptable. Section 6.4 looks in more detail at the terminology of independent variables in continuum thermodynamics with internal variables. Second, the style for introduction of Equations (5.29) and (5.30) may lead to confusion about the idea of modelling materials. When the theory of internal variables is used for preparation of constitutive equations, one does not assume any equations to be similar to Equations (5.29) and (5.30) but introduces explicit forms for two functions, namely, the specific Helmholtz free energy $\psi$ and the specific dissipation function $\varphi$. The stress tensor $\boldsymbol{\sigma}$, for example, is obtained as a partial derivative of the specific Helmholtz free energy $\psi$ with respect to the strain tensor $\boldsymbol{\varepsilon}$. This gives a state equation as described by Equation (5.29). The introduction of the explicit form for the specific dissipation function $\varphi$ applied to the normality rule yields evolution equations for internal variables, such as Equation (5.30).

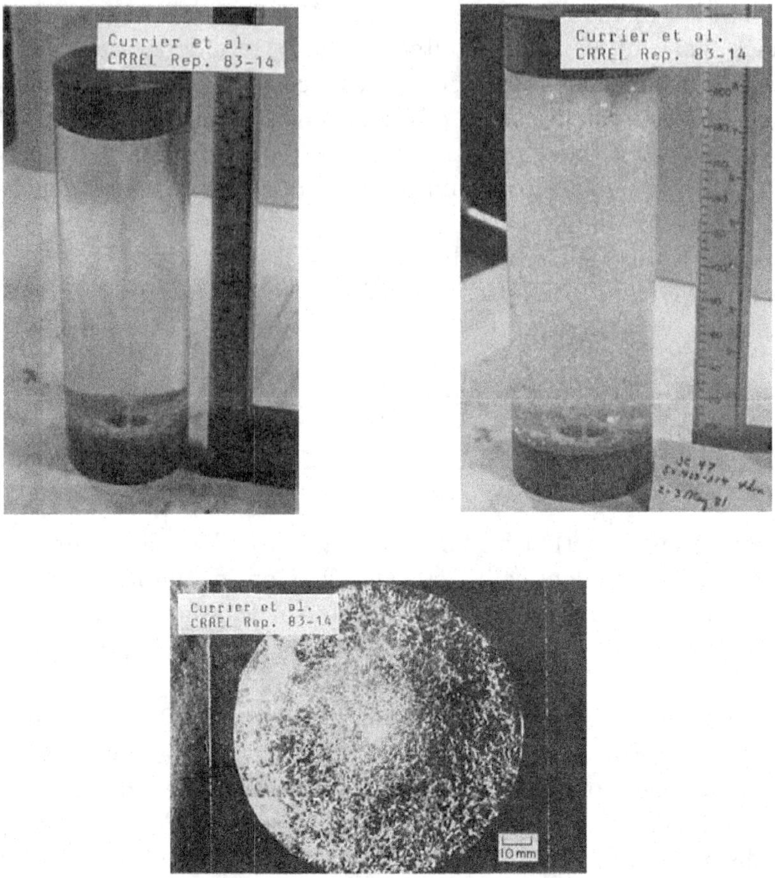


**Figure 5.12** Visible microcracks (i.e., internal variable) in ice. A fine-grained compression test specimen (a) before and (b, c) after the test. See Currier et al. ([19], Figures 10, 31, and 32).

As already mentioned, continuum thermodynamics with internal variables is the simplest generalization of the classical thermodynamic theory of irreversible processes CIT. According to Maugin and Muschik ([61], p. 226), it is the most commonly accepted viewpoint, which consists in replacing the axiom of the local equilibrium state of classical thermodynamic theory of irreversible processes with a somewhat straightforward generalization known under the name of axiom of the local accompanying state (for short L.A.S.). The authors stress that the preceding axioms are concepts for introduction of the formalism and results of thermostatics, also for processes outside the thermodynamic equilibrium. This viewpoint is also followed here.

The introduction of internal variables has several advantages of great practical importance. First, the history dependence of deformation is described by the internal variables, and the obtained evolution equation is a set of differential equations, as given by Expression (5.30). Rational thermodynamics RT, however, uses for description of the history-dependent deformation models the functional form,

$$\epsilon^{\mathrm{v}} = \int_0^t \ldots \mathrm{d}t \,. \tag{5.31}$$

The classical theory of viscoelasticity (see, e.g., Flügge [22]) is a good example of functional constitutive equations. For a varying state of stress, this theory gives formulations that are difficult to solve, since they use computing time excessively and require a considerable amount of computer memory, as observed by Santaoja [87] and Santaoja ([90], Sections 4.4–4.9 and Applications 3 and 4). Strictly speaking, in general, performing a structural analysis with a history-dependent material model by using a finite difference or a finite element method is not possible if the constitutive equation has a functional appearance.

Second, the classical thermodynamic theory of irreversible processes CIT cannot describe processes with memory. Plastic yield is a path-dependent process and therefore displays memory effects. Internal variables are introduced for modelling of history dependence. This is the major power of continuum thermodynamics with internal variables over CIT.

Finally, the foundation of the classical thermodynamic theory of irreversible processes is the assumption that the (generalized) forces and fluxes have a linear mutual relationship. The celebrated work of Onsager [73, 74] showed that the coefficients of the aforementioned relation satisfy the Onsager reciprocal relations. This assumption has not been adopted into continuum thermodynamics with internal variables but is replaced by introduction of the dissipation potential and the principle of maximum dissipation. This approach allows the formulation of several constitutive equations that do not fit CIT.

Little criticism has been raised against the continuum thermodynamics with internal variables. Lebon et al. ([46], p. 50) pointed out the following: in the thermodynamics of internal variables, the internal variables do not appear in the balance equations of momentum and energy. Moreover, since the selection of internal variables is not regulated by strict rules, the same class of materials can be described by several formalisms.

## 5.8 Summary

Since thermostatics forms a natural path from thermodynamics to continuum thermodynamics, the present chapter first looked at thermostatics. Thermostatics actually does not study processes, but compares two states of systems

$m^b$ being at thermodynamic equilibrium. Furthermore, thermostatics assumes that the state of the system $m^b$ is homogeneous. This means, for example, that the absolute temperature $T$ is the same everywhere within the system $m^b$. Since these assumptions are too restrictive for the validation purposes of the material models, thermodynamics was introduced. Thermodynamics studies processes outside the thermodynamic equilibrium in non-homogeneous systems $m^b$.

Three examples were presented for derivation of the state equations that play an important role in the validation of the material models. State equations will be studied in more details in Sections 6.4.4 and 6.4.5.

Figure 5.9 introduced fundamental quantities for thermodynamics, and they are repeated in Figure 5.13. As Figure 5.13 shows, thermodynamics introduces one new fundamental unit, namely, the kelvin K. This set of fundamental quantities also applies for continuum thermodynamics. The fundamental quantities for continuum mechanics are given in Figure 5.8 and repeated in Figure 5.14.

| Specific entropy | Heat flux vector | Heat source per unit mass |
|---|---|---|
| $s$ | $\vec{q}$ | $r$ |
| [Nm / K kg] | [Nm / s m$^2$] | [Nm / s kg] |

**Figure 5.13** Fundamental quantities and associated units of thermodynamics beyond continuum mechanics.

| Density $\underline{\rho}$ | Length $\underline{\ell}$ | Time $t$ | Traction | Body force |
|---|---|---|---|---|
| kilogram /m$^3$ | meter [m] | second [s] | vector $\vec{\underline{t}}$ | vector $\vec{\underline{b}}$ |
| [kg/meter$^3$] | | | [N/m$^2$] | [N/kg] |

**Figure 5.14** Fundamental quantities and associated units for continuum mechanics.

This book utilizes continuum thermodynamics with internal variables. The benefit of the internal variables was briefly discussed in Section 5.7.

# 6 Axioms of Thermodynamics and Their Consequences

## 6.1 Introduction

This chapter introduces the axioms of thermostatics and thermodynamics in the forms that apply to continuum thermodynamics. These axioms are: the first law of thermodynamics, the second law of thermodynamics, the axiom of caloric equation of state, the axiom of local accompanying state, and the principle of maximum dissipation. The local forms of the first and second laws of thermodynamics are derived for solids in current material coordinates $\underline{x}_i(t)$ and for fluids using the spatial description. The axiom of caloric equation of state is introduced to express that the specific internal energy $u$ is in a form in which the set of state variables has one thermal variable, which in this book is the specific entropy. For solids, the set of state variables includes internal variables, which are a vital part of material modelling, as discussed in Section 6.4.2. The axiom of local accompanying state is applied in Section 6.4.4 to extend the derived state equations to cover the new set of state variables. The heat equation and the Clausius–Duhem inequality are derived for solids and fluids. The Clausius–Duhem inequality is the validation tool for material models. Finally, the principle of maximum dissipation is used in formulating the normality rule for different cases.

## 6.2 First Law of Thermodynamics

The mathematical investigation of the axioms of continuum thermodynamics follows the same concepts that were used when continuum mechanics was investigated. In this section, the first law of thermodynamics is considered, and its local form, the energy equation, is derived. In Section 6.2.1, the local form of the first law of thermodynamics for solids is formulated, and in Section 6.2.3, the same is developed for fluids. Section 6.2.2 studies the local form for the current material description. The first law of thermodynamics is the same for all materials. The difference is in the local form of the first law of thermodynamics. For example, instead of the strain rate tensor $\dot{\boldsymbol{\varepsilon}}$ applied for solids, fluids prefer the velocity gradient $\vec{\nabla}\vec{v}$ in the local form of the first law of thermodynamics.

According to Narasimhan ([66], p. 321), the principle of conservation of energy, also referred to as the first law of thermodynamics, can be stated as follows: the time rate of change of the sum total of the kinetic energy $\underline{K}$ and the internal energy $\underline{U}$ in the body is equal to the sum of the rates of work done by the surface and body loads in producing the deformation (or flow) together with heat energy that may leave or enter the body at a certain rate. Thus is obtained

**AX 3**
$$\frac{\mathrm{D}}{\mathrm{D}t}(\underline{K}+\underline{U}) = \underline{P}^{\mathrm{ext}} + \underline{Q}\,. \tag{6.1}$$

In Axiom (6.1) [**AX 3**], $\underline{P}^{\mathrm{ext}}$ is the power input of the external forces, and $\underline{Q}$ is the heat input rate. It is worth noting that Form (6.1) is valid for thermomechanical materials where other influences, such as electrical and chemical are neglected compared to the thermomechanical effects. Form (6.1) also assumes a closed subsystem $\underline{v}(t)$ (see Kestin [37], Vol. 1, Chapter 6).

The total energy of the subsystem is considered here as the sum of two parts, the kinetic energy $\underline{K}$ and the internal energy $\underline{U}$. By kinetic energy $\underline{K}$ is meant the macroscopic kinetic energy associated with the usual macroscopically observable velocity of the continuum. The kinetic energy of the random thermal motions of molecules, associated with temperature measurements instead of velocity measurements, is considered part of the internal energy $\underline{U}$. The internal energy $\underline{U}$ also includes stored elastic energy and possibly other forms of energy not specified explicitly. (Malvern [57], p. 230).

As already mentioned in Section 4.4, the basic laws and axioms of continuum mechanics and continuum thermodynamics have the form of the current material description, and they are written for an arbitrarily selected but then nonchanging entity of matter $m$. Therefore the preceding variables were underlined and the current material coordinates $\underline{\vec{x}}_i(t)$ are used. Figure 6.1 shows the subsystem $m$ and the independent quantities that are present in the first law of thermodynamics.

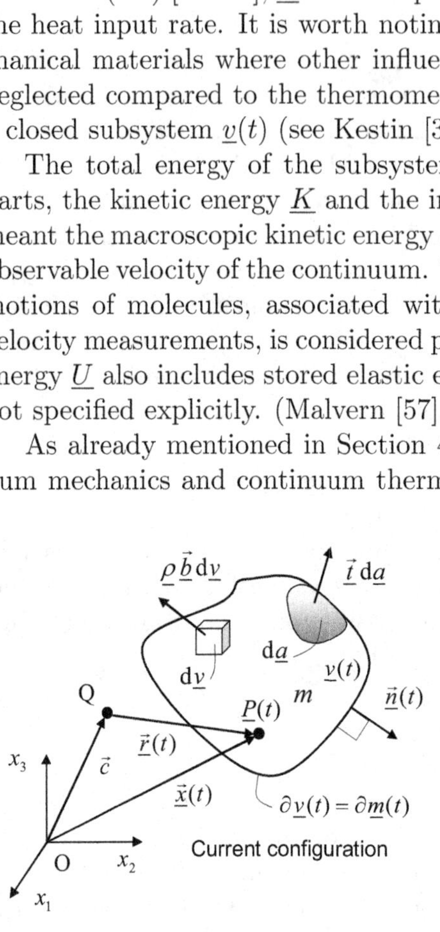


**Figure 6.1** Subsystem $m$ in the current configuration $\underline{v}^{\mathrm{b}}(t)$.

The kinetic energy $\underline{K}$ is defined by

$$\underline{K} := \int_m \tfrac{1}{2}\,\underline{\vec{v}}(\underline{\vec{x}}(t),t)\cdot\underline{\vec{v}}(\underline{\vec{x}}(t),t)\,\mathrm{d}m\,. \tag{6.2}$$

Since $\mathrm{d}m = \underline{\rho}(\underline{\vec{x}}(t),t)\,\mathrm{d}\underline{v}(t)$, Definition (6.2) yields

$$\underline{K} = \int_{\underline{v}(t)} \tfrac{1}{2}\,\underline{\rho}(\underline{\vec{x}}(t),t)\,\underline{\vec{v}}(\underline{\vec{x}}(t),t)\cdot\underline{\vec{v}}(\underline{\vec{x}}(t),t)\,\mathrm{d}\underline{v}\,. \tag{6.3}$$

Equation $(5.25)_2$ is recalled:

$$\underline{U} = \int_{\underline{v}(t)} \underline{\rho}(\underline{\vec{x}}(t),t)\, \underline{u}(\underline{\vec{x}}(t),t)\, \mathrm{d}\underline{v}. \tag{6.4}$$

The power input $\underline{P}^{\text{ext}}$ for the subdomain $m$ has the form

$$\underline{P}^{\text{ext}} := \oint_{\partial\underline{m}(t)} \underline{\vec{t}}(\underline{\vec{x}}(t),t,\underline{\vec{n}}) \cdot \underline{\vec{v}}(\underline{\vec{x}}(t),t)\, \mathrm{d}(\partial m) + \int_m \underline{\vec{b}}(\underline{\vec{x}}(t),t) \cdot \underline{\vec{v}}(\underline{\vec{x}}(t))\, \mathrm{d}m\,, \tag{6.5}$$

where the first term on the right side is the power caused by the (surface) traction vector $\underline{\vec{t}}$ and the second term is the power caused by the body force vector $\underline{\vec{b}}$. By taking the identities $\partial\underline{m}(t) = \partial\underline{v}(t)$ and $\mathrm{d}m = \underline{\rho}\,(\underline{\vec{x}}(t),t)\,\mathrm{d}\underline{v}(t)$ into account, Definition (6.5) yields

$$\underline{P}^{\text{ext}} = \oint_{\partial\underline{v}(t)} \underline{\vec{t}}(\underline{\vec{x}}(t),t,\underline{\vec{n}}) \cdot \underline{\vec{v}}(\underline{\vec{x}}(t),t)\, \mathrm{d}\underline{a} + \int_{\underline{v}(t)} \underline{\rho}(\underline{\vec{x}}(t),t)\, \underline{\vec{b}}(\underline{\vec{x}}(t),t) \cdot \underline{\vec{v}}(\underline{\vec{x}}(t),t)\, \mathrm{d}\underline{v}\,. \tag{6.6}$$

The expression of the heat input rate $\underline{Q}$, Expression (5.27), reads

$$\underline{Q} = -\oint_{\partial\underline{v}(t)} \underline{\vec{n}}(\underline{\vec{x}}(t),t) \cdot \underline{\vec{q}}^{\,\text{c}}(\underline{\vec{x}}(t),t)\, \mathrm{d}\underline{a} + \int_{\underline{v}(t)} \underline{\rho}(\underline{\vec{x}}(t),t)\, \underline{r}(\underline{\vec{x}}(t),t)\, \mathrm{d}\underline{v}, \tag{6.7}$$

where $\underline{\vec{q}}^{\,\text{c}}$ is the Cauchy heat flux vector and $\underline{r}$ is the heat source per unit mass. The negative sign in front of the first term on the right side of Expression (6.7) is due to $\underline{\vec{n}}$ being defined as the outward vector, and the heat input rate $\underline{Q}$ is positive into the interior of the subsystem $\underline{v}(t)$. This is sketched in Figure 6.2. The heat source $\underline{r}$ describes the influence of the radiation, as well as other heat sources not specified here.

## 6.2.1 First Law of Thermodynamics for Solids

This section derives the local form of the first law of thermodynamics for solid material. Small deformations and rotations are assumed. In order to make it easier to follow the derivation, independent variables are not always shown in the argument list.

In order to derive the local form of the first law of thermodynamics for solids, Expressions (6.3), (6.4), (6.6), and (6.7) have to be converted into the expressions for the IG-CtP configuration $\mathrm{V}_0^{\mathrm{b}}$. Thus initial material coordinates $X_i$ are applied.

In order to make the derivation clearer, the first law of thermodynamics, given by Axiom (6.1) [**AX 3**], is rewritten in the form

$$\dot{\underline{U}} - (\underline{P}^{\text{ext}} - \dot{\underline{K}}) - \underline{Q} = 0\,. \tag{6.8}$$

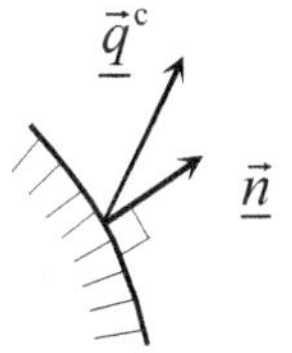


**Figure 6.2** Positive outward unit vector $\underline{\vec{n}}$ and Cauchy heat flux vector $\underline{\vec{q}}^{\,\text{c}}$.

Equation (4.75) is recalled:

$$\int_{\underline{v}(t)} \underline{\rho}(\underline{\vec{x}}(t),t)\, \underline{\theta}(\underline{\vec{x}}(t),t)\, \mathrm{d}\underline{v} = \int_{V_0} \rho_0(\vec{X})\, \theta(\vec{X},t)\, \mathrm{d}V\,. \tag{6.9}$$

Reynolds Transport Theorem (4.80) is

$$\frac{\mathrm{D}}{\mathrm{D}t}\int_{V_0} \rho_0\,(\vec{X})\, \theta(\vec{X},t)\, \mathrm{d}V = \int_{V_0} \rho_0(\vec{X})\, \frac{\mathrm{D}\theta(\vec{X},t)}{\mathrm{D}t}\, \mathrm{d}V\,. \tag{6.10}$$

Reynolds Transport Theorem (4.81) reads

$$\frac{\mathrm{D}}{\mathrm{D}t}\int_{\underline{v}(t)} \underline{\rho}(\underline{\vec{x}}(t),t)\, \underline{\theta}(\underline{\vec{x}}(t),t)\, \mathrm{d}\underline{v} = \int_{\underline{v}(t)} \underline{\rho}(\underline{\vec{x}}(t),t)\, \frac{\mathrm{D}\theta(\underline{\vec{x}}(t),t)}{\mathrm{D}t}\, \mathrm{d}\underline{v}\,. \tag{6.11}$$

Based on Expressions (6.4), (6.9), and (6.10), the rate of internal energy $\mathrm{D}U/\mathrm{D}t$ takes the following form:

$$\frac{\mathrm{D}U}{\mathrm{D}t} = \frac{\mathrm{D}}{\mathrm{D}t}\int_{\underline{v}(t)} \underline{\rho}(\underline{\vec{x}}(t),t)\, \underline{u}(\underline{\vec{x}}(t),t)\, \mathrm{d}\underline{v} = \frac{\mathrm{D}}{\mathrm{D}t}\int_{V_0} \rho_0(\vec{X})\, u(\vec{X},t)\, \mathrm{d}V = \int_{V_0} \rho_0\, \dot{u}\, \mathrm{d}V. \tag{6.12}$$

The quantity $\mathrm{D}U/\mathrm{D}t \equiv \dot{U}$ is not underlined, since it is expressed in the initial material coordinates $X_i$, as the right side of Expression (6.12) shows.

The definition of the Cauchy stress tensor $\underline{\boldsymbol{\sigma}}^{\mathrm{c}}$ from Definition (4.90) is recalled:

$$\underline{\vec{n}}(\underline{\vec{x}}(t),t)\cdot \underline{\boldsymbol{\sigma}}^{\mathrm{c}}(\underline{\vec{x}}(t),t) := \underline{\vec{t}}\,(\underline{\vec{x}}(t),t,\vec{n}) \qquad \text{or} \qquad \underline{\vec{n}}\cdot\underline{\boldsymbol{\sigma}}^{\mathrm{c}} := \underline{\vec{t}}\,. \tag{6.13}$$

Substitution of Definition (6.13) into the expression for the power input $P^{\mathrm{ext}}$, Expression (6.6), gives

$$\begin{aligned} \underline{P}^{\mathrm{ext}} &:= \oint_{\partial\underline{v}(t)} \underline{\vec{t}}(\underline{\vec{x}}(t),t,\underline{\vec{n}})\cdot \underline{\vec{v}}(\underline{\vec{x}}(t),t)\, \mathrm{d}\underline{a} + \int_{\underline{v}(t)} \underline{\rho}(\underline{\vec{x}}(t),t)\, \underline{\vec{b}}(\underline{\vec{x}}(t),t)\cdot\underline{\vec{v}}(\underline{\vec{x}}(t),t)\, \mathrm{d}\underline{v} \\ &= \oint_{\partial\underline{v}(t)} \underline{\vec{n}}\cdot\underline{\boldsymbol{\sigma}}^{\mathrm{c}}\cdot\underline{\vec{v}}\, \mathrm{d}\underline{a} + \int_{\underline{v}(t)} \underline{\rho}\,\underline{\vec{b}}\cdot\underline{\vec{v}}\, \mathrm{d}\underline{v}\,. \end{aligned} \tag{6.14}$$

Generalized Gauss's Theorem $(2.170)_1$ reads

$$\int_V \vec{\nabla} * f \mathrm{d}V = \oint_{\partial V} \vec{n} * f \mathrm{d}A\,. \tag{6.15}$$

By replacing the volume $V$ with the volume in the current configuration $\underline{v}(t)$, the star product $*$ with the dot product $\cdot$ and the outward unit normal $\vec{n}$ with the outward unit normal $\underline{\vec{n}}(\underline{\vec{x}}(t),t)$, and denoting $f = \underline{\boldsymbol{\sigma}}^{\mathrm{c}}(\underline{\vec{x}}(t),t)\cdot\underline{\vec{v}}(\underline{\vec{x}}(t),t)$ as Generalized Gauss's Theorem (6.15), yields

$$\oint\limits_{\partial\underline{v}(t)} \underline{\vec{n}}(\underline{\vec{x}}(t),t)\cdot\underline{\boldsymbol{\sigma}}^{\mathrm{c}}(\underline{\vec{x}}(t),t)\cdot\underline{\vec{v}}(\underline{\vec{x}}(t),t)\, \mathrm{d}\underline{a} = \int\limits_{\underline{v}(t)} \underline{\vec{\nabla}}(\underline{\vec{x}})\cdot\big[\,\underline{\boldsymbol{\sigma}}^{\mathrm{c}}(\underline{\vec{x}}(t),t)\cdot\underline{\vec{v}}(\underline{\vec{x}}(t),t)\,\big]\, \mathrm{d}\underline{v} \tag{6.16}$$

or

$$\oint_{\partial \underline{v}(t)} \underline{\vec{n}} \cdot \underline{\boldsymbol{\sigma}}^{\mathrm{c}} \cdot \underline{\vec{v}}\, \mathrm{d}\underline{a} = \int_{\underline{v}(t)} \underline{\vec{\nabla}} \cdot (\underline{\boldsymbol{\sigma}}^{\mathrm{c}} \cdot \underline{\vec{v}})\, \mathrm{d}\underline{v}\,. \tag{6.17}$$

Substitution of Expression (6.17) [or Expression (6.16)] into the first term of the third line of Equation (6.14) gives

$$\begin{aligned} \underline{P}^{\mathrm{ext}} &= \oint_{\partial \underline{v}(t)} \underline{\vec{n}} \cdot \underline{\boldsymbol{\sigma}}^{\mathrm{c}} \cdot \underline{\vec{v}}\, \mathrm{d}\underline{a} + \int_{\underline{v}(t)} \underline{\rho}\, \underline{\vec{b}} \cdot \underline{\vec{v}}\, \mathrm{d}\underline{v} = \int_{\underline{v}(t)} \underline{\vec{\nabla}} \cdot (\underline{\boldsymbol{\sigma}}^{\mathrm{c}} \cdot \underline{\vec{v}})\, \mathrm{d}\underline{v} + \int_{\underline{v}(t)} \underline{\rho}\, \underline{\vec{b}} \cdot \underline{\vec{v}}\, \mathrm{d}\underline{v} \\ &= \int_{\underline{v}(t)} \left[ \underline{\vec{\nabla}} \cdot (\underline{\boldsymbol{\sigma}}^{\mathrm{c}} \cdot \underline{\vec{v}}) + \underline{\rho}\, \underline{\vec{b}} \cdot \underline{\vec{v}} \right] \mathrm{d}\underline{v}\,. \end{aligned} \tag{6.18}$$

Problem 2.23 is to show that the following holds:

$$\vec{\nabla} \cdot (\mathbf{h} \cdot \vec{e}) = (\vec{\nabla} \cdot \mathbf{h}) \cdot \vec{e} + \mathbf{h} : \vec{\nabla} \vec{e}, \tag{6.19}$$

where $\mathbf{h}$ is a second-order tensor and $\vec{\nabla}\vec{e}$ is the open product of vectors $\vec{\nabla}$ and $\vec{e}$. By replacing the tensor $\mathbf{h}$ with the Cauchy stress tensor $\underline{\boldsymbol{\sigma}}^{\mathrm{c}}$ and the vector $\vec{e}$ with the velocity vector $\underline{\vec{v}}$, Equation (6.19) takes the following form:

$$\underline{\vec{\nabla}} \cdot (\underline{\boldsymbol{\sigma}}^{\mathrm{c}} \cdot \underline{\vec{v}}) = (\underline{\vec{\nabla}} \cdot \underline{\boldsymbol{\sigma}}^{\mathrm{c}}) \cdot \underline{\vec{v}} + \underline{\boldsymbol{\sigma}}^{\mathrm{c}} : \underline{\vec{\nabla}}\, \underline{\vec{v}}\,. \tag{6.20}$$

Cauchy Equations of Motion $(4.123)_1$ read

$$\underline{\vec{\nabla}} \cdot \underline{\boldsymbol{\sigma}}^{\mathrm{c}} + \underline{\rho}\, \underline{\vec{b}} = \underline{\rho}\, \frac{\mathrm{D}\underline{\vec{v}}}{\mathrm{D}t} \qquad \text{or} \qquad \underline{\vec{\nabla}} \cdot \underline{\boldsymbol{\sigma}}^{\mathrm{c}} + \underline{\rho}\, \underline{\vec{b}} = \underline{\rho}\, \dot{\underline{\vec{v}}}\,. \tag{6.21}$$

Multiplication of Equation (6.21) by $\cdot\, \underline{\vec{v}}$ from the right leads to

$$(\underline{\vec{\nabla}} \cdot \underline{\boldsymbol{\sigma}}^{\mathrm{c}}) \cdot \underline{\vec{v}} + \underline{\rho}\, \underline{\vec{b}} \cdot \underline{\vec{v}} = \underline{\rho}\, \dot{\underline{\vec{v}}} \cdot \underline{\vec{v}} \quad \Rightarrow \quad (\underline{\vec{\nabla}} \cdot \underline{\boldsymbol{\sigma}}^{\mathrm{c}}) \cdot \underline{\vec{v}} = \underline{\rho}\, \dot{\underline{\vec{v}}} \cdot \underline{\vec{v}} - \underline{\rho}\, \underline{\vec{b}} \cdot \underline{\vec{v}}\,. \tag{6.22}$$

Substitution of Expression $(6.22)_2$ into Equation (6.20) yields

$$\underline{\vec{\nabla}} \cdot (\underline{\boldsymbol{\sigma}}^{\mathrm{c}} \cdot \underline{\vec{v}}) = \underline{\rho}\, \dot{\underline{\vec{v}}} \cdot \underline{\vec{v}} - \underline{\rho}\, \underline{\vec{b}} \cdot \underline{\vec{v}} + \underline{\boldsymbol{\sigma}}^{\mathrm{c}} : \underline{\vec{\nabla}}\, \underline{\vec{v}}\,. \tag{6.23}$$

Problem 4.3 is to show that the following holds:

$$\underline{\boldsymbol{\sigma}}^{\mathrm{c}} : \underline{\vec{\nabla}}\, \underline{\vec{v}} = \underline{\boldsymbol{\sigma}}^{\mathrm{c}} : \underline{\mathbf{d}}\,, \tag{6.24}$$

where $\underline{\mathbf{d}}$ is the rate of deformation tensor. Relation (6.24) allows Expression (6.23) to take the form

$$\underline{\vec{\nabla}} \cdot (\underline{\boldsymbol{\sigma}}^{\mathrm{c}} \cdot \underline{\vec{v}}) = \underline{\rho}\, \dot{\underline{\vec{v}}} \cdot \underline{\vec{v}} - \underline{\rho}\, \underline{\vec{b}} \cdot \underline{\vec{v}} + \underline{\boldsymbol{\sigma}}^{\mathrm{c}} : \underline{\mathbf{d}}\,. \tag{6.25}$$

When substituting Equation (6.25) into Equation (6.18), the latter takes the following form:

$$\begin{aligned} \underline{P}^{\mathrm{ext}} &= \int_{\underline{v}(t)} \left[ \underline{\vec{\nabla}} \cdot (\underline{\boldsymbol{\sigma}}^{\mathrm{c}} \cdot \underline{\vec{v}}) + \underline{\rho}\, \underline{\vec{b}} \cdot \underline{\vec{v}} \right] \mathrm{d}\underline{v} \\ &= \int_{\underline{v}(t)} \left( \underline{\rho}\, \dot{\underline{\vec{v}}} \cdot \underline{\vec{v}} - \underline{\rho}\, \underline{\vec{b}} \cdot \underline{\vec{v}} + \underline{\boldsymbol{\sigma}}^{\mathrm{c}} : \underline{\mathbf{d}} + \underline{\rho}\, \underline{\vec{b}} \cdot \underline{\vec{v}} \right) \mathrm{d}\underline{v} = \int_{\underline{v}(t)} \left( \underline{\rho}\, \dot{\underline{\vec{v}}} \cdot \underline{\vec{v}} + \underline{\boldsymbol{\sigma}}^{\mathrm{c}} : \underline{\mathbf{d}} \right) \mathrm{d}\underline{v}\,. \end{aligned} \tag{6.26}$$

Based on Expressions (6.3) and (6.11), the rate of the kinetic energy $\dot{\underline{K}}$ takes the appearance

$$\begin{aligned}\dot{\underline{K}} = \frac{\mathrm{D}}{\mathrm{D}t}\underline{K} &= \frac{\mathrm{D}}{\mathrm{D}t}\int_{\underline{v}(t)} \tfrac{1}{2}\,\underline{\rho}(\vec{\underline{x}}(t),t)\,\vec{\underline{v}}(\vec{\underline{x}}(t),t)\cdot\vec{\underline{v}}(\vec{\underline{x}}(t),t)\,\mathrm{d}\underline{v}\\
&= \int_{\underline{v}(t)} \tfrac{1}{2}\,\underline{\rho}(\vec{\underline{x}}(t),t)\,\frac{\mathrm{D}}{\mathrm{D}t}\left[\vec{\underline{v}}(\vec{\underline{x}}(t),t)\cdot\vec{\underline{v}}(\vec{\underline{x}}(t),t)\right]\mathrm{d}\underline{v}\\
&= \int_{\underline{v}(t)} \tfrac{1}{2}\,\underline{\rho}(\vec{\underline{x}}(t),t)\left[\vec{\underline{v}}(\vec{\underline{x}}(t),t)\cdot\dot{\vec{\underline{v}}}(\vec{\underline{x}}(t),t)+\dot{\vec{\underline{v}}}(\vec{\underline{x}}(t))\cdot\vec{\underline{v}}(\vec{\underline{x}}(t),t)\right]\mathrm{d}\underline{v}\\
&= \int_{\underline{v}(t)} \underline{\rho}(\vec{\underline{x}}(t),t)\,\dot{\vec{\underline{v}}}(\vec{\underline{x}}(t),t)\cdot\vec{\underline{v}}(\vec{\underline{x}}(t),t)\,\mathrm{d}\underline{v} = \int_{\underline{v}(t)} \underline{\rho}\,\dot{\vec{\underline{v}}}\cdot\vec{\underline{v}}\,\mathrm{d}\underline{v}\,. \qquad (6.27)\end{aligned}$$

Equations (6.26) and (6.27) give

$$\underline{P}^{\mathrm{ext}} - \dot{\underline{K}} = \int_{\underline{v}(t)} \left(\underline{\rho}\,\dot{\vec{\underline{v}}}\cdot\vec{\underline{v}} + \underline{\boldsymbol{\sigma}}^{\mathrm{c}}:\underline{\mathbf{d}}\right)\mathrm{d}\underline{v} - \int_{\underline{v}(t)} \underline{\rho}\,\dot{\vec{\underline{v}}}\cdot\vec{\underline{v}}\,\mathrm{d}\underline{v} = \int_{\underline{v}(t)} \underline{\boldsymbol{\sigma}}^{\mathrm{c}}:\underline{\mathbf{d}}\,\mathrm{d}\underline{v}\,. \qquad (6.28)$$

The term $\underline{\boldsymbol{\sigma}}^{\mathrm{c}}:\underline{\mathbf{d}}$ in expression (6.28) is called the stress power (per unit volume). For the initial material description, the corresponding term reads $\mathbf{S}:\dot{\mathbf{E}}$ [cf. Holzapfel [29], Eqs. (4.100) and (4.107)]. Thus Equation (6.28) gives

$$P^{\mathrm{ext}} - \dot{K} = \int\limits_{\underline{v}(t)} \underline{\boldsymbol{\sigma}}^{\mathrm{c}}:\underline{\mathbf{d}}\,\mathrm{d}\underline{v} = \int\limits_{\underline{v}(t)} \underline{\boldsymbol{\sigma}}^{\mathrm{c}}(\vec{\underline{x}}(t),t):\underline{\mathbf{d}}(\vec{\underline{x}}(t),t)\,\mathrm{d}\underline{v} = \int\limits_{V_0} \mathbf{S}(\vec{X},t):\dot{\mathbf{E}}(\vec{X},t)\,\mathrm{d}V\,, \qquad (6.29)$$

where $\mathbf{S}$ is the second Piola–Kirchhoff stress tensor and $\dot{\mathbf{E}}$ is the Green–Lagrange strain rate tensor. The second Piola–Kirchhoff stress tensor $\mathbf{S}$ is not defined in this book. It is not necessary, because the second Piola–Kirchhoff stress tensor $\mathbf{S}$ is not used elsewhere in this book and will soon be replaced by the small deformation and rotation stress tensor $\boldsymbol{\sigma}$.

The heat input rate $\underline{Q}$, Expression (6.7), reads

$$\underline{Q} = -\oint_{\partial\underline{v}(t)} \vec{\underline{n}}(\vec{\underline{x}}(t),t)\cdot\vec{\underline{q}}^{\mathrm{c}}(\vec{\underline{x}}(t),t)\,\mathrm{d}\underline{a} + \int_{\underline{v}(t)} \underline{\rho}(\vec{\underline{x}}(t),t)\,\underline{r}(\vec{\underline{x}}(t),t)\,\mathrm{d}\underline{v}\,, \qquad (6.30)$$

The Piola–Kirchhoff type of a heat flux vector $\vec{q}^{\,\square}$ is expressed in the initial material coordinates $X_i$ and is defined to be

$$\vec{N}(\vec{X})\cdot\vec{q}^{\,\square}(\vec{X},t) := \vec{\underline{n}}(\vec{\underline{x}}(t),t)\cdot\vec{\underline{q}}^{\mathrm{c}}(\vec{\underline{x}}(t),t)\,. \qquad (6.31)$$

The key point in Definition (6.31) is that the outward unit vector $\vec{N}(\vec{X})$ is not the vector $\vec{\underline{n}}(\vec{\underline{x}}(t),t)$ expressed in different coordinates but a totally different one, as discussed regarding the definition of the first Piola–Kirchhoff stress tensor $\boldsymbol{\sigma}^{\square}$ in Definition (4.91). Since the vector $\vec{N}(\vec{X})$ is uniquely defined, Expression

(6.31) is the definition for the Piola–Kirchhoff type of heat flux vector $\vec{q}^{\square}$. Both sides of Definition (6.31) are integrated over the surface of the subsystem $m$, and we arrive at the following:

$$\oint_{\partial V_0} \vec{N}(\vec{X}) \cdot \vec{q}^{\square}(\vec{X}, t)\, \mathrm{d}A = \oint_{\partial \underline{v}(t)} \underline{\vec{n}}(\underline{\vec{x}}(t), t) \cdot \underline{\vec{q}}^{\mathrm{c}}(\underline{\vec{x}}(t), t)\, \mathrm{d}\underline{a}\,. \tag{6.32}$$

Generalized Gauss's Theorem (6.15) is applied to the left side of Expression (6.32), which leads to

$$\int_{V_0} \vec{\nabla}(\vec{X}) \cdot \vec{q}^{\square}(\vec{X}, t)\, \mathrm{d}V = \oint_{\partial \underline{v}(t)} \underline{\vec{n}}(\underline{\vec{x}}(t), t) \cdot \underline{\vec{q}}^{\mathrm{c}}(\underline{\vec{x}}(t), t)\, \mathrm{d}\underline{a}\,. \tag{6.33}$$

Equation (6.9) applies to the second term on the right side of Expression (6.30). Thus, based on Expression (6.9) and Result (6.33), Expression (6.30) can be expressed as

$$\begin{aligned} Q &= -\oint_{\partial \underline{v}(t)} \underline{\vec{n}}(\underline{\vec{x}}(t), t) \cdot \underline{\vec{q}}^{\mathrm{c}}(\underline{\vec{x}}(t), t)\, \mathrm{d}\underline{a} + \int_{\underline{v}(t)} \underline{\rho}(\underline{\vec{x}}(t), t)\, \underline{r}(\underline{\vec{x}}(t), t)\, \mathrm{d}\underline{v} \\ &= \int_{V_0} \Big[ -\vec{\nabla}(\vec{X}) \cdot \vec{q}^{\square}(\vec{X}, t) + \rho_0(\vec{X})\, r(\vec{X}, t) \Big]\, \mathrm{d}V = \int_{V_0} \Big( -\vec{\nabla} \cdot \vec{q}^{\square} + \rho_0\, r \Big)\, \mathrm{d}V\,. \end{aligned} \tag{6.34}$$

Next, the study reverts to the equation describing the first law of thermodynamics. Substitution of Equations (6.12), (6.29), and (6.34) into Expression (6.8), representing the first law of thermodynamics, yields

$$\int_{V_0} (\rho_0\, \dot{u} - \mathbf{S} : \dot{\mathbf{E}} + \vec{\nabla} \cdot \vec{q}^{\square} - \rho_0\, r)\, \mathrm{d}V = 0\,. \tag{6.35}$$

Since the volume $V_0$ is an arbitrary part of the volume $\mathrm{V}_0^{\mathrm{b}}$, the integrand of Integrand (6.35) must vanish for every $V_0$ within the volume of the system in the initial geometry $\mathrm{V}_0^{\mathrm{b}}$. Thus the term in parentheses vanishes at every point of the material in the initial configuration $\mathrm{V}_0^{\mathrm{b}}$, and the following local form is achieved (see a more detailed study in Section 4.7):

$$\rho_0\, \dot{u} = \mathrm{S} : \dot{\mathbf{E}} + \rho_0\, r - \vec{\nabla} \cdot \vec{q}^{\square}\,. \tag{6.36}$$

When small deformations and rotations are studied, the scalar product of the second Piola–Kirchhoff stress tensor $\mathbf{S}$ and the Green–Lagrange strain rate tensor $\dot{\mathbf{E}}$, that is, $\mathbf{S} : \dot{\mathbf{E}}$ reduce to the scalar product of the small deformation stress tensor $\boldsymbol{\sigma}$ and the small deformation strain rate tensor $\dot{\boldsymbol{\varepsilon}}$, that is, $\boldsymbol{\sigma} : \dot{\boldsymbol{\varepsilon}}$. The heat flux vector of the Piola–Kirchhoff type $\vec{q}^{\square}$ reduces to the heat flux vector $\vec{q}$. Thus, for small deformations and rotations, Expression (6.36) reduces to

$$\rho_0\, \dot{u} = \boldsymbol{\sigma} : \dot{\boldsymbol{\varepsilon}} + \rho_0\, r - \vec{\nabla} \cdot \vec{q}\,. \tag{6.37}$$

Equation (6.37) is called the energy equation (in a non-polar case) or the equation of balance of energy (Fung [24], p. 347; Malvern [57], p. 230). Equation (6.37) is the local form for the first law of thermodynamics for solids when small deformations and rotations are assumed.

With the introduction of independent variables, Expression (6.37) can be written with the following appearance:

$$\rho_0(\vec{X})\,\dot{u}(\vec{X},t) = \boldsymbol{\sigma}(\vec{X},t) : \dot{\boldsymbol{\varepsilon}}(\vec{X},t) + \rho_0(\vec{X})\,r(\vec{X},t) - \vec{\nabla}(\vec{X})\cdot\vec{q}(\vec{X},t)\,. \tag{6.38}$$

### 6.2.2 First Law of Thermodynamics for Current Material Coordinates $\underline{x}_i(t)$

This section derives the local form of the first law of thermodynamics for the current material coordinates $\underline{x}_i(t)$. In order to make it easier to follow the derivation, independent variables are not always shown.

As done in the previous section, the first law of thermodynamics [**AX 3**], Expression (6.1), is rewritten in the form

$$\dot{\underline{U}} - (\underline{P}^{\text{ext}} - \dot{\underline{K}}) - \underline{Q} = 0\,. \tag{6.39}$$

Reynolds Transport Theorem (4.81) reads

$$\frac{\mathrm{D}}{\mathrm{D}t}\int_{\underline{v}(t)} \underline{\rho}(\vec{\underline{x}}(t),t)\,\underline{\theta}(\vec{\underline{x}}(t),t)\,\mathrm{d}\underline{v} = \int_{\underline{v}(t)} \underline{\rho}(\vec{\underline{x}}(t),t)\,\frac{\mathrm{D}\theta(\vec{\underline{x}}(t),t)}{\mathrm{D}t}\,\mathrm{d}\underline{v}\,. \tag{6.40}$$

Equation (6.4) gives

$$\dot{\underline{U}} = \frac{\mathrm{D}\underline{U}}{\mathrm{D}t} = \frac{\mathrm{D}}{\mathrm{D}t}\int_{\underline{v}(t)} \underline{\rho}(\vec{\underline{x}}(t),t)\,\underline{u}(\vec{\underline{x}}(t),t)\,\mathrm{d}\underline{v}. \tag{6.41}$$

Applying Reynolds Transport Theorem (6.40), Rate (6.41) yields

$$\begin{aligned}\dot{\underline{U}} = \frac{\mathrm{D}}{\mathrm{D}t}\int_{\underline{v}(t)} \underline{\rho}(\vec{\underline{x}}(t),t)\,\underline{u}(\vec{\underline{x}}(t),t)\,\mathrm{d}\underline{v} &= \int_{\underline{v}(t)} \underline{\rho}(\vec{\underline{x}}(t),t)\,\frac{\mathrm{D}\underline{u}(\vec{\underline{x}}(t),t)}{\mathrm{D}t}\,\mathrm{d}\underline{v} \\ &= \int_{\underline{v}(t)} \underline{\rho}(\vec{\underline{x}}(t),t)\,\dot{\underline{u}}(\vec{\underline{x}}(t),t)\,\mathrm{d}\underline{v}\,.\end{aligned} \tag{6.42}$$

Equation (6.29) gives

$$\underline{P}^{\text{ext}} - \dot{\underline{K}} = \int_{\underline{v}(t)} \underline{\boldsymbol{\sigma}}^{\mathrm{c}} : \underline{\mathbf{d}}\,\mathrm{d}\underline{v} = \int_{\underline{v}(t)} \underline{\boldsymbol{\sigma}}^{\mathrm{c}}(\vec{\underline{x}}(t),t) : \underline{\mathbf{d}}(\vec{\underline{x}}(t),t)\,\mathrm{d}\underline{v}\,. \tag{6.43}$$

Equation (6.7) is

$$\underline{Q} = -\oint_{\partial\underline{v}(t)} \vec{\underline{n}}(\vec{\underline{x}}(t),t)\cdot\vec{\underline{q}}^{\mathrm{c}}(\vec{\underline{x}}(t),t)\,\mathrm{d}\underline{a} + \int_{\underline{v}(t)} \underline{\rho}(\vec{\underline{x}}(t),t)\,\underline{r}(\vec{\underline{x}}(t),t)\,\mathrm{d}\underline{v}\,. \tag{6.44}$$

Generalized Gauss's Theorem (6.15) is applied to the first term on the right side of Expression (6.44). This act gives

$$\underline{Q} = \int_{\underline{v}(t)} \Big[ -\vec{\underline{\nabla}}(\vec{\underline{x}}) \cdot \vec{\underline{q}}^{\,\mathrm{c}}(\vec{\underline{x}}(t),t) + \underline{\rho}(\vec{\underline{x}}(t),t)\, \underline{r}(\vec{\underline{x}}(t),t) \Big] \,\mathrm{d}\underline{v}\,. \tag{6.45}$$

Substitution of Terms (6.42), (6.43), and (6.45) into Form (6.39) yields

$$\int_{\underline{v}(t)} \big[ \underline{\rho}(\vec{\underline{x}}(t),t)\, \underline{\dot{u}}(\vec{\underline{x}}(t),t) - \underline{\boldsymbol{\sigma}}^{\mathrm{c}}(\vec{\underline{x}}(t),t) : \underline{\mathbf{d}}(\vec{\underline{x}}(t),t)$$
$$+ \vec{\underline{\nabla}}(\vec{\underline{x}}) \cdot \vec{\underline{q}}^{\,\mathrm{c}}(\vec{\underline{x}}(t),t) - \underline{\rho}(\vec{\underline{x}}(t),t)\, \underline{r}(\vec{\underline{x}}(t),t) \,\big] \,\mathrm{d}\underline{v} = 0\,, \tag{6.46}$$

or without the independent variables,

$$\int_{\underline{v}(t)} \Big( \underline{\rho}\,\underline{\dot{u}} - \underline{\boldsymbol{\sigma}}^{\mathrm{c}} : \underline{\mathbf{d}} + \vec{\underline{\nabla}} \cdot \vec{\underline{q}}^{\,\mathrm{c}} - \underline{\rho}\,\underline{r} \Big) \,\mathrm{d}\underline{v} = 0\,. \tag{6.47}$$

Since the volume $\underline{v}(t)$ is an arbitrary part of the volume $\underline{v}^{\mathrm{b}}(t)$, Integrand (6.47) must vanish for every $\underline{v}(t)$ within the volume of the system $\mathrm{m}^{\mathrm{b}}$ in the current configuration $\underline{v}^{\mathrm{b}}(t)$. Thus the term in square brackets vanishes at every point of the material in the current configuration $\underline{v}^{\mathrm{b}}(t)$ and the following local form is achieved (see a more detailed study in Section 4.7):

$$\underline{\rho}\,\underline{\dot{u}} = \underline{\boldsymbol{\sigma}}^{\mathrm{c}} : \underline{\mathbf{d}} + \underline{\rho}\,\underline{r} - \vec{\underline{\nabla}} \cdot \vec{\underline{q}}^{\,\mathrm{c}}\,. \tag{6.48}$$

Equation (6.48) is called the energy equation (in the nonpolar case) or the equation of balance of energy (Fung [24], p. 347; Malvern [57], p. 230). Equation (6.48) is the local form for the first law of thermodynamics for a matter expressed by the current material coordinates $\underline{x}_i(t)$.

With introduction of the independent variables, Expression (6.48) can take the following appearance:

$$\underline{\rho}(\vec{\underline{x}}(t),t)\underline{\dot{u}}(\vec{\underline{x}}(t),t) = \underline{\boldsymbol{\sigma}}^{\mathrm{c}}(\vec{\underline{x}}(t),t){:}\underline{\mathbf{d}}(\vec{\underline{x}}(t),t) + \underline{\rho}(\vec{\underline{x}}(t),t)\, \underline{r}(\vec{\underline{x}}(t),t) - \vec{\underline{\nabla}}(\vec{\underline{x}}) \cdot \vec{\underline{q}}^{\,\mathrm{c}}(\vec{\underline{x}}(t),t). \tag{6.49}$$

### 6.2.3 First Law of Thermodynamics for Fluids: Spatial Description

This section derives the local form of the first law of thermodynamics for spatial description, which is often used for fluids. Naturally, the first law of thermodynamics has the same form for solids and fluids. The energy equation, which is the local form of the first law of thermodynamics, however, takes different appearances. This is because the set of state variables for fluids differs from that for solids.

Since fluid mechanics utilizes spatial coordinates $x_i$ and the basic laws and axioms are expressed in the current material coordinates $\underline{x}_i(t)$, a coordinate transformation is needed. The derivation follows the same concept as that introduced in Section 4.18. The keen reader is encouraged to study Section 4.18 for a more detailed understanding.

The first law of thermodynamics, Axiom (6.1) [**AX 3**], is recalled. It has the form

**AX 3**
$$\frac{\mathrm{D}}{\mathrm{D}t}(\underline{K}+\underline{U}) = \underline{P}^{\mathrm{ext}} + \underline{Q}\,. \tag{6.50}$$

Equations (6.3), (6.4), (6.6), and (6.7) give the expressions for the quantities in Axiom (6.50) [**AX 3**]. They are

$$\begin{aligned}\underline{K} &= \int_{\underline{v}(t)} \tfrac{1}{2}\,\underline{\rho}(\underline{\vec{x}}(t),t)\,\underline{\vec{v}}(\underline{\vec{x}}(t),t)\cdot\underline{\vec{v}}(\underline{\vec{x}}(t),t)\,\mathrm{d}\underline{v}\\ \underline{U} &= \int_{\underline{v}(t)} \underline{\rho}(\underline{\vec{x}}(t),t)\,\underline{u}(\underline{\vec{x}}(t),t)\,\mathrm{d}\underline{v}\,,\end{aligned} \tag{6.51}$$

and furthermore,

$$\underline{P}^{\mathrm{ext}} = \oint_{\partial\underline{v}(t)} \underline{\vec{t}}(\underline{\vec{x}}(t),t,\underline{\vec{n}})\cdot\underline{\vec{v}}(\underline{\vec{x}}(t),t)\,\mathrm{d}\underline{a} + \int_{\underline{v}(t)} \underline{\rho}(\underline{\vec{x}}(t),t)\,\underline{\vec{b}}(\underline{\vec{x}}(t),t)\cdot\underline{\vec{v}}(\underline{\vec{x}}(t),t)\,\mathrm{d}\underline{v}\,, \tag{6.52}$$

and, finally,

$$\underline{Q} = -\oint_{\partial\underline{v}(t)} \underline{\vec{n}}(\underline{\vec{x}}(t),t)\cdot\underline{\vec{q}}^{\,\mathrm{c}}(\underline{\vec{x}}(t),t)\,\mathrm{d}\underline{a} + \int_{\underline{v}(t)} \underline{\rho}(\underline{\vec{x}}(t),t)\,\underline{r}(\underline{\vec{x}}(t),t)\,\mathrm{d}\underline{v}\,. \tag{6.53}$$

Reynolds Transport Theorem (4.89) reads

$$\frac{\mathrm{D}}{\mathrm{D}t}\int_{\underline{v}(t)} \underline{\rho}(\underline{\vec{x}}(t),t)\,\underline{\theta}(\underline{\vec{x}}(t),t)\,\mathrm{d}\underline{v} = \int_{\mathrm{V^{cv}}} \rho(\vec{x},t)\,\frac{\mathrm{D}\theta(\vec{x},t)}{\mathrm{D}t}\,\mathrm{d}V\,. \tag{6.54}$$

Substitution of $\theta = \vec{v}\cdot\vec{v}$ into Theorem (6.54) allows the kinetic energy rate $\dot{\underline{K}}$ present in Axiom (6.50) [**AX 3**] to take the following appearance:

$$\begin{aligned}\frac{\mathrm{D}\underline{K}}{\mathrm{D}t} &= \frac{\mathrm{D}}{\mathrm{D}t}\int_{\underline{v}(t)} \tfrac{1}{2}\,\underline{\rho}(\underline{\vec{x}}(t),t)\,\underline{\vec{v}}(\underline{\vec{x}}(t),t)\cdot\underline{\vec{v}}(\underline{\vec{x}}(t),t)\,\mathrm{d}\underline{v}\\ &= \frac{1}{2}\int_{\mathrm{V^{cv}}} \rho(\vec{x},t)\,\frac{\mathrm{D}}{\mathrm{D}t}\left[\vec{v}(\vec{x},t)\cdot\vec{v}(\vec{x},t)\right]\mathrm{d}V\\ &= \frac{1}{2}\int_{\mathrm{V^{cv}}} \rho(\vec{x},t)\left[\vec{v}(\vec{x},t)\cdot\dot{\vec{v}}(\vec{x},t) + \dot{\vec{v}}(\vec{x},t)\cdot\vec{v}(\vec{x},t)\right]\mathrm{d}V\\ &= \int_{\mathrm{V^{cv}}} \rho(\vec{x},t)\,\dot{\vec{v}}(\vec{x},t)\cdot\vec{v}(\vec{x},t)\,\mathrm{d}V = \int_{\mathrm{V^{cv}}} \rho\,\dot{\vec{v}}\cdot\vec{v}\,\mathrm{d}V\,.\end{aligned} \tag{6.55}$$

Substitution of $\theta = u$ into Theorem (6.54) allows the internal energy rate $\dot{\underline{U}}$ present in Axiom (6.50) [**AX 3**] to take the following appearance:

$$\dot{U} = \frac{\mathrm{D}U}{\mathrm{D}t} = \frac{\mathrm{D}}{\mathrm{D}t}\int_{\underline{v}(t)} \underline{\rho}(\underline{\vec{x}}(t),t)\,\underline{u}(\underline{\vec{x}}(t),t)\,\mathrm{d}\underline{v} = \int_{\mathrm{V^{cv}}} \rho(\vec{x},t)\,\frac{\mathrm{D}u(\vec{x},t)}{\mathrm{D}t}\,\mathrm{d}V$$
$$= \int_{\mathrm{V^{cv}}} \rho(\vec{x},t)\,\dot{u}(\vec{x},t)\,\mathrm{d}V = \int_{\mathrm{V^{cv}}} \rho\,\dot{u}\,\mathrm{d}V\,. \tag{6.56}$$

The quantities $\dot{K}$ and $\dot{U}$ are not underlined, since the far-left terms in Equations (6.55) and (6.56) are expressed in the spatial coordinates $x_i$.

It is assumed that the volume $\underline{v}(t_2)$ of the subsystem $m$ occupies exactly the control volume $\mathrm{V^{cv}}$ at time $t_2$. It is possible to select another control volume $\mathrm{V^{cv}}$ in such a way that the volume $\underline{v}(t_3)$ of the subsystem $m$ occupies exactly the control volume $\mathrm{V^{cv}}$ at time $t_3$. Thus it is possible to introduce a control volume $\mathrm{V^{cv}}$ for any volume $\underline{v}(t)$ of the subsystem $m$ at any moment $t$.

The value of a quantity $\theta$ is independent of the coordinate system. For the present case, it means that

$$\underline{\theta}(\underline{\vec{x}}(t),t) = \theta(\vec{x},t)\,. \tag{6.57}$$

The foregoing discussion allows the first term on the right side of Axiom (6.50) [**AX 3**] to take the form

$$P^{\mathrm{ext}} = \oint_{\partial\underline{v}(t)} \underline{\vec{t}}(\underline{\vec{x}}(t),t,\underline{\vec{n}})\cdot\underline{\vec{v}}(\underline{\vec{x}}(t),t)\,\mathrm{d}\underline{a} + \int_{\underline{v}(t)} \underline{\rho}(\underline{\vec{x}}(t),t)\,\underline{\vec{b}}(\underline{\vec{x}}(t),t)\cdot\underline{\vec{v}}(\underline{\vec{x}}(t),t)\,\mathrm{d}\underline{v}$$
$$= \oint_{\partial\mathrm{V^{cv}}} \vec{t}(\vec{x},t,\vec{n})\cdot\vec{v}(\vec{x},t)\mathrm{d}A + \int_{\mathrm{V^{cv}}} \rho(\vec{x},t)\,\vec{b}(\vec{x},t)\cdot\vec{v}(\vec{x},t)\,\mathrm{d}V\,. \tag{6.58}$$

Definition (4.95) for the stress tensor $\boldsymbol{\sigma}^{\mathrm{f}}$, that is, $\vec{n}\cdot\boldsymbol{\sigma}^{\mathrm{f}} := \vec{t}$, allows Expression (6.58) to take the form

$$P^{\mathrm{ext}} = \oint_{\partial\mathrm{V^{cv}}} \vec{n}(\vec{x},t)\cdot\boldsymbol{\sigma}^{\mathrm{f}}(\vec{x},t)\cdot\vec{v}(\vec{x},t)\,\mathrm{d}A + \int_{\mathrm{V^{cv}}} \rho(\vec{x},t)\,\vec{b}(\vec{x},t)\cdot\vec{v}(\vec{x},t)\,\mathrm{d}V$$
$$= \oint_{\partial\mathrm{V^{cv}}} \vec{n}\cdot\boldsymbol{\sigma}^{\mathrm{f}}\cdot\vec{v}\,\mathrm{d}A + \int_{\mathrm{V^{cv}}} \rho\,\vec{b}\cdot\vec{v}\,\mathrm{d}V = \oint_{\partial\mathrm{V^{cv}}} \vec{n}\cdot(\boldsymbol{\sigma}^{\mathrm{f}}\cdot\vec{v})\,\mathrm{d}A + \int_{\mathrm{V^{cv}}} \rho\,\vec{b}\cdot\vec{v}\,\mathrm{d}V\,. \tag{6.59}$$

The second term on the right side of Axiom (6.1) [**AX 3**] reads

$$Q = -\oint_{\partial\underline{v}(t)} \underline{\vec{n}}(\underline{\vec{x}}(t),t)\cdot\underline{\vec{q}}^{\,\mathrm{c}}(\underline{\vec{x}}(t),t)\,\mathrm{d}\underline{a} + \int_{\underline{v}(t)} \underline{\rho}(\underline{\vec{x}}(t),t)\,\underline{r}(\underline{\vec{x}}(t),t)\,\mathrm{d}\underline{v}$$
$$= -\oint_{\partial\mathrm{V^{cv}}} \vec{n}(\vec{x},t)\cdot\vec{q}(\vec{x},t)\,\mathrm{d}A + \int_{\mathrm{V^{cv}}} \rho(\vec{x},t)\,r(\vec{x},t)\,\mathrm{d}V$$
$$= -\oint_{\partial\mathrm{V^{cv}}} \vec{n}\cdot\vec{q}\,\mathrm{d}A + \int_{\mathrm{V^{cv}}} \rho\,r\,\mathrm{d}V\,. \tag{6.60}$$

Generalized Gauss's Theorem $(2.170)_1$ reads

$$\int_V \vec{\nabla} * \, \mathrm{d}V = \oint_{\partial V} \vec{n} * f \, \mathrm{d}A \,. \tag{6.61}$$

Substituting $f = \boldsymbol{\sigma}^{\mathrm{f}} \cdot \vec{v}$ and $f = \vec{q}$ into Theorem (6.61) and replacing the star product $*$ with the dot product $\cdot$ and the volume $V$ with the volume $\mathrm{V}^{\mathrm{cv}}$ gives the following expressions:

$$\int_{\mathrm{V}^{cv}} \vec{\nabla} \cdot (\boldsymbol{\sigma}^{\mathrm{f}} \cdot \vec{v}) \, \mathrm{d}V = \oint_{\partial \mathrm{V}^{\mathrm{cv}}} \vec{n} \cdot (\boldsymbol{\sigma}^{\mathrm{f}} \cdot \vec{v}) \, \mathrm{d}A \quad \text{and} \quad \int_{\mathrm{V}^{\mathrm{cv}}} \vec{\nabla} \cdot \vec{q} \, \mathrm{d}V = \oint_{\partial \mathrm{V}^{\mathrm{cv}}} \vec{n} \cdot \vec{q} \, \mathrm{d}A \,. \tag{6.62}$$

Substitution of Theorems (6.62) into Expressions (6.59) and (6.60) leads to

$$P^{\mathrm{ext}} = \oint_{\partial \mathrm{V}^{\mathrm{cv}}} \vec{n} \cdot (\boldsymbol{\sigma}^{\mathrm{f}} \cdot \vec{v}) \, \mathrm{d}A + \int_{\mathrm{V}^{\mathrm{cv}}} \rho \vec{b} \cdot \vec{v} \, \mathrm{d}V = \int_{\mathrm{V}^{\mathrm{cv}}} \left[ \vec{\nabla} \cdot (\boldsymbol{\sigma}^{\mathrm{f}} \cdot \vec{v}) + \rho \vec{b} \cdot \vec{v} \right] \mathrm{d}V \tag{6.63}$$

and

$$Q = - \oint_{\partial \mathrm{V}^{\mathrm{cv}}} \vec{n} \cdot \vec{q} \, \mathrm{d}A + \int_{\mathrm{V}^{\mathrm{cv}}} \rho \, r \, \mathrm{d}V = \int_{\mathrm{V}^{\mathrm{cv}}} \left( -\vec{\nabla} \cdot \vec{q} + \rho \, r \right) \mathrm{d}V \,. \tag{6.64}$$

Problem 2.23 is to show that the following holds:

$$\vec{\nabla} \cdot (\mathbf{h} \cdot \vec{e}) = (\vec{\nabla} \cdot \mathbf{h}) \cdot \vec{e} + \mathbf{h} : \vec{\nabla} \vec{e}, \tag{6.65}$$

where $\mathbf{h}$ is a second-order tensor and $\vec{\nabla} \vec{e}$ is the open product of vectors $\vec{\nabla}$ and $\vec{e}$. By replacing the tensor $\mathbf{h}$ with the fluid stress tensor $\boldsymbol{\sigma}^{\mathrm{f}}$ and the vector $\vec{e}$ with the velocity vector $\vec{v}$, Equation (6.65) takes the following form:

$$\vec{\nabla} \cdot (\boldsymbol{\sigma}^{\mathrm{f}} \cdot \vec{v}) = (\vec{\nabla} \cdot \boldsymbol{\sigma}^{\mathrm{f}}) \cdot \vec{v} + \boldsymbol{\sigma}^{\mathrm{f}} : \vec{\nabla} \vec{v} \,. \tag{6.66}$$

The equations of motion for fluids, Expression $(4.135)_1$, read

$$\vec{\nabla} \cdot \boldsymbol{\sigma}^{\mathrm{f}} + \rho \vec{b} = \rho \frac{\mathrm{D} \vec{v}}{\mathrm{D} t}, \qquad \text{that is,} \qquad \vec{\nabla} \cdot \boldsymbol{\sigma}^{\mathrm{f}} + \rho \vec{b} = \rho \dot{\vec{v}}. \tag{6.67}$$

Equation $(6.67)_2$ is multiplied by $\cdot \vec{v}$ from the right, and the obtained expression is substituted into Equation (6.66). The following is achieved:

$$\vec{\nabla} \cdot (\boldsymbol{\sigma}^{\mathrm{f}} \cdot \vec{v}) = \rho \dot{\vec{v}} \cdot \vec{v} - \rho \vec{b} \cdot \vec{v} + \boldsymbol{\sigma}^{\mathrm{f}} : \vec{\nabla} \vec{v} \,. \tag{6.68}$$

Substitution of Term (6.68) into the first term on the right side of Expression (6.63) yields

$$\begin{aligned} P^{\mathrm{ext}} &= \int_{\mathrm{V}^{\mathrm{cv}}} \left[ \vec{\nabla} \cdot (\boldsymbol{\sigma}^{\mathrm{f}} \cdot \vec{v}) + \rho \vec{b} \cdot \vec{v} \right] \mathrm{d}V \\ &= \int_{\mathrm{V}^{\mathrm{cv}}} \left[ \rho \dot{\vec{v}} \cdot \vec{v} - \rho \vec{b} \cdot \vec{v} + \boldsymbol{\sigma}^{\mathrm{f}} : \vec{\nabla} \vec{v} + \rho \vec{b} \cdot \vec{v} \right] \mathrm{d}V \,. \end{aligned} \tag{6.69}$$

Equation (6.69) reduces to

$$P^{\mathrm{ext}} = \int_{\mathrm{V^{cv}}} \left( \vec{\nabla} \cdot (\boldsymbol{\sigma}^{\mathrm{f}} \cdot \vec{v}) + \rho \vec{b} \cdot \vec{v} \right) \mathrm{d}V = \int_{\mathrm{V^{cv}}} \left( \rho \dot{\vec{v}} \cdot \vec{v} + \boldsymbol{\sigma}^{\mathrm{f}} : \vec{\nabla} \vec{v} \right) \mathrm{d}V \,. \tag{6.70}$$

Terms (6.55), (6.56), (6.64), and (6.70) are substituted into the first law of thermodynamics, Axiom (6.50) [**AX 3**]. This gives

$$\int_{\mathrm{V^{cv}}} \left( \rho \dot{\vec{v}} \cdot \vec{v} + \rho \dot{u} \right) \mathrm{d}V = \int_{\mathrm{V^{cv}}} \left( \rho \dot{\vec{v}} \cdot \vec{v} + \boldsymbol{\sigma}^{\mathrm{f}} : \vec{\nabla} \vec{v} - \vec{\nabla} \cdot \vec{q} + \rho r \right) \mathrm{d}V \,. \tag{6.71}$$

The terms in Expression (6.71) are collected under one integral operator on the left side of the expression. This gives

$$\int_{\mathrm{V^{cv}}} \left( \rho \dot{u} - \boldsymbol{\sigma}^{\mathrm{f}} : \vec{\nabla} \vec{v} + \vec{\nabla} \cdot \vec{q} - \rho r \right) \mathrm{d}V = 0 \,. \tag{6.72}$$

Since the control volume $\mathrm{V^{cv}}$ is an arbitrary part of the system $\underline{v}^{\mathrm{b}}(t)$, Integrand (6.72) must vanish for every $\mathrm{V^{cv}}$ within the system $\underline{v}^{\mathrm{b}}(t)$ (Section 4.7 gives a detailed study). Thus the term in parentheses vanishes at every point of the material, and the following local form is achieved:

$$\rho \dot{u} = \boldsymbol{\sigma}^{\mathrm{f}} : \vec{\nabla} \vec{v} + \rho r - \vec{\nabla} \cdot \vec{q} \,, \tag{6.73}$$

or with independent variables,

$$\rho(\vec{x}, t) \, \dot{u}(\vec{x}, t) = \boldsymbol{\sigma}^{\mathrm{f}}(\vec{x}, t) : \vec{\nabla}(\vec{x}) \, \vec{v}(\vec{x}, t) + \rho(\vec{x}, t) \, r(\vec{x}, t) - \vec{\nabla}(\vec{x}) \cdot \vec{q}(\vec{x}, t) \,. \tag{6.74}$$

As already discussed in Section 4.18, in fluid mechanics, it is common practice to separate the fluid stress tensor $\boldsymbol{\sigma}^{\mathrm{f}}$ into two parts, as follows [see Equation (4.136)]:

$$\boldsymbol{\sigma}^{\mathrm{f}} = -p \, \mathbf{1} + \boldsymbol{\sigma}^{\mathrm{d}} \,, \tag{6.75}$$

where $p$ is the thermodynamic pressure and $\boldsymbol{\sigma}^{\mathrm{d}}$ is the viscous stress tensor or dissipative stress tensor. It is important to note that the thermodynamic pressure $p$ is not the same as the mechanical pressure $p_{\mathrm{mech}}$, defined by Definition (2.117) as follows:

$$p_{\mathrm{mech}} := -\tfrac{1}{3} \, \mathbf{1} : \boldsymbol{\sigma} \,. \tag{6.76}$$

Form (6.73) for the fluid stress tensor $\boldsymbol{\sigma}^{\mathrm{f}}$ fulfills the requirements set by the surface integrals present in the basic laws, as can be seen by studying Section 4.22.

Substitution of Separation (6.75) into Expression (6.73) gives

$$\rho \dot{u} = -p \, \mathbf{1} : \vec{\nabla} \vec{v} + \boldsymbol{\sigma}^{\mathrm{d}} : \vec{\nabla} \vec{v} + \rho r - \vec{\nabla} \cdot \vec{q} \,. \tag{6.77}$$

In order to obtain an acceptable state variable for the energy equation, the term $\mathbf{1} : \vec{\nabla} \vec{v}$ must be recast in a new appearance.

Problem 2.21 is to show that the following holds:

$$\mathbf{1} : \vec{\nabla}\vec{v} = \vec{\nabla}\cdot\vec{v}\,. \tag{6.78}$$

Continuity Equation (4.69) is

$$\frac{\mathrm{D}\rho(\vec{x},t)}{\mathrm{D}t} + \rho\,(\vec{x},t)\,\vec{\nabla}(\vec{x})\cdot\vec{v}(\vec{x},t) = 0\,, \qquad \text{which yields} \qquad \vec{\nabla}\cdot\vec{v} = -\frac{\dot{\rho}}{\rho}\,. \tag{6.79}$$

Equations (6.78) and $(6.79)_2$ give the following result:

$$\mathbf{1} : \vec{\nabla}\vec{v} = -\frac{\dot{\rho}}{\rho}\,. \tag{6.80}$$

Based on Result (6.80), Equation (6.77) can be written as

$$\rho\,\dot{u} = p\,\frac{\dot{\rho}}{\rho} + \boldsymbol{\sigma}^{\mathrm{d}} : \vec{\nabla}\vec{v} + \rho\,r - \vec{\nabla}\cdot\vec{q}\,. \tag{6.81}$$

Equation (6.81) is the energy equation for fluids when expressed with the spatial description.

By introduction of the specific enthalpy denoted by $h$ and defined by

$$h := u + \frac{p}{\rho} \qquad \Rightarrow \qquad \dot{u} = \dot{h} - \frac{\dot{p}}{\rho} + \frac{p}{\rho^2}\,\dot{\rho}\,, \tag{6.82}$$

Energy Equation (6.81) can be written in the form

$$\rho\,\dot{h} = \dot{p} + \boldsymbol{\sigma}^{\mathrm{d}} : \vec{\nabla}\,\vec{v} + \rho\,r - \vec{\nabla}\cdot\vec{q}\,. \tag{6.83}$$

Form (6.83) equals that given by White [[116], Eq. (2.43)], although White did not introduce the heat source term $\rho\,r$ and assumed in the last term Fourier's law of heat conduction (see Section 8.2),

$$\vec{q} = -\gamma\,\vec{\nabla}T\,. \tag{6.84}$$

The fluid strain rate tensor $\mathring{\boldsymbol{\varepsilon}}$ and the fluid vorticity tensor $\mathring{\boldsymbol{\omega}}$ are introduced in Definitions (3.123) and (3.124). The definitions are

$$\mathring{\boldsymbol{\varepsilon}}(\vec{x},t) := \frac{1}{2}\left[\vec{v}(\vec{x},t)\overleftarrow{\nabla}(\vec{x}) + \vec{\nabla}(\vec{x})\vec{v}(\vec{x},t)\right] \quad \text{or} \quad \mathring{\boldsymbol{\varepsilon}} := \frac{1}{2}\left(\vec{v}\overleftarrow{\nabla} + \vec{\nabla}\vec{v}\right) \tag{6.85}$$

and

$$\mathring{\boldsymbol{\omega}}(\vec{x},t) := \frac{1}{2}\left[\vec{v}(\vec{x},t)\overleftarrow{\nabla}(\vec{x}) - \vec{\nabla}(\vec{x})\vec{u}(\vec{x},t)\right] \quad \text{or} \quad \mathring{\boldsymbol{\omega}} := \frac{1}{2}\left(\vec{v}\overleftarrow{\nabla} - \vec{\nabla}\vec{v}\right). \tag{6.86}$$

Based on Forms $(6.85)_2$ and $(6.86)_2$, the following relation is obtained:

$$\vec{\nabla}\vec{v} = \mathring{\boldsymbol{\varepsilon}} - \mathring{\boldsymbol{\omega}}\,. \tag{6.87}$$

Since the viscous stress tensor $\boldsymbol{\sigma}^{\mathrm{d}}$ is a symmetric second-order tensor and the fluid vorticity tensor $\overset{\circ}{\boldsymbol{\omega}}$ is a skew-symmetric second-order tensor, their scalar product vanishes, that is, $\boldsymbol{\sigma}^{\mathrm{d}} : \overset{\circ}{\boldsymbol{\omega}} = 0$ (see **Example 2.9**), and the following relation is achieved:

$$\boldsymbol{\sigma}^{\mathrm{d}} : \vec{\nabla}\vec{v} = \boldsymbol{\sigma}^{\mathrm{d}} : (\overset{\circ}{\boldsymbol{\varepsilon}} - \overset{\circ}{\boldsymbol{\omega}}) = \boldsymbol{\sigma}^{\mathrm{d}} : \overset{\circ}{\boldsymbol{\varepsilon}} . \tag{6.88}$$

Result (6.88) allows Energy Equation (6.81) to be written as

$$\rho\,\dot{u} = p\,\frac{\dot{\rho}}{\rho} + \boldsymbol{\sigma}^{\mathrm{d}} : \overset{\circ}{\boldsymbol{\varepsilon}} + \rho\,r - \vec{\nabla}\cdot\vec{q} . \tag{6.89}$$

Form (6.89) of the energy equation is used in the present book.

In fluid mechanics, the function $\Phi^{\mathrm{fluid}}$ is introduced. It is defined by

$$\Phi^{\mathrm{fluid}} := \boldsymbol{\sigma}^{\mathrm{d}} : \vec{\nabla}\vec{v} = \boldsymbol{\sigma}^{\mathrm{d}} : \overset{\circ}{\boldsymbol{\varepsilon}} . \tag{6.90}$$

The quantity $\Phi^{\mathrm{fluid}}$ is referred to as the fluid dissipation function.

### 6.2.4 First Law of Thermodynamics: Comparison of Expressions for Current Material Description and the Spatial Description

This section compares the local forms of the first law of thermodynamics written for the current material description and for the spatial description. Equation (6.73) gives the local form of the first law of thermodynamics expressed by the spatial coordinates $x_i$:

$$\rho\,\dot{u} = \boldsymbol{\sigma}^{\mathrm{f}} : \vec{\nabla}\vec{v} + \rho\,r - \vec{\nabla}\cdot\vec{q} . \tag{6.91}$$

Equation (6.87) is recalled:

$$\vec{\nabla}\vec{v} = \overset{\circ}{\boldsymbol{\varepsilon}} - \overset{\circ}{\boldsymbol{\omega}} . \tag{6.92}$$

Since the fluid stress tensor $\boldsymbol{\sigma}^{\mathrm{f}}$ is symmetric and the fluid vorticity tensor $\overset{\circ}{\boldsymbol{\omega}}$ is skew symmetric, the following holds: $\boldsymbol{\sigma}^{\mathrm{f}} : \overset{\circ}{\boldsymbol{\omega}} = 0$. Thus we arrive at the following:

$$\boldsymbol{\sigma}^{\mathrm{f}} : \vec{\nabla}\vec{v} = \boldsymbol{\sigma}^{\mathrm{f}} : (\overset{\circ}{\boldsymbol{\varepsilon}} - \overset{\circ}{\boldsymbol{\omega}}) = \boldsymbol{\sigma}^{\mathrm{f}} : \overset{\circ}{\boldsymbol{\varepsilon}} . \tag{6.93}$$

Substitution of Term (6.93) into Expression (6.91) gives

$$\rho\,\dot{u} = \boldsymbol{\sigma}^{\mathrm{f}} : \overset{\circ}{\boldsymbol{\varepsilon}} + \rho\,r - \vec{\nabla}\cdot\vec{q} . \tag{6.94}$$

Equation (6.91) was not the final form for the local form of the first law of thermodynamics for spatial description, but Expression (6.91) was recast further, and finally, Expression (6.89) was derived. It reads

$$\rho\,\dot{u} = p\,\frac{\dot{\rho}}{\rho} + \boldsymbol{\sigma}^{\mathrm{d}} : \overset{\circ}{\boldsymbol{\varepsilon}} + \rho\,r - \vec{\nabla}\cdot\vec{q} . \qquad \text{For spatial description.} \tag{6.95}$$

When expressed by the current material coordinates $\underline{x}_i(t)$, Form (6.48) of the energy equation was derived. It is

$$\underline{\rho}\,\dot{\underline{u}} = \underline{\boldsymbol{\sigma}}^{\mathrm{c}} : \underline{\mathbf{d}} + \underline{\rho}\,\underline{r} - \underline{\vec{\nabla}}\cdot\underline{\vec{q}}^{\,\mathrm{c}} . \tag{6.96}$$

Equations (6.94) and (6.96) have almost identical appearances. Therefore the derivation carried out for the spatial description at the end of Section 6.2.3 can be copied and tuned to the current material description with small differences. Thus the following expression can be written without any detailed derivation:

$$\underline{\rho}\,\dot{\underline{u}} = \underline{p}\,\frac{\dot{\underline{\rho}}}{\underline{\rho}} + \underline{\boldsymbol{\sigma}}^{\mathrm{dc}} : \underline{\mathbf{d}} + \underline{\rho}\,\underline{r} - \underline{\vec{\nabla}}\cdot\underline{\vec{q}}^{\,\mathrm{c}}, \quad \text{For current material description.} \tag{6.97}$$

where $\underline{\boldsymbol{\sigma}}^{\mathrm{dc}}$ is the dissipative Cauchy stress tensor $\underline{\boldsymbol{\sigma}}^{\mathrm{c}}$ given by expression [cf. Expression (6.75)]

$$\boldsymbol{\sigma}^{\mathrm{c}} = -p\,\mathbf{1} + \boldsymbol{\sigma}^{\mathrm{dc}} . \tag{6.98}$$

Local Forms (6.95) and (6.97) have similar appearances. This may be why there are cases in the literature in which the derivation of the local form of the first law of thermodynamics is done for the current material coordinates $\underline{x}_i(t)$ and the result is interpreted to be derived for the spatial coordinates $x_i$.

## 6.3 Second Law of Thermodynamics

This section looks at the second law of thermodynamics and its local forms. The second law of thermodynamics reads

$$\textbf{AX 4} \quad \dot{\underline{S}}(t) \geq -\oint_{\partial \underline{m}(t)} \frac{\underline{\vec{n}}(\underline{x}(t),t)\cdot\underline{\vec{q}}^{\,\mathrm{c}}(\underline{\vec{x}}(t),t)}{\underline{T}(\underline{\vec{x}}(t),t)}\,\mathrm{d}(\partial m) + \int_m \frac{\underline{r}(\underline{\vec{x}}(t),t)}{\underline{T}(\underline{\vec{x}}(t),t)}\,\mathrm{d}m . \tag{6.99}$$

The surface around the arbitrary subsystem $\partial\underline{m}(t)$ equals the surface of the volume $\underline{v}(t)$, that is, $\partial\underline{m}(t) = \partial\underline{v}(t)$. The differential mass element $\mathrm{d}m$ can be expressed by $\mathrm{d}m = \underline{\rho}\,(\underline{\vec{x}}(t),t)\,\mathrm{d}\underline{v}(t)$. Substitution of these two things into Axiom (6.99) [**AX 4**] gives

$$\textbf{AX 4} \quad \dot{\underline{S}}(t) \geq -\oint_{\partial \underline{v}(t)} \frac{\underline{\vec{n}}(\underline{x}(t),t)\cdot\underline{\vec{q}}^{\,\mathrm{c}}(\underline{\vec{x}}(t),t)}{\underline{T}(\underline{\vec{x}}(t),t)}\,\mathrm{d}\underline{a} + \int_{\underline{v}(t)} \underline{\rho}(\underline{\vec{x}}(t),t)\,\frac{\underline{r}(\underline{\vec{x}}(t),t)}{\underline{T}(\underline{\vec{x}}(t),t)}\,\mathrm{d}\underline{v} \tag{6.100}$$

or

$$\textbf{AX 4} \qquad \dot{\underline{S}}(t) \geq -\oint_{\partial \underline{v}(t)} \frac{\underline{\vec{n}}\cdot\underline{\vec{q}}^{\,\mathrm{c}}}{\underline{T}}\,\mathrm{d}\underline{a} + \int_{\underline{v}(t)} \underline{\rho}\,\frac{\underline{r}}{\underline{T}}\,\mathrm{d}\underline{v}, \tag{6.101}$$

where $\dot{\underline{S}}(t)$ is the entropy rate.

When expressed by the integrals over the volume $\underline{v}(t)$, entropy $\underline{S}(t)$ and entropy rate $\dot{\underline{S}}(t)$ take the forms given by Expressions $(5.24)_2$ and $(5.28)_1$:

$$\underline{S}(t) = \int_{\underline{v}(t)} \underline{\rho}(\vec{\underline{x}}(t),t)\, \underline{s}(\vec{\underline{x}}(t),t)\,\mathrm{d}\underline{v} \quad \text{and} \quad \dot{\underline{S}}(t) = \int_{\underline{v}(t)} \underline{\rho}(\vec{\underline{x}}(t),t)\, \dot{\underline{s}}(\vec{\underline{x}}(t),t)\,\mathrm{d}\underline{v}\,. \tag{6.102}$$

### 6.3.1 Second Law of Thermodynamics for Solids

This section derives the local form of the second law of thermodynamics for solids. Equation (6.31) gives the following definition for the Piola–Kirchhoff type of heat flux vector $\vec{q}^{\,\square}$:

$$\vec{N}(\vec{X})\cdot\vec{q}^{\,\square}(\vec{X},t) := \vec{\underline{n}}(\vec{\underline{x}}(t),t)\cdot\vec{\underline{q}}^{\,\mathrm{c}}(\vec{\underline{x}}(t),t)\,. \tag{6.103}$$

As Definition (6.103) shows, the Piola–Kirchhoff type of a heat flux vector $\vec{q}^{\,\square}$ is expressed in the initial material coordinates $X_i$.

Both sides of Definition (6.103) are divided by the absolute temperature $T$, which leads to

$$\frac{\vec{N}(\vec{X})\cdot\vec{q}^{\,\square}(\vec{X},t)}{T(\vec{X},t)} = \frac{\vec{\underline{n}}(\underline{x}(t),t)\cdot\vec{\underline{q}}^{\,\mathrm{c}}(\vec{\underline{x}}(t),t)}{\underline{T}(\vec{\underline{x}}(t),t)}\,. \tag{6.104}$$

Since the value of the absolute temperature $T$ is independent of the coordinate system, that is, $T(\vec{X},t) = \underline{T}(\vec{\underline{x}}(t),t)$ [cf. Expression (4.76)], Equation (6.103) is correct. Both sides of Expression (6.104) are integrated over the surface of the subdomain $m$, which gives

$$\oint_{\partial V_0} \frac{\vec{N}(\vec{X})\cdot\vec{q}^{\,\square}(\vec{X},t)}{T(\vec{X},t)}\,\mathrm{d}A = \oint_{\partial\underline{v}(t)} \frac{\vec{\underline{n}}(\underline{x}(t),t)\cdot\vec{\underline{q}}^{\,\mathrm{c}}(\vec{\underline{x}}(t),t)}{\underline{T}(\vec{\underline{x}}(t),t)}\,\mathrm{d}\underline{a}\,. \tag{6.105}$$

Equation (4.75) is recalled. It is

$$\int_{\underline{v}(t)} \underline{\rho}(\vec{\underline{x}}(t),t)\,\underline{\theta}(\vec{\underline{x}}(t),t)\,\mathrm{d}\underline{v} = \int_{V_0} \rho_0(\vec{X})\,\theta(\vec{X},t)\,\mathrm{d}V\,. \tag{6.106}$$

Equations (6.105) and (6.106) allow the second law of thermodynamics [**AX 4**], Inequality (6.100), to be expressed as follows:

$$\textbf{AX 4} \quad \dot{\underline{S}}(t) \geq -\oint_{\partial\underline{v}(t)} \frac{\vec{\underline{n}}(\vec{\underline{x}}(t),t)\cdot\vec{\underline{q}}^{\,\mathrm{c}}(\vec{\underline{x}}(t),t)}{\underline{T}(\vec{\underline{x}}(t),t)}\,\mathrm{d}\underline{a} + \int_{\underline{v}(t)} \underline{\rho}(\vec{\underline{x}}(t),t)\,\frac{\underline{r}(\vec{\underline{x}}(t),t)}{\underline{T}(\vec{\underline{x}}(t),t)}\,\mathrm{d}\underline{v}$$

$$\dot{S}(t) \geq -\oint_{\partial V_0} \frac{\vec{N}(\vec{X})\cdot\vec{q}^{\,\square}(\vec{X},t)}{T(\vec{X},t)}\,\mathrm{d}A + \int_{V_0} \rho_0(\vec{X})\,\frac{r(\vec{X},t)}{T(\vec{X},t)}\,\mathrm{d}V\,. \tag{6.107}$$

Without the independent variables, the second line of Inequality (6.107) takes the following appearance:

$$\dot{S}(t) \geq -\oint_{\partial V_0} \frac{\vec{N}\cdot\vec{q}^{\,\square}}{T}\,\mathrm{d}A + \int_{V_0} \rho_0\,\frac{r}{T}\,\mathrm{d}V\,. \tag{6.108}$$

Generalized Gauss's Theorem $(2.170)_1$ reads

$$\int_V \vec{\nabla} * f \mathrm{d}V = \oint_{\partial V} \vec{n} * f \mathrm{d}A \,. \tag{6.109}$$

The volume $V$ is replaced with the volume $V_0$, the star product $*$ is replaced with the dot product $\cdot$, the outward unit normal $\vec{n}$ is replaced with the outward unit normal $\vec{N}$, and the function $f$ is replaced with $\vec{q}^{\square}/T$. Thus Generalized Gauss's Theorem (6.109) takes the appearance

$$\oint_{\partial V_0} \frac{\vec{N} \cdot \vec{q}^{\square}}{T} \mathrm{d}A = \int_{V_0} \vec{\nabla} \cdot \left( \frac{\vec{q}^{\square}}{T} \right) \mathrm{d}V \,. \tag{6.110}$$

Substitution of Expression (6.110) into Equation (6.108) gives

$$\dot{S}(t) \geq - \oint_{\partial V_0} \frac{\vec{N} \cdot \vec{q}^{\square}}{T} \mathrm{d}A + \int_{V_0} \rho_0 \frac{r}{T} \mathrm{d}V = - \int_{V_0} \vec{\nabla} \cdot \left( \frac{\vec{q}^{\square}}{T} \right) \mathrm{d}V + \int_{V_0} \rho_0 \frac{r}{T} \mathrm{d}V \,, \tag{6.111}$$

which yields

$$\dot{S} \geq \int_{V_0} \left( -\frac{1}{T} \vec{\nabla} \cdot \vec{q}^{\square} + \frac{1}{T^2} \vec{q}^{\square} \cdot \vec{\nabla} T + \frac{1}{T} \rho_0 \, r \right) \mathrm{d}V \,. \tag{6.112}$$

Based on Expression (6.106), Rate $(6.102)_2$ yields

$$\dot{S}(t) = \int_{\underline{v}(t)} \underline{\rho}(\vec{\underline{x}}(t), t) \, \underline{\dot{s}}(\vec{\underline{x}}(t), t) \, \mathrm{d}v = \int_{V_0} \rho_0(\vec{X}) \, \dot{s}(\vec{X}, t) \, \mathrm{d}V \,. \tag{6.113}$$

The entropy rate $\dot{S}(t)$ in Equation (6.113) is substituted into Expression (6.111) and the obtained terms are rearranged. This gives

$$\int_{V_0} \left( \rho_0 \, \dot{s} + \frac{1}{T} \vec{\nabla} \cdot \vec{q}^{\square} - \frac{1}{T^2} \vec{q}^{\square} \cdot \vec{\nabla} T - \frac{1}{T} \rho_0 \, r \right) \mathrm{d}V \geq 0 \,. \tag{6.114}$$

Since the volume $V_0$ is an arbitrary part of volume of the system $\mathrm{V}_0^{\mathrm{b}}$, Integrand (6.114) must be nonnegative for every $V_0$ within the volume system in the initial geometry $\mathrm{V}_0^{\mathrm{b}}$. Thus the term in parentheses is nonnegative at every point of the material, and the following local form is achieved (see a more detailed study in Section 4.7):

$$\rho_0 \, \dot{s} + \frac{1}{T} \vec{\nabla} \cdot \vec{q}^{\square} - \frac{1}{T^2} \vec{q}^{\square} \cdot \vec{\nabla} T - \frac{1}{T} \rho_0 \, r \geq 0 \,. \tag{6.115}$$

If the terms of Inequality (6.115) are multiplied by the absolute temperature $T$, the following expression is achieved:

$$\rho_0 \, T \, \dot{s} + \vec{\nabla} \cdot \vec{q}^{\square} - \frac{\vec{\nabla} T}{T} \cdot \vec{q}^{\square} - \rho_0 \, r \geq 0 \,. \tag{6.116}$$

When small deformations and rotations are examined, the Piola–Kirchhoff type of heat flux vector $\vec{q}^{\square}$ reduces to the heat flux vector $\vec{q}$ and Expression (6.116) reduces to

$$\rho_0 \, T \, \dot{s} + \vec{\nabla} \cdot \vec{q} - \frac{\vec{\nabla} T}{T} \cdot \vec{q} - \rho_0 \, r \geq 0 \, . \tag{6.117}$$

Inequality (6.117) is the local form of the second law of thermodynamics for solids when small deformations and rotations are studied.

When the independent variables are introduced, Local Form (6.117) takes the appearance

$$\rho_0(\vec{X}) \, T(\vec{X},t) \, \dot{s}(\vec{X},t) + \vec{\nabla}(\vec{X}) \cdot \vec{q}(\vec{X},t) - \frac{\vec{\nabla}(\vec{X}) T(\vec{X},t)}{T(\vec{X},t)} \cdot \vec{q}(\vec{X},t) \\ - \rho_0(\vec{X}) \, r(\vec{X},t) \geq 0 \, . \tag{6.118}$$

The concept of the dissipation $\Phi$ is introduced. It is defined by [cf. Inequality (6.117)]

$$\Phi := \rho_0 \, T \, \dot{s} + \vec{\nabla} \cdot \vec{q} - \frac{\vec{\nabla} T}{T} \cdot \vec{q} - \rho_0 \, r \quad (\geq 0) \, . \tag{6.119}$$

Based on Inequality (6.119), the dissipation $\Phi$ is nonnegative:

$$\Phi \geq 0 \, . \tag{6.120}$$

### 6.3.2 Second Law of Thermodynamics for Current Material Coordinates $\underline{x}_i(t)$

This section derives the local form of the second law of thermodynamics for the current material coordinates $\underline{x}_i(t)$. The second law of thermodynamics [**AX 4**], Expression (6.100), reads

$$\textbf{AX4} \quad \underline{\dot{S}}(t) \geq - \oint_{\partial \underline{v}(t)} \frac{\underline{\vec{n}}(\vec{\underline{x}}(t),t) \cdot \underline{\vec{q}}^{\,\mathrm{c}}(\vec{\underline{x}}(t),t)}{\underline{T}(\vec{\underline{x}}(t),t)} \, \mathrm{d}\underline{a} + \int_{\underline{v}(t)} \underline{\rho}(\vec{\underline{x}}(t),t) \, \frac{\underline{r}(\vec{\underline{x}}(t),t)}{\underline{T}(\vec{\underline{x}}(t),t)} \, \mathrm{d}\underline{v} \, , \tag{6.121}$$

or without the independent variables [i.e., Expression (6.101)],

$$\textbf{AX4} \qquad \underline{\dot{S}}(t) \geq - \oint_{\partial \underline{v}(t)} \frac{\underline{\vec{n}} \cdot \underline{\vec{q}}^{\,\mathrm{c}}}{\underline{T}} \, \mathrm{d}\underline{a} + \int_{\underline{v}(t)} \underline{\rho} \, \frac{\underline{r}}{\underline{T}} \, \mathrm{d}\underline{v} \, . \tag{6.122}$$

Generalized Gauss's Theorem $(2.170)_1$ reads

$$\int_V \vec{\nabla} * f \, \mathrm{d}V = \oint_{\partial V} \vec{n} * f \, \mathrm{d}A \, . \tag{6.123}$$

The volume $V$ is replaced with the volume $\underline{v}(t)$, the vector operator $\vec{\nabla}$ is replaced with the vector operator $\underline{\vec{\nabla}}$, the star product $*$ is replaced with the dot product

$\cdot$, the outward unit normal $\vec{n}$ is replaced with the outward unit normal $\underline{\vec{n}}$, and the function $f$ is replaced with $\underline{\vec{n}}/\underline{T}$. Thus Generalized Gauss's Theorem (6.123) takes the appearance

$$\oint_{\partial \underline{v}(t)} \underline{\vec{n}} \cdot \left( \frac{\underline{\vec{q}}^{\mathrm{c}}}{\underline{T}} \right) \mathrm{d}\underline{a} = \int_{\underline{v}(t)} \underline{\vec{\nabla}} \cdot \left( \frac{\underline{\vec{q}}^{\mathrm{c}}}{\underline{T}} \right) \mathrm{d}\underline{v} \,. \tag{6.124}$$

Theorem (6.124) allows Expression (6.122) to take the form

$$\underline{\dot{S}}(t) \geq - \oint_{\partial \underline{v}(t)} \frac{\underline{\vec{n}} \cdot \underline{\vec{q}}^{\mathrm{c}}}{\underline{T}} \mathrm{d}\underline{a} + \int_{\underline{v}(t)} \underline{\rho} \, \frac{\underline{r}}{\underline{T}} \, \mathrm{d}\underline{v} = \int_{\underline{v}(t)} \left[ -\underline{\vec{\nabla}} \cdot \left( \frac{\underline{\vec{q}}^{\mathrm{c}}}{\underline{T}} \right) + \underline{\rho} \, \frac{\underline{r}}{\underline{T}} \right] \mathrm{d}\underline{v} \,. \tag{6.125}$$

Equation $(6.102)_2$ yields

$$\underline{\dot{S}}(t) = \int_{\underline{v}(t)} \underline{\rho}(\underline{\vec{x}}(t), t) \, \underline{\dot{s}}(\underline{\vec{x}}(t), t) \, \mathrm{d}\underline{v} = \int_{\underline{v}(t)} \underline{\rho} \, \underline{\dot{s}} \, \mathrm{d}\underline{v} \,. \tag{6.126}$$

Substituting Rate (6.126) into Inequality (6.125) and rearranging the obtained the terms, the following relation is obtained:

$$\int_{\underline{v}(t)} \left[ \underline{\rho} \, \underline{\dot{s}} + \underline{\vec{\nabla}} \cdot \left( \frac{\underline{\vec{q}}^{\mathrm{c}}}{\underline{T}} \right) - \underline{\rho} \, \frac{\underline{r}}{\underline{T}} \right] \mathrm{d}\underline{v} \geq 0 \,. \tag{6.127}$$

Since the volume $\underline{v}(t)$ is an arbitrary part of the volume $\underline{v}^{\mathrm{b}}(t)$, Integrand (6.127) must be nonnegative for every $\underline{v}(t)$ within the volume of the system in the current configuration $\underline{v}^{\mathrm{b}}(t)$. Thus the term in parentheses is nonnegative at every point of the material in the current configuration $\underline{v}^{\mathrm{b}}(t)$, and the following local form is achieved (the proof follows the concept taken in Section 4.7):

$$\underline{\rho} \, \underline{\dot{s}} + \underline{\vec{\nabla}} \cdot \left( \frac{\underline{\vec{q}}^{\mathrm{c}}}{\underline{T}} \right) - \underline{\rho} \, \frac{\underline{r}}{\underline{T}} \geq 0 \,. \tag{6.128}$$

Inequality (6.128) gives

$$\underline{\rho} \, \underline{\dot{s}} + \frac{1}{\underline{T}} \underline{\vec{\nabla}} \cdot \underline{\vec{q}}^{\mathrm{c}} - \frac{\underline{\vec{\nabla}} \underline{T}}{\underline{T}^2} \cdot \underline{\vec{q}}^{\mathrm{c}} - \underline{\rho} \, \frac{\underline{r}}{\underline{T}} \geq 0 \,. \tag{6.129}$$

Inequality (6.129) is multiplied by $\underline{T}$, which yields

$$\underline{\rho} \, \underline{\dot{s}} \, \underline{T} + \underline{\vec{\nabla}} \cdot \underline{\vec{q}}^{\mathrm{c}} - \frac{\underline{\vec{\nabla}} \underline{T}}{\underline{T}} \cdot \underline{\vec{q}}^{\mathrm{c}} - \underline{\rho} \, \underline{r} \geq 0 \,. \tag{6.130}$$

Inequality (6.130) is the local form of the second law of thermodynamics for the current material description.

When the independent variables are introduced, Local Form (6.130) takes the appearance

$$\underline{\rho}(\underline{\vec{x}}(t),t)\,\underline{\dot{s}}(\underline{\vec{x}}(t),t)\,\underline{T}(\underline{\vec{x}}(t),t)+\underline{\vec{\nabla}}(\underline{\vec{x}}(t))\cdot\underline{\vec{q}}^{\,\mathrm{c}}(\underline{\vec{x}}(t),t)$$

$$-\frac{\underline{\vec{\nabla}}(\underline{\vec{x}}(t))}{\underline{T}(\underline{\vec{x}}(t),t)}\cdot\underline{\vec{q}}^{\,\mathrm{c}}(\underline{\vec{x}}(t),t)-\underline{\rho}(\underline{\vec{x}}(t),t)\,\underline{r}(\underline{\vec{x}}(t),t)\geq 0\,. \tag{6.131}$$

The concept of dissipation $\underline{\Phi}$ is introduced. It is defined by [cf. Inequality (6.130)]

$$\underline{\Phi}=\underline{\rho}\,\underline{\dot{s}}\,\underline{T}+\underline{\vec{\nabla}}\cdot\underline{\vec{q}}^{\,\mathrm{c}}-\frac{\underline{\vec{\nabla}}\underline{T}}{\underline{T}}\cdot\underline{\vec{q}}^{\,\mathrm{c}}-\underline{\rho}\,\underline{r}\quad(\geq 0\,)\,. \tag{6.132}$$

Based on Inequality (6.132), the dissipation $\underline{\phi}$ is nonnegative:

$$\underline{\Phi}\geq 0\,. \tag{6.133}$$

### 6.3.3 Second Law of Thermodynamics for Fluids: Spatial Description

This section derives the local form of the second law of thermodynamics for fluids. Since fluid mechanics utilizes spatial coordinates $x_i$ and the basic laws and axioms are expressed in the current material coordinates $\underline{x}_i(t)$, a coordinate transformation is needed. The derivation follows the same concept as that introduced in Section 4.18. Here the concept is discussed briefly; the interested reader is encouraged to study Section 4.18 for more detailed understanding.

According to Equation (4.76), the value of a quantity $\theta$ is independent of the coordinate system. For the present frames, this means that

$$\theta(\underline{\vec{x}}(t),t)=\theta(\vec{x},t)\,. \tag{6.134}$$

Furthermore, it is assumed that the current volume $\underline{v}(t)$ of the subsystem $m$ occupies exactly the control volume $\mathrm{V}^{\mathrm{cv}}$ at time $t$. Based on Expression (6.134) and the preceding assumption, the right side of Axiom (6.100) [**AX 4**] takes the following appearance:

$$\dot{S}(t)\geq-\oint_{\partial\mathrm{V}^{\mathrm{cv}}}\frac{\vec{n}(\vec{x},t)\cdot\vec{q}(\vec{x},t)}{T(\vec{x},t)}\,\mathrm{d}A+\int_{\mathrm{V}^{\mathrm{cv}}}\rho(\vec{x},t)\,\frac{r(\vec{x},t)}{T(\vec{x},t)}\,\mathrm{d}V\,. \tag{6.135}$$

Reynolds Transport Theorem (4.89) reads

$$\frac{\mathrm{D}}{\mathrm{D}t}\int_{\underline{v}(t)}\rho(\underline{\vec{x}}(t),t)\,\theta(\underline{\vec{x}}(t),t)\mathrm{d}\underline{v}=\int_{\mathrm{V}^{\mathrm{cv}}}\rho(\vec{x},t)\,\frac{\mathrm{D}\theta(\vec{x},t)}{\mathrm{D}t}\,\mathrm{d}V\,. \tag{6.136}$$

Substitution of $\theta=s$ into Theorem (6.136) gives

$$\dot{S}(t)=\frac{\mathrm{D}S(t)}{\mathrm{D}t}=\frac{\mathrm{D}}{\mathrm{D}t}\int_{\underline{v}(t)}\rho(\underline{\vec{x}}(t),t)\,s(\underline{\vec{x}}(t),t)\,\mathrm{d}\underline{v}=\int_{\mathrm{V}^{\mathrm{cv}}}\rho(\vec{x},t)\,\dot{s}(\vec{x},t)\,\mathrm{d}V. \tag{6.137}$$

When the independent variables $\vec{x}$ and $t$ are dropped, Expressions (6.135) and (6.137) give

$$\int_{\mathrm{V^{cv}}} \rho\,\dot{s}\,\mathrm{d}V \geq -\oint_{\partial\mathrm{V^{cv}}} \frac{\vec{n}\cdot\vec{q}}{T}\,\mathrm{d}A + \int_{\mathrm{V^{cv}}} \rho\,\frac{r}{T}\,\mathrm{d}V\,. \tag{6.138}$$

Generalized Gauss's Theorem $(2.170)_1$ reads

$$\int_V \vec{\nabla} * f\,\mathrm{d}V = \oint_{\partial V} \vec{n} * f\,\mathrm{d}A\,. \tag{6.139}$$

The volume $V$ is replaced with the volume $\mathrm{V^{cv}}$, the star product $*$ is replaced with the dot product $\cdot$, and the function $f$ is replaced with $\vec{n}/T$. Thus Generalized Gauss's Theorem (6.139) takes the appearance

$$\oint_{\partial\mathrm{V^{cv}}} \vec{n}\cdot\left(\frac{\vec{q}}{T}\right)\mathrm{d}A = \int_{\mathrm{V^{cv}}} \vec{\nabla}\cdot\left(\frac{\vec{q}}{T}\right)\mathrm{d}V = \int_{\mathrm{V^{cv}}} \left(\frac{1}{T}\,\vec{\nabla}\cdot\vec{q} - \frac{\vec{\nabla}T}{T^2}\cdot\vec{q}\right)\mathrm{d}V. \tag{6.140}$$

Theorem (6.140) is substituted into Expression (6.138) and the terms of the obtained expression are rearranged, leading to

$$\int_{\mathrm{V^{cv}}} \left(\rho\,\dot{s} + \frac{1}{T}\,\vec{\nabla}\cdot\vec{q} - \frac{\vec{\nabla}T}{T^2}\cdot\vec{q} - \rho\,\frac{r}{T}\right)\mathrm{d}V \geq 0\,. \tag{6.141}$$

Since the control volume $\mathrm{V^{cv}}$ is an arbitrary part of the system $\underline{v}^{\mathrm{b}}(t)$, Integrand (6.141) must vanish for every $\mathrm{V^{cv}}$ within the system $\underline{v}^{\mathrm{b}}(t)$ (Section 4.7 gives a detailed study). Thus the term in parentheses vanishes at every point of the material and the following local form is achieved:

$$\rho\,\dot{s} + \frac{1}{T}\,\vec{\nabla}\cdot\vec{q} - \frac{\vec{\nabla}T}{T^2}\cdot\vec{q} - \rho\,\frac{r}{T} \geq 0\,. \tag{6.142}$$

Expression (6.142) is multiplied by the absolute temperature $T$ to give

$$\rho\,T\,\dot{s} + \vec{\nabla}\cdot\vec{q} - \frac{\vec{\nabla}T}{T}\cdot\vec{q} - \rho\,r \geq 0\,. \tag{6.143}$$

Inequality (6.143) is the local form of the second law of thermodynamics for fluids.

When independent variables are introduced, Local Form (6.143) takes the appearance

$$\begin{aligned}\rho(\vec{x},t)\,T(\vec{x},t)\,\dot{s}(\vec{x},t) + \vec{\nabla}(\vec{x})\cdot\vec{q}(\vec{x},t) - \frac{\vec{\nabla}(\vec{x})T(\vec{x},t)}{T(\vec{x},t)}\cdot\vec{q}(\vec{x},t)&\\ - \rho(\vec{x},t)\,r(\vec{x},t) \geq 0.&\end{aligned} \tag{6.144}$$

The concept of the dissipation $\Phi$ is introduced. It is defined by [cf. Inequality (6.144)]

$$\Phi := \rho T \dot{s} + \vec{\nabla} \cdot \vec{q} - \frac{\vec{\nabla} T}{T} \cdot \vec{q} - \rho r \quad (\geq 0) . \tag{6.145}$$

Based on Inequality (6.145), the dissipation $\phi$ is nonnegative:

$$\Phi \geq 0 . \tag{6.146}$$

## 6.4 State Variables and State Equations

The aim of this section is to introduce the independent variables for description of the modelled processes. They are called state variables. Since the local form of the first law of thermodynamics takes a different appearance for solids, liquids, and gases, the state variables for these materials are different. Therefore, they are introduced in two sections. Finally, the state equations outside the thermodynamical equilibrium are introduced by utilizing the axiom of local accompanying state. The caloric equation of state utilizing the introduced state variables is expressed for each material category separately.

### 6.4.1 On State Variables

There are two types of state variables when processes within the framework of continuum thermodynamics with internal variables are described. These two categories of state variables are: controllable variables and internal variables. The independent variables present in the basic laws and axioms of continuum thermodynamics are controllable variables. This division is supported by the discussion with Kukudžanov [40] and the text by Lebon et al. ([46], p. 50), which expresses that the internal variables (they call them hidden variables) do not appear in the balance equations of momentum and energy.

According to Section 6.4.2, the strain rate tensor $\dot{\boldsymbol{\varepsilon}}$ is a controllable variable when response of solid material is modelled. This fact is used next, when the role of controllable variables is discussed.

It is important to note that controllable variables are only controllable to some extent. Thus, for example, the value for the strain tensor $\boldsymbol{\varepsilon}$ is controllable only (if ever) at the boundaries of the system. A uniaxial tension/compression is investigated as an example. If the researcher had an extremely strong and stiff loading machine, they could control the strain $\varepsilon$ of the specimen on the surfaces where the specimen is attached to the loading device. The preceding means that the adjective *controllable* cannot agree fully with the quantities called controllable variables. The phenomenon *controllable* is not the key concept for separation of the variables into two categories.

In conclusion, the following definition is expressed:

*The independent variables present in the basic laws and axioms of continuum thermodynamics are called controllable variables. The other independent variables are called internal variables.*

The internal variables and their form, are determined by the material model under consideration.

### 6.4.2 State Variables for Solids

This section introduces the state variables for solid material. The local forms of the first law and of the second law of thermodynamics for solid material are given by Equations (6.37) and (6.117). These equations are

$$\rho_0 \, \dot{u} = \boldsymbol{\sigma} : \dot{\boldsymbol{\varepsilon}} + \rho_0 \, r - \vec{\nabla} \cdot \vec{q} \tag{6.147}$$

and

$$\rho_0 \, T \, \dot{s} + \vec{\nabla} \cdot \vec{q} - \frac{\vec{\nabla} T}{T} \cdot \vec{q} - \rho_0 \, r \geq 0 \, . \tag{6.148}$$

Based on Local Forms (6.147) and (6.148), the controllable state variables for thermomechanical processes in deformable solids are as follows.

$\boldsymbol{\varepsilon}$ The strain tensor. It is a second-order tensor describing both mechanical and thermal deformation.

$s$ The specific entropy.

Gibbs Equation (5.19) clarifies the selection of controllable variables. Gibbs Equation (5.19) is

$$\mathrm{d}u = \frac{1}{\rho_0} \, \sigma \, \mathrm{d}\varepsilon + T \, \mathrm{d}s \, , \qquad \text{which gives} \qquad \dot{u} = \frac{1}{\rho_0} \, \sigma \, \dot{\varepsilon} + T \, \dot{s} \, . \tag{6.149}$$

According to Expression $(6.149)_2$, the change of the values of the state variables $\varepsilon$ and $s$ leads to the change of the state function $u$.

As mentioned, the set of internal variables and their forms are determined by the material model under consideration. For the sake of simplicity, this part of the work studies material models that (also) model elastic deformation. Thus, the present formulation is for cases in which the response of the solid consists of elastic and inelastic deformation or pure elastic deformation. Also thermal expansion can be simulated, for example. These types of material models are here called "material models with elastic deformation," in contrast to the others, called "material models without elastic deformation" in Section 7.4. Material models with elastic deformation have the irreversible strain tensor $\boldsymbol{\varepsilon}^{\mathrm{i}}$ in the set of internal variables. The term *irreversible strain tensor* $\boldsymbol{\varepsilon}^{\mathrm{i}}$ is a continuum thermodynamic expression. In material science, it is known as the inelastic strain, which term is used here. If pure elastic deformation is modelled, the terms containing inelastic strain tensor $\boldsymbol{\varepsilon}^{\mathrm{i}}$ are neglected. Since in material models without elastic deformation the inelastic strain tensor $\boldsymbol{\varepsilon}^{\mathrm{i}}$ equals the strain tensor $\boldsymbol{\varepsilon}$, these models do not have the inelastic strain tensor $\boldsymbol{\varepsilon}^{\mathrm{i}}$ in the set of state variables. The set of other internal variables is denoted here by a second-order tensor $\boldsymbol{\alpha}$. The results derived later for $\boldsymbol{\alpha}$ are generally valid and can therefore easily be extended by the reader to their own set of independent variables. In some special cases, for example, when the gradient theory is used, a slightly different approach may be necessary. Gradient theory is considered in Section 7.5. As assumed by Astarita ([8], Chapter 2), the set of state variables may also contain rate terms (see Malvern [57], p. 257). The key problem in the mathematical modelling of material behavior is the choice of a set of independent variables. In light of the preceding, the number of state variables may be considerably high, which makes continuum thermodynamics with internal variables more complicated and – we can hope – more capable. On the other hand, use of an unnecessary large set of variables may lead to an overly complicated theory with many (physically unclear) variables. A theory with too few variables is not capable of describing the event. The "correct" set is dependent on the processes modelled and on the desired accuracy of the constitutive equation. Thus no generally valid set of independent variables can be given. The choice of a representative set of state variables is a constitutive assumption and shows – in the case in which it is done correctly – the professionalism of the physicist.

The internal variables for material models showing elastic deformation or pure elastic deformation are (does not hold, e.g., for gradient theory) as follows.

$\boldsymbol{\varepsilon}^{\mathrm{i}}$ The strain tensor for description of inelastic deformation. It is a second-order tensor describing dissipative deformation mechanisms.

$\boldsymbol{\alpha}$ This represents the other internal variables. It can be a scalar, vector, or tensor of any order. In derivations, $\boldsymbol{\alpha}$ is usually a second-order tensor.

The only explicitly given independent field variable in Definition (5.10) for the caloric equation of state [**AX 2**] for a solid material is the specific entropy $s$. Now, based on the foregoing discussion, the axiom of caloric equation of state [**AX 2**] for a solid material takes the following appearance:

**AX 2**
$$u := u(\boldsymbol{\varepsilon}, \boldsymbol{\varepsilon}^{\mathrm{i}}, \boldsymbol{\alpha}, s, h(\vec{X})) \,, \tag{6.150}$$

where the notation $h(\vec{X})$ indicates that the system $\mathrm{m^b}$ may be thermodynamically inhomogeneous. If the specific internal energy $u(\vec{X}, t)$ (i.e., internal energy per unit mass) does not depend on $h(\vec{X})$, the system $\mathrm{m^b}$ is said to be thermodynamically homogeneous; that is, the mechanical as well as thermal properties are the same everywhere within the system $\mathrm{m^b}$. This terminology originates from thermostatics, in which only reversible deformation (elastic deformation) was evaluated. Inelastic properties are determined by the values of the internal variables ($\boldsymbol{\alpha}$ can describe, e.g., strain hardening), and their values can vary from point to point within the system $\mathrm{m^b}$ independently of value of the function

$h(\vec{X})$. This means that the quantity $h(\vec{X})$ is related to the homogeneity of the reversible deformation.

Equation (6.150) and the preceding notation $u(\vec{X}, t)$ show two different sets of independent variables for the specific internal energy $u$. Both are correct, of course. The latter form is a consequence of Expression (6.150), since the independent variables $\boldsymbol{\varepsilon}$, $\boldsymbol{\varepsilon}^{\mathrm{i}}$, $\boldsymbol{\alpha}$, and $s$ are dependent on $\vec{X}$ and $t$. Furthermore, the independent variables do not necessarily have to be mutually independent. The terminology comes from pure mathematics, where from the point of view of exactness the independent variables are assumed to be mutually independent. However, mathematical operations are valid also in the case of mutual dependence. The variables the values of which are dependent on independent variables are called dependent variables. In this case, the specific internal energy $u$ is a dependent variable.

It is important to note the similarity of the two caloric equations of state, namely, Expressions (5.1) and (6.150). The internal variables $\boldsymbol{\varepsilon}^{\mathrm{i}}$ and $\boldsymbol{\alpha}$ are the extension due to the introduction of continuum thermodynamics with internal variables. Of course, the theory of thermostatics could be extended by the introduction of internal variables. Since the traditional form of thermostatics does not contain internal variables, they are not adopted here.

At the risk of sounding repetitive, the merits of using continuum thermodynamics with internal variables, and the requirements for these internal variables, are discussed briefly next.

(1) An internal variable is a variable describing the internal response of a material. The value of an internal variable can be measured, but its evolution cannot be directly controlled. In other words, an internal variable is measurable but not controllable. Although internal variables are sometimes called hidden variables, they can be visible, as Figure 5.12 illustrates.

(2) By introduction of a set of internal variables, the constitutive equations can be expressed in rate form, in which no time functionals appear. This is a major advantage over the traditional theory of viscoelasticity (see, e.g., Flügge [22]) or the rational thermodynamics of Coleman [18], Truesdell and Noll [111], and Truesdell [110], because numerical analysis with time functionals may require a great amount of computer capacity (see, e.g., Santaoja [87], Table 3 or Santaoja [90], Table 4).

(3) The authors cannot agree with the comment by Muschik that an internal variable may or may not be a state variable (Muschik [65], p. 48). In his studies, Muschik may have used an incorrect set of internal variables, which led him to this erroneous conclusion. All the internal variables present in this work are of course also state variables.

(4) Internal variables need a model or an (microscopic or molecular) interpretation (Muschik [65], p. 53).

Item (4), requiring that an internal variable needs a microscopic or molecular interpretation, is difficult to fulfill within the gradient theory, which uses gradient of damage as a variable to prevent localization.

The authors suggest the terminology shown in Figure 6.3 for the independent variables for solid material within the framework of continuum thermodynamics with internal variables. The following pages show how the specific entropy $s$ is replaced by the absolute temperature $T$.

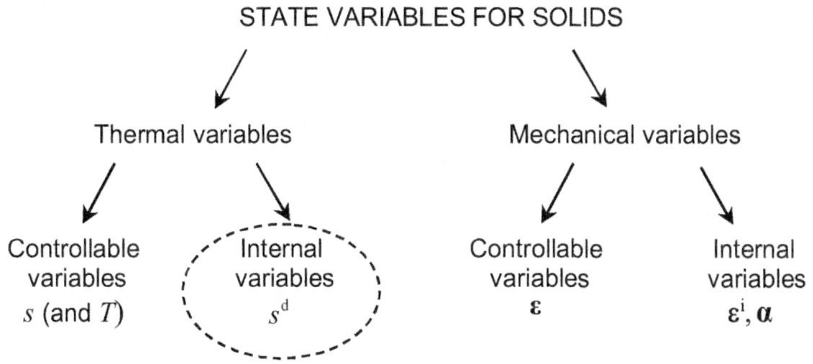


**Figure 6.3** Different types of independent variables for solids in continuum thermodynamics with internal variables. The thermal internal variable called specific dissipative entropy and denoted by $s^{\mathrm{d}}$ is present only in the gradient theory (see Section 7.5.2).

### 6.4.3 State Variables for Fluids

This section introduces state variables for fluids. The local forms of the first law and of the second law of thermodynamics for fluids are given by Expressions (6.89) and (6.143). These equations are

$$\rho\,\dot{u} = p\,\frac{\dot{\rho}}{\rho} + \boldsymbol{\sigma}^{\mathrm{d}} : \overset{\circ}{\boldsymbol{\varepsilon}} + \rho\,r - \vec{\nabla}\cdot\vec{q} \tag{6.151}$$

and

$$\rho\,T\,\dot{s} + \vec{\nabla}\cdot\vec{q} - \frac{\vec{\nabla}T}{T}\cdot\vec{q} - \rho\,r \geq 0\,. \tag{6.152}$$

Based on Local Forms (6.151) and (6.152), the controllable state variables for thermodynamical processes in fluids are as follows:

$\rho$ The density
$s$ The specific entropy

Gibbs Equation (5.15) clarifies the selection of controllable variables. It is

$$\mathrm{d}u = \frac{1}{\rho^2}\,p\,\mathrm{d}\rho + T\,\mathrm{d}s\,, \qquad \text{which gives} \qquad \dot{u} = \frac{1}{\rho^2}\,p\,\dot{\rho} + T\,\dot{s}\,. \tag{6.153}$$

According to Expression $(6.153)_2$, the change of values of the state variables $\rho$ and $s$ leads to a change of the state function $u$.

The axiom of caloric equation of state [**AX 2**] for fluids takes the following appearance:

$$\textbf{AX 2} \qquad u := u(\rho, s)\,. \tag{6.154}$$

### 6.4.4 State Equations for Solids

This section introduces state equations for solid material. The axiom of caloric equation of state [**AX 2**] for material modelling of solids is given by Axiom (6.150) [**AX 2**],

$$\textbf{AX 2} \qquad u := u(\boldsymbol{\varepsilon}, \boldsymbol{\varepsilon}^{\mathrm{i}}, \boldsymbol{\alpha}, s, h(\vec{X}))\,. \tag{6.155}$$

State Equations (5.21) are

$$\sigma = \rho_0 \frac{\partial u(\varepsilon, s)}{\partial \varepsilon} \qquad \text{and} \qquad T = \frac{\partial u(\varepsilon, s)}{\partial s}\,. \tag{6.156}$$

State Equations (6.156) are derived for thermostatic processes, that is, for processes that occur at thermodynamical equilibrium. The axiom of local accompanying state [**AX 5**] expresses that State Equation (6.156) also holds for nonequilibrium processes. Therefore the following state equations can be written:

$$\boldsymbol{\sigma} = \rho_0 \frac{\partial u(\boldsymbol{\varepsilon}, \boldsymbol{\varepsilon}^{\mathrm{i}}, \boldsymbol{\alpha}, s, h(\vec{X}))}{\partial \boldsymbol{\varepsilon}} \qquad \text{and} \qquad T = \frac{\partial u(\boldsymbol{\varepsilon}, \boldsymbol{\varepsilon}^{\mathrm{i}}, \boldsymbol{\alpha}, s, h(\vec{X}))}{\partial s}\,. \tag{6.157}$$

According to Figure 6.4, forces $\boldsymbol{\sigma}$ and $T$ are calculated at the equilibrium process PZ$(t)$ and utilized for the nonequilibrium process Z$(t)$.

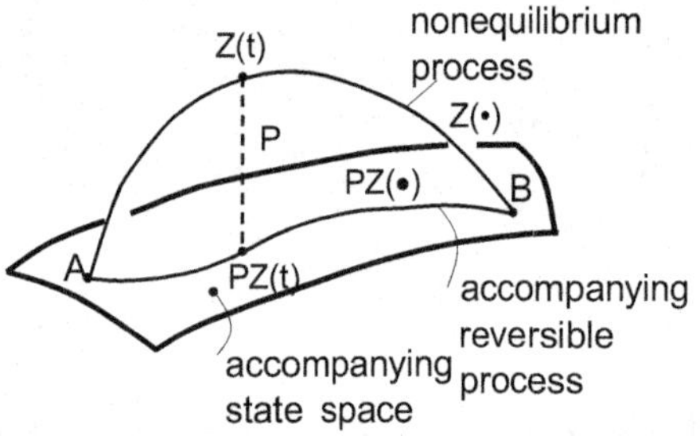


**Figure 6.4** Projection P maps the (nonequilibrium) process Z$(\cdot)$ point by point onto the equilibrium subspace represented as a hypersurface in the state space. Modified from Muschik ([65], Figure 1.16).

State Equation $(6.157)_1$ is written into a multiaxial case, whereas State Equation $(6.156)_1$ is a uniaxial one. It is worth noting that the axiom of local accompanying state [**AX 5**] does not extend the uniaxial state equation into a multiaxial case but expresses that the state equations derived at thermodynamical equilibrium are also valid for nonequilibrium processes.

The local forms of the first law and of the second law of thermodynamics for solid material are given by Equations (6.37) and (6.117). These equations are:

$$\rho_0\, \dot{u} = \boldsymbol{\sigma} : \dot{\boldsymbol{\varepsilon}} + \rho_0\, r - \vec{\nabla} \cdot \vec{q} \tag{6.158}$$

and

$$\rho_0\, T\, \dot{s} + \vec{\nabla}\cdot\vec{q} - \frac{\vec{\nabla} T}{T}\cdot\vec{q} - \rho_0\, r \geq 0\,. \tag{6.159}$$

Based on the Local Forms (6.158) and (6.159) and State Equations (6.157), there are two pairs of variables in Local Forms (6.158) and (6.159): $(\boldsymbol{\sigma}, \boldsymbol{\varepsilon})$ and $(T, s)$. The second component of the pair is a controllable state variable, and the first component is the corresponding state function the value of which is obtained from the state equation.

Following the concept of State Equations (6.157), the internal forces $\boldsymbol{\upsilon}$ and $\boldsymbol{\beta}$ are defined by

$$\boldsymbol{\upsilon} := -\,\rho_0\, \frac{\partial u(\boldsymbol{\varepsilon}, \boldsymbol{\varepsilon}^{\mathrm{i}}, \boldsymbol{\alpha}, s, h(\vec{X}))}{\partial \boldsymbol{\varepsilon}^{\mathrm{i}}} \tag{6.160}$$

and

$$\boldsymbol{\beta} := -\,\rho_0\, \frac{\partial u(\boldsymbol{\varepsilon}, \boldsymbol{\varepsilon}^{\mathrm{i}}, \boldsymbol{\alpha}, s, h(\vec{X}))}{\partial \boldsymbol{\alpha}}\,. \tag{6.161}$$

Equations (6.157), (6.160), and (6.161) are state equations. The quantities $\boldsymbol{\sigma}$, $T$, $\boldsymbol{\upsilon}$, and $\boldsymbol{\beta}$ are called state functions. The minus sign on the right side of State Equations (6.160) and (6.161) is due to the definition of positive value for internal force $\boldsymbol{\beta}$. A different definition is of course possible. The idea behind the preceding sign definition is that it leads to a more aesthetic form for the heat equation and for the Clausius–Duhem inequality, as Sections 6.5 and 6.6 show. State Equations (6.157), (6.160), and (6.161) are field equations, which means that their quantities are dependent on the position of the studied material point $P(t)$.

The present book evaluates mainly material models that model elastic and inelastic deformation or pure elastic deformation. For those material models, the specific internal energy $u(\vec{X}, t)$ takes the following appearance:

$$u = u(\boldsymbol{\varepsilon} - \boldsymbol{\varepsilon}^{\mathrm{i}}, \boldsymbol{\alpha}, s, h(\vec{X}))\,. \tag{6.162}$$

Equation (6.162) displays the fact that for the material models with elastic deformation, the difference $\boldsymbol{\varepsilon} - \boldsymbol{\varepsilon}^{\mathrm{i}}$ belongs to the description of state. Section 7.4 studies material models without elastic deformation. For Material Model (6.162), State Equations $(6.157)_1$ and (6.160) are modified. This can be done by the following two manipulations. By following the concept of Expression (2.163), the first one is found to take the following appearance:

$$\begin{aligned}\frac{\partial u(\boldsymbol{\varepsilon} - \boldsymbol{\varepsilon}^{\mathrm{i}}, \boldsymbol{\alpha}, s, h(\vec{X}))}{\partial \boldsymbol{\varepsilon}} &= \frac{\partial(\boldsymbol{\varepsilon} - \boldsymbol{\varepsilon}^{\mathrm{i}})}{\partial \boldsymbol{\varepsilon}} : \frac{\partial u(\boldsymbol{\varepsilon} - \boldsymbol{\varepsilon}^{\mathrm{i}}, \boldsymbol{\alpha}, s, h(\vec{X}))}{\partial(\boldsymbol{\varepsilon} - \boldsymbol{\varepsilon}^{\mathrm{i}})} \\ &= \mathbf{I} : \frac{\partial u(\boldsymbol{\varepsilon} - \boldsymbol{\varepsilon}^{\mathrm{i}}, \boldsymbol{\alpha}, s, h(\vec{X}))}{\partial(\boldsymbol{\varepsilon} - \boldsymbol{\varepsilon}^{\mathrm{i}})} = \frac{\partial u(\boldsymbol{\varepsilon} - \boldsymbol{\varepsilon}^{\mathrm{i}}, \boldsymbol{\alpha}, s, h(\vec{X}))}{\partial(\boldsymbol{\varepsilon} - \boldsymbol{\varepsilon}^{\mathrm{i}})}\,.\end{aligned} \tag{6.163}$$

In arriving at Manipulation (6.163), the following three facts were exploited: First, the derivative of a second-order tensor with respect to itself is a fourth-order identity tensor $\mathbf{I}$. Second, the derivative of a scalar (here it is $u$) with respect to a second-order tensor (here it is $\boldsymbol{\varepsilon} - \boldsymbol{\varepsilon}^{\mathrm{i}}$) is a second-order tensor. Finally, a double-dot product between an arbitrary second-order tensor $\mathbf{g}$ and a fourth-order identity tensor $\mathbf{I}$ gives the second-order tensor $\mathbf{g}$. The results are given by Expressions (2.149), $(2.172)_1$, and (2.69). The second manipulation is

$$\begin{aligned}\frac{\partial u(\boldsymbol{\varepsilon} - \boldsymbol{\varepsilon}^{\mathrm{i}}, \boldsymbol{\alpha}, s, h(\vec{X}))}{\partial \boldsymbol{\varepsilon}^{\mathrm{i}}} &= \frac{\partial(\boldsymbol{\varepsilon} - \boldsymbol{\varepsilon}^{\mathrm{i}})}{\partial \boldsymbol{\varepsilon}^{\mathrm{i}}} : \frac{\partial u(\boldsymbol{\varepsilon} - \boldsymbol{\varepsilon}^{\mathrm{i}}, \boldsymbol{\alpha}, s, h(\vec{X}))}{\partial(\boldsymbol{\varepsilon} - \boldsymbol{\varepsilon}^{\mathrm{i}})} \\ &= (-\mathbf{I}) : \frac{\partial u(\boldsymbol{\varepsilon} - \boldsymbol{\varepsilon}^{\mathrm{i}}, \boldsymbol{\alpha}, s, h(\vec{X}))}{\partial(\boldsymbol{\varepsilon} - \boldsymbol{\varepsilon}^{\mathrm{i}})} \\ &= -\frac{\partial u(\boldsymbol{\varepsilon} - \boldsymbol{\varepsilon}^{\mathrm{i}}, \boldsymbol{\alpha}, s, h(\vec{X}))}{\partial(\boldsymbol{\varepsilon} - \boldsymbol{\varepsilon}^{\mathrm{i}})} .\end{aligned} \tag{6.164}$$

Due to Material Model (6.162) and Manipulations (6.163) and (6.164), State Equations (6.157), (6.160), and (6.161) are replaced with

$$\boldsymbol{\sigma} = \rho_0 \frac{\partial u(\boldsymbol{\varepsilon} - \boldsymbol{\varepsilon}^{\mathrm{i}}, \boldsymbol{\alpha}, s, h(\vec{X}))}{\partial \boldsymbol{\varepsilon}} = \rho_0 \frac{\partial u(\boldsymbol{\varepsilon} - \boldsymbol{\varepsilon}^{\mathrm{i}}, \boldsymbol{\alpha}, s, h(\vec{X}))}{\partial(\boldsymbol{\varepsilon} - \boldsymbol{\varepsilon}^{\mathrm{i}})} \tag{6.165}$$

and

$$T = \frac{\partial u(\boldsymbol{\varepsilon} - \boldsymbol{\varepsilon}^{\mathrm{i}}, \boldsymbol{\alpha}, s, h(\vec{X}))}{\partial s} , \tag{6.166}$$

and furthermore,

$$\boldsymbol{\upsilon} := -\rho_0 \frac{\partial u(\boldsymbol{\varepsilon} - \boldsymbol{\varepsilon}^{\mathrm{i}}, \boldsymbol{\alpha}, s, h(\vec{X}))}{\partial \boldsymbol{\varepsilon}^{\mathrm{i}}} = \rho_0 \frac{\partial u(\boldsymbol{\varepsilon} - \boldsymbol{\varepsilon}^{\mathrm{i}}, \boldsymbol{\alpha}, s, h(\vec{X}))}{\partial(\boldsymbol{\varepsilon} - \boldsymbol{\varepsilon}^{\mathrm{i}})} , \tag{6.167}$$

and, finally,

$$\boldsymbol{\beta} := -\rho_0 \frac{\partial u(\boldsymbol{\varepsilon} - \boldsymbol{\varepsilon}^{\mathrm{i}}, \boldsymbol{\alpha}, s, h(\vec{X}))}{\partial \boldsymbol{\alpha}} . \tag{6.168}$$

State Equations (6.165) and (6.167) show that the state functions $\boldsymbol{\sigma}$ and $\boldsymbol{\upsilon}$ coincide. Therefore the state function $\boldsymbol{\upsilon}$ can be dismissed.

For the future purposes the material derivative of the specific internal free energy $\dot{u}$ is given. Based on Expression (6.162) and the concept given by Expression (2.167) and by multiplying the obtained expression by $\rho_0$ the following is obtained:

$$\begin{aligned}\rho_0 \, \dot{u}(\boldsymbol{\varepsilon} - \boldsymbol{\varepsilon}^{\mathrm{i}}, \boldsymbol{\alpha}, s, h(\vec{X})) &= \rho_0 \frac{\partial u(\ldots)}{\partial(\boldsymbol{\varepsilon} - \boldsymbol{\varepsilon}^{\mathrm{i}})} : (\dot{\boldsymbol{\varepsilon}} - \dot{\boldsymbol{\varepsilon}}^{\mathrm{i}}) + \rho_0 \frac{\partial u(\ldots)}{\partial \boldsymbol{\alpha}} : \dot{\boldsymbol{\alpha}} + \rho_0 \frac{\partial u(\ldots)}{\partial s} \, \dot{s} \\ &\quad + \rho_0 \frac{\partial u(\ldots)}{\partial h} \, \dot{h} .\end{aligned} \tag{6.169}$$

Substitution of State Equations (6.165), (6.166), and (6.168) into Derivative (6.169) yields

$$\rho_0\,\dot{u}\big(\boldsymbol{\varepsilon}-\boldsymbol{\varepsilon}^{\mathrm{i}},\boldsymbol{\alpha},s,h(\vec{X})\big)=\boldsymbol{\sigma}:(\dot{\boldsymbol{\varepsilon}}-\dot{\boldsymbol{\varepsilon}}^{\mathrm{i}})-\boldsymbol{\beta}:\dot{\boldsymbol{\alpha}}+\rho_0 T\,\dot{s}\,. \tag{6.170}$$

Since the quantity $h(\vec{X})$ does not depend on time, its material derivative vanishes. This is exploited in Derivative (6.170).

Instead of the specific internal energy $u$, the state of solids is usually described by the specific Helmholtz free energy $\psi$ which is a Legendre partial transformation of the specific internal energy $u$ (see **Appendix E**). This is done because writing a material model using the specific internal energy $u$ is very difficult, given that the specific entropy $s$ is an argument of the specific internal energy $u$. It is very difficult to construct a constitutive model as a function of the specific entropy $s$. In the formulation of the specific Helmholtz free energy $\psi$ the specific entropy $s$ is replaced with the absolute temperature $T$. Since the absolute temperature $T$ is a well-known quantity for a human being, writing a material model using the specific Helmholtz free energy $\psi$ is much simpler than doing so with the specific internal energy $u$. The specific Helmholtz free energy $\psi$ is defined by

$$\begin{aligned}\psi(\boldsymbol{\epsilon}-\boldsymbol{\varepsilon}^{\mathrm{i}},\boldsymbol{\alpha},T,h(\vec{X}))&\overset{\text{Legendre}}{=}u(\boldsymbol{\epsilon}-\boldsymbol{\varepsilon}^{\mathrm{i}},\boldsymbol{\alpha},s,h(\vec{X}))-T\,s\\ \text{with}\quad T&=\frac{\partial u(\boldsymbol{\varepsilon}-\boldsymbol{\varepsilon}^{\mathrm{i}},\boldsymbol{\alpha},s,h(\vec{X}))}{\partial s}\,.\end{aligned} \tag{6.171}$$

It is worth noting that the specific Helmholtz free energy $\psi$ does not contain any new information, because it is represented by the two previously introduced state functions $u$ and $T$ and a state variable $s$. When the state is expressed by the specific internal energy $u$, temperature $T$ is a state function, as State Equation (6.166) shows, whereas in the specific Helmholtz free energy $\psi$ temperature $T$ is a state variable. State Equation $(6.171)_2$ [i.e., State Equation (6.166)] is a vital part of the Legendre transformation; without it no Legendre transformation can be performed. Thus the Legendre (partial) transformation consists of two parts. They are Expression $(6.171)_1$ and Derivative $(6.171)_2$ [i.e., State Equation (6.166)].

State equations carry the information between the independent variables of the state functions, above $\psi$ and $u$, on the left and right sides of the Legendre (partial) transformation. Other kinds of state functions, such as the Gibbs free energy $g$, can be introduced. Gibbs free energy $g$ is studied later in this chapter.

According to **Appendices E** and **F** the specific Helmholtz free energy $\psi$ is a Legendre partial transformation of the specific internal energy $u$. Comparison of Definition (6.171) and Transformation (E.11) of **Appendix E** gives $b=-1$ and $a=1$, as well as the fact that the specific entropy $s$ is the active variable of the transformation whereas the quantities $\boldsymbol{\varepsilon}-\boldsymbol{\varepsilon}^{\mathrm{i}}$, $\boldsymbol{\alpha}$, and $h(\vec{X})$ are passive ones.

Based on Results (E.14) and (E.15) in **Appendix E**, the following is achieved:

$$s = -\frac{\partial\psi(\boldsymbol{\varepsilon} - \boldsymbol{\varepsilon}^{\mathrm{i}}, \boldsymbol{\alpha}, T, h(\vec{X}))}{\partial T} \tag{6.172}$$

and

$$\frac{\partial u(\ldots, s, \ldots)}{\partial(\boldsymbol{\varepsilon} - \boldsymbol{\varepsilon}^{\mathrm{i}})} = \frac{\partial\psi(\ldots, T, \ldots)}{\partial(\boldsymbol{\varepsilon} - \boldsymbol{\varepsilon}^{\mathrm{i}})} \quad \text{and} \quad \frac{\partial u(\ldots, s, \ldots)}{\partial \boldsymbol{\alpha}} = \frac{\partial\psi(\ldots, T, \ldots)}{\partial \boldsymbol{\alpha}}. \tag{6.173}$$

Substitution of Expressions (6.173) into State Equations (6.165) and (6.168) yields

$$\boldsymbol{\sigma} = \rho_0 \frac{\partial\psi(\boldsymbol{\varepsilon} - \boldsymbol{\varepsilon}^{\mathrm{i}}, \boldsymbol{\alpha}, T, h(\vec{X}))}{\partial(\boldsymbol{\varepsilon} - \boldsymbol{\varepsilon}^{\mathrm{i}})} \quad \text{and} \quad \boldsymbol{\beta} = -\rho_0 \frac{\partial\psi(\boldsymbol{\varepsilon} - \boldsymbol{\varepsilon}^{\mathrm{i}}, \boldsymbol{\alpha}, T, h(\vec{X}))}{\partial \boldsymbol{\alpha}}. \tag{6.174}$$

State Equations (6.172) and (6.174) are the state equations often used in this book. As Equations (6.173) and (6.174) show, Legendre Partial Transformation (6.171) is not enough but Equations (6.165) and (6.168) are needed for State Equations. (6.174).

In case of many internal variables, the specific Helmholtz free energy may, for example, take the following appearance:

$$\psi = \psi(\boldsymbol{\varepsilon} - \boldsymbol{\varepsilon}^{\mathrm{i}}, \boldsymbol{\alpha}, \vec{\alpha}^{\,1} \alpha^2, T, h(\vec{X})). \tag{6.175}$$

In this case the following two state functions and state equations beyond those given by State Equations (6.172) and (6.174) are obtained:

$$\vec{\beta}^{\,1} = -\rho_0 \frac{\partial\psi(\boldsymbol{\varepsilon} - \boldsymbol{\varepsilon}^{\mathrm{i}}, \boldsymbol{\alpha}, \vec{\alpha}^{\,1}, \alpha^2 T, h(\vec{X}))}{\partial \vec{\alpha}^{\,1}} \tag{6.176}$$

and

$$\beta^2 = -\rho_0 \frac{\partial\psi(\boldsymbol{\varepsilon} - \boldsymbol{\varepsilon}^{\mathrm{i}}, \boldsymbol{\alpha}, \vec{\alpha}^{\,1}, \alpha^2 T, h(\vec{X}))}{\partial \alpha^2}. \tag{6.177}$$

In this case, the specific Helmholtz free energy $\psi$ in State Equations (6.172) and (6.174) is given in Equation (6.175). The origin of State Equations (6.176) and (6.177) is the specific internal energy $u = u(\boldsymbol{\varepsilon} - \boldsymbol{\varepsilon}^{\mathrm{i}}, \boldsymbol{\alpha}, \vec{\alpha}^1, \alpha^2, T, h(\vec{X}))$ and the state equations similar to State Equations (6.168). Good examples of material models with many internal variables are the models for thermoplasticity (see Section 8.4), and the model for creep by Le Gac and Duval [48] (see Section 9.6).

Sometimes the material model is expressed by the specific Gibbs free energy $g$ instead of the specific Helmholtz free energy $\psi$. The specific Gibbs free energy $g$ is a Legendre transformation of the specific Helmholtz free energy $\psi$. Depending on which state variables are replaced by their associative state functions $g$ also can be a Legendre partial transformation. As an example the specific Gibbs free energy $g$ is assumed to have the following appearance:

$$g = g(\boldsymbol{\sigma}, \boldsymbol{\alpha}, T, h(\vec{X}))\,. \tag{6.178}$$

This means that the difference of the strain tensors [ i.e., $(\boldsymbol{\varepsilon} - \boldsymbol{\varepsilon}^{\mathrm{i}})$ ] is replaced by its associative force $\boldsymbol{\sigma}$ whereas the other state variables remain unchanged. Thus the following Legendre partial transformation is made:

$$\begin{aligned} \rho_0\, g(\boldsymbol{\sigma}, \boldsymbol{\alpha}, T, h(\vec{X})) &\overset{\text{Legendre}}{=} \boldsymbol{\sigma} : (\boldsymbol{\varepsilon} - \boldsymbol{\varepsilon}^{\mathrm{i}}) - \rho_0\, \psi(\boldsymbol{\varepsilon} - \boldsymbol{\varepsilon}^{\mathrm{i}}, \boldsymbol{\alpha}, T, h(\vec{X})) \\ \text{with} \quad \boldsymbol{\sigma} &= \rho_0\, \frac{\partial \psi(\boldsymbol{\varepsilon} - \boldsymbol{\varepsilon}^{\mathrm{i}}, \boldsymbol{\alpha}, T, h(\vec{X}))}{\partial (\boldsymbol{\varepsilon} - \boldsymbol{\varepsilon}^{\mathrm{i}})}\,. \end{aligned} \tag{6.179}$$

Comparison of Transformation (6.179) and State Equations (6.172) and (6.174) with the corresponding equations in **Appendix E** [Equation (E.10) and Transformation (E.11)] gives $a = \rho_0$ and $b = \rho_0$. Based on Expressions (E.14) and (E.15), State Equations (6.172) and (6.174) are now replaced by

$$s = \frac{\partial g(\boldsymbol{\sigma}, \boldsymbol{\alpha}, T, h(\vec{X}))}{\partial T} \tag{6.180}$$

and

$$\boldsymbol{\varepsilon} - \boldsymbol{\varepsilon}^{\mathrm{i}} = \rho_0\, \frac{\partial g(\boldsymbol{\sigma}, \boldsymbol{\alpha}, T, h(\vec{X}))}{\partial \boldsymbol{\sigma}} \quad \text{and} \quad \boldsymbol{\beta} = \rho_0 \frac{\partial g(\boldsymbol{\sigma}, \boldsymbol{\alpha}, T, h(\vec{X}))}{\partial \boldsymbol{\alpha}}\,. \tag{6.181}$$

Often the specific Gibbs free energy $g$ is defined to take the opposite sign [see, e.g., Holzapfel [29], Eq. (7.33); Malvern [57], Table 5-1; Reddy [84], Eq. (6.8.2)]. Since the present definition leads to the positive-valued energy function $g$, Definition (6.179) is used in this book.

Similar to the preceding, in case of many internal variables the specific Gibbs free energy $g$ may, for example, take the following appearance:

$$g = g(\boldsymbol{\sigma}, \boldsymbol{\alpha}, \vec{\alpha}^{1}, \alpha^2, T, h(\vec{X}))\,. \tag{6.182}$$

In this case the following two state functions and state equations beyond those given by Expressions (6.180) and (6.181) are obtained:

$$\vec{\beta}^{1} = \rho_0\, \frac{\partial g(\boldsymbol{\varepsilon} - \boldsymbol{\varepsilon}^{\mathrm{i}}, \boldsymbol{\alpha}, \vec{\alpha}^{1}, \alpha^2, T, h(\vec{X}))}{\partial \vec{\alpha}^{1}} \tag{6.183}$$

and

$$\beta^2 = \rho_0\, \frac{\partial g(\boldsymbol{\sigma}, \boldsymbol{\alpha}, \vec{\alpha}^{1}, \alpha^2, T, h(\vec{X}))}{\partial \alpha^2}\,. \tag{6.184}$$

In this case, the specific Gibbs free energy $g$ in State Equations (6.180) and (6.181) is given in Equation (6.178).

### 6.4.5 State Equations for Fluids

This section introduces state equations for fluids. The local forms of the first law of thermodynamics and of the second law of thermodynamics for fluids are given by Equations (6.89) and (6.143). These equations are:

$$\rho\,\dot{u} = p\,\frac{\dot{\rho}}{\rho} + \boldsymbol{\sigma}^{\mathrm{d}} : \overset{\circ}{\boldsymbol{\varepsilon}} + \rho\,r - \vec{\nabla}\cdot\vec{q} \tag{6.185}$$

and

$$\rho\,T\,\dot{s} + \vec{\nabla}\cdot\vec{q} - \frac{\vec{\nabla}T}{T}\cdot\vec{q} - \rho\,r \geq 0\,. \tag{6.186}$$

The axiom of caloric equation of state [**AX 2**] for material modelling of fluids is given by Axiom (6.154) [**AX 2**]:

**AX 2**
$$u \;:= u(\rho, s)\,. \tag{6.187}$$

Based on Expressions (6.185)–(6.187) and the investigation carried out in the previous section, the state function, state variable pairs are; $(p, \rho)$ and $(T, s)$.

Based on State Equations (5.17) and on the axiom of local accompanying state [**AX 5**], the state equations for nonhomogeneous processes outside thermodynamic equilibrium are

$$p = \rho^2\,\frac{\partial u(\rho, s)}{\partial \rho} \qquad \text{and} \qquad T = \frac{\partial u(\rho, s)}{\partial s}\,. \tag{6.188}$$

Based on Expression (6.187) [**AX 2**], the following is achieved (the chain rule):

$$\rho\,\dot{u}(\rho, s) = \rho\,\frac{\partial u}{\partial \rho}\,\dot{\rho} + \rho\,\frac{\partial u}{\partial s}\,\dot{s}\,. \tag{6.189}$$

Substitution of State Equations (6.188) into Rate (6.189) gives

$$\rho\,\dot{u}(\rho, s) = \frac{1}{\rho}\,p\,\dot{\rho} + \rho\,T\,\dot{s}\,. \tag{6.190}$$

The specific Helmholtz free energy $\psi$ is defined by the following Legendre partial transformation:

$$\psi(\rho, T) \overset{\text{Legendre}}{=} u(\rho, s) - T\,s\,, \qquad \text{with} \qquad T = \frac{\partial u(\rho, s)}{\partial s}. \tag{6.191}$$

According to **Appendix E**, the specific Helmholtz free energy $\psi$ is a Legendre partial transformation of the specific internal energy $u$. Comparison of Definition (6.189) and Transformation (E.11) in **Appendix E** gives $b = -1$ and $a = 1$, and shows the fact that the specific entropy $s$ is the active variable of the transformation whereas the quantity $\rho$ is a passive one. Based on Results (E.14) and (E.15) in **Appendix E**, the following result is obtained:

$$s = -\frac{\partial \psi(\rho, T)}{\partial T} \qquad \text{and} \qquad \frac{\partial u(\rho, s)}{\partial \rho} = \frac{\partial \psi(\rho, T)}{\partial \rho}\,. \tag{6.192}$$

Substitution of Expression $(6.192)_2$ into State Equation $(6.188)_1$ yields

$$p = \rho^2 \frac{\partial \psi(\rho, T)}{\partial \rho}. \tag{6.193}$$

## 6.5 Heat Equation

This section takes a close look at the heat equation. As pointed out by Maugin ([59], p. 178), the heat equation is none other than the energy equation in disguise. The appearance of the heat equation depends on the set of state variables. Since the state variables are different for solids and fluids, the heat equations for these states of material are derived in separate sections.

### 6.5.1 Heat Equation for Solids

This section derives heat equation for a solid material. Energy Equation (6.37) is

$$\rho_0 \dot{u} = \boldsymbol{\sigma} : \dot{\boldsymbol{\varepsilon}} + \rho_0 r - \vec{\nabla} \cdot \vec{q}. \tag{6.194}$$

Again, the definition for the specific Helmholtz free energy $\psi$, Definition (6.171), is

$$\begin{aligned} \psi(\boldsymbol{\varepsilon} - \boldsymbol{\varepsilon}^{\mathrm{i}}, \boldsymbol{\alpha}, T, h(\vec{X})) &\overset{\text{Legendre}}{=} u(\boldsymbol{\varepsilon} - \boldsymbol{\varepsilon}^{\mathrm{i}}, \boldsymbol{\alpha}, s, h(\vec{X})) - T\,s \\ \text{with} \quad T &= \frac{\partial u(\boldsymbol{\varepsilon} - \boldsymbol{\varepsilon}^{\mathrm{i}}, \boldsymbol{\alpha}, s, h(\vec{X}))}{\partial s}. \end{aligned} \tag{6.195}$$

Transformation $(6.195)_1$ gives

$$\dot{u} = \dot{\psi} + \dot{T}\,s + T\,\dot{s}. \tag{6.196}$$

Substitution of Derivative (6.196) into Energy Equation (6.194) yields

$$\rho_0 \dot{\psi} + \rho_0 \dot{T}\,s + \rho_0 T\,\dot{s} - \boldsymbol{\sigma} : \dot{\boldsymbol{\varepsilon}} - \rho_0 r + \vec{\nabla} \cdot \vec{q} = 0. \tag{6.197}$$

Since the specific Helmholtz free energy $\psi$ has the appearance $\psi(\boldsymbol{\varepsilon} - \boldsymbol{\varepsilon}^{\mathrm{i}}, \boldsymbol{\alpha}, T, h)$, we arrive at the following [cf. Equation (2.167)]:

$$\dot{\psi}(\boldsymbol{\varepsilon} - \boldsymbol{\varepsilon}^{\mathrm{i}}, \boldsymbol{\alpha}, T, h(\vec{X})) = \frac{\partial \psi}{\partial(\boldsymbol{\varepsilon} - \boldsymbol{\varepsilon}^{\mathrm{i}})} : (\dot{\boldsymbol{\varepsilon}} - \dot{\boldsymbol{\varepsilon}}^{\mathrm{i}}) + \frac{\partial \psi}{\partial \boldsymbol{\alpha}} : \dot{\boldsymbol{\alpha}} + \frac{\partial \psi}{\partial T}\,\dot{T}. \tag{6.198}$$

Equation (6.198) exploits the fact that $\dot{h}(\vec{X})$ vanishes. State Equations Equations (6.172) and (6.174) are

$$\boldsymbol{\sigma} = \rho_0 \frac{\partial \psi(\ldots)}{\partial(\boldsymbol{\varepsilon} - \boldsymbol{\varepsilon}^{\mathrm{i}})}, \qquad \boldsymbol{\beta} = -\rho_0 \frac{\partial \psi(\ldots)}{\partial \boldsymbol{\alpha}}, \qquad \text{and} \qquad s = -\frac{\partial \psi(\ldots)}{\partial T}. \tag{6.199}$$

By multiplying Derivative (6.198) by the density $\rho_0$ and by taking State Equations (6.199) into consideration, Derivative (6.198) gives

$$\rho_0 \dot{\psi}(\boldsymbol{\varepsilon} - \boldsymbol{\varepsilon}^{\mathrm{i}}, \boldsymbol{\alpha}, T, h) = \boldsymbol{\sigma} : (\dot{\boldsymbol{\varepsilon}} - \dot{\boldsymbol{\varepsilon}}^{\mathrm{i}}) - \boldsymbol{\beta} : \dot{\boldsymbol{\alpha}} - \rho_0\,s\,\dot{T}. \tag{6.200}$$

State Equation $(6.199)_3$ and Derivative (6.198) give

$$\dot{s} = -\frac{\partial \dot{\psi}(\ldots)}{\partial T} = -\frac{\partial^2 \psi}{\partial T\, \partial(\boldsymbol{\varepsilon} - \boldsymbol{\varepsilon}^{\mathrm{i}})} : (\dot{\boldsymbol{\varepsilon}} - \dot{\boldsymbol{\varepsilon}}^{\mathrm{i}}) - \frac{\partial^2 \psi}{\partial T\, \partial \boldsymbol{\alpha}} : \dot{\boldsymbol{\alpha}} - \frac{\partial^2 \psi}{\partial T^2}\, \dot{T}\,. \tag{6.201}$$

Substitution of Derivatives (6.200) and (6.201) into Equation (6.197) gives

$$\begin{aligned} &\boldsymbol{\sigma} : (\dot{\boldsymbol{\varepsilon}} - \dot{\boldsymbol{\varepsilon}}^{\mathrm{i}}) - \boldsymbol{\beta} : \dot{\boldsymbol{\alpha}} - \rho_0\, s\, \dot{T} + \rho_0\, \dot{T}\, s \\ &\quad - \rho_0\, T \left[ \frac{\partial^2 \psi}{\partial T\, \partial(\boldsymbol{\varepsilon} - \boldsymbol{\varepsilon}^{\mathrm{i}})} : (\dot{\boldsymbol{\varepsilon}} - \dot{\boldsymbol{\varepsilon}}^{\mathrm{i}}) + \frac{\partial^2 \psi}{\partial T\, \partial \boldsymbol{\alpha}} : \dot{\boldsymbol{\alpha}} + \frac{\partial^2 \psi}{\partial T^2}\, \dot{T} \right] \\ &\quad - \boldsymbol{\sigma} : \dot{\boldsymbol{\varepsilon}} - \rho_0\, r + \vec{\nabla} \cdot \vec{q} = 0\,. \end{aligned} \tag{6.202}$$

By rearranging the terms, Expression (6.202) can be cast into the following form:

$$\vec{\nabla} \cdot \vec{q} = \boldsymbol{\sigma} : \dot{\boldsymbol{\varepsilon}}^{\mathrm{i}} + \boldsymbol{\beta} : \dot{\boldsymbol{\alpha}} + \rho_0\, r + \rho_0\, T \left[ \frac{\partial^2 \psi}{\partial T\, \partial(\boldsymbol{\varepsilon} - \boldsymbol{\varepsilon}^{\mathrm{i}})} : (\dot{\boldsymbol{\varepsilon}} - \dot{\boldsymbol{\varepsilon}}^{\mathrm{i}}) + \frac{\partial^2 \psi}{\partial T \partial \boldsymbol{\alpha}} : \dot{\boldsymbol{\alpha}} + \frac{\partial^2 \psi}{\partial T^2} \dot{T} \right]. \tag{6.203}$$

Equation (6.203) is the heat equation for solids for a case when the material model is expressed by the specific Helmholtz free energy $\psi$.

In a case of pure reversible (elastic) deformation, that is, when $\dot{\boldsymbol{\varepsilon}}^{\mathrm{i}} = \dot{\boldsymbol{\alpha}} = \mathbf{0}$, $\boldsymbol{\varepsilon}$ reduces to $\boldsymbol{\varepsilon}^{\mathrm{e}}$ and $\psi(\boldsymbol{\varepsilon} - \boldsymbol{\varepsilon}^{\mathrm{i}}, \boldsymbol{\alpha}, T, h)$ reduces to $\psi^{\mathrm{r}}(\boldsymbol{\varepsilon}^{\mathrm{e}}, T, h)$ where the superscripts e and r refer to elastic and reversible process, respectively, Heat Equation (6.203) reduces to

$$\vec{\nabla} \cdot \vec{q} = \rho_0\, T \left[ \frac{\partial^2 \psi^{\mathrm{r}}}{\partial T\, \partial \boldsymbol{\varepsilon}^{\mathrm{e}}} : \dot{\boldsymbol{\varepsilon}}^{\mathrm{e}} + \frac{\partial^2 \psi^{\mathrm{r}}}{\partial T^2}\, \dot{T} \right] + \rho_0\, r\,. \tag{6.204}$$

as derived by Parkus [[75], Eq. (5.28)].

Following Maugin [[59], Definition $(4.4)_1$] the specific heat capacity denoted by $c$ is defined by

$$c := -T\, \frac{\partial^2 \psi(\, \boldsymbol{\varepsilon} - \boldsymbol{\varepsilon}^{\mathrm{i}}, \boldsymbol{\alpha}, T, h(\vec{X}))}{\partial T^2} \qquad (\geq 0)\,. \tag{6.205}$$

Definition (6.205) is evaluated in Section 6.5.2. Definition (6.205) allows Heat Equation (6.203) to be written in the form

$$\vec{\nabla} \cdot \vec{q} = \boldsymbol{\sigma} : \dot{\boldsymbol{\varepsilon}}^{\mathrm{i}} + \boldsymbol{\beta} : \dot{\boldsymbol{\alpha}} + \rho_0\, r + \rho_0\, T \left[ \frac{\partial^2 \psi}{\partial T\, \partial(\boldsymbol{\varepsilon} - \boldsymbol{\varepsilon}^{\mathrm{i}})} : (\dot{\boldsymbol{\varepsilon}} - \dot{\boldsymbol{\varepsilon}}^{\mathrm{i}}) + \frac{\partial^2 \psi}{\partial T\, \partial \boldsymbol{\alpha}} : \dot{\boldsymbol{\alpha}} \right] - \rho_0\, c\, \dot{T}. \tag{6.206}$$

**Appendix F** contains a derivation of the heat equation for a case when the material model is expressed by the specific Gibbs free energy $g$. The following form, Expression (F.12), is obtained:

$$\vec{\nabla} \cdot \vec{q} = \boldsymbol{\sigma} : \dot{\boldsymbol{\varepsilon}}^{\mathrm{i}} + \boldsymbol{\beta} : \dot{\boldsymbol{\alpha}} + \rho_0\, r - \rho_0\, T \left[ \frac{\partial^2 g}{\partial T\, \partial \boldsymbol{\sigma}} : \dot{\boldsymbol{\sigma}} + \frac{\partial^2 g}{\partial T\, \partial \boldsymbol{\alpha}} : \dot{\boldsymbol{\alpha}} + \frac{\partial^2 g}{\partial T^2}\, \dot{T} \right], \tag{6.207}$$

where $g = g(\boldsymbol{\sigma}, \boldsymbol{\alpha}, T, h(\vec{X}))$. Based on the text above State Equation (6.180), the specific Gibbs free energy $g$ was obtained from the specific Helmholtz free energy $\psi$ by the Legendre partial transformation with the values $a = \rho_0$ and $b = \rho_0$. Based on the values for $a$ and $b$, Equation (E.15) of **Appendix E** gives

$$\rho_0 \frac{\partial \psi(\boldsymbol{\varepsilon} - \boldsymbol{\varepsilon}^{\mathrm{i}}, \boldsymbol{\alpha}, T, h(\vec{X}))}{\partial T} = -\rho_0 \frac{g(\boldsymbol{\sigma}, \boldsymbol{\alpha}, T, h(\vec{X}))}{\partial T}. \tag{6.208}$$

Equation (6.208) gives

$$\frac{\partial^2 \psi(\boldsymbol{\varepsilon} - \boldsymbol{\varepsilon}^{\mathrm{i}}, \boldsymbol{\alpha}, T, h(\vec{X}))}{\partial T^2} = -\frac{\partial^2 g(\boldsymbol{\sigma}, \boldsymbol{\alpha}, T, h(\vec{X}))}{\partial T^2}. \tag{6.209}$$

Definition (6.205) and Equality (6.209) give

$$c = T \frac{\partial^2 g(\boldsymbol{\sigma}, \boldsymbol{\alpha}, T, h(\vec{X}))}{\partial T^2}. \tag{6.210}$$

Substitution of the specific heat capacity $c$ from Equation (6.210) into Equation (6.207) leads to

$$\vec{\nabla} \cdot \vec{q} = \boldsymbol{\sigma} : \dot{\boldsymbol{\varepsilon}}^{\mathrm{i}} + \boldsymbol{\beta} : \dot{\boldsymbol{\alpha}} + \rho_0 \, r - \rho_0 \, T \left[ \frac{\partial^2 g}{\partial T \, \partial \boldsymbol{\sigma}} : \dot{\boldsymbol{\sigma}} + \frac{\partial^2 g}{\partial T \, \partial \boldsymbol{\alpha}} : \dot{\boldsymbol{\alpha}} \right] - \rho_0 \, c \, \dot{T}. \tag{6.211}$$

Equation (6.211) is the heat equation for solids for a case when the material model is expressed by the specific Gibbs free energy $g$.

## 6.5.2 Specific Heat Capacity $c$

In this section provide the definition for the specific heat capacity $c$ and compare it with that for the specific heat capacity at constant volume $c_{\mathrm{V}}$ in thermostatics. According to Young and Freedman [[119], Eq. (18.24)], the expression for the amount of heat $\mathrm{d}Q^*$ required to change the temperature by $\mathrm{d}T$ of the system $\mathrm{m}^{\mathrm{b}}$ reads

$$\mathrm{d}Q^* = n \, C_{\mathrm{V}} \, \mathrm{d}T, \tag{6.212}$$

where $n$ is the number of moles and $C_{\mathrm{V}}$ is the heat capacity at constant volume. Young and Freedman [119] provide, in Equations (17.19) and (17.16), the following expressions:

$$C_{\mathrm{V}} = M \, c_{\mathrm{V}} \quad \text{and} \quad \mathrm{m}^{\mathrm{b}} = n \, M \quad \Rightarrow \quad n = \frac{\mathrm{m}^{\mathrm{b}}}{M}, \tag{6.213}$$

where $M$ is the molar mass of the matter and $c_{\mathrm{V}}$ is the specific heat at constant volume. Substitution of Expressions $(6.213)_1$ and $(6.213)_3$ into Equation (6.212) yields

$$\mathrm{d}Q^* = n \, C_{\mathrm{V}} \, \mathrm{d}T = \frac{\mathrm{m}^{\mathrm{b}}}{M} M \, c_{\mathrm{V}} \, \mathrm{d}T = \mathrm{m}^{\mathrm{b}} \, c_{\mathrm{V}} \, \mathrm{d}T. \tag{6.214}$$

An arbitrary subsystem $m$ in the IG-CtP configuration $\mathrm{V}_0^{\mathrm{b}}$ is studied. In the foregoing study, the (specific) heat capacity at constant volume were introduced. A corresponding assumption for solid material is that the subsystem $m$ is rigid. This means that the mechanical response of the subsystem $m$ can be neglected as a small quantity. Since the mechanical response is neglected, Heat Equations (6.206) and (6.211) reduce to

$$\vec{\nabla}\cdot\vec{q} = \rho_0\, r - \rho_0\, c\,\dot{T}\,. \tag{6.215}$$

Equation (6.215) is multiplied by $-1$, the first term on the right side of the equals sign is moved to the left side and the obtained equation is integrated over the volume $V_0$ of the subsystem $m$. This yields

$$\int_{V_0}\left(-\vec{\nabla}\cdot\vec{q} + \rho_0\, r\right)\,\mathrm{d}V = \int_{V_0}\rho_0\, c\,\dot{T}\,\mathrm{d}V\,. \tag{6.216}$$

Equation (6.34) gives the following expression for the heat input rate $Q$:

$$Q = \int_{V_0}\left(-\vec{\nabla}\cdot\vec{q}^{\,\square} + \rho_0\, r\right)\,\mathrm{d}V\,, \tag{6.217}$$

where $\vec{q}^{\,\square}$ is the Piola–Kirchhoff type of a heat flux vector. When small deformations and rotations are studied, the Piola–Kirchhoff type of heat flux vector $\vec{q}^{\,\square}$ reduces to the heat flux vector $\vec{q}$ and therefore Expression (6.217) reduces to

$$Q = \int_{V_0}\left(-\vec{\nabla}\cdot\vec{q} + \rho_0\, r\right)\,\mathrm{d}V\,. \tag{6.218}$$

Substitution of Expression (6.218) into Equation (6.216) yields

$$Q = \int_{V_0}\rho_0\, c\,\dot{T}\,\mathrm{d}V\,. \tag{6.219}$$

The subsystem $m$ is assumed to be homogeneous in such a way that the values for the density $\rho_0$ and specific heat capacity $c$ are the same everywhere within the subsystem $m$. Furthermore, the value of the specific heat capacity $c$ is assumed to be independent of the absolute temperature $T$. Thus Equation (6.219) can be manipulated as follows:

$$Q = \int_{V_0}\rho_0\, c\,\dot{T}\,\mathrm{d}V = \rho_0\, c\int_{V_0}\dot{T}\,\mathrm{d}V = \rho_0\, c\, V_0\,\frac{1}{V_0}\int_{V_0}\dot{T}\,\mathrm{d}V = \rho_0\, c\, V_0\,\dot{T}^{\mathrm{aver}}\,, \tag{6.220}$$

where $\dot{T}^{\mathrm{aver}}$ is the average of the temperature rate $\dot{T}$ within the subsystem $m$. The mass of the subsystem $m$ in the IG-CtP configuration $\mathrm{V}_0^{\mathrm{b}}$ reads

$$m = \rho_0\, V_0\,. \tag{6.221}$$

Substitution of Expression (6.221) into the right side of Equation (6.220) yields

$$Q = \mathrm{m}\, c\, \dot{T}^{\mathrm{aver}} \,. \tag{6.222}$$

Equation (6.222) has the rate form, is based on field quantities and it is written for an arbitrary subsystem $m$, whereas Equation (6.214) is for the whole system $\mathrm{m^b}$ and is for comparison of two states of equilibrium. However, Comparison of Expressions (6.214) and (6.222) shows that Definition (6.205) for the specific heat capacity $c$ has the same physical background as the specific heat capacity at constant volume $c_\mathrm{V}$ in Equation (6.214) for thermostatics. It is important to note that the unit for the amount of heat $\mathrm{d}Q^*$ is Nm, whereas the unit for heat input rate $Q$ is Nm/s. Besides the evaluation of the specific heat capacity $c$, the following approach can be taken.

When thermostatics in Section 5.3 was studied, the second law of thermodynamics was written in the form of Axiom (5.3) [**AX 4**]:

**AX 4**
$$đQ^* = T\, \mathrm{d}S \,. \tag{6.223}$$

Equation (6.223) is for a homogeneous system $\mathrm{m^b}$. Comparison of Expressions (6.214) and (6.223) yields

$$\mathrm{m^b}\, c_\mathrm{V}\, \mathrm{d}T = T\, \mathrm{d}S \quad \Rightarrow \quad c_\mathrm{V} = \frac{T}{\mathrm{m^b}} \frac{\mathrm{d}S}{\mathrm{d}T} = T \frac{\mathrm{d}s}{\mathrm{d}T} \,, \tag{6.224}$$

where the specific entropy $s = S/\mathrm{m^b}$ [cf. Equation (5.12)] is introduced. The preceding manipulation utilized Expression (6.214), which was written for a constant volume. When gas within a cylinder was studied in **Example 5.1**, the internal energy $U$ was shown to depend on the volume of the gas $V$ and the entropy $S$. State Equation $(5.9)_2$ creates an expression for the relationship between the absolute temperature $T$, the volume of the gas $V$ and the entropy $S$. Since here case of constant volume is studied, the specific entropy $s$ in Manipulation $(6.224)_2$ depends only on the temperature $T$, and therefore the rightmost term of Manipulation $(6.224)_2$ can be replaced with the partial derivative $\partial s/\partial T$. Furthermore, Manipulation $(6.224)_2$ is extended to cover processes in general and for arbitrary nonhomogeneous subsystems $m$. Therefore, the following is written:

$$c = T \frac{\partial s}{\partial T} \,. \tag{6.225}$$

It is worth noting that the preceding "derivation" is not exact physical and mathematical proof, but it motivates the definition of the specific heat capacity $c$ as given in Definition (6.205).

State Equations (6.172) and $(6.192)_1$ for solids and for fluids read

$$s = -\frac{\partial \psi(\boldsymbol{\varepsilon} - \boldsymbol{\varepsilon}^\mathrm{i},\, \boldsymbol{\alpha}, T, h(\vec{X}))}{\partial T} \quad \text{and} \quad s = -\frac{\partial \psi(\rho, T)}{\partial T} \,. \tag{6.226}$$

Based on Expressions (6.225) and (6.226), we arrive at the following:

$$c = -T\,\frac{\partial^2\psi}{\partial T^2}\,. \tag{6.227}$$

Result (6.227) has the form at which we are aiming, as it covers solids and fluids and it equals Definition (6.205) for the specific heat capacity $c$.

### 6.5.3 Heat Equation for Fluids

In this section we derive the heat equation for fluids. Energy Equation (6.89) is

$$\rho\,\dot{u} = p\,\frac{\dot{\rho}}{\rho} + \boldsymbol{\sigma}^{\mathrm{d}} : \overset{\circ}{\boldsymbol{\varepsilon}} + \rho\,r - \vec{\nabla}\cdot\vec{q}\,. \tag{6.228}$$

Definition for the specific Helmholtz free energy $\psi$, Definition (6.191), is

$$\psi(\rho,T) \overset{\text{Legendre}}{=} u(\rho,s) - T\,s, \qquad \text{with} \qquad T = \frac{\partial u(\rho,s)}{\partial s}\,. \tag{6.229}$$

Transformation $(6.229)_1$ gives

$$\dot{u} = \dot{\psi} + \dot{T}\,s + T\,\dot{s}\,. \tag{6.230}$$

Substitution of Derivative (6.230) into Energy Equation (6.228) yields

$$\rho\,\dot{\psi} + \rho\,\dot{T}\,s + \rho\,T\,\dot{s} - p\,\frac{\dot{\rho}}{\rho} - \boldsymbol{\sigma}^{\mathrm{d}} : \overset{\circ}{\boldsymbol{\varepsilon}} - \rho\,r + \vec{\nabla}\cdot\vec{q} = 0\,. \tag{6.231}$$

Since the specific Helmholtz free energy $\psi$ has the appearance $\psi(\rho,T)$ [cf. Expression (6.189)], we arrive at the following:

$$\dot{\psi}(\rho,T) = \frac{\partial\psi(\ldots)}{\partial\rho}\,\dot{\rho} + \frac{\partial\psi(\ldots)}{\partial T}\,\dot{T}\,. \tag{6.232}$$

State Equations (6.193) and $(6.192)_1$ are

$$p = \rho^2\,\frac{\partial\psi(\rho,T)}{\partial\rho} \qquad \text{and} \qquad s = -\,\frac{\partial\psi(\rho,T)}{\partial T}\,. \tag{6.233}$$

By multiplying Derivative (6.232) by the density $\rho$ and by taking State Equations (6.233) into consideration, Derivative (6.232) yields

$$\rho\,\dot{\psi}(\rho,T) = \frac{1}{\rho}\,p\,\dot{\rho} - \rho\,s\,\dot{T}\,. \tag{6.234}$$

State Equation $(6.233)_2$ and Derivative (6.232) give

$$\dot{s} = -\,\frac{\partial\dot{\psi}}{\partial T} = -\,\frac{\partial^2\psi}{\partial T\,\partial\rho}\,\dot{\rho} - \frac{\partial^2\,\psi}{\partial T^2}\,\dot{T}\,. \tag{6.235}$$

Substitution of Derivatives (6.234) and (6.235) into Equation (6.231) gives

$$\frac{p}{\rho}\,\dot{\rho} - \rho\, s\dot{T} + \rho\dot{T}\, s - \rho T\left[\frac{\partial^2\psi}{\partial T\,\partial\rho}\,\dot{\rho} + \frac{\partial^2\psi}{\partial T^2}\,\dot{T}\right] - p\,\frac{\dot{\rho}}{\rho} - \boldsymbol{\sigma}^{\mathrm{d}} : \overset{\circ}{\boldsymbol{\varepsilon}} - \rho\, r + \vec{\nabla}\cdot\vec{q} = 0\,. \tag{6.236}$$

By rearranging the terms, Expression (6.236) can be cast into the following appearance:

$$\vec{\nabla}\cdot\vec{q} = \boldsymbol{\sigma}^{\mathrm{d}} : \overset{\circ}{\boldsymbol{\varepsilon}} + \rho\, r + \rho\, T\left[\frac{\partial^2\psi}{\partial T\,\partial\rho}\,\dot{\rho} + \frac{\partial^2\psi}{\partial T^2}\,\dot{T}\right]\,. \tag{6.237}$$

Equation (6.237) is the heat equation for fluids for a case when the material model is expressed by the specific Helmholtz free energy $\psi$.

## 6.6 Clausius–Duhem Inequality

In this section we derive the Clausius–Duhem inequality, which is a combination of the first and second laws of thermodynamics. Since the local forms of the first law of thermodynamics take different appearances for different states of material, the Clausius–Duhem inequality for different material behavior is derived in separate sections.

### 6.6.1 Clausius–Duhem Inequality for Solids

Here we derive the Clausius–Duhem inequality for a solid material. The local forms of the first and second laws of thermodynamics are recalled. They are [see Expressions (6.37) and (6.119)]

$$\rho_0\,\dot{u} = \boldsymbol{\sigma} : \dot{\boldsymbol{\varepsilon}} + \rho_0\, r - \vec{\nabla}\cdot\vec{q} \tag{6.238}$$

and

$$\Phi := \rho_0\, T\,\dot{s} + \vec{\nabla}\cdot\vec{q} - \frac{\vec{\nabla}T}{T}\cdot\vec{q} - \rho_0\, r \quad (\geq 0)\,. \tag{6.239}$$

Substitution of $\vec{\nabla}\cdot\vec{q}$ evaluated from Equation (6.238) into Definition (6.239) gives

$$\Phi = \boldsymbol{\sigma} : \dot{\boldsymbol{\varepsilon}} - \rho_0\,\dot{u} + \rho_0\, T\,\dot{s} - \frac{\vec{\nabla}T}{T}\cdot\vec{q} \quad (\geq 0)\,. \tag{6.240}$$

Inequality (6.240) reads

$$\boldsymbol{\sigma} : \dot{\boldsymbol{\varepsilon}} - \rho_0\,\dot{u} + \rho_0\, T\,\dot{s} - \frac{\vec{\nabla}T}{T}\cdot\vec{q} \geq 0\,. \tag{6.241}$$

Equation (6.241) is the Clausius–Duhem inequality for solids. The material model has to obey Clausius–Duhem Inequality (6.241). The validation can be done either numerically during the computation procedure or analytically, the latter of course being much more convenient. In both cases, Form (6.241) is not very useful and must therefore be recast into a new appearance. This recasting procedure follows below.

Following the concept by Truesdell and Noll ([111], p. 295), the dissipation $\Phi$ is assumed to be separable into the mechanical dissipation, denoted by $\Phi_{\text{mech}}$, and into the thermal dissipation, denoted by $\Phi_{\text{ther}}$, as follows:

$$\Phi = \Phi_{\text{mech}} + \Phi_{\text{ther}}\,, \tag{6.242}$$

where

$$\Phi_{\text{mech}} := \boldsymbol{\sigma} : \dot{\boldsymbol{\varepsilon}} - \rho_0\,\dot{u} + \rho_0\,T\,\dot{s} \tag{6.243}$$

and

$$\Phi_{\text{ther}} := -\frac{\vec{\nabla}T}{T}\cdot\vec{q}\,. \tag{6.244}$$

They also proposed a stronger assumption, requiring separately

$$\Phi_{\text{mech}} \geq 0 \qquad \text{and} \qquad \Phi_{\text{ther}} \geq 0\,. \tag{6.245}$$

The first of these inequalities corresponds to the physical observation that a substance at uniform temperature free from sources of heat may consume mechanical energy but cannot give it out. The second inequality corresponds to the fact that heat does not flow spontaneously from the colder to the hotter parts of the system (body). Postulate (6.245) is a special case of the Clausius–Duhem inequality or the principle of dissipation. See Truesdell and Noll ([111], p. 295).

Rate (6.170) reads

$$\rho_0\,\dot{u}(\,\boldsymbol{\varepsilon} - \boldsymbol{\varepsilon}^{\text{i}}, \boldsymbol{\alpha}, s, h(\vec{X})\,) \;= \boldsymbol{\sigma} : (\,\dot{\boldsymbol{\varepsilon}} - \dot{\boldsymbol{\varepsilon}}^{\text{i}}\,) - \;\boldsymbol{\beta} : \dot{\boldsymbol{\alpha}} + \rho_0\,T\,\dot{s}\,. \tag{6.246}$$

Substitution of Rate (6.246) into Expression (6.240) gives

$$\Phi = \boldsymbol{\sigma} : \dot{\boldsymbol{\varepsilon}} - \boldsymbol{\sigma} : (\,\dot{\boldsymbol{\varepsilon}} - \dot{\boldsymbol{\varepsilon}}^{\text{i}}\,) + \boldsymbol{\beta} : \dot{\boldsymbol{\alpha}} - \rho_0\,T\,\dot{s} + \rho_0\,T\,\dot{s} - \frac{\vec{\nabla}T}{T}\cdot\vec{q} \qquad (\geq 0)\,, \tag{6.247}$$

which yields

$$\Phi = \boldsymbol{\sigma} : \dot{\boldsymbol{\varepsilon}}^{\text{i}} + \boldsymbol{\beta} : \dot{\boldsymbol{\alpha}} - \frac{\vec{\nabla}T}{T}\cdot\vec{q} \qquad (\geq 0)\,. \tag{6.248}$$

Inequality (6.248) is

$$\boldsymbol{\sigma} : \dot{\boldsymbol{\varepsilon}}^{\text{i}} + \boldsymbol{\beta} : \dot{\boldsymbol{\alpha}} - \frac{\vec{\nabla}T}{T}\cdot\vec{q} \geq 0\,, \tag{6.249}$$

which is the Clausius–Duhem inequality.

The preceding derivation of the Clausius–Duhem inequality assumed that the material model is expressed by the specific internal energy $u$. Inequality (6.249) is also obtained when the material model is expressed by the specific

Helmholtz free energy $\psi$, as Expression (G.8) of Appendix G shows. It is reasonable to assume that the same result holds when the material model is expressed by the specific Gibbs free energy and, therefore, Inequality (6.249) is a general one.

Forms (6.241) and (6.249) are two different formulations for the Clausius–Duhem inequality. The difference between them is that the latter introduces a certain set of internal variables ($\boldsymbol{\varepsilon}^{\mathrm{i}}$, $\boldsymbol{\alpha}$). Form (6.249) is used with the principle of maximum dissipation, as Section 6.7 shows. By following Separation (6.242) one will obtain

$$\Phi_{\mathrm{mech}} = \boldsymbol{\sigma} : \dot{\boldsymbol{\varepsilon}}^{\mathrm{i}} + \boldsymbol{\beta} : \dot{\boldsymbol{\alpha}} \quad (\geq 0) \tag{6.250}$$

and

$$\Phi_{\mathrm{ther}} = -\frac{\vec{\nabla} T}{T} \cdot \vec{q} \quad (\geq 0)\,. \tag{6.251}$$

According to Expressions (6.248)–(6.251), the terms in the Clausius–Duhem inequality have the form "force times flux." In continuum thermodynamics the rates of the state variables are called fluxes, whereas the corresponding state functions are called forces. In the mechanical part of the Clausius–Duhem inequality the rates of the internal variables are present. Thus the explicit form of Inequality (6.250) depends on the set of internal variables. The material model under consideration determines the set of internal variables. The thermal part of the Clausius–Duhem inequality [Inequality (6.251)] always has the same appearance. The form of the Clausius–Duhem inequality is discussed in more detail at the end of Section 7.6.1. Therefore, if the set of internal variables also contained the variables $\vec{\alpha}^{1}$ and $\alpha^2$, Clausius–Duhem Inequality (6.249) would be replaced with

$$\boldsymbol{\sigma} : \dot{\boldsymbol{\varepsilon}}^{\mathrm{i}} + \boldsymbol{\beta} : \dot{\boldsymbol{\alpha}} + \vec{\beta}^{1} \cdot \dot{\vec{\alpha}}^{1} + \beta^2\, \dot{\alpha}^2 - \frac{\vec{\nabla} T}{T} \cdot \vec{q} \geq 0\,. \tag{6.252}$$

### 6.6.2 Clausius–Duhem Inequality for Fluids

In this section we derive the Clausius–Duhem inequality for fluids. The local forms of the first and second laws of thermodynamics are recalled. They are [see Expressions (6.89) and (6.145)]

$$\rho\, \dot{u} = p\, \frac{\dot{\rho}}{\rho} + \boldsymbol{\sigma}^{\mathrm{d}} : \overset{\circ}{\boldsymbol{\varepsilon}} + \rho\, r - \vec{\nabla} \cdot \vec{q} \tag{6.253}$$

and

$$\Phi := \rho\, T\, \dot{s} + \vec{\nabla} \cdot \vec{q} - \frac{\vec{\nabla} T}{T} \cdot \vec{q} - \rho\, r \quad (\geq 0\,)\,. \tag{6.254}$$

Substitution of $\vec{\nabla} \cdot \vec{q}$ evaluated from Equation (6.253) into Definition (6.254) gives

$$\Phi = p\, \frac{\dot{\rho}}{\rho} + \boldsymbol{\sigma}^{\mathrm{d}} : \overset{\circ}{\boldsymbol{\varepsilon}} - \rho\, \dot{u} + \rho\, T\, \dot{s} - \frac{\vec{\nabla} T}{T} \cdot \vec{q} \quad (\geq 0)\,. \tag{6.255}$$

Inequality (6.255) reads

$$p\,\frac{\dot{\rho}}{\rho} + \boldsymbol{\sigma}^{\mathrm{d}} : \overset{\circ}{\boldsymbol{\varepsilon}} - \rho\,\dot{u} + \rho\,T\,\dot{s} - \frac{\vec{\nabla}T}{T}\cdot\vec{q} \geq 0\,. \tag{6.256}$$

Inequality (6.256) is the Clausius–Duhem inequality for fluids.

The expression for the rate of the specific internal energy $\dot{u}$, Expression (6.190), is

$$\rho\,\dot{u}(\rho, s) = \frac{1}{\rho}\,p\,\dot{\rho} + \rho\,T\,\dot{s}\,. \tag{6.257}$$

Substitution of Derivative (6.257) into Expression (6.255) leads to

$$\Phi = p\,\frac{\dot{\rho}}{\rho} + \boldsymbol{\sigma}^{\mathrm{d}} : \overset{\circ}{\boldsymbol{\varepsilon}} - \frac{1}{\rho}\,p\,\dot{\rho} - \rho\,T\,\dot{s} + \rho\,T\,\dot{s} - \frac{\vec{\nabla}T}{T}\cdot\vec{q} \quad (\geq 0)\,, \tag{6.258}$$

which yields

$$\Phi = \boldsymbol{\sigma}^{\mathrm{d}} : \overset{\circ}{\boldsymbol{\varepsilon}} - \frac{\vec{\nabla}T}{T}\cdot\vec{q} \quad (\geq 0)\,. \tag{6.259}$$

Inequality (6.259) reads

$$\boldsymbol{\sigma}^{\mathrm{d}} : \overset{\circ}{\boldsymbol{\varepsilon}} - \frac{\vec{\nabla}T}{T}\cdot\vec{q} \geq 0\,, \tag{6.260}$$

which is another form for the Clausius–Duhem inequality for fluids. As discussed in the previous section, Truesdell and Noll ([111], p. 295) made a stronger assumption than Inequality (6.259). They assumed that the mechanical part and the thermal part of Inequality (6.259) are separately nonnegative. This is

$$\Phi_{\mathrm{mech}} = \boldsymbol{\sigma}^{\mathrm{d}} : \overset{\circ}{\boldsymbol{\varepsilon}} \quad (\geq 0) \tag{6.261}$$

and

$$\Phi_{\mathrm{ther}} = -\frac{\vec{\nabla}T}{T}\cdot\vec{q} \quad (\geq 0)\,. \tag{6.262}$$

## 6.7 Principle of Maximum Dissipation Provides the Normality Rule

This section studies the principle of maximum dissipation and derives its consequence: the normality rule. The principle of maximum dissipation is first formulated for the mechanical response of a solid material. Investigations of a thermal problem and of fluids are brief and are based on the study of solid material. The reader who is interested in thermal problems or fluids is referred first to Sections 6.7.3 and 6.7.7. The present chapter follows the concept given by Santaoja [95].

Section 6.7.1 studies the time-dependent deformation of a solid material when the material model is expressed in terms of fluxes. Section 6.7.2 studies the time-dependent deformation of a solid material when the material model is expressed in terms of forces. Section 6.7.3 gives the normality rule for a thermal problem. Section 6.7.4 introduces the normality rule for time-independent thermoplasticity. Sections 6.7.5 and 6.7.6 make some comments on the principle of maximum dissipation. Section 6.7.7 studies the normality rule for fluids.

The principle of maximum dissipation is based on the principle of maximal rate of entropy production proposed by Ziegler ([120], p. 134). It should be pointed out that this principle as proposed by Ziegler (see, e.g., [121], pp. 271 and 272) is not (yet?) a basic law of physics, contrary to those discussed in the previous chapters. In Ziegler's own words, the principle of maximal rate of entropy production is quite general (Ziegler and Wehrli [122], p. 186).

According to Ziegler ([121], p. 272), the physical foundation of this principle is as follows: from the physical point of view, this principle is particularly appealing, since it may be considered as an extension of the second fundamental law. In fact, if a closed system tends towards its state of maximal entropy, it seems reasonable that the rate of entropy increase (the specific entropy rate) under prescribed forces would take a maximum value, that is, the system should approach its final state along the fastest (shortest) possible path.

The phrase "the principle of maximal rate of entropy production" is replaced nowadays by an expression referring to maximum dissipation. Therefore the phrase "the principle of maximum dissipation" is adopted here.

## 6.7.1 Normality Rule for Time-Dependent Deformation When Material Model $\varphi = \varphi(\dot{\boldsymbol{\varepsilon}}^{\mathrm{i}}, \dot{\boldsymbol{\alpha}}, \ldots)$ Is Expressed by Fluxes

This section derives the normality rule for time-dependent deformation when the material model is expressed by the fluxes. Here the separation by Truesdell and Noll ([111], p. 295) is assumed to hold [see Separation (6.245)].

Equations (6.248), (6.250), and (6.251) are recalled:

$$\Phi = \boldsymbol{\sigma} : \dot{\boldsymbol{\varepsilon}}^{\mathrm{i}} + \boldsymbol{\beta} : \dot{\boldsymbol{\alpha}} - \frac{\vec{\nabla} T}{T} \cdot \vec{q} \quad (\geq 0) \tag{6.263}$$

and

$$\Phi_{\mathrm{mech}} = \boldsymbol{\sigma} : \dot{\boldsymbol{\varepsilon}}^{\mathrm{i}} + \boldsymbol{\beta} : \dot{\boldsymbol{\alpha}} \quad (\geq 0) \quad \text{and} \quad \Phi_{\mathrm{ther}} = -\frac{\vec{\nabla} T}{T} \cdot \vec{q} \quad (\geq 0)\,. \tag{6.264}$$

Investigation of the expression for the dissipation $\Phi$ in Equation (6.263) shows that $\Phi$ is dependent upon both forces and fluxes (processes). For example, in the expression $\boldsymbol{\sigma} : \dot{\boldsymbol{\varepsilon}}^{\mathrm{i}}$ the term $\boldsymbol{\sigma}$ represents the force and $\dot{\boldsymbol{\varepsilon}}^{\mathrm{i}}$ describes the flux. Furthermore, Ziegler ([120], p. 129) assumes the existence of the specific dissipation function (see also Ziegler [121], p. 76)

$$\varphi = \varphi(\dot{\boldsymbol{\varepsilon}}^{\mathrm{i}}, \dot{\boldsymbol{\alpha}}, \vec{q}; \boldsymbol{\varepsilon} - \boldsymbol{\varepsilon}^{\mathrm{i}}, \boldsymbol{\alpha}, T, h)\,. \tag{6.265}$$

The concept of the specific dissipation function $\varphi$ is that in case of an actual process (i.e., when the maximum is present), it contains the same information (except for the density $\rho$ ) about the state and the process as the dissipation $\Phi$, but the arguments of $\varphi$ are only fluxes $(\dot{\boldsymbol{\varepsilon}}^{\mathrm{i}}, \dot{\boldsymbol{\alpha}}, \vec{q})$ [and state $(\boldsymbol{\varepsilon}-\boldsymbol{\varepsilon}^{\mathrm{i}}, \boldsymbol{\alpha}, T, h)$], whereas the expression for $\Phi$ contains also the conjugate variables $[\boldsymbol{\sigma}, \boldsymbol{\beta}, -(\vec{\nabla}T)/T]$. This can be seen in Equations (6.263) and (6.265). The main concept of introducing the specific dissipation function $\varphi$ is that it is the explicit form of the material model. The specific dissipation function $\varphi$ is defined by the following [see Ziegler [120], Eq. (4.3); [121], Eq. (5.1)]:

$$\text{For an actual process} \qquad \varphi := \frac{1}{\rho_0}\,\Phi \quad \Rightarrow \quad \rho_0\,\varphi - \Phi = 0\,. \tag{6.266}$$

Ziegler did not cast his earlier-presented concept into an exact mathematical framework. The authors therefore propose the following formulation for the principle of maximum dissipation (the principle of maximal rate of entropy production): the process is investigated at a certain state $(\boldsymbol{\varepsilon}-\boldsymbol{\varepsilon}^{\mathrm{i}}, \boldsymbol{\alpha}, T, h)$ and the values for the fluxes $(\dot{\boldsymbol{\varepsilon}}^{\mathrm{i}}, \dot{\boldsymbol{\alpha}}, \vec{q})$ have to be determined in order to maximize the dissipation $\phi$. The state gives the values for the state functions $[\boldsymbol{\sigma}, \boldsymbol{\beta}, -(\vec{\nabla}T)/T]$ as follows: State Equations (6.174) give the values for the forces $\boldsymbol{\sigma}$ and $\boldsymbol{\beta}$. Since the state $(\boldsymbol{\varepsilon}-\boldsymbol{\varepsilon}^{\mathrm{i}}, \boldsymbol{\alpha}, T, h)$ is known, the temperature $T$ is known, and therefore the state function $-(\vec{\nabla}T)/T$ is also known. Thus the values for the state functions $[\boldsymbol{\sigma}, \boldsymbol{\beta}, -(\vec{\nabla}T)/T]$ are known.

Based on the foregoing discussion, the problem can be expressed in the following way: it is assumed that the state is known; that is, the set $(\boldsymbol{\varepsilon}-\boldsymbol{\varepsilon}^{\mathrm{i}}, \boldsymbol{\alpha}, T, h)$ is assumed to be known. This implies that the values for the forces $[\boldsymbol{\sigma}, \boldsymbol{\beta}, -(\vec{\nabla}T)/T]$ are known. The question is, What are the magnitudes of the fluxes $(\dot{\boldsymbol{\varepsilon}}^{\mathrm{i}}, \dot{\boldsymbol{\alpha}}, \vec{q})$ that maximize the dissipation $\Phi$? At the same time also Definition $(6.266)_1$ has to be satisfied.

To make the evaluation shorter, General Problem (6.263) is not evaluated [it is done in Section 6.7.5], but the Mechanical $(6.264)_1$ and Thermal $(6.264)_2$ parts are studied separately. The mechanical part of the problem [Expression $(6.264)_1$] is considered first. Based on the preceding the principle of maximum dissipation is written in the following mathematical form:

maximise with respect to the fluxes $(\boldsymbol{\varepsilon}^{\mathrm{i}}, \boldsymbol{\alpha})$,

$$\Phi_{\text{mech}}(\dot{\boldsymbol{\varepsilon}}^{\mathrm{i}}, \dot{\boldsymbol{\alpha}}, \boldsymbol{\sigma}, \boldsymbol{\beta}) = \boldsymbol{\sigma} : \dot{\boldsymbol{\varepsilon}}^{\mathrm{i}} + \boldsymbol{\beta} : \dot{\boldsymbol{\alpha}}\,, \tag{6.267}$$

subject to

$$\tau_{\text{mech}} = \rho_0\,\varphi_{\text{mech}}(\dot{\boldsymbol{\varepsilon}}^{\mathrm{i}}, \dot{\boldsymbol{\alpha}}; \boldsymbol{\varepsilon}-\boldsymbol{\varepsilon}^{\mathrm{i}}, \boldsymbol{\alpha}, T, h) - \Phi_{\text{mech}}(\dot{\boldsymbol{\varepsilon}}^{\mathrm{i}}, \dot{\boldsymbol{\alpha}}, \boldsymbol{\sigma}, \boldsymbol{\beta}) = 0\,, \tag{6.268}$$

where $\tau_{\text{mech}} = 0$ is a constraint and $\varphi_{\text{mech}}$ is the specific dissipation function for mechanical behavior. It is worth noting that the specific dissipation function $\varphi$ is dependent on the fluxes present in the Clausius–Duhem inequality [see Inequality $(6.264)_1$] and on the state variables [see Equation (6.162)]. In this case this means that the function $\varphi_{\text{mech}}$ is dependent on sets $(\dot{\boldsymbol{\varepsilon}}^{\text{i}}, \dot{\boldsymbol{\alpha}})$ and $(\boldsymbol{\varepsilon} - \boldsymbol{\epsilon}^{\text{i}}, \boldsymbol{\alpha}, T, h)$. The general maximization problem (where the dissipation $\Phi$ is not separated) follows the preceding concept.

Both $\Phi_{\text{mech}}$ and $\tau_{\text{mech}}$ are assumed to have at least continuous second partial derivatives with respect to the arguments $(\dot{\boldsymbol{\varepsilon}}^{\text{i}}, \dot{\boldsymbol{\alpha}})$. It should be pointed out that also Inequality $(6.264)_1$ must be satisfied. Applying Luenberger ([55], p. 225), the first-order sufficient condition for the point $(\dot{\boldsymbol{\varepsilon}}^{\text{i}}, \dot{\boldsymbol{\alpha}})$ to be a local maximum is

$$\frac{\partial}{\partial \dot{\boldsymbol{\varepsilon}}^{\text{i}}} (\Phi_{\text{mech}} + \lambda\, \tau_{\text{mech}}) = \mathbf{0}\,, \qquad \frac{\partial}{\partial \dot{\boldsymbol{\alpha}}} (\Phi_{\text{mech}} + \lambda\, \tau_{\text{mech}}) = \mathbf{0} \quad \text{and} \quad \tau_{\text{mech}} = 0\,, \tag{6.269}$$

where $\lambda$ is the Lagrange multiplier. As mentioned by Arfken ([7], p. 946), the method based on Lagrange multipliers will fail if in Expressions $(6.269)_1$ and $(6.269)_2$ the coefficients of $\lambda$ vanish at the extremum (i.e., $\tau_{\text{mech}} = 0$). Therefore, also special points where

$$\frac{\partial \tau_{\text{mech}}}{\partial \dot{\boldsymbol{\varepsilon}}^{\text{i}}} = \mathbf{0} \qquad \text{and} \qquad \frac{\partial \tau_{\text{mech}}}{\partial \dot{\boldsymbol{\alpha}}} = \mathbf{0} \tag{6.270}$$

must be studied. The aforementioned coefficients of $\lambda$ are

$$\frac{\partial}{\partial \dot{\boldsymbol{\varepsilon}}^{\text{i}}} \tau_{\text{mech}} \qquad \text{and} \qquad \frac{\partial}{\partial \dot{\boldsymbol{\alpha}}} \tau_{\text{mech}}\,. \tag{6.271}$$

The preceding indicates that there are two different cases for evaluation of the local maximum; utilization of Expression (6.269) referred to as Case A, and the special case described by Expression (6.270) referred to as Case B.

<u>Starting with Case A:</u>

Substitution of Equations (6.267) and (6.268) into Equation $(6.269)_1$ gives

$$\frac{\partial}{\partial \dot{\boldsymbol{\varepsilon}}^{\text{i}}} \Big\{ (\boldsymbol{\sigma} : \dot{\boldsymbol{\varepsilon}}^{\text{i}} + \boldsymbol{\beta} : \dot{\boldsymbol{\alpha}}) + \lambda \Big[ \rho_0\, \varphi_{\text{mech}} - (\boldsymbol{\sigma} : \dot{\boldsymbol{\varepsilon}}^{\text{i}} + \boldsymbol{\beta} : \dot{\boldsymbol{\alpha}}) \Big] \Big\} = 0\,. \tag{6.272}$$

Equation (6.272) leads to

$$\boldsymbol{\sigma} + \lambda \left[ \rho_0 \frac{\partial \varphi_{\text{mech}}}{\partial \dot{\boldsymbol{\varepsilon}}^{\text{i}}} - \boldsymbol{\sigma} \right] = \mathbf{0}\,, \qquad \text{which gives} \qquad \boldsymbol{\sigma} = \frac{\lambda}{\lambda - 1}\, \rho_0 \frac{\partial \varphi_{\text{mech}}}{\partial \dot{\boldsymbol{\varepsilon}}^{\text{i}}}\,. \tag{6.273}$$

Similarly, Equation $(6.269)_2$ gives

$$\boldsymbol{\beta} = \frac{\lambda}{\lambda - 1}\, \rho_0 \frac{\partial \varphi_{\text{mech}}}{\partial \dot{\boldsymbol{\alpha}}}\,. \tag{6.274}$$

By substituting the results in Equations $(6.273)_2$ and (6.274) into Equation $(6.269)_3$ and reordering the obtained equation, the following result is obtained:

$$\varphi_{\text{mech}}(\dot{\boldsymbol{\varepsilon}}^{\text{i}}, \dot{\boldsymbol{\alpha}}; \boldsymbol{\varepsilon} - \boldsymbol{\varepsilon}^{\text{i}}, \boldsymbol{\alpha}, T, h) = \frac{\lambda}{\lambda - 1} \left( \frac{\partial \varphi_{\text{mech}}}{\partial \dot{\boldsymbol{\varepsilon}}^{\text{i}}} : \dot{\boldsymbol{\varepsilon}}^{\text{i}} + \frac{\partial \varphi_{\text{mech}}}{\partial \dot{\boldsymbol{\alpha}}} : \dot{\boldsymbol{\alpha}} \right) . \qquad (6.275)$$

It is worth noting that the value of the Lagrange multiplier $\lambda$ is dependent on the set $(\boldsymbol{\varepsilon} - \boldsymbol{\varepsilon}^{\text{i}}, \boldsymbol{\alpha}, T, h)$. This is based on the definition of the maximization problem, which assumed that the value of the set $(\boldsymbol{\varepsilon} - \boldsymbol{\varepsilon}^{\text{i}}, \boldsymbol{\alpha}, T, h)$ is known and that the values for the corresponding fluxes $(\dot{\boldsymbol{\varepsilon}}^{\text{i}}, \dot{\boldsymbol{\alpha}})$ have to be determined. This implies that for a certain set $(\boldsymbol{\varepsilon} - \boldsymbol{\varepsilon}^{\text{i}}, \boldsymbol{\alpha}, T, h)$, a unique value for $\lambda$ is obtained. Thus, the following holds: $\lambda = \lambda(\boldsymbol{\varepsilon} - \boldsymbol{\varepsilon}^{\text{i}}, \boldsymbol{\alpha}, T, h)$.

By extending the definition for homogeneous functions given, for example, by Widder ([117], pp. 19 and 20), the following is achieved: a function $\theta(x, y, z, u, v)$ is homogeneous of degree $\omega$ in variables $(x,\ y,\ z)$ in a region $R$, if and only if, for $(x,\ y,\ z)$ in $R$ and for every positive value of $k$, the following holds:

$$\theta(k\,x, k\,y, k\,z, u, v) := k^{\omega}\,\theta(x, y, z, u, v) . \qquad (6.276)$$

Sometimes the definition is assumed to hold for every real $k$, and if the values of $k$ are restricted to being positive, the function $\theta(x, y, z, u, v)$ is said to be a positive homogeneous function.

Euler's theorem on homogeneous functions (see original form in Widder [117], p. 20) for the preceding extended definition reads

$$\omega\,\theta(x, y, z, u, v) = \frac{\partial \theta}{\partial x}\,x + \frac{\partial \theta}{\partial y}\,y + \frac{\partial \theta}{\partial z}\,z . \qquad (6.277)$$

In the special case that the Lagrange multiplier $\lambda$ is a constant, the extended Euler's theorem for homogeneous functions [Theorem (6.277)] and Result (6.275) indicate that the specific dissipation function $\varphi_{\text{mech}}$ is a homogeneous function of degree $(\lambda - 1)/\lambda$ in the variables $(\dot{\boldsymbol{\varepsilon}}^{\text{i}}, \dot{\boldsymbol{\alpha}})$.

The following notation is introduced:

$$\mu := \frac{\lambda}{\lambda - 1}, \qquad \text{which gives} \qquad \lambda = \frac{\mu}{\mu - 1} . \qquad (6.278)$$

Equation $(6.278)_2$ shows that the multiplier $\mu = \mu(\boldsymbol{\varepsilon} - \boldsymbol{\varepsilon}^{\text{i}}, \boldsymbol{\alpha}, T, h)$ can be any real number excluding $\mu = 1$. This means that in Case A, the specific dissipation function $\varphi_{\text{mech}}$ cannot be a homogeneous function of $1/\mu = 1$.

Continuing with Case B:

The candidates for the extremum points defined by Equation (6.270) are investigated next. Substitution of Equation (6.268) into Equation $(6.270)_1$ gives

$$\rho_0\,\frac{\partial \varphi_{\text{mech}}}{\partial \dot{\boldsymbol{\varepsilon}}^{\text{i}}} - \boldsymbol{\sigma} = 0 , \qquad \text{which gives} \qquad \boldsymbol{\sigma} = \rho_0\,\frac{\partial \varphi_{\text{mech}}}{\partial \dot{\boldsymbol{\varepsilon}}^{\text{i}}} . \qquad (6.279)$$

Correspondingly, the following is obtained:

$$\boldsymbol{\beta} = \rho_0 \frac{\partial \varphi_{\text{mech}}}{\partial \dot{\boldsymbol{\alpha}}} \, . \tag{6.280}$$

Instead of Expression (6.275), Case B gives

$$\varphi_{\text{mech}}(\dot{\boldsymbol{\varepsilon}}^{\text{i}}, \dot{\boldsymbol{\alpha}}; \boldsymbol{\varepsilon} - \boldsymbol{\varepsilon}^{\text{i}}, \boldsymbol{\alpha}, T, h) = \frac{\partial \varphi_{\text{mech}}}{\partial \dot{\boldsymbol{\varepsilon}}^{\text{i}}} : \dot{\boldsymbol{\varepsilon}}^{\text{i}} + \frac{\partial \varphi_{\text{mech}}}{\partial \dot{\boldsymbol{\alpha}}} : \dot{\boldsymbol{\alpha}} \, . \tag{6.281}$$

Comparison of Equations $(6.279)_2$, (6.280), and (6.281) with Equations $(6.273)_2$, (6.274), and (6.275) shows that the special points defined by Case B give the same solution as Case A, except that $\varphi_{\text{mech}}$ is a homogeneous function of degree $1 \; (= 1/\mu)$.

Concluding from Cases A and B, the following can be said:

At the start of this chapter, the following problem was set: the state is known, which means that the set $(\boldsymbol{\varepsilon} - \boldsymbol{\varepsilon}^{\text{i}}, \boldsymbol{\alpha}, T, h)$ is known. Due to Equations $(6.174)_1$ and $(6.174)_2$, this implies that the values for the forces $(\boldsymbol{\sigma}, \boldsymbol{\beta})$ are known. The magnitudes of the fluxes $(\dot{\boldsymbol{\varepsilon}}^{\text{i}}, \dot{\boldsymbol{\alpha}})$ have to be determined in order for the dissipation $\Phi$ to be maximized. At the same time also Definition $(6.266)_1$ has to be satisfied.

As a result the following normality rule was achieved:

$$\boldsymbol{\sigma} = \mu \, \rho_0 \frac{\partial \varphi_{\text{mech}}(\dot{\boldsymbol{\varepsilon}}^{\text{i}}, \dot{\boldsymbol{\alpha}}; \boldsymbol{\varepsilon} - \boldsymbol{\varepsilon}^{\text{i}}, \boldsymbol{\alpha}, T, h)}{\partial \dot{\boldsymbol{\varepsilon}}^{\text{i}}} \tag{6.282}$$

and

$$\boldsymbol{\beta} = \mu \, \rho_0 \frac{\partial \varphi_{\text{mech}}(\dot{\boldsymbol{\varepsilon}}^{\text{i}}, \dot{\boldsymbol{\alpha}}; \boldsymbol{\varepsilon} - \boldsymbol{\varepsilon}^{\text{i}}, \boldsymbol{\alpha}, T, h)}{\partial \dot{\boldsymbol{\alpha}}} \, . \tag{6.283}$$

The specific dissipation function $\varphi_{\text{mech}}$ has to obey the following condition:

$$\varphi_{\text{mech}}(\dot{\boldsymbol{\varepsilon}}^{\text{i}}, \dot{\boldsymbol{\alpha}}; \boldsymbol{\varepsilon} - \boldsymbol{\varepsilon}^{\text{i}}, \boldsymbol{\alpha}, T, h) = \mu \left( \frac{\partial \varphi_{\text{mech}}}{\partial \dot{\boldsymbol{\varepsilon}}^{\text{i}}} : \dot{\boldsymbol{\varepsilon}}^{\text{i}} + \frac{\partial \varphi_{\text{mech}}}{\partial \dot{\boldsymbol{\alpha}}} : \dot{\boldsymbol{\alpha}} \right) . \tag{6.284}$$

The first-order sufficient condition for the point $(\dot{\boldsymbol{\varepsilon}}^{\text{i}}, \dot{\boldsymbol{\alpha}}^{\text{i}})$ to be a local maximum is that Equations (6.282) and (6.283) hold and that the specific dissipation function $\varphi_{\text{mech}}$ is a homogeneous function of degree $1/\mu$. The latter property is obtained if the coefficient $\mu$ in Expression (6.284) is a constant. If the multiplier $\mu$ is not a constant but $\mu = \mu(\boldsymbol{\varepsilon} - \boldsymbol{\varepsilon}^{\text{i}}, \boldsymbol{\alpha}, T, h)$, the specific dissipation function $\varphi_{\text{mech}}$ is not a homogeneous function, and the value for $\mu$ is obtained from Equation (6.284). Equations (6.266), (6.282), and (6.283) show that the specific dissipation function $\varphi_{\text{mech}}$ is a scalar potential, and it is therefore also called the dissipation potential.

The second-order both necessary and sufficient conditions for a local maximum lead to matrices so extensive (see Luenberger [55], pp. 226 and 227) that investigating them is very complicated and it hardly provides any practical results.

However, in practice, when constitutive models are evaluated, the explicit form for the specific dissipation function $\varphi_{\mathrm{mech}}$ is assumed, and the forces $(\boldsymbol{\sigma}, \boldsymbol{\beta})$ are determined by Normality Rules (6.282) and (6.283).

If a material model contained more than two internal variables, as sketched in the Clausius–Duhem Inequality (6.252), a new normality rule similar to that of Expression (6.283) would be obtained for every internal variable–force pair. In such a case, the right side of Equation (6.284) would have new terms.

### 6.7.2 Normality Rule for Time-Dependent Deformation When Material Model $\varphi^{\mathrm{c}} = \varphi^{\mathrm{c}}(\boldsymbol{\sigma}, \boldsymbol{\beta}, \ldots)$ Is Expressed by Forces

This section gives the normality rule for time-dependent deformation when the material model is expressed as a function of state functions, that is, forces instead of the rates of internal variables, that is, fluxes.

Normality Rules (6.282) and (6.283) assume that the material model is expressed by the fluxes $(\dot{\boldsymbol{\varepsilon}}^{\mathrm{i}}, \dot{\boldsymbol{\alpha}})$. Very often, however, the conjugate forces $(\sigma, \boldsymbol{\beta})$ are the arguments of the material model. Therefore the mechanical part of the specific complementary dissipation function $\varphi^{\mathrm{c}}_{\mathrm{mech}}$ is introduced. It is a Legendre partial transformation of the specific dissipation function $\varphi_{\mathrm{mech}}$. The transformation is defined by

$$\begin{aligned}
\rho_0\, \varphi^{\mathrm{c}}_{\mathrm{mech}}(\boldsymbol{\sigma}, \boldsymbol{\beta}; \boldsymbol{\varepsilon} - \boldsymbol{\varepsilon}^{\mathrm{i}}, \boldsymbol{\alpha}, T, h) &\overset{\text{Legendre}}{=} \boldsymbol{\sigma} : \dot{\boldsymbol{\varepsilon}}^{\mathrm{i}} + \boldsymbol{\beta} : \dot{\boldsymbol{\alpha}} \\
&\qquad - \mu\, \rho_0\, \varphi_{\mathrm{mech}}(\dot{\boldsymbol{\varepsilon}}^{\mathrm{i}}, \dot{\boldsymbol{\alpha}}; \boldsymbol{\varepsilon} - \boldsymbol{\varepsilon}^{\mathrm{i}}, \boldsymbol{\alpha}, T, h) \\
\text{with} \quad \boldsymbol{\sigma} &= \mu\, \rho_0 \frac{\partial \varphi_{\mathrm{mech}}(\dot{\boldsymbol{\varepsilon}}^{\mathrm{i}}, \dot{\boldsymbol{\alpha}}; \boldsymbol{\varepsilon} - \boldsymbol{\varepsilon}^{\mathrm{i}}, \boldsymbol{\alpha}, T, h)}{\partial \dot{\boldsymbol{\varepsilon}}^{\mathrm{i}}} \qquad (6.285) \\
\text{and} \quad \boldsymbol{\beta} &= \mu\, \rho_0 \frac{\partial \varphi_{\mathrm{mech}}(\dot{\boldsymbol{\varepsilon}}^{\mathrm{i}}, \dot{\boldsymbol{\alpha}}; \boldsymbol{\varepsilon} - \boldsymbol{\varepsilon}^{\mathrm{i}}, \boldsymbol{\alpha}, T, h)}{\partial \dot{\boldsymbol{\alpha}}},
\end{aligned}$$

where superscript c in $\varphi^{\mathrm{c}}_{\mathrm{mech}}$ refers to the complementary function.

Based on Normality Rules (6.282) and (6.283) and Definition (6.285), the coefficient in Expression (E.11) of Appendix E take the values $a = \mu\, \rho_0$ and $b = \rho_0$. Thus, Expression (E.14) of **Appendix E** gives the following normality rule:

$$\dot{\boldsymbol{\varepsilon}}^{\mathrm{i}} = \rho_0 \frac{\partial \varphi^{\mathrm{c}}_{\mathrm{mech}}(\boldsymbol{\sigma}, \boldsymbol{\beta}; \boldsymbol{\varepsilon} - \boldsymbol{\varepsilon}^{\mathrm{i}}, \boldsymbol{\alpha}, T, h)}{\partial \boldsymbol{\sigma}} \qquad (6.286)$$

and

$$\dot{\boldsymbol{\alpha}} = \rho_0 \frac{\partial \varphi^{\mathrm{c}}_{\mathrm{mech}}(\boldsymbol{\sigma}, \boldsymbol{\beta}; \boldsymbol{\varepsilon} - \boldsymbol{\varepsilon}^{\mathrm{i}}, \boldsymbol{\alpha}, T, h)}{\partial \boldsymbol{\beta}}. \qquad (6.287)$$

Substitution of Transformation (6.285) and Normality Rules (6.282), (6.283), (6.286), and (6.287) into Expression (6.284) gives

$$\varphi^{\mathrm{c}}_{\mathrm{mech}}(\boldsymbol{\sigma}, \boldsymbol{\beta}; \boldsymbol{\varepsilon} - \boldsymbol{\varepsilon}^{\mathrm{i}}, \boldsymbol{\alpha}, T, h) = (1 - \mu) \left[ \frac{\partial \varphi^{\mathrm{c}}_{\mathrm{mech}}}{\partial \boldsymbol{\sigma}} : \boldsymbol{\sigma} + \frac{\partial \varphi^{\mathrm{c}}_{\mathrm{mech}}}{\partial \boldsymbol{\beta}} : \boldsymbol{\beta} \right]. \qquad (6.288)$$

The first-order sufficient condition for the point $(\boldsymbol{\sigma}, \boldsymbol{\beta})$ to be a local maximum is that Equations (6.286) and (6.287) hold and that the specific complementary dissipation function $\varphi^{\mathrm{c}}_{\mathrm{mech}}$ is a homogeneous function of degree $1/(1-\mu)$. The latter property is obtained if the coefficient $\mu$ in Expression (6.288) is a constant. If the multiplier $\mu$ is not a constant but $\mu = \mu(\boldsymbol{\varepsilon} - \boldsymbol{\varepsilon}^{\mathrm{i}}, \boldsymbol{\alpha}, T, h)$, the specific complementary dissipation function $\varphi^{\mathrm{c}}_{\mathrm{mech}}$ is not a homogeneous function and the value for $\mu$ is obtained from Equation (6.288).

It is important to note that the specific complementary dissipation function $\varphi^{\mathrm{c}}_{\mathrm{mech}}$ cannot be a homogeneous function of degree 1 in the variables $(\boldsymbol{\sigma}, \boldsymbol{\beta})$, since, according to Expression (6.288), in such a case, the quantity $\mu$ would be zero and Transformation (6.285) would vanish. This means that if the specific complementary dissipation function $\varphi^{\mathrm{c}}_{\mathrm{mech}}$ is a homogeneous function of degree 1, Transformation (6.285) vanishes, and it must be replaced with a different transformation. This new transformation is used when time-independent processes are modelled in Section 6.7.4.

Equation (E.31) of **Appendix E** shows that if the specific dissipation function $\varphi_{\mathrm{mech}}$ is a homogeneous function of degree $1/\mu$ $(\neq 1)$, the specific complementary dissipation function $\varphi^{\mathrm{c}}_{\mathrm{mech}}$ is a homogeneous function of degree $1/(1-\mu)$.

In the case of more than two internal variables, new members for the normality rule given by Expressions (6.286) and (6.287) are obtained. In such a case, the right side of Expression (6.288) would take more terms as well. This case is demonstrated in Section 9.6 with the model for creep proposed by Le Gac and Duval [48].

### 6.7.3 Normality Rule for a Thermal Problem

This section gives the normality rule for a thermal problem. The thermal problem [see Equation $(6.264)_2$] is considered next. The mathematical derivation of the thermal problem follows the same path as that taken earlier. The result is given without further elaboration. Corresponding to Expressions (6.282) and (6.283), the normality rule takes the form

$$-\frac{\vec{\nabla} T}{T} = \mu_{\mathrm{ther}}\,\rho_0\,\frac{\partial \varphi_{\mathrm{ther}}(\vec{q};\, \boldsymbol{\varepsilon} - \boldsymbol{\varepsilon}^{\mathrm{i}}, \boldsymbol{\alpha}, T, h)}{\partial \vec{q}}\,. \tag{6.289}$$

The specific dissipation function $\varphi_{\mathrm{ther}}(\vec{q};\, \boldsymbol{\varepsilon} - \boldsymbol{\varepsilon}^{\mathrm{i}}, \boldsymbol{\alpha}, T, h)$ must satisfy [cf. Eq. (6.284)]

$$\varphi_{\mathrm{ther}}(\vec{q};\ \boldsymbol{\varepsilon} - \boldsymbol{\varepsilon}^{\mathrm{i}},\ \boldsymbol{\alpha}, T, h) = \mu_{\mathrm{ther}}\,\frac{\partial \varphi_{\mathrm{ther}}}{\partial \vec{q}} \cdot \vec{q}\,. \tag{6.290}$$

If the multiplier $\mu_{\mathrm{ther}}$ is a constant, $\varphi_{\mathrm{ther}}(\vec{q};\, \boldsymbol{\varepsilon} - \boldsymbol{\varepsilon}^{\mathrm{i}}, \boldsymbol{\alpha}, T, h)$ is a homogeneous function of any degree $(= 1/\mu_{\mathrm{ther}})$ in the variable $\vec{q}$. If the coefficient $\mu_{\mathrm{ther}}$ is not a constant but $\mu_{\mathrm{ther}} = \mu_{\mathrm{ther}}(\boldsymbol{\varepsilon} - \epsilon^{\mathrm{i}},\ \boldsymbol{\alpha},\ T,\ h)$, the specific dissipation function $\varphi_{\mathrm{ther}}$ is not a homogeneous function, and the value for $\mu_{\mathrm{ther}}$ can be determined by Equation (6.290).

### 6.7.4 Normality Rule for Time-Independent Deformation Applied to Thermoplasticity

The aim of the present section is to evaluate the normality rule for thermoplastic material behavior. It is a special case, since the thermoplastic response is modelled as a time-independent process, whereas the previous sections are for modelling time-dependent processes. Of course, all the real processes are time-dependent. However, sometimes the response of the material is so fast that it follows the application of the load almost without any delay. In such cases, the time-independent model for material response is accurate enough and can provide some mathematical and/or numerical benefits beyond time-dependent material models. In this section, the plastic strain rate $\dot{\boldsymbol{\varepsilon}}^{\mathrm{p}}$ replaces the inelastic strain rate $\dot{\boldsymbol{\varepsilon}}^{\mathrm{i}}$. This section utilizes the results obtained in Sections 6.7.1, 6.7.2, and 6.7.3.

As in Section 6.7.1, it is assumed that the fluxes $(\dot{\boldsymbol{\varepsilon}}^{\mathrm{p}}, \dot{\boldsymbol{\alpha}})$ are independent of the heat flux vector $\vec{q}$, which means that the specific dissipation function for thermoplasticity $\varphi^{\mathrm{p}}$ is separable as follows:

$$\varphi^{\mathrm{p}}(\dot{\boldsymbol{\varepsilon}}^{\mathrm{p}}, \dot{\boldsymbol{\alpha}}, \vec{q}; \ldots) = \varphi^{\mathrm{p}}_{\mathrm{mech}}(\dot{\boldsymbol{\varepsilon}}^{\mathrm{p}}, \dot{\boldsymbol{\alpha}}; \ldots) + \varphi^{\mathrm{p}}_{\mathrm{ther}}(\vec{q}; \ldots)\,, \tag{6.291}$$

where the superscript $p$ refers to plasticity. With Separation (6.291), the maximization problem is separated into two different ones. The heat problem related to $\varphi^{\mathrm{p}}_{\mathrm{ther}}(\vec{q}; \ldots)$ leads to the results given in Section 6.7.3 and is thus not evaluated here. The mechanical part of the problem which is related to term $\varphi^{\mathrm{p}}_{\mathrm{mech}}(\dot{\boldsymbol{\varepsilon}}^{\mathrm{p}}, \dot{\boldsymbol{\alpha}}; \ldots)$ requires a deeper analysis, which follows.

The principle of maximum dissipation described at the start of Section 6.7.1 is written next. It is assumed that the state is known; that is, the set $(\boldsymbol{\varepsilon} - \boldsymbol{\varepsilon}^{\mathrm{i}}, \boldsymbol{\alpha}, T, h)$ is assumed to be known. This implies that the values for the forces $[\boldsymbol{\sigma}, \boldsymbol{\beta}, -(\vec{\nabla}T)/T]$ are known. The question is, what are the magnitudes of the fluxes $(\dot{\boldsymbol{\varepsilon}}^{\mathrm{i}}, \dot{\boldsymbol{\alpha}}, \vec{q}\,)$ which maximize the dissipation $\phi$? At the same time also Definition $(6.264)_1$ has to be satisfied.

The preceding formulation is capable describing of several different material responses. However, it is not for thermoplasticity, since the amount of plastic yield is dependent on state $(\boldsymbol{\varepsilon} - \boldsymbol{\varepsilon}^{\mathrm{p}}, \boldsymbol{\alpha}, T, h)$ and on the rate of loading. The state on the yield surface is not a sufficient condition for plastic yield, but the loading direction has to be toward the outside region of the yield locus. In the theory of plasticity, this is called the loading/unloading condition. Rate of loading means the rates of controllable state variables, that is, the strain rate tensor $\dot{\boldsymbol{\varepsilon}}$ and the rate of the temperature $\dot{T}$. In the theory of plasticity, these rates of loading are often replaced by the corresponding differentials. The values for the internal variables $\dot{\boldsymbol{\varepsilon}}^{\mathrm{p}}$, $\boldsymbol{\alpha}$, and so on (if others exist) cannot be adjusted (controlled) by an external agent, as discussed in Section 6.4.1. Based on the preceding, for the theory of thermoplasticity, Form (6.265) is replaced with

$$\varphi^{\mathrm{p}}_{\mathrm{mech}} = \varphi^{\mathrm{p}}_{\mathrm{mech}}(\dot{\boldsymbol{\varepsilon}}^{\mathrm{p}}, \dot{\boldsymbol{\alpha}}; \boldsymbol{\varepsilon} - \boldsymbol{\varepsilon}^{\mathrm{p}}, \boldsymbol{\alpha}, T, h, \dot{\boldsymbol{\varepsilon}}, \dot{T})\,. \tag{6.292}$$

Based on Expressions (6.267) and (6.268), it is written as follows:

maximize with respect to the fluxes $(\dot{\boldsymbol{\varepsilon}}^{\mathrm{p}}, \dot{\boldsymbol{\alpha}})$,

$$\Phi_{\text{mech}}(\dot{\boldsymbol{\varepsilon}}^{\mathrm{p}}, \dot{\boldsymbol{\alpha}}, \boldsymbol{\sigma}, \boldsymbol{\beta}) = \boldsymbol{\sigma} : \dot{\boldsymbol{\varepsilon}}^{\mathrm{p}} + \boldsymbol{\beta} : \dot{\boldsymbol{\alpha}}\,, \tag{6.293}$$

subject to

$$\tau_{\text{mech}} = \rho_0\, \varphi^{\mathrm{p}}_{\text{mech}}(\dot{\boldsymbol{\varepsilon}}^{\mathrm{p}}, \dot{\boldsymbol{\alpha}}; \boldsymbol{\varepsilon} - \boldsymbol{\varepsilon}^{\mathrm{p}}, \boldsymbol{\alpha}, T, h, \dot{\boldsymbol{\varepsilon}}, \dot{T}) - (\boldsymbol{\sigma} : \dot{\boldsymbol{\varepsilon}}^{\mathrm{p}} + \boldsymbol{\beta} : \dot{\boldsymbol{\alpha}}) = 0\,, \tag{6.294}$$

where $\tau_{mech} = 0$ is a constraint. Both $\Phi_{\text{mech}}$ and $\tau_{\text{mech}}$ are assumed to have at least continuous second partial derivatives with respect to the arguments $(\dot{\boldsymbol{\varepsilon}}^{\mathrm{p}}, \dot{\boldsymbol{\alpha}})$.

It is convenient to assume that the mechanical part of the specific dissipation function $\varphi^{\mathrm{p}}_{\text{mech}}(\dot{\boldsymbol{\varepsilon}}^{\mathrm{p}}, \dot{\boldsymbol{\alpha}}; \ldots)$ for time-independent processes is a homogeneous function of degree 1. The reason is shown later in this section. Based on Normality Rules $(6.279)_2$ and (6.280), the following is achieved:

$$\boldsymbol{\sigma} = \rho_0 \frac{\partial \varphi^{\mathrm{p}}_{\text{mech}}}{\partial \dot{\boldsymbol{\varepsilon}}^{\mathrm{p}}} \qquad \text{and} \qquad \boldsymbol{\beta} = \rho_0 \frac{\partial \varphi^{\mathrm{p}}_{\text{mech}}}{\partial \dot{\boldsymbol{\alpha}}}\,. \tag{6.295}$$

In the theory of thermoplasticity, the mechanical part of the specific dissipation function $\varphi^{\mathrm{p}}_{\text{mech}}(\dot{\boldsymbol{\varepsilon}}^{\mathrm{p}}, \dot{\boldsymbol{\alpha}}; \ldots)$ is usually replaced by the yield function $F$ the arguments of which are the forces $(\boldsymbol{\sigma}, \boldsymbol{\beta})$ instead of the fluxes $(\dot{\boldsymbol{\varepsilon}}^{\mathrm{p}}, \dot{\boldsymbol{\alpha}})$. Furthermore, the traditions of thermoplasticity give the following property for the yield function $F$:

$$F := 0 \quad \text{during plastic flow}\,, \tag{6.296}$$

which gives

$$\dot{F} = 0 \quad \text{during plastic flow}\,. \tag{6.297}$$

Condition (6.297) is the consistency condition.

Based on the preceding, the following Legendre partial transformation is made:

$$\begin{aligned}
\overset{\circ}{\lambda}\, F(\boldsymbol{\sigma}, \boldsymbol{\beta}; \boldsymbol{\varepsilon} - \boldsymbol{\varepsilon}^{\mathrm{p}}, \boldsymbol{\alpha}, T, h) &\overset{\text{Legendre}}{=} \boldsymbol{\sigma} : \dot{\boldsymbol{\varepsilon}}^{\mathrm{p}} + \boldsymbol{\beta} : \dot{\boldsymbol{\alpha}} \\
&\qquad - \rho_0\, \varphi^{\mathrm{p}}_{\text{mech}}(\dot{\boldsymbol{\varepsilon}}^{\mathrm{p}}, \dot{\boldsymbol{\alpha}}; \boldsymbol{\varepsilon} - \boldsymbol{\varepsilon}^{\mathrm{p}}, \boldsymbol{\alpha}, T, h, \dot{\boldsymbol{\varepsilon}}, \dot{T})\,, \\
\text{with} \quad \boldsymbol{\sigma} &= \rho_0 \frac{\partial \varphi^{\mathrm{p}}_{\text{mech}}(\dot{\boldsymbol{\varepsilon}}^{\mathrm{p}}, \dot{\boldsymbol{\alpha}}; \boldsymbol{\varepsilon} - \boldsymbol{\varepsilon}^{\mathrm{p}}, \boldsymbol{\alpha}, T, h, \dot{\boldsymbol{\varepsilon}}, \dot{T})}{\partial \dot{\boldsymbol{\varepsilon}}^{\mathrm{p}}} \\
\text{and} \quad \boldsymbol{\beta} &= \rho_0 \frac{\partial \varphi^{\mathrm{p}}_{\text{mech}}(\dot{\boldsymbol{\varepsilon}}^{\mathrm{p}}, \dot{\boldsymbol{\alpha}}; \boldsymbol{\varepsilon} - \boldsymbol{\varepsilon}^{\mathrm{p}}, \boldsymbol{\alpha}, T, h, \dot{\boldsymbol{\varepsilon}}, \dot{T})}{\partial \dot{\boldsymbol{\alpha}}}\,,
\end{aligned} \tag{6.298}$$

where the plasticity multiplier $\overset{\circ}{\lambda}$ is defined by

$$\overset{\circ}{\lambda} := \overset{\circ}{\lambda}(\boldsymbol{\varepsilon} - \boldsymbol{\varepsilon}^{\mathrm{p}}, \boldsymbol{\alpha}, T, h, \dot{\boldsymbol{\varepsilon}}, \dot{T}). \tag{6.299}$$

Thus the plasticity multiplier $\overset{\circ}{\lambda}$ is dependent on the state $(\boldsymbol{\varepsilon} - \boldsymbol{\varepsilon}^{\mathrm{p}}, \boldsymbol{\alpha}, T, h)$ and on the rate of loading $(\dot{\boldsymbol{\varepsilon}}, \dot{T})$. The notation $\circ$ above the quantity $\lambda$ indicates that the plasticity multiplier $\overset{\circ}{\lambda}$ is not a material derivative but a function also having material derivatives of quantities as independent variables.

Based on Normality Rules (6.295) and Transformation (6.298), the coefficients in Expression (E.11) of **Appendix E** take the values $a = \rho_0$ and $b = \overset{\circ}{\lambda}$. Thus, Expression (E.14) of **Appendix E** gives the following normality rule:

$$\dot{\boldsymbol{\varepsilon}}^{\mathrm{p}} = \overset{\circ}{\lambda} \frac{\partial F(\boldsymbol{\sigma}, \boldsymbol{\beta}; \boldsymbol{\varepsilon} - \boldsymbol{\varepsilon}^{\mathrm{p}}, \boldsymbol{\alpha}, T, h)}{\partial \boldsymbol{\sigma}} \tag{6.300}$$

and

$$\dot{\boldsymbol{\alpha}} = \overset{\circ}{\lambda} \frac{\partial F(\boldsymbol{\sigma}, \boldsymbol{\beta}; \boldsymbol{\varepsilon} - \boldsymbol{\varepsilon}^{\mathrm{p}}, \boldsymbol{\alpha}, T, h)}{\partial \boldsymbol{\beta}}. \tag{6.301}$$

Two further topics are considered: first, the background to the fact that for time-independent processes the specific dissipation function $\varphi^{\mathrm{p}}_{\mathrm{mech}}(\dot{\boldsymbol{\varepsilon}}^{\mathrm{p}}, \dot{\boldsymbol{\alpha}}; \ldots)$ is a homogeneous function of degree 1; second, demonstration that the homogeneity (of degree 1) of the specific dissipation function $\varphi^{\mathrm{p}}_{\mathrm{mech}}(\dot{\boldsymbol{\varepsilon}}^{\mathrm{p}}, \dot{\boldsymbol{\alpha}}; \ldots)$ has no homogeneity implications for the yield function $F(\boldsymbol{\sigma}, \boldsymbol{\beta}; \ldots)$.

The following manipulation is needed:

$$\begin{aligned}
\frac{\partial \varphi^{\mathrm{p}}_{\mathrm{mech}}(\dot{\boldsymbol{\varepsilon}}^{\mathrm{p}}, \dot{\boldsymbol{\alpha}}; \boldsymbol{\varepsilon} - \boldsymbol{\varepsilon}^{\mathrm{p}}, \boldsymbol{\alpha}, \ldots)}{\partial \dot{\boldsymbol{\varepsilon}}^{\mathrm{p}}} &= \frac{k}{k} \frac{\partial \varphi^{\mathrm{p}}_{\mathrm{mech}}(\dot{\boldsymbol{\varepsilon}}^{\mathrm{p}}, \dot{\boldsymbol{\alpha}}; \boldsymbol{\varepsilon} - \boldsymbol{\varepsilon}^{\mathrm{p}}, \boldsymbol{\alpha}, \ldots)}{\partial \dot{\boldsymbol{\varepsilon}}^{\mathrm{p}}} \\
&= \frac{1}{k} \frac{\partial \varphi^{\mathrm{p}}_{\mathrm{mech}}(k\dot{\boldsymbol{\varepsilon}}^{\mathrm{p}}, k\dot{\boldsymbol{\alpha}}; \boldsymbol{\varepsilon} - \boldsymbol{\varepsilon}^{\mathrm{p}}, \boldsymbol{\alpha}, \ldots)}{\partial \dot{\boldsymbol{\varepsilon}}^{\mathrm{p}}} = \frac{1}{k} \frac{\partial \varphi^{\mathrm{p}}_{\mathrm{mech}}(k\dot{\boldsymbol{\varepsilon}}^{\mathrm{p}}, k\dot{\boldsymbol{\alpha}}; \boldsymbol{\varepsilon} - \boldsymbol{\varepsilon}^{\mathrm{p}}, \boldsymbol{\alpha}, \ldots)}{\partial (k\dot{\boldsymbol{\varepsilon}}^{\mathrm{p}})} : \frac{\partial (k\dot{\boldsymbol{\varepsilon}}^{\mathrm{p}})}{\partial \dot{\boldsymbol{\varepsilon}}^{\mathrm{p}}} \\
&= \frac{1}{k} \frac{\partial \varphi^{\mathrm{p}}_{\mathrm{mech}}(k\,\dot{\boldsymbol{\varepsilon}}^{\mathrm{p}}, k\,\dot{\boldsymbol{\alpha}}; \boldsymbol{\varepsilon} - \boldsymbol{\varepsilon}^{\mathrm{p}}, \boldsymbol{\alpha}, \ldots)}{\partial (k\,\dot{\boldsymbol{\varepsilon}}^{\mathrm{p}})} : k\,\mathbf{I} = \frac{\partial \varphi^{\mathrm{p}}_{\mathrm{mech}}(k\,\dot{\boldsymbol{\varepsilon}}^{\mathrm{p}}, k\,\dot{\boldsymbol{\alpha}}; \boldsymbol{\varepsilon} - \boldsymbol{\varepsilon}^{\mathrm{p}}, \boldsymbol{\alpha}, \ldots)}{\partial (k\,\dot{\boldsymbol{\varepsilon}}^{\mathrm{p}})},
\end{aligned} \tag{6.302}$$

where the following facts are exploited: (a) Expression (2.69), that is, $\mathbf{c}:\mathbf{I} = \mathbf{c}$, (b) the mechanical part of the specific dissipation function for thermoplasticity $\varphi^{\mathrm{p}}_{\mathrm{mech}}(\dot{\boldsymbol{\varepsilon}}^{\mathrm{p}}, \dot{\boldsymbol{\alpha}}; \ldots)$ is a homogeneous function of degree 1 in the variables $(\dot{\boldsymbol{\varepsilon}}^{\mathrm{p}}, \dot{\boldsymbol{\alpha}})$, and (c) Definition (6.276), that is, $\theta(k\,x, k\,y, k\,z, u, v) := k^{\omega}\theta(x, y, z, u, v)$, where $\omega$ is the degree of homogeneity.

Again, Normality Rule $(6.295)_1$ is

$$\boldsymbol{\sigma}(\dot{\boldsymbol{\varepsilon}}^{\mathrm{p}}, \dot{\boldsymbol{\alpha}}; \boldsymbol{\varepsilon} - \boldsymbol{\varepsilon}^{\mathrm{p}}, \boldsymbol{\alpha}, \ldots) = \rho_0 \frac{\partial \varphi^{\mathrm{p}}_{\mathrm{mech}}(\dot{\boldsymbol{\varepsilon}}^{\mathrm{p}}, \dot{\boldsymbol{\alpha}}; \boldsymbol{\varepsilon} - \boldsymbol{\varepsilon}^{\mathrm{p}}, \boldsymbol{\alpha}, \ldots)}{\partial \dot{\boldsymbol{\varepsilon}}^{\mathrm{p}}}, \tag{6.303}$$

which exploits the fact that a derivative of a function with respect to one of its arguments is a function having no more than the same arguments. Based on Normality Rule (6.303), one can achieve

$$\boldsymbol{\sigma}(k\,\dot{\boldsymbol{\varepsilon}}^{\mathrm{p}}, k\,\dot{\boldsymbol{\alpha}}; \boldsymbol{\varepsilon} - \boldsymbol{\varepsilon}^{\mathrm{p}}, \boldsymbol{\alpha}, \ldots) = \rho_0 \frac{\partial \varphi^{\mathrm{p}}_{\mathrm{mech}}(k\,\dot{\boldsymbol{\varepsilon}}^{\mathrm{p}}, k\,\dot{\boldsymbol{\alpha}}; \boldsymbol{\varepsilon} - \boldsymbol{\varepsilon}^{\mathrm{p}}, \boldsymbol{\alpha}, \ldots)}{\partial (k\,\dot{\boldsymbol{\varepsilon}}^{\mathrm{p}})}. \tag{6.304}$$

According to Manipulation (6.302), the right sides of Normality Rules (6.303) and (6.304) coincide. Therefore the left sides are also equal. Since two arbitrarily different sets of fluxes, that is, $(\dot{\boldsymbol{\varepsilon}}^{\mathrm{p}}, \dot{\boldsymbol{\alpha}})$ and $(k\dot{\boldsymbol{\varepsilon}}^{\mathrm{p}}, k\dot{\boldsymbol{\alpha}})$, give the same value for the stress tensor $\boldsymbol{\sigma}$, the value of the stress tensor $\boldsymbol{\sigma}$ cannot depend on fluxes $(\dot{\boldsymbol{\varepsilon}}^{\mathrm{p}}, \dot{\boldsymbol{\alpha}})$. This means that the fluxes $(\dot{\boldsymbol{\varepsilon}}^{\mathrm{p}}, \dot{\boldsymbol{\alpha}})$, that is, a time-dependent process, have no influence on the forces $(\boldsymbol{\sigma}, \boldsymbol{\beta})$ and through State Equations (6.174) do not affect the state $(\boldsymbol{\varepsilon} - \boldsymbol{\varepsilon}^{\mathrm{p}}, \boldsymbol{\alpha}, T, h)$. Thus modelling the specific dissipation function $\varphi^{\mathrm{p}}_{\mathrm{mech}}(\dot{\boldsymbol{\varepsilon}}^{\mathrm{p}}, \dot{\boldsymbol{\alpha}}; \ldots)$ by a homogeneous function of degree 1 in fluxes gives a time-independent process. This is a sufficient condition for time-independence.

It is shown next that although the mechanical part of the specific dissipation function for thermoplasticity $\varphi^{\mathrm{p}}_{\mathrm{mech}}(\dot{\boldsymbol{\varepsilon}}^{\mathrm{p}}, \dot{\boldsymbol{\alpha}}; \ldots)$ is a homogeneous function of degree 1, the yield function $F$ can be nonhomogeneous. This is not obvious since usually the Legendre transformation of a homogeneous function is a homogeneous function, as proved in **Appendix E**. The only exception is the case in which the original function is a homogeneous function of degree 1.

Definition (6.276) and Theorem (6.277) for homogeneous functions are recalled and applied to the function $\varphi^{\mathrm{p}}_{\mathrm{mech}}(\dot{\boldsymbol{\varepsilon}}^{\mathrm{p}}, \dot{\boldsymbol{\alpha}}; \ldots)$, which is a homogeneous function of degree 1 in the fluxes $(\dot{\boldsymbol{\varepsilon}}^{\mathrm{p}}, \dot{\boldsymbol{\alpha}})$. Obtained are

$$\varphi^{\mathrm{p}}_{\mathrm{mech}}(k\,\dot{\boldsymbol{\varepsilon}}^{\mathrm{p}}, k\,\dot{\boldsymbol{\alpha}}; \boldsymbol{\varepsilon} - \boldsymbol{\varepsilon}^{\mathrm{p}}, \boldsymbol{\alpha}, \ldots) = k\,\varphi^{\mathrm{p}}_{\mathrm{mech}}(\dot{\boldsymbol{\varepsilon}}^{\mathrm{p}}, \dot{\boldsymbol{\alpha}}; \boldsymbol{\varepsilon} - \boldsymbol{\varepsilon}^{\mathrm{p}}, \boldsymbol{\alpha}, \ldots) \tag{6.305}$$

and

$$\begin{aligned}\varphi^{\mathrm{p}}_{\mathrm{mech}}&(\dot{\boldsymbol{\varepsilon}}^{\mathrm{p}}, \dot{\boldsymbol{\alpha}}; \boldsymbol{\varepsilon} - \boldsymbol{\varepsilon}^{\mathrm{p}}, \boldsymbol{\alpha}, \ldots)\\ &= \frac{\partial \varphi^{\mathrm{p}}_{\mathrm{mech}}(\dot{\boldsymbol{\varepsilon}}^{\mathrm{p}}, \dot{\boldsymbol{\alpha}}; \boldsymbol{\varepsilon} - \boldsymbol{\varepsilon}^{\mathrm{p}}, \boldsymbol{\alpha}, \ldots)}{\partial \dot{\boldsymbol{\varepsilon}}^{\mathrm{p}}} : \dot{\boldsymbol{\varepsilon}}^{\mathrm{p}} + \frac{\partial \varphi^{\mathrm{p}}_{\mathrm{mech}}(\dot{\boldsymbol{\varepsilon}}^{\mathrm{p}}, \dot{\boldsymbol{\alpha}}; \boldsymbol{\varepsilon} - \boldsymbol{\varepsilon}^{\mathrm{p}}, \boldsymbol{\alpha}, \ldots)}{\partial \dot{\boldsymbol{\alpha}}} : \dot{\boldsymbol{\alpha}}\,.\end{aligned} \tag{6.306}$$

Multiplying both sides of Transformation $(6.298)_1$ by a positive-valued scalar $k$ gives

$$\begin{aligned}k\,\overset{\circ}{\lambda}\,F(\boldsymbol{\sigma}, \boldsymbol{\beta}; \boldsymbol{\varepsilon} - \boldsymbol{\varepsilon}^{\mathrm{p}}, \boldsymbol{\alpha}, T, h) &= \boldsymbol{\sigma} : (k\,\dot{\boldsymbol{\varepsilon}}^{\mathrm{p}}) + \boldsymbol{\beta} : (k\,\dot{\boldsymbol{\alpha}})\\ &\quad - k\,\rho_0\,\varphi^{\mathrm{p}}_{\mathrm{mech}}(\dot{\boldsymbol{\varepsilon}}^{\mathrm{p}}, \dot{\boldsymbol{\alpha}}; \boldsymbol{\varepsilon} - \boldsymbol{\varepsilon}^{\mathrm{p}}, \boldsymbol{\alpha}, T, h, \dot{\boldsymbol{\sigma}}, \dot{T})\,.\end{aligned} \tag{6.307}$$

Utilization of Normality Rule (6.304), the corresponding normality rule for the internal force $\boldsymbol{\beta}$, and Property (6.305) in Expression (6.307) gives

$$\begin{aligned}k\,\overset{\circ}{\lambda}\,(\boldsymbol{\sigma}, \boldsymbol{\beta}; \boldsymbol{\varepsilon} - \boldsymbol{\varepsilon}^{\mathrm{p}}, \boldsymbol{\alpha}, T, h) &= \rho_0\,\frac{\partial \varphi^{\mathrm{p}}_{\mathrm{mech}}(k\,\dot{\boldsymbol{\varepsilon}}^{\mathrm{p}}, k\,\dot{\boldsymbol{\alpha}}; \boldsymbol{\varepsilon} - \boldsymbol{\varepsilon}^{\mathrm{p}}, \boldsymbol{\alpha}, \ldots)}{\partial (k\,\dot{\boldsymbol{\varepsilon}}^{\mathrm{p}})} : (k\,\dot{\boldsymbol{\varepsilon}}^{\mathrm{p}})\\ &\quad + \rho_0\,\frac{\partial \varphi^{\mathrm{p}}_{\mathrm{mech}}(k\,\dot{\boldsymbol{\varepsilon}}^{\mathrm{p}}, k\,\dot{\boldsymbol{\alpha}}; \boldsymbol{\varepsilon} - \boldsymbol{\varepsilon}^{\mathrm{p}}, \boldsymbol{\alpha}, \ldots)}{\partial (k\,\dot{\boldsymbol{\alpha}})} : (k\,\dot{\boldsymbol{\alpha}})\\ &\quad - \rho_0\,\varphi^{\mathrm{p}}_{\mathrm{mech}}(k\,\dot{\boldsymbol{\varepsilon}}^{\mathrm{p}}, k\,\dot{\boldsymbol{\alpha}}; \boldsymbol{\varepsilon} - \boldsymbol{\varepsilon}^{\mathrm{p}}, \boldsymbol{\alpha}, T, h, \dot{\boldsymbol{\sigma}}, \dot{T})\,.\end{aligned} \tag{6.308}$$

Based on Property (6.306), the right side of Expression (6.308) vanishes and Expression (6.308) reduces to

$$k \overset{\circ}{\lambda} F(\boldsymbol{\sigma}, \boldsymbol{\beta};\, \boldsymbol{\varepsilon} - \boldsymbol{\varepsilon}^{\mathrm{p}}, \boldsymbol{\alpha}, T, h) = 0\,, \qquad \text{which yields} \quad F = 0\,. \tag{6.309}$$

This is the condition aimed at, since the theory of thermoplasticity requires that the value for the yield function $F$ vanishes during plastic flow, as indicated by Definition (6.296).

Substitution of Normality Rule (6.295) into Transformation $(6.298)_1$ leads to

$$\begin{aligned} \overset{\circ}{\lambda} F(\boldsymbol{\sigma}, \boldsymbol{\beta}; \boldsymbol{\varepsilon} - \boldsymbol{\varepsilon}^{\mathrm{p}}, \boldsymbol{\alpha}, T, h) = \rho_0\, & \frac{\partial \varphi^{\mathrm{p}}_{\mathrm{mech}}(\dot{\boldsymbol{\varepsilon}}^{\mathrm{p}}, \dot{\boldsymbol{\alpha}}; \boldsymbol{\varepsilon} - \boldsymbol{\varepsilon}^{\mathrm{p}}, \boldsymbol{\alpha}, \ldots)}{\partial \dot{\boldsymbol{\varepsilon}}^{\mathrm{p}}} : \dot{\boldsymbol{\varepsilon}}^{\mathrm{p}} \\ + \rho_0\, & \frac{\partial \varphi^{\mathrm{p}}_{\mathrm{mech}}(\dot{\boldsymbol{\varepsilon}}^{\mathrm{p}}, \dot{\boldsymbol{\alpha}}; \boldsymbol{\varepsilon} - \boldsymbol{\varepsilon}^{\mathrm{p}}, \boldsymbol{\alpha}, \ldots)}{\partial \dot{\boldsymbol{\alpha}}} : \dot{\boldsymbol{\alpha}} \\ - \rho_0\, & \varphi^{\mathrm{p}}_{\mathrm{mech}}(\dot{\boldsymbol{\varepsilon}}^{\mathrm{p}}, \dot{\boldsymbol{\alpha}}; \boldsymbol{\varepsilon} - \boldsymbol{\varepsilon}^{\mathrm{p}}, \boldsymbol{\alpha}, T, h, \dot{\boldsymbol{\sigma}}, \dot{T})\,. \end{aligned} \tag{6.310}$$

Once again, Expression (6.306) ensures that the right side of Expression (6.310) vanishes and therefore $F = 0$ during plastic flow.

The preceding indicates that although the mechanical part of the specific dissipation function for thermoplasticity $\varphi^{\mathrm{p}}_{\mathrm{mech}}(\dot{\boldsymbol{\varepsilon}}^{\mathrm{p}}, \dot{\boldsymbol{\alpha}}; \ldots)$ is a homogeneous function of degree 1, the yield function $F$ can be non-homogeneous.

### 6.7.5 Normality Rule for Nonseparated Dissipation $\Phi$

This section gives the normality for a case where the dissipation $\Phi$ is not separated into mechanical part $\Phi_{\mathrm{mech}}$ and into thermal part $\Phi_{\mathrm{ther}}$. The consequences of the separation are evaluated from the material modelling point of view.

Section **H.1** of **Appendix H** evaluates the principle of maximum dissipation $\Phi$ in case the dissipation $\Phi$ is not separated into mechanical part $\Phi_{\mathrm{mech}}$ and thermal part $\Phi_{\mathrm{ther}}$. The evaluation starts from the full form of the Clausius–Duhem Inequality (6.263):

$$\Phi = \boldsymbol{\sigma} : \dot{\boldsymbol{\varepsilon}}^{\mathrm{i}} + \boldsymbol{\beta} : \dot{\boldsymbol{\alpha}} - \frac{\vec{\nabla} T}{T} \cdot \vec{q} \qquad (\geq 0)\,. \tag{6.311}$$

According to Equations (H.7)–(H.9) of **Appendix H**, the normality rule takes the form

$$\boldsymbol{\sigma} = \mu\, \rho_0\, \frac{\partial \varphi(\dot{\boldsymbol{\varepsilon}}^{\mathrm{i}}, \dot{\boldsymbol{\alpha}}, \vec{q}; \boldsymbol{\varepsilon} - \boldsymbol{\varepsilon}^{\mathrm{i}}, \boldsymbol{\alpha}, T, h)}{\partial \dot{\boldsymbol{\varepsilon}}^{\mathrm{i}}} \tag{6.312}$$

and

$$\boldsymbol{\beta} = \mu\, \rho_0\, \frac{\partial \varphi(\dot{\boldsymbol{\varepsilon}}^{\mathrm{i}}, \dot{\boldsymbol{\alpha}}, \vec{q}; \boldsymbol{\varepsilon} - \boldsymbol{\varepsilon}^{\mathrm{i}}, \boldsymbol{\alpha}, T, h)}{\partial \dot{\boldsymbol{\alpha}}}\,, \tag{6.313}$$

and, finally,

$$-\frac{\vec{\nabla} T}{T} = \mu\, \rho_0\, \frac{\partial \varphi(\dot{\boldsymbol{\varepsilon}}^{\mathrm{i}}, \dot{\boldsymbol{\alpha}}, \vec{q}; \boldsymbol{\varepsilon} - \boldsymbol{\varepsilon}^{\mathrm{i}}, \boldsymbol{\alpha}, T, h)}{\partial \vec{q}}\,. \tag{6.314}$$

According to Expression (H.10) of **Appendix H**, the specific dissipation function $\varphi$ has to satisfy the following condition:

$$\varphi(\dot{\boldsymbol{\varepsilon}}^{\mathrm{i}}, \dot{\boldsymbol{\alpha}}, \vec{q}; \boldsymbol{\varepsilon} - \boldsymbol{\varepsilon}^{\mathrm{i}}, \boldsymbol{\alpha}, T, h) = \mu \left( \frac{\partial \varphi}{\partial \dot{\boldsymbol{\varepsilon}}^{\mathrm{i}}} : \dot{\boldsymbol{\varepsilon}}^{\mathrm{i}} + \frac{\partial \varphi}{\partial \dot{\boldsymbol{\alpha}}} : \dot{\boldsymbol{\alpha}} + \frac{\partial \varphi}{\partial \vec{q}} \cdot \vec{q} \right) . \qquad (6.315)$$

Sections 6.7.1 and 6.7.3 evaluated a case where Clausius–Duhem Inequality (6.311) was separated into a mechanical part $\Phi_{\mathrm{mech}}$ and a thermal part $\Phi_{\mathrm{ther}}$,

$$\Phi_{\mathrm{mech}} = \boldsymbol{\sigma} : \dot{\boldsymbol{\varepsilon}}^{\mathrm{i}} + \boldsymbol{\beta} : \dot{\boldsymbol{\alpha}} \;\; (\geq 0) \quad \text{and} \quad \Phi_{\mathrm{ther}} = -\frac{\vec{\nabla} T}{T} \cdot \vec{q} \;\; (\geq 0) \qquad (6.316)$$

and created two mutually independent maximization problems. According to Equations (6.282), (6.283), and (6.289) the normality rule takes the form

$$\boldsymbol{\sigma} = \mu \, \rho_0 \, \frac{\partial \varphi_{\mathrm{mech}}(\dot{\boldsymbol{\varepsilon}}^{\mathrm{i}}, \dot{\boldsymbol{\alpha}}; \boldsymbol{\varepsilon} - \boldsymbol{\varepsilon}^{\mathrm{i}}, \boldsymbol{\alpha}, T, h)}{\partial \dot{\boldsymbol{\varepsilon}}^{\mathrm{i}}} \qquad (6.317)$$

and

$$\boldsymbol{\beta} = \mu \, \rho_0 \, \frac{\partial \varphi_{\mathrm{mech}}(\dot{\boldsymbol{\varepsilon}}^{\mathrm{i}}, \dot{\boldsymbol{\alpha}}; \boldsymbol{\varepsilon} - \boldsymbol{\varepsilon}^{\mathrm{i}}, \boldsymbol{\alpha}, T, h)}{\partial \dot{\boldsymbol{\alpha}}} , \qquad (6.318)$$

and, finally,

$$-\frac{\vec{\nabla} T}{T} = \mu_{\mathrm{ther}} \, \rho_0 \, \frac{\partial \varphi_{\mathrm{ther}}(\vec{q}; \boldsymbol{\varepsilon} - \boldsymbol{\varepsilon}^{\mathrm{i}}, \boldsymbol{\alpha}, T, h)}{\partial \vec{q}} . \qquad (6.319)$$

For this latter case, the specific dissipation functions $\varphi_{\mathrm{mech}}$ and $\varphi_{\mathrm{ther}}$ have to satisfy the following conditions [see Equations (6.284) and (6.290)]:

$$\varphi_{\mathrm{mech}} = \mu \left( \frac{\partial \varphi_{\mathrm{mech}}}{\partial \dot{\boldsymbol{\varepsilon}}^{\mathrm{i}}} : \dot{\boldsymbol{\varepsilon}}^{\mathrm{i}} + \frac{\partial \varphi_{\mathrm{mech}}}{\partial \dot{\boldsymbol{\alpha}}} : \dot{\boldsymbol{\alpha}} \right) \quad \text{and} \quad \varphi_{\mathrm{ther}} = \mu_{\mathrm{ther}} \frac{\partial \varphi_{\mathrm{ther}}}{\partial \vec{q}} \cdot \vec{q} . \qquad (6.320)$$

From a material modelling point of view, there is a remarkable difference between Normality Rules (6.312)–(6.314) and Normality Rules (6.317)–(6.319). The first set, that is, Expressions (6.312)–(6.314), have the same multipliers in front of the partial derivatives whereas the second set, Equations (6.317)–(6.319), does not. The same difference holds when Conditions (6.315) and (6.320) are compared. The fact that the thermal problem has a different multiplier, that is, $\mu_{\mathrm{ther}}$, from that of the mechanical problem, which is $\mu$, makes the latter approach more flexible and therefore more capable for description of material behavior.

The same conclusion can be drawn when the material models are expressed by the specific complementary dissipation function $\varphi^{\mathrm{c}}_{***}$ (where $***$ can be mech, ther, or an empty space).

### 6.7.6 Two Different Ways to Define the Maximization Problem

In this section we evaluate the difference between two mathematical formulations of the principle of maximum dissipation when the dissipation $\Phi$ is separated into two parts. **Appendix H** studies the principle of maximum dissipation in case the Clausius–Duhem inequality is separated into two different inequalities and therefore the dissipation $\Phi$ is separated into two parts. This separation is shown in Expression (6.316).

By following Separation (6.316) also the specific dissipation function $\varphi$ is separated. Thus the following is obtained:

$$\varphi(\dot{\boldsymbol{\varepsilon}}^{\mathrm{i}}, \dot{\boldsymbol{\alpha}}, \vec{q}; \boldsymbol{\varepsilon} - \boldsymbol{\varepsilon}^{\mathrm{i}}, \boldsymbol{\alpha}, T, h) = \varphi_{\mathrm{mech}}(\dot{\boldsymbol{\varepsilon}}^{\mathrm{i}}, \dot{\boldsymbol{\alpha}}; \boldsymbol{\varepsilon} - \boldsymbol{\varepsilon}^{\mathrm{i}}, \boldsymbol{\alpha}, T, h) + \varphi_{\mathrm{ther}}(\vec{q}; \boldsymbol{\varepsilon} - \boldsymbol{\varepsilon}^{\mathrm{i}}, \boldsymbol{\alpha}, T, h). \tag{6.321}$$

The principle of maximum dissipation is written in the following two mathematical forms:

Case A: Consider two different problems:

maximize with respect to the fluxes $(\dot{\boldsymbol{\varepsilon}}^{\mathrm{i}}, \dot{\boldsymbol{\alpha}})$,

$$\Phi_{\mathrm{mech}}(\dot{\boldsymbol{\varepsilon}}^{\mathrm{i}}, \dot{\boldsymbol{\alpha}}, \boldsymbol{\sigma}, \boldsymbol{\beta}) = \boldsymbol{\sigma} : \dot{\boldsymbol{\varepsilon}}^{\mathrm{i}} + \boldsymbol{\beta} : \dot{\boldsymbol{\alpha}}\,, \tag{6.322}$$

subject to

$$\tau_{\mathrm{mech}} = \rho_0\, \varphi_{\mathrm{mech}}(\dot{\boldsymbol{\varepsilon}}^{\mathrm{i}}, \dot{\boldsymbol{\alpha}}; \boldsymbol{\varepsilon} - \boldsymbol{\varepsilon}^{\mathrm{i}}, \boldsymbol{\alpha}, T, h) - \Phi_{\mathrm{mech}}(\dot{\boldsymbol{\varepsilon}}^{\mathrm{i}}, \dot{\boldsymbol{\alpha}}, \boldsymbol{\sigma}, \boldsymbol{\beta}) = 0\,, \tag{6.323}$$

where $\tau_{\mathrm{mech}} = 0$ is a constraint,

and

maximize with respect to the flux $(\vec{q})$,

$$\Phi_{\mathrm{ther}}(\vec{q}, \vec{\nabla} T/T) = \frac{\vec{\nabla} T}{T} \cdot \vec{q}, \tag{6.324}$$

subject to

$$\tau_{\mathrm{ther}} = \rho_0\, \varphi_{\mathrm{ther}}(\vec{q}; \boldsymbol{\varepsilon} - \boldsymbol{\varepsilon}^{\mathrm{i}}, \boldsymbol{\alpha}, T, h) - \Phi_{\mathrm{ther}}(\vec{q}, \vec{\nabla} T/T) = 0\,, \tag{6.325}$$

where $\tau_{\mathrm{ther}} = 0$ is a constraint.

Case B: Consider one problem:

maximize with respect to the fluxes $(\dot{\boldsymbol{\varepsilon}}^{\mathrm{i}}, \dot{\boldsymbol{\alpha}}, \vec{q})$

$$\Phi = \Phi_{\mathrm{mech}}(\dot{\boldsymbol{\varepsilon}}^{\mathrm{i}}, \dot{\boldsymbol{\alpha}}, \boldsymbol{\sigma}, \boldsymbol{\beta}) + \Phi_{\mathrm{ther}}(\vec{q}, \vec{\nabla} T/T) = \boldsymbol{\sigma} : \dot{\boldsymbol{\varepsilon}}^{\mathrm{i}} + \boldsymbol{\beta} : \dot{\boldsymbol{\alpha}} - \frac{\vec{\nabla} T}{T} \cdot \vec{q}, \tag{6.326}$$

subject to the constraints:

$$\tau_{\mathrm{mech}} = \rho_0\, \varphi_{\mathrm{mech}}\,(\dot{\boldsymbol{\varepsilon}}^{\mathrm{i}}, \dot{\boldsymbol{\alpha}}; \boldsymbol{\varepsilon} - \boldsymbol{\varepsilon}^{\mathrm{i}}, \boldsymbol{\alpha}, T, h) - \Phi_{\mathrm{mech}}(\dot{\boldsymbol{\varepsilon}}^{\mathrm{i}}, \dot{\boldsymbol{\alpha}}, \boldsymbol{\sigma}, \boldsymbol{\beta}) = 0 \tag{6.327}$$

and

$$\tau_{\mathrm{ther}} = \rho_0\, \varphi_{\mathrm{ther}}(\vec{q}; \boldsymbol{\varepsilon} - \boldsymbol{\varepsilon}^{\mathrm{i}}, \boldsymbol{\alpha}, T, h) - \Phi_{\mathrm{ther}}(\vec{q}, \vec{\nabla} T/T) = 0\,. \tag{6.328}$$

Case A was studied in Sections 6.7.1 and 6.7.3, and case B is studied in **Appendix H**. It was shown in **Appendix H** that both approaches lead to the same result.

### 6.7.7 Normality Rule or Fluids

This section gives the normality rule for fluids. Equations (6.261) and (6.262) are

$$\Phi_{\text{mech}} = \boldsymbol{\sigma}^{\text{d}} : \overset{\circ}{\boldsymbol{\varepsilon}} \qquad (\geq 0) \tag{6.329}$$

and

$$\Phi_{\text{ther}} = -\frac{\vec{\nabla} T}{T} \cdot \vec{q} \qquad (\geq 0)\,. \tag{6.330}$$

By repeating the steps taken in Section 6.7.1, the following normality rule is obtained:

$$\boldsymbol{\sigma}^{\text{d}} = \mu\, \rho\, \frac{\partial \varphi_{\text{mech}}(\overset{\circ}{\boldsymbol{\varepsilon}}; \rho, T)}{\partial \overset{\circ}{\boldsymbol{\varepsilon}}}\,. \tag{6.331}$$

The specific dissipation function $\varphi_{\text{mech}}$ has to obey the following condition:

$$\varphi_{\text{mech}}(\overset{\circ}{\boldsymbol{\varepsilon}}; \rho, T) = \mu\, \frac{\partial \varphi_{\text{mech}}}{\partial \overset{\circ}{\boldsymbol{\varepsilon}}} : \overset{\circ}{\boldsymbol{\varepsilon}}\,. \tag{6.332}$$

The first-order sufficient condition for the point $\overset{\circ}{\boldsymbol{\varepsilon}}$ to be a local maximum is that Equations (6.331) and (6.332) hold and that the specific dissipation function $\varphi_{\text{mech}}$ is a homogeneous function of degree $1/\mu$. The latter property is obtained if the coefficient $\mu$ in Expression (6.332) is a constant. If the multiplier $\mu$ is not a constant but $\mu = \mu(\rho, T)$, the specific dissipation function $\varphi_{\text{mech}}$ is not a homogeneous function, and the value for $\mu$ is obtained from Equation (6.332).

Based on Section 6.7.3 and Expression (6.330), the normality rule for a thermal problem takes the following appearance:

$$-\frac{\vec{\nabla} T}{T} = \mu_{\text{ther}}\, \rho\, \frac{\partial \varphi_{\text{ther}}(\vec{q}; \rho, T)}{\partial \vec{q}}\,. \tag{6.333}$$

The specific dissipation function $\varphi_{\text{ther}}$ must satisfy

$$\varphi_{\text{ther}}(\vec{q}; \rho, T) = \mu_{\text{ther}}\, \frac{\partial \varphi_{\text{ther}}}{\partial \vec{q}} \cdot \vec{q}. \tag{6.334}$$

If the multiplier $\mu_{\text{ther}}$ is a constant, $\varphi_{\text{ther}}(\vec{q}; \rho, T)$ is a homogeneous function of any degree ($= 1/\mu_{\text{ther}}$) in the variable $\vec{q}$. If the coefficient $\mu_{\text{ther}}$ is not a constant but $\mu_{\text{ther}} = \mu_{\text{ther}}(\rho, T)$, the specific dissipation function $\varphi_{\text{ther}}$ is not a homogeneous function, and the value for $\mu_{\text{ther}}$ can be determined by Equation (6.334) .

## 6.8 Summary

This chapter derived the local forms of the first and second laws of thermodynamics. From the point of view of material modelling, the local form of the first law of thermodynamics is important, since its different appearance is the heat equation, which plays an important role in material modelling, as Chapter 8 will show. Sections 6.4.1–6.4.3 discussed state variables. These sections are important, since preparation of the reader's own material model may need the introduction of new internal variables being peculiar just for the material model at which they are aiming. Introduction of new internal variables implies introduction of new state equations. State equations are discussed in Sections 6.4.4 and 6.4.5. The validation tool for material models is the Clausius–Duhem inequality, which is the topic of Section 6.6. As Figure 5.1 shows, the Clausius–Duhem inequality contains information of all the other basic laws and axioms, except the principle of maximum dissipation. Usually the material models are prepared by writing an expression for the specific Helmholtz free energy $\psi$ or for the specific Gibbs free energy $g$ and the specific dissipation function $\varphi$ or the specific complementary dissipation function $\varphi^{\mathrm{c}}$. Often the mechanical and thermal parts of the problem are separated, and therefore the specific dissipation functions for the mechanical parts are $\varphi_{\mathrm{mech}}$ or $\varphi^{\mathrm{c}}_{\mathrm{mech}}$. For thermoplasticity, the specific dissipation functions are replaced by the yield function $F$. More detailed discussion on material modelling is carried out in Section 7.6.

# Problems

**6.1** The present problem studies the determination of the principal normal stresses and corresponding principal planes.

Although the value of the stress tensor $\boldsymbol{\sigma}$ does not depend on the coordinate system, the scalar components take different values when expressed in different coordinate systems. The present problem derives expressions to obtain the maximum value of the scalar components of the stress tensor $\boldsymbol{\sigma}$ and gives the plane which the maximum normal stress is acting on. The stress tensor $\boldsymbol{\sigma}$ is assumed to be symmetric. This assumption does not hold for all stress measures. The first Piola–Kirchhoff stress tensor $\boldsymbol{\sigma}^{\square}$ is not a symmetric tensor.

The scalar component of the stress tensor $\boldsymbol{\sigma}$ on the plane, the normal of which is $\vec{N}$, is given by the expression $\vec{N} \cdot \boldsymbol{\sigma} \cdot \vec{N}$. Thus the determination of the maximum value for the scalar component of the stress tensor means that the value for the expression $\vec{N} \cdot \boldsymbol{\sigma} \cdot \vec{N}$ is maximized. At the same time, it should be kept in mind that the length of the unit normal $\vec{N}$ is 1. By following the concept of the Lagrange multiplier introduced in Section 6.7.1, the following maximization problem can be written:

maximize with respect to the unit normal vector $\vec{N}$,

$$\sigma^{\mathrm{N}} := \vec{N}\cdot\boldsymbol{\sigma}\cdot\vec{N}\,, \tag{1}$$

subject to

$$\tau = -(\,\vec{N}\cdot\vec{N} - 1\,) = 0\,, \tag{2}$$

where $\tau = 0$ is a constraint. The minus sign in Constraint (2) is for convenience, as the forthcoming rows will show.

You need the following results:

$$\frac{\partial(\vec{N}\cdot\vec{N})}{\partial\vec{N}} = 2\,\vec{N} \tag{3}$$

and

$$\frac{\partial(\vec{N}\cdot\boldsymbol{\sigma}\cdot\vec{N})}{\partial\vec{N}} = \boldsymbol{\sigma}\cdot\vec{N} + \vec{N}\cdot\boldsymbol{\sigma}\,. \tag{4}$$

Since the stress tensor $\boldsymbol{\sigma}$ is symmetric, Derivative (4) can be written in the following two forms:

$$\frac{\partial(\vec{N}\cdot\boldsymbol{\sigma}\cdot\vec{N})}{\partial\vec{N}} = 2\,\boldsymbol{\sigma}\cdot\vec{N} \quad \text{and} \quad \frac{\partial(\vec{N}\cdot\boldsymbol{\sigma}\cdot\vec{N})}{\partial\vec{N}} = 2\,\vec{N}\cdot\boldsymbol{\sigma}\,. \tag{5}$$

Take the following steps:

**(a)** Apply Expression (6.269) and Results (4) and (5).

**(b)** Write the equations obtained in step (a) in the index form where the indices are $s$ and $t$, for example.

**(c)** Apply the Einstein summation convention, and let the indices $s$ and $t$ take the numerical values 1, 2, and 3.

**(d)** Write the obtained equations in the matrix format.

**(e)** Formally solve the matrix equation, and denote the solutions by $\lambda^{①}$, $\lambda^{②}$, etc. (give the missing variables that are denoted by "etc.").

**(f)** Every quantity $\lambda^{①}$, $\lambda^{②}$, etc. has a corresponding unit normal vector $\vec{N}^{①}$, $\vec{N}^{②}$ etc. Give the expression(s) for calculating the value for $\vec{N}^{①}$.

**(g)** Study the physical meaning of the quantities $\lambda^{①}$, $\lambda^{②}$, etc. by applying the expression obtained in Step (a). Study the quantity $\lambda^{①}$, for example. Use the equation(s) you obtained in Step (a).

**(h)** Substitute the definition of the stress tensor, $\vec{T}^{①} = \vec{N}^{①}\cdot\boldsymbol{\sigma}$, into the equation obtained in Step (a). Interpret the result.

**6.2** Small deformations and rotations are studied. When material models for solid materials are evaluated, the strain tensor $\boldsymbol{\varepsilon}$ is assumed to be separable into different terms, each of which describes a particular deformation mechanism. For fluids, on the other hand, the stress tensor $\boldsymbol{\sigma}$ is assumed to be separable into two terms.

The concept of this study is to study a material model that assumes that both strain tensor $\boldsymbol{\varepsilon}$ and stress tensor $\boldsymbol{\sigma}$ are separable, and to write the Clausius–Duhem inequality for such materials and assume that they are solids. They cannot be fluids, since the formulation for fluids does not have strain tensor $\boldsymbol{\varepsilon}$.

Therefore the specific internal energy $u$ is assumed to take the form

$$u = u(\boldsymbol{\varepsilon} - \boldsymbol{\varepsilon}^{\mathrm{i}},\, \boldsymbol{\alpha}, s)\,. \tag{1}$$

The stress tensor $\boldsymbol{\sigma}$ is assumed to be separable, as follows:

$$\boldsymbol{\sigma} = \boldsymbol{\sigma}^{\mathrm{q}} + \boldsymbol{\sigma}^{\mathrm{d}}\,. \tag{2}$$

Assume that one of the state equations reads [cf. Equation (9.26)]

$$\boldsymbol{\sigma}^{\mathrm{q}} = \rho_0 \,\frac{\partial u(\boldsymbol{\varepsilon} - \boldsymbol{\varepsilon}^{\mathrm{i}}, \boldsymbol{\alpha}, s, h(\vec{X}))}{\partial(\boldsymbol{\varepsilon} - \boldsymbol{\varepsilon}^{\mathrm{i}})}\,. \tag{3}$$

The other state equations follow those given in Section 6.4.4.

Derive the Clausius–Duhem inequality in terms of state variables and functions. Use the first and second laws of thermodynamics for solid materials.

**Hint.** Follow the concept given in Section 6.6. Start by writing the local forms of the first and second laws of thermodynamics. Be careful, since the derivation is not the same as carried out in Section 6.6; there are small differences.

**6.3** Small deformations and rotations are studied. When material models for solid materials are evaluated, the strain tensor $\boldsymbol{\varepsilon}$ is assumed to be separable into different terms, each of which describes a particular deformation mechanism. For fluids, on the other hand, the stress tensor $\boldsymbol{\sigma}$ is assumed to be separable into two terms.

The concept of this study is to study a material model that assumes that the explicit form for the specific internal energy $u$ is not known but the explicit form for the specific Helmholtz free energy $\psi$ is known. So, the specific Helmholtz free energy $\psi$ is assumed to take the form

$$\psi = \psi(\boldsymbol{\varepsilon} - \boldsymbol{\varepsilon}^{\mathrm{i}}, \boldsymbol{\alpha}, T, h(\vec{X}))\,. \tag{1}$$

Derive the Clausius–Duhem inequality for solids in the form "force times flux" [cf. Expression (6.249)]. Use the first and second laws of thermodynamics. Write the Clausius–Duhem inequality for such materials.

**Hint.** Follow the concept given in Section 6.6. Start by writing the local forms of the first and second laws of thermodynamics. After combining the first and second laws of thermodynamics, replace the specific internal energy $u$ with the specific Helmholtz free energy $\psi$. You do not need the explicit form for the specific Helmholtz free energy $\psi$. Be careful, since the derivation is not the same as carried out in Section 6.6; there are small differences.

# 7 Special Cases of Continuum Thermodynamics and Validation of Material Models

## 7.1 Introduction

This chapter studies special cases of continuum thermodynamics and gives step by step instructions for the validation of a material model. Pay special attention to Section 7.2, which shows that if the material model is formulated in such a way that the dissipation potential is a nonnegative homogeneous function, the Clausius–Duhem inequality is automatically satisfied. Section 7.6 gives rigorous instructions on the material model validation procedure, with an illustrative example.

## 7.2 Nonnegative Homogeneous Dissipation Potentials Satisfy the Clausius–Duhem Inequality

This section shows that the Clausius–Duhem inequality (CDI) is satisfied if the specific dissipation function $\varphi$ is a nonnegative homogeneous function and if the normality rule is assumed to hold. Solids are studied, but the results also hold for fluids.

Expressions (6.250) and (6.251) give the Clausius–Duhem inequalities for the separated problems. They are

$$\boldsymbol{\sigma} : \dot{\boldsymbol{\varepsilon}}^{\mathrm{i}} + \boldsymbol{\beta} : \dot{\boldsymbol{\alpha}} \geq 0, \qquad \text{and} \qquad -\frac{\vec{\nabla} T}{T} \cdot \vec{q} \geq 0 \,. \tag{7.1}$$

According to Equations (6.282), (6.283), and (6.289), the normality rule takes the form

$$\boldsymbol{\sigma} = \mu\, \rho_0 \, \frac{\partial \varphi_{\mathrm{mech}}(\dot{\boldsymbol{\varepsilon}}^{\mathrm{i}}, \dot{\boldsymbol{\alpha}}; \boldsymbol{\varepsilon} - \boldsymbol{\varepsilon}^{\mathrm{i}}, \boldsymbol{\alpha}, T, h)}{\partial \dot{\boldsymbol{\varepsilon}}^{\mathrm{i}}} \tag{7.2}$$

and

$$\boldsymbol{\beta} = \mu\, \rho_0 \, \frac{\partial \varphi_{\mathrm{mech}}(\dot{\boldsymbol{\varepsilon}}^{\mathrm{i}}, \dot{\boldsymbol{\alpha}}; \boldsymbol{\varepsilon} - \boldsymbol{\varepsilon}^{\mathrm{i}}, \boldsymbol{\alpha}, T, h)}{\partial \dot{\boldsymbol{\alpha}}}\,, \tag{7.3}$$

and finally,

$$-\frac{\vec{\nabla} T}{T} = \mu_{\mathrm{ther}}\, \rho_0 \, \frac{\partial \varphi_{\mathrm{ther}}(\vec{q}; \boldsymbol{\varepsilon} - \boldsymbol{\varepsilon}^{\mathrm{i}}, \boldsymbol{\alpha}, T, h)}{\partial \vec{q}}\,. \tag{7.4}$$

Substitution of Normality Rules (7.2)–(7.4) into the two parts of Clausius–Duhem Equality (7.1) gives the following:

$$\mu\,\rho_0\left[\frac{\partial\varphi_{\rm mech}}{\partial\dot{\boldsymbol{\varepsilon}}^{\rm i}}:\dot{\boldsymbol{\varepsilon}}^{\rm i}+\frac{\partial\varphi_{\rm mech}}{\partial\dot{\boldsymbol{\alpha}}}:\dot{\boldsymbol{\alpha}}\right]\geq 0 \quad\text{and}\quad \mu_{\rm ther}\,\rho_0\,\frac{\partial\varphi_{\rm ther}}{\partial\vec{q}}\cdot\vec{q}\geq 0\,. \tag{7.5}$$

Based on the discussions at the end of Sections 6.7.1 and 6.7.3, the following can be argued: if the specific dissipation functions $\varphi_{\rm mech}(\dot{\boldsymbol{\varepsilon}}^{\rm i},\dot{\boldsymbol{\alpha}};\boldsymbol{\varepsilon}-\boldsymbol{\varepsilon}^{\rm i},\boldsymbol{\alpha},T,h)$ and $\varphi_{\rm ther}(\vec{q};\boldsymbol{\varepsilon}-\boldsymbol{\varepsilon}^{\rm i},\boldsymbol{\alpha},T,h)$ are homogeneous functions, they are homogeneous functions of degree $1/\mu$ and $1/\mu_{\rm ther}$, respectively.

Equation (2.208) gives Euler's theorem on homogeneous functions for the extended definition formulated in Expression (2.207). It reads

$$\omega\,\theta(x,y,z,u,v)=\frac{\partial\theta}{\partial x}x+\frac{\partial\theta}{\partial y}y+\frac{\partial\theta}{\partial z}z\,, \tag{7.6}$$

where the function $\theta(x,y,z,u,v)$ is a homogeneous function of degree $\omega$ in variables $x$, $y$, and $z$.

Now, it is assumed that the specific dissipation functions $\varphi_{\rm mech}$ and $\varphi_{\rm ther}$ are homogeneous functions. Thus, for the present dissipation functions $\varphi_{\rm mech}(\dot{\boldsymbol{\varepsilon}}^{\rm i},\dot{\boldsymbol{\alpha}};\boldsymbol{\varepsilon}-\boldsymbol{\varepsilon}^{\rm i},\boldsymbol{\alpha},T,h)$ and $\varphi_{\rm ther}(\vec{q};\boldsymbol{\varepsilon}-\boldsymbol{\varepsilon}^{\rm i},\boldsymbol{\alpha},T,h)$, we arrive at the following:

$$\omega=\frac{1}{\mu}\quad\text{and}\quad\omega_{\rm ther}=\frac{1}{\mu_{\rm ther}}\,. \tag{7.7}$$

Applying Euler's Theorem (7.6) and Expressions (7.7) into Inequalities (7.5) yields

$$\mu\,\rho_0\left[\omega\,\varphi_{\rm mech}\right]\geq 0\quad\Rightarrow\quad\mu\rho_0\left[\frac{1}{\mu}\,\varphi_{\rm mech}\right]\geq 0\quad\Rightarrow\quad\rho_0\,\varphi_{\rm mech}\geq 0 \tag{7.8}$$

and

$$\mu_{\rm ther}\,\rho_0\,\frac{1}{\mu_{\rm ther}}\,\varphi_{\rm ther}\geq 0\quad\Rightarrow\quad\rho_0\,\varphi_{\rm ther}\geq 0\,. \tag{7.9}$$

Inequalities $(7.8)_2$ and $(7.9)_2$ show that if $\varphi_{\rm mech}(\dot{\boldsymbol{\varepsilon}}^{\rm i},\dot{\boldsymbol{\alpha}};\boldsymbol{\varepsilon}-\boldsymbol{\varepsilon}^{\rm i},\boldsymbol{\alpha},T,h)$ and $\varphi_{\rm ther}(\vec{q};\boldsymbol{\varepsilon}-\boldsymbol{\varepsilon}^{\rm i},\boldsymbol{\alpha},T,h)$ are nonnegative functions, Inequalities $(7.8)_2$ and $(7.9)_2$ hold and therefore Clausius–Duhem Inequalities (7.1) are satisfied. A similar investigation can also be made for the nonseparated specific dissipation function $\varphi(\dot{\boldsymbol{\varepsilon}}^{\rm i},\dot{\boldsymbol{\alpha}},\vec{q};\boldsymbol{\varepsilon}-\boldsymbol{\varepsilon}^{\rm i},\boldsymbol{\alpha},T,h)$.

The result obtained here has great practical importance. If a physicist selects a material model in which the specific dissipation function $\varphi$ is a nonnegative homogeneous function, and if they introduce the normality rule, they do not have to be unsure of the satisfaction of the Clausius–Duhem inequality. It is satisfied, no validation is necessary.

By repeating the aforementioned steps, it can be shown that material models described by the specific complementary dissipation functions $\varphi^{\rm c}_{\rm mech}(\boldsymbol{\sigma},\boldsymbol{\beta};\boldsymbol{\varepsilon}-$

$\boldsymbol{\varepsilon}^{\mathrm{i}}, \boldsymbol{\alpha}, T, h)$ and $\varphi^{\mathrm{c}}_{\mathrm{ther}}(-\nabla T/T; \boldsymbol{\varepsilon} - \boldsymbol{\varepsilon}^{\mathrm{i}}, \boldsymbol{\alpha}, T, h)$, which are nonnegative homogeneous functions, satisfy the Clausius–Duhem inequality.

The dissipative part of the material model is sometimes expressed as a sum of two dissipation potentials. Instead of the specific dissipation function $\varphi_{\mathrm{mech}}(\dot{\boldsymbol{\varepsilon}}^{\mathrm{i}}, \dot{\boldsymbol{\alpha}}; \boldsymbol{\varepsilon} - \boldsymbol{\varepsilon}^{\mathrm{i}}, \boldsymbol{\alpha}, T, h)$, the material model is assumed to be expressed by the specific complementary dissipation function $\varphi^{\mathrm{c}}_{\mathrm{mech}}(\boldsymbol{\sigma}, \boldsymbol{\beta}; \boldsymbol{\varepsilon} - \boldsymbol{\varepsilon}^{\mathrm{i}}, \boldsymbol{\alpha}, T, h)$. Thus, the following model is introduced:

$$\varphi^{\mathrm{c}}_{\mathrm{mech}}(\boldsymbol{\sigma}, \boldsymbol{\beta}; \boldsymbol{\varepsilon}-\boldsymbol{\varepsilon}^{\mathrm{i}}, \boldsymbol{\alpha}, T, h) := \varphi^{\mathrm{c1}}_{\mathrm{mech}}(\boldsymbol{\sigma}, \boldsymbol{\beta}; \boldsymbol{\varepsilon}-\boldsymbol{\varepsilon}^{\mathrm{i}}, \boldsymbol{\alpha}, T, h) + \varphi^{\mathrm{c2}}_{\mathrm{mech}}(\boldsymbol{\beta}; \boldsymbol{\varepsilon}-\boldsymbol{\varepsilon}^{\mathrm{i}}, \boldsymbol{\alpha}, T, h). \tag{7.10}$$

Substitution of Separation (7.10) into Normality Rules (6.286) and (6.287) gives

$$\dot{\boldsymbol{\varepsilon}}^{\mathrm{i}} = \rho_0 \frac{\partial \varphi^{\mathrm{c}}_{\mathrm{mech}}(\boldsymbol{\sigma}, \boldsymbol{\beta}; \boldsymbol{\varepsilon} - \boldsymbol{\varepsilon}^{\mathrm{i}}, \boldsymbol{\alpha}, T, h)}{\partial \boldsymbol{\sigma}} = \rho_0 \frac{\partial \varphi^{\mathrm{c1}}_{\mathrm{mech}}(\boldsymbol{\sigma}, \boldsymbol{\beta}; \mathit{state})}{\partial \boldsymbol{\sigma}} \tag{7.11}$$

and

$$\begin{aligned} \dot{\boldsymbol{\alpha}} &= \rho_0 \frac{\partial \varphi^{\mathrm{c}}_{\mathrm{mech}}(\boldsymbol{\sigma}, \boldsymbol{\beta}; \boldsymbol{\varepsilon} - \boldsymbol{\varepsilon}^{\mathrm{i}}, \boldsymbol{\alpha}, T, h)}{\partial \boldsymbol{\beta}} \\ &= \rho_0 \frac{\partial \varphi^{\mathrm{c1}}_{\mathrm{mech}}(\boldsymbol{\sigma}, \boldsymbol{\beta}; \mathit{state})}{\partial \boldsymbol{\beta}} + \rho_0 \frac{\partial \varphi^{\mathrm{c2}}_{\mathrm{mech}}(\boldsymbol{\beta}; \mathit{state})}{\partial \boldsymbol{\beta}} . \end{aligned} \tag{7.12}$$

In Normality Rules (7.11) and (7.12), the state variables $(\boldsymbol{\varepsilon} - \boldsymbol{\varepsilon}^{\mathrm{i}}, \boldsymbol{\alpha}, T, h)$ are expressed by a phrase *state*.

Substitution of the rates $(\dot{\boldsymbol{\varepsilon}}^{\mathrm{i}}, \dot{\boldsymbol{\alpha}})$ obtained from Normality Rules (7.11) and (7.12) into Clausius–Duhem Inequality $(7.1)_1$ gives the following:

$$\begin{aligned} \rho_0 \Bigg[ & \frac{\partial \varphi^{\mathrm{c1}}_{\mathrm{mech}}(\boldsymbol{\sigma}, \boldsymbol{\beta}; \mathit{state})}{\partial \boldsymbol{\sigma}} : \boldsymbol{\sigma} + \frac{\partial \varphi^{\mathrm{c1}}_{\mathrm{mech}}(\boldsymbol{\sigma}, \boldsymbol{\beta}; \mathit{state})}{\partial \boldsymbol{\beta}} : \boldsymbol{\beta} \\ & + \frac{\partial \varphi^{\mathrm{c2}}_{\mathrm{mech}}(\boldsymbol{\beta}; \mathit{state})}{\partial \boldsymbol{\beta}} : \boldsymbol{\beta} \Bigg] \geq 0 . \end{aligned} \tag{7.13}$$

If the function $\varphi^{\mathrm{c1}}_{\mathrm{mech}}(\boldsymbol{\sigma}, \boldsymbol{\beta}; \mathit{state})$ is a homogeneous function of degree $\omega^1$ in the variables $(\boldsymbol{\sigma}, \boldsymbol{\beta})$ and if the function $\varphi^{\mathrm{c2}}_{\mathrm{mech}}(\boldsymbol{\beta}; \mathit{state})$ is a homogeneous function of degree $\omega^2$ in the variable $(\boldsymbol{\beta})$, Property (7.6) can be applied and Inequality (7.13) reduces to

$$\omega^1 \rho_0 \varphi^{\mathrm{c1}}_{\mathrm{mech}} + \omega^2 \rho_0 \varphi^{\mathrm{c2}}_{\mathrm{mech}} \geq 0 . \tag{7.14}$$

If functions $\varphi^{\mathrm{c1}}_{\mathrm{mech}}$ and $\varphi^{\mathrm{c2}}_{\mathrm{mech}}$ are nonnegative functions, Inequality (7.14) is satisfied and therefore Clausius–Duhem Inequality $(7.1)_1$ is satisfied. This result can be extended to more complex dissipation potentials than that given by Model (7.10).

By repeating the aforementioned steps, the following can be shown: if the specific dissipation function $\varphi_{\mathrm{mech}}$ is separated into parts, which are homogeneous functions, the Clausius–Duhem inequality is satisfied.

## 7.3 Normality Rule by the French School of Thermodynamics

The French school of thermodynamics does not use the principle of maximum dissipation (or the principle of the maximal rate of entropy production) but adopts a different approach. This chapter evaluates the approach à la French school of thermodynamics. Solids are studied.

The Clausius–Duhem Inequality (6.249) is

$$\boldsymbol{\sigma} : \dot{\boldsymbol{\varepsilon}}^{\mathrm{i}} + \boldsymbol{\beta} : \dot{\boldsymbol{\alpha}} - \frac{\vec{\nabla} T}{T} \cdot \vec{q} \geq 0 \,. \tag{7.15}$$

The following presentation is based on Germain et al. ([25], pp. 1015–1016), Lemaitre [50], Chaboche ([14], pp. 60–63), Maugin ([59], pp. 177–180), Maugin ([60], pp. 38, 44, and 45), and Maugin and Muschik ([61], p. 221).

The left side of Inequality (7.15) is the dissipation $\Phi$ (power per unit volume) and can be written in the form

$$\Phi := \boldsymbol{\sigma} : \dot{\boldsymbol{\varepsilon}}^{\mathrm{i}} + \boldsymbol{\beta} : \dot{\boldsymbol{\alpha}} - \frac{\vec{\nabla} T}{T} \cdot \vec{q} \quad (\geq 0) \,. \tag{7.16}$$

We postulate the existence of a dissipation potential $\hat{\varphi}$ expressed as **(a)** a continuous and **(b)** a convex scalar valued function of the flux variables $(\dot{\boldsymbol{\varepsilon}}^{\mathrm{i}}, \dot{\boldsymbol{\alpha}}, \vec{q})$, in which the state variables may appear as parameters:

$$\hat{\varphi} = \hat{\varphi}(\dot{\boldsymbol{\varepsilon}}^{\mathrm{i}}, \dot{\boldsymbol{\alpha}}, \vec{q}; \boldsymbol{\varepsilon} - \boldsymbol{\varepsilon}^{\mathrm{i}}, \boldsymbol{\alpha}, T, h) \,. \tag{7.17}$$

This potential $\hat{\varphi}$ is **(c)** a nonnegative, **(d)** homogeneous function of a certain degree **(e)** with a zero value at the origin of the space of the flux variables $(\dot{\boldsymbol{\varepsilon}}^{\mathrm{i}}, \dot{\boldsymbol{\alpha}}, \vec{q})$. The complementary laws are then expressed by **(f)** the normality property

$$\boldsymbol{\sigma} = \frac{\partial \hat{\varphi}}{\partial \dot{\boldsymbol{\varepsilon}}^{\mathrm{i}}} \,, \qquad \boldsymbol{\beta} = \frac{\partial \hat{\varphi}}{\partial \dot{\boldsymbol{\alpha}}} , \qquad \text{and} \qquad - \frac{\vec{\nabla} T}{T} = \frac{\partial \hat{\varphi}}{\partial \vec{q}} \,. \tag{7.18}$$

In the foregoing evaluation, the dissipation $\Phi$ was not separated into mechanical and thermal parts, as is usually done by a researcher following the French style. It is equivalent to what was done in Sections 6.6 and 6.7.

The notations **(a)**–**(f)** indicating the properties of the volumetric dissipation function $\hat{\varphi}$ are by the authors. The term *volumetric* in front of $\hat{\varphi}$ indicates that $\hat{\varphi}$ is the power per unit volume. The specific dissipation function $\varphi$ is the power per unit mass.

The authors make the following comments on the preceding.

If the volumetric dissipation function $\hat{\varphi}$ is a convex function, it is necessarily a continuous function. This means that Property **(a)** is a consequence of Property **(b)**.

Normality Rules (7.18) **(f)** are substituted into Definition (7.16), and the following is achieved:

$$\Phi = \frac{\partial \hat{\varphi}}{\partial \dot{\boldsymbol{\varepsilon}}^{\mathrm{i}}} : \dot{\boldsymbol{\varepsilon}}^{\mathrm{i}} + \frac{\partial \hat{\varphi}}{\partial \dot{\boldsymbol{\alpha}}} : \dot{\boldsymbol{\alpha}} + \frac{\partial \hat{\varphi}}{\partial \vec{q}} \cdot \vec{q} \qquad (\geq 0)\,. \tag{7.19}$$

Next it will be shown that the dissipation $\Phi$ expressed by Equation (7.19) takes nonnegative values.

Since, according to Item **(d)**, the dissipation potential $\hat{\varphi}$ is a homogeneous function of a certain degree (let us say, of degree $\omega$) in the variables $(\dot{\boldsymbol{\varepsilon}}^{\mathrm{i}}, \dot{\boldsymbol{\alpha}}, \vec{q})$, according to Euler's theorem on homogeneous functions [Theorem (2.208)], the following holds:

$$\omega\,\hat{\phi} = \frac{\partial \hat{\varphi}}{\partial \dot{\boldsymbol{\varepsilon}}^{\mathrm{i}}} : \dot{\boldsymbol{\varepsilon}}^{\mathrm{i}} + \frac{\partial \hat{\varphi}}{\partial \dot{\boldsymbol{\alpha}}} : \dot{\boldsymbol{\alpha}} + \frac{\partial \hat{\varphi}}{\partial \vec{q}} \cdot \vec{q}\,. \tag{7.20}$$

Comparison of Expressions (7.19) and (7.20) gives

$$\omega\,\hat{\varphi} = \Phi\,. \tag{7.21}$$

In Expression (7.21), the function $\hat{\varphi}$ is nonnegative, since Property **(c)** assumed it to be nonnegative. In order to ensure the nonnegativeness of the dissipation $\Phi$, the homogeneity of the dissipation potential $\hat{\varphi}$ has to be nonnegative, that is, $\omega \;\geq 0$. This is not a general property, since according to Widder ([117], p. 20) a function can be homogeneous of degree $n$, where $n$ can be positive, negative, or zero, and it need not be an integer. Later in this chapter, it will be shown that the volumetric dissipation function $\hat{\varphi}$ cannot be a homogeneous function of a degree less than unity.

Thus, Normality Rule (7.18) leads to the satisfaction of Clausius–Duhem Inequality (7.15) if the dissipation potential $\hat{\varphi}$ is a homogeneous nonnegative function of a degree larger than zero. A corresponding result was already obtained in the previous section.

Next it is shown that the approach by the French school of thermodynamics leads to a maximization problem similar to the principle of maximum dissipation [principle of maximal rate of entropy production by Ziegler ([120], p. 134)].

Above the dissipation potential $\hat{\varphi}$ is said to be **(b)** a convex function. This implies that the domain (denoted by $\Omega$) bounded by the dissipation potential $\hat{\varphi}$ is a convex domain. The domain $\Omega$ is called an epigraph. Now the following investigation can be made.

According to Luenberger ([55], p. 116), the following propositions holds. Let $f \in C^1$. Then $f$ is convex over a convex set $\Omega$ if and only if

$$f(y) \geq f(x) + \nabla f(x)\,(y - x) \tag{7.22}$$

for all $x, y \in \omega$.

In the foregoing proposition, the notation $f \in C^1$ indicates that the first derivative of $f$ is a continuous function and that $\nabla$ is a gradient operator.

Inequality (7.22) is multiplied by $-1$ and reorganized. This gives

$$\nabla f(x)\,(x-y) \geq f(x) - f(y)\,. \tag{7.23}$$

In Inequality (7.23), the function $f$ is replaced with the dissipation potential $\hat{\varphi}$. The variable $x$ is replaced with the set actual fluxes $(\dot{\boldsymbol{\varepsilon}}^{\mathrm{i}}_{\mathrm{a}}, \dot{\boldsymbol{\alpha}}_{\mathrm{a}}, \vec{q}_{\mathrm{a}})$, where the subscript a refers to an actual process. The variable $y$ is an arbitrary set of fluxes $(\dot{\boldsymbol{\varepsilon}}^{\mathrm{i}}, \dot{\boldsymbol{\alpha}}, \vec{q})$ being within the domain $\Omega$ but not being the actual set of fluxes. Thus Inequality (7.23) yields

$$\frac{\partial \hat{\varphi}_{\mathrm{a}}}{\partial \dot{\boldsymbol{\varepsilon}}^{\mathrm{i}}} : (\dot{\boldsymbol{\varepsilon}}^{\mathrm{i}}_{\mathrm{a}} - \dot{\boldsymbol{\varepsilon}}^{\mathrm{i}}) + \frac{\partial \hat{\varphi}_{\mathrm{a}}}{\partial \dot{\boldsymbol{\alpha}}} : (\dot{\boldsymbol{\alpha}}_{\mathrm{a}} - \dot{\boldsymbol{\alpha}}) + \frac{\partial \hat{\varphi}_{\mathrm{a}}}{\partial \vec{q}} \cdot (\vec{q}_{\mathrm{a}} - \vec{q}) \geq \hat{\varphi}_{a} - \hat{\varphi}\,, \tag{7.24}$$

where

$$\hat{\varphi}_{\mathrm{a}} := \hat{\varphi}(\dot{\boldsymbol{\varepsilon}}^{\mathrm{i}}_{\mathrm{a}}, \dot{\boldsymbol{\alpha}}_{\mathrm{a}}, \vec{q}_{\mathrm{a}}; \boldsymbol{\varepsilon} - \boldsymbol{\varepsilon}^{\mathrm{i}}, \boldsymbol{\alpha}, T, h)\,. \tag{7.25}$$

Since Normality Rule (7.18) holds for an actual process, it allows Inequality (7.24) to be written in the form

$$\boldsymbol{\sigma} : (\dot{\boldsymbol{\varepsilon}}^{\mathrm{i}}_{\mathrm{a}} - \dot{\boldsymbol{\varepsilon}}^{\mathrm{i}}) + \boldsymbol{\beta} : (\dot{\boldsymbol{\alpha}}_{\mathrm{a}} - \dot{\boldsymbol{\alpha}}) - \frac{\vec{\nabla} T}{T} \cdot (\vec{q}_{\mathrm{a}} - \vec{q}) \geq \hat{\varphi}_{\mathrm{a}} - \hat{\varphi}\,. \tag{7.26}$$

Inequality (7.26) can be written in the form

$$\boldsymbol{\sigma} : \dot{\boldsymbol{\varepsilon}}^{\mathrm{i}}_{\mathrm{a}} + \boldsymbol{\beta} : \dot{\boldsymbol{\alpha}}_{\mathrm{a}} - \frac{\vec{\nabla} T}{T} \cdot \vec{q}_{\mathrm{a}} - \boldsymbol{\sigma} : \dot{\boldsymbol{\varepsilon}}^{\mathrm{i}} - \boldsymbol{\beta} : \dot{\boldsymbol{\alpha}} + \frac{\vec{\nabla} T}{T} \cdot \vec{q}_{\mathrm{a}} \geq \hat{\varphi}_{a} - \hat{\varphi}\,. \tag{7.27}$$

Comparison of the terms on the left side of Inequality (7.27) with Definition (7.16) gives

$$\Phi(\dot{\boldsymbol{\varepsilon}}^{\mathrm{i}}_{\mathrm{a}}, \dot{\boldsymbol{\alpha}}_{\mathrm{a}}, \vec{q}_{\mathrm{a}}; \boldsymbol{\varepsilon} - \boldsymbol{\varepsilon}^{\mathrm{i}}, \boldsymbol{\alpha}, T, h) - \Phi(\dot{\boldsymbol{\varepsilon}}^{\mathrm{i}}, \dot{\boldsymbol{\alpha}}, \vec{q}; \boldsymbol{\varepsilon} - \boldsymbol{\varepsilon}^{\mathrm{i}}, \boldsymbol{\alpha}, T, h) \geq \hat{\varphi}_{\mathrm{a}} - \hat{\varphi}\,. \tag{7.28}$$

If the dissipation potential $\varphi$ is a homogeneous function of degree $\omega$, Result (7.21) holds and Inequality (7.28) takes the form

$$\Phi_{\mathrm{a}} - \Phi \geq \frac{1}{\omega}\,\Phi_{\mathrm{a}} - \frac{1}{\omega}\,\Phi\,, \qquad \text{which yields} \quad (\omega - 1)\,(\Phi_{\mathrm{a}} - \Phi) \geq 0\,. \tag{7.29}$$

If the homogeneity of the dissipation potential $\hat{\varphi}$ is higher than 1, that is, $\omega > 1$, Inequality $(7.29)_2$ gives

$$\Phi_{\mathrm{a}} - \Phi \geq 0\,. \tag{7.30}$$

According to Widder ([117], p. 23), if $f$ is a homogeneous function of degree $\omega$, any derivative of order $p$ is a homogeneous function of degree $\omega - p$.

It is assumed that the volumetric dissipation function $\hat{\varphi}$ is a homogeneous function of degree $\omega < 1$. According to Normality Rule $(7.18)_1$, the stress tensor $\boldsymbol{\sigma}$ is obtained as a first derivative of the volumetric dissipation function $\varphi$. Based on the theorem by Widder, the stress tensor $\boldsymbol{\sigma}$ is a homogeneous function of degree $\omega - 1 < 0$. The degree of homogeneity of the stress tensor $\boldsymbol{\sigma}$ is denoted by $-\kappa$, where $\kappa$ is a positive constant. The extended definition for the homogeneous functions was given by Definition (2.207),

$$\theta(kx, ky, kz, u, v) := k^{\omega}\, \theta(x, y, z, u, v)\,, \tag{7.31}$$

where $\theta$ is a homogeneous function of degree $\omega$ in the variables $x$, $y$, and $z$. Based on the foregoing discussion, Definition (7.31), and Expression (7.17), the following is achieved:

$$\boldsymbol{\sigma}(k\,\dot{\boldsymbol{\varepsilon}}^{\mathrm{i}}, k\,\dot{\boldsymbol{\alpha}}, k\vec{q}; \boldsymbol{\varepsilon} - \boldsymbol{\varepsilon}^{\mathrm{i}}, \boldsymbol{\alpha}, T, h) = k^{-\kappa}\boldsymbol{\sigma}(\dot{\boldsymbol{\varepsilon}}^{\mathrm{i}}, \dot{\boldsymbol{\alpha}}, \vec{q}; \boldsymbol{\varepsilon} - \boldsymbol{\varepsilon}^{\mathrm{i}}, \boldsymbol{\alpha}, T, h). \tag{7.32}$$

Equation (7.32) exploits the fact that the derivative of a function $f$ has the same arguments and parameters as the function $f$ itself. Of course, the inelastic strain rate $\dot{\boldsymbol{\varepsilon}}^{\mathrm{i}}$ can vanish, because the stress tensor $\boldsymbol{\sigma}$ is obtained from the derivative of the volumetric dissipation function $\hat{\varphi}$ with respect to the inelastic strain rate $\dot{\boldsymbol{\varepsilon}}^{\mathrm{i}}$. Since $k$ takes only positive values, the information of Expression (7.32) is as follows: if the value of the applied stress $\boldsymbol{\sigma}$ is decreased by dividing it by a factor $k^{\kappa}$, the values of the fluxes $(\dot{\boldsymbol{\varepsilon}}^{\mathrm{i}}, \dot{\boldsymbol{\alpha}}, \vec{q})$ grow by a positive-valued factor $k$. Creep rate $\dot{\boldsymbol{\varepsilon}}^{\mathrm{v}}$, for example, does not grow when the applied stress $\boldsymbol{\sigma}$ is decreased. Thus, from the physics point of view, the volumetric dissipation function $\hat{\varphi}$ cannot be a homogeneous function of degree $\omega < 1$.

Result (7.30) and the foregoing discussion indicate that if the volumetric dissipation function $\hat{\varphi} \in C^1$, the actual process follows a path where the dissipation $\Phi$ takes its maximum value. This result is equivalent to Ziegler's principle of maximal rate of entropy production (or the principle of maximum dissipation) evaluated in Section 6.7.

According to Property **(d)**, the dissipation potential $\hat{\varphi}$ has to take zero value at the origin of the space of the flux variables $(\dot{\boldsymbol{\varepsilon}}^{\mathrm{i}}, \dot{\boldsymbol{\alpha}}, \vec{q})$. According to Equation (7.21), $\omega\,\hat{\varphi} = \Phi$, volumetric dissipation function $\hat{\varphi}$ is linearly related to the dissipation of the material $\Phi$. When dissipative fluxes $(\dot{\boldsymbol{\varepsilon}}^{\mathrm{i}}, \dot{\boldsymbol{\alpha}}, \vec{q})$ vanish, dissipation $\Phi$ also vanishes. Therefore volumetric dissipation function $\hat{\varphi}$ has to vanish at the origin of the space of the flux variables $(\dot{\boldsymbol{\varepsilon}}^{\mathrm{i}}, \dot{\boldsymbol{\alpha}}, \vec{q})$.

According to Nguyen [69], the volumetric dissipation function $\hat{\varphi}$ does not have to be a homogeneous function. Instead, the derivative of the dissipation potential $\hat{\varphi}$ vanishes at the origin.

This is not a new property, as demonstrated by Kanervo [36]. According to Properties **(c)** and **(e)**, $\hat{\varphi} \geq 0$ and $\hat{\varphi}(0) = 0$. If also $\hat{\varphi} \in C^1$, the derivative of the volumetric dissipation function $\hat{\varphi}$ exists and vanishes at the origin.

The authors have the following comments on the preceding.

First, a mathematical result is derived. Property (7.22) is

$$f(y) \geq f(x) + \nabla f(x)\,(y - x)\,. \tag{7.33}$$

By interchanging points $y$ and $x$, Property (7.33) takes the form

$$f(x) \geq f(y) + \nabla f(y)\,(x - y)\,. \tag{7.34}$$

By adding Forms (7.33) and (7.34) together, we arrive at the following:

$$f(x) + f(y) \geq f(x) + f(y) + \nabla f(x)\,(y - x) + \nabla f(y)\,(x - y)\,, \tag{7.35}$$

which gives

$$[\,\nabla f(x) - \nabla f(y)\,]\,(x - y) \geq 0\,. \tag{7.36}$$

The condition given by Nguyen [69] can be expressed as follows:

$$\frac{\partial \hat{\varphi}(\mathbf{0}, \mathbf{0}, \vec{0}; \boldsymbol{\varepsilon} - \boldsymbol{\varepsilon}^{\mathrm{i}}, \boldsymbol{\alpha}, T, h)}{\partial\, \dot{\boldsymbol{\varepsilon}}^{\mathrm{i}}} = \mathbf{0} \quad \text{and} \quad \frac{\partial \hat{\varphi}(\mathbf{0}, \mathbf{0}, \vec{0}; \boldsymbol{\varepsilon} - \boldsymbol{\varepsilon}^{\mathrm{i}}, \boldsymbol{\alpha}, T, h)}{\partial \dot{\boldsymbol{\alpha}}} = \mathbf{0}\,, \tag{7.37}$$

and, finally,

$$\frac{\partial \hat{\varphi}(\mathbf{0}, \mathbf{0}, \vec{0}; \boldsymbol{\varepsilon} - \boldsymbol{\varepsilon}^{\mathrm{i}}, \boldsymbol{\alpha}, T, h)}{\partial \vec{q}} = \vec{0}\,. \tag{7.38}$$

In Inequality (7.36), the function $f$ is replaced by the dissipation potential $\hat{\varphi}$. The variable $x$ is replaced by the set actual fluxes $(\dot{\boldsymbol{\varepsilon}}^{\mathrm{i}}_{\mathrm{a}}, \dot{\boldsymbol{\alpha}}_{\mathrm{a}}, \vec{q}_{\mathrm{a}})$, where the subscript a refers to an actual process. The variable $y$ is an arbitrary set of fluxes $(\dot{\boldsymbol{\varepsilon}}^{\mathrm{i}}, \dot{\boldsymbol{\alpha}}, \vec{q}\,)$ being within the domain $\Omega$ but not being the actual set of fluxes. Thus Inequality (7.36) yields

$$\left[\frac{\partial \hat{\varphi}_{\mathrm{a}}}{\partial \dot{\boldsymbol{\varepsilon}}^{\mathrm{i}}} - \frac{\partial \hat{\varphi}}{\partial \dot{\boldsymbol{\varepsilon}}^{\mathrm{i}}}\right] : (\dot{\boldsymbol{\varepsilon}}^{\mathrm{i}}_{\mathrm{a}} - \dot{\boldsymbol{\varepsilon}}^{\mathrm{i}}) + \left[\frac{\partial \hat{\varphi}_{\mathrm{a}}}{\partial \dot{\boldsymbol{\alpha}}} - \frac{\partial \hat{\varphi}}{\partial \dot{\boldsymbol{\alpha}}}\right] : (\dot{\boldsymbol{\alpha}}_{\mathrm{a}} - \dot{\boldsymbol{\alpha}}) \\ + \left[\frac{\partial \hat{\varphi}_{\mathrm{a}}}{\partial \vec{q}} - \frac{\partial \hat{\varphi}}{\partial \vec{q}}\right] \cdot (\vec{q}_{\mathrm{a}} - \vec{q}\,) \geq 0. \tag{7.39}$$

Now it is assumed that the point $(\dot{\boldsymbol{\varepsilon}}^{\mathrm{i}}, \dot{\boldsymbol{\alpha}}, \vec{q}\,)$ is the origin. This is the point

$$\dot{\boldsymbol{\varepsilon}}^{\mathrm{i}} = \mathbf{0}\,, \qquad \dot{\boldsymbol{\alpha}} = \mathbf{0}\,, \qquad \text{and} \qquad \vec{q} = \vec{0}\,. \tag{7.40}$$

This means that the derivatives $\partial\hat{\varphi}/\partial\dot{\boldsymbol{\varepsilon}}^{\mathrm{i}}$, $\partial\hat{\varphi}/\partial\dot{\boldsymbol{\alpha}}$, and $\partial\hat{\varphi}/\partial\vec{q}$ in Inequality (7.39) are calculated at the origin. Thus, based on Expressions (7.37), (7.38), and (7.40), Inequality (7.39) yields

$$\frac{\partial \hat{\varphi}_{\mathrm{a}}}{\partial \dot{\boldsymbol{\varepsilon}}^{\mathrm{i}}} : \dot{\boldsymbol{\varepsilon}}^{\mathrm{i}}_{\mathrm{a}} + \frac{\partial \hat{\varphi}_{\mathrm{a}}}{\partial \dot{\boldsymbol{\alpha}}} : \dot{\boldsymbol{\alpha}}_{\mathrm{a}} + \frac{\partial \hat{\varphi}_{\mathrm{a}}}{\partial \vec{q}} \cdot \vec{q}_{\mathrm{a}} \geq 0\,. \tag{7.41}$$

Normality Rule (7.18) holds for an actual process. By taking into account Normality Rules (7.18) and **(f)**, Inequality (7.41) takes the form

$$\boldsymbol{\sigma} : \dot{\boldsymbol{\varepsilon}}^{\mathrm{i}}_{\mathrm{a}} + \boldsymbol{\beta} : \dot{\boldsymbol{\alpha}}_{\mathrm{a}} + \frac{\vec{\nabla} T}{T} \cdot \vec{q}_{\mathrm{a}} \geq 0\,. \tag{7.42}$$

Result (7.42) means that Clausius–Duhem Inequality (7.15) is satisfied.

## 7.4 Material Models without Elastic Deformation

This section evaluates material models for solids without elastic deformation. It utilizes the ideas introduced in the previous chapters. The set of internal variables and state functions are different. Therefore the investigation is brief, and the reader is encouraged to study previous chapters before evaluating this one. A Kelvin–Voigt solid is a material without elastic deformation.

In contrast to Expression (6.162), here the state is assumed to be expressed by the following caloric equation of state:

$$u = u(\boldsymbol{\varepsilon}, \boldsymbol{\alpha}, s, h(\vec{X}))\,. \tag{7.43}$$

Comparing Expression (7.43) with Expressions (6.162) and (6.171), the specific Helmholtz free energy $\psi$ takes the form

$$\begin{aligned} \psi(\varepsilon, \boldsymbol{\alpha}, T, h(\vec{X})) &\overset{\text{Legendre}}{=} u(\varepsilon, \boldsymbol{\alpha}, s, h(\vec{X})) - T\,s \\ \text{with} \quad T &= \frac{\partial u(\boldsymbol{\varepsilon}, \boldsymbol{\alpha}, s, h(\vec{X}))}{\partial s}. \end{aligned} \tag{7.44}$$

State Equations (6.172) and (6.174) are now replaced with

$$s = -\frac{\partial \psi(\boldsymbol{\varepsilon}, \boldsymbol{\alpha}, T, h(\vec{X}))}{\partial T} \tag{7.45}$$

and

$$\boldsymbol{\sigma}^{\mathrm{q}} = \rho_0 \frac{\partial \psi(\boldsymbol{\varepsilon}, \boldsymbol{\alpha}, s, h(\vec{X}))}{\partial \boldsymbol{\varepsilon}} \quad \text{and} \quad \boldsymbol{\beta} = -\rho_0 \frac{\partial \psi(\boldsymbol{\varepsilon}, \boldsymbol{\alpha}, s, h(\vec{X}))}{\partial \boldsymbol{\alpha}}\,. \tag{7.46}$$

In State Equation $(7.46)_1$, the notation $\boldsymbol{\sigma}^{\mathrm{q}}$ stands for the quasi-conservative stress tensor (due to Ziegler [121], p. 61). The notion of the quasi-conservative stress tensor $\boldsymbol{\sigma}^{\mathrm{q}}$ is the second deviation from the formulation for solid materials prepared in the previous chapters.

The material derivative of the specific Helmholtz free energy $\dot{\psi}$ is given. Based on Definition (7.44), the following is achieved:

$$\rho_0\,\dot{\psi}(\boldsymbol{\varepsilon}, \boldsymbol{\alpha}, T, h(\vec{X})) = \rho_0 \frac{\partial \psi(\ldots)}{\partial \boldsymbol{\varepsilon}}) : \dot{\boldsymbol{\varepsilon}} + \rho_0 \frac{\partial \psi(\ldots)}{\partial \boldsymbol{\alpha}} : \dot{\boldsymbol{\alpha}} + \rho_0 \frac{\partial \psi(\ldots)}{\partial T} \dot{T}\,. \tag{7.47}$$

Since the quantity $h(\vec{X})$ does not depend on time, its material derivative vanishes. This is exploited in Derivative (7.47). Substitution of State Equations (7.45) and (7.46) into Derivative (7.47) yields

$$\rho_0\,\dot{\psi}(\boldsymbol{\varepsilon}, \boldsymbol{\alpha}, T, h) = \boldsymbol{\sigma}^{\mathrm{q}} : \dot{\boldsymbol{\varepsilon}} - \boldsymbol{\beta} : \dot{\boldsymbol{\alpha}} - \rho_0\, s\,\dot{T}\,. \tag{7.48}$$

Study of Section 6.2.1 shows that the introduction of the quasi-conservative stress tensor $\boldsymbol{\sigma}^{\mathrm{q}}$ does not change the appearance of Energy Equation (6.37), which is

$$\rho_0\,\dot{u} = \boldsymbol{\sigma} : \dot{\boldsymbol{\varepsilon}} + \rho_0\, r - \vec{\nabla} \cdot \vec{q}\,. \tag{7.49}$$

Expression $(7.44)_1$ is

$$\psi = u - T s\,, \qquad \text{and it gives} \qquad \dot{u} = \dot{\psi} + \dot{T}\, s + T\, \dot{s}\,. \tag{7.50}$$

Substitution of Derivative $(7.50)_2$ into Energy Equation (7.49) yields

$$\rho_0\, \dot{\psi} + \rho_0\, \dot{T}\, s + \rho_0\, T\, \dot{s} - \boldsymbol{\sigma} : \dot{\boldsymbol{\varepsilon}} - \rho_0\, r + \vec{\nabla} \cdot \vec{q} = 0\,. \tag{7.51}$$

Substitution of the derivative $\rho\, \dot{\psi}$ from Equation (7.48) into Expression (7.51) leads to

$$\vec{\nabla} \cdot \vec{q} = (\boldsymbol{\sigma} - \boldsymbol{\sigma}^{\mathrm{q}}) : \dot{\boldsymbol{\varepsilon}} + \boldsymbol{\beta} : \dot{\boldsymbol{\alpha}} - \rho_0\, T\, \dot{s} + \rho_0\, r\,. \tag{7.52}$$

Following Ziegler ([121], p. 60), the dissipative stress tensor $\boldsymbol{\sigma}^{\mathrm{d}}$ is defined by [see also Equation (6.75)]

$$\boldsymbol{\sigma}^{\mathrm{d}} := \boldsymbol{\sigma} - \boldsymbol{\sigma}^{\mathrm{q}}\,. \tag{7.53}$$

State Equation (7.45) is

$$s = -\frac{\partial \psi}{\partial T}\,, \qquad \text{which gives} \qquad \dot{s} = -\frac{\partial \dot{\psi}}{\partial T}\,. \tag{7.54}$$

Substitution of the rate of the specific Helmholtz free energy $\dot{\psi}$ from Derivative (7.47) into Rate $(7.54)_2$ and the obtained result further into Expression (7.52) gives

$$\vec{\nabla} \cdot \vec{q} = \boldsymbol{\sigma}^{\mathrm{d}} : \dot{\boldsymbol{\varepsilon}} + \boldsymbol{\beta} : \dot{\boldsymbol{\alpha}} + \rho_0\, r + \rho_0\, T \left( \frac{\partial^2 \psi}{\partial T\, \partial \boldsymbol{\varepsilon}} : \dot{\boldsymbol{\varepsilon}} + \frac{\partial^2 \psi}{\partial T\, \partial \boldsymbol{\alpha}} : \dot{\boldsymbol{\alpha}} + \frac{\partial^2 \psi}{\partial T^2}\, \dot{T} \right), \tag{7.55}$$

where Definition (7.53) is taken into account. Expression (7.55) is the heat equation. In the case of pure reversible (elastic) deformation, that is, when $\boldsymbol{\sigma}^{\mathrm{d}} = \mathbf{0}$, $\dot{\boldsymbol{\alpha}} = \mathbf{0}$, $\boldsymbol{\varepsilon}$ reduces to $\boldsymbol{\varepsilon}^{\mathrm{e}}$ and $\psi(\boldsymbol{\varepsilon}, \boldsymbol{\alpha}, T, h)$ reduces to $\psi^{\mathrm{r}}(\boldsymbol{\varepsilon}^{\mathrm{e}}, T, h)$, where the superscripts e and r refer to elastic and reversible process, respectively. Heat Equation (7.55) reduces to

$$\vec{\nabla} \cdot \vec{q} = \rho_0\, T \left[ \frac{\partial^2 \psi^{\mathrm{r}}}{\partial T\, \partial \boldsymbol{\varepsilon}^{\mathrm{e}}} : \dot{\boldsymbol{\varepsilon}}^{\mathrm{e}} + \frac{\partial^2 \psi^{\mathrm{r}}}{\partial T^2}\, \dot{T} \right] + \rho_0\, r \tag{7.56}$$

as derived by Parkus [[75], Eq. (5.28)].

Definition (6.205) for the specific heat capacity $c$ reads

$$c := -T\, \frac{\partial \psi^2(\boldsymbol{\varepsilon}, \boldsymbol{\alpha}, T, h(\vec{x}))}{\partial T^2} \qquad (\geq 0)\,. \tag{7.57}$$

Definition (7.57) allows Heat Equation (7.55) to be written in the form

$$\vec{\nabla} \cdot \vec{q} = \boldsymbol{\sigma}^{\mathrm{d}} : \dot{\boldsymbol{\varepsilon}} + \boldsymbol{\beta} : \dot{\boldsymbol{\alpha}} + \rho_0\, r + \rho_0\, T \left[ \frac{\partial^2 \psi}{\partial T\, \partial \boldsymbol{\varepsilon}} : \dot{\boldsymbol{\varepsilon}} + \frac{\partial^2 \psi}{\partial T\, \partial \boldsymbol{\alpha}} : \dot{\boldsymbol{\alpha}} \right] - \rho_0\, c\, \dot{T}\,. \tag{7.58}$$

The local form of the second law of thermodynamics takes the same form as obtained in Expression (6.119). Thus the following is written:

$$\Phi := \rho_0 \, T \, \dot{s} + \vec{\nabla} \cdot \vec{q} - \frac{\vec{\nabla} T}{T} \cdot \vec{q} - \rho_0 \, r \qquad (\geq 0) \, . \tag{7.59}$$

Substitution of $\vec{\nabla} \cdot \vec{q}$ evaluated from Energy Equation (7.49) into Definition (7.59) gives

$$\Phi = \boldsymbol{\sigma} : \dot{\boldsymbol{\varepsilon}} - \rho_0 \, \dot{u} + \rho_0 \, T \, \dot{s} - \frac{\vec{\nabla} T}{T} \cdot \vec{q} \qquad (\geq 0) \, . \tag{7.60}$$

Substitution of Equation $(7.50)_2$ into Equation (7.60) yields

$$\Phi = \boldsymbol{\sigma} : \dot{\boldsymbol{\varepsilon}} - \rho_0 \, \dot{\psi} - \rho_0 \, \dot{T} \, s - \frac{\vec{\nabla} T}{T} \cdot \vec{q} \qquad (\geq 0) \, . \tag{7.61}$$

Inequality (7.61) is

$$\boldsymbol{\sigma} : \dot{\boldsymbol{\varepsilon}} - \rho_0 \, \dot{\psi} - \rho_0 \, \dot{T} \, s - \frac{\vec{\nabla} T}{T} \cdot \vec{q} \geq 0 \, , \tag{7.62}$$

which is the Clausius–Duhem inequality. The material model has to obey Clausius–Duhem Inequality (7.62). The validation can be done either numerically during the computation procedure or analytically, the latter of course being much more convenient. In both cases Form (7.62) is not very useful and must therefore be recast into a new appearance. This recasting procedure follows.

Substitution of derivative $\rho_0 \, \dot{\psi}$ from Equation (7.48) into Equation (7.62) leads to

$$\Phi = \boldsymbol{\sigma}^{\mathrm{d}} : \dot{\boldsymbol{\varepsilon}} + \boldsymbol{\beta} : \dot{\boldsymbol{\alpha}} - \frac{\vec{\nabla} T}{T} \cdot \vec{q} \qquad (\geq 0) \, , \tag{7.63}$$

where Definition (7.53) for the dissipative stress tensor $\boldsymbol{\sigma}^{\mathrm{d}}$ is exploited. Inequality (7.63) is

$$\boldsymbol{\sigma}^{\mathrm{d}} : \dot{\boldsymbol{\varepsilon}} + \boldsymbol{\beta} : \dot{\boldsymbol{\alpha}} - \frac{\vec{\nabla} T}{T} \cdot \vec{q} \geq 0 \, . \tag{7.64}$$

Forms (7.62) and (7.64) are two different formulations of the Clausius–Duhem inequality. The difference between them is that the latter introduces an internal variable $\boldsymbol{\alpha}$. Form (7.64) is used with the principle of maximum dissipation, as shown in Section 6.7.

By following Separation (6.242), Expression (7.63) gives

$$\Phi_{\mathrm{mech}} = \boldsymbol{\sigma}^{\mathrm{d}} : \dot{\boldsymbol{\varepsilon}} + \boldsymbol{\beta} : \dot{\boldsymbol{\alpha}} \qquad (\geq 0) \tag{7.65}$$

and

$$\Phi_{\mathrm{ther}} = - \frac{\vec{\nabla} T}{T} \cdot \vec{q} \qquad (\geq 0) \, . \tag{7.66}$$

The mechanical part, Equation (7.65), is studied first.

If the material model is expressed by fluxes, $\varphi_{\text{mech}} = \varphi_{\text{mech}}(\dot{\boldsymbol{\varepsilon}}, \dot{\boldsymbol{\alpha}}; \boldsymbol{\varepsilon}, \boldsymbol{\alpha}, T, h)$, the results obtained in Expressions (6.282)–(6.284) provide the following:

$$\boldsymbol{\sigma}^{\text{d}} = \mu\,\rho_0\,\frac{\partial\varphi_{\text{mech}}(\dot{\boldsymbol{\varepsilon}}, \dot{\boldsymbol{\alpha}}; \boldsymbol{\varepsilon}, \boldsymbol{\alpha}, T, h)}{\partial\dot{\boldsymbol{\varepsilon}}} \tag{7.67}$$

and

$$\boldsymbol{\beta} = \mu\,\rho_0\,\frac{\partial\varphi_{\text{mech}}(\dot{\boldsymbol{\varepsilon}}, \dot{\boldsymbol{\alpha}}; \boldsymbol{\varepsilon}, \boldsymbol{\alpha}, T, h)}{\partial\dot{\boldsymbol{\alpha}}}\,. \tag{7.68}$$

The specific dissipation function $\varphi_{\text{mech}}$ has to obey the following condition:

$$\varphi_{\text{mech}}(\dot{\boldsymbol{\varepsilon}}, \dot{\boldsymbol{\alpha}}; \boldsymbol{\varepsilon}, \boldsymbol{\alpha}, T, h) = \mu\left(\frac{\partial\varphi_{\text{mech}}}{\partial\dot{\boldsymbol{\varepsilon}}} : \dot{\boldsymbol{\varepsilon}} + \frac{\partial\varphi_{\text{mech}}}{\partial\dot{\boldsymbol{\alpha}}} : \dot{\boldsymbol{\alpha}}\right). \tag{7.69}$$

The first-order sufficient condition for the point $(\dot{\boldsymbol{\varepsilon}}, \dot{\boldsymbol{\alpha}})$ to be a local maximum is that Equations (7.67) and (7.68) hold and that the specific dissipation function $\varphi_{\text{mech}}$ is a homogeneous function of degree $1/\mu$. The latter property is obtained if the coefficient $\mu$ in Expression (7.69) is a constant. If the multiplier $\mu$ is not a constant but $\mu = \mu(\boldsymbol{\varepsilon}, \boldsymbol{\alpha}, T, h)$, the specific dissipation function $\varphi_{\text{mech}}$ is not a homogeneous function and the value for $\mu$ is obtained from Equation (7.69).

If the material model is expressed by forces, $\varphi^{\text{c}}_{\text{mech}} = \varphi^{\text{c}}_{\text{mech}}(\boldsymbol{\sigma}^{\text{d}}, \boldsymbol{\beta}; \boldsymbol{\varepsilon}, \boldsymbol{\alpha}, T, h)$, the results obtained in Expressions (6.286)–(6.288) provide the following:

$$\dot{\boldsymbol{\varepsilon}} = \rho_0\,\frac{\partial\varphi^{\text{c}}_{\text{mech}}(\boldsymbol{\sigma}^{\text{d}}, \boldsymbol{\beta}; \boldsymbol{\varepsilon}, \boldsymbol{\alpha}, T, h)}{\partial\boldsymbol{\sigma}^{\text{d}}} \tag{7.70}$$

and

$$\dot{\boldsymbol{\alpha}} = \rho_0\,\frac{\partial\varphi^{\text{c}}_{\text{mech}}(\boldsymbol{\sigma}^{\text{d}}, \boldsymbol{\beta}; \boldsymbol{\varepsilon}, \boldsymbol{\alpha}, T, h)}{\partial\boldsymbol{\beta}}\,. \tag{7.71}$$

The specific complementary dissipation function $\varphi^{\text{c}}_{\text{mech}}$ has to obey the following condition:

$$\varphi^{\text{c}}_{\text{mech}}(\boldsymbol{\sigma}^{\text{d}}, \boldsymbol{\beta}; \boldsymbol{\varepsilon}, \boldsymbol{\alpha}, T, h) = (1-\mu)\left(\frac{\partial\varphi^{\text{c}}_{\text{mech}}}{\partial\boldsymbol{\sigma}^{\text{d}}} : \boldsymbol{\sigma}^{\text{d}} + \frac{\partial\varphi^{\text{c}}_{\text{mech}}}{\partial\boldsymbol{\beta}} : \boldsymbol{\beta}\right). \tag{7.72}$$

The first-order sufficient conditions for the point $\boldsymbol{\beta}$ to be a local maximum are that Equations (7.70) and (7.71) hold and that the specific complementary dissipation function $\varphi^{\text{c}}_{\text{mech}}$ be a homogeneous function of degree $1/(1-\mu)$. The latter property is obtained if the coefficient $\mu$ in Expression (7.72) is a constant. If the multiplier $\mu$ is not a constant but $\mu = \mu(\boldsymbol{\varepsilon}, \boldsymbol{\alpha}, T, h)$, the specific complementary dissipation function $\varphi^{\text{c}}_{\text{mech}}$ is not a homogeneous function, and the value for $\mu$ is obtained from Equation (7.72). The specific complementary dissipation function $\varphi^{\text{c}}_{\text{mech}}$ cannot be a homogeneous function of degree 1 in the variables $(\boldsymbol{\sigma}^{\text{d}}, \boldsymbol{\beta})$.

The normality rule for the thermal response equals that given in Section 6.3.3. It is recommended that the reader reverts to Section 6.3.3.

## 7.5 Gradient Theory

Strain softening has been observed in many materials, such as concrete, rocks, ice, and metallic materials at elevated temperatures. In order to model strain softening, usually a new variable called damage and denoted by $D$ is introduced. Damage $D$ does not model any specific deformation mechanism but is a general concept for strain softening. In the high-temperature materials of power plant components, where the service temperature is around 550 °C, damage $D$ may be nucleation, growth and coalescence of voids on grain boundaries, or carbide coarsening. In ice, by comparison, damage is related to microcracking.

The standard form of continuum mechanics is a local theory. This means that the functions are dependent only on the values of the variables at the same material points. For Hooke's law, this is as follows: stress $\sigma$ depends only on the value of the strain $\varepsilon$ at that specific point, that is, $\sigma = E\,\varepsilon$. The problem with local theories is that when strain softening is modelled, they lead to stability problems and to mesh dependency in finite element analysis. Nonlocal theories solve these problems. In nonlocal theories, (some) functions are dependent on the values of the variables at that particular point but also on the values of the variables in the neighborhood of the investigated point. As mentioned by Bažant ([10], p. 593), finite element analysis of distributed strain-softening damage, including its final localization into a sharp fracture, requires the use of some type of non-local continuum.

Nonlocal continuum theories have a material length scale. According to Ramaswamy and Aravas ([83], pp. 11 and 12), variety of methods have introduced a length scale. In micropolar continuum theories, the rotational degrees of freedom are added to the conventional translational degrees of freedom. In integral-type constitutive models, the evolution of certain internal variables is expressed by means of integral equations. In the gradient type of plasticity models, for example, the Laplacian value of the effective plastic strain $\varepsilon^{\mathrm{ef}}$ can be a vital part of the yield function. A counterpart gradient type of a constitutive equation for Hooke's law might be $\sigma = E\,\varepsilon + k\,\nabla^2\,\varepsilon$, where $\nabla^2$ stands for the Laplacian operator.

This chapter evaluates material models within the framework of continuum thermodynamics with internal variables. It has an ingredient beyond the standard formulation: the gradient theory. Usually, internal variables describe the mechanical part of the response of a material, that is, deformation. Internal variables are scalars or tensors of any order. In the gradient theory also the gradients of internal variable(s) are a vital part of the set of internal variables. This chapter demonstrates how gradients can be implemented into the continuum thermodynamics with internal variables. This is done by introducing a thermal internal variable called the specific dissipative entropy.

### 7.5.1 Continuum Thermodynamics with Internal Variables

The present section evaluates the problems that originate when the classical approach to continuum thermodynamics with internal variables is used with a gradient term in the list of internal variables. This chapter follows the work by Santaoja [93, 96] (see also Santaoja [92], Chapter 3), in which a Laplacian of an internal variable is used in the list of state variables.

For the sake of simplicity, this chapter studies material models with elastic deformation, which include the inelastic strain tensor $\boldsymbol{\varepsilon}^{\mathrm{i}}$ in the set of internal variables and also assume that the difference $\boldsymbol{\varepsilon} - \boldsymbol{\varepsilon}^{\mathrm{i}}$ describes the state of a material point. The present chapter studies a special case of gradient theory. Traditionally, when strain softening has been studied with the theory of plasticity, the higher-order gradients of inelastic strain tensor $\boldsymbol{\varepsilon}^{\mathrm{i}}$ have been used in the yield functions (see, e.g., Aifantis [3]). This has been done to avoid localization (see, e.g., Kukudžanov et al. [39], pp. 2–6). Here a different approach is used, in which the gradient operator $\vec{\nabla}$ is acting on the internal variable. This work follows the popular notation and assumes that this particular variable is a scalar-valued quantity denoted by $D$. Its gradient is $\vec{\nabla} D$. The quantity $D$ is only an example and can therefore be replaced by a vector or a tensor of any order.

The state $\mathcal{E}$ is expressed by

$$u = u(\boldsymbol{\varepsilon} - \boldsymbol{\varepsilon}^{\mathrm{i}}, D, \vec{\nabla} D, \boldsymbol{\alpha}, s), \tag{7.73}$$

where $\boldsymbol{\alpha}$ represents the set of other mechanical internal variables. The set $\boldsymbol{\alpha}$ can contain scalars, vectors, or tensors of any order. As before, instead of the specific internal energy $u$, the state of solids is usually described by the specific Helmholtz free energy $\psi$, which is a Legendre partial transformation of the specific internal energy $u$.

State functions are obtained as partial derivatives of the specific internal energy $u$ with respect to the state variables. Due to the introduction of the specific Helmholtz free energy $\psi$, state equations take the forms [cf. Equations (6.172) and (6.174)]

$$\boldsymbol{\sigma} = \rho_0 \frac{\partial \psi(\ldots)}{\partial(\boldsymbol{\varepsilon} - \boldsymbol{\varepsilon}^{\mathrm{i}})} \qquad \text{and} \qquad \upsilon := -\rho_0 \frac{\partial \psi(\ldots)}{\partial D}, \tag{7.74}$$

and furthermore,

$$\vec{\zeta} := -\rho_0 \frac{\partial \psi(\ldots)}{\partial \vec{\nabla} D} \qquad \text{and} \qquad \boldsymbol{\beta} := -\rho_0 \frac{\partial \psi(\ldots)}{\partial \boldsymbol{\alpha}}, \tag{7.75}$$

and finally,

$$s = -\frac{\partial \psi(\ldots)}{\partial T}, \qquad \text{where} \qquad \psi = \psi(\boldsymbol{\varepsilon} - \boldsymbol{\varepsilon}^{\mathrm{i}}, D, \vec{\nabla} D, \boldsymbol{\alpha}, T). \tag{7.76}$$

By combining the local forms of the first and second laws of thermodynamics, the Clausius–Duhem inequality is obtained. This was done in Section 6.6. For the present set of state variables [see Expression $(7.76)_2$], it takes the following form:

$$\boldsymbol{\sigma} : \dot{\boldsymbol{\varepsilon}}^{\mathrm{i}} + v\,\dot{D} + \vec{\zeta}\cdot\vec{\nabla}\dot{D} + \boldsymbol{\beta} : \dot{\boldsymbol{\alpha}} - \frac{\vec{\nabla}T}{T}\cdot\vec{q} \geq 0\,. \tag{7.77}$$

Based on Clausius–Duhem Inequality (7.77), the dissipation $\Phi$ is introduced. It is defined by

$$\Phi := \boldsymbol{\sigma} : \dot{\boldsymbol{\varepsilon}}^{\mathrm{i}} + v\,\dot{D} + \vec{\zeta}\cdot\vec{\nabla}\dot{D} + \boldsymbol{\beta} : \dot{\boldsymbol{\alpha}} - \frac{\vec{\nabla}T}{T}\cdot\vec{q} \qquad (\geq 0)\,. \tag{7.78}$$

The principle of maximum dissipation [or the principle of the maximal rate of entropy production by Ziegler ([120], Chapter 4)] is used to obtain the normality rule. The mechanical part of the problem [i.e., of Expression (7.78)] is considered. Based on Ziegler's principle and Normality Rules (6.286) and (6.287), the mechanical part of the normality rules take the forms

$$\dot{\boldsymbol{\varepsilon}}^{\mathrm{i}} = \rho_0\,\frac{\partial\varphi^{\mathrm{c}}_{\mathrm{mech}}(\ldots)}{\partial\boldsymbol{\sigma}} \qquad \text{and} \qquad \dot{D} = \rho_0\,\frac{\partial\varphi^{\mathrm{c}}_{\mathrm{mech}}(\ldots)}{\partial v} \tag{7.79}$$

and

$$\vec{\nabla}\dot{D} = \rho_0\,\frac{\partial\varphi^{\mathrm{c}}_{\mathrm{mech}}(\ldots)}{\partial\vec{\zeta}} \qquad \text{and} \qquad \dot{\boldsymbol{\alpha}} = \rho_0\,\frac{\partial\varphi^{\mathrm{c}}_{\mathrm{mech}}(\ldots)}{\partial\boldsymbol{\beta}}\,, \tag{7.80}$$

where

$$\varphi^{\mathrm{c}}_{\mathrm{mech}} = \varphi^{\mathrm{c}}_{\mathrm{mech}}(\boldsymbol{\sigma}, v, \vec{\zeta}, \boldsymbol{\beta}; \boldsymbol{\varepsilon} - \boldsymbol{\varepsilon}^{\mathrm{i}}, D, \vec{\nabla}D, \boldsymbol{\alpha}, T)\,. \tag{7.81}$$

There is a serious problem with Form (7.77) of the Clausius–Duhem inequality, and this is discussed next. If the gradient operator $\vec{\nabla}$ acted on Equation $(7.79)_2$ from the left, the left sides of Equations $(7.79)_2$ and $(7.80)_1$ would be identical. This means that also the right sides of Equations $(7.79)_2$ and $(7.80)_1$ should be the same. This condition is

$$\frac{1}{\rho_0}\,\vec{\nabla}\dot{D} = \vec{\nabla}\,\frac{\partial\varphi^{\mathrm{c}}_{\mathrm{mech}}(\ldots)}{\partial v} = \frac{\partial\varphi^{\mathrm{c}}_{\mathrm{mech}}(\ldots)}{\partial\vec{\zeta}}\,. \tag{7.82}$$

Condition (7.82) is too restrictive for preparation of a material model expressed by potential $\varphi^{\mathrm{c}}_{\mathrm{mech}}(\boldsymbol{\sigma}, v, \vec{\zeta}, \boldsymbol{\beta}; \boldsymbol{\varepsilon} - \boldsymbol{\varepsilon}^{\mathrm{i}}, D, \vec{\nabla}D, \boldsymbol{\alpha}, T)$. It is most difficult, even impossible, to write any explicit form for the specific complementary dissipation function $\varphi^{\mathrm{c}}_{\mathrm{mech}}(\boldsymbol{\sigma}, v, \vec{\zeta}, \boldsymbol{\beta}; \boldsymbol{\varepsilon} - \boldsymbol{\varepsilon}^{\mathrm{i}}, D, \vec{\nabla}D, \boldsymbol{\alpha}, T)$ that would both satisfy Condition (7.82) and provide an acceptable material model. Therefore Clausius–Duhem Inequality (7.77) [actually Expression (7.78)] has to be rewritten in a form in which both $\dot{D}$ and $\vec{\nabla}\dot{D}$ are not present. This is done in the following section.

### 7.5.2 Continuum Thermodynamics with Internal Variables with Gradients

This section shows that the introduction of the specific dissipative entropy $s^{\mathrm{d}}$ solves the problems caused by the gradient term in the set of internal variables.

In order to avoid the aforementioned problems when the gradient term of an internal variable is introduced, a new internal (state) variable $s^{\mathrm{d}}$ is introduced. The notation $s^{\mathrm{d}}$ stands for the specific dissipative entropy. It is the difference between the values of the specific entropy in nonequilibrium and equilibrium. It is a thermal internal variable, as Figure 7.1 shows (Figure 6.3 is reproduced here for the reader's convenience).

Instead of Expression (7.73), the state $\mathcal{E}$ is now expressed by

$$u = u(\boldsymbol{\varepsilon} - \boldsymbol{\varepsilon}^{\mathrm{i}}, D, \vec{\nabla} D, \boldsymbol{\alpha}, s - s^{\mathrm{d}})\,. \tag{7.83}$$

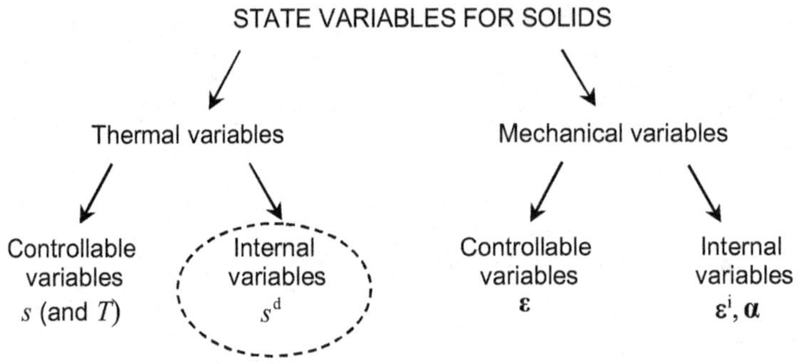


**Figure 7.1** Different types of independent variables for solids in continuum thermodynamics with internal variables. The thermal internal variable called specific dissipative entropy and denoted by $s^{\mathrm{d}}$ is present only in the gradient theory.

The specific Helmholtz free energy $\psi$ is defined by two expressions [see Definition (6.171)]. For the present case, the first expression is

$$\psi(\boldsymbol{\varepsilon} - \boldsymbol{\varepsilon}^{\mathrm{i}}, D, \vec{\nabla} D, \boldsymbol{\alpha}, T) \overset{\text{Legendre}}{=} u(\boldsymbol{\varepsilon} - \boldsymbol{\varepsilon}^{\mathrm{i}}, D, \vec{\nabla} D, \boldsymbol{\alpha}, s - s^{\mathrm{d}}) - T\,(s - s^{\mathrm{d}})\,. \tag{7.84}$$

The second expression is the state equation (or more than one). State Equation $(7.76)_1$ is now replaced by

$$s - s^{\mathrm{d}} := -\frac{\partial \psi(\ldots)}{\partial T}\,. \tag{7.85}$$

The other state equations and the form of the specific Helmholtz free energy $\psi$ keep their Forms (7.74), (7.75), and $(7.76)_2$.

The investigation of Sections 6.2 and 6.3 shows that the local forms of the first and second laws of thermodynamics keep their original forms. Therefore

Expression (6.240) holds. It is

$$\Phi = \boldsymbol{\sigma} : \dot{\boldsymbol{\varepsilon}} - \rho_0\, \dot{u} + \rho_0\, T\, \dot{s} - \frac{\vec{\nabla} T}{T} \cdot \vec{q} \qquad (\geq 0)\,. \tag{7.86}$$

Based on Expression (7.84), the following can be written:

$$\psi := u - T\,(s - s^{\mathrm{d}}) \qquad \text{and} \qquad \dot{u} = \dot{\psi} + \dot{T}\,(s - s^{\mathrm{d}}) + T\,(\dot{s} - \dot{s}^{\mathrm{d}})\,. \tag{7.87}$$

Substitution of Derivative $(7.87)_2$ into Expression (7.86) yields

$$\Phi = \boldsymbol{\sigma} : \dot{\boldsymbol{\varepsilon}} - \rho_0\, \dot{\psi} - \rho_0\, \dot{T}\,(s - s^{\mathrm{d}}) + \rho_0\, T\, \dot{s}^{\mathrm{d}} - \frac{\vec{\nabla} T}{T} \cdot \vec{q} \qquad (\geq 0)\,. \tag{7.88}$$

Based on Expression (7.84), the rate of the specific Helmholtz free energy $\dot{\psi}$ takes the form

$$\begin{aligned}\rho_0\, \dot{\psi}(\boldsymbol{\varepsilon} - \boldsymbol{\varepsilon}^{\mathrm{i}}, D, \vec{\nabla} D, \boldsymbol{\alpha}, T) &= \rho_0 \frac{\partial \psi(\dots)}{\partial(\boldsymbol{\varepsilon} - \boldsymbol{\varepsilon}^{\mathrm{i}})} : (\dot{\boldsymbol{\varepsilon}} - \dot{\boldsymbol{\varepsilon}}^{\mathrm{i}}) + \rho_0\, \frac{\partial \psi(\dots)}{\partial D}\, \dot{D} \\ &+ \rho_0\, \frac{\partial \psi(\dots)}{\partial \vec{\nabla} D} \cdot \vec{\nabla} \dot{D} + \rho_0\, \frac{\partial \psi(\dots)}{\partial \boldsymbol{\alpha}} : \dot{\boldsymbol{\alpha}} + \rho_0\, \frac{\partial \psi(\dots)}{\partial T}\, \dot{T}\,.\end{aligned} \tag{7.89}$$

Substitution of State Equations (7.74), (7.75), and (7.85) into Derivative (7.89) gives

$$\rho_0\, \dot{\psi}(\boldsymbol{\varepsilon} - \boldsymbol{\varepsilon}^{\mathrm{i}}, D, \vec{\nabla} D, \boldsymbol{\alpha}, T) = \boldsymbol{\sigma} : (\dot{\boldsymbol{\varepsilon}} - \dot{\boldsymbol{\varepsilon}}^{\mathrm{i}}) - \upsilon\, \dot{D} - \vec{\zeta} \cdot \vec{\nabla} \dot{D} - \boldsymbol{\beta} : \dot{\boldsymbol{\alpha}} - \rho_0\,(s - s^{\mathrm{d}})\, \dot{T}\,. \tag{7.90}$$

Substitution of the rate of the specific Helmholtz free energy $\dot{\psi}$ obtained from Equation (7.90) into Expression (7.88) yields

$$\Phi = \boldsymbol{\sigma} : \dot{\boldsymbol{\varepsilon}}^{\mathrm{i}} + \upsilon\, \dot{D} + \vec{\zeta} \cdot \vec{\nabla} \dot{D} + \boldsymbol{\beta} : \dot{\boldsymbol{\alpha}} + \rho_0\, T\, \dot{s}^{\mathrm{d}} - \frac{\vec{\nabla} T}{T} \cdot \vec{q} \qquad (\geq 0)\,. \tag{7.91}$$

Expression (7.91) gives the following form for the Clausius–Duhem inequality:

$$\boldsymbol{\sigma} : \dot{\boldsymbol{\varepsilon}}^{\mathrm{i}} + \upsilon\, \dot{D} + \vec{\zeta} \cdot \vec{\nabla} \dot{D} + \boldsymbol{\beta} : \dot{\boldsymbol{\alpha}} + \rho_0\, T\, \dot{s}^{\mathrm{d}} - \frac{\vec{\nabla} T}{T} \cdot \vec{q} \geq 0\,. \tag{7.92}$$

Following a concept with similar elements to those introduced by Maugin ([59], Eq.(3.5)), the following can be written:

$$\vec{\nabla} \cdot (\vec{\zeta}\, \dot{D}) = (\vec{\nabla} \cdot \vec{\zeta})\, \dot{D} + \vec{\zeta} \cdot \vec{\nabla} \dot{D}\,. \tag{7.93}$$

Substitution of Manipulation (7.93) into Clausius–Duhem Inequality (7.92) gives

$$\boldsymbol{\sigma} : \dot{\boldsymbol{\varepsilon}}^{\mathrm{i}} + \upsilon\, \dot{D} - (\vec{\nabla} \cdot \vec{\zeta})\, \dot{D} + \vec{\nabla} \cdot (\vec{\zeta}\, \dot{D}) + \boldsymbol{\beta} : \dot{\boldsymbol{\alpha}} + \rho_0\, T\, \dot{s}^{\mathrm{d}} - \frac{\vec{\nabla} T}{T} \cdot \vec{q} \geq 0\,. \tag{7.94}$$

The following notation and definition are introduced:

$$\eta := \langle \upsilon - \vec{\nabla}\cdot\vec{\zeta}\,\rangle \qquad \text{and} \qquad \rho_0\, T\, \dot{s}^{\mathrm{d}} := -\vec{\nabla}\cdot(\vec{\zeta}\,\dot{D})\,. \tag{7.95}$$

The Macaulay brackets guarantee that the amount of damage cannot decrease. If this is not the case, the Macaulay brackets can of course be dropped out. The consequence of the Macaulay brackets is shown in Section 10.6, where this theory is applied to a material model.

Notation $(7.95)_1$ and Definition $(7.95)_2$ allow Expression (7.94) to be rewritten in the form

$$\boldsymbol{\sigma} : \dot{\boldsymbol{\varepsilon}}^{\mathrm{i}} + \eta\,\dot{D} + \boldsymbol{\beta} : \dot{\boldsymbol{\alpha}} - \frac{\vec{\nabla} T}{T}\cdot\vec{q} \geq 0\,. \tag{7.96}$$

Clausius–Duhem Inequality (7.96) has the form at which we were aiming. It is the sum of force times flux pairs, and it does not contain a restriction on the description of the material model described in Equation (7.82). The normality rule takes the form

$$\dot{\boldsymbol{\varepsilon}}^{\mathrm{i}} = \rho_0\,\frac{\partial \varphi^{\mathrm{c}}_{\mathrm{mech}}}{\partial \boldsymbol{\sigma}} \qquad \text{and} \qquad \dot{D} = \rho_0\,\frac{\partial \varphi^{\mathrm{c}}_{\mathrm{mech}}}{\partial \eta}\,, \tag{7.97}$$

and furthermore,

$$\dot{\boldsymbol{\alpha}} = \rho_0\,\frac{\partial \varphi^{\mathrm{c}}_{\mathrm{mech}}}{\partial \boldsymbol{\beta}}\,, \tag{7.98a}$$

where

$$\varphi^{\mathrm{c}}_{\mathrm{mech}} = \varphi^{\mathrm{c}}_{\mathrm{mech}}(\boldsymbol{\sigma}, \eta, \boldsymbol{\beta}; \boldsymbol{\varepsilon} - \boldsymbol{\varepsilon}^{\mathrm{i}}, D, \vec{\nabla} D, \boldsymbol{\alpha}, T)\,. \tag{7.98b}$$

## 7.6 Validation of a Material Model by Continuum Thermodynamics

In this section we summarize the results of the previous chapters and sections by giving the steps to prepare a validation of a material model from a continuum thermodynamics point of view. The validation is carried out by studying whether the constitutive equation satisfies the Clausius–Duhem inequality in order to be thermodynamically consistent. Also, a simple uniaxial material model is studied as an example in order to clarify the validation process. Although this chapter is written for solid materials, it can be easily modified for fluids.

Based on a microscopic investigation of the response of a material under mechanical and thermal loading, a macroscopic material model is written. This model consists of a set of differential and algebraic equations. The idea of the thermodynamic validation is to test that the constitutive equation satisfies the basic laws and axioms of continuum thermodynamics. If the material model fails the test, it has to be forgotten, since that kind of model cannot exist. The satisfaction of the basic laws and axioms of continuum thermodynamics by a material model is as important as the satisfaction of the equilibrium equations by a beam when beam bending is analyzed. Actually, equilibrium equations are a reduced form of equations of motion (see Sections 4.15, 4.17, and 4.18), the latter being one of the basic laws of continuum thermodynamics. The validation is carried out by ensuring that the material model satisfies the Clausius–Duhem inequality, which is a combination of two basic laws and five axioms listed in Figure 7.2, which is a reproduction of Figure 5.1.

The preceding means that the information of one axiom of continuum thermodynamics is missing from the information of the Clausius–Duhem inequality. The missing axiom reads **AX 6** *Principle of maximum dissipation* (being present in Figure 5.1). A well-set question is: does the Clausius–Duhem inequality represent the basic laws and axioms of continuum thermodynamics so that the Clausius–Duhem inequality contains the information of all the basic laws and axioms necessary for satisfaction of a material model?

**BL 1** Law of balance of momentum
**BL 2** Law of balance of moment of momentum
**AX 1** Conservation of mass

**AX 2** Axiom of caloric equation of state
**AX 3** First law of thermodynamics
**AX 4** Second law of thermodynamics

**AX 5** Axiom of local accompanying state

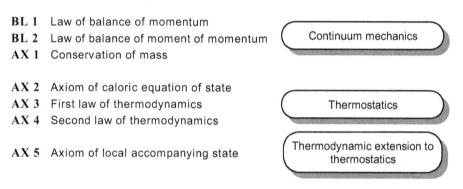

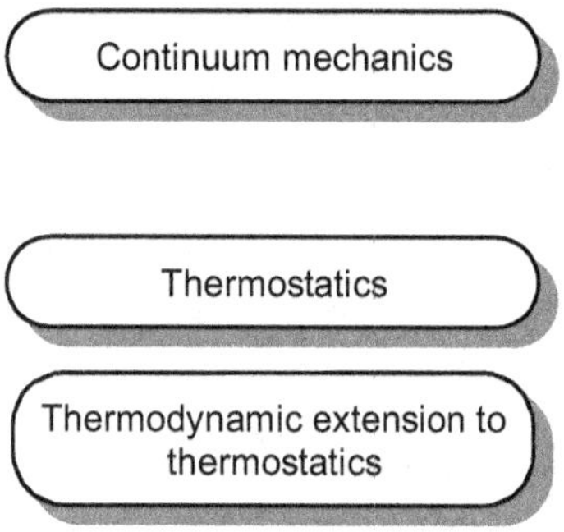


**Figure 7.2** Two basic laws and five axioms of continuum thermodynamics.

The following procedure is written for solids also showing (instantaneous) elastic deformation. This type of formulation of solids is in this book referred to as the standard formulation for solids. However, the following procedure can be applied for solids without elastic deformation (see Section 7.4) and for fluids (see Section 6.6.2) with minor modifications. As everywhere in this book, the notations $\boldsymbol{\alpha}$ and $\boldsymbol{\beta}$ represent sets of internal variables, for example, $\alpha^1$, $\vec{\alpha}^2$, $\boldsymbol{\alpha}^3$, and their associative internal forces $\beta^1$, $\vec{\beta}^2$, $\boldsymbol{\beta}^3$.

The validation of a material model may have the following steps:

(1) Select the set of state variables for this particular material model.

The selection is based on the material model that is usually obtained by studying the micromechanical processes of deformation. It could be, for example, the set $(\boldsymbol{\varepsilon} - \boldsymbol{\varepsilon}^{\mathrm{i}}, \boldsymbol{\alpha}, T)$, where the difference $\boldsymbol{\varepsilon} - \boldsymbol{\varepsilon}^{\mathrm{i}}$ shows that the model also describes the (instantaneous) elastic deformation. Sometimes the material model asks to introduce the set $(\boldsymbol{\sigma}, \boldsymbol{\alpha}, T)$ or $(\boldsymbol{\sigma}, \boldsymbol{\beta}, T)$.

**(2)** Write the explicit form for a potential describing the relationships between the sets $(\boldsymbol{\varepsilon} - \boldsymbol{\varepsilon}^{\mathrm{i}}, \boldsymbol{\alpha}, T)$ and $(\boldsymbol{\sigma}, \boldsymbol{\beta}, T)$ for the material model.

The potential is the specific Helmholtz free energy $\psi$ or the specific Gibbs free energy $g$. The specific Helmholtz free energy $\psi$ is an extension to the strain-energy density $w(\boldsymbol{\varepsilon}^{\mathrm{e}})$, whereas the specific Gibbs free energy $g$ is related to the complementary strain-energy density $w^{\mathrm{c}}(\boldsymbol{\sigma})$. Although the role of these two functions is to create expressions between the sets $(\boldsymbol{\varepsilon} - \boldsymbol{\varepsilon}^{\mathrm{i}}, \boldsymbol{\alpha}, T)$ and $(\boldsymbol{\sigma}, \boldsymbol{\beta}, T)$, the example in the next section shows that a relation between the rates $\dot{\boldsymbol{\alpha}}$ and $\dot{\boldsymbol{\beta}}$ may be needed, and sometimes this relation may not be a part of the final constitutive equation. Hooke's law, Hooke's law enriched by the damage description, and thermal expansion are the fundamental deformation mechanisms modelled by the specific Helmholtz free energy $\psi$ or the specific Gibbs free energy $g$.

Thus, the explicit form for one of the following two expressions must be written:

$$\psi = \psi(\varepsilon - \boldsymbol{\varepsilon}^{\mathrm{i}}, \boldsymbol{\alpha}, T, h(\vec{X})) \tag{7.99}$$

or

$$g = g(\boldsymbol{\sigma}, \boldsymbol{\alpha}, T, h(\vec{X}))\,, \tag{7.100}$$

where the role of the function $h(\vec{X})$ is to model inhomogeneous elastic deformation.

**(3)** Apply state equations.

In a case in which the material model is written in terms of the specific Helmholtz free energy $\psi$, apply State Equations (6.172) and (6.174):

$$\boldsymbol{\sigma} = \rho_0 \frac{\partial \psi(\boldsymbol{\varepsilon} - \boldsymbol{\varepsilon}^{\mathrm{i}}, \boldsymbol{\alpha}, T, h(\vec{X}))}{\partial(\boldsymbol{\varepsilon} - \boldsymbol{\varepsilon}^{\mathrm{i}})} \quad \text{and} \quad \boldsymbol{\beta} = -\rho_0 \frac{\partial \psi(\boldsymbol{\varepsilon} - \boldsymbol{\varepsilon}^{\mathrm{i}}, \boldsymbol{\alpha}, T, h(\vec{X}))}{\partial \boldsymbol{\alpha}}\,, \tag{7.101}$$

and so on. If the material model requires introduction of the specific Gibbs free energy $g$ instead of the specific Helmholtz free energy $\psi$, State Equations (7.101) are replaced with State Equations (6.181), viz.

$$\boldsymbol{\varepsilon} - \boldsymbol{\varepsilon}^{\mathrm{i}} = \rho_0 \frac{\partial g(\sigma, \boldsymbol{\alpha}, T, h(\vec{X}))}{\partial \boldsymbol{\sigma}} \quad \text{and} \quad \boldsymbol{\beta} = \rho_0 \frac{\partial g(\sigma, \boldsymbol{\alpha}, T, h(\vec{X}))}{\partial \boldsymbol{\alpha}}\,, \tag{7.102}$$

and so on.

**(4)** Based on step 3, write the (total) strain tensor $\boldsymbol{\varepsilon}$ in terms of the other strain tensors in the form $\boldsymbol{\varepsilon} = \boldsymbol{\varepsilon}^{\mathrm{e}} + \ldots$.

**(5)** Write the Clausius–Duhem inequalities for this particular material model.

Based on the foregoing steps, the Clausius–Duhem inequalities for this particular material model are given by Expressions (6.250) and (6.251), viz.

$$(\Phi_{\text{mech}} =) \qquad \boldsymbol{\sigma} : \dot{\boldsymbol{\varepsilon}}^{\text{i}} + \boldsymbol{\beta} : \dot{\boldsymbol{\alpha}} \geq 0 \tag{7.103}$$

and

$$(\Phi_{\text{ther}} =) \qquad -\frac{\vec{\nabla} T}{T} \cdot \vec{q} \geq 0 \,, \tag{7.104}$$

where $\Phi_{\text{mech}}$ and $\Phi_{\text{ther}}$ are the mechanical dissipation and thermal dissipation, respectively.

In a case in which the material model consists of the sets $\alpha^1$, $\vec{\alpha}^2$, $\boldsymbol{\alpha}^3$, and $\beta^1$, $\vec{\beta}^2$, $\boldsymbol{\beta}^3$, the mechanical part of the Clausius–Duhem inequality takes the following appearance:

$$(\Phi_{\text{mech}} =) \qquad \boldsymbol{\sigma} : \dot{\boldsymbol{\varepsilon}}^{\text{i}} + \beta^1 \dot{\alpha}^1 + \vec{\beta}^2 \cdot \dot{\vec{\alpha}}^2 + \boldsymbol{\beta}^3 : \dot{\boldsymbol{\alpha}}^3 \geq 0 \,. \tag{7.105}$$

As the preceding inequality shows, the Clausius–Duhem inequality contains the terms in which the rate of the internal variable (called flux) is multiplied by the associative force. The first term of Inequalities (7.103) and (7.105) is "stress times the inelastic strain rate" and is always present when the standardized formulation for solids is applied. The exceptions to this rule are material models for fluids (see Section 6.6.2) and for solids without elastic deformation (see Section 7.4).

There are two ways to continue, denoted as a and b. The letter a indicates that a second potential for the material model exists, and b that it does not. Thus the following steps will be labelled 6a, 7a, 7b, 8a, and 8b.

**(6a)** Write the explicit form of the second potential for the material model.

The second potential is the specific dissipation function $\varphi$, the specific complementary dissipation function $\varphi^{\text{c}}$, or the yield function $F$. The role of the second potential is to create expressions between the internal variable rates $\dot{\boldsymbol{\varepsilon}}^{\text{i}}$ and $\dot{\boldsymbol{\alpha}}$ and the forces $\boldsymbol{\sigma}$ and $\boldsymbol{\beta}$.

As discussed in Section 6.7.1, the rates $\dot{\boldsymbol{\varepsilon}}^{\text{i}}$ and $\dot{\boldsymbol{\alpha}}$ in Clausius–Duhem Inequality (7.103) are the independent variables of the mechanical part of the specific dissipation function $\varphi_{\text{mech}}$, whereas the state variables $(\boldsymbol{\varepsilon} - \boldsymbol{\varepsilon}^{\text{i}})$, $\boldsymbol{\alpha}$ and $T$, are the parameters of the specific dissipation function $\varphi_{\text{mech}}$. Thus the following (see details in Section 6.7.1):

$$\varphi = \varphi_{\text{mech}}(\dot{\boldsymbol{\varepsilon}}^{\text{i}}, \dot{\boldsymbol{\alpha}}; \boldsymbol{\varepsilon} - \boldsymbol{\varepsilon}^{\text{i}}, \boldsymbol{\alpha}, T, h) + \varphi_{\text{ther}}(\vec{q}; \boldsymbol{\varepsilon} - \boldsymbol{\varepsilon}^{\text{i}}, \boldsymbol{\alpha}, T, h) \,. \tag{7.106}$$

However, any of the internal variables $\boldsymbol{\alpha}$ can be replaced with its associative internal force $\boldsymbol{\beta}$ given by State Equations (7.101).

The difference between the specific dissipation function $\varphi$ and the specific complementary dissipation function $\varphi^{\mathrm{c}}$ is the variables. The variables for the mechanical part of the specific complementary dissipation function $\varphi^{\mathrm{c}}_{\mathrm{mech}}$ are the forces $\boldsymbol{\sigma}$ and $\boldsymbol{\beta}$ and, as earlier, the state variables $(\boldsymbol{\varepsilon} - \boldsymbol{\varepsilon}^{\mathrm{i}})$, $\boldsymbol{\alpha}$ and $T$, are the parameters, as discussed in Section 6.7.2. This leads to the following form:

$$\varphi^{\mathrm{c}}_{\mathrm{mech}} = \varphi^{\mathrm{c}}_{\mathrm{mech}}(\boldsymbol{\sigma}, \boldsymbol{\beta}; \boldsymbol{\varepsilon} - \boldsymbol{\varepsilon}^{\mathrm{i}}, \boldsymbol{\alpha}, T, h)\,. \tag{7.107}$$

As earlier, any of the internal variables $\boldsymbol{\alpha}$ can be replaced by its associative internal force $\boldsymbol{\beta}$ given by State Equations (7.101).

Modelling of (instantaneous) thermoplasticity requires the yield function $F$ to be written. It has the following appearance:

$$F = F(\boldsymbol{\sigma}, \boldsymbol{\beta}; \boldsymbol{\varepsilon} - \boldsymbol{\varepsilon}^{\mathrm{p}}, \boldsymbol{\alpha}, T, h)\,. \tag{7.108}$$

**(7a)** Apply normality rules.

The normality rule for the mechanical part of the problem is given by Expressions (6.282) and (6.283), viz.

$$\boldsymbol{\sigma} = \mu\,\rho_0\,\frac{\partial \varphi_{\mathrm{mech}}(\dot{\boldsymbol{\varepsilon}}^{\mathrm{i}}, \dot{\boldsymbol{\alpha}}; \boldsymbol{\varepsilon} - \boldsymbol{\varepsilon}^{\mathrm{i}}, \boldsymbol{\alpha}, T, h)}{\partial \dot{\boldsymbol{\varepsilon}}^{\mathrm{i}}}\,, \qquad \text{and so on.} \tag{7.109}$$

Satisfaction of the principle of maximum dissipation requires that the mechanical part of the specific dissipation function $\varphi_{\mathrm{mech}}$ obey Property (6.284),

$$\varphi_{\mathrm{mech}}(\dot{\boldsymbol{\varepsilon}}^{\mathrm{i}}, \dot{\boldsymbol{\alpha}}; \boldsymbol{\varepsilon} - \boldsymbol{\varepsilon}^{\mathrm{i}}, \boldsymbol{\alpha}, T, h) = \mu\left(\frac{\partial \varphi_{\mathrm{mech}}}{\partial \dot{\boldsymbol{\varepsilon}}^{\mathrm{i}}} : \dot{\boldsymbol{\varepsilon}}^{\mathrm{i}} + \frac{\partial \varphi_{\mathrm{mech}}}{\partial \dot{\boldsymbol{\alpha}}} : \dot{\boldsymbol{\alpha}}\right). \tag{7.110}$$

If the specific dissipation function $\varphi_{\mathrm{mech}}$ is a nonnegative homogeneous function of degree $\omega$ in the variables $\dot{\boldsymbol{\varepsilon}}^{\mathrm{i}}$ and $\dot{\boldsymbol{\alpha}}$, the multiplier $\mu$ takes the constant value $\mu = 1/\omega$, which satisfies Expression (7.110). This stems from the form of Expression (7.110), which equals the form of Euler's theorem on homogeneous functions given in Expression (2.208). In such a case, the constant value for the multiplier $\mu$ can easily be applied in Normality Rule (7.109). Form (7.110) may not be the most convenient way to study whether the specific dissipation function $\varphi_{\mathrm{mech}}$ is a homogeneous function; a more convenient approach is given in step 8a.

If the specific dissipation function $\varphi_{\mathrm{mech}}$ is not a homogeneous function, Equation (7.110) is used in determining the value for the multiplier $\mu$, which in this case is dependent on the quantities of the specific dissipation function $\varphi_{\mathrm{mech}}$.

If the material model requires the introduction of the specific complementary dissipation function $\varphi^{\mathrm{c}}_{\mathrm{mech}}$ instead of the specific dissipation function $\varphi_{\mathrm{mech}}$, the normality rule takes Forms (6.286) and (6.287), which are

$$\dot{\boldsymbol{\varepsilon}}^{\mathrm{i}} = \rho_0 \, \frac{\partial \varphi^{\mathrm{c}}_{\mathrm{mech}}(\boldsymbol{\sigma}, \boldsymbol{\beta}; \boldsymbol{\varepsilon} - \boldsymbol{\varepsilon}^{\mathrm{i}}, \boldsymbol{\alpha}, T, h)}{\partial \boldsymbol{\sigma}} \,, \qquad \text{and so on.} \qquad (7.111)$$

Satisfaction of the principle of maximum dissipation requires that the specific complementary dissipation function $\varphi^{\mathrm{c}}_{\mathrm{mech}}$ obey Property (6.288), viz.

$$\varphi^{\mathrm{c}}_{\mathrm{mech}}(\boldsymbol{\sigma}, \boldsymbol{\beta}; \boldsymbol{\varepsilon} - \boldsymbol{\varepsilon}^{\mathrm{i}}, \boldsymbol{\alpha}, T, h) = (1 - \mu) \left( \frac{\partial \varphi^{\mathrm{c}}_{\mathrm{mech}}}{\partial \boldsymbol{\sigma}} : \boldsymbol{\sigma} + \frac{\partial \varphi^{\mathrm{c}}_{\mathrm{mech}}}{\partial \boldsymbol{\beta}} : \boldsymbol{\beta} \right) . \qquad (7.112)$$

If the specific complementary dissipation function $\varphi^{\mathrm{c}}_{\mathrm{mech}}$ is a nonnegative homogeneous function, Equation (7.112) is satisfied as discussed earlier.

If the specific complementary dissipation function $\varphi^{\mathrm{c}}_{\mathrm{mech}}$ is not a homogeneous function, the value for the multiplier $\mu$ is not needed, since Normality Rules (7.111) does not utilize the multiplier $\mu$.

If the specific dissipation function $\varphi$ is not separated, the normality rule takes Forms (6.312)–(6.314). If the material model describes processes indepedent of time, consult Section 6.7.4. In both cases, the instructions given earlier can be applied.

**(7b)** If the material model cannot be formulated by the second potential, apply a different approach.

Based on micromechanical investigation, for example, write the explicit form for the "rate part" of the material model, that is, write, for example,

$$\dot{\boldsymbol{\varepsilon}}^{\mathrm{i}} = \ldots \,, \qquad \dot{\alpha}^1 = \ldots \,, \qquad \dot{\vec{\alpha}}^2 = \ldots \,, \qquad \text{and so on.} \qquad (7.113)$$

**(8)** Validate the material model by satisfaction of the Clausius–Duhem inequality.

The separated form of the Clausius–Duhem inequality is obtained from Inequalites (7.103) and (7.104), which give

$$\boldsymbol{\sigma} : \dot{\boldsymbol{\varepsilon}}^{\mathrm{i}} + \boldsymbol{\beta} : \dot{\boldsymbol{\alpha}} \geq 0 \qquad \text{and} \qquad -\frac{\vec{\nabla} T}{T} \cdot \vec{q} \geq 0 \,. \qquad (7.114)$$

As discussed at the end of Section 6.6.1, the Clausius–Duhem inequality for the heat problem, Inequality $(7.114)_2$, always takes the same appearance. On the other hand, the Clausius–Duhem inequality for the mechanical problem, Inequality $(7.114)_1$, is dependent on the material model under consideration. The mechanical part of Clausius–Duhem Inequality $(7.114)_1$ is studied.

There are two ways to evaluate whether the present constitutive equation satisfies Clausius–Duhem Inequality $(7.114)_1$. The first is based on evaluating whether the dissipation potential is a homogeneous function. The second is to substitute the material model into the Clausius–Duhem inequality and evaluate analytically or numerically the satisfaction of the obtained inequality.

**(8a)** Dissipation potential is a nonnegative homogeneous function.

If the mechanical part of the specific dissipation function $\varphi_{\text{mech}}$ or of the specific complementary dissipation function $\varphi^{\text{c}}_{\text{mech}}$ is a nonnegative homogeneous function, according to Section 7.2, the material model satisfies the Clausius–Duhem inequality. Since the mechanical part of the specific dissipation function has the form $\varphi_{\text{mech}}(\dot{\boldsymbol{\varepsilon}}^{\text{i}}, \dot{\boldsymbol{\alpha}}; \boldsymbol{\varepsilon} - \boldsymbol{\varepsilon}^{\text{i}}, \boldsymbol{\alpha}, T, h)$, it has to be a homogeneous function in the variables $\dot{\boldsymbol{\varepsilon}}^{\text{i}}$ and $\dot{\boldsymbol{\alpha}}$. The quantities after ";" are the parameters. In the dissipation potential $\varphi^{\text{c}}_{\text{mech}}$, there can be another set of variables, in which case the variables of the homogeneity are different. The same rule applies to $\varphi^{\text{c}}_{\text{mech}}$.

Expressions (7.110) and (7.112) may not be good candidates for evaluating $\varphi_{\text{mech}}$ or $\varphi^{\text{c}}_{\text{mech}}$ is a homogeneous function. The definition for homogeneous functions, Definition (2.207), is a better tool for this purpose. Definition (2.207) reads that function $\theta(x, y, z, u, v)$ is homogeneous of degree $\omega$ in variables $x$, $y$, and $z$ in a region $R$, if and only if, for $x$, $y$, and $z$ in $R$ and for every positive value of $k$, the following holds:

$$\theta(k\,x, k\,y, k\,z, u, v) := k^{\omega}\,\theta(x, y, z, u, v)\,. \tag{7.115}$$

**8a, 8b** Dissipation potential is not a nonnegative homogeneous function or no dissipation potential exists.

If the specific dissipation function $\varphi_{\text{mech}}$ or if the specific complementary dissipation function $\varphi^{\text{c}}_{\text{mech}}$ is not a homogeneous function, evaluation of the satisfaction of the Clausius–Duhem inequality needs more extensive study.

There are two procedures for investigating the satisfaction of Clausius–Duhem Inequality $(7.114)_1$. Validation of the material model can be done either analytically or numerically. An analytical validation is proposed, of course. The analytical evaluation can be carried out by substitution of the material model into Inequality $(7.114)_1$ and by checking if Inequality $(7.114)_1$ is satisfied. If analytical validation is not possible, the validation has to be carried out numerically during computation. This procedure is not proposed, since numerical validation requires computer resources that should be used for the "real" problem.

### 7.6.1 Example

Next, a simple uniaxial example is studied. In order to make the material model as simple as possible, the constitutive equation is written for constant tension only.

The response of a high-temperature power plant material 10CrMo910 at a service temperature of 550 °C is studied. The material model is assumed to consist of three different strain terms: Hooke's law, thermal expansion and creep. Figure 7.3 sketches a strain-time relationship of the aforementioned power plant material at 550 °C under constant stress $\sigma$ (or load). In order to simplify the investigation further, only primary and secondary stages of creep are modelled. Thus the (total) tensile strain $\varepsilon$ is separable as

$$\varepsilon = \varepsilon^{\mathrm{e}} + \varepsilon^{\mathrm{th}} + \varepsilon^{\mathrm{v}} . \tag{7.116}$$

In this example, the inelastic strain $\varepsilon^{\mathrm{i}}$ is the viscous strain $\varepsilon^{\mathrm{v}}$. In Equation (7.116), the Hookean deformation is modelled by

$$\sigma = E\,\varepsilon^{\mathrm{e}} , \tag{7.117}$$

where $E$ is Young's modulus. Thermal strain $\varepsilon^{\mathrm{Th}}$ is modelled as

$$\varepsilon^{\mathrm{Th}} = \hat{\alpha}\,(T - T_{\mathrm{r}}) , \tag{7.118}$$

where $\hat{\alpha}$ is the linear coefficient of thermal expansion and $T_{\mathrm{r}}$ is the reference temperature. Creep strain $\varepsilon^{\mathrm{v}}$ is modelled by a set of differential equations having the form

$$\dot{\varepsilon}^{\mathrm{v}} = \overset{\circ}{\varepsilon}_{\mathrm{re}} \left( \frac{\sigma - \beta}{\sigma_{\mathrm{re}}} \right)^{n} \quad \text{and} \quad \dot{\beta} = e\,\dot{\varepsilon}^{\mathrm{v}} - b \left( \frac{\beta}{\sigma_{\mathrm{re}}} \right)^{m} e^{\frac{E_{\mathrm{a}}}{R}\left(\frac{1}{T_{\mathrm{r}}} - \frac{1}{T}\right)} . \tag{7.119}$$

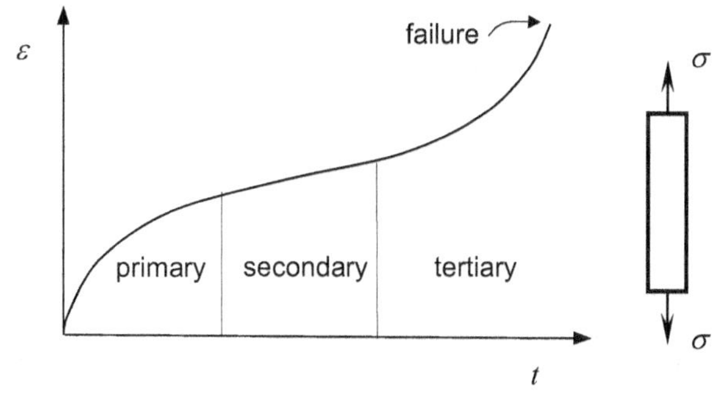


**Figure 7.3** Creep curve.

The notations $\overset{\circ}{\varepsilon}_{\mathrm{re}}, \sigma_{\mathrm{re}}, n, e, b$, and $m$ are material parameters, $E_{\mathrm{a}}$ is the activation energy, and $R$ is the universal gas constant. The deformation described in Separation (7.116) is obtained as follows: the test specimen shown in Figure 7.3 is attached to the testing machine and the temperature is raised to 550 °C. This process will cause the thermal strain $\varepsilon^{\mathrm{Th}}$. Then the testing machine is started and the tensile stress $\sigma$ that yields Hookean strain $\varepsilon^{\mathrm{e}}$ is generated. During the continuation of loading the creep strain $\varepsilon^{\mathrm{v}}$ grows monotonously. Actually, the driving force for dislocation creep is the shear stress $\tau$. Thus, instead of $\sigma - \beta$, the numerator in Material Model $(7.119)_1$ should be $\tau - \beta$. Unfortunately, this latter notation does not fit with the following state equation and the Clausius–Duhem inequality, for example. This is one of the limitations of the uniaxial model.

Now the aforementioned steps are taken one by one.

**(1)** Select the set of state variables for this particular material model.

Based on the material model given by Expressions (7.116)–(7.119), the material model has one internal force, denoted by $\beta$. The associative internal variables read $\alpha$. Thus the state variables for the present model are $(\varepsilon, \varepsilon^{\mathrm{v}}, \alpha, T)$. In uniaxial cases, $\boldsymbol{\alpha}$ is replaced with a scalar $\alpha$.

**(2)** Write the explicit form for a potential describing the relationships between the sets $(\boldsymbol{\varepsilon} - \boldsymbol{\varepsilon}^{\mathrm{i}}, \boldsymbol{\alpha}, T)$ and $(\boldsymbol{\sigma}, \boldsymbol{\beta}, T)$ for the material model.

The specific Helmholtz free energy $\psi$ takes the following appearance:

$$\psi(\varepsilon - \varepsilon^{\mathrm{v}}, \alpha, T) = \frac{1}{2\,\rho_0}\,E\,(\varepsilon - \varepsilon^{\mathrm{v}})^2 + \frac{e}{2\,\rho_0}\,\alpha^2 - \frac{E\,\hat{\alpha}}{\rho_0}\,(\varepsilon - \varepsilon^{\mathrm{v}})\,(T - T_{\mathrm{r}}) - c_{\mathrm{r}}\,[T\,\ln(T/T_{\mathrm{r}}) - T]\,. \tag{7.120}$$

**(3)** Apply state equations.

State Equations $(6.174)_1$ and $(6.174)_2$ are applied. They give

$$\sigma = \rho_0\,\frac{\partial\psi(\varepsilon - \varepsilon^{\mathrm{v}}, \alpha, T)}{\partial(\varepsilon - \varepsilon^{\mathrm{v}})} = E\,(\varepsilon - \varepsilon^{\mathrm{v}}) - E\,\hat{\alpha}\,(T - T_{\mathrm{r}}) \tag{7.121}$$

and

$$\beta = -\rho_0\,\frac{\partial\psi(\varepsilon - \varepsilon^{\mathrm{v}}, \alpha, T))}{\partial\alpha} = -\,e\,\alpha \quad\Rightarrow\quad \dot{\beta} = -\,e\,\dot{\alpha}\,. \tag{7.122}$$

**(4)** Based on step 3, write the (total) strain tensor $\boldsymbol{\varepsilon}$ in terms of the other strain tensors in the form $\boldsymbol{\varepsilon} = \boldsymbol{\varepsilon}^{\mathrm{e}} + \cdots$ .

Equation (7.121) gives

$$(\varepsilon - \varepsilon^{\mathrm{v}}) = \frac{\sigma}{E} + \hat{\alpha}\,(T - T_{\mathrm{r}})\,, \tag{7.123}$$

which, according to Separation (7.116), yields

$$\varepsilon - \varepsilon^{\mathrm{v}} = \varepsilon^{\mathrm{e}} + \varepsilon^{\mathrm{th}} \qquad \Rightarrow \qquad \varepsilon = \varepsilon^{\mathrm{e}} + \varepsilon^{\mathrm{th}} + \varepsilon^{\mathrm{v}}\,, \tag{7.124}$$

where

$$\varepsilon^{\mathrm{e}} = \frac{\sigma}{E} \qquad \text{and} \qquad \varepsilon^{\mathrm{th}} = \hat{\alpha}\,(T - T_{\mathrm{r}})\,. \tag{7.125}$$

**(5)** Write the Clausius–Duhem inequalities for this particular material model.

As pointed out in step 1, the set of state variables for the present material model is $(\varepsilon, \varepsilon^{\mathrm{v}}, \alpha, T)$. Thus the viscous strain $\varepsilon^{\mathrm{v}}$ and internal variable $\alpha$ are the internal variables for the present constitutive equation. Therefore, based on Inequality (7.103), the mechanical part of the Clausius–Duhem inequality for the present material model takes the form

$$\sigma\,\dot{\varepsilon}^{\mathrm{v}} + \beta\,\dot{\alpha} \geq 0\,. \tag{7.126}$$

**(6a)** Write the explicit form of the second potential for the material model.

The specific complementary dissipation function $\varphi^{\mathrm{c}}_{\mathrm{mech}}$ is

$$\varphi^{\mathrm{c}}_{\mathrm{mech}}(\sigma, \beta; T) = \frac{\overset{\circ}{\varepsilon}_{\mathrm{re}}\,\sigma_{\mathrm{re}}}{\rho_0\,(n+1)}\left(\frac{\sigma - \beta}{\sigma_{\mathrm{re}}}\right)^{n+1} + \frac{b\,\sigma_{\mathrm{re}}}{\rho_0\,c\,(m+1)}\left(\frac{\beta}{\sigma_{\mathrm{re}}}\right)^{m+1} e^{\frac{E_{\mathrm{a}}}{R}\left(\frac{1}{T_{\mathrm{r}}} - \frac{1}{T}\right)}. \tag{7.127}$$

**(7a)** Apply normality rules.

Normality Rules (6.286) and (6.287) are applied, giving

$$\begin{aligned} \dot{\varepsilon}^{\mathrm{v}} &= \rho_0\,\frac{\partial \varphi^{\mathrm{c}}_{\mathrm{mech}}(\sigma, \beta; T)}{\partial \sigma} = \rho_0\,\frac{\partial}{\partial \sigma}\left[\frac{\overset{\circ}{\varepsilon}_{\mathrm{re}}\,\sigma_{\mathrm{re}}}{\rho_0\,(n+1)}\left(\frac{\sigma - \beta}{\sigma_{\mathrm{re}}}\right)^{n+1}\right] \\ &= \overset{\circ}{\varepsilon}_{\mathrm{re}}\left(\frac{\sigma - \beta}{\sigma_{\mathrm{re}}}\right)^{n} \end{aligned} \tag{7.128}$$

and

$$\begin{aligned} \dot{\alpha} = \rho_0\,\frac{\partial \varphi^{\mathrm{c}}_{\mathrm{mech}}(\sigma, \beta; T)}{\partial \beta} &= -\overset{\circ}{\varepsilon}_{\mathrm{re}}\left(\frac{\sigma - \beta}{\sigma_{\mathrm{re}}}\right)^{n} + \frac{b}{e}\left(\frac{\beta}{\sigma_{\mathrm{re}}}\right)^{m} e^{\frac{E_{\mathrm{a}}}{R}\left(\frac{1}{T_{\mathrm{r}}} - \frac{1}{T}\right)} \\ &= -\dot{\varepsilon}^{\mathrm{v}} + \frac{b}{e}\left(\frac{\beta}{\sigma_{\mathrm{re}}}\right)^{m} e^{\frac{E_{\mathrm{a}}}{R}\left(\frac{1}{T_{\mathrm{r}}} - \frac{1}{T}\right)}\,. \end{aligned} \tag{7.129}$$

Substitution of Derivative $(7.122)_2$ into Evolution Equation (7.129) gives

$$-\frac{1}{e}\,\dot{\beta} = -\dot{\varepsilon}^{\mathrm{v}} + \frac{b}{e}\left(\frac{\beta}{\sigma_{\mathrm{re}}}\right)^{m} e^{\frac{E_{\mathrm{a}}}{R}\left(\frac{1}{T_{\mathrm{r}}} - \frac{1}{T}\right)}, \tag{7.130}$$

which yields

$$\dot{\beta} = e\,\dot{\varepsilon}^{\mathrm{v}} - b\left(\frac{\beta}{\sigma_{\mathrm{re}}}\right)^{m} e^{\frac{E_{\mathrm{a}}}{R}\left(\frac{1}{T_{\mathrm{r}}}-\frac{1}{T}\right)}\,. \tag{7.131}$$

Comparison of Results $(7.125)_1$, $(7.125)_2$, (7.128), and (7.131) with the Target Material Model (7.116)–(7.119) proves that Potentials (7.120) and (7.127) are the correct ones.

**(8a, 8b)** Validate the material model by satisfaction of the Clausius–Duhem inequality.

**(8a)** Dissipation potential is a nonnegative homogeneous function.

Based on Section 7.2, the following can be argued: if the specific complementary dissipation function $\varphi^{\mathrm{c}}_{\mathrm{mech}}$ is a sum of nonnegative homogeneous functions of different degrees in forces [in forces $(\sigma, \beta)$ in this case], the Clausius–Duhem inequality is automatically satisfied. Material Model (7.127) gives

$$\varphi^{\mathrm{c1}}_{\mathrm{mech}}(\sigma,\beta) = \frac{\overset{\circ}{\varepsilon}_{\mathrm{re}}\,\sigma_{\mathrm{re}}}{\rho_0\,(n+1)}\left(\frac{\sigma-\beta}{\sigma_{\mathrm{re}}}\right)^{n+1} \tag{7.132}$$

and

$$\varphi^{\mathrm{c2}}_{\mathrm{mech}}(\beta;T) = \frac{b\,\sigma_{\mathrm{re}}}{\rho_0\,c\,(m+1)}\left(\frac{\beta}{\sigma_{\mathrm{re}}}\right)^{m+1} e^{\frac{E_{\mathrm{a}}}{R}\left(\frac{1}{T_{\mathrm{r}}}-\frac{1}{T}\right)}\,. \tag{7.133}$$

Functions (7.132) and (7.133) give

$$\varphi^{\mathrm{c1}}_{\mathrm{mech}}(k\,\sigma, k\,\beta) = k^{n+1}\,\varphi^{\mathrm{c1}}_{\mathrm{mech}}(\sigma,\beta) \tag{7.134}$$

and

$$\varphi^{\mathrm{c2}}_{\mathrm{mech}}(k\,\beta;T) = k^{m+1}\,\varphi^{\mathrm{c2}}_{\mathrm{mech}}(\beta;T)\,. \tag{7.135}$$

Based on Expressions (7.134) and (7.135) [see also Definition (2.207) or Definition (7.115)], functions $\varphi^{\mathrm{c1}}_{\mathrm{mech}}(\sigma,\beta)$ and $\varphi^{\mathrm{c2}}_{\mathrm{mech}}(\beta)$ are homogeneous functions of degree $n+1$ and $m+1$ in forces $\sigma$ and $\beta$ , respectively. Furthermore, functions $\varphi^{\mathrm{c1}}_{\mathrm{mech}}(\sigma,\beta)$ and $\varphi^{\mathrm{c2}}_{\mathrm{mech}}(\beta;T)$ are nonnegative. Thus Clausius–Duhem Inequality (7.126) is satisfied.

**(8a, 8b)** Dissipation potential is not a nonnegative homogeneous function, or no dissipation potential exists.

Next, the approach using substitution of the material model into the Clausius–Duhem inequality is considered.

Based on Rate Equations (7.128) and (7.129), the following forms for the terms in Inequality (7.126) are obtained:

$$\sigma\,\dot{\varepsilon}^{\mathrm{v}} = \overset{\circ}{\varepsilon}_{\mathrm{re}}\left(\frac{\sigma-\beta}{\sigma_{\mathrm{re}}}\right)^{n}\sigma \tag{7.136}$$

and

$$\beta\,\dot{\alpha} = \left[ -\overset{\circ}{\varepsilon}_{\mathrm{re}} \left( \frac{\sigma - \beta}{\sigma_{\mathrm{re}}} \right)^{n} + \frac{b}{e} \left( \frac{\beta}{\sigma_{\mathrm{re}}} \right)^{m} e^{\frac{E_{\mathrm{a}}}{R}\left(\frac{1}{T_{\mathrm{r}}} - \frac{1}{T}\right)} \right] \beta\,. \tag{7.137}$$

Substitution of Terms (7.136) and (7.137) into Inequality (7.126) gives

$$\sigma\,\dot{\varepsilon}^{\mathrm{v}} + \beta\,\dot{\alpha} = \overset{\circ}{\varepsilon}_{\mathrm{re}} \left( \frac{\sigma - \beta}{\sigma_{\mathrm{re}}} \right)^{n} (\sigma - \beta) + \frac{b}{e} \left( \frac{\beta}{\sigma_{\mathrm{re}}} \right)^{m} e^{\frac{E_{\mathrm{a}}}{R}\left(\frac{1}{T_{\mathrm{r}}} - \frac{1}{T}\right)} \beta\,. \tag{7.138}$$

It is common practice to formulate material models in such a way that the values for material parameters are nonnegative. This implies that in Equation (7.138), the parameters $\overset{\circ}{\varepsilon}_{\mathrm{re}}, \sigma_{\mathrm{re}}, n$, and so on are nonnegative.

The term $\sigma - \beta$ is nonnegative. This originates from the fact that if the value of $\beta$ approaches the value of $\sigma$, according to Rate Equation (7.128), the value of $\dot{\varepsilon}^{\mathrm{v}}$ tends to zero, and based on Material Model (7.131), the value of $\beta$ no longer grows. Thus the first term on the right side of Expression (7.138) is nonnegative.

Since the creep strain rate $\dot{\varepsilon}^{\mathrm{v}}$ takes only nonnegative values, the first term on the right side of Evolution Equation (7.131) is nonnegative. If the value for the second term on the right side of Evolution Equation (7.131) approaches the value of the first term, according to Evolution Equation (7.131), the rate of internal force $\dot{\beta}$ diminishes and therefore the second term on the right side no longer grows. The natural exponential term $e^{\cdot}$ is always nonnegative. Thus the second term on the right side of Expression (7.138) is nonnegative.

Based on the preceding, Expression (7.138) is nonnegative and therefore Clausius–Duhem Inequality (7.126) is satisfied.

The uniaxial material model is a simplified example, but it gives good insight into the procedure for validating material models with the satisfaction of the Clausius–Duhem inequality. Evaluation carried out in Expressions (7.136)–(7.138) is the only exception. As evaluation of the multiaxial counterpart for the present model, that is, the material model proposed by Le Gac and Duval, shows, that the satisfaction of the multiaxial expression by substitution of the terms into the Clausius–Duhem inequality is much clearer.

## 7.7 Example of Legendre Partial Transformation

This section gives an example of the Legendre partial transformation of a material model, which is written with the specific Helmholtz free energy $\psi$ and transformed into the form of specific Gibbs free energy $g$. To make the example easy to follow, a uniaxial material model is studied; thus the example can be expressed in scalar-valued quantities.

Definition (6.179) gives the relation between the specific Gibbs free energy $g$ and the specific Helmholtz free energy $\psi$. It reads

$$\rho_0\, g(\sigma, \boldsymbol{\alpha}, T, h(\vec{X})) \overset{\text{Legendre}}{=} \boldsymbol{\sigma} : (\boldsymbol{\varepsilon} - \boldsymbol{\varepsilon}^{\mathrm{i}}) - \rho_0\, \psi(\varepsilon - \boldsymbol{\varepsilon}^{\mathrm{i}}, \boldsymbol{\alpha}, T, h(\vec{X}))$$
$$\text{with} \quad \boldsymbol{\sigma} = \rho_0 \frac{\partial \psi(\varepsilon - \boldsymbol{\varepsilon}^{\mathrm{i}}, \boldsymbol{\alpha}, T, h(\vec{X}))}{\partial(\boldsymbol{\varepsilon} - \boldsymbol{\varepsilon}^{\mathrm{i}})} \,. \tag{7.139}$$

The response of the material is assumed to have three mutually independent deformation mechanisms, which are modelled as follows:

$$\varepsilon = \varepsilon^{\mathrm{e}} + \varepsilon^{\mathrm{th}} + \varepsilon^{\mathrm{v}} \quad \Rightarrow \quad \varepsilon = \varepsilon^{\mathrm{eT}} + \varepsilon^{\mathrm{v}}, \quad \text{where} \quad \varepsilon^{\mathrm{eT}} = \varepsilon^{\mathrm{e}} + \varepsilon^{\mathrm{th}} \,. \tag{7.140}$$

In Material Model (7.140), the quantity $\varepsilon^{\mathrm{e}}$ is the elastic strain, $\varepsilon^{\mathrm{th}}$ is the thermal strain, $\varepsilon^{\mathrm{v}}$ is the viscous strain, and $\varepsilon^{\mathrm{eT}}$ is the thermoelastic strain. Based on Expression $(7.140)_2$, we arrive at the following:

$$\varepsilon - \varepsilon^{\mathrm{v}} = \varepsilon^{\mathrm{eT}} \,. \tag{7.141}$$

The specific Helmholtz free energy $\psi$ describing the uniaxial response is given by the following material model:

$$\psi(\varepsilon - \varepsilon^{\mathrm{v}}, \alpha, T) = \frac{1}{2\,\rho_0} E\,(\varepsilon - \varepsilon^{\mathrm{v}})^2 + \frac{e}{2\,\rho_0} \alpha^2 - \frac{E\,\hat{\alpha}}{\rho_0} (\varepsilon - \varepsilon^{\mathrm{v}})(T - T_{\mathrm{r}})$$
$$- c_{\mathrm{r}}\,[T \ln(T/T_{\mathrm{r}}) - T] \,, \tag{7.142}$$

where $\alpha$ is an internal variable related to viscous deformation, $\hat{\alpha}$ is the linear coefficient of thermal expansion, and $E$ is Young's modulus. The natural logarithm is denoted by $\ln(\cdot)$. The roles of the parameters $T_{\mathrm{r}}$ and $c_{\mathrm{r}}$ and quantity $h(\vec{X})$ are studied in more detail with the material models studied in Chapter 8.

Since the viscous deformation is the only inelastic response of the material, the inelastic strain $\varepsilon^{\mathrm{i}}$ is replaced with the viscous strain $\varepsilon^{\mathrm{v}}$ in Expressions (7.139) and their uniaxial counterparts take the form

$$\rho_0\, g(\sigma, \alpha, T, h(\vec{X})) \overset{\text{Legendre}}{=} \sigma\,(\varepsilon - \varepsilon^{\mathrm{v}}) - \rho_0\, \psi(\varepsilon - \varepsilon^{\mathrm{v}}, \alpha, T, h(\vec{X}))$$
$$\text{with} \quad \sigma = \rho_0 \frac{\partial \psi(\varepsilon - \varepsilon^{\mathrm{v}}, \alpha, T, h(\vec{X}))}{\partial(\varepsilon - \varepsilon^{\mathrm{v}})} \,. \tag{7.143}$$

Substitution of Material Model (7.142) into State Equation $(7.143)_2$ gives

$$\sigma = E\,(\varepsilon - \varepsilon^{\mathrm{v}}) - E\,\hat{\alpha}\,(T - T_{\mathrm{r}}) \quad \Rightarrow \quad (\varepsilon - \varepsilon^{\mathrm{v}}) = \frac{\sigma}{E} + \hat{\alpha}\,(T - T_{\mathrm{r}}) \,. \tag{7.144}$$

Substitution of Material Model (7.142) into Legendre Partial Transformation $(7.143)_1$ yields

$$\rho_0\, g(\sigma,\alpha,T,h(\vec{X})) = \sigma\,(\varepsilon-\varepsilon^{\mathrm{v}}) - \frac{1}{2}\,E\,(\varepsilon-\varepsilon^{\mathrm{v}})^2 - \frac{e}{2}\,\alpha^2 + E\,\hat{\alpha}\,(\varepsilon-\varepsilon^{\mathrm{v}})\,(T-T_{\mathrm{r}}) \\ + \rho_0\, c_{\mathrm{r}}\,[T\,\ln(T/T_{\mathrm{r}}) - T]\,. \tag{7.145}$$

Substitution of the difference $(\varepsilon-\varepsilon^{\mathrm{v}})$ from Equation $(7.144)_2$ into Expression (7.145) gives

$$\rho_0\, g(\sigma,\alpha,T,h(\vec{X})) = \sigma\,\Big[\frac{\sigma}{E} + \hat{\alpha}\,(T-T_{\mathrm{r}})\Big] - \frac{E}{2}\,\Big[\frac{\sigma}{E} + \hat{\alpha}\,(T-T_{\mathrm{r}})\Big]^2 - \frac{e}{2}\,\alpha^2 \\ + E\,\hat{\alpha}\,\Big[\frac{\sigma}{E} + \hat{\alpha}\,(T-T_{\mathrm{r}})\Big]\,(T-T_{\mathrm{r}}) + \rho_0\, c_{\mathrm{r}}\,[T\,\ln(T/T_{\mathrm{r}}) - T]\,. \tag{7.146}$$

Expression (7.146) reduces to

$$\rho_0\, g(\sigma,\alpha,T,h(\vec{X})) = \frac{\sigma^2}{2\,E} - \frac{e}{2}\,\alpha^2 + \hat{\alpha}\,\sigma\,(T-T_{\mathrm{r}}) + \frac{E}{2}\,\hat{\alpha}^2\,(T-T_{\mathrm{r}})^2 \\ + \rho_0\, c_{\mathrm{r}}\,[T\,\ln(T/T_{\mathrm{r}}) - T]\,, \tag{7.147}$$

which leads to

$$g(\sigma,\alpha,T,h(\vec{X})) = \frac{\sigma^2}{2\,\rho_0\,E} - \frac{e}{2\,\rho_0}\,\alpha^2 + \frac{\hat{\alpha}\,\sigma}{\rho_0}\,(T-T_{\mathrm{r}}) + \frac{E}{2\,\rho_0}\,\hat{\alpha}^2\,(T-T_{\mathrm{r}})^2 \\ + c_{\mathrm{r}}\,[T\,\ln(T/T_{\mathrm{r}}) - T]\,. \tag{7.148}$$

Expression (7.148) is the Legendre partial transformation of Expression (7.142). The strain difference $\varepsilon-\varepsilon^{\mathrm{v}}$ in the specific Helmholtz free energy $\psi$ is replaced with the stress $\sigma$ in the expression of the specific Gibbs free energy $g$. The earlier study shows that besides the relation between the energy functions $\psi$ and $g$ [here Expression $(7.143)_1$], state equations are also needed. Therefore, in Expression $(7.143)_1$, the text "Legendre" was added above the equals sign. Since in this case only the strain difference $\varepsilon-\varepsilon^{\mathrm{v}}$ is replaced with the stress $\sigma$, one state equation is enough.

To verify Result (7.148), it is checked that the specific Helmholtz free energy $\psi$ given in Material Model (7.142) and the specific Gibbs free energy $g$ given in Material Model (7.148) lead to the same material behavior, although the variables are different.

State Equations $(6.174)_2$ and (6.172) are

$$\boldsymbol{\beta} = -\,\rho_0\,\frac{\partial\psi(\boldsymbol{\varepsilon}-\boldsymbol{\varepsilon}^{\mathrm{i}},\boldsymbol{\alpha},T,h(\vec{X}))}{\partial\boldsymbol{\alpha}} \qquad \text{and} \qquad s = -\,\frac{\partial\psi(\boldsymbol{\varepsilon}-\boldsymbol{\varepsilon}^{\mathrm{i}},\boldsymbol{\alpha},T,h(\vec{X}))}{\partial T}\,. \tag{7.149}$$

Since the viscous strain $\varepsilon^{\mathrm{v}}$ is the only inelastic strain $\varepsilon^{\mathrm{i}}$ in this material model, the uniaxial counterparts for State Equation (7.149) are

$$\beta = -\rho_0 \frac{\partial \psi(\varepsilon - \varepsilon^{\mathrm{v}}, \alpha, T, h(\vec{X}))}{\partial \alpha} \quad \text{and} \quad s = -\frac{\partial \psi(\varepsilon - \varepsilon^{\mathrm{v}}, \alpha, T, h(\vec{X}))}{\partial T}. \tag{7.150}$$

Substitution of Material Model (7.142) into State Equations (7.150) gives

$$\beta = -\rho_0 \frac{\partial \psi(\varepsilon - \varepsilon^{\mathrm{v}}, \alpha, T, h(\vec{X}))}{\partial \alpha} = -e\,\alpha \tag{7.151}$$

and

$$\begin{aligned} \rho_0\, s = -\rho_0 \frac{\partial \psi(\varepsilon - \varepsilon^{\mathrm{v}}, \alpha, T, h(\vec{X}))}{\partial T} &= E\,\hat{\alpha}\,(\varepsilon - \varepsilon^{\mathrm{v}}) + \rho_0\, c_{\mathrm{r}}\,[\ln(T/T_{\mathrm{r}}) + 1 - 1] \\ &= E\,\hat{\alpha}\,(\varepsilon - \varepsilon^{\mathrm{v}}) + \rho_0\, c_{\mathrm{r}} \ln(T/T_{\mathrm{r}}). \end{aligned} \tag{7.152}$$

State Equations (6.181) and (6.180) are

$$\boldsymbol{\varepsilon} - \boldsymbol{\varepsilon}^{\mathrm{i}} = \rho_0 \frac{\partial g(\boldsymbol{\sigma}, \boldsymbol{\alpha}, T, h(\vec{X}))}{\partial \boldsymbol{\sigma}} \quad \text{and} \quad \boldsymbol{\beta} = \rho_0 \frac{\partial g(\boldsymbol{\sigma}, \boldsymbol{\alpha}, T, h(\vec{X}))}{\partial \boldsymbol{\alpha}}, \tag{7.153}$$

and furthermore,

$$s = \frac{\partial g(\boldsymbol{\sigma}, \boldsymbol{\alpha}, T, h(\vec{X}))}{\partial T}. \tag{7.154}$$

Since the viscous strain $\varepsilon^{\mathrm{v}}$ is the only inelastic strain $\varepsilon^{\mathrm{i}}$ in this material model, the uniaxial counterparts for State Equations (7.153) and (7.154) are

$$\varepsilon - \varepsilon^{\mathrm{v}} = \rho_0 \frac{\partial g(\sigma, \alpha, T, h(\vec{X}))}{\partial \sigma} \quad \text{and} \quad \beta = \rho_0 \frac{\partial g(\sigma, \alpha, T, h(\vec{X}))}{\partial \alpha}, \tag{7.155}$$

and furthermore,

$$s = \frac{\partial g(\sigma, \alpha, T, h(\vec{X}))}{\partial T}. \tag{7.156}$$

Substitution of Material Model (7.148) into State Equations (7.155) and (7.156) yields

$$\varepsilon - \varepsilon^{\mathrm{v}} = \rho_0 \frac{\partial g(\sigma, \alpha, T, h(\vec{X}))}{\partial \sigma} = \frac{\sigma}{E} + \hat{\alpha}\,(T - T_{\mathrm{r}}), \tag{7.157}$$

and

$$\beta = \rho_0 \frac{\partial g(\sigma, \alpha, T, h(\vec{X}))}{\partial \alpha} = -e\,\alpha, \tag{7.158}$$

and furthermore,

$$\begin{aligned} \rho_0\, s = \rho_0 \frac{\partial g(\sigma, \alpha, T, h(\vec{X}))}{\partial T} &= \hat{\alpha}\,\sigma + E\,\hat{\alpha}^2\,(T - T_{\mathrm{r}}) + \rho_0\, c_{\mathrm{r}}\,[\ln(T/T_{\mathrm{r}}) + 1 - 1] \\ &= E\,\hat{\alpha}\left[\frac{\sigma}{E} + \hat{\alpha}\,(T - T_{\mathrm{r}})\right] + \rho_0\, c_{\mathrm{r}} \ln(T/T_{\mathrm{r}}). \end{aligned} \tag{7.159}$$

Based on Material Model (7.157), Expression (7.159) yields

$$\rho_0\, s = \rho_0\, \frac{\partial g(\sigma, \alpha, T, h(\vec{X})}{\partial T} = E\, \hat{\alpha} \left[ \frac{\sigma}{E} + \hat{\alpha}\, (T - T_{\mathrm{r}}) \right] + \rho_0\, c_{\mathrm{r}}\, \ln(T/T_{\mathrm{r}})$$
$$= E\, \hat{\alpha}\, (\varepsilon - \varepsilon^{\mathrm{v}}) + \rho_0\, c_{\mathrm{r}}\, \ln(T/T_{\mathrm{r}})\,. \tag{7.160}$$

Comparison of Material Models (7.157), (7.158), and (7.160) with Material Models $(7.144)_2$, (7.151), and (7.152) shows that they are equal. Thus, the specific Helmholtz free energy $\psi$ given in Expression (7.142) and the specific Gibbs free energy $g$ given in Expression (7.148) give the same material behavior as the necessary result. This means that the person preparing the material model can create the same material response by writing the specific Helmholtz free energy $\psi$ or the specific Gibbs free energy $g$. The choice between these two functions is up to the modeller.

## 7.8 Summary

There are some sections that are very important for material modelling. First, Section 7.2 showed that the nonnegative homogeneous dissipation potentials satisfy the Clausius–Duhem inequality. This is important, because in many material model, the specific dissipation function (aka dissipation potential) is a nonnegative homogeneous function. Therefore no other validation procedure is needed. Sometimes there is a material model that does not describe elastic deformation. Such a case requires a slightly different formulation compared to those elsewhere in this book. Section 7.4 showed how to prepare material models that do not show elastic deformation. Material models describing material damage may have a localization problem. Section 7.5 introduces gradient theory, which is one way to avoid this problem. Section 7.6 introduced the validation procedure for a material model and provided an illustrative example. Finally, Section 7.7 demonstrated the way to transfer a material model written in the specific Helmholtz free energy $\psi$ into the specific Gibbs free energy $g$ and showed that these two functions give the same material response.

# 8 Material Models for Heat Conduction and Time-Independent Deformation

## 8.1 Introduction

In this chapter, material models for heat conduction and time-independent deformation are formulated. The idea behind validating material models using continuum thermodynamics is to introduce one or two energy functions, then use the state equations normality rule to derive the material model, and finally validate the obtained constitutive equation(s) using the Clausius–Duhem inequality. Sometimes only one energy function is needed, as can be seen in Section 8.2, where the thermal part of the specific dissipation function $\varphi_{\text{ther}}$ is enough to describe the material behavior. In such a case, state equations are not needed. If the specific dissipation function is a nonnegative homogeneous function, no Clausius–Duhem inequality is necessary to introduce, since the Clausius–Duhem inequality is automatically satisfied. Section 8.2 is a good example of that.

## 8.2 Fourier's Law of Heat Conduction

In this section, the equation of heat transport according to Fourier's law of heat conduction is derived. The heat conduction equation has the following appearance:

$$\rho_0\, c\, \dot{T} = \rho_0\, r + \vec{\nabla} \cdot [\gamma(T)\, \vec{\nabla} T]\,, \tag{8.1}$$

where, according to Definition (6.205), $c$ is the specific heat capacity. The notation $\gamma(T)$ is a temperature-dependent positive scalar called thermal conductivity. The first term on the right side of Expression (8.1) is the heat source term.

Since no deformation is considered, no mechanical state variables are needed. Thus, according to Figure 6.3 and Section 6.5.1, the only state variable is the absolute temperature $T$.

Applying Ziegler and Wehrli [[122], Eq. (3.10)], the specific dissipation function $\varphi_{\text{ther}}$ for the linear isotropic material (in the sense of heat conduction) can be written in the form

$$\varphi_{\text{ther}}(\vec{q};T) = \frac{\vec{q}\cdot\vec{q}}{2\,\mu_{\text{ther}}\,\rho_0\,\gamma(T)\,T}\,, \tag{8.2}$$

where $\vec{q}$ is the heat flux vector. The right side of Material Model (8.2) shows that the influence of the mechanical state variables $\boldsymbol{\varepsilon}$, $\boldsymbol{\varepsilon}^{\text{i}}$, and $\boldsymbol{\alpha}$ is neglected. Obviously, it is obtained that

$$\varphi_{\text{ther}}(k\,\vec{q};T) = k^2\,\varphi_{\text{ther}}(\vec{q};T)\,, \tag{8.3}$$

which implies that $\varphi_{\text{con}}$ is a homogeneous function of degree $1/\mu_{\text{ther}} = 2$ in the heat flux vector $\vec{q}$, as can be seen from Equation (2.207) and after Equation (6.290). Material Model (8.2) shows that the specific dissipation function $\varphi_{\text{ther}}$ is nonnegative. Since it is also a homogeneous function, according to Section 7.2, the Clausius–Duhem inequality is satisfied.

Normality Rule (6.289) reads

$$-\frac{\vec{\nabla}T}{T} = \mu_{\text{ther}}\,\rho_0\,\frac{\partial\varphi_{\text{ther}}(\vec{q};\boldsymbol{\varepsilon}-\boldsymbol{\varepsilon}^{\text{i}},\boldsymbol{\alpha},T,h)}{\partial\vec{q}}\,. \tag{8.4}$$

Substitution of Material Model (8.2) into Normality Rule (8.4) gives

$$-\frac{\vec{\nabla}T}{T} = \mu_{\text{ther}}\,\rho_0\,\frac{\partial}{\partial\vec{q}}\left(\frac{\vec{q}\cdot\vec{q}}{2\,\mu_{\text{ther}}\,\rho_0\,\gamma(T)\,T}\right) = \frac{1}{\gamma(T)\,T}\,\frac{\partial}{\partial\vec{q}}\left(\tfrac{1}{2}\,\vec{q}\cdot\vec{q}\right)\,. \tag{8.5}$$

Problem 8.1 is to show that for an arbitrary vector $\vec{a}$ and for an arbitrary symmetric second-order tensor $\mathbf{c}$, the following holds:

$$\frac{\partial}{\partial\vec{a}}\left[\tfrac{1}{2}\,\vec{a}\cdot\mathbf{c}\cdot\vec{a}\right] = \vec{a}\cdot\mathbf{c}\,. \tag{8.6}$$

Since $\vec{q}\cdot\vec{q} = \vec{q}\cdot\mathbf{1}\cdot\vec{q}$, Equation (8.6) gives

$$\frac{\partial}{\partial\vec{q}}\left[\tfrac{1}{2}\,\vec{q}\cdot\vec{q}\right] = \frac{\partial}{\partial\vec{q}}\left[\tfrac{1}{2}\,\vec{q}\cdot\mathbf{1}\cdot\vec{q}\right] = \vec{q}\cdot\mathbf{1} = \vec{q}\,. \tag{8.7}$$

Derivative (8.7) is substituted into Equation (8.5). This leads to

$$-\frac{\vec{\nabla}T}{T} = \frac{1}{\gamma(T)\,T}\,\frac{\partial}{\partial\vec{q}}\left(\tfrac{1}{2}\,\vec{q}\cdot\vec{q}\right) = \frac{\vec{q}}{\gamma(T)\,T}\,, \quad \text{which gives} \quad \vec{q} = -\gamma(T)\,\vec{\nabla}T\,. \tag{8.8}$$

Expression $(8.8)_2$ is Fourier's law of heat conduction [see, e.g., Malvern [57], Eq. (7.1.6)]. According to Isachenko et al. ([33], p. 22), it has been shown experimentally that with accuracy sufficient for practical applications, the dependence of thermal conductivity $\gamma(T)$ on temperature can be assumed to be linear for many materials, viz.

$$\gamma(T) = \gamma_0 \left[ 1 + \gamma_1 (T - T_0) \right], \tag{8.9}$$

where $\gamma_0$ is the thermal conductivity at the temperature $T_0$ and $\gamma_1$ is a material constant.

It has already been shown that Material Model (8.2) satisfies the Clausius–Duhem inequality. Now the proof is carried out in a more traditional way. Recalling the Clausius–Duhem inequality for heat conduction [Inequality $(6.264)_2$]:

$$(\Phi_{\text{ther}} =) \qquad -\frac{\vec{\nabla} T}{T} \cdot \vec{q} \geq 0 \,. \tag{8.10}$$

Substitution of $-\vec{\nabla} T / T$, obtained from Fourier's Law $(8.8)_2$, into Clausius–Duhem Inequality (8.10) yields

$$\frac{\vec{q} \cdot \vec{q}}{\gamma(T)\, T} \geq 0 \,. \tag{8.11}$$

Since thermal conductivity $\gamma(T)$ and absolute temperature $T$ are positive scalars, Inequality (8.11) holds true and the Clausius–Duhem inequality is satisfied.

Equation $(8.8)_2$ reads

$$\vec{q} = -\gamma(T) \vec{\nabla} T \qquad \Rightarrow \qquad \vec{\nabla} \cdot \vec{q} = -\vec{\nabla} \cdot \left[ \gamma(T)\, \vec{\nabla} T \right]. \tag{8.12}$$

Heat Equation (6.206) is

$$\vec{\nabla} \cdot \vec{q} = \boldsymbol{\sigma} : \dot{\boldsymbol{\varepsilon}}^{\mathrm{i}} + \boldsymbol{\beta} : \dot{\boldsymbol{\alpha}} + \rho_0\, r + \rho_0\, T \left[ \frac{\partial^2 \psi}{\partial T\, \partial(\boldsymbol{\varepsilon} - \boldsymbol{\varepsilon}^{\mathrm{i}})} : (\dot{\boldsymbol{\varepsilon}} - \dot{\boldsymbol{\varepsilon}}^{\mathrm{i}}) + \frac{\partial^2 \psi}{\partial T\, \partial \boldsymbol{\alpha}} : \dot{\boldsymbol{\alpha}} \right] - \rho_0\, c\, \dot{T} \,. \tag{8.13}$$

Substitution of $\vec{\nabla} \cdot \vec{q}$, obtained from Equation $(8.12)_2$, into Heat Equation (8.13) gives

$$\begin{aligned} \rho_0\, c \dot{T} &= \boldsymbol{\sigma} : \dot{\boldsymbol{\varepsilon}}^{\mathrm{i}} + \boldsymbol{\beta} : \dot{\boldsymbol{\alpha}} + \rho_0\, r \\ &\quad + \rho_0\, T \left[ \frac{\partial^2 \psi}{\partial T\, \partial(\boldsymbol{\varepsilon} - \boldsymbol{\varepsilon}^{\mathrm{i}})} : (\dot{\boldsymbol{\varepsilon}} - \dot{\boldsymbol{\varepsilon}}^{\mathrm{i}}) + \frac{\partial^2 \psi}{\partial T\, \partial \boldsymbol{\alpha}} : \dot{\boldsymbol{\alpha}} \right] + \vec{\nabla} \cdot \left[ \gamma(T) \vec{\nabla} T \right]. \end{aligned} \tag{8.14}$$

If only the thermal part of the problem is considered, Heat Conduction Equation (8.14) reduces to

$$\rho_0\, c \dot{T} = \rho_0\, r + \vec{\nabla} \cdot \left[ \gamma(T)\, \vec{\nabla} T \right]. \tag{8.15}$$

The meaning of the phrase "only thermal part of the problem is considered" requires more careful study. As can be seen in Section 8.3.4, Expression (8.15) is the heat conduction for a rigid material.

Certainly the best-known model for heat conduction is Fourier's law of heat conduction, Expression $(8.8)_2$. However, some criticism is worth making here. According to Lebon et al. ([47], p. 182), "although this equation [Equation (8.15) without the source term] is well tested for most practical problems, it fails to describe the transient temperature field in situations involving short time, high frequencies, and small wavelengths." Lebon et al. ([47], p. 182) argued that according to Fourier's law, a sudden application of temperature difference gives instantaneous rise to a heat flux everywhere in the system. In other words, any temperature disturbance will propagate at infinite velocity.

In 1948, Cattaneo [13] proposed the following expression to replace Fourier's law of heat conduction:

$$\vec{q} + \tau_0 \frac{\partial \vec{q}}{\partial t} = -\gamma(T)\, \vec{\nabla} T\,, \tag{8.16}$$

where $\tau_0$ is the thermal relaxation characteristic time. According to Lebon et al. ([47], p. 182), when the relaxation time $\tau_0$ of the heat flux is negligible or the time variation of the heat flux is slow, Cattaneo Equation (8.16) reduces to Fourier's Law $(8.8)_2$.

Assuming that the specific heat capacity $c$ and the thermal conductivity $\gamma$ take constant values, and neglecting the heat source, Cattaneo Equation (8.16) gives the following form for the heat conduction equation [see Lebon et al. [47], Eq. (7.5)]:

$$\tau_0 \frac{\partial^2 T}{\partial t^2} + \frac{\partial T}{\partial t} = \frac{\gamma}{\rho_0\, c} \nabla^2 T\,. \tag{8.17}$$

The following example shows the difference in simulated responses to the heat conduction problem when applying Fourier's law and the expression proposed by Cattaneo [13]. According to Lebon et al. ([47], p. 182), Fourier's Law $(8.8)_2$ fails to describe the transient temperature field in situations involving a short time, high frequencies, and small wavelengths. Therefore the comparison is made by studying the heat response due to the delta pulse of the temperature of a one-dimensional rod. The temperature distributions by Lebon et al. ([47], Figure 7.1) are redrawn in Figure 8.1. It is interesting to see that the Cattaneo equation predicts a finite speed of information, shown by the sudden temperature drop to zero, whereas the response calculation with Fourier's law of heat conduction leads to a decay of the temperature. The difference between the Cattaneo equation and Fourier's law is that they lead to different forms of the heat conduction equation. The Cattaneo equation leads to a hyperbolic differential equation, whereas with Fourier's law, it is a parabolic differential equation.

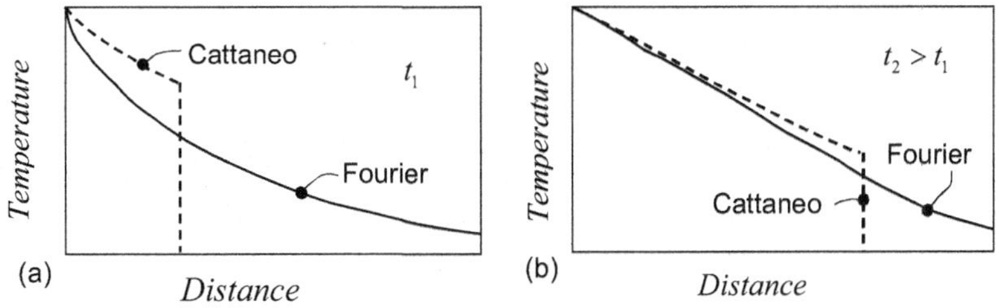


**Figure 8.1** Temperature distributions determined by the heat conduction equation based on Fourier's law and the Cattaneo equation. Adapted from: Lebon et al. ([47], Figure 7.1).

The mathematical form of the solution for Heat Conduction Equation (8.15) (Fourier's law) is studied with the following two simple examples. The first is a modified copy from the book by Boas [11] (see Chapter 13, Section 9, Example 1). Since the example by Boas [11] studies a uniaxial case where the thermal conductivity $\gamma(T)$ takes a constant value, that is, $\gamma(T) = \gamma$, and where the heat source is given by a boundary condition, Equation (8.15) reduces to

$$\rho_0 \, c \, \frac{\partial T}{\partial t} = \gamma \, \frac{\partial^2 T}{\partial x^2} \qquad \Rightarrow \qquad \frac{\partial T}{\partial t} = \omega^2 \, \frac{\partial^2 T}{\partial x^2} \, , \quad \text{where } \omega = \sqrt{\frac{\gamma}{\rho_0 \, c}} \, . \tag{8.18}$$

**Example 8.1:** A semi-infinite (rigid) bar (extending from $x = 0$ to $x = \infty$), with insulated sides, is initially at the uniform temperature $T$. At time $t = 0$, the end $x = 0$ is brought to $T + \Delta T$ and held there. Find the temperature distribution in the bar as a function of $x$ and $t$. Apply Fourier's law of heat conduction, that is, Equation $(8.8)_2$.

**Solution:** Boas [[11], p. 660, Eq. (9.7)] gives the following solution:

$$T(x,t) = \Delta T \left[ 1 - \operatorname{erf}\left( \frac{x}{2\,\omega\,\sqrt{t}} \right) \right] , \tag{8.19}$$

where the error function $\operatorname{erf}(\xi)$ is defined to be [Boas [11], p. 548, Eq. (9.7)]

$$\operatorname{erf}(\xi) := \frac{2}{\sqrt{\pi}} \int_{\kappa=0}^{\xi} e^{-\kappa^2} \, d\kappa \, . \tag{8.20}$$

The complementary error function $\operatorname{cerf}(\xi)$ reads [Boas [11], p. 547, Eq. (9.3a)]

$$\operatorname{erfc}(\xi) := \frac{2}{\sqrt{\pi}} \int_{\kappa=\xi}^{\infty} e^{-\kappa^2} \, d\kappa = 1 - \operatorname{erf}(\xi) \, . \tag{8.21}$$

Based on Expression (8.21), Solution (8.19) takes the form

$$T(x,t) = \Delta T \operatorname{erfc}\left( \frac{x}{2\,\omega\,\sqrt{t}} \right) . \tag{8.22}$$

■

**Example 8.2:** The conditions are the same as in **Example 8.1**. What is the temperature in the rod at time $t = 1$ s at a distance of 0.25 m from the end of the rod? The material of the rod is pure aluminium, with the following values of the material parameters at 300 K (see Incropera and DeWitt [32], p. 51): $\rho_0 = 2702$ kg/m$^3$, $c_{\varepsilon\alpha} = 903$ J/(kg K), and $\gamma = 237$ W/(m K). Apply Fourier's law of heat conduction, that is, Equation $(8.8)_2$.

**Solution**: The preceding values for the material parameters $\rho_0$, $c_{\varepsilon\alpha}$, and $\gamma$ yield

$$\omega = \sqrt{\frac{\gamma}{\rho_0\, c_{\varepsilon\alpha}}} = \sqrt{\frac{237}{2702 \times 903}} = \sqrt{97.135 \cdot 10^{-6}} = 9.8557 \cdot 10^{-3} \mathrm{m}/\sqrt{\mathrm{s}}\,. \quad (8.23)$$

The following notation is introduced:

$$\xi = \frac{x}{2\,\omega\,\sqrt{t}} \qquad \Rightarrow \qquad T(\xi) = \Delta T\,\mathrm{erfc}(\xi)\,. \quad (8.24)$$

Based on Equations (8.20), (8.22), and (8.24) and the given values for the quantities $x$ and $t$, we arrive at the following:

$$\xi = \frac{x}{2\,\omega\,\sqrt{t}} = \frac{0.25}{2 \times 9.8557 \cdot 10^{-3}\sqrt{1}} = 12.683\,. \quad (8.25)$$

Boas [[11], p. 549, Eq. 10.4)] gives a series approximation for the complementary error function erfc($\xi$) when the absolute value for the independent variable takes large values, $\mid \xi \mid \gg 1$. It reads

$$\mathrm{erfc}(\xi) \approx \frac{e^{-\xi^2}}{\xi\,\sqrt{\pi}}\left[1 - \frac{1}{2\,\xi^2} + \frac{1\times 3}{(2\,\xi^2)^2} - \frac{1 \times 3 \times 5}{(2\,\xi^2)^3} + \ldots\right], \quad \text{for} \mid \xi \mid \;\gg 1\,. \quad (8.26)$$

Although the value for the quantity $\mid \xi \mid$ is not much larger than unity, as assumed in Approximation (8.26), Approximation (8.26) has to be utilized. The increase of the value for the distance $x$ would make the quantity $|\xi|$ larger, but this is not possible, as the following figures show.

Substitution of the value for the quantity $\xi$ from Expression (8.25) into Series (8.26) (where only the first term between the parentheses needs to be taken into account) and the obtained value further into Solution (8.22) yields

$$T(x = 0.25\mathrm{m}, t = 1\mathrm{s}) = \Delta T\,\mathrm{erfc}(\xi = 12.683)$$

$$\approx \Delta T\,\frac{e^{-(12.683)^2}}{12.683\,\sqrt{\pi}}\left[1 - \frac{1}{2 \times 12.683^2} + \frac{1 \times 3}{(2 \times 12.683^2)^2} - \frac{1 \times 3 \times 5}{(2 \times 12.683^2)^3} + \ldots\right]. \quad (8.27)$$

Equation (8.27) yields

$$T(x = 0.25\mathrm{m}, t = 1\mathrm{s}) = \Delta T\,\mathrm{erfc}(\xi = 12.683)$$

$$\approx 6.1411 \cdot 10^{-72}(1 - 3.108322 \cdot 10^{-3} + 2.8985 \cdot 10^{-5} - 4.5047 \cdot 10^{-7} + \ldots)\,\Delta T\,. \quad (8.28)$$

As Result (8.28) shows, the error in the heat conduction equation caused by Fourier's law is minimal in most applications. ■

The preceding study was prepared for a rigid body, that is, thermal expansion was neglected. This is the case in many books on physics. This was the reason for replacing in Equations (8.17) and (8.18) the material time derivative of temperature $\dot{T}$ with the partial time derivative $\partial T/\partial t$.

## 8.3 Elasticity Combined with Heat Equation

The present section studies hyperelasticity and thermoelasticity combined with heat conduction. Section 8.3.1 examines hyperelastic deformation. Section 8.3.2 derives limits for the values of Young's modulus $E$ and Poisson's ratio $\nu$. In Section 8.3.3, thermoelastic deformation is considered, and Section 8.3.4 evaluates Fourier's law of heat conduction coupled with thermoelastic deformation. Section 8.3.6 studies thermodynamically inhomogeneous Hookean material.

### 8.3.1 Hyperelastic Deformation

This section studies hyperelastic deformation. Hyperelasticity means that the value for the specific Helmholtz free energy $\psi$ (or strain-energy density $w$) depends only on the strain tensors and therefore neglects, for example, temperature effects on the deformation of a body. Deformation is assumed to be linear elastic; that is, it follows Hooke's law. The material is assumed to be isotropic.

The specific Helmholtz free energy $\psi$ for an isotropic linear hyperelastic deformation takes the following appearance:

$$\psi^{\mathrm{e}}(\boldsymbol{\varepsilon}-\boldsymbol{\varepsilon}^{\mathrm{i}}) = \psi^{\mathrm{e}}_{\mathrm{r}} + \frac{1}{2\,\rho_0}\,(\boldsymbol{\varepsilon}-\boldsymbol{\varepsilon}^{\mathrm{i}}):\mathbf{C}:(\boldsymbol{\varepsilon}-\boldsymbol{\varepsilon}^{\mathrm{i}})\,, \tag{8.29}$$

where the subscript r denotes a value in the reference state (initial configuration). The fourth-order constitutive tensor $\mathbf{C}$ for a Hookean deformation (i.e., for Hooke's law) is defined by

$$\mathbf{C} := \lambda\,\mathbf{1}\,\mathbf{1} + 2\,\mu\,\mathbf{I}^{\mathrm{s}}\,, \tag{8.30}$$

where $\lambda$ and $\mu$ are Lamé elastic constants defined by

$$\lambda := \frac{\nu\,E}{(1+\nu)\,(1-2\,\nu)} \qquad \text{and} \qquad \mu := \frac{E}{2\,(1+\nu)}\,. \tag{8.31}$$

In Definitions (8.31), $E$ and $\nu$ are Young's modulus and Poisson's ratio, respectively. The second-order identity tensor $\mathbf{1}$ is given by Definition (2.63), and the fourth-order symmetric identity tensor $\mathbf{I}^{\mathrm{s}}$ is given by Definition (2.106).

Material Model (8.29) and State Equation $(6.174)_1$ give [see Result $(2.131)_2$]

$$\boldsymbol{\sigma} = \rho_0 \frac{\partial \psi^{\mathrm{e}}(\boldsymbol{\varepsilon} - \boldsymbol{\varepsilon}^{\mathrm{i}})}{\partial (\boldsymbol{\varepsilon} - \boldsymbol{\varepsilon}^{\mathrm{i}})} = \mathbf{C} : (\boldsymbol{\varepsilon} - \boldsymbol{\varepsilon}^{\mathrm{i}}) \quad \Rightarrow \quad \boldsymbol{\sigma} = \mathbf{C} : \boldsymbol{\varepsilon}^{\mathrm{e}} , \tag{8.32}$$

where the major symmetry of the constitutive tensor $\mathbf{C}$, that is, $\mathrm{C}_{ijkl} = \mathrm{C}_{klij}$, is exploited. The symmetry of the constitutive tensor $\mathbf{C}$ is discussed in more detail later in this section. Hooke's Law $(8.32)_2$ introduced the elastic strain tensor, denoted by $\boldsymbol{\varepsilon}^{\mathrm{e}}$ and defined to be

$$\boldsymbol{\varepsilon}^{\mathrm{e}} := \boldsymbol{\varepsilon} - \boldsymbol{\varepsilon}^{\mathrm{i}} . \qquad \text{For hyperelastic deformation.} \tag{8.33}$$

It has to be kept in mind that the comment "for hyperelastic deformation" is crucial, since there are cases in which the strain tensor difference $\boldsymbol{\varepsilon} - \boldsymbol{\varepsilon}^{\mathrm{i}}$ does not equal the elastic strain tensor $\boldsymbol{\varepsilon}^{\mathrm{e}}$. Such cases are evaluated later in this chapter under thermoelastic deformation and in Sections 10.3 and 10.4, in which expressions for continuum damage mechanics are derived.

Substitution of Definition (8.30) into Hooke's Law $(8.32)_2$ gives

$$\boldsymbol{\sigma} = \lambda \mathbf{1}\mathbf{1} : \boldsymbol{\varepsilon}^{\mathrm{e}} + 2\mu \mathbf{I}^{\mathrm{s}} : \boldsymbol{\varepsilon}^{\mathrm{e}} . \tag{8.34}$$

Next a relation for the elastic strain tensor $\boldsymbol{\varepsilon}^{\mathrm{e}}$ in terms of the stress tensor $\boldsymbol{\sigma}$ is formulated. Expression (8.34) is multiplied by $\mathbf{1}:$ from the left. This gives

$$\mathbf{1} : \boldsymbol{\sigma} = \lambda \mathbf{1} : \mathbf{1}\mathbf{1} : \boldsymbol{\varepsilon}^{\mathrm{e}} + 2\mu \mathbf{1} : \mathbf{I}^{\mathrm{s}} : \boldsymbol{\varepsilon}^{\mathrm{e}} . \tag{8.35}$$

Since $\mathbf{1}:\mathbf{1} = 3$ [Equation $(2.65)_2$] and $\mathbf{1}:\mathbf{I}^{\mathrm{s}} = \frac{1}{2}(\mathbf{1} + \mathbf{1}^{\mathrm{T}}) = \mathbf{1}$, [see Definition (2.106)], Expression (8.35) reduces to

$$\mathbf{1} : \boldsymbol{\sigma} = 3\lambda \mathbf{1} : \boldsymbol{\varepsilon}^{\mathrm{e}} + 2\mu \mathbf{1} : \boldsymbol{\varepsilon}^{\mathrm{e}} = (3\lambda + 2\mu) \mathbf{1} : \boldsymbol{\varepsilon}^{\mathrm{e}} . \tag{8.36}$$

Substitution of $\mathbf{1} : \boldsymbol{\varepsilon}^{\mathrm{e}}$, obtained from Equation (8.36), into Hooke's Law (8.34) yields

$$\boldsymbol{\sigma} = \frac{\lambda}{(3\lambda + 2\mu)} \mathbf{1}\mathbf{1} : \boldsymbol{\sigma} + 2\mu \mathbf{I}^{\mathrm{s}} : \boldsymbol{\varepsilon}^{\mathrm{e}} , \tag{8.37}$$

which for the symmetric elastic strain tensor $\boldsymbol{\varepsilon}^{\mathrm{e}}$ leads to [see Definition (2.106) and note that $\mathbf{I}^{\mathrm{s}} : \boldsymbol{\varepsilon}^{\mathrm{e}} = [\boldsymbol{\varepsilon}^{\mathrm{e}} + (\boldsymbol{\varepsilon}^{\mathrm{e}})^{\mathrm{T}}]/2 = \boldsymbol{\varepsilon}^{\mathrm{e}}$]

$$\boldsymbol{\varepsilon}^{\mathrm{e}} = -\frac{\lambda}{2\mu(3\lambda + 2\mu)} \mathbf{1}\mathbf{1} : \boldsymbol{\sigma} + \frac{1}{2\mu} \boldsymbol{\sigma} \tag{8.38}$$

or

$$\boldsymbol{\varepsilon}^{\mathrm{e}} = -\frac{\lambda}{2\mu(3\lambda + 2\mu)} \mathbf{1}\mathbf{1} : \boldsymbol{\sigma} + \frac{1}{2\mu} \mathbf{I}^{\mathrm{s}} : \boldsymbol{\sigma} , \tag{8.39}$$

where the symmetry of the stress tensor $\boldsymbol{\sigma}$ is exploited. Substitution of the Lamé elastic constants $\lambda$ and $\mu$ into Expression (8.39) yields

$$\boldsymbol{\varepsilon}^{\mathrm{e}} = -\frac{\nu}{E} \mathbf{1}\mathbf{1} : \boldsymbol{\sigma} + \frac{1+\nu}{E} \mathbf{I}^{\mathrm{s}} : \boldsymbol{\sigma} \quad \text{or} \quad \boldsymbol{\varepsilon}^{\mathrm{e}} = -\frac{\nu}{E} \mathbf{1}\mathbf{1} : \boldsymbol{\sigma} + \frac{1+\nu}{E} \boldsymbol{\sigma} . \tag{8.40}$$

In Expression $(8.40)_2$, Definition (2.106) and the symmetry of the stress tensor $\boldsymbol{\sigma}$ were exploited; that is, $\mathbf{I}^{\mathrm{s}}:\boldsymbol{\sigma} = (\boldsymbol{\sigma} + \boldsymbol{\sigma}^{\mathrm{T}})/2 = \boldsymbol{\sigma}$. Hooke's Law $(8.40)_1$ can be further recast in the forms

$$\boldsymbol{\varepsilon}^{\mathrm{e}} = \mathbf{S}:\boldsymbol{\sigma} \qquad \text{or} \qquad \boldsymbol{\varepsilon} - \boldsymbol{\varepsilon}^{\mathrm{i}} = \mathbf{S}:\boldsymbol{\sigma}, \tag{8.41}$$

where the compliance tensor for a Hookean deformation $\mathbf{S}$ is defined by

$$\mathbf{S} := -\frac{\nu}{E}\mathbf{1}\mathbf{1} + \frac{1+\nu}{E}\mathbf{I}^{\mathrm{s}}. \tag{8.42}$$

Constitutive tensor $\mathbf{C}$ and compliance tensor $\mathbf{S}$ obey the following property:

$$\mathbf{C}:\mathbf{S} = \mathbf{S}:\mathbf{C} = \mathbf{I}^{\mathrm{s}}. \tag{8.43}$$

Theorem 2 of **Appendix I** shows that Expressions (8.42) and (8.43) for the compliance tensor $\mathbf{S}$ are equal.

According to Theorem 3 of **Appendix I**, the constitutive tensor for a Hookean deformation $\mathbf{C}$ is major symmetric,

$$C_{ijkl} = C_{klij}. \tag{8.44}$$

According to Theorem 4 of **Appendix I**, the constitutive tensor $\mathbf{C}$ is minor symmetric,

$$C_{ijkl} = C_{jikl} \qquad \text{and} \qquad C_{ijkl} = C_{ijlk}. \tag{8.45}$$

It can be shown that Properties (8.44) and (8.45) are also valid for the compliance tensor for a Hookean deformation $\mathbf{S}$.

From the physics point of view, it is reasonable to assume that straining of a body requires work. This implies that the second term of Expression (8.29) is positive. With Definition (8.33), this gives

$$\boldsymbol{\varepsilon}^{\mathrm{e}}:\mathbf{C}:\boldsymbol{\varepsilon}^{\mathrm{e}} > 0 \quad \text{if} \quad \boldsymbol{\varepsilon}^{\mathrm{e}} \neq \mathbf{0}, \quad \text{for hyperelastic deformation}, \tag{8.46}$$

where the fact that the term $1/2\,\rho_0$ is positive is exploited. Inequality (8.46) is the definition for a tensor $\mathbf{C}$ to be positive-definite. Thus the constitutive tensor $\mathbf{C}$ is a positive-definite tensor. Starting from the expression for the specific Gibbs free energy $g$, the compliance tensor $\mathbf{S}$ can be shown to be positive-definite, as well.

### 8.3.2 Restrictions for Values of Young's Modulus and Poisson's Ratio

The positive-definiteness of the constitutive tensor $\mathbf{C}$ imposes restrictions on the values of Young's modulus $E$ and Poisson's ratio $\nu$. These restrictions are studied next.

Substitution of the constitutive tensor $\mathbf{C}$ from Equation (8.30) into Inequality (8.46) gives

$$\begin{aligned}\boldsymbol{\varepsilon}^{\mathrm{e}}:\mathbf{C}:\boldsymbol{\varepsilon}^{\mathrm{e}} &= \boldsymbol{\varepsilon}^{\mathrm{e}}:\left[\lambda\,\mathbf{1}\,\mathbf{1}+2\,\mu\,\mathbf{I}^{\mathrm{s}}\right]:\boldsymbol{\varepsilon}^{\mathrm{e}} = \lambda\,\boldsymbol{\varepsilon}^{\mathrm{e}}:\mathbf{1}\,\mathbf{1}:\boldsymbol{\varepsilon}^{\mathrm{e}}+2\,\mu\,\boldsymbol{\varepsilon}^{\mathrm{e}}:\mathbf{I}^{\mathrm{s}}:\boldsymbol{\varepsilon}^{\mathrm{e}}\\ &= \lambda\,\boldsymbol{\varepsilon}^{\mathrm{e}}:\mathbf{1}\,\mathbf{1}:\boldsymbol{\varepsilon}^{\mathrm{e}}\\ &\quad +2\,\mu\left[(\mathbf{I}-\tfrac{1}{3}\mathbf{1}\,\mathbf{1}+\tfrac{1}{3}\mathbf{1}\,\mathbf{1}):\boldsymbol{\varepsilon}^{\mathrm{e}}\right]:\mathbf{I}^{\mathrm{s}}:\left[\boldsymbol{\varepsilon}^{\mathrm{e}}:(\mathbf{I}-\tfrac{1}{3}\mathbf{1}\,\mathbf{1}+\tfrac{1}{3}\mathbf{1}\,\mathbf{1})\right]>0\,.\end{aligned} \tag{8.47}$$

The concept of the third line of Manipulation (8.47) is to divide the elastic strain tensor $\boldsymbol{\varepsilon}^{\mathrm{e}}$ into a spherical part and a deviatoric part. Since $\mathbf{I}:\boldsymbol{\varepsilon}^{\mathrm{e}}=\boldsymbol{\varepsilon}^{\mathrm{e}}:\mathbf{I}=\boldsymbol{\varepsilon}^{\mathrm{e}}$ [see Definition (2.69)], the terms in parentheses in the third line of Manipulation (8.47) do not change the information given by Equation (8.47).

Based on Equations $(2.108)_2$ and (2.120), the following holds:

$$(\mathbf{I}-\tfrac{1}{3}\mathbf{1}\,\mathbf{1}):\boldsymbol{\varepsilon}^{\mathrm{e}}=\mathbf{e}^{\mathrm{e}} \qquad \Rightarrow \qquad \boldsymbol{\varepsilon}^{\mathrm{e}}:(\mathbf{I}-\tfrac{1}{3}\mathbf{1}\,\mathbf{1})=\mathbf{e}^{\mathrm{e}}\,, \tag{8.48}$$

where $\mathbf{e}^{\mathrm{e}}$ is the deviatoric elastic strain tensor. Since the deviatoric elastic strain tensor $\mathbf{e}^{\mathrm{e}}$ is symmetric, Definition (2.106) yields

$$\mathbf{I}^{\mathrm{s}}:\mathbf{e}^{\mathrm{e}}=\tfrac{1}{2}\left[\mathbf{e}^{\mathrm{e}}+(\mathbf{e}^{\mathrm{e}})^{\mathrm{T}}\right]=\mathbf{e}^{\mathrm{e}}\,. \tag{8.49}$$

Application of Equations (8.48) and (8.49) to Expression (8.47) leads to

$$\begin{aligned}\boldsymbol{\varepsilon}^{\mathrm{e}}:\mathbf{C}:\boldsymbol{\varepsilon}^{\mathrm{e}} &= \lambda\,\boldsymbol{\varepsilon}^{\mathrm{e}}:\mathbf{1}\,\mathbf{1}:\boldsymbol{\varepsilon}^{\mathrm{e}}\\ &\quad +2\,\mu\left[(\mathbf{I}-\tfrac{1}{3}\mathbf{1}\,\mathbf{1}+\tfrac{1}{3}\mathbf{1}\,\mathbf{1}):\boldsymbol{\varepsilon}^{\mathrm{e}}\right]:\mathbf{I}^{\mathrm{s}}:\left[\boldsymbol{\varepsilon}^{\mathrm{e}}:(\mathbf{I}-\tfrac{1}{3}\mathbf{1}\,\mathbf{1}+\tfrac{1}{3}\mathbf{1}\,\mathbf{1})\right]\\ &= \lambda\,\boldsymbol{\varepsilon}^{\mathrm{e}}:\mathbf{1}\,\mathbf{1}:\boldsymbol{\varepsilon}^{\mathrm{e}}+2\,\mu\left[\mathbf{e}^{\mathrm{e}}+\tfrac{1}{3}\mathbf{1}\,\mathbf{1}:\boldsymbol{\varepsilon}^{\mathrm{e}}\right]:\mathbf{I}^{\mathrm{s}}:\left[\mathbf{e}^{\mathrm{e}}+\tfrac{1}{3}\mathbf{1}\,\mathbf{1}:\boldsymbol{\varepsilon}^{\mathrm{e}}\right]\\ &= \lambda\,\boldsymbol{\varepsilon}^{\mathrm{e}}:\mathbf{1}\,\mathbf{1}:\boldsymbol{\varepsilon}^{\mathrm{e}}+2\,\mu\left[\mathbf{e}^{\mathrm{e}}+\tfrac{1}{3}\mathbf{1}\,\mathbf{1}:\boldsymbol{\varepsilon}^{\mathrm{e}}\right]:\left[\mathbf{e}^{\mathrm{e}}+\tfrac{1}{3}\mathbf{1}\,\mathbf{1}:\boldsymbol{\varepsilon}^{\mathrm{e}}\right]>0\,.\end{aligned} \tag{8.50}$$

Based on Equation $(2.116)_2$, the following holds:

$$\mathbf{1}:\mathbf{e}^{\mathrm{e}}=\mathbf{e}^{\mathrm{e}}:\mathbf{1}=0\,. \tag{8.51}$$

Next, the double-dot product between the terms in square brackets is carried out. By utilizing Expression (8.51), the following is achieved:

$$\begin{aligned}\boldsymbol{\varepsilon}^{\mathrm{e}}:\mathbf{C}:\boldsymbol{\varepsilon}^{\mathrm{e}} &= \lambda\,\boldsymbol{\varepsilon}^{\mathrm{e}}:\mathbf{1}\,\mathbf{1}:\boldsymbol{\varepsilon}^{\mathrm{e}}+2\,\mu\left[\mathbf{e}^{\mathrm{e}}+\tfrac{1}{3}\mathbf{1}\,\mathbf{1}:\boldsymbol{\varepsilon}^{\mathrm{e}}\right]:\left[\mathbf{e}^{\mathrm{e}}+\tfrac{1}{3}\mathbf{1}\,\mathbf{1}:\boldsymbol{\varepsilon}^{\mathrm{e}}\right]\\ &= \lambda\,\boldsymbol{\varepsilon}^{\mathrm{e}}:\mathbf{1}\,\mathbf{1}:\boldsymbol{\varepsilon}^{\mathrm{e}}+2\,\mu\,\mathbf{e}^{\mathrm{e}}:\mathbf{e}^{\mathrm{e}}+2\,\mu\,\tfrac{1}{9}\,\mathbf{1}:\boldsymbol{\varepsilon}^{\mathrm{e}}\,\mathbf{1}:\mathbf{1}\,\mathbf{1}:\boldsymbol{\varepsilon}^{\mathrm{e}}\\ &= \lambda\,\boldsymbol{\varepsilon}^{\mathrm{e}}:\mathbf{1}\,\mathbf{1}:\boldsymbol{\varepsilon}^{\mathrm{e}}+2\,\mu\,\mathbf{e}^{\mathrm{e}}:\mathbf{e}^{\mathrm{e}}+2\,\mu\,\tfrac{1}{3}\,\mathbf{1}:\boldsymbol{\varepsilon}^{\mathrm{e}}\,\mathbf{1}:\boldsymbol{\varepsilon}^{\mathrm{e}}>0\,.\end{aligned} \tag{8.52}$$

On the last line of Expression (8.52), Expression $(2.65)_2$, that is, $\mathbf{1}:\mathbf{1}=3$, is exploited. Expression (8.52) can be recast in the form

$$\begin{aligned}\boldsymbol{\varepsilon}^{\mathrm{e}}:\mathbf{C}:\boldsymbol{\varepsilon}^{\mathrm{e}} &= \lambda\,\boldsymbol{\varepsilon}^{\mathrm{e}}:\mathbf{1}\,\mathbf{1}:\boldsymbol{\varepsilon}^{\mathrm{e}}+2\,\mu\,\mathbf{e}^{\mathrm{e}}:\mathbf{e}^{\mathrm{e}}+2\,\mu\,\tfrac{1}{3}\,\mathbf{1}:\boldsymbol{\varepsilon}^{\mathrm{e}}\,\mathbf{1}:\boldsymbol{\varepsilon}^{\mathrm{e}}\\ &= (\lambda+\tfrac{2}{3}\,\mu)\,\boldsymbol{\varepsilon}^{\mathrm{e}}:\mathbf{1}\,\mathbf{1}:\boldsymbol{\varepsilon}^{\mathrm{e}}+2\,\mu\,\mathbf{e}^{\mathrm{e}}:\mathbf{e}^{\mathrm{e}}>0\,.\end{aligned} \tag{8.53}$$

The next step is to show that the two terms on the second line of Expression (8.53) have to take zero values independently. Component forms of the quantities present in Expression (8.53) help perform this examination. Based on Equations $(2.65)_1$, $(2.108)_2$, (2.120), and (2.69), we arrive at the following:

$$\boldsymbol{\varepsilon}^{\mathrm{e}} : \mathbf{1} = \mathbf{1} : \boldsymbol{\varepsilon}^{\mathrm{e}} = \varepsilon^{\mathrm{e}}_{ii} = \varepsilon^{\mathrm{e}}_{11} + \varepsilon^{\mathrm{e}}_{22} + \varepsilon^{\mathrm{e}}_{33} \,. \tag{8.54}$$

and

$$\mathbf{e}^{\mathrm{e}} := \mathbf{K} : \boldsymbol{\varepsilon}^{\mathrm{e}} = (\mathbf{I} - \tfrac{1}{3}\,\mathbf{1}\,\mathbf{1}) : \boldsymbol{\varepsilon}^{\mathrm{e}} = \boldsymbol{\varepsilon}^{\mathrm{e}} - \tfrac{1}{3}\,\mathbf{1}\,\mathbf{1} : \boldsymbol{\varepsilon}^{\mathrm{e}} = \boldsymbol{\varepsilon}^{\mathrm{e}} - \tfrac{1}{3}\,\mathbf{1}\varepsilon^{\mathrm{e}}_{ss} \,. \tag{8.55}$$

Based on Equation (2.64), that is, $\mathbf{1} = \delta_{ij}\,\vec{i}_i\,\vec{i}_j$ , Equation (8.55) yields

$$e^{\mathrm{e}}_{ij} = \varepsilon^{\mathrm{e}}_{ij} - \tfrac{1}{3}\,\delta_{ij}\,\varepsilon^{\mathrm{e}}_{ss} \,. \tag{8.56}$$

Expression (8.56) gives

$$e^{\mathrm{e}}_{11} = \varepsilon^{\mathrm{e}}_{11} - \tfrac{1}{3}(\varepsilon^{\mathrm{e}}_{11} + \varepsilon^{\mathrm{e}}_{22} + \varepsilon^{\mathrm{e}}_{33}) = \frac{2\,\varepsilon^{\mathrm{e}}_{11} - \varepsilon^{\mathrm{e}}_{22} - \varepsilon^{\mathrm{e}}_{33}}{3} \,, \tag{8.57}$$

and

$$e^{\mathrm{e}}_{22} = \frac{2\,\varepsilon^{\mathrm{e}}_{22} - \varepsilon^{\mathrm{e}}_{33} - \varepsilon^{\mathrm{e}}_{11}}{3} \qquad \text{and} \qquad e^{\mathrm{e}}_{33} = \frac{2\,\varepsilon^{\mathrm{e}}_{33} - \varepsilon^{\mathrm{e}}_{11} - \varepsilon^{\mathrm{e}}_{22}}{3} \,, \tag{8.58}$$

and furthermore,

$$e^{\mathrm{e}}_{12} = \varepsilon^{\mathrm{e}}_{12} \,, \qquad e^{\mathrm{e}}_{13} = \varepsilon^{\mathrm{e}}_{13} \,, \qquad e^{\mathrm{e}}_{21} = \varepsilon^{\mathrm{e}}_{21} \,, \qquad \text{and so on.} \tag{8.59}$$

First, the straining of type A is applied. The material is strained by a combination of elastic shear strains only. Based on Equation (8.54), the first term on the second line of Equation (8.53) vanishes, whereas according to Equations (8.59), the second term is nonzero. Since the term $\boldsymbol{\varepsilon}^{\mathrm{e}} : \mathbf{C} : \boldsymbol{\varepsilon}^{\mathrm{e}}$ has to be positive, straining type A leads to the following inequality:

$$\mu > 0 \qquad \Rightarrow \qquad \frac{E}{2\,(1+\nu)} > 0 \,, \tag{8.60}$$

where Definition $(8.31)_2$ is exploited.

Second, the straining of type B is applied. The material is strained so that the normal strain components take equal values, that is, $\varepsilon^{\mathrm{e}}_{11} = \varepsilon^{\mathrm{e}}_{22} = \varepsilon^{\mathrm{e}}_{33}$ whereas the shear strain components are zero, that is, $\varepsilon^{\mathrm{e}}_{12} = \varepsilon^{\mathrm{e}}_{13} = \varepsilon^{\mathrm{e}}_{21} = \ldots = 0$. Based on Expression (8.54), this gives the first term on the second line of Expression (8.53) as nonzero, whereas Expressions (8.57)–(8.59) show that the second term vanishes. Since the term $\boldsymbol{\varepsilon}^{\mathrm{e}} : \mathbf{C} : \boldsymbol{\varepsilon}^{\mathrm{e}}$ has to be positive, type B straining leads to the following inequality:

$$\lambda + \tfrac{2}{3}\,\mu > 0 \qquad \Rightarrow \qquad \frac{E\,(1+\nu)}{3\,(1+\nu)\,(1-2\,\nu)} > 0 \,, \tag{8.61}$$

where Definitions (8.31) were exploited.

If Young's modulus $E$ is negative, according to Inequality $(8.60)_2$, the term $(1+\nu)$ has to be negative, but in such a case, Inequality $(8.61)_2$ does not hold, since the term $(1-2\nu)$ is positive. Thus Young's modulus $E$ is positive, and therefore, based on Inequality $(8.60)_2$, the term $(1+\nu)$ is positive. This is

$$E > 0 \qquad \text{and} \qquad 1+\nu > 0 \qquad \Rightarrow \qquad \nu > -1\,. \tag{8.62}$$

Based on Inequalities $(8.61)_2$ and (8.62), the following result holds:

$$1 - 2\,\nu > 0 \qquad \Rightarrow \qquad \nu < \tfrac{1}{2}\,. \tag{8.63}$$

Combination of Results (8.62) and (8.63) is

$$E > 0 \qquad \text{and} \qquad -1 < \nu < \tfrac{1}{2}\,. \tag{8.64}$$

The limits for Young's modulus $E$ and Poisson's ratio $\nu$ given by Expression (8.64) are not only valid for linear hyperelasticity but also cover other deformations. The preceding results are valid for material responses where linear elasticity is combined with other deformation mechanisms that are independent of the Hookean deformation. A good example of such a response is the thermoelastic deformation, which is studied in the following section, and microvoid- or microcrack-assisted damage, covered in Sections 10.4 and 10.5.

### 8.3.3 Thermoelastic Deformation

This section provides the formulation for thermoelastic deformation in an isotropic material. The thermoelastic strain tensor $\boldsymbol{\varepsilon}^{\mathrm{eT}}$ includes pure elastic as well as thermal expansion. The material is assumed to be isotropic.

The (total) deformation is assumed to be the sum of elastic, thermal, and inelastic deformation, which are assumed to be mutually independent deformation mechanisms. Thus the following model for the response of the material can be written:

$$\boldsymbol{\varepsilon} = \boldsymbol{\varepsilon}^{\mathrm{e}} + \boldsymbol{\varepsilon}^{\mathrm{Th}} + \boldsymbol{\varepsilon}^{\mathrm{i}} \qquad \Rightarrow \qquad \boldsymbol{\varepsilon} = \boldsymbol{\varepsilon}^{\mathrm{eT}} + \boldsymbol{\varepsilon}^{\mathrm{i}} \qquad \Rightarrow \qquad \boldsymbol{\varepsilon} - \boldsymbol{\varepsilon}^{\mathrm{i}} = \boldsymbol{\varepsilon}^{\mathrm{eT}}\,,$$
$$\text{where} \qquad \boldsymbol{\varepsilon}^{\mathrm{eT}} := \boldsymbol{\varepsilon}^{\mathrm{e}} + \boldsymbol{\varepsilon}^{\mathrm{Th}}\,. \tag{8.65}$$

The thermoelastic part of the specific Helmholtz free energy $\psi$ for an isotropic material when the response is linear thermoelastic can be written as follows:

$$\begin{aligned}\psi^{\mathrm{eT}}(\boldsymbol{\varepsilon}-\boldsymbol{\varepsilon}^{\mathrm{i}}, T) = {} & \psi_{\mathrm{r}}^{\mathrm{eT}} - s_{\mathrm{r}}\,(T - T_{\mathrm{r}}) + \frac{1}{2\,\rho_0}\,(\boldsymbol{\varepsilon}-\boldsymbol{\varepsilon}^{\mathrm{i}}):\mathbf{C}:(\boldsymbol{\varepsilon}-\boldsymbol{\varepsilon}^{\mathrm{i}}) \\ & - \frac{E\,\hat{\alpha}}{\rho_0\,(1-2\,\nu)}\,\mathbf{1}:(\boldsymbol{\varepsilon}-\boldsymbol{\varepsilon}^{\mathrm{i}})\,(T-T_{\mathrm{r}}) + \frac{1}{\rho_0}\,\boldsymbol{\sigma}_{\mathrm{r}}:(\boldsymbol{\varepsilon}-\boldsymbol{\varepsilon}^{\mathrm{i}}) \\ & - c_{\mathrm{r}}\,[\,T\,\ln(T/T_{\mathrm{r}}) - T\,]\,.\end{aligned} \tag{8.66}$$

In Material Model (8.66), the subscript $r$ denotes a value in the reference state $\mathcal{E}_{\mathrm{r}}$ (initial configuration), the linear coefficient of thermal expansion is denoted by $\hat{\alpha}$, and $c_{\mathrm{r}}$ is the specific heat capacity in the reference state $\mathcal{E}_{\mathrm{r}}$ [see Definition (6.205)]. Thus the reference stress tensor, that is, the state of stress in the reference state $\mathcal{E}_{\mathrm{r}}$, and the reference temperature, are denoted by $\boldsymbol{\sigma}_{\mathrm{r}}$ and $T_{\mathrm{r}}$, respectively. The notation $\ln(\cdot)$ refers to the natural logarithm. The specific heat capacity $c_{\mathrm{r}}$ in the last term on the right side of Material Model (8.66) shows that this term is associated with Heat Equation (6.203). The role of the last term on right side of Material Model (8.66) is discussed in more detail in Section 8.3.5.

If the response of the material shows inelastic deformation as well, the specific Helmholtz free energy $\psi$ is the sum of the earlier-given $\psi^{\mathrm{eT}}$ and $\psi^{\mathrm{i}}$, where the latter energy function gives a model for inelastic deformation. However, inelasticity is outside the scope of the present section, which studies thermoelasticity.

Material Model (8.66) and State Equations $(6.174)_1$ give [see Material Models (8.29), (8.32), Equation (2.149) and Definition (2.69)]

$$\boldsymbol{\sigma} = \rho_0 \frac{\partial \psi^{\mathrm{eT}}(\boldsymbol{\varepsilon} - \boldsymbol{\varepsilon}^{\mathrm{i}}, T)}{\partial(\boldsymbol{\varepsilon} - \boldsymbol{\varepsilon}^{\mathrm{i}})} = \mathbf{C} : (\boldsymbol{\varepsilon} - \boldsymbol{\varepsilon}^{\mathrm{i}}) - \frac{E\,\hat{\alpha}}{1 - 2\nu}(T - T_{\mathrm{r}})\,\mathbf{1} + \boldsymbol{\sigma}_{\mathrm{r}}\,. \tag{8.67}$$

Based on Equation $(8.65)_3$, Equation (8.67) can be cast in the following appearance:

$$\boldsymbol{\sigma} = \rho_0 \frac{\partial \psi^{\mathrm{eT}}(\boldsymbol{\varepsilon} - \boldsymbol{\varepsilon}^{\mathrm{i}}, T)}{\partial(\boldsymbol{\varepsilon} - \boldsymbol{\varepsilon}^{\mathrm{i}})} = \mathbf{C} : \boldsymbol{\varepsilon}^{\mathrm{eT}} - \frac{E\,\hat{\alpha}}{1 - 2\nu}(T - T_{\mathrm{r}})\,\mathbf{1} + \boldsymbol{\sigma}_{\mathrm{r}}\,, \tag{8.68}$$

which is the extended Duhamel–Neumann form of Hooke's law. The expression of the Duhamel–Neumann form of Hooke's law does not have the reference stress tensor $\boldsymbol{\sigma}_{\mathrm{r}}$. Based on Material Model (8.68), when the material is not loaded mechanically or thermally, the reference stress tensor $\boldsymbol{\sigma}_{\mathrm{r}}$ gives the state of stress in the material. The Duhamel–Neuman equation is valid for the cases in which the response of the material contains other deformation mechanisms as well.

It is worth noting that the first term on the right side of Equation (8.68) is not Hooke's law, since the thermal part, that is, $\boldsymbol{\varepsilon}^{\mathrm{T}}$, of the thermoelastic strain tensor $\boldsymbol{\varepsilon}^{\mathrm{eT}}$ does not follow Hooke's law. Therefore the stress state must be "corrected" with the second term on the right side of Equation (8.68).

Expression $(\mathrm{I}.4)_2$ of **Appendix I**, Definition (2.106), and Expression $(2.65)_2$ are recalled. They are

$$\mathbf{S} : \mathbf{C} = \mathbf{I}^{\mathrm{s}}\,, \quad \mathbf{I}^{\mathrm{s}} : \mathbf{c} = \mathbf{c} : \mathbf{I}^{\mathrm{s}} = \frac{1}{2}(\mathbf{c} + \mathbf{c}^{\mathrm{T}})\,, \quad \text{and} \quad \mathbf{1} : \mathbf{1} = \delta_{ii} = 3\,, \tag{8.69}$$

where $\mathbf{c}$ is an arbitrary second-order tensor.

Since the thermoelastic strain tensor $\boldsymbol{\varepsilon}^{\mathrm{eT}}$ and the second-order identity tensor $\mathbf{1}$ are symmetric tensors, Expression $(8.69)_2$ gives

$$\mathbf{I}^{\mathrm{s}} : \boldsymbol{\varepsilon}^{\mathrm{eT}} = \tfrac{1}{2}\left[\, \boldsymbol{\varepsilon}^{\mathrm{eT}} + (\boldsymbol{\varepsilon}^{\mathrm{eT}})^{\mathrm{T}} \,\right] = \boldsymbol{\varepsilon}^{\mathrm{eT}} \quad \text{and} \quad \mathbf{I}^{\mathrm{s}} : \mathbf{1} = \tfrac{1}{2}\,(\,\mathbf{1} + \mathbf{1}^{\mathrm{T}}\,) = \mathbf{1}\,. \tag{8.70}$$

Based on Expressions $(8.69)_1$ and (8.70), the following result is obtained:

$$\mathbf{S} : \mathbf{C} : \boldsymbol{\varepsilon}^{\mathrm{eT}} = \mathbf{I}^{\mathrm{s}} : \boldsymbol{\varepsilon}^{\mathrm{eT}} = \boldsymbol{\varepsilon}^{\mathrm{eT}} \qquad \text{and} \qquad \mathbf{S} : \mathbf{C} : \mathbf{1} = \mathbf{I}^{\mathrm{s}} : \mathbf{1} = \mathbf{1}\,. \tag{8.71}$$

Definition (8.42) reads

$$\mathbf{S} := -\frac{\nu}{E}\,\mathbf{1}\,\mathbf{1} + \frac{1+\nu}{E}\,\mathbf{I}^{\mathrm{s}}\,. \tag{8.72}$$

Definition (8.72) is multiplied by $:\mathbf{1}$ from the right, and Expressions $(8.69)_3$ and $(8.70)_2$ are exploited. This gives

$$\mathbf{S} : \mathbf{1} = -\frac{\nu}{E}\,\mathbf{1}\,\mathbf{1} : \mathbf{1} + \frac{1+\nu}{E}\,\mathbf{I}^{\mathrm{s}} : \mathbf{1} = -\frac{3\,\nu}{E}\,\mathbf{1} + \frac{1+\nu}{E}\,\mathbf{1} = \frac{(1-2\,\nu)}{E}\,\mathbf{1}\,. \tag{8.73}$$

Material Model (8.68) is multiplied by $\mathbf{S}:$ from the left. It gives

$$\mathbf{S} : \boldsymbol{\sigma} = \mathbf{S} : \mathbf{C} : \boldsymbol{\varepsilon}^{\mathrm{eT}} - \mathbf{S} : \mathbf{1}\,\frac{E\,\hat{\alpha}}{1-2\,\nu}\,(\,T - T_{\mathrm{r}}\,) + \mathbf{S} : \boldsymbol{\sigma}_{\mathrm{r}}\,. \tag{8.74}$$

Expressions (8.71) and (8.73) allow Equation (8.74) to take the form

$$\begin{aligned} \mathbf{S} : \boldsymbol{\sigma} &= \boldsymbol{\varepsilon}^{\mathrm{eT}} - \frac{(1-2\,\nu)}{E}\,\mathbf{1}\,\frac{E\,\hat{\alpha}}{(1-2\,\nu)}\,(\,T - T_{\mathrm{r}}\,) + \mathbf{S} : \boldsymbol{\sigma}_{\mathrm{r}} \\ &= \boldsymbol{\varepsilon}^{\mathrm{eT}} - \hat{\alpha}\,(\,T - T_{\mathrm{r}}\,)\,\mathbf{1} + \mathbf{S} : \boldsymbol{\sigma}_{\mathrm{r}}\,. \end{aligned} \tag{8.75}$$

Equation (8.75) gives

$$\boldsymbol{\varepsilon}^{\mathrm{eT}} = \mathbf{S} : \boldsymbol{\sigma} + \hat{\alpha}\,(\,T - T_{\mathrm{r}}\,)\,\mathbf{1}\,. \tag{8.76}$$

In Expression (8.76) the value of the stress tensor in the reference state $\mathcal{E}_{\mathrm{r}}$, that is, $\boldsymbol{\sigma}_{\mathrm{r}}$, is neglected as a small quantity. Based on Definition (8.72) of the compliance tensor $\mathbf{S}$, Constitutive Equation (8.76) can be cast in the form

$$\boldsymbol{\varepsilon}^{\mathrm{eT}} = \mathbf{S} : \boldsymbol{\sigma} + \hat{\alpha}\,(\,T - T_{\mathrm{r}}\,)\,\mathbf{1} = -\frac{\nu}{E}\,\mathbf{1}\,\mathbf{1} : \boldsymbol{\sigma} + \frac{1+\nu}{E}\,\mathbf{I}^{\mathrm{s}} : \boldsymbol{\sigma} + \hat{\alpha}\,(\,T - T_{\mathrm{r}}\,)\,\mathbf{1}\,. \tag{8.77}$$

Based on Constitutive Equation (8.77), the thermoelastic strain tensor $\boldsymbol{\varepsilon}^{\mathrm{eT}}$ can be separated into two parts, as follows:

$$\boldsymbol{\varepsilon}^{\mathrm{eT}} = \boldsymbol{\varepsilon}^{\mathrm{e}} + \boldsymbol{\varepsilon}^{\mathrm{Th}}\,, \tag{8.78}$$

where the strain tensors for linear elastic deformation (i.e., for Hooke's law) $\boldsymbol{\varepsilon}^{\mathrm{e}}$ and the thermal expansion $\boldsymbol{\varepsilon}^{\mathrm{Th}}$ take the forms

$$\boldsymbol{\varepsilon}^{\mathrm{e}} = \mathbf{S} : \boldsymbol{\sigma} = -\frac{\nu}{E}\,\mathbf{1}\,\mathbf{1} : \boldsymbol{\sigma} + \frac{1+\nu}{E}\,\mathbf{I}^{\mathrm{s}} : \boldsymbol{\sigma} \quad \text{and} \quad \boldsymbol{\varepsilon}^{\mathrm{Th}} = \hat{\alpha}\,(\,T - T_{\mathrm{r}}\,)\,\mathbf{1}\,. \tag{8.79}$$

Equation $(8.79)_2$ shows that the thermal strain tensor $\boldsymbol{\varepsilon}^{\mathrm{Th}}$ vanishes at the reference temperature $T_{\mathrm{r}}$.

### 8.3.4 Fourier's Law of Heat Conduction Coupled with Thermoelastic Deformation

This section evaluates Fourier's law of heat conduction coupled with thermoelastic deformation. Heat Equation (6.203) is

$$\vec{\nabla}\cdot\vec{q} = \boldsymbol{\sigma}:\dot{\boldsymbol{\varepsilon}}^{\mathrm{i}} + \boldsymbol{\beta}:\dot{\boldsymbol{\alpha}} + \rho_0\, r + \rho_0\, T\left[\frac{\partial^2\psi}{\partial T\,\partial(\boldsymbol{\varepsilon}-\boldsymbol{\varepsilon}^{\mathrm{i}})}:(\dot{\boldsymbol{\varepsilon}}-\dot{\boldsymbol{\varepsilon}}^{\mathrm{i}}) + \frac{\partial^2\psi}{\partial T\,\partial\boldsymbol{\alpha}}:\dot{\boldsymbol{\alpha}} + \frac{\partial^2\psi}{\partial T^2}\,\dot{T}\right]. \tag{8.80}$$

Fourier's Law $(8.8)_2$ is extended to an anisotropic material (in the sense of heat conduction) by introducing the following specific dissipation function:

$$\varphi_{\text{ther}}(\vec{q};T) = \frac{1}{2\,\mu_{\text{ther}}\,\rho_0\,T}\,\vec{q}\cdot[\boldsymbol{\gamma}(T)]^{-1}\cdot\vec{q}, \qquad \text{where} \qquad \boldsymbol{\gamma} = \boldsymbol{\gamma}^{\mathrm{T}}. \tag{8.81}$$

The second-order tensor $\boldsymbol{\gamma}(T)$ is the thermal conductivity tensor. Constitutive Equation $(8.81)_1$ yields

$$\varphi_{\text{ther}}(k\,\vec{q};T) = k^2\,\varphi_{\text{ther}}(\vec{q};T), \tag{8.82}$$

which implies that $\varphi_{\text{ther}}$ is a homogeneous function of degree $1/\mu_{\text{ther}} = 2$ in the heat flux vector $\vec{q}$. Material Model (8.81) shows that the specific dissipation function $\varphi_{\text{ther}}$ is nonnegative. Since it is also a homogeneous function, according to Section 7.2, the Clausius–Duhem inequality is satisfied. The satisfaction of the Clausius–Duhem inequality requires that no other dissipation potentials be introduced, of course.

Normality Rule (6.289) reads

$$-\frac{\vec{\nabla}T}{T} = \mu_{\text{ther}}\,\rho_0\,\frac{\partial\varphi_{\text{ther}}(\vec{q};\boldsymbol{\varepsilon}-\boldsymbol{\varepsilon}^{\mathrm{i}},\boldsymbol{\alpha},T,h)}{\partial\vec{q}}. \tag{8.83}$$

Substitution of Material Model $(8.81)_1$ into Normality Rule (8.83) gives

$$-\frac{\vec{\nabla}T}{T} = \mu_{\text{ther}}\,\rho_0\,\frac{\partial}{\partial\vec{q}}\left\{\frac{1}{2\,\mu_{\text{ther}}\,\rho_0\,T}\,\vec{q}\cdot[\boldsymbol{\gamma}(T)]^{-1}\cdot\vec{q}\right\} = \frac{1}{2T}\,\frac{\partial}{\partial\vec{q}}\left\{\vec{q}\cdot[\boldsymbol{\gamma}(T)]^{-1}\cdot\vec{q}\right\}. \tag{8.84}$$

Problem 8.1 is to show that for an arbitrary vector $\vec{a}$ and for an arbitrary symmetric second-order tensor $\mathbf{c}$, the following holds:

$$\frac{1}{2}\,\frac{\partial}{\partial\vec{a}}\,[\vec{a}\cdot\mathbf{c}\cdot\vec{a}] = \mathbf{c}\cdot\vec{a}. \tag{8.85}$$

Assuming that $\vec{a} = \vec{q}$ and $\mathbf{c} = [\boldsymbol{\gamma}(T)]^{-1}$ in Expression (8.85), we arrive at the following:

$$-\frac{\vec{\nabla}T}{T} = \frac{1}{2T}\,\frac{\partial}{\partial\vec{q}}\{\vec{q}\cdot[\boldsymbol{\gamma}(T)]^{-1}\cdot\vec{q}\} = \frac{1}{T}\,[\boldsymbol{\gamma}(T)]^{-1}\cdot\vec{q}. \tag{8.86}$$

Expression (8.86) is multiplied by the absolute temperature $T$ and by $[\gamma(T)]\cdot$ from the left. This gives

$$\vec{q} = -\gamma(T)\cdot\vec{\nabla}T \quad \Rightarrow \quad \vec{\nabla}\cdot\vec{q} = -\vec{\nabla}\cdot\left[\gamma(T)\cdot\vec{\nabla}T\right]. \tag{8.87}$$

Substitution of $\vec{\nabla}\cdot\vec{q}$ obtained from Expression $(8.87)_2$ into Heat Equation (8.80) gives

$$\begin{aligned} -\vec{\nabla}\cdot\left[\gamma(T)\cdot\vec{\nabla}T\right] &= \boldsymbol{\sigma} : \dot{\boldsymbol{\varepsilon}}^{\mathrm{i}} + \boldsymbol{\beta} : \dot{\boldsymbol{\alpha}} + \rho_0\, r \\ &\quad + \rho_0\, T\left[\frac{\partial^2\psi}{\partial T\,\partial(\boldsymbol{\varepsilon}-\boldsymbol{\varepsilon}^{\mathrm{i}})} : (\dot{\boldsymbol{\varepsilon}} - \dot{\boldsymbol{\varepsilon}}^{\mathrm{i}}) + \frac{\partial^2\psi}{\partial T\,\partial\boldsymbol{\alpha}} : \dot{\boldsymbol{\alpha}} + \frac{\partial^2\psi}{\partial T^2}\,\dot{T}\right]. \end{aligned} \tag{8.88}$$

Material Model (8.66) gives

$$\frac{\partial\psi^{\mathrm{eT}}}{\partial T} = -s_{\mathrm{r}} - \frac{E\,\hat{\alpha}}{\rho_0\,(1-2\,\nu)}\,\mathbf{1} : (\boldsymbol{\varepsilon} - \boldsymbol{\varepsilon}^{\mathrm{i}}) - c_{\mathrm{r}}\,\ln(T/T_{\mathrm{r}}), \tag{8.89}$$

which yields

$$\frac{\partial^2\psi^{\mathrm{eT}}}{\partial T^2} = -c_{\mathrm{r}}\,\frac{1}{T}, \qquad \frac{\partial^2\psi^{\mathrm{eT}}}{\partial T\,\partial(\boldsymbol{\varepsilon}-\boldsymbol{\varepsilon}^{\mathrm{i}})} = -\frac{E\,\hat{\alpha}}{\rho_0\,(1-2\,\nu)}\,\mathbf{1}, \quad \text{and} \quad \frac{\partial^2\psi^{\mathrm{eT}}}{\partial T\,\partial\boldsymbol{\alpha}} = \mathbf{0}. \tag{8.90}$$

Substitution of Derivatives (8.90) into Heat Conduction Equation (8.88) gives

$$-\vec{\nabla}\cdot[\gamma(T)\cdot\vec{\nabla}T] = \boldsymbol{\sigma} : \dot{\boldsymbol{\varepsilon}}^{\mathrm{i}} + \boldsymbol{\beta} : \dot{\boldsymbol{\alpha}} + \rho_0\, r - \frac{E\,\hat{\alpha}}{1-2\,\nu}\,T\,\mathbf{1} : (\dot{\boldsymbol{\varepsilon}} - \dot{\boldsymbol{\varepsilon}}^{\mathrm{i}}) - \rho_0\, c_{\mathrm{r}}\,\dot{T}. \tag{8.91}$$

Taking into account that no inelastic deformation occurs, that is, $\dot{\boldsymbol{\varepsilon}}^{\mathrm{i}} = \dot{\boldsymbol{\alpha}} \equiv \mathbf{0}$ and $\dot{\boldsymbol{\varepsilon}} = \dot{\boldsymbol{\varepsilon}}^{\mathrm{eT}}$, Expression (8.91) yields

$$\rho_0\, c_{\mathrm{r}}\,\dot{T} = \rho_0\, r - \frac{E\,\hat{\alpha}}{1-2\,\nu}\,T\,\mathbf{1} : \dot{\boldsymbol{\varepsilon}}^{\mathrm{eT}} + \vec{\nabla}\cdot\left[\gamma(T)\cdot\vec{\nabla}T\right], \tag{8.92}$$

which is the differential equation of heat conduction in the material having thermoelastic response. It is worth noting that Equation (8.92) assumes that there is no heat exchange between the body (system) through the surfaces of the body. The only heat source or sink is the quantity $r$, but it is the rate of heat energy (energy over second) distributed in the material. Next, some special cases of Equation (8.92) are studied.

If the heat source term $\rho_0\, r$ vanishes, the material parameter $\gamma$ is a scalar-valued constant, that is, $\boldsymbol{\gamma} = \gamma\,\mathbf{1}$, and if the temperature $T$ is approximated by the reference temperature $T_{\mathrm{r}}$ in the second term on the right side of Expression (8.92), it reduces to

$$\rho_0\, c_{\mathrm{r}}\,\dot{T} + \frac{E\,\hat{\alpha}}{1-2\,\nu}\,T_{\mathrm{r}}\,\mathbf{1} : \dot{\boldsymbol{\varepsilon}}^{\mathrm{eT}} = \gamma\,\nabla^2 T, \tag{8.93}$$

which is Duhamel's equation of heat conduction.

If the second term on the right side of Equation (8.92) is neglected, it reduces to

$$\rho_0 \, c_{\mathrm{r}} \, \dot{T} = \rho_0 \, r + \vec{\nabla} \cdot \left[ \gamma(T) \vec{\nabla} T \right] , \tag{8.94}$$

which is the differential equation of heat conduction in a rigid material.

It is assumed that there is no heat source within the body (system). This means that the first term on the right side of Equation (8.92) vanishes. It is further assumed that there is a homogeneous temperature field in the body. This means that the quantity $\vec{\nabla} T$ is zero and therefore the last term on the right side of Equation (8.92) vanishes. Thus Equation (8.92) reduces to

$$\rho_0 \, c_{\mathrm{r}} \, \dot{T} = - \frac{E \, \hat{\alpha}}{1 - 2 \, \nu} \, T \, \mathbf{1} : \dot{\boldsymbol{\varepsilon}}^{\mathrm{eT}} \, . \tag{8.95}$$

Equation (8.95) is Kelvin's formula for the change of temperature of an insulated body. The term $\mathbf{1} : \dot{\boldsymbol{\varepsilon}}^{\mathrm{eT}} = \dot{\varepsilon}^{\mathrm{eT}}_{11} + \dot{\varepsilon}^{\mathrm{eT}}_{22} + \dot{\varepsilon}^{\mathrm{eT}}_{33}$ gives the rate of volume change due to elastic deformation and thermal expansion. Thus, Equation (8.95) expresses the amount of temperature $T$ decrease due to the volume increase of the body. Temperature $T$ decreases, since the same amount of thermal energy is in a larger body, and vice versa.

### 8.3.5 Terms of the Specific Helmholtz Free Energy $\psi^{\mathrm{et}}$ Given by Material Model (8.65) Having Temperature $T$

This section discusses the form of the specific Helmholtz free energy $\psi^{\mathrm{eT}}$ introduced in Material Model (8.66) and focuses on the terms that have the absolute temperature $T$. Material Model (8.66) reads

$$\begin{aligned} \psi^{\mathrm{eT}}(\boldsymbol{\varepsilon} - \boldsymbol{\varepsilon}^{\mathrm{i}}, T) = {} & \psi^{\mathrm{eT}}_{\mathrm{r}} - s_{\mathrm{r}} \, (T - T_{\mathrm{r}}) + \frac{1}{2 \, \rho_0} \, (\boldsymbol{\varepsilon} - \boldsymbol{\varepsilon}^{\mathrm{i}}) : \mathbf{C} : (\boldsymbol{\varepsilon} - \boldsymbol{\varepsilon}^{\mathrm{i}}) \\ & - \frac{E \, \hat{\alpha}}{\rho_0 \, (1 - 2 \nu)} \, \mathbf{1} : (\boldsymbol{\varepsilon} - \boldsymbol{\varepsilon}^{\mathrm{i}}) \, (T - T_{\mathrm{r}}) + \frac{1}{\rho_0} \, \boldsymbol{\sigma}_{\mathrm{r}} : (\boldsymbol{\varepsilon} - \boldsymbol{\varepsilon}^{\mathrm{i}}) \\ & - c_{\mathrm{r}} \left[ T \, \ln(T/T_{\mathrm{r}}) - T \right] . \end{aligned} \tag{8.96}$$

In Constitutive Equation (8.96), three terms have the absolute temperature $T$. They are studied next.

Equation (8.67) is

$$\boldsymbol{\sigma} = \rho_0 \, \frac{\partial \psi^{\mathrm{eT}}(\boldsymbol{\varepsilon} - \boldsymbol{\varepsilon}^{\mathrm{i}}, T)}{\partial (\boldsymbol{\varepsilon} - \boldsymbol{\varepsilon}^{\mathrm{i}})} = \mathbf{C} : (\boldsymbol{\varepsilon} - \boldsymbol{\varepsilon}^{\mathrm{i}}) - \frac{E \, \hat{\alpha}}{1 - 2 \, \nu} \, (T - T_{\mathrm{r}}) \, \mathbf{1} + \boldsymbol{\sigma}_{\mathrm{r}} \, , \tag{8.97}$$

Comparison of Expressions (8.96) and (8.97) shows that the only temperature-dependent term in Material Model (8.96) for the state of stress is the first one on the second line of Expression (8.96). In fact, it was the only temperature-dependent term utilized in Section 8.3.3 that only studied the mechanical response of the matter.

State Equation (6.172) and Derivative (8.89) give the expression for the specific entropy $s$. They are

$$s = -\frac{\partial \psi}{\partial T} \quad \text{and} \quad \frac{\partial \psi^{\mathrm{eT}}}{\partial T} = -s_{\mathrm{r}} - \frac{E\,\hat{\alpha}}{\rho_0\,(1-2\,\nu)}\,\mathbf{1}:(\boldsymbol{\varepsilon}-\boldsymbol{\varepsilon}^{\mathrm{i}}) - c_{\mathrm{r}}\,\ln(T/T_{\mathrm{r}})\,. \quad (8.98)$$

Expressions (8.98) give

$$s = -\frac{\partial \psi}{\partial T} = s_{\mathrm{r}} + \frac{E\,\hat{\alpha}}{\rho_0\,(1-2\,\nu)}\,\mathbf{1}:(\boldsymbol{\varepsilon}-\boldsymbol{\varepsilon}^{\mathrm{i}}) + c_{\mathrm{r}}\,\ln(T/T_{\mathrm{r}})\,. \qquad (8.99)$$

Equation (8.99) shows that at the reference state $\mathcal{E}_{\mathrm{r}}$, the specific entropy $s$ takes the value $s_{\mathrm{r}}$, which is the correct value. Derivation of Equation (8.99) for the specific entropy $s$ requires all three temperature-dependent terms from Material Model (8.65) [i.e. Material Model (8.96)].

Although the heat equation makes use of Equation (8.99), the influence of the first temperature-dependent term in Material Model (8.96) [i.e. $s_{\mathrm{r}}\,(T-T_{\mathrm{r}})$] and Expression (8.99) vanishes due to the further derivatives of Equation (8.99). Since material modelling does not exploit the values of the specific Helmholz free energy $\psi$ and the specific entropy $s$ at the reference state $\mathcal{E}_{\mathrm{r}}$, the quantities $\psi_{\mathrm{r}}$ and $s_{\mathrm{r}}$ are not usually included in the material models in this book.

### 8.3.6 Inhomogeneous Hookean Deformation

This section evaluates a material that has a thermodynamically inhomogeneous Hookean deformation, that is, Hooke's law, where the constitutive tensor $\mathbf{C}$ varies from point to point within the material. The specific Helmholtz free energy $\psi$ for a linear isotropic inhomogeneous elastic material can be written

$$\psi^{\mathrm{e}}(\boldsymbol{\varepsilon}-\boldsymbol{\varepsilon}^{\mathrm{i}}, h(\vec{X})) = \psi^{\mathrm{e}}_{\mathrm{r}} + \frac{1}{2\,\rho_0}\,(\boldsymbol{\varepsilon}-\boldsymbol{\varepsilon}^{\mathrm{i}}):\mathbf{h}(\vec{X}):(\boldsymbol{\varepsilon}-\boldsymbol{\varepsilon}^{\mathrm{i}})\,, \qquad (8.100)$$

where the function $h(\vec{X})$ is assumed to be a fourth-order major-symmetric tensor $\mathbf{h}(\vec{X})$ .

Material Model (8.100) and State Equations $(6.174)_1$ give [see Equation $(2.131)_2$]

$$\boldsymbol{\sigma} = \rho_0\,\frac{\partial \psi^{\mathrm{eT}}(\boldsymbol{\varepsilon}-\boldsymbol{\varepsilon}^{\mathrm{i}}, T)}{\partial(\boldsymbol{\varepsilon}-\boldsymbol{\varepsilon}^{\mathrm{i}})} = \mathbf{h}(\vec{X}):(\boldsymbol{\varepsilon}-\boldsymbol{\varepsilon}^{\mathrm{i}})\,. \qquad (8.101)$$

If it is defined that

$$\mathbf{h}(\vec{X}) = \mathbf{C}(\vec{X})\,, \qquad (8.102)$$

then Hooke's Law (8.101) can be written as

$$\boldsymbol{\sigma} = \mathbf{C}(\vec{X}):(\boldsymbol{\varepsilon}-\boldsymbol{\varepsilon}^{\mathrm{i}})\,. \qquad (8.103)$$

Expression (8.103) is Hooke's law, where the value for the constitutive tensor $\mathbf{C}(\vec{X})$ varies from point to point within the material.

## 8.4 Thermoplasticity

This section evaluates thermoplasticity. A careful reader may have noticed a certain difference between the traditional expressions for modelling of plastic flow and the equations derived here. Usually in the theory of plasticity, for example, the normality rule is formulated in the differential form, whereas this work writes equations in the rate form. This is because the rate form fits well with the continuum thermodynamic framework. The timescale in plastic yield is not dependent on the rate of the material response, since deformation is modelled as instantaneous. The timescale is a consequence of the rate of loading. Thus the minimal difference between these two approaches depends only on the presentation of loading and can be neglected by a simple time integration.

### 8.4.1 Prerequisites for Thermoplasticity

This section assumes the following set of internal variables: $(\boldsymbol{\varepsilon}^{\mathrm{p}}, \boldsymbol{\alpha}^1, \alpha^2)$.

#### 8.4.1.1 Rates of Internal Forces $\dot{\boldsymbol{\beta}}^1$ and $\dot{\beta}^2$

This section derives expressions for the rates of internal forces $\dot{\boldsymbol{\beta}}^1$ and $\dot{\beta}^2$. It is assumed that the internal variables $\boldsymbol{\alpha}^1$ and $\alpha^2$ have no influence on the relationship between the stress tensor $\boldsymbol{\sigma}$ and the thermoelastic strain tensor $\boldsymbol{\varepsilon}^{\mathrm{eT}}$. Therefore the following can be formulated:

$$\boldsymbol{\varepsilon} = \boldsymbol{\varepsilon}^{\mathrm{eT}}(\boldsymbol{\sigma}, T, h(\vec{X})) + \boldsymbol{\varepsilon}^{\mathrm{p}}(\boldsymbol{\alpha}^1, \alpha^2, T)\,. \tag{8.104}$$

Modifying Lubliner ([53], Chapter 4) (see also Lubliner [54], p. 63), the following can be argued: if decomposition of strain tensor $\boldsymbol{\varepsilon}$ into elastic and plastic parts is assumed to take Form (8.104), then such a decomposition is compatible with existence of the specific Helmholtz free energy $\psi(\boldsymbol{\varepsilon} - \boldsymbol{\varepsilon}^{\mathrm{p}}, \boldsymbol{\alpha}^1, \alpha^2, h(\vec{X}), T)$ if and only if $\psi$ can be decomposed as

$$\psi(\boldsymbol{\varepsilon} - \boldsymbol{\varepsilon}^{\mathrm{p}}, \boldsymbol{\alpha}^1, \alpha^2, T, h(\vec{X})) = \psi^{\mathrm{eT}}(\boldsymbol{\varepsilon} - \boldsymbol{\varepsilon}^{\mathrm{p}}, T, h(\vec{X})) + \psi^{\mathrm{p}}(\boldsymbol{\alpha}^1, \alpha^2, T)\,. \tag{8.105}$$

In Decomposition (8.105), the notation $\boldsymbol{\varepsilon} - \boldsymbol{\varepsilon}^{\mathrm{p}}$ is used instead of $\boldsymbol{\varepsilon}^{\mathrm{eT}}$ to show that the variables $\boldsymbol{\varepsilon}$ and $\boldsymbol{\varepsilon}^{\mathrm{p}}$ are the state variables, not the thermoelastic strain tensor $\boldsymbol{\varepsilon}^{\mathrm{eT}}$.

Based on Decomposition (8.105) and the fact that the order of the partial derivative operators $\partial/\partial(\boldsymbol{\varepsilon} - \boldsymbol{\varepsilon}^{\mathrm{p}})$ and $\partial/\partial\boldsymbol{\alpha}^1$ can be changed, the following manipulation can be made:

$$\frac{\partial^2 \psi}{\partial \boldsymbol{\alpha}^1 \, \partial(\boldsymbol{\varepsilon} - \boldsymbol{\varepsilon}^{\mathrm{p}})} = \frac{\partial}{\partial \boldsymbol{\alpha}^1} \left( \frac{\partial \psi}{\partial(\boldsymbol{\varepsilon} - \boldsymbol{\varepsilon}^{\mathrm{p}})} \right) = \frac{\partial}{\partial \boldsymbol{\alpha}^1} \left( \frac{\partial \psi^{eT}}{\partial(\boldsymbol{\varepsilon} - \boldsymbol{\varepsilon}^{\mathrm{p}})} \right)$$
$$= \frac{\partial}{\partial(\boldsymbol{\varepsilon} - \boldsymbol{\varepsilon}^{\mathrm{p}})} \left( \frac{\partial \psi^{eT}}{\partial \boldsymbol{\alpha}^1} \right) = \frac{\partial(\mathbf{0})}{\partial(\boldsymbol{\varepsilon} - \boldsymbol{\varepsilon}^{\mathrm{p}})} = \mathbf{O}\,, \tag{8.106}$$

where $\mathbf{0}$ and $\mathbf{O}$ are the second-order and fourth-order zero tensors, respectively. The components of the zero tensors $\mathbf{0}$ and $\mathbf{O}$ vanish.

Correspondingly, it is obtained that

$$\frac{\partial^2 \psi}{\partial \alpha^2 \, \partial(\boldsymbol{\varepsilon} - \boldsymbol{\varepsilon}^{\mathrm{p}})} = \mathbf{0}\,. \tag{8.107}$$

For the current set of state variables, State Equation $(6.174)_2$ gives

$$\boldsymbol{\beta}^1 = -\rho_0 \frac{\partial \psi(\boldsymbol{\varepsilon} - \boldsymbol{\varepsilon}^{\mathrm{p}}, \boldsymbol{\alpha}^1, \alpha^2, T, h(\vec{X}))}{\partial \boldsymbol{\alpha}^1}\,. \tag{8.108}$$

Based on State Equation (8.108), the rate of the internal force $\dot{\boldsymbol{\beta}}^1$ reads

$$\dot{\boldsymbol{\beta}}^1 = -\rho_0 \frac{\partial^2 \psi}{\partial \boldsymbol{\alpha}^1 \, \partial(\boldsymbol{\varepsilon} - \boldsymbol{\varepsilon}^{\mathrm{p}})} : (\dot{\boldsymbol{\varepsilon}} - \dot{\boldsymbol{\varepsilon}}^{\mathrm{p}}) - \rho_0 \frac{\partial^2 \psi}{\partial \boldsymbol{\alpha}^1 \, \partial \boldsymbol{\alpha}^1} : \dot{\boldsymbol{\alpha}}^1 - \rho_0 \frac{\partial^2 \psi}{\partial \boldsymbol{\alpha}^1 \, \partial \alpha^2} \dot{\alpha}^2$$
$$- \rho_0 \frac{\partial^2 \psi}{\partial \boldsymbol{\alpha}^1 \, \partial T} \dot{T}\,. \tag{8.109}$$

For the current set of state variables, Normality Rules (6.300) and (6.301) give

$$\dot{\boldsymbol{\varepsilon}}^{\mathrm{p}} = \overset{\circ}{\lambda} \frac{\partial F}{\partial \boldsymbol{\sigma}}\,, \qquad \dot{\boldsymbol{\alpha}}^1 = \overset{\circ}{\lambda} \frac{\partial F}{\partial \boldsymbol{\beta}^1}\,, \qquad \text{and} \qquad \dot{\alpha}^2 = \overset{\circ}{\lambda} \frac{\partial F}{\partial \beta^2}\,, \tag{8.110}$$

where the yield function $F$ is for the present case,

$$F = F(\boldsymbol{\sigma}, \boldsymbol{\beta}^1, \beta^2; T)\,. \tag{8.111}$$

Substitution of Result (8.106) and Normality Rules $(8.110)_2$ and $(8.110)_3$ into Rate (8.109) yields

$$\dot{\boldsymbol{\beta}}^1 = -\rho_0 \overset{\circ}{\lambda} \left[ \frac{\partial^2 \psi}{\partial \boldsymbol{\alpha}^1 \, \partial \boldsymbol{\alpha}^1} : \frac{\partial F}{\partial \boldsymbol{\beta}^1} + \frac{\partial^2 \psi}{\partial \boldsymbol{\alpha}^1 \, \partial \alpha^2} \frac{\partial F}{\partial \beta^2} \right] - \rho_0 \frac{\partial^2 \psi}{\partial \boldsymbol{\alpha}^1 \, \partial T} \dot{T}\,. \tag{8.112}$$

Correspondingly, it is obtained that

$$\dot{\beta}^2 = -\rho_0 \overset{\circ}{\lambda} \left[ \frac{\partial^2 \psi}{\partial \alpha^2 \, \partial \boldsymbol{\alpha}^1} : \frac{\partial F}{\partial \boldsymbol{\beta}^1} + \frac{\partial^2 \psi}{\partial \alpha^2 \, \partial \alpha^2} \frac{\partial F}{\partial \beta^2} \right] - \rho_0 \frac{\partial^2 \psi}{\partial \alpha^2 \, \partial T} \dot{T}\,. \tag{8.113}$$

#### 8.4.1.2 Plasticity Multiplier $\overset{\circ}{\lambda}$

This section studies the consistency condition and derives the expression for the plasticity multiplier $\overset{\circ}{\lambda}$ of the normality rule. Recalling Consistency Condition (6.297) and introducing the variables needed in Section 8.4.1.1, the following is achieved [see also Equation (6.296)]:

$$\dot{F}(\boldsymbol{\sigma}, \boldsymbol{\beta}^1, \beta^2; T, h) = 0\,. \tag{8.114}$$

By using the chain rule, the rate of the yield function $F$ takes the form

$$\dot{F}(\boldsymbol{\sigma}, \boldsymbol{\beta}^1, \beta^2; T, h) = \frac{\partial F}{\partial \boldsymbol{\sigma}} : \dot{\boldsymbol{\sigma}} + \frac{\partial F}{\partial \boldsymbol{\beta}^1} : \dot{\boldsymbol{\beta}}^1 + \frac{\partial F}{\partial \beta^2} \dot{\beta}^2 + \frac{\partial F}{\partial T} \dot{T}\,. \tag{8.115}$$

Substitution of Expression (8.115) into Consistency Condition (8.114), and subsequently of Results (8.112) and (8.113) into the obtained condition gives

$$\begin{aligned}\frac{\partial F}{\partial \boldsymbol{\sigma}} : \dot{\boldsymbol{\sigma}} - \rho_0 \overset{\circ}{\lambda} \left[ \frac{\partial F}{\partial \boldsymbol{\beta}^1} : \frac{\partial^2 \psi}{\partial \boldsymbol{\alpha}^1 \partial \boldsymbol{\alpha}^1} : \frac{\partial F}{\partial \boldsymbol{\beta}^1} + 2 \frac{\partial F}{\partial \boldsymbol{\beta}^1} : \frac{\partial^2 \psi}{\partial \boldsymbol{\alpha}^1 \partial \alpha^2} \frac{\partial F}{\partial \beta^2} + \frac{\partial F}{\partial \beta^2} \frac{\partial^2 \psi}{\partial \alpha^2 \partial \alpha^2} \frac{\partial F}{\partial \beta^2} \right] \\ - \left[ \rho_0 \left( \frac{\partial F}{\partial \boldsymbol{\beta}^1} : \frac{\partial^2 \psi}{\partial \boldsymbol{\alpha}^1 \partial T} + \frac{\partial F}{\partial \beta^2} \frac{\partial^2 \psi}{\partial \alpha^2 \partial T} \right) - \frac{\partial F}{\partial T} \right] \dot{T} = 0\,.\end{aligned} \tag{8.116}$$

Derivation of Expression (8.116) exploits the fact that for second-order tensors $\mathbf{a}$ and $\mathbf{b}$, $\mathbf{a} : \mathbf{b} = \mathbf{b} : \mathbf{a}$ holds. This fact was utilized in the second term within the first square brackets.

Expression (8.116) gives

$$\begin{aligned}\overset{\circ}{\lambda} = \mathrm{H}(F) \left\langle \frac{\partial F}{\partial \boldsymbol{\sigma}} : \dot{\boldsymbol{\sigma}} - \left[ \rho_0 \left( \frac{\partial F}{\partial \boldsymbol{\beta}^1} : \frac{\partial^2 \psi}{\partial \boldsymbol{\alpha}^1 \partial T} + \frac{\partial F}{\partial \beta^2} \frac{\partial^2 \psi}{\partial \alpha^2 \partial T} \right) - \frac{\partial F}{\partial T} \right] \dot{T} \right\rangle \\ \Big/ \rho_0 \left[ \frac{\partial F}{\partial \boldsymbol{\beta}^1} : \frac{\partial^2 \psi}{\partial \boldsymbol{\alpha}^1 \partial \boldsymbol{\alpha}^1} : \frac{\partial F}{\partial \boldsymbol{\beta}^1} + 2 \frac{\partial F}{\partial \boldsymbol{\beta}^1} : \frac{\partial^2 \psi}{\partial \boldsymbol{\alpha}^1 \partial \alpha^2} \frac{\partial F}{\partial \beta^2} + \frac{\partial F}{\partial \beta^2} \frac{\partial^2 \psi}{\partial \alpha^2 \partial \alpha^2} \frac{\partial F}{\partial \beta^2} \right],\end{aligned} \tag{8.117}$$

where on the right side, the first row is the numerator and the second row is the denominator. The Heaviside function $H(F)$ [see Definition (2.2)] is added to guarantee that the plasticity multiplier $\overset{\circ}{\lambda}$ vanishes in the pure elastic region (cf. Lemaitre [50]; Chaboche [14], p. 194). The Macaulay brackets $\langle\ \rangle$ [see Definition (2.1)] are adopted to imply that the plasticity multiplier $\overset{\circ}{\lambda}$ takes only positive values [see Nguyen and Bui [70], Eq. (7)].

#### 8.4.1.3 Clausius–Duhem Inequality

This section evaluates the restrictions for satisfaction of the Clausius–Duhem inequality. For the current set of state variables, the mechanical part of Clausius–Duhem Inequality (6.250) takes the form

$$\boldsymbol{\sigma} : \dot{\boldsymbol{\varepsilon}}^{\mathrm{p}} + \boldsymbol{\beta}^1 : \dot{\boldsymbol{\alpha}}^1 + \beta^2 \dot{\alpha}^2 \geq 0\,. \tag{8.118}$$

Substitution of Normality Rules (8.110) into Inequality (8.118) gives

$$\overset{\circ}{\lambda} \left[ \frac{\partial F}{\partial \boldsymbol{\sigma}} : \boldsymbol{\sigma} + \frac{\partial F}{\partial \boldsymbol{\beta}^1} : \boldsymbol{\beta}^1 + \frac{\partial F}{\partial \beta^2} \beta^2 \right] \geq 0 , \tag{8.119}$$

where the order of the terms in square brackets has been changed.

#### 8.4.1.4 Strain Property for Deformation without Volume Change

This section derives a property for a strain tensor term related to deformation that does not show volume changes. Based on Definitions (2.120) and $(2.108)_2$, the deviatoric plastic strain rate $\dot{\mathbf{e}}^{\mathrm{p}}$ is

$$\dot{\mathbf{e}}^{\mathrm{p}} := \mathbf{K} : \dot{\boldsymbol{\varepsilon}}^{\mathrm{p}} \qquad \Rightarrow \qquad \dot{\mathbf{e}}^{\mathrm{p}} = \mathbf{I} : \dot{\boldsymbol{\varepsilon}}^{\mathrm{p}} - \tfrac{1}{3} \mathbf{1}\,\mathbf{1} : \dot{\boldsymbol{\varepsilon}}^{\mathrm{p}} . \tag{8.120}$$

According to Definition (2.69) and Equation $(2.65)_1$, the following result is obtained:

$$\mathbf{I} : \dot{\boldsymbol{\varepsilon}}^{\mathrm{p}} = \dot{\boldsymbol{\varepsilon}}^{\mathrm{p}} \qquad \text{and} \qquad \mathbf{1} : \dot{\boldsymbol{\varepsilon}}^{\mathrm{p}} = \dot{\varepsilon}^{\mathrm{p}}_{ii} = \dot{\varepsilon}^{\mathrm{p}}_{11} + \dot{\varepsilon}^{\mathrm{p}}_{22} + \dot{\varepsilon}^{\mathrm{p}}_{33} . \tag{8.121}$$

According to Lin ([52], p. 22), the volumetric strain rate is $(\dot{\varepsilon}_{11} + \dot{\varepsilon}_{22} + \dot{\varepsilon}_{33})/3$. Pure plastic strain tensor $\boldsymbol{\varepsilon}^{\mathrm{p}}$ is associated with deformation mechanisms that do not show volume changes. Therefore strain rate tensor $\dot{\varepsilon}^{\mathrm{p}}_{ii}$ vanishes since no volume change is present. This and Equation (8.121) together with Definition (8.120) lead to

$$\dot{\mathbf{e}}^{\mathrm{p}} = \dot{\boldsymbol{\varepsilon}}^{\mathrm{p}} , \qquad \text{since no volume change is related to plastic deformation.} \tag{8.122}$$

The integration of Equality (8.122) from $t = 0$ to $t = t_{\mathrm{curr}}$, where the notation "curr" refers to the current configuration $\underline{v}^{\mathrm{b}}(t)$, gives

$$\mathbf{e}^{\mathrm{p}} = \boldsymbol{\varepsilon}^{\mathrm{p}} , \qquad \text{since no volume change is related to plastic deformation.} \tag{8.123}$$

Result (8.123) holds also for the viscous strain tensor $\boldsymbol{\varepsilon}^{\mathrm{v}}$. It is noteworthy that when a deformation consists of two parts, such as $\boldsymbol{\varepsilon} = \boldsymbol{\varepsilon}^{\mathrm{eT}} + \boldsymbol{\varepsilon}^{\mathrm{p}}$, the elastic deformation includes a volume change and therefore the (total) strain tensor $\boldsymbol{\varepsilon}$ shows a volume change. Sometimes void nucleation is connected to plastic deformation (or viscous flow) and the combined deformation is modelled by plastic strain tensor $\boldsymbol{\varepsilon}^{\mathrm{p}}$ (or by viscous strain tensor $\boldsymbol{\varepsilon}^{\mathrm{v}}$). In those models, plastic strain tensor $\boldsymbol{\varepsilon}^{\mathrm{p}}$ (or viscous strain tensor $\boldsymbol{\varepsilon}^{\mathrm{v}}$) includes volume changes.

### 8.4.2 Model for Isotropic and Kinematic Hardening

This section evaluates the traditional models for plastic yield: the model for isotropic hardening and kinematic hardening associated with the von Mises yield condition. The following yield function showing isotropic and kinematic hardening is introduced:

$$F(\boldsymbol{\sigma}, \boldsymbol{\beta}^1, \beta^2; T) = \frac{2}{3} J_{\rm vM}(\boldsymbol{\sigma} - \boldsymbol{\beta}^1) - \beta^2 - \sigma^{\rm Y}, \tag{8.124}$$

where $J_{\rm vM}(\boldsymbol{\sigma} - \boldsymbol{\beta}^1)$ is the von Mises value of the tensorial difference $\boldsymbol{\sigma} - \boldsymbol{\beta}^1$ [see Definition (2.124)]. In Material Model (8.124), the term $\sigma^{\rm Y}$ is the (initial) yield stress. According to Material Model (8.124), the internal force $\boldsymbol{\beta}^1$ shows kinematic hardening, whereas the internal $\beta^2$ force displays isotropic hardening. Since, during plastic yield, the yield function $F$ vanishes, the increasing values of the internal forces $\boldsymbol{\beta}^1$ and $\beta^2$ lead to a higher value of the stress tensor $\boldsymbol{\sigma}$ during plastic yield.

Based on Equations (A.14) and (A.15) of **Appendix A**, Material Model (8.124) gives

$$\frac{\partial F}{\partial \boldsymbol{\sigma}} = \frac{\mathbf{s} - \mathbf{b}^1}{J_{\rm vM}(\boldsymbol{\sigma} - \boldsymbol{\beta}^1)} \quad \text{and} \quad \frac{\partial F}{\partial \boldsymbol{\beta}^1} = -2\frac{\mathbf{s} - \mathbf{b}^1}{J_{\rm vM}(\boldsymbol{\sigma} - \boldsymbol{\beta}^1)}. \tag{8.125}$$

In Derivatives (8.125), the notations $\mathbf{s}$ and $\mathbf{b}^1$ are deviatoric tensors of $\boldsymbol{\sigma}$ and $\boldsymbol{\beta}^1$, respectively. The definition for a deviatoric tensor is given by Definitions (2.120) and $(2.108)_2$. Material Model (8.124) gives also

$$\frac{\partial F}{\partial \beta^2} = -1. \tag{8.126}$$

Substitution of Derivatives (8.125) and (8.126) into Normality Rule (8.110) yields

$$\dot{\boldsymbol{\varepsilon}}^{\rm p} = \mathring{\lambda}\frac{\partial F}{\partial \boldsymbol{\sigma}} = \mathring{\lambda}\frac{\mathbf{s} - \mathbf{b}^1}{J_{\rm vM}(\boldsymbol{\sigma} - \boldsymbol{\beta}^1)} \tag{8.127}$$

and

$$\dot{\boldsymbol{\alpha}}^1 = \mathring{\lambda}\frac{\partial F}{\partial \boldsymbol{\beta}^1} = -\mathring{\lambda}\frac{\mathbf{s} - \mathbf{b}^1}{J_{\rm vM}(\boldsymbol{\sigma} - \boldsymbol{\beta}^1)} = -\dot{\boldsymbol{\varepsilon}}^{\rm p} \tag{8.128}$$

and, finally,

$$\dot{\alpha}^2 = \mathring{\lambda}\frac{\partial F}{\partial \beta^2} = -\mathring{\lambda}. \tag{8.129}$$

By combining Normality Rule (8.127) and Result (8.122) [$\dot{\mathbf{e}}^{\rm p} = \dot{\boldsymbol{\varepsilon}}^{\rm p}$] with the definition of the von Mises operator [Definition (2.124)], the following expression is obtained:

$$\begin{aligned}[J_{\rm vM}(\dot{\boldsymbol{\varepsilon}}^{\rm p})]^2 &= \frac{3}{2}\dot{\mathbf{e}}^{\rm p} : \dot{\mathbf{e}}^{\rm p} = \frac{3}{2}\dot{\boldsymbol{\varepsilon}}^{\rm p} : \dot{\boldsymbol{\varepsilon}}^{\rm p} = \frac{3}{2}[\mathring{\lambda}]^2\frac{(\mathbf{s} - \mathbf{b}^1) : (\mathbf{s} - \mathbf{b}^1)}{[J_{\rm vM}(\boldsymbol{\sigma} - \boldsymbol{\beta}^1)]^2} \\ &= [\mathring{\lambda}]^2\frac{[J_{\rm vM}(\boldsymbol{\sigma} - \boldsymbol{\beta}^1)]^2}{[J_{\rm vM}(\boldsymbol{\sigma} - \boldsymbol{\beta}^1)]^2} = [\mathring{\lambda}]^2.\end{aligned} \tag{8.130}$$

Substitution of Result (8.130) into Flux (8.129) yields

$$\dot{\alpha}^2 = -\,J_{\mathrm{vM}}(\dot{\boldsymbol{\varepsilon}}^{\mathrm{p}})\,. \tag{8.131}$$

By extending State Equations (6.174) for the current set of state variables having two internal variables, the following can be written:

$$\boldsymbol{\sigma} = \rho_0\,\frac{\partial\psi}{\partial(\boldsymbol{\varepsilon}-\boldsymbol{\varepsilon}^{\mathrm{p}})} \quad \text{and} \quad \boldsymbol{\beta}^1 = -\,\rho_0\,\frac{\partial\psi}{\partial\boldsymbol{\alpha}^1} \quad \text{and} \quad \beta^2 = -\,\rho_0\,\frac{\partial\psi}{\partial\alpha^2}\,. \tag{8.132}$$

Next, three special cases, that is, (a) isotropic hardening and (b, c) kinematic hardening, are evaluated. Combination of these two cases is of course possible.

**(a) Isotropic Hardening with von Mises Yield Criterion [114]**

Kinematic hardening is assumed to be negligible, which implies that all the components of the internal force $\boldsymbol{\beta}^1$ vanish. The plastic part of the specific Helmholtz free energy $\psi^{\mathrm{p}}$ is modelled as follows:

$$\psi^{\mathrm{p}}(\alpha^2,T) = \frac{k}{2\,\rho_0}\,(\alpha^2)^2\,, \tag{8.133}$$

where the scalar-valued quantity $k$ is a material parameter. Substitution of Material Model (8.133) into State Equation $(8.132)_3$ gives

$$\beta^2 = -\,\rho_0\,\frac{\partial\psi^{\mathrm{p}}}{\partial\alpha^2} = -\,k\,\alpha^2\,, \quad \text{which yields} \quad \dot{\beta}^2 = -\,k\,\dot{\alpha}^2\,. \tag{8.134}$$

By substitution of Expression (8.131) into Material Derivative $(8.134)_2$, the following is obtained:

$$\dot{\beta} = k\,J_{\mathrm{vM}}(\dot{\boldsymbol{\varepsilon}}^{\mathrm{p}})\,. \tag{8.135}$$

Assuming that $\beta^2$ vanishes in the initial configuration, Expression (8.135) gives

$$\beta^2 = k\int_0^t J_{\mathrm{vM}}(\dot{\boldsymbol{\varepsilon}}^{\mathrm{p}})\,\mathrm{d}t = \tfrac{3}{2}\,k\int_0^t \left[\tfrac{2}{3}\,\dot{\boldsymbol{\varepsilon}}^{\mathrm{p}}:\dot{\boldsymbol{\varepsilon}}^{\mathrm{p}}\right]^{1/2}\mathrm{d}t\,, \tag{8.136}$$

where the latter integral is called cumulative plastic strain, effective inelastic strain, or accumulated plastic strain. The latter integral without the multiplier 2/3 is called, according to Odqvist [71], the Odqvist parameter. Equation (8.136) shows that the hardening $\beta^2$ is related to the von Mises criterion. Expressions (8.133), $(8.134)_1$, and (8.136) give

$$\psi^{\mathrm{p}}(\alpha^2,T) = \frac{1}{2\,\rho_0\,k}(\beta^2)^2 = \frac{k}{2\,\rho_0}\left[\int_0^t J_{\mathrm{vM}}\,(\dot{\boldsymbol{\varepsilon}}^{\mathrm{p}})\,\mathrm{d}t\right]^2\,. \tag{8.137}$$

In this case, Normality Rule (8.127) collapses to

$$\dot{\boldsymbol{\varepsilon}}^{\mathrm{p}} = \overset{\circ}{\lambda}\,\frac{\mathbf{s}}{J_{\mathrm{vM}}(\boldsymbol{\sigma})}\,. \tag{8.138}$$

The preceding was for a case in which the hardening and the plastic part of the specific Helmholtz free energy $\psi^{\mathrm{p}}$ [cf. Equation (8.137)] are related to the accumulated plastic strain or to the Odqvist parameter.

The appearance of Expression (8.138) may give the illusion that the internal force $\beta^2$ and therefore the accumulated plastic strain have no influence on the plastic strain rate $\dot{\boldsymbol{\varepsilon}}^{\mathrm{p}}$. This is not the case, since the plasticity multiplier $\overset{\circ}{\lambda}$ is dependent on the internal force $\beta^2$, as Section 8.4.3 will show.

**(b) Kinematic Hardening**

Isotropic hardening is assumed to be negligible, which implies that the internal force $\beta^2$ vanishes. The plastic part of the specific Helmholtz free energy $\psi^{\mathrm{p}}$ is modelled as follows:

$$\psi^{\mathrm{p}}(\boldsymbol{\alpha}^1, T) = \frac{1}{2\,\rho_0}\,\boldsymbol{\alpha}^1 : \mathbf{D} : \boldsymbol{\alpha}^1\,, \qquad \text{where} \qquad D_{ijkl} = D_{klij}\,. \tag{8.139}$$

According to Property $(8.139)_2$, the constitutive tensor $\mathbf{D}$ is major symmetric. The fourth-order tensor $\mathbf{D}$ contains material parameters having in Property $(8.139)_2$ the described symmetry. Substitution of Material Model (8.139) into State Equation $(8.132)_2$ gives [see Equation $(2.131)_2$]

$$\boldsymbol{\beta}^1 = -\,\rho_0\,\frac{\partial \psi}{\partial \boldsymbol{\alpha}^1} = -\,\mathbf{D} : \boldsymbol{\alpha}^1\,, \qquad \text{which yields} \qquad \dot{\boldsymbol{\beta}}^1 = -\,\mathbf{D} : \dot{\boldsymbol{\alpha}}^1\,. \tag{8.140}$$

By substituting Expression (8.128) into Derivative $(8.140)_2$, the following is obtained:

$$\dot{\boldsymbol{\beta}}^1 = \mathbf{D} : \dot{\boldsymbol{\varepsilon}}^{\mathrm{p}}\,. \tag{8.141}$$

Assume that, at time $t = 0$, $\boldsymbol{\varepsilon}^{\mathrm{p}} = \mathbf{0}$, $\boldsymbol{\alpha}^1 = \mathbf{0}$, and $\boldsymbol{\beta}^1 = \mathbf{0}$ hold in Rate Equations (8.128) and (8.141). This yields

$$\boldsymbol{\alpha}^1 = -\,\boldsymbol{\varepsilon}^{\mathrm{p}} \qquad \text{and} \qquad \boldsymbol{\beta}^1 = \mathbf{D} : \boldsymbol{\varepsilon}^{\mathrm{p}}\,. \tag{8.142}$$

Substitution of Equation $(8.142)_1$ into Material Model (8.139) yields

$$\psi^{\mathrm{p}}(\boldsymbol{\alpha}^1, T) = \frac{1}{2\,\rho_0}\,\boldsymbol{\varepsilon}^{\mathrm{p}} : \mathbf{D} : \boldsymbol{\varepsilon}^{\mathrm{p}}\,. \tag{8.143}$$

In kinematic hardening, the hardening is related to the plastic strain tensor $\boldsymbol{\varepsilon}^{\mathrm{p}}$, as Equation $(8.142)_2$ shows.

In this case, Normality Rule (8.127) becomes

$$\dot{\boldsymbol{\varepsilon}}^{\mathrm{p}} = \overset{\circ}{\lambda}\,\frac{\partial F}{\partial \boldsymbol{\sigma}} = \overset{\circ}{\lambda}\,\frac{\mathbf{s} - \mathbf{b}^1}{J_{\mathrm{vM}}(\boldsymbol{\sigma} - \boldsymbol{\beta}^1)}\,. \tag{8.144}$$

**(c) Kinematic Hardening with the Prager Yield Criterion**

Although the material model presented here is usually attributed to Prager [77, 78], it should be remembered that he did not write this model but investigated plastic flow by drawing movements of yield surfaces due to hardening. The present model is a mathematical formulation of his drawings. It is not known to the authors who was first to write the mathematical formulation for the following constitutive equation.

It is assumed that

$$\mathbf{D} := c\,\mathbf{I}\,, \tag{8.145}$$

where $c$ is a material parameter. Due to Assumption (8.145), Material Model (8.139) reduces to

$$\psi^{\mathrm{p}}(\boldsymbol{\alpha}^1, T) = \frac{c}{2\,\rho_0}\,\boldsymbol{\alpha}^1 : \boldsymbol{\alpha}^1\,. \tag{8.146}$$

Based on Material Model (8.145), Equation (8.141) yields

$$\dot{\boldsymbol{\beta}}^1 = \mathbf{D} : \dot{\boldsymbol{\varepsilon}}^{\mathrm{p}} \qquad\Rightarrow\qquad \boldsymbol{\beta}^1 = c\,\boldsymbol{\varepsilon}^{\mathrm{p}}\,. \tag{8.147}$$

Based on Property (8.123) (i.e., $\mathbf{e}^{\mathrm{p}} = \boldsymbol{\varepsilon}^{\mathrm{p}}$ ), Constitutive Equation $(8.147)_2$ takes the form

$$\boldsymbol{\beta}^1 = c\,\mathbf{e}^{\mathrm{p}}\,. \tag{8.148}$$

Following Definition (2.120), the deviatoric plastic strain $\mathbf{e}^{\mathrm{p}}$ and the deviatoric part of the internal force $\mathbf{b}^1$ are obtained from

$$\mathbf{e}^{\mathrm{p}} := \mathbf{K} : \boldsymbol{\varepsilon}^{p} \qquad \text{and} \qquad \mathbf{b}^1 = \mathbf{K} : \boldsymbol{\beta}^1\,. \tag{8.149}$$

Multiplying Hardening Model $(8.147)_2$ by $\mathbf{K}:$ from the left and by taking Definition $(8.149)_1$ into account, the following is achieved:

$$\mathbf{K} : \boldsymbol{\beta}^1 = c\,\mathbf{K} : \boldsymbol{\varepsilon}^{\mathrm{p}} = c\,\mathbf{e}^{\mathrm{p}}\,. \tag{8.150}$$

Substitution of Expression (8.150) into Equation (8.149) gives

$$\mathbf{b}^1 = c\,\mathbf{e}^{\mathrm{p}} \qquad\Rightarrow\qquad \mathbf{b}^1 = \boldsymbol{\beta}^1\,, \tag{8.151}$$

where Hardening Models (8.148) and $(8.151)_1$ are compared.

Substitution of Result $(8.151)_2$ into Normality Rule (8.144) gives

$$\dot{\boldsymbol{\varepsilon}}^{\mathrm{p}} = \overset{\circ}{\lambda}\,\frac{\partial F}{\partial \boldsymbol{\sigma}} = \overset{\circ}{\lambda}\,\frac{\mathbf{s} - \boldsymbol{\beta}^1}{J_{\mathrm{vM}}(\boldsymbol{\sigma} - \boldsymbol{\beta}^1)}\,. \tag{8.152}$$

### 8.4.3 Value for the Plasticity Multiplier $\overset{\circ}{\lambda}$

This section derives the value for the plasticity multiplier $\overset{\circ}{\lambda}$ for the material model showing isotropic and kinematic hardening evaluated in Section 8.4.2.

The elastic deformation is assumed to follow the generalized Duhamel–Neuman form of Hooke's law.

Material Model (8.105) is

$$\psi(\boldsymbol{\varepsilon}-\boldsymbol{\varepsilon}^{\mathrm{p}},\boldsymbol{\alpha}^1,\alpha^2,h(\vec{X}),T)=\psi^{\mathrm{eT}}(\boldsymbol{\varepsilon}-\boldsymbol{\varepsilon}^{\mathrm{p}},h(\vec{X}),T)+\psi^{\mathrm{p}}(\boldsymbol{\alpha}^1,\alpha^2,T)\,. \tag{8.153}$$

The thermoelastic part of the specific Helmholtz free energy $\psi^{\mathrm{eT}}$ [see Material Model (8.153)] is assumed to follow Material Model (8.66),

$$\begin{aligned}\psi^{\mathrm{eT}}(\boldsymbol{\varepsilon}-\boldsymbol{\varepsilon}^{\mathrm{i}},T)=\psi_{\mathrm{r}}^{\mathrm{eT}}&-s_{\mathrm{r}}\,(T-T_{\mathrm{r}})\;+\frac{1}{2\,\rho_0}\,(\boldsymbol{\varepsilon}-\boldsymbol{\varepsilon}^{\mathrm{i}}):\mathbf{C}:(\boldsymbol{\varepsilon}-\boldsymbol{\varepsilon}^{\mathrm{i}})\\&-\frac{E\,\hat{\alpha}}{\rho_0\,(1-2\,\nu)}\,\mathbf{1}:(\boldsymbol{\varepsilon}-\boldsymbol{\varepsilon}^{\mathrm{i}})\,(T-T_{\mathrm{r}})+\frac{1}{\rho_0}\,\boldsymbol{\sigma}_{\mathrm{r}}:(\boldsymbol{\varepsilon}-\boldsymbol{\varepsilon}^{\mathrm{i}})\\&-c_{\mathrm{r}}\,[\,T\,\ln(T/T_{\mathrm{r}})-T\,]\,.\end{aligned} \tag{8.154}$$

Based on Equation (8.67), Material Model (8.154) leads to the following constitutive equation:

$$\begin{aligned}\boldsymbol{\sigma}=\rho_0\,\frac{\partial\psi^{\mathrm{eT}}(\boldsymbol{\varepsilon}-\boldsymbol{\varepsilon}^{\mathrm{i}},T)}{\partial(\boldsymbol{\varepsilon}-\boldsymbol{\varepsilon}^{\mathrm{i}})}&=\mathbf{C}:(\boldsymbol{\varepsilon}-\boldsymbol{\varepsilon}^{\mathrm{i}})-\frac{E\,\hat{\alpha}}{(1-2\,\nu)}\,(T-T_{\mathrm{r}})\,\mathbf{1}+\boldsymbol{\sigma}_{\mathrm{r}}\\&=\mathbf{C}:(\boldsymbol{\varepsilon}-\boldsymbol{\varepsilon}^{\mathrm{p}})-\frac{E\,\hat{\alpha}}{(1-2\,\nu)}\,(T-T_{\mathrm{r}})\,\mathbf{1}+\boldsymbol{\sigma}_{\mathrm{r}}\,.\end{aligned} \tag{8.155}$$

Constitutive Equation (8.155) gives

$$\dot{\boldsymbol{\sigma}}=\mathbf{C}:(\dot{\boldsymbol{\varepsilon}}-\dot{\boldsymbol{\varepsilon}}^{\mathrm{p}})-\frac{E\,\hat{\alpha}}{(1-2\,\nu)}\,\dot{T}\,\mathbf{1}\,. \tag{8.156}$$

Normality Rule (8.127) is

$$\dot{\boldsymbol{\varepsilon}}^{\mathrm{p}}=\overset{\circ}{\lambda}\,\frac{\partial F}{\partial\boldsymbol{\sigma}}\qquad\left[=\overset{\circ}{\lambda}\,\frac{\mathbf{s}-\mathbf{b}^1}{J_{\mathrm{vM}}(\boldsymbol{\sigma}-\boldsymbol{\beta}^1)}\right]. \tag{8.157}$$

Substitution of Rate (8.157) into Derivative (8.156) yields

$$\dot{\boldsymbol{\sigma}}=\mathbf{C}:\dot{\boldsymbol{\varepsilon}}-\overset{\circ}{\lambda}\,\mathbf{C}:\frac{\partial F}{\partial\boldsymbol{\sigma}}-\frac{E\,\hat{\alpha}}{(1-2\,\nu)}\,\dot{T}\;\mathbf{1}\,. \tag{8.158}$$

Equation (8.117) is

$$\begin{aligned}\overset{\circ}{\lambda}=&H(F)\left\langle\frac{\partial F}{\partial\boldsymbol{\sigma}}:\dot{\boldsymbol{\sigma}}-\left[\rho_0\left(\frac{\partial F}{\partial\boldsymbol{\beta}^1}:\frac{\partial^2\psi}{\partial\boldsymbol{\alpha}^1\,\partial T}+\frac{\partial F}{\partial\beta^2}\,\frac{\partial^2\psi}{\partial\alpha^2\,\partial T}\right)-\frac{\partial F}{\partial T}\right]\dot{T}\right\rangle\\&\Big/\rho_0\left[\frac{\partial F}{\partial\boldsymbol{\beta}^1}:\frac{\partial^2\psi}{\partial\boldsymbol{\alpha}^1\,\partial\boldsymbol{\alpha}^1}:\frac{\partial F}{\partial\boldsymbol{\beta}^1}+2\,\frac{\partial F}{\partial\boldsymbol{\beta}^1}:\frac{\partial^2\psi}{\partial\boldsymbol{\alpha}^1\,\partial\alpha^2}\,\frac{\partial F}{\partial\beta^2}+\frac{\partial F}{\partial\beta^2}\,\frac{\partial^2\psi}{\partial\alpha^2\,\partial\alpha^2}\,\frac{\partial F}{\partial\beta^2}\right].\end{aligned} \tag{8.159}$$

Stress rate $\dot{\boldsymbol{\sigma}}$ from Equation (8.158) is substituted into Equation (8.159), and the value for the plasticity multiplier $\overset{\circ}{\lambda}$ is solved. This process gives

$$\overset{\circ}{\lambda} = H(F)\Big\langle \frac{\partial F}{\partial \boldsymbol{\sigma}} : \mathbf{C} : \dot{\boldsymbol{\varepsilon}} - \frac{E\,\hat{\alpha}}{(1-2\,\nu)}\,\dot{T}\,\frac{\partial F}{\partial \boldsymbol{\sigma}} : \mathbf{1} - \rho_0\,\frac{\partial F}{\partial \boldsymbol{\beta}^1} : \frac{\partial^2 \psi}{\partial \boldsymbol{\alpha}^1\,\partial T}\,\dot{T}$$
$$+ \rho_0\,\frac{\partial F}{\partial \beta^2}\,\frac{\partial^2 \psi}{\partial \alpha^2\,\partial T}\,\dot{T} + \frac{\partial F}{\partial T}\,\dot{T}\Big\rangle$$
$$\Big/ \rho_0 \left( \frac{\partial F}{\partial \boldsymbol{\beta}^1} : \frac{\partial^2 \psi}{\partial \boldsymbol{\alpha}^1\,\partial \boldsymbol{\alpha}^1} : \frac{\partial F}{\partial \boldsymbol{\beta}^1} + 2\,\frac{\partial F}{\partial \boldsymbol{\beta}^1} : \frac{\partial^2 \psi}{\partial \boldsymbol{\alpha}^1\,\partial \alpha^2}\,\frac{\partial F}{\partial \beta^2} + \frac{\partial F}{\partial \beta^2}\,\frac{\partial^2 \psi}{\partial \alpha^2\,\partial \alpha^2}\,\frac{\partial F}{\partial \beta^2} \right)$$
$$+ \frac{\partial F}{\partial \boldsymbol{\sigma}} : \mathbf{C} : \frac{\partial F}{\partial \boldsymbol{\sigma}}\,. \tag{8.160}$$

In Expression (8.160), the first expression on the first two lines after the equals sign is the numerator, and the expression on the third and fourth lines is the denominator.

Definition (6.299) shows that the value of the plasticity multiplier $\overset{\circ}{\lambda}$ depends on state and on the rate of loading, as follows:

$$\overset{\circ}{\lambda} := \overset{\circ}{\lambda}(\boldsymbol{\varepsilon} - \boldsymbol{\varepsilon}^{\mathrm{p}}, \boldsymbol{\alpha}^1, \alpha^2, T, h, \dot{\boldsymbol{\varepsilon}}, \dot{T})\,. \tag{8.161}$$

Expression (8.160) supports this. Since the yield function $F$ is expressed by forces $(\boldsymbol{\sigma}, \boldsymbol{\beta}^1, \beta^2)$, many of the derivatives in Expression (8.160) are functions of forces. This does not violate the foregoing statement, since the state equations create a link between the state variables $(\boldsymbol{\varepsilon} - \boldsymbol{\varepsilon}^{\mathrm{p}}, \boldsymbol{\alpha}^1, \alpha^2)$ and the state functions $(\boldsymbol{\sigma}, \boldsymbol{\beta}^1, \beta^2)$. Thus, according to Equations (8.157) and (8.160), plastic yield takes place when the state $(\boldsymbol{\varepsilon} - \boldsymbol{\varepsilon}^{\mathrm{p}}, \boldsymbol{\alpha}^1, \alpha^2)$ is on the yield surface and the rate of loading $(\dot{\boldsymbol{\varepsilon}}, \dot{T})$ [i.e., material derivatives of the controllable variables $(\boldsymbol{\varepsilon}, T)$] takes a value outward from the yield surface.

Casting Material Models (8.133) and (8.139) together gives the following:

$$\psi^{\mathrm{p}}(\boldsymbol{\alpha}^1, \alpha^2, T) \equiv \psi^{\mathrm{p}}(\boldsymbol{\alpha}^1, \alpha^2) = \frac{k}{2\,\rho_0}\,(\alpha^2)^2 + \frac{1}{2\,\rho_0}\,\boldsymbol{\alpha}^1 : \mathbf{D} : \boldsymbol{\alpha}^1\,. \tag{8.162}$$

Based on Property $(8.139)_2$, material property tensor $\mathbf{D}$ is major symmetric; that is, it obeys the following symmetry condition $D_{ijkl} = D_{klij}$. Based on Material Models (8.153), (8.154), and (8.162) and Expressions $(2.131)_2$ and (2.133), the following result

$$\frac{\partial \psi}{\partial \boldsymbol{\alpha}^1} = \frac{1}{\rho_0}\,\mathbf{D} : \boldsymbol{\alpha}^1 \quad \Rightarrow \quad \frac{\partial^2 \psi}{\partial \boldsymbol{\alpha}^1\,\partial \boldsymbol{\alpha}^1} = \frac{1}{\rho_0}\,\mathbf{D} \quad \text{and} \quad \frac{\partial^2 \psi}{\partial \boldsymbol{\alpha}^1\,\partial \alpha^2} = \mathbf{0}\,. \tag{8.163}$$

Based on Material Models (8.153), (8.154), and (8.162), the following is achieved:

$$\frac{\partial \psi}{\partial \alpha^2} = \frac{k}{\rho_0}\,\alpha^2 \quad \Rightarrow \quad \frac{\partial^2 \psi}{\partial \alpha^2\,\partial \alpha^2} = \frac{k}{\rho_0}\,. \tag{8.164}$$

Derivatives (8.125) and (8.126) are

$$\frac{\partial F}{\partial \boldsymbol{\sigma}} = \frac{\mathbf{s} - \mathbf{b}^1}{J_{\mathrm{vM}}(\boldsymbol{\sigma} - \boldsymbol{\beta}^1)}, \quad \frac{\partial F}{\partial \boldsymbol{\beta}^1} = -\frac{\mathbf{s} - \mathbf{b}^1}{J_{\mathrm{vM}}(\boldsymbol{\sigma} - \boldsymbol{\beta}^1)}, \quad \text{and} \quad \frac{\partial F}{\partial \beta^2} = -1. \tag{8.165}$$

Material Model (8.124) is

$$F(\boldsymbol{\sigma}, \boldsymbol{\beta}^1, \beta^2; T) = \frac{2}{3} J_{\mathrm{vM}}(\boldsymbol{\sigma} - \boldsymbol{\beta}^1) - \beta^2 - \sigma^{\mathrm{Y}} \quad \Rightarrow \quad \frac{\partial F}{\partial T} = 0. \tag{8.166}$$

Based on Derivatives $(8.163)_1$ and $(8.164)_1$

$$\frac{\partial^2 \psi}{\partial \boldsymbol{\alpha}^1 \, \partial T} = \mathbf{0} \quad \text{and} \quad \frac{\partial^2 \psi}{\partial \alpha^2 \, \partial T} = 0. \tag{8.167}$$

Thus, based on Results (8.163)–(8.167), Equation (8.160) takes the form

$$\overset{\circ}{\lambda} = H(F) \left\langle \frac{\mathbf{s} - \mathbf{b}^1}{J_{\mathrm{vM}}(\boldsymbol{\sigma} - \boldsymbol{\beta}^1)} : \mathbf{C} : \dot{\boldsymbol{\varepsilon}} - \frac{E\,\hat{\alpha}}{(1 - 2\,\nu)} \dot{T} \frac{\mathbf{s} - \mathbf{b}^1}{J_{\mathrm{vM}}(\boldsymbol{\sigma} - \boldsymbol{\beta}^1)} : \mathbf{1} \right\rangle$$
$$\Big/ \left\{ \left[ \frac{\mathbf{s} - \mathbf{b}^1}{J_{\mathrm{vM}}(\boldsymbol{\sigma} - \boldsymbol{\beta}^1)} : \mathbf{D} : \frac{\mathbf{s} - \mathbf{b}^1}{J_{\mathrm{vM}}(\boldsymbol{\sigma} - \boldsymbol{\beta}^1)} + k \right] + \frac{\mathbf{s} - \mathbf{b}^1}{J_{\mathrm{vM}}(\boldsymbol{\sigma} - \boldsymbol{\beta}^1)} : \mathbf{C} : \frac{\mathbf{s} - \mathbf{b}^1}{J_{\mathrm{vM}}(\boldsymbol{\sigma} - \boldsymbol{\beta}^1)} \right\}. \tag{8.168}$$

Based on Property (2.116), $\mathbf{s} : \mathbf{1} \equiv 0$ and $\mathbf{b}^1 : \mathbf{1} \equiv 0$ hold, and therefore Expression (8.168) reduces to

$$\overset{\circ}{\lambda} = H(F) \left\langle \frac{\mathbf{s} - \mathbf{b}^1}{J_{\mathrm{vM}}(\boldsymbol{\sigma} - \boldsymbol{\beta}^1)} : \mathbf{C} : \dot{\boldsymbol{\varepsilon}} \right\rangle$$
$$\Big/ \left\{ \left[ \frac{\mathbf{s} - \mathbf{b}^1}{J_{\mathrm{vM}}(\boldsymbol{\sigma} - \boldsymbol{\beta}^1)} : \mathbf{D} : \frac{\mathbf{s} - \mathbf{b}^1}{J_{\mathrm{vM}}(\boldsymbol{\sigma} - \boldsymbol{\beta}^1)} + k \right] + \frac{\mathbf{s} - \mathbf{b}^1}{J_{\mathrm{vM}}(\boldsymbol{\sigma} - \boldsymbol{\beta}^1)} : \mathbf{C} : \frac{\mathbf{s} - \mathbf{b}^1}{J_{\mathrm{vM}}(\boldsymbol{\sigma} - \boldsymbol{\beta}^1)} \right\}. \tag{8.169}$$

## 8.4.4 Satisfaction of the Clausius–Duhem Inequality

This section shows that the model for isotropic and kinematic hardening satisfies the Clausius–Duhem inequality. The study starts from the following form of Clausius–Duhem Inequality (8.118):

$$\boldsymbol{\sigma} : \dot{\boldsymbol{\varepsilon}}^{\mathrm{p}} + \boldsymbol{\beta}^1 : \dot{\boldsymbol{\alpha}}^1 + \beta^2 \, \dot{\alpha}^2 \geq 0. \tag{8.170}$$

Rate Equations (8.127), (8.128), and (8.129) are

$$\dot{\boldsymbol{\varepsilon}}^{\mathrm{p}} = \overset{\circ}{\lambda} \frac{\partial F}{\partial \boldsymbol{\sigma}} = \overset{\circ}{\lambda} \frac{\mathbf{s} - \mathbf{b}^1}{J_{\mathrm{vM}}(\boldsymbol{\sigma} - \boldsymbol{\beta}^1)} \tag{8.171}$$

and

$$\dot{\boldsymbol{\alpha}}^1 = \overset{\circ}{\lambda}\,\frac{\partial F}{\partial \boldsymbol{\beta}^1} = -\,\overset{\circ}{\lambda}\,\frac{\mathbf{s} - \mathbf{b}^1}{J_{\mathrm{vM}}(\boldsymbol{\sigma} - \boldsymbol{\beta}^1)} = -\,\dot{\boldsymbol{\varepsilon}}^{\mathrm{p}} \tag{8.172}$$

and, finally,

$$\dot{\alpha}^2 = \overset{\circ}{\lambda}\,\frac{\partial F}{\partial \beta^2} = -\,\overset{\circ}{\lambda}\,. \tag{8.173}$$

Substitution of Material Derivatives (8.171)–(8.173) into Inequality (8.170) gives

$$\overset{\circ}{\lambda}\left[\frac{\mathbf{s} - \mathbf{b}^1}{J_{\mathrm{vM}}(\boldsymbol{\sigma} - \boldsymbol{\beta}^1)} : (\boldsymbol{\sigma} - \boldsymbol{\beta}^1) - \beta^2\right] \geq 0\,. \tag{8.174}$$

Problem 8.2 is to show that the following holds:

$$(\boldsymbol{\sigma} - \boldsymbol{\beta}^1) : (\mathbf{s} - \mathbf{b}^1) = (\mathbf{s} - \mathbf{b}^1) : (\mathbf{s} - \mathbf{b}^1)\,. \tag{8.175}$$

Based on Equality (8.175), Inequality (8.174) changes to

$$\overset{\circ}{\lambda}\left[\frac{(\mathbf{s} - \mathbf{b}^1) : (\mathbf{s} - \mathbf{b}^1)}{J_{\mathrm{vM}}(\boldsymbol{\sigma} - \boldsymbol{\beta}^1)} - \beta^2\right] \geq 0\,. \tag{8.176}$$

Expression (2.124) gives the definition for the von Mises operator $J_{\mathrm{vM}}(\cdot)$. For the difference $(\boldsymbol{\sigma} - \boldsymbol{\beta}^1)$, it gives

$$J_{\mathrm{vM}}(\boldsymbol{\sigma} - \boldsymbol{\beta}^1) = \sqrt{\tfrac{3}{2}\,(\mathbf{s} - \mathbf{b}^1) : (\mathbf{s} - \mathbf{b}^1)}\,. \tag{8.177}$$

Expression (8.177) allows Inequality (8.176) to be written in the form

$$\overset{\circ}{\lambda}\left[\frac{\tfrac{2}{3}\left[J_{\mathrm{vM}}(\boldsymbol{\sigma} - \boldsymbol{\beta}^1)\right]^2}{J_{\mathrm{vM}}(\boldsymbol{\sigma} - \boldsymbol{\beta}^1)} - \beta^2\right] \geq 0\,, \tag{8.178}$$

which yields

$$\overset{\circ}{\lambda}\left[\tfrac{2}{3}\,J_{\mathrm{vM}}(\boldsymbol{\sigma} - \boldsymbol{\beta}^1) - \beta^2\right] \geq 0\,. \tag{8.179}$$

According to Material Model (8.124), the yield function $F$ is

$$F(\boldsymbol{\sigma}, \boldsymbol{\beta}^1, \beta^2; T) = \frac{2}{3}\,J_{\mathrm{vM}}(\boldsymbol{\sigma} - \boldsymbol{\beta}^1) - \beta^2 - \sigma^{\mathrm{Y}}\,. \tag{8.180}$$

According to the definition of the yield function $F$, it is nonnegative, that is, $F \geq 0$. Since the (initial) yield stress $\sigma^{\mathrm{Y}}$ is nonnegative, Material Model (8.180) ensures the nonnegativeness of the term between the square brackets in Inequality (8.179). Since the plasticity multiplier $\overset{\circ}{\lambda}$ is also nonnegative [see the discussion under Equation (8.117)], Clausius–Duhem Inequality (8.179) is satisfied.

## 8.5 Summary

This chapter studied material models for heat conduction and time-independent deformation. The validation procedure introduced in this book was used to show that the studied material models do not violate the basic laws and axioms of continuum thermodynamics. An interesting result may be that according to Fourier's law of heat conduction, any temperature disturbance will propagate at infinite velocity. As discussed in Section 8.2 and shown in Figure 8.1, the Cattaneo Equation (8.16) solves the problem with Fourier's Heat Conduction Equation (6.203). However, Fourier's law of heat conduction is fully applicable to most engineering applications. Section 8.3.2 derived the limits for Poisson's ratio $\nu$ [see Equation (8.64)]. These limits are good to remember.

## Problems

**8.1** Show that for an arbitrary vector $\vec{a}$ and for an arbitrary symmetric second-order tensor $\mathbf{c}$, the following holds:

$$\frac{\partial}{\partial \vec{a}} \left[ \tfrac{1}{2}\, \vec{a} \cdot \mathbf{C} \cdot \vec{a} \right] = \vec{a} \cdot \mathbf{C} \,. \tag{1}$$

**8.2** Show that the following expression holds:

$$(\boldsymbol{\sigma} - \boldsymbol{\beta}) : (\mathbf{s} - \mathbf{b}) = (\mathbf{s} - \mathbf{b}) : (\mathbf{s} - \mathbf{b}) \,, \tag{1}$$

where $\mathbf{s}$ is the deviatoric stress tensor and $\mathbf{b}$ is the deviatoric part of the tensor $\boldsymbol{\beta}$.

**8.3** The fourth-order constitutive tensor of the Hookean deformation $\mathbf{C}$ reads

$$\mathbf{C} := \lambda\, \mathbf{1}\, \mathbf{1} + 2\, \mu\, \mathbf{I}^{\mathrm{s}} \,, \tag{1}$$

where $\lambda$ and $\mu$ are Lamé elastic constants defined by

$$\lambda := \frac{\nu\, E}{(1 + \nu)\,(1 - 2\,\nu)} \qquad \text{and} \qquad \mu := \frac{E}{2\,(1 + \nu)} \,. \tag{2}$$

The fourth-order compliance tensor $\mathbf{S}$ for a Hookean deformation is defined by

$$\mathbf{S} := \frac{1 + \nu}{E}\, \mathbf{I}^{\mathrm{s}} - \frac{\nu}{E}\, \mathbf{1}\, \mathbf{1} \,. \tag{3}$$

Calculate $\mathbf{C} : \mathbf{S}$.

**Hint.** Do **not** use the index notation, but apply

$$\mathbf{I}^{\mathrm{s}} : \mathbf{I}^{\mathrm{s}} = \mathbf{I}^{\mathrm{s}} \,. \tag{4}$$

# 9 Material Models for Creep

## 9.1 Introduction

This chapter studies material models for viscoelastic and viscoplastic deformation. They are time-dependent processes and therefore are usually modelled by a (set of) differential equation(s). Figure 9.1 compares viscoelastic and viscoplastic behavior under a simple loading shown by Figure 9.1(a). The main difference between these two time-dependent constitutive relations is that for viscoelasticity, the deformation is fully recoverable, whereas viscoplastic deformation is a permanent response. Of course, a mixture of these two responses is possible, as Figure 9.1(d) shows.

The first few sections study some simple examples that provide basic information on viscoelastic and viscoplastic material models. Subsequent sections discuss the models for specific material behaviors.

The tradition of studying time-dependent deformation of materials at elevated temperatures goes back to the beginning of the 20th century. In 1910, Costa Andrade initiated a systematic study of viscous deformation and introduced the idea of dividing the time-dependent deformation curve into different stages. Nowadays, those stages are called the primary, secondary, and tertiary stages of the time-dependent deformation. The instantaneous part of deformation is included in Figure 9.2. Usually it is not presented, because it is negligible compared with time-

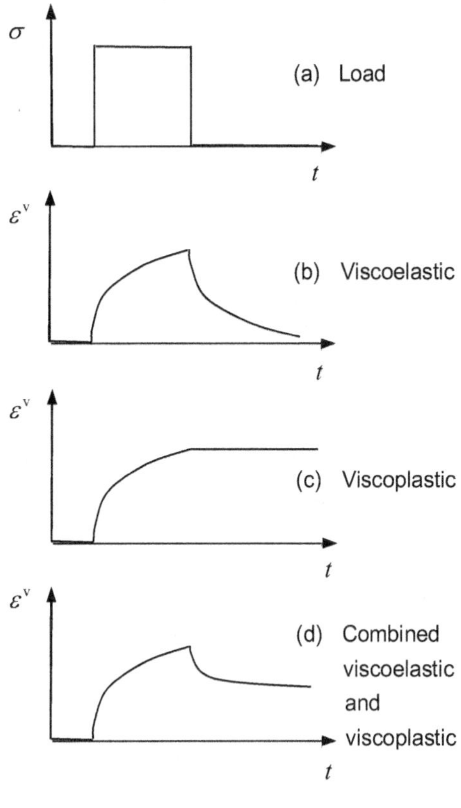


**Figure 9.1** (a) Applied load, (b) viscoelastic deformation, (c) viscoplastic deformation, and (d) combination of viscoelastic and viscoplastic deformation.

dependent strain terms.

The primary stage is often called transient creep, and the secondary stage is steady state creep. The three stages can be described as follows:

- *Primary creep.* During primary, creep the strain rate decreases.
- *Secondary creep.* The strain rate takes a constant value.
- *Tertiary creep.* The strain rate increases. Usually tertiary creep is related to the damage of the material, but this depends on the definition of damage.

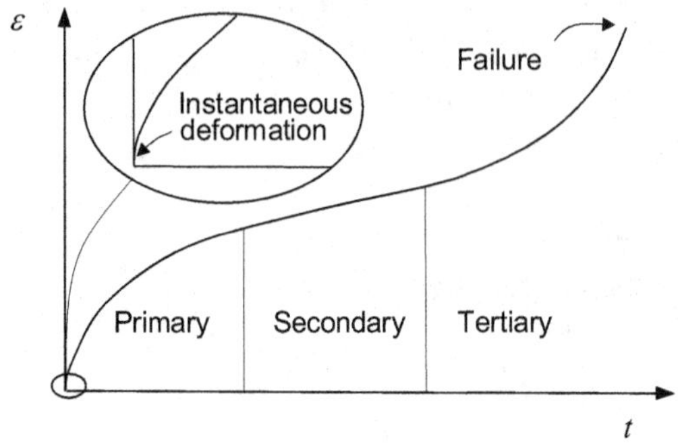


**Figure 9.2** Typical strain-time curve for time-dependent deformation of a body at elevated temperatures under constant load and temperature. The curve is divided into (a) instantaneous deformation, (b) primary (transient) creep, (c) secondary (steady state) creep, and (d) tertiary creep.

During the past ninety years, many different formulae have been proposed for predicting time-dependent strain. Many of the early expressions were collected and published by Lin ([52], pp. 62–64). Very few of the expressions are based on a micromechanical study of the material. Instead, the curve-fitting approach has usually been applied to find a formula that best fits the measured strain–time curve. The formulae have almost always been developed in order to predict the uniaxial strain–time response of a body under constant stress and temperature.

First two simple material models are studied. These models are the Maxwell solid and Kelvin–Voigt solid, which are traditionally introduced in textbooks for basic courses. The Maxwell solid describes viscoplastic deformation, whereas the Kelvin solid is for viscoelasticity. In a uniaxial case, both models can be described by a combination of springs and dashpots. Springs and dashpots may have some pedagogical value, but it has to be kept in mind that real materials do not contain springs and dashpots, which means that a sketch with springs and dashpots does not prove any material model to be correct or false.

Traditional power plants play an important role in modern societies, since they are used in energy production. Estimating their life-span requires knowledge of high-temperature creep in the component materials of power plants. Therefore, this chapter studies material models for high-temperature creep in metallic material. The brief introduction is followed by studies of specific material models later in this section.

High-temperature creep is a time-dependent, thermally activated deformation that occurs in metallic materials above temperatures 30%–40% (in kelvins) of the melting point of the material. There are three mechanisms of high-temperature creep: diffusional creep, grain boundary sliding, and dislocation creep. Diffusional creep is dominant when the temperature is close to the melting point and when the stress level is low. Grain boundary sliding is dominant when the temperature of the material is very close to its melting point. There are rare cases in which grain boundary sliding has engineering importance. However, at slow strain rates, grain boundary sliding is the dominant deformation mechanism of sea ice. Dislocation creep is dominant when the temperature is lower and the stress level is higher than in the case of diffusional creep. Modern power plants operate at temperatures around 550 °C, and the critical components are under a stress level for dislocation creep. Thus, in this chapter the term *high-temperature creep* refers to dislocation creep.

When using commercial or university-based computer programs such as the Abaqus, there are two different ways to utilize a material model. The user can select a suitable model from the existing library of constitutive equations in the codes or the user can prepare a model of their own. In the case of high-temperature creep, there are already several more or less acceptable models for primary and secondary stages of creep. Often these models are already implemented into the structural analysis codes. The complete constitutive equation also needs a model for tertiary creep. Modelling tertiary creep with a description of damage is a more complicated task, and only a few (if any) reliable constitutive equations exist that also cover the tertiary stage. Furthermore, localization of damage is a severe problem for all material models showing softening. Localization is a process in the numerical solution where, during time-stepping, damage tends to concentrate on a small volume of the body. This behavior is unrealistic for high-temperature creep damage. Thus an analysis of high-temperature creep damage requires the preparation of a customized material model. The gradient theory discussed in Section 7.5 is prepared for avoidance of localization. The concepts introduced in Sections 7.5 and 10.5 are valid also when an acceptable damage model for high-temperature creep is prepared.

Whether the researcher is selecting or deriving a material model, they will face the following questions: What is the appearance of an acceptable material model? What are the features of a reliable constitutive equation for creep? How can one check the quality of the model? How can one change this model to be better suited for computer simulation? This is the topic of the following section, which discusses two frequently used and discussed material models for primary and secondary creep. Although these models are often expressed as equivalent, they are hugely different from the material science, from the mathematical and continuum thermodynamic point of view. These differences are discussed in the Section 9.5.

## 9.2 Maxwell Solid

This section evaluates a traditional and simple model for viscoplastic behavior. This model is called the Maxwell solid. A more well known material model is the Maxwell fluid, which has the same spring–dashpot appearance as the Maxwell solid. Actually it is odd to refer to the spring–dashpot combination shown in Figure 9.3 as a model for fluid, since the spring and the dashpot are in series and therefore the total strain $\varepsilon$ is a sum of the strain over the spring $\varepsilon^{\mathrm{e}}$ and over the dashpot $\varepsilon^{\mathrm{v}}$, as shown in Figure 9.3. Usually the response of solids is modelled in this way, whereas fluids prefer summation of the stress $\sigma$, as shown in Expression (6.75).

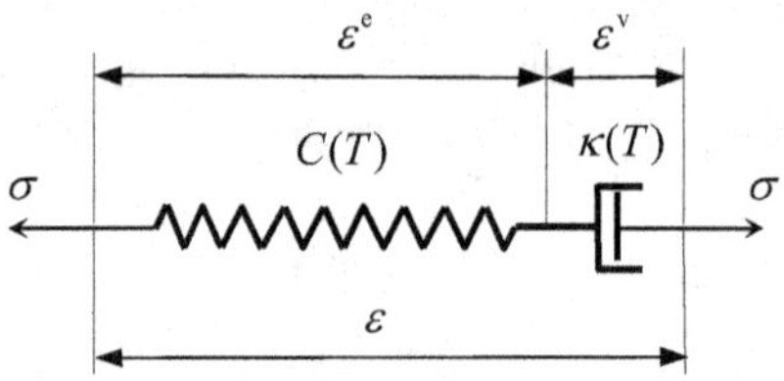


**Figure 9.3** Maxwell material model (also known as the Maxwell solid model).

The uniaxial form of the Maxwell material models can be described by the spring and dashpot combination presented in Figure 9.3. For the Maxwell solid, the specific Helmholtz free energy $\psi$ takes the form

$$\psi(\boldsymbol{\varepsilon}-\boldsymbol{\varepsilon}^{\mathrm{v}},T)=\frac{1}{2\,\rho_0}\,(\,\boldsymbol{\varepsilon}-\boldsymbol{\varepsilon}^{\mathrm{v}}\,):\mathbf{C}(T):(\,\boldsymbol{\varepsilon}-\boldsymbol{\varepsilon}^{\mathrm{v}}\,)\,, \tag{9.1}$$

where $\boldsymbol{\varepsilon}^{\mathrm{v}}$ is the viscous strain tensor describing time-dependent deformation, which can be viscoelastic, viscoplastic (as in this section), or a combination of these two deformation mechanisms. Since, in the Maxwell model, Material Model (9.1), the fourth-order material property tensor $\mathbf{C}(T)$ is the constitutive tensor for a Hookean deformation, it satisfies the condition $C_{ijkl}(T)=C_{klij}(T)$. It is worth noting that Material Model (9.1) for the specific Helmholtz free energy $\psi$ does not describe thermal expansion, as will be discussed next.

Substitution of Material Model (9.1) into State Equations (6.172) and $(6.174)_1$ gives

$$s=-\frac{\partial\psi(\boldsymbol{\varepsilon}-\boldsymbol{\varepsilon}^{\mathrm{v}},T)}{\partial T}=-\frac{1}{2\,\rho_0}\,(\,\boldsymbol{\varepsilon}-\boldsymbol{\varepsilon}^{\mathrm{v}}\,):\frac{\partial\mathbf{C}(T)}{\partial T}:(\,\boldsymbol{\varepsilon}-\boldsymbol{\varepsilon}^{\mathrm{v}}\,) \tag{9.2}$$

and [see Equation (2.131)]

$$\boldsymbol{\sigma}=\rho_0\,\frac{\partial\psi(\boldsymbol{\varepsilon}-\boldsymbol{\varepsilon}^{\mathrm{v}},T)}{\partial(\boldsymbol{\varepsilon}-\boldsymbol{\varepsilon}^{\mathrm{v}})}=\mathbf{C}(T):(\,\boldsymbol{\varepsilon}-\boldsymbol{\varepsilon}^{\mathrm{v}}\,)\,. \tag{9.3}$$

As Result (9.3) shows, the stress–strain relationship depends on temperature $T$, but the form of the dependency does not have the standard form of the thermal expansion being independent of the stress tensor $\boldsymbol{\sigma}$. Thermal expansion can be included in the model by adding to the expression for the specific Helmholtz free energy $\psi$, Material Model (9.1), the first term on the second line of Material Model (8.66). This would replace Constitutive Relation (9.3) with Material Model (8.67) without the reference stress tensor $\boldsymbol{\sigma}_{\mathrm{r}}$. In this case, the expression for the specific entropy $s$ would take a different form as well.

The evaluation continues by studying a case in which the driving force for creep is the normal stress tensor $\boldsymbol{\sigma}$. The specific dissipation function $\varphi$ is modelled as separable into a mechanical part and a thermal part, as follows:

$$\varphi(\dot{\boldsymbol{\varepsilon}}^{\mathrm{V}}, \vec{q}; \boldsymbol{\varepsilon}-\boldsymbol{\varepsilon}^{\mathrm{V}}, T) = \varphi_{\mathrm{mech}}(\dot{\boldsymbol{\varepsilon}}^{\mathrm{V}}; \boldsymbol{\varepsilon}-\boldsymbol{\varepsilon}^{\mathrm{V}}, T) + \varphi_{\mathrm{ther}}(\vec{q}; \boldsymbol{\varepsilon}-\boldsymbol{\varepsilon}^{\mathrm{V}}, T) \,. \tag{9.4}$$

The mechanical part of the problem is modelled by the specific complementary dissipation function $\varphi^{\mathrm{c}}_{\mathrm{mech}}$. It has the following appearance:

$$\varphi^{\mathrm{c}}_{\mathrm{mech}}(\boldsymbol{\sigma}; T) = \frac{1}{2\,\rho_0}\boldsymbol{\sigma} : \boldsymbol{\kappa}(T) : \boldsymbol{\sigma} \,, \tag{9.5}$$

where $\boldsymbol{\kappa}(T)$ is the fourth-order constitutive tensor, having the symmetry property $\kappa_{ijkl}(T) = \kappa_{klij}(T)$. By substituting Material Model (9.5) into Normality Rule (6.286), we arrive at the following result [see Equation (2.131)]:

$$\dot{\boldsymbol{\varepsilon}}^{\mathrm{V}} = \rho_0 \, \frac{\partial \varphi^{\mathrm{c}}_{\mathrm{mech}}(\boldsymbol{\sigma}; T)}{\partial \boldsymbol{\sigma}} = \boldsymbol{\kappa}(T) : \boldsymbol{\sigma} \,. \tag{9.6}$$

If the value for the constitutive tensor for a Hookean deformation $\mathbf{C}(T)$ is constant, that is, $\mathbf{C}(T) = \mathbf{C}$, Expression (9.3) yields

$$\boldsymbol{\sigma} = \mathbf{C} : (\boldsymbol{\varepsilon} - \boldsymbol{\varepsilon}^{\mathrm{V}}) \,, \qquad \text{which gives} \qquad \dot{\boldsymbol{\sigma}} = \mathbf{C} : (\dot{\boldsymbol{\varepsilon}} - \dot{\boldsymbol{\varepsilon}}^{\mathrm{V}}) \,. \tag{9.7}$$

Equation $(9.7)_2$ was obtained by taking a derivative of Equation $(9.7)_1$ with respect to time $t$. Equation (8.43) and Definition (2.106) are recalled:

$$\mathbf{C} : \mathbf{S} = \mathbf{S} : \mathbf{C} = \mathbf{I}^{\mathrm{s}} \quad \text{and} \quad \mathbf{Def} \quad \mathbf{I}^{\mathrm{s}} : \mathbf{c} = \mathbf{c} : \mathbf{I}^{\mathrm{s}} = \frac{1}{2}(\mathbf{c} + \mathbf{c}^{\mathrm{T}}) \,, \tag{9.8}$$

where $\mathbf{c}$ is an arbitrary second-order tensor. Equation $(9.7)_2$ is multiplied by compliance tensor for a Hookean deformation $\mathbf{S}:$ from the left, and thus the following is achieved:

$$\mathbf{S} : \dot{\boldsymbol{\sigma}} = \mathbf{S} : \mathbf{C} : (\dot{\boldsymbol{\varepsilon}} - \dot{\boldsymbol{\varepsilon}}^{\mathrm{V}}) \,, \qquad \text{which gives} \qquad \mathbf{S} : \dot{\boldsymbol{\sigma}} = \dot{\boldsymbol{\varepsilon}} - \dot{\boldsymbol{\varepsilon}}^{\mathrm{V}} \,, \tag{9.9}$$

where Expressions (9.8) and the symmetry of the strain rate tensors $\dot{\boldsymbol{\varepsilon}}$ and $\dot{\boldsymbol{\varepsilon}}^{\mathrm{V}}$, that is, $(\dot{\boldsymbol{\varepsilon}})^{\mathrm{T}} = \dot{\boldsymbol{\varepsilon}}$, and $(\dot{\boldsymbol{\varepsilon}}^{\mathrm{V}})^{\mathrm{T}} = \dot{\boldsymbol{\varepsilon}}^{\mathrm{V}}$, are exploited.

Substitution of Rate (9.6) into Expression $(9.9)_2$ leads to

$$\dot{\boldsymbol{\varepsilon}} = \mathbf{S} : \dot{\boldsymbol{\sigma}} + \boldsymbol{\kappa}(T) : \boldsymbol{\sigma} \,. \tag{9.10}$$

Rate Equation (9.10) is a candidate for the three-dimensional counterpart to the Maxwell spring–dashpot assembly. It describes Hookean deformation and viscoplastic creep. The normal stress $\sigma$ is the driving force for the volumetric part of the creep strain rate $\dot{\varepsilon}_{ii}$. This feature is obtained by setting the values of the components of the constitutive tensor $\boldsymbol{\kappa}(T)$ properly.

In a uniaxial case, Material Model (9.10) takes the form [if the values of the components of the constitutive tensor $\boldsymbol{\kappa}(T)$ are set properly]

$$\dot{\varepsilon} = S\,\dot{\sigma} + \kappa(T)\,\sigma\,. \tag{9.11}$$

Equation (9.11) defines the Maxwell model (cf. Flügge ([22], pp. 6 and 22; Reddy [84], Section 9.2).

Material Model (9.5) fulfills the following relationship:

$$\varphi^{\mathrm{c}}_{\mathrm{mech}}(k\,\boldsymbol{\sigma}; T) = k^2\,\varphi^{\mathrm{c}}_{\mathrm{mech}}(\boldsymbol{\sigma}; T)\,. \tag{9.12}$$

Result (9.12) implies that the specific complementary dissipation function $\varphi^{\mathrm{c}}_{\mathrm{mech}}$ is a homogeneous function of degree 2. From the physics point of view, the energy has to be nonnegative, and therefore the components of the tensor $\boldsymbol{\kappa}(T)$ have to take values that lead to the nonnegative energy. This means that the specific complementary dissipation function $\varphi^{\mathrm{c}}_{\mathrm{mech}}$ is a nonnegative function. Thus, according to Section 7.2, the Clausius–Duhem inequality is satisfied.

Although the response of the material obeying Constitutive Relation (9.10) shows viscoplastic behavior, the preceding material model is not applicable to creep of metallic materials of which the driving force is shear stress $\tau$. The model derived next fulfills this requirement. Nor is the material model in Material Model (9.10) applicable to polymers, since polymers show viscoelastic creep behavior.

For the latter type of creep deformation, where the driving force for creep is the deviatoric stress tensor $\mathbf{s}$; that is, the shear stress $\tau$ is the driving force for creep. The mechanical part of the specific complementary dissipation function $\varphi^{\mathrm{c}}_{\mathrm{mech}}$ is assumed to take the following appearance:

$$\varphi^{\mathrm{c}}_{\mathrm{mech}}(\boldsymbol{\sigma}; T) = \frac{\kappa(T)}{2\,\rho_0}\,\mathbf{s}:\mathbf{s} = \frac{\kappa(T)}{2\,\rho_0}\,(\mathbf{K}:\boldsymbol{\sigma}):(\mathbf{K}:\boldsymbol{\sigma})\,, \tag{9.13}$$

where $\kappa(T)$ is a scalar-valued material parameter for the viscous part of the response. Equation (9.13) exploits Definition (2.108) of the deviatoric stress tensor $\mathbf{s}$,

$$\mathbf{s} := \mathbf{K}:\boldsymbol{\sigma}\,, \quad \text{where the fourth-order tensor} \quad \mathbf{K} : = \mathbf{I} - \tfrac{1}{3}\,\mathbf{1}\,\mathbf{1}\,. \tag{9.14}$$

Derivative (2.142) is recalled:

$$\frac{\partial}{\partial\boldsymbol{\sigma}}\{\mathbf{s}:\mathbf{s}\} = 2\,\mathbf{s}\,. \tag{9.15}$$

By substituting Material Model (9.13) into Normality Rule (6.286) and by taking Derivative (9.15) into account, the following result is obtained:

$$\dot{\boldsymbol{\varepsilon}}^{\mathrm{v}} = \rho_0 \, \frac{\partial \varphi^{\mathrm{c}}_{\mathrm{mech}}(\boldsymbol{\sigma};T)}{\partial \boldsymbol{\sigma}} = \rho_0 \, \frac{\partial}{\partial \boldsymbol{\sigma}} \left[ \frac{\kappa(T)}{2\,\rho_0} \, \mathbf{s}:\mathbf{s} \right] = \frac{\kappa(T)}{2} \, \frac{\partial}{\partial \boldsymbol{\sigma}} \{\mathbf{s}:\mathbf{s}\} = \kappa(T)\, \mathbf{s}\,. \tag{9.16}$$

By introducing the notation $\kappa(T) := \overset{\circ}{\varepsilon}_{\mathrm{re}}(T)/\sigma_{\mathrm{re}}$, Rate (9.16) takes the form

$$\dot{\boldsymbol{\varepsilon}}^{\mathrm{v}} = \overset{\circ}{\varepsilon}_{\mathrm{re}}(T) \, \frac{\mathbf{s}}{\sigma_{\mathrm{re}}}\,, \tag{9.17}$$

where $\overset{\circ}{\varepsilon}_{\mathrm{re}}(T)$ is the reference strain rate and $\sigma_{\mathrm{re}}$ is the reference stress making the term $\mathbf{s}/\sigma_{\mathrm{re}}$ dimensionless. The role of the quantity $\sigma_{\mathrm{re}}$ is discussed in more detail in Section 9.4, when more realistic material models are evaluated. Expression (9.17) has the correct form for creep. The reference strain rate $\overset{\circ}{\varepsilon}_{\mathrm{re}}(T)$ has the correct dimension, and the driving force for creep is the deviatoric stress tensor $\mathbf{s}$, which is the three-dimensional counterpart for the shear stress $\tau$.

If the value for the constitutive tensor for a Hookean deformation $\mathbf{C}(T)$ is constant, that is, $\mathbf{C}(T) = \mathbf{C}$, Expression (9.3) yields

$$\boldsymbol{\sigma} = \mathbf{C}:(\boldsymbol{\varepsilon} - \boldsymbol{\varepsilon}^{\mathrm{v}})\,, \qquad \text{which gives} \qquad \dot{\boldsymbol{\sigma}} = \mathbf{C}:(\dot{\boldsymbol{\varepsilon}} - \dot{\boldsymbol{\varepsilon}}^{\mathrm{v}})\,. \tag{9.18}$$

In the manipulation from Equation $(9.18)_1$ to Equation $(9.18)_2$, a derivative with respect to time $t$ is created. Equation (8.43) and Definition (2.106) are recalled:

$$\mathbf{C}:\mathbf{S} = \mathbf{S}:\mathbf{C} = \mathbf{I}^{\mathrm{s}} \quad \text{and} \quad \mathbf{Def} \quad \mathbf{I}^{\mathrm{s}}:\mathbf{c} = \mathbf{c}:\mathbf{I}^{\mathrm{s}} = \frac{1}{2}\,(\mathbf{c} + \mathbf{c}^{\mathrm{T}})\,, \tag{9.19}$$

where $\mathbf{c}$ is an arbitrary second-order tensor. Equation $(9.18)_2$ is multiplied by compliance tensor for a Hookean deformation $\mathbf{S}:$ from the left, and thus the following is achieved:

$$\mathbf{S}:\dot{\boldsymbol{\sigma}} = \mathbf{S}:\mathbf{C}:(\dot{\boldsymbol{\varepsilon}} - \dot{\boldsymbol{\varepsilon}}^{\mathrm{v}})\,, \qquad \text{which gives} \qquad \mathbf{S}:\dot{\boldsymbol{\sigma}} = \dot{\boldsymbol{\varepsilon}} - \dot{\boldsymbol{\varepsilon}}^{\mathrm{v}}\,, \tag{9.20}$$

where Expressions (9.19) and the symmetry of the strain rate tensors $\dot{\boldsymbol{\varepsilon}}$ and $\dot{\boldsymbol{\varepsilon}}^{\mathrm{v}}$, that is, $(\dot{\boldsymbol{\varepsilon}})^{\mathrm{T}} = \dot{\boldsymbol{\varepsilon}}$, and $(\dot{\boldsymbol{\varepsilon}}^{\mathrm{v}})^{\mathrm{T}} = \dot{\boldsymbol{\varepsilon}}^{\mathrm{v}}$, are exploited.

Substitution of Rate (9.17) into Material Model $(9.20)_2$ leads to

$$\dot{\boldsymbol{\varepsilon}} = \mathbf{S}:\dot{\boldsymbol{\sigma}} + \overset{\circ}{\varepsilon}_{\mathrm{re}}(T) \, \frac{\mathbf{s}}{\sigma_{\mathrm{re}}}\,. \tag{9.21}$$

In the uniaxial case, Material Model (9.21) takes the form [cf. Equation (2.115)]

$$\dot{\varepsilon} = S\,\dot{\sigma} + \overset{\circ}{\varepsilon}_{\mathrm{re}}(T) \, \frac{\tau}{\sigma_{\mathrm{re}}}\,. \tag{9.22}$$

Expression (9.22) defines the Maxwell model (cf. Flügge [22], pp. 6 and 22).

Section 9.8 will show that, when the driving force for creep is the deviatoric stress tensor $\mathbf{s}$, creep deformation (creep rate $\dot{\boldsymbol{\varepsilon}}^{\mathrm{v}}$ ) does not include a volumetric change of matter. This means that the deviatoric stress tensor $\mathbf{s}$ has a pure shear character. This was the reason in Equation (9.22) for reducing the tensor $\mathbf{s}$ to the shear stress $\tau$. The viscous deformation in Material Model (9.21) describes a permanent, that is, viscoplastic, response. Thus, Material Model (9.21) is the multiaxial form of the Maxwell solid for Hookean deformation and dislocation creep in metallic materials.

Material Model (9.13) fulfills the following relationship:

$$\varphi^{\mathrm{c}}_{\mathrm{mech}}(k\,\boldsymbol{\sigma};T) = k^2\,\varphi^{\mathrm{c}}_{\mathrm{mech}}(\boldsymbol{\sigma};T)\,. \tag{9.23}$$

Result (9.23) implies that the specific complementary dissipation function $\varphi^{\mathrm{c}}_{\mathrm{mech}}$ is a homogeneous function of degree 2. From the physics point of view, the energy has to be nonnegative, and therefore the components of the material parameter $\kappa(T)$ have to be positive, which leads to the nonnegative energy. This means that the specific complementary dissipation function $\varphi^{\mathrm{c}}_{\mathrm{mech}}$ is a nonnegative function. Thus, according to Section 7.2, Clausius–Duhem inequality is satisfied.

## 9.3 Kelvin–Voigt Solid

This section evaluates another traditional and simple model for viscous behavior. This model is called the Kelvin–Voigt solid. The Kelvin–Voigt solid can be described by the spring and dashpot combination presented in Figure 9.4.

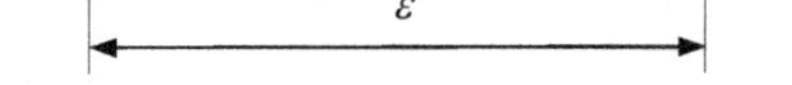

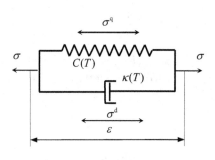

**Figure 9.4** Kelvin–Voigt solid.

As Figure 9.4 shows, the "total" stress tensor $\boldsymbol{\sigma}$ is the sum of two stress tensors. According to Section 7.4, these stress tensors are called the quasi-conservative stress tensor and the dissipative stress tensor, denoted, respectively, by $\boldsymbol{\sigma}^{\mathrm{q}}$ and $\boldsymbol{\sigma}^{\mathrm{d}}$:

$$\boldsymbol{\sigma} = \boldsymbol{\sigma}^{\mathrm{q}} + \boldsymbol{\sigma}^{\mathrm{d}}\,. \tag{9.24}$$

For the Kelvin–Voigt solid, the specific Helmholtz free energy $\psi$ takes the form

$$\psi(\boldsymbol{\varepsilon},T) = \frac{1}{2\,\rho_0}\,\boldsymbol{\varepsilon} : \mathbf{C}(T) : \boldsymbol{\varepsilon}\,, \tag{9.25}$$

where $\mathbf{C}(T)$ is the fourth-order constitutive tensor for a Hookean deformation having the major-symmetry property $C_{ijkl}(T) = C_{klij}(T)$ [cf. Expression (2.93)]. It is worth noting that Form (9.1) for the specific Helmholtz free energy

$\psi$ does not describe thermal expansion. See a more detailed discussion after Equation (9.3).

Substitution of Material Model (9.25) into State Equation $(7.46)_1$ gives

$$\boldsymbol{\sigma}^{\mathrm{q}} = \rho_0 \, \frac{\partial \psi(\boldsymbol{\varepsilon}, \boldsymbol{\alpha}, s, h(\vec{X}))}{\partial \boldsymbol{\varepsilon}} = \mathbf{C}(T) : \boldsymbol{\varepsilon} \, . \tag{9.26}$$

Clausius–Duhem Inequality (7.64) becomes

$$\boldsymbol{\sigma}^{\mathrm{d}} : \dot{\boldsymbol{\varepsilon}} + \boldsymbol{\beta} : \dot{\boldsymbol{\alpha}} - \frac{\vec{\nabla} T}{T} \cdot \vec{q} \, \geq \, 0 \, . \tag{9.27}$$

For the present material model, Clausius–Duhem Inequality (9.27) reduces to

$$\boldsymbol{\sigma}^{\mathrm{d}} : \dot{\boldsymbol{\varepsilon}} - \frac{\vec{\nabla} T}{T} \cdot \vec{q} \geq 0 \, . \tag{9.28}$$

The specific dissipation function $\varphi$ is modelled as separable into a mechanical part and a thermal part, that is,

$$\varphi(\dot{\boldsymbol{\varepsilon}}, \vec{q}; \boldsymbol{\varepsilon}, T) = \varphi_{\mathrm{mech}}(\dot{\boldsymbol{\varepsilon}}; \boldsymbol{\varepsilon}, T) + \varphi_{\mathrm{ther}}(\vec{q}; \boldsymbol{\varepsilon}, T) \, . \tag{9.29}$$

The present model does not include any thermal part, as Figure 9.4 illustrates. This is obvious, since heat conduction, for example, is not possible to express by a spring/dashpot model. Therefore only the mechanical part is evaluated.

In the first case, the driving force for creep is modelled to be the normal dissipative stress tensor $\boldsymbol{\sigma}^{\mathrm{d}}$. The mechanical part of the problem is modelled by the specific complementary dissipation function $\varphi^{\mathrm{c}}_{\mathrm{mech}}$. It has the following appearance:

$$\varphi^{\mathrm{c}}_{\mathrm{mech}}(\boldsymbol{\sigma}^{\mathrm{d}}; T) = \frac{1}{2\,\rho_0} \, \boldsymbol{\sigma}^{\mathrm{d}} : \boldsymbol{\kappa}(T) : \boldsymbol{\sigma}^{\mathrm{d}} \, , \tag{9.30}$$

where $\boldsymbol{\kappa}(T)$ is the fourth-order constitutive tensor having the major symmetry property, that is, $\kappa_{ijkl}(T) = \kappa_{klij}(T)$. By substituting Material Model (9.30) into Normality Rule (7.70), the following result is obtained:

$$\dot{\boldsymbol{\varepsilon}} = \rho_0 \, \frac{\partial \varphi^{\mathrm{c}}_{\mathrm{mech}}(\boldsymbol{\sigma}^{\mathrm{d}}; \boldsymbol{\varepsilon}, T)}{\partial \boldsymbol{\sigma}^{\mathrm{d}}} = \boldsymbol{\kappa}(T) : \boldsymbol{\sigma}^{\mathrm{d}} \, , \tag{9.31}$$

where Result (2.131) is exploited. Substitution of quasi-conservative stress tensor $\boldsymbol{\sigma}^{\mathrm{q}}$ from Equation (9.26) and dissipative stress tensor $\boldsymbol{\sigma}^{\mathrm{d}}$ obtained from Equation (9.31) into Separation (9.24) yields

$$\boldsymbol{\sigma} = \mathbf{C}(T) : \boldsymbol{\varepsilon} + [\boldsymbol{\kappa}(T)]^{-1} : \dot{\boldsymbol{\varepsilon}} \, . \tag{9.32}$$

The uniaxial counterpart for Constitutive Equation (9.32) is

$$\sigma = C(T) \, \varepsilon + \frac{1}{\kappa(T)} \, \dot{\varepsilon} \, . \tag{9.33}$$

Expression (9.33) defines the traditionally expressed form for the Kelvin–Voigt solid (see, e.g., Flügge [22], pp. 9 and 22).

Material Model (9.32) is a good candidate for a constitutive model for creep response of amorphous polymers. It describes a response that is the sum of Hookean and viscoelastic deformation. According to Hertzberg et al. ([28], pp. 17 and 229), "the elastic response of elastomers is approximately linear up to about 1%" and "amorphous high-molecular-weight polymer chains are highly kinked in the unloaded state. When a chain is straightened under load, there is a strong entropic driving force to rekink it once the load is removed. This provides a driving force for viscoelastic strain recovery in amorphous polymers." Based on Hertzberg et al. ([28], Figure 4.37), there is remarkable volume change during the creep process of polymer materials. These are the vital features of Material Model (9.33).

The present model has not the standard character of the material models for solid material, since it assumes stress tensor separation, as expressed in Equation (9.24). Viscoelastic response is possible to be modelled with material models having traditional appearance. Good candidates for description of viscoelastic response are models having the form of the model by Le Gac and Duval [48], although here it is used for modelling viscoplastic creep of high-temperature power plant material.

Material Model (9.30) fulfills the following relationship:

$$\varphi^{\mathrm{c}}_{\mathrm{mech}}(k\,\boldsymbol{\sigma}^{\mathrm{d}};\boldsymbol{\varepsilon},T) = k^2\,\varphi^{\mathrm{c}}_{\mathrm{mech}}(\boldsymbol{\sigma}^{\mathrm{d}};\boldsymbol{\varepsilon},T)\,. \tag{9.34}$$

Result (9.34) implies that the specific complementary dissipation function $\varphi^{\mathrm{c}}_{\mathrm{mech}}$ is a homogeneous function of degree 2. From the physics point of view, the energy has to be nonnegative and therefore the components of the tensor $\boldsymbol{\kappa}(T)$ have to take values that lead to the nonnegative energy. This means that the specific complementary dissipation function $\varphi^{\mathrm{c}}_{\mathrm{mech}}$ is a nonnegative function. Thus, according to Section 7.2, the Clausius–Duhem inequality is satisfied. A Kelvin–Voigt solid is a material response without elastic deformation described in Section 7.4.

In the second case, the driving force for creep is modelled to be the deviatoric dissipative stress tensor $\mathbf{s}^{\mathrm{d}}$. The mechanical part of the problem is modelled by the specific complementary dissipation function $\varphi^{\mathrm{c}}_{\mathrm{mech}}$ having the form

$$\varphi^{\mathrm{c}}_{\mathrm{mech}}(\boldsymbol{\sigma}^{\mathrm{d}};T) = \frac{\overset{\circ}{\varepsilon}_{\mathrm{re}}\,\sigma_{\mathrm{re}}}{\rho_0\,(n+1)}\left[\frac{J_{\mathrm{vM}}(\boldsymbol{\sigma}^{\mathrm{d}})}{\sigma_{\mathrm{re}}}\right]^{(n+1)}, \tag{9.35}$$

where $\overset{\circ}{\varepsilon}_{\mathrm{re}}$ is a material parameter and the role of the material parameter $\sigma_{\mathrm{re}}$ is to make the quantity within the parentheses dimensionless. The operator $J_{\mathrm{vM}}(\cdot)$ is the von Mises operator given in Definition (2.124). When applied to the dissipative stress tensor $\boldsymbol{\sigma}^{\mathrm{d}}$, it takes the following appearance:

$$J_{\mathrm{vM}}(\boldsymbol{\sigma}^{\mathrm{d}}) := \sqrt{\tfrac{3}{2}\,\mathbf{s}^{\mathrm{d}} : \mathbf{s}^{\mathrm{d}}}\,. \tag{9.36}$$

By substituting Material Model (9.35) into Normality Rule (7.70), the following is obtained:

$$\dot{\boldsymbol{\varepsilon}} = \rho_0 \frac{\partial \varphi^{\mathrm{c}}_{\mathrm{mech}}(\boldsymbol{\sigma}^{\mathrm{d}}; \boldsymbol{\varepsilon}, T)}{\partial \boldsymbol{\sigma}^{\mathrm{d}}} = \overset{\circ}{\varepsilon}_{\mathrm{re}} \left[ \frac{J_{\mathrm{vM}}(\boldsymbol{\sigma}^{\mathrm{d}})}{\sigma_{\mathrm{re}}} \right]^n \frac{\mathbf{s}^{\mathrm{d}}}{J_{\mathrm{vM}}(\boldsymbol{\sigma}^{\mathrm{d}})}, \tag{9.37}$$

where Expression (2.152) is exploited.

Equation (9.24) and Result (9.26) are recalled:

$$\boldsymbol{\sigma} = \boldsymbol{\sigma}^{\mathrm{q}} + \boldsymbol{\sigma}^{\mathrm{d}} \qquad \text{and} \qquad \boldsymbol{\sigma}^{\mathrm{q}} = \mathbf{C}(T) : \boldsymbol{\varepsilon}\,. \tag{9.38}$$

Equations (9.37) and (9.38) form the material model.

Based on Expression (9.35) and Definition (9.36), the specific complementary dissipation function $\varphi^{\mathrm{c}}_{\mathrm{mech}}$ is a homogeneous function of degree $n+1$, that is, the following holds: $\varphi^{\mathrm{c}}_{\mathrm{mech}}(k\,\boldsymbol{\sigma}^{\mathrm{d}}; \boldsymbol{\varepsilon}, T) = k^{(n+1)}\,\varphi^{\mathrm{c}}_{\mathrm{mech}}(\boldsymbol{\sigma}^{\mathrm{d}}; \boldsymbol{\varepsilon}, T)$. It is a nonnegative function as well. Therefore, according to Section 7.2, the Clausius–Duhem inequality is satisfied.

The application of Material Model (9.37) and (9.38) requires significant computer capacity. If the solution is known at time $t$ and is sought for the time $t + \Delta t$, Abaqus finite element code, for example, provides, besides other information, the time increment $\Delta t$, the value of the strain tensor $\boldsymbol{\varepsilon}$ at time $t$, and the approximation of the strain tensor increment $\Delta\boldsymbol{\varepsilon}$ for use in the material subroutine, and the task for the subroutine is to compute the value of the stress tensor $\boldsymbol{\sigma}$ at time $t+\Delta t$. Based on Material Model $(9.38)_2$, the value of the quasi-conservative stress tensor $\boldsymbol{\sigma}^{\mathrm{q}}$ at the end of increment $\Delta t$ can be computed. The strain rate $\dot{\boldsymbol{\varepsilon}}$ can be computed using expression $\dot{\boldsymbol{\varepsilon}} = \Delta\boldsymbol{\varepsilon}/\Delta t$. The value of the dissipative stress tensor $\boldsymbol{\sigma}^{\mathrm{d}}$ is now obtained from Equation (9.37), but an iterative process is required. When the value of the dissipative stress tensor $\boldsymbol{\sigma}^{\mathrm{d}}$ is computed, Material Model $(9.38)_1$ is used to compute the value for the stress tensor $\boldsymbol{\sigma}$ at time $t+\Delta t$. The problem with Material Model (9.37) and (9.38) is that they require the earlier described iteration, which takes computing time. Thus, instead of Material Model (9.37) and (9.38), that is, Kelvin–Voigt solid, the model by Le Gac and Duval [48] described in Section 9.6 is proposed.

## 9.4 Norton's Law

This section studies Norton's law, presented by Bailey [9] and Norton [68]. Unfortunately, often Norton's law is written in the form

$$\dot{\varepsilon}^{\mathrm{v}} = \overset{\circ}{\varepsilon}_{\mathrm{re}}\,\sigma^n\,. \tag{9.39}$$

Material Model (9.39) is not acceptable, as will be shown next.

If Material Model (9.39) were used, the parameter $\overset{\circ}{\varepsilon}_{\mathrm{re}}$ would take strange units that would be dependent on the value of the exponent $n$. The following simple example displays the problem. The exponent $n$ is assumed to have the value 3.427, and the stress $\sigma$ is represented in MPa. This leads to the conclusion that the unit for the parameter $\overset{\circ}{\varepsilon}_{\mathrm{re}}$ should be $(1/\mathrm{MPa})^{3.427}1/\mathrm{s}$. If the value for the parameter $n$ was known within a tolerance (as it should be), what would be the unit for the parameter $\overset{\circ}{\varepsilon}_{\mathrm{re}}$? This implies that Form (9.39) is not acceptable.

The correct form for the uniaxial constitutive equation called Norton's law can be written as

$$\dot{\varepsilon}^{\mathrm{v}} = \overset{\circ}{\varepsilon}_{\mathrm{re}} \left(\frac{\sigma}{\sigma_{\mathrm{re}}}\right)^{n}. \tag{9.40}$$

The notations $\overset{\circ}{\varepsilon}_{\mathrm{re}}$, $\sigma_{\mathrm{re}}$, and $n$ are material parameters. In Material Model (9.40), the reference stress $\sigma_{\mathrm{re}}$ is a parameter the value of which can be fixed before determination of the values for the reference strain rate $\overset{\circ}{\varepsilon}_{\mathrm{re}}$ and stress exponent $n$. It is important to note that Form (9.40) does not have more parameters to fit by a curve fitting procedure than Material Model (9.39). In principle, the value for $\sigma_{\mathrm{re}}$ is arbitrary. The role of $\sigma_{\mathrm{re}}$ is to make the quantity $\sigma/\sigma_{\mathrm{re}}$ dimensionless. However, in order to reduce the error caused by the inaccuracy of the value for the exponent $n$, the value of $\sigma_{\mathrm{re}}$ should be chosen so that the ratio $\sigma/\sigma_{\mathrm{re}}$ takes values close to unity. This means that if there were three data curves with the stress values 40 MPa, 60 MPa, and 80 MPa, a recommended value for $\sigma_{\mathrm{re}}$ would be 60 MPa. It is worth noting that the parameters $\overset{\circ}{\varepsilon}_{\mathrm{re}}$ and $n$ have a similar effect on the viscous strain rate $\dot{\varepsilon}^{\mathrm{v}}$. Increasing values for $\overset{\circ}{\varepsilon}_{\mathrm{re}}$ and $n$ lead to an increasing value for the viscous strain rate $\dot{\varepsilon}^{\mathrm{v}}$. Only results obtained by different values of the stress $\sigma$ can display the different roles of parameters $\overset{\circ}{\varepsilon}_{\mathrm{re}}$ and $n$ in Creep Law (9.40).

It is necessary to show the dependence of the parameters on the variables in the model. Therefore, if the values for the parameters $\overset{\circ}{\varepsilon}_{\mathrm{re}}$ and $n$ were obtained as dependent on the stress $\sigma$ (i.e., they take different values for the different stresses $\sigma$), Material Model (9.40) should be replaced by the constitutive equation

$$\dot{\varepsilon}^{\mathrm{v}} = \overset{\circ}{\varepsilon}_{\mathrm{re}}(\sigma) \left(\frac{\sigma}{\sigma_{\mathrm{re}}}\right)^{n(\sigma)}, \tag{9.41}$$

where the notations $\overset{\circ}{\varepsilon}_{\mathrm{re}}(\sigma)$ and $n(\sigma)$ show the dependence on the stress $\sigma$. It should be kept in mind that Material Models (9.40) and (9.41) represent two different constitutive equations.

On the other hand, the parameters $\overset{\circ}{\varepsilon}_{\mathrm{re}}$ and $n$ in Material Model (9.40) may be dependent on temperature $T$. Since temperature $T$ is not a "visible" variable in the model, the temperature dependence of the parameters $\overset{\circ}{\varepsilon}_{\mathrm{re}}$ and $n$ does not have to be shown. Of course, if the researcher wishes to stress the temperature dependence, they can replace Material Model (9.40) by the constitutive equation having the appearance

$$\dot{\varepsilon}^{\mathrm{v}} = \overset{\circ}{\varepsilon}_{\mathrm{re}}(T) \left(\frac{\sigma}{\sigma_{\mathrm{re}}}\right)^{n(T)}. \tag{9.42}$$

In contrast to Material Models (9.40) and (9.41), Material Models (9.40) and (9.42) can be equal. Material parameters are sometimes called (material) constants. This is acceptable terminology. However, it should be remembered that those constants are constant with respect to the variables of the model but can be dependent on other variables, as already discussed. This means that an existing model can adopt new variables and the obtained model can be an extended model from the present one.

Later, Odqvist [72] extended Uniaxial Material Model (9.40) to multiaxial cases by writing

$$\dot{\boldsymbol{\varepsilon}}^{\mathrm{v}} = \frac{3}{2} \overset{\circ}{\varepsilon}_{\mathrm{re}} \left[ \frac{J_{\mathrm{vM}}(\boldsymbol{\sigma})}{\sigma_{\mathrm{re}}} \right]^{n} \frac{\mathbf{s}}{J_{\mathrm{vM}}(\boldsymbol{\sigma})} . \tag{9.43}$$

In this study, the thermoelastic strain $\boldsymbol{\varepsilon}^{\mathrm{eT}}$ is added into the material model; therefore, the (total) strain $\boldsymbol{\varepsilon}$ is obtained from

$$\boldsymbol{\varepsilon} = \boldsymbol{\varepsilon}^{\mathrm{eT}} + \boldsymbol{\varepsilon}^{\mathrm{v}} . \tag{9.44}$$

In this section, the thermoelastic strain tensor $\boldsymbol{\varepsilon}^{\mathrm{eT}}$ is not studied, but that work was carried out in Section 8.3.3 on thermoelastic deformation. Should the reader wish to apply Material Model (9.44), they can copy the form for the thermoelastic strain tensor $\boldsymbol{\varepsilon}^{\mathrm{eT}}$ from there. The state is described by the state variables $(\boldsymbol{\varepsilon}, \boldsymbol{\varepsilon}^{\mathrm{v}}, T)$.

The following specific Helmholtz free energy $\psi$ and the specific complementary dissipation function $\varphi^{\mathrm{c}}_{\mathrm{mech}}$ are written

$$\psi(\boldsymbol{\varepsilon}, \boldsymbol{\varepsilon}^{\mathrm{v}}, T) = \psi^{\mathrm{eT}}(\boldsymbol{\varepsilon} - \boldsymbol{\varepsilon}^{\mathrm{v}}, T) \tag{9.45}$$

and

$$\varphi^{\mathrm{c}}_{\mathrm{mech}}(\boldsymbol{\sigma}) = \frac{\overset{\circ}{\varepsilon}_{\mathrm{re}}\, \sigma_{\mathrm{re}}}{\rho_0\, (n+1)} \left[ \frac{J_{\mathrm{vM}}(\boldsymbol{\sigma})}{\sigma_{\mathrm{re}}} \right]^{(n+1)} . \tag{9.46}$$

Substitution of Material Model (9.46) into Normality Rule (6.286) yields

$$\dot{\boldsymbol{\varepsilon}}^{\mathrm{v}} = \rho_0 \frac{\partial \varphi^{\mathrm{c}}_{\mathrm{mech}}(\boldsymbol{\sigma})}{\partial \boldsymbol{\sigma}} = \frac{3}{2} \overset{\circ}{\varepsilon}_{\mathrm{re}} \left[ \frac{J_{\mathrm{vM}}(\boldsymbol{\sigma})}{\sigma_{\mathrm{re}}} \right]^{n} \frac{\mathbf{s}}{J_{\mathrm{vM}}(\boldsymbol{\sigma})} . \tag{9.47}$$

Derivation (9.47) exploits Result (2.152).

The specific complementary dissipation function $\varphi^{\mathrm{c}}_{\mathrm{mech}}$ given by Material Model (9.46) obeys the following property:

$$\varphi^{\mathrm{c}}_{\mathrm{mech}}(k\, \boldsymbol{\sigma}) = k^{(n+1)}\, \varphi^{\mathrm{c}}_{\mathrm{mech}}(\boldsymbol{\sigma}) . \tag{9.48}$$

This means that the specific complementary dissipation function $\varphi^{\mathrm{c}}_{\mathrm{mech}}$ is a homogeneous function of degree $(n+1)$. It also is nonnegative. Thus, based on Section 7.2, the Clausius–Duhem inequality is satisfied.

## 9.5 Time-Hardening and Strain-Hardening Models for Primary Creep

The creep model introduced by Costa Andrade ([5], p. 11) is studied next. The original appearance, viz.

$$\dot{\varepsilon}^{\mathrm{v}} = A\,t^{-2/3} + \text{ other terms}, \tag{9.49}$$

is replaced by

$$\dot{\varepsilon}^{\mathrm{v}} = A\left(\frac{t}{t_0}\right)^{-2/3} + \text{ other terms}. \tag{9.50}$$

In Equation (9.50), $t_0$ is a parameter the value of which can be fixed before determination of the values for the parameter $A$. This is the same as was discussed when evaluating Norton's law in Section 9.4.

Material Model (9.50) can be generalized by writing (without "other terms")

$$\dot{\varepsilon}^{\mathrm{v}} = A(\sigma)\left(\frac{t}{t_0}\right)^{-k}, \tag{9.51}$$

where $k$ is a positive number less than 1 and the other terms are dropped out. The modifications of Form (9.51) are widely used today.

One of the problems in Material Model (9.51) is that at the moment $t = 0$, the value for the viscous strain rate $\dot{\varepsilon}^{\mathrm{v}}$ tends towards infinity. However, the quantity $(t/t_0)^{-k}$ $(0 < k < 1)$ has a weak singularity at the point $t = 0$. This means that $(t/t_0)^{-k}$ is integrable and therefore that the viscous strain $\varepsilon^{\mathrm{v}}$ is bounded (i.e., it has a finite value). The problem arises when finite element or finite difference codes are used. Such programs carry out the time integration numerically. Usually numerical methods utilize the derivative of the integrand at the beginning of the time step. For the first time step (starting from $t = 0$), this derivative is unbounded, and therefore no numerical integration is possible. This problem can be solved through careful modelling.

The second problem of Form (9.51) is the starting shot to measure time $t$, that is, the definition of the moment $t = 0$. The natural time scale is the one that started with the Big Bang. It is not for creep models. The other possibility is to measure time from the moment the load was applied. The difficulties related to this interpretation will be discussed next.

Because of the nature of the creep process, the function $A(\sigma)$ is a monotonically increasing function. This means that if the stress $\sigma$ tends toward zero, the value of the function $A(\sigma)$ approaches zero and, according to Form (9.51), the viscous strain rate $\dot{\varepsilon}^{\mathrm{v}}$ approaches zero as well. Two different loadings denoted by 1 and 2, shown by Figure 9.5(a), are studied. Load 1 immediately takes the value $\sigma_1$, whereas load 2 is extremely low up to the moment $t = t_1$ and then takes the value $\sigma_1$. Time-Hardening Material Model (9.51) gives strain–time curves with the forms sketched in Figure 9.5(b). The paradox in

Figure 9.5(b) is that the shapes of the strain–time curves differ from each other. They should be equal, since the extremely low loading "equals" no loading at all, therefore it can have virtually no effect on the strain–time relationship. This paradox can be solved by the following interpretation: Material Model (9.51) is prepared for constant stress cases, that is, for $\sigma =$ constant, and $t = 0$ is the moment of application of the constant stress. This means that the varying stress has to be described by a sequence of constant stresses, that is, by a staircase function, as shown in Figure 9.6. It is not a big problem for computational mechanics, since in the numerical integration procedures, the stress–time curve has to be approximated in any case. On the other hand, the interpretation of Material Model (9.51) for varying stress–time dependence is a difficult task. This problem is discussed next.

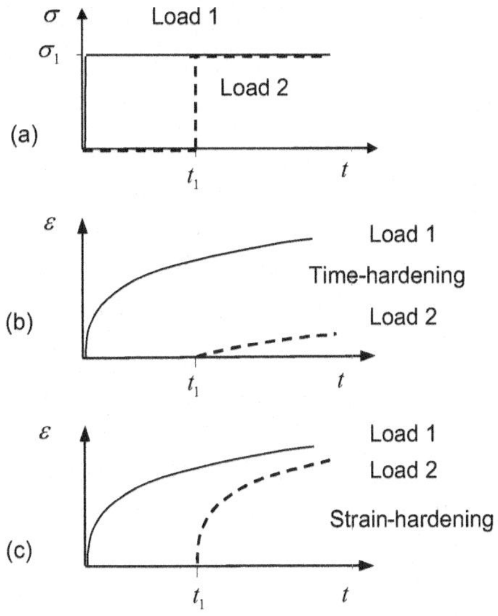


**Figure 9.5** (a) Two different loadings and the response of the material according to (b) Time-Hardening Model (9.51) and (c) Strain-Hardening Model (9.53).

The response of creeping materials can display a viscoelastic and/or viscoplastic character. Both deformations are time-dependent, but viscoelastic strain is recoverable, whereas viscoplastic strain is permanent. The theory of viscoelasticity (see, e.g., Flügge [22] or the powerful computational procedure of Santaoja [87] and Santaoja [90], Sections 4.4–4.9 and Appendices 3 and 4) provides methods for analyzing the material response, if Material Model (9.51) simulates viscoelastic deformation. However, the dislocation creep studied here is mainly permanent deformation, and therefore the strain rate $\dot{\varepsilon}^{\mathrm{v}}$ has to be interpreted as viscoplastic strain rate. This makes Material Model (9.51) very complicated, or even impossible, to solve with satisfaction for varying stress cases. Therefore a modification for the time-hardening creep model, Material Model (9.51), is sought. It can be obtained by assuming that the stress $\sigma$ takes a constant value and by integrating Material Model (9.51) over time $t$. This yields

$$\varepsilon^{\mathrm{v}} = A(\sigma)\,\frac{t_0}{1-k}\left(\frac{t}{t_0}\right)^{1-k}. \tag{9.52}$$

Elimination of time $t$ between Material Models (9.51) and (9.52) results in the strain-hardening creep law,

$$\dot{\varepsilon}^{\mathrm{v}} = [A(\sigma)]^{\frac{1}{1-k}} \left(\frac{t_0}{1-k}\right)^{\frac{k}{1-k}} (\varepsilon^{\mathrm{v}})^{\frac{-k}{1-k}} . \tag{9.53}$$

Although Material Models (9.51) and (9.53) give an equal response under a constant stress condition, they yield different results for varying stress, as shown in Figure 9.5. Strain-Hardening Model (9.53) gives the expected shape for hardening, as shown by Figure 9.5(c). Time-Hardening Material Model (9.51), however, gives an incorrect response, as already discussed. This is a remarkable phenomenon, since also a constant load on a structure usually causes varying stress in the material. Therefore Strain-Hardening Model (9.53) is proposed here instead of Time-Hardening Model (9.51). Strictly speaking, Material Model (9.51) should be forgotten. Thermodynamics supports this fact. Material Model (9.51) cannot be derived by continuum thermodynamics with internal variables that adopt the specific Helmholtz free energy and the specific dissipation function. On the other hand, Material Model (9.53) is compatible with internal variable theory, as the following study demonstrates. Also, the terminology indicates the excellence of Form (9.53) over Material Model (9.51). Time hardening does not exist, because the flow of time is not a cause and therefore cannot have any effect. On the other hand, strain hardening exists, since creep strain is a consequence of dislocation movement in the material and hardening is associated with the increase of resistance to dislocation motion. Also, Hult ([31], p. 32) preferred Material Model (9.53) in his book, although he did not mention so drastically the problem with Material Model (9.51), as expressed in Figure 9.5.

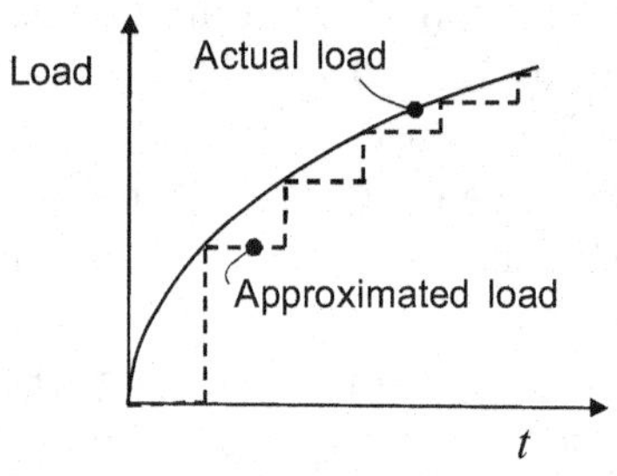


**Figure 9.6** The load approximated by a staircase function.

The authors propose the following strategy: never use the time-hardening model. If the values of the material parameters are determined for the time-hardening model using a constant stress $\sigma$, Material Models (9.51) and (9.53) can be used for determination of the values of the material parameters for the strain-hardening model. Do this conversion procedure and use the strain-hardening model. If the conversion is not possible to carry out analytically, create artificial creep curves using the time-hardening model and determine the values of the strain-hardening material model using a curve-fitting technique. The latter case also allows the introduction of other constitutive equations. The artificial creep curves should be created using same loading as in the original measurements. In the case of Strain-Hardening Material Model (9.53), stress $\sigma$ is the only load, but for creep temperature $T$ is also a load.

If the strain-hardening is modelled by

$$\dot{\alpha} := \tfrac{2}{3}\, J_{\mathrm{vM}}(\dot{\boldsymbol{\varepsilon}}^{\mathrm{v}})\,, \tag{9.54}$$

the expression for the strain-hardening creep can be written in the following multiaxial form:

$$\dot{\boldsymbol{\varepsilon}}^{\mathrm{v}} = \frac{\overset{\circ}{\varepsilon}_{\mathrm{re}}}{\alpha^m}\left[\frac{J_{\mathrm{vM}}(\boldsymbol{\sigma})}{\sigma_{\mathrm{re}}}\right]^n \frac{\mathbf{s}}{J_{\mathrm{vM}}(\boldsymbol{\sigma})} \qquad \Rightarrow \qquad \dot{\varepsilon}^{\mathrm{v}} = \frac{2}{3}\,\frac{\overset{\circ}{\varepsilon}_{\mathrm{re}}}{(\varepsilon^{\mathrm{v}})^m}\left[\frac{\sigma}{\sigma_{\mathrm{re}}}\right]^n. \tag{9.55}$$

Constitutive Equation $(9.55)_2$ is the uniaxial counterpart to the Multiaxial Strain-Hardening Creep Law $(9.55)_1$. Material Models (9.53) and $(9.55)_2$ have similar forms.

The (total) strain tensor $\boldsymbol{\varepsilon}$ is assumed to be separable as

$$\boldsymbol{\varepsilon} = \boldsymbol{\varepsilon}^{\mathrm{eT}} + \boldsymbol{\varepsilon}^{\mathrm{v}}\,. \tag{9.56}$$

This section does not deal with the thermoelastic strain tensor $\boldsymbol{\varepsilon}^{\mathrm{eT}}$, as it is covered in Sections 9.6 and 8.3.

State variables for Material Model (9.54)–(9.56) are $(\boldsymbol{\varepsilon}, \boldsymbol{\varepsilon}^{\mathrm{v}}, \alpha, T)$. The specific Helmholtz free energy $\psi$ and the specific complementary dissipation function $\varphi^{\mathrm{c}}_{\mathrm{mech}}$ are

$$\psi(\boldsymbol{\varepsilon} - \boldsymbol{\varepsilon}^{\mathrm{v}}, \alpha, T) = \psi^{\mathrm{eT}}(\boldsymbol{\varepsilon} - \boldsymbol{\varepsilon}^{\mathrm{v}}, T) + \frac{1}{\rho_0}\, f(\alpha) \tag{9.57}$$

and

$$\varphi^{\mathrm{c}}_{\mathrm{mech}}(\boldsymbol{\sigma}, \beta; \boldsymbol{\varepsilon} - \boldsymbol{\varepsilon}^{\mathrm{v}}, \alpha) = \frac{2}{3}\,\frac{\overset{\circ}{\varepsilon}_{\mathrm{re}}\,\sigma_{\mathrm{re}}}{\rho_0\,(n+1)\,\alpha^m}\left[\frac{\langle J_{\mathrm{vM}}(\boldsymbol{\sigma}) + \beta + \frac{\partial f(\alpha)}{\partial \alpha} - \sigma_{\mathrm{tr}}\rangle}{\sigma_{\mathrm{re}}}\right]^{(n+1)}. \tag{9.58}$$

In Material Models (9.57) and (9.58), the function $f(\alpha)$ is a continuously differentiable function specified more exactly later. The quantity $\sigma_{\mathrm{tr}}$ is the threshold value for creep. State Equation $(6.174)_2$ and Material Model (9.57) give

$$\beta = -\,\rho_0\,\frac{\partial \psi(\boldsymbol{\varepsilon} - \boldsymbol{\varepsilon}^{\mathrm{v}}, \alpha, T)}{\partial \alpha} = -\,\frac{\partial f(\alpha)}{\partial \alpha}\,. \tag{9.59}$$

Based on Results (2.152), the derivative $\partial J_{\mathrm{vM}}(\boldsymbol{\sigma})/\partial \boldsymbol{\sigma}$ is

$$\frac{\partial J_{\mathrm{vM}}(\boldsymbol{\sigma})}{\partial \boldsymbol{\sigma}} = \frac{3}{2}\,\frac{\mathbf{s}}{J_{\mathrm{vM}}(\boldsymbol{\sigma})}\,. \tag{9.60}$$

Substitution of Material Model (9.58) into Normality Rule (6.286) gives

$$\dot{\boldsymbol{\varepsilon}}^{\mathrm{v}} = \rho_0\,\frac{\partial \varphi^{\mathrm{c}}_{\mathrm{mech}}(\boldsymbol{\sigma}, \beta; \alpha)}{\partial \boldsymbol{\sigma}} = \frac{2}{3}\,\frac{\overset{\circ}{\varepsilon}_{\mathrm{re}}}{\alpha^m}\left[\frac{\langle J_{\mathrm{vM}}(\boldsymbol{\sigma}) + \beta + \frac{\partial f(\alpha)}{\partial \alpha} - \sigma_{\mathrm{tr}}\rangle}{\sigma_{\mathrm{re}}}\right]^n \frac{\partial J_{\mathrm{vM}}(\boldsymbol{\sigma})}{\partial \boldsymbol{\sigma}}\,. \tag{9.61}$$

Substitution of derivative $\partial J_{\mathrm{vM}}(\boldsymbol{\sigma})/\partial\boldsymbol{\sigma}$ obtained from Equation (9.60) and internal force $\beta$ obtained from Equation (9.59) into Expression (9.61) leads to

$$\dot{\boldsymbol{\varepsilon}}^{\mathrm{v}} = \rho_0\,\frac{\partial\varphi^{\mathrm{c}}_{\mathrm{mech}}(\boldsymbol{\sigma},\beta;\alpha)}{\partial\boldsymbol{\sigma}} = \frac{\overset{\circ}{\varepsilon}_{\mathrm{re}}}{\alpha^m}\left[\frac{\langle J_{\mathrm{vM}}(\boldsymbol{\sigma})-\sigma_{\mathrm{tr}}\rangle}{\sigma_{\mathrm{re}}}\right]^n \frac{\mathbf{s}}{J_{\mathrm{vM}}(\boldsymbol{\sigma})}\,. \tag{9.62}$$

Substitution of Material Model (9.58) into Normality Rule (6.287) yields

$$\begin{aligned}\dot{\alpha} = \rho_0\,\frac{\partial\varphi^{\mathrm{c}}_{\mathrm{mech}}(\boldsymbol{\sigma},\beta;\alpha)}{\partial\beta} &= \frac{2}{3}\,\frac{\overset{\circ}{\varepsilon}_{\mathrm{re}}}{\alpha^m}\left[\frac{\langle J_{\mathrm{vM}}(\boldsymbol{\sigma})+\beta+\frac{\partial f(\alpha)}{\partial\alpha}-\sigma_{\mathrm{tr}}\rangle}{\sigma_{\mathrm{re}}}\right]^n\\ &= \frac{2}{3}\,\frac{\overset{\circ}{\varepsilon}_{\mathrm{re}}}{\alpha^m}\left[\frac{\langle J_{\mathrm{vM}}(\boldsymbol{\sigma})-\sigma_{\mathrm{tr}}\rangle}{\sigma_{\mathrm{re}}}\right]^n.\end{aligned} \tag{9.63}$$

Following Result (8.122), $\dot{\mathbf{e}}^{\mathrm{v}} = \dot{\boldsymbol{\varepsilon}}^{\mathrm{v}}$, since no volume change is related to creep (viscous) deformation. This fact, Definition (2.124) for the von Mises operator $J_{\mathrm{vM}}(\cdot)$, and Evolution Equation (9.62) allow the following study to be carried out:

$$\begin{aligned}[J_{\mathrm{vM}}(\dot{\boldsymbol{\varepsilon}}^{\mathrm{v}})]^2 = \tfrac{2}{3}\,\dot{\mathbf{e}}^{\mathrm{v}}:\dot{\mathbf{e}}^{\mathrm{v}} = \tfrac{2}{3}\,\dot{\boldsymbol{\varepsilon}}^{\mathrm{v}}:\dot{\boldsymbol{\varepsilon}}^{\mathrm{v}} &= \frac{2}{3}\left(\frac{\overset{\circ}{\varepsilon}_{\mathrm{re}}}{\alpha^m}\right)^2\left[\frac{\langle J_{\mathrm{vM}}(\boldsymbol{\sigma})-\sigma_{\mathrm{tr}}\rangle}{\sigma_{\mathrm{re}}}\right]^{2n}\frac{\mathbf{s}:\mathbf{s}}{[J_{\mathrm{vM}}(\boldsymbol{\sigma})]^2}\\ &= \left(\frac{\overset{\circ}{\varepsilon}_{\mathrm{re}}}{\alpha^m}\right)^2\left[\frac{\langle J_{\mathrm{vM}}(\boldsymbol{\sigma})-\sigma_{\mathrm{tr}}\rangle}{\sigma_{\mathrm{re}}}\right]^{2n}\frac{[J_{\mathrm{vM}}(\boldsymbol{\sigma})]^2}{[J_{\mathrm{vM}}(\boldsymbol{\sigma})]^2}\\ &= \left(\frac{\overset{\circ}{\varepsilon}_{\mathrm{re}}}{\alpha^m}\right)^2\left[\frac{\langle J_{\mathrm{vM}}(\boldsymbol{\sigma})-\sigma_{\mathrm{tr}}\rangle}{\sigma_{\mathrm{re}}}\right]^{2n},\end{aligned} \tag{9.64}$$

which gives

$$J_{\mathrm{vM}}(\dot{\boldsymbol{\varepsilon}}^{\mathrm{v}}) = \frac{\overset{\circ}{\varepsilon}_{\mathrm{re}}}{\alpha^m}\left[\frac{\langle J_{\mathrm{vM}}(\boldsymbol{\sigma})-\sigma_{\mathrm{tr}}\rangle}{\sigma_{\mathrm{re}}}\right]^n. \tag{9.65}$$

Comparison of Expressions (9.63) and (9.65) yields

$$\dot{\alpha} = \tfrac{2}{3}\,J_{\mathrm{vM}}(\dot{\boldsymbol{\varepsilon}}^{\mathrm{v}})\,. \tag{9.66}$$

Comparison of Rate Equations (9.62) and (9.66) with Material Model (9.54) and (9.55) proves that Potentials (9.57) and (9.58) have the correct forms.

The specific complementary dissipation function $\varphi^{\mathrm{c}}_{\mathrm{mech}}$, Material Model (9.58), is not a homogeneous function in the variables $\boldsymbol{\sigma}$ and $\beta$. Thus the satisfaction of the Clausius–Duhem inequality is evaluated by studying the inequality itself. The mechanical part of Clausius–Duhem Inequality (6.250) is

$$\boldsymbol{\sigma}:\dot{\boldsymbol{\varepsilon}}^{\mathrm{i}} + \boldsymbol{\beta}:\dot{\alpha} \,\geq\, 0\,. \tag{9.67}$$

For the present set of the state variables, Clausius–Duhem Inequality (9.67) takes the form

$$\boldsymbol{\sigma} : \dot{\boldsymbol{\varepsilon}}^{\mathrm{v}} + \beta\,\dot{\alpha} \geq 0\,. \tag{9.68}$$

Based on Problem 8.2, the following holds: $\boldsymbol{\sigma}:\mathbf{s} = \mathbf{s}:\mathbf{s}$. This and Rate Equation (9.62) allow the first term of Inequality (9.68) to be cast in the form

$$\boldsymbol{\sigma} : \dot{\boldsymbol{\varepsilon}}^{\mathrm{v}} = \frac{\overset{\circ}{\varepsilon}_{\mathrm{re}}}{\alpha^m}\left[\frac{\langle J_{\mathrm{vM}}(\boldsymbol{\sigma}) - \sigma_{\mathrm{tr}}\rangle}{\sigma_{\mathrm{re}}}\right]^n \frac{\mathbf{s}:\mathbf{s}}{J_{\mathrm{vM}}(\boldsymbol{\sigma})} = \frac{2}{3}\frac{\overset{\circ}{\varepsilon}_{\mathrm{re}}}{\alpha^m}\left[\frac{\langle J_{\mathrm{vM}}(\boldsymbol{\sigma}) - \sigma_{\mathrm{tr}}\rangle}{\sigma_{\mathrm{re}}}\right]^n J_{\mathrm{vM}}(\boldsymbol{\sigma}). \tag{9.69}$$

In Expression (9.69), Definition (2.124) of the von Mises operator $J_{\mathrm{vM}}(\cdot)$ is utilized. Term (9.69) is nonnegative.

Evaluation of the second term of Inequality (9.68) is more complicated. Based on Rate Equation (9.63), the following result is obtained:

$$\beta\,\dot{\alpha} = \frac{2}{3}\frac{\overset{\circ}{\varepsilon}_{\mathrm{re}}}{\alpha^m}\left[\frac{\langle J_{\mathrm{vM}}(\boldsymbol{\sigma}) - \sigma_{\mathrm{tr}}\rangle}{\sigma_{\mathrm{re}}}\right]^n \beta\,. \tag{9.70}$$

The sign of Term (9.70) depends on the sign of the internal force $\beta$.

Combining Terms (9.69) and (9.70) gives

$$\boldsymbol{\sigma} : \dot{\boldsymbol{\varepsilon}}^{\mathrm{v}} + \beta\,\dot{\alpha} = \frac{2}{3}\frac{\overset{\circ}{\varepsilon}_{\mathrm{re}}}{\alpha^m}\left[\frac{\langle J_{\mathrm{vM}}(\boldsymbol{\sigma}) - \sigma_{\mathrm{tr}}\rangle}{\sigma_{\mathrm{re}}}\right]^n \left[J_{\mathrm{vM}}(\boldsymbol{\sigma}) + \beta\right]. \tag{9.71}$$

It is not clear that Expression (9.71) is nonnegative, since the variable $\beta$ takes negative values. This stems from the fact that it is physically necessary that the function $f(\alpha)$ present in the expression for the specific Helmholtz free energy $\psi$, Expression (9.57), takes zero value when $\alpha = 0$ and that the function $f(\alpha)$ be an increasing function with increasing values for the internal variable $\alpha$. Thus the derivative of the function $f(\alpha)$ with respect to $\alpha$ is positive. This fact and Equation (9.59) show that the internal force $\beta$ takes negative values. This leads to the conclusion that with low values of the stress tensor $\boldsymbol{\sigma}$, the term within the latter square brackets tends to be negative, and therefore Clausius–Duhem Inequality (9.68) is not satisfied. To avoid this problem, the threshold value for creep $\sigma_{\mathrm{tr}}$ was introduced. The concept of the threshold value $\sigma_{\mathrm{tr}}$ is to make the term between the Macaulay brackets $\langle\cdot\rangle$ negative if the term $J_{\mathrm{vM}}(\boldsymbol{\sigma}) + \beta$ tends to take a negative value. This implies that no creep occurs [see Equation (9.62)] and the Clausius–Duhem inequality takes zero value; that is, it is not violated.

The preceding verbal description is not enough and requires mathematical proof. The function $f(\alpha)$ is assumed to have the form

$$f(\alpha) = \sigma_{tr}\left(1 - e^{-\alpha}\right). \tag{9.72}$$

Based on Expression (9.72), function $f(\alpha)$ takes zero value when $\alpha = 0$, and it is an increasing positive-valued function, as shown in Figure 9.7. Material

Model (9.72) gives the following expression for the internal force $\beta$:

$$\beta = -\frac{\partial f(\alpha)}{\partial \alpha} = -\sigma_{\mathrm{tr}}\, e^{-\alpha}\,. \tag{9.73}$$

Expression (9.73) shows that the value for $\beta$ does not fall so much that it will be smaller than $-\sigma_{\mathrm{tr}}$. This ensures that Expression (9.71) is nonnegative and Clausius–Duhem Inequality (9.68) is satisfied.

Comparison of the material model introduced in this text and given by Equations (9.62) and (9.66) with the strain-hardening creep law given by Expressions (9.54) and (9.55) shows that they are not equal, but Constitutive Equation (9.62) introduces the threshold value for creep $\sigma_{\mathrm{tr}}$. This does not make a great difference between these two constitutive equations, since $\sigma_{\mathrm{tr}}$ can take very small values. Furthermore, from a material mechanics point of view, the introduction of the threshold $\sigma_{\mathrm{tr}}$ may be a good concept.

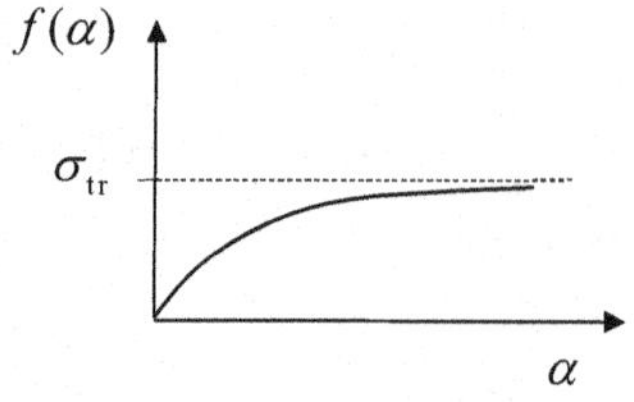

**Figure 9.7** Sketch for the function $f(\alpha)$ given by Model (9.57).

## 9.6 Material Model of Le Gac and Duval

This section derives a model for primary and secondary creep. The response of these deformation mechanisms is expressed by the viscous strain tensor $\boldsymbol{\varepsilon}^{\mathrm{v}}$. This model for creep is taken from the work by Le Gac and Duval [48], who introduced this model for the description of ice creep. However, the model is discussed here since the authors have applied it for simulation of creep in high-temperature component materials used in conventional power plants.

According to Le Gac and Duval [48], creep strain $\boldsymbol{\varepsilon}^{\mathrm{v}}$ is obtained from

$$\dot{\boldsymbol{\varepsilon}}^{\mathrm{v}} = \overset{\circ}{\varepsilon}_{\mathrm{re}} \left[ \frac{\left\langle J_{\mathrm{vM}}(\boldsymbol{\sigma} - \boldsymbol{\beta}^1) - \beta^2 - \sigma_{\mathrm{tr}} \right\rangle}{\sigma_{\mathrm{re}}} \right]^n \frac{\mathbf{s} - \mathbf{b}^1}{J_{\mathrm{vM}}(\boldsymbol{\sigma} - \boldsymbol{\beta}^1)}\,. \tag{9.74}$$

In Constitutive Equation (9.74), the notations $\overset{\circ}{\varepsilon}_{\mathrm{re}}$ and $\sigma_{\mathrm{re}}$ stand for the reference strain rate and for the reference stress, respectively. The quantity $\sigma_{\mathrm{tr}}$ refers to the onset of creep. As already discussed with Norton's law in Section 9.4, the reference stress $\sigma_{\mathrm{re}}$ is not a "real" material parameter. Its value can be set before determination of the values for the other material parameters. A good choice is to set the value for the reference stress $\sigma_{\mathrm{re}}$ so that the term between the square brackets in Rate (9.74) takes values close to unity.

The value for the kinematic hardening $\boldsymbol{\beta}^1$ is obtained from

$$\dot{\boldsymbol{\beta}}^1 = m^1\, \dot{\boldsymbol{\varepsilon}}^{\mathrm{v}} - m^2 \left[ \frac{J_{\mathrm{vM}}(\boldsymbol{\beta}^1)}{\sigma_{\mathrm{re}}} \right]^k \frac{\mathbf{b}^1}{J_{\mathrm{vM}}(\boldsymbol{\beta}^1)}\,, \tag{9.75}$$

where the notations $m^1$, $m^2$, and $k$ refer to material parameters associated with kinematic hardening.

The evolution equation for isotropic hardening $\beta^2$ has a similar appearance to that for kinematic hardening $\boldsymbol{\beta}^1$. It is

$$\dot{\beta}^2 = n^1\, J_{\mathrm{vM}}(\dot{\boldsymbol{\varepsilon}}^{\mathrm{v}}) - n^2 \left(\frac{\beta^2}{\sigma_{\mathrm{re}}}\right)^m , \tag{9.76}$$

where the material parameters $n^1$, $n^2$, and $m$ are associated with isotropic hardening.

In this section, it is assumed that the stress level is below the yield point $\sigma^{\mathrm{Y}}$, which implies that it is not necessary to model plastic deformation. Since traditional power plants operate at 550 °C and since during maintenance they are cooled to room temperature, the strain associated with thermal expansion is remarkable, and therefore it must be included in the material model.

Thus the (total) strain tensor $\boldsymbol{\varepsilon}$ consists of three parts, as follows:

$$\boldsymbol{\varepsilon} = \boldsymbol{\varepsilon}^{\mathrm{e}} + \boldsymbol{\varepsilon}^{\mathrm{Th}} + \boldsymbol{\varepsilon}^{\mathrm{v}} , \tag{9.77}$$

where the elastic strain tensor $\boldsymbol{\varepsilon}^{\mathrm{e}}$ is modelled by Hooke's law and thermal strain tensor $\boldsymbol{\varepsilon}^{\mathrm{Th}}$ by a linear relationship,

$$\boldsymbol{\sigma} = \mathbf{C} : \boldsymbol{\varepsilon}^{\mathrm{e}} \qquad \text{and} \qquad \boldsymbol{\varepsilon}^{\mathrm{Th}} = \hat{\alpha}\,(T - T_{\mathrm{r}})\,\mathbf{1} , \tag{9.78}$$

where $\mathbf{C}$ is the constitutive tensor for a Hookean deformation, $\hat{\alpha}$ is the linear coefficient of the thermal expansion, and $T_{\mathrm{r}}$ is the reference temperature. At reference temperature, no thermal deformation exists.

The formulation within the framework of continuum thermodynamics for the preceding material models follows. The first step of the continuum thermodynamic evaluation of the material model is the determination of the set of state variables.

The controllable variables are the strain tensor $\boldsymbol{\varepsilon}$ and the specific entropy $s$. Since the foregoing material model shows elastic deformation, the inelastic strain tensor $\boldsymbol{\varepsilon}^{\mathrm{i}}$ belongs to the set of internal variables. In the present material model, the inelastic strain tensor $\boldsymbol{\varepsilon}^{\mathrm{i}}$ is the creep strain tensor denoted by $\boldsymbol{\varepsilon}^{\mathrm{v}}$. Furthermore, two internal variables $\boldsymbol{\alpha}^1$ and $\alpha^2$ associated with kinematic and isotropic hardening are introduced. Thus the set of state variables is $(\boldsymbol{\varepsilon}, \boldsymbol{\varepsilon}^{\mathrm{v}}, \boldsymbol{\alpha}^1, \alpha^2, s)$. As discussed in Section 6.4.4, in the presence of elastic deformation, the difference $\boldsymbol{\varepsilon} - \boldsymbol{\varepsilon}^{\mathrm{v}}$ belongs to the description of state. Since the state $\mathcal{E}$ is expressed by the specific Helmholtz free energy $\psi$, the state variable specific entropy $s$ is replaced by the absolute temperature $T$. Thus the specific Helmholtz free energy $\psi$ has the following appearance:

$$\psi = \psi(\boldsymbol{\varepsilon} - \boldsymbol{\varepsilon}^{\mathrm{v}}, \boldsymbol{\alpha}^1, \alpha^2, T) . \tag{9.79}$$

The specific Helmholtz free energy $\psi$ is separable as follows:

$$\psi(\boldsymbol{\varepsilon} - \boldsymbol{\varepsilon}^{\mathrm{v}}, \boldsymbol{\alpha}^1, \alpha^2, T) = \psi^{\mathrm{eT}}(\boldsymbol{\varepsilon} - \boldsymbol{\varepsilon}^{\mathrm{v}}, T) + \psi^{\mathrm{v1}}(\boldsymbol{\alpha}^1) + \psi^{\mathrm{v2}}(\alpha^2) . \tag{9.80}$$

The specific Helmholtz free energy $\psi$ for a linear isotropic thermoelastic material is obtained from Material Model (8.66),

$$\psi^{\mathrm{eT}}(\boldsymbol{\varepsilon}-\boldsymbol{\varepsilon}^{\mathrm{i}},T)=\psi_{\mathrm{r}}^{\mathrm{eT}}-s_{\mathrm{r}}\,(T-T_{\mathrm{r}})+\frac{1}{2\,\rho_0}\,(\boldsymbol{\varepsilon}-\boldsymbol{\varepsilon}^{\mathrm{i}}):\mathbf{C}:(\boldsymbol{\varepsilon}-\boldsymbol{\varepsilon}^{\mathrm{i}})$$
$$-\frac{E\,\hat{\alpha}}{\rho_0\,(1-2\,\nu)}\,\mathbf{1}:(\boldsymbol{\varepsilon}-\boldsymbol{\varepsilon}^{\mathrm{i}})\,(T-T_{\mathrm{r}})+\frac{1}{\rho_0}\,\boldsymbol{\sigma}_{\mathrm{r}}:(\boldsymbol{\varepsilon}-\boldsymbol{\varepsilon}^{\mathrm{i}})$$
$$-\,c_{\mathrm{r}}\,[T\,\ln(T/T_{\mathrm{r}})-T\,]\,. \tag{9.81}$$

By following Section 8.3, we arrive at the following [cf. Eqs (8.67) and (8.68)]:

$$\boldsymbol{\sigma}=\rho_0\,\frac{\partial 1\psi(\boldsymbol{\varepsilon}-\boldsymbol{\varepsilon}^{\mathrm{i}},T)}{\partial(\boldsymbol{\varepsilon}-\boldsymbol{\varepsilon}^{\mathrm{i}})}=\rho_0\,\frac{\partial\psi(\boldsymbol{\varepsilon}-\boldsymbol{\varepsilon}^{\mathrm{v}},T)}{\partial(\boldsymbol{\varepsilon}-\boldsymbol{\varepsilon}^{\mathrm{v}})}=\rho_0\,\frac{\partial\psi^{\mathrm{eT}}(\boldsymbol{\varepsilon}-\boldsymbol{\varepsilon}^{\mathrm{v}},T)}{\partial(\boldsymbol{\varepsilon}-\boldsymbol{\varepsilon}^{\mathrm{v}})}$$
$$=\mathbf{C}:(\boldsymbol{\varepsilon}-\boldsymbol{\varepsilon}^{\mathrm{v}})-\frac{E\,\hat{\alpha}}{(1-2\,\nu)}\,(T-T_{\mathrm{r}})\,\mathbf{1}+\boldsymbol{\sigma}_{\mathrm{r}}\,, \tag{9.82}$$

where $\hat{\alpha}$ is the linear coefficient of thermal expansion and $\boldsymbol{\sigma}_{\mathrm{r}}$ is the stress in the reference state. Equation (9.82) utilizes the fact that the only inelastic deformation (modelled with the inelastic strain tensor $\boldsymbol{\varepsilon}^{\mathrm{i}}$ ) is the viscous deformation (modelled with the viscous strain tensor $\boldsymbol{\varepsilon}^{\mathrm{v}}$). Equation (9.82) can take the following appearance:

$$\boldsymbol{\sigma}=\rho_0\,\frac{\partial\psi(\boldsymbol{\varepsilon}-\boldsymbol{\varepsilon}^{\mathrm{v}},T)}{\partial(\boldsymbol{\varepsilon}-\boldsymbol{\varepsilon}^{\mathrm{v}})}=\rho_0\,\frac{\partial\psi^{\mathrm{eT}}(\boldsymbol{\varepsilon}-\boldsymbol{\varepsilon}^{\mathrm{v}},T)}{\partial(\boldsymbol{\varepsilon}-\boldsymbol{\varepsilon}^{\mathrm{v}})}$$
$$=\mathbf{C}:(\boldsymbol{\varepsilon}-\boldsymbol{\varepsilon}^{\mathrm{v}})-(3\,\lambda+2\,\mu)\,\hat{\alpha}\,(T-T_{\mathrm{r}})\,\mathbf{1}+\boldsymbol{\sigma}_{\mathrm{r}}\,, \tag{9.83}$$

where $E/(1-2\nu)=(3\lambda+2\mu)$ is used in arriving at the last step. It is assumed that the value of the stress tensor in the reference state $\boldsymbol{\sigma}_{\mathrm{r}}$ vanishes. Therefore, Constitutive Equation (9.82) can be cast in the following form, given in Equation (8.77):

$$\boldsymbol{\varepsilon}^{\mathrm{eT}}=\mathbf{S}:\boldsymbol{\sigma}+\hat{\alpha}\,(T-T_{\mathrm{r}})\,\mathbf{1}=-\frac{\nu}{E}\,\mathbf{1}\,\mathbf{1}:\boldsymbol{\sigma}+\frac{1+\nu}{E}\,\mathbf{I}^{\mathrm{s}}:\boldsymbol{\sigma}+\hat{\alpha}\,(T-T_{\mathrm{r}})\,\mathbf{1}\,. \tag{9.84}$$

Equation (9.84) introduced the notation

$$\boldsymbol{\varepsilon}^{\mathrm{eT}}:=\boldsymbol{\varepsilon}-\boldsymbol{\varepsilon}^{\mathrm{v}}\,. \qquad \text{for thermoelastic deformation.} \tag{9.85}$$

Definition (9.85) is similar to Definition $(8.65)_3$, but the inelastic strain tensor $\boldsymbol{\varepsilon}^{\mathrm{i}}$ is replaced by the viscous strain tensor $\boldsymbol{\varepsilon}^{\mathrm{v}}$. This is because, in the present model, the viscous strain tensor $\boldsymbol{\varepsilon}^{\mathrm{v}}$ equals the inelastic strain tensor $\boldsymbol{\varepsilon}^{\mathrm{i}}$, since no other inelastic deformations are present. This is seen in Expression (9.77).

Based on Constitutive Equation (9.84), the thermoelastic strain tensor $\boldsymbol{\varepsilon}^{\mathrm{eT}}$ can be separated into two parts, as follows:

$$\boldsymbol{\varepsilon}^{\mathrm{eT}}:=\boldsymbol{\varepsilon}^{\mathrm{e}}+\boldsymbol{\varepsilon}^{\mathrm{Th}}\,, \tag{9.86}$$

where the strain tensor for linear elastic deformation (i.e., for Hooke's law) $\boldsymbol{\varepsilon}^{\mathrm{e}}$ and the thermal strain $\boldsymbol{\varepsilon}^{\mathrm{Th}}$ take the forms

$$\boldsymbol{\varepsilon}^{\mathrm{e}} = -\frac{\nu}{E}\,\mathbf{1}\,\mathbf{1} : \boldsymbol{\sigma} + \frac{1+\nu}{E}\,\mathbf{I}^{\mathrm{s}} : \boldsymbol{\sigma} \qquad \text{and} \qquad \boldsymbol{\varepsilon}^{\mathrm{Th}} = \hat{\alpha}\,(T - T_{\mathrm{r}})\,\mathbf{1}\,. \tag{9.87}$$

As given in Expressions $(8.41)_1$ and (8.42), Hooke's Law $(9.87)_1$ can be written as

$$\boldsymbol{\varepsilon}^{\mathrm{e}} = \mathbf{S} : \boldsymbol{\sigma}\,, \qquad \text{where} \qquad \mathbf{S} := -\frac{\nu}{E}\,\mathbf{1}\,\mathbf{1} + \frac{1+\nu}{E}\,\mathbf{I}^{\mathrm{s}}\,. \tag{9.88}$$

Multiplication of Material Model $(9.88)_1$ by $\mathbf{C}:$ from the left gives [see Equation $(8.32)_2$ ]

$$\boldsymbol{\sigma} = \mathbf{C} : \boldsymbol{\varepsilon}^{\mathrm{e}}\,. \tag{9.89}$$

Constitutive Equations $(9.87)_2$ and (9.89) have the forms aimed at Material Models (9.78).

The last two terms for the specific Helmholtz free energy $\psi$ [Material Model (9.80)] have the following appearances:

$$\psi^{\mathrm{v1}}(\boldsymbol{\alpha}^1) = \frac{1}{2\,\rho_0}\,\boldsymbol{\alpha}^1 : \mathbf{C}^1 : \boldsymbol{\alpha}^1 \qquad \text{and} \qquad \psi^{\mathrm{v2}}(\alpha^2) = \frac{3\,n^1}{4\,\rho_0}\,\alpha^2\,\alpha^2\,, \tag{9.90}$$

where $\mathbf{C}^1$ is a constant tensor. The term *constant tensor* means a tensor whose components have constant values. The fourth-order tensor $\mathbf{C}^1$ is assumed to have the following symmetry properties:

$$C^1_{ijkl} = C^1_{jikl} = C^1_{ijlk} = C^1_{jilk} = C^1_{klij}\,. \tag{9.91}$$

Expression (9.91) implies that the tensor $\mathbf{C}^1$ has a minor and major symmetry (see Theorems 3 and 4 of **Appendix I**). Forms $(9.90)_1$ and $(9.90)_2$ are taken from Santaoja [[90], Eqs. (272) and (273)].

With regard to Material Modes (9.80), (9.81), $(9.90)_1$, and $(9.90)_2$, State Equation $(6.174)_2$ gives

$$\boldsymbol{\beta}^1 = -\,\rho_0\,\frac{\partial\psi}{\partial\boldsymbol{\alpha}^1} = -\,\mathbf{C}^1 : \boldsymbol{\alpha}^1\,, \qquad \text{which yields} \qquad \dot{\boldsymbol{\beta}}^1 = -\,\mathbf{C}^1 : \dot{\boldsymbol{\alpha}}^1\,, \tag{9.92}$$

and

$$\beta^2 = -\,\rho_0\,\frac{\partial\psi}{\partial\alpha^2} = -\,\tfrac{3}{2}\,n^1\,\alpha^2\,, \qquad \text{which yields} \qquad \dot{\beta}^2 = -\,\tfrac{3}{2}\,n^1\,\dot{\alpha}^2\,. \tag{9.93}$$

Derivative $(9.92)_1$ exploits Expression $(2.131)_2$.

The specific complementary dissipation function $\varphi^{\mathrm{c}}_{\mathrm{mech}}$ is separable as follows:

$$\varphi^{\mathrm{c}}_{\mathrm{mech}}(\boldsymbol{\sigma}, \boldsymbol{\beta}^1, \beta^2; \boldsymbol{\varepsilon} - \boldsymbol{\varepsilon}^{\mathrm{v}}, \boldsymbol{\alpha}^1, \alpha^2, T) = \varphi^{\mathrm{c1}}_{\mathrm{mech}}(\boldsymbol{\sigma}, \boldsymbol{\beta}^1, \beta^2) + \varphi^{\mathrm{c2}}_{\mathrm{mech}}(\boldsymbol{\beta}^1) + \varphi^{\mathrm{c3}}_{\mathrm{mech}}(\beta^2). \tag{9.94}$$

The explicit forms for the terms $\varphi^{c1}_{mech}$ through $\varphi^{c3}_{mech}$ are given next. Following the work by Santaoja [[91], Models (218), (221) and (222)], these three terms take the appearance

$$\varphi^{c1}_{mech}(\boldsymbol{\sigma}, \boldsymbol{\beta}^1, \beta^2) = \frac{2\,\overset{\circ}{\varepsilon}_{re}\,\sigma_{re}}{3\,\rho_0\,(n+1)} \left[ \frac{\langle\, J_{vM}(\boldsymbol{\sigma} - \boldsymbol{\beta}^1) - \beta^2 - \sigma_{tr}\,\rangle}{\sigma_{re}} \right]^{(n+1)} \tag{9.95}$$

and

$$\varphi^{c2}_{mech}(\boldsymbol{\beta}^1) = \frac{2\,\tilde{C}^{11}\,\sigma_{re}}{3\,\rho_0(k+1)} \left[ \frac{J_{vM}(\boldsymbol{\beta}^1)}{\sigma_{re}} \right]^{(k+1)}, \tag{9.96}$$

and, finally,

$$\varphi^{c3}_{mech}(\beta^2) = \frac{2\,C^{22}\,\sigma_{re}}{3\,\rho_0\,(m+1)} \left( \frac{\beta^2}{\sigma_{re}} \right)^{(m+1)}. \tag{9.97}$$

In Material Models (9.96) and (9.97), $\tilde{C}^{11}$ and $C^{22}$ are nonnegative material parameters for kinematic and isotropic recovery.

Derivatives (2.152) and (2.153) are

$$\frac{\partial J_{vM}(\boldsymbol{\sigma} - \boldsymbol{\beta}^1)}{\partial \boldsymbol{\sigma}} = \frac{3}{2}\,\frac{\mathbf{s} - \mathbf{b}^1}{J_{vM}(\boldsymbol{\sigma} - \boldsymbol{\beta}^1)} \tag{9.98}$$

and

$$\frac{\partial J_{vM}(\boldsymbol{\sigma} - \boldsymbol{\beta}^1)}{\partial \boldsymbol{\beta}^1} = -\,\frac{3}{2}\,\frac{\mathbf{s} - \mathbf{b}^1}{J_{vM}(\boldsymbol{\sigma} - \boldsymbol{\beta}^1)}. \tag{9.99}$$

Since the specific complementary dissipation function $\varphi^{c1}_{mech}$ is the only part of the specific complementary dissipation function $\varphi^{c}_{mech}$ to have the stress tensor $\boldsymbol{\sigma}$ as a variable, Normality Rule (6.286) yields

$$\dot{\boldsymbol{\varepsilon}}^{v} = \rho_0\,\frac{\partial \varphi^{c}_{mech}}{\partial \boldsymbol{\sigma}} = \overset{\circ}{\varepsilon}_{re} \left[ \frac{\langle\, J_{vM}(\boldsymbol{\sigma} - \boldsymbol{\beta}^1) - \beta^2 - \sigma_{tr}\,\rangle}{\sigma_{re}} \right]^{n} \frac{\mathbf{s} - \mathbf{b}^1}{J_{vM}(\boldsymbol{\sigma} - \boldsymbol{\beta}^1)}. \tag{9.100}$$

The rate equation exploits Derivative (9.98). Material Model (9.100) has the same appearance as Material Model (9.74).

The specific complementary dissipation functions $\varphi^{c1}_{mech}$ and $\varphi^{c2}_{mech}$ have the internal force $\boldsymbol{\beta}^1$ as an independent variable. Therefore Normality Rule (6.287) provides

$$\dot{\boldsymbol{\alpha}}^1 = \rho_0\,\frac{\partial \varphi^{c}_{mech}}{\partial \boldsymbol{\beta}^1} = -\,\dot{\boldsymbol{\varepsilon}}^{v} + \tilde{C}^{11} \left[ \frac{J_{vM}(\boldsymbol{\beta}^1)}{\sigma_{re}} \right]^{k} \frac{\mathbf{b}^1}{J_{vM}(\boldsymbol{\beta}^1)}. \tag{9.101}$$

The first term on the right side of Expression (9.101) utilizes the fact that Derivatives (9.98) and (9.99) take the same appearance, except the sign. Substitution of $\dot{\boldsymbol{\alpha}}^1$ obtained from Equation $(9.92)_2$ into Equation (9.101) leads to

$$\dot{\boldsymbol{\beta}}^1 = m^1\,\dot{\boldsymbol{\varepsilon}}^{v} - m^2 \left[ \frac{J_{vM}(\boldsymbol{\beta}^1)}{\sigma_{re}} \right]^{k} \frac{\mathbf{b}^1}{J_{vM}(\boldsymbol{\beta}^1)}, \tag{9.102}$$

where the following were assumed:

$$\mathbf{C}^1 = m^1\,\mathbf{I} \qquad \text{and} \qquad \tilde{C}^{11} = \frac{m^2}{m^1}\,. \tag{9.103}$$

Evolution Equation (9.102) is that given by Rate Equation (9.75).

The specific dissipation functions $\varphi^{c1}_{\text{mech}}$ and $\varphi^{c3}_{\text{mech}}$ have the internal force $\beta^2$ as an independent variable. Therefore Normality Rule (6.287) provides

$$\dot{\alpha}^2 = \rho_0\,\frac{\partial\varphi^{c}_{\text{mech}}}{\partial\beta^2} = -\frac{2}{3}\,\overset{\circ}{\varepsilon}_{\text{re}}\left[\frac{\langle J_{\text{vM}}(\boldsymbol{\sigma}-\boldsymbol{\beta}^1)-\beta^2-\sigma_{\text{tr}}\rangle}{\sigma_{\text{re}}}\right]^n + \frac{2}{3}\,C^{22}\left(\frac{\beta^2}{\sigma_{\text{re}}}\right)^m. \tag{9.104}$$

Definition (2.124), Result (8.122) (for this case, $\dot{\mathbf{e}}^{\text{v}} = \dot{\boldsymbol{\varepsilon}}^{\text{v}}$ ), and Rate (9.100) provide the following manipulation:

$$\begin{aligned}
[\,J_{\text{vM}}(\dot{\boldsymbol{\varepsilon}}^{\text{v}})\,]^2 &= \tfrac{3}{2}\,\dot{\mathbf{e}}^{\text{v}}:\dot{\mathbf{e}}^{\text{v}} = \tfrac{3}{2}\,\dot{\boldsymbol{\varepsilon}}^{\text{v}}:\dot{\boldsymbol{\varepsilon}}^{\text{v}}\\
&= (\overset{\circ}{\varepsilon}_{\text{re}})^2\left[\frac{\langle J_{\text{vM}}(\boldsymbol{\sigma}-\boldsymbol{\beta}^1)-\beta^2-\sigma_{\text{tr}}\rangle}{\sigma_{\text{re}}}\right]^{2n}\frac{\tfrac{3}{2}\,(\mathbf{s}-\mathbf{b}^1):(\mathbf{s}-\mathbf{b}^1)}{[\,J_{\text{vM}}(\boldsymbol{\sigma}-\boldsymbol{\beta}^1)\,]^2}\\
&= (\overset{\circ}{\varepsilon}_{\text{re}})^2\left[\frac{\langle J_{\text{vM}}(\boldsymbol{\sigma}-\boldsymbol{\beta}^1)-\beta^2-\sigma_{\text{tr}}\rangle}{\sigma_{\text{re}}}\right]^{2n}\frac{[\,J_{\text{vM}}(\boldsymbol{\sigma}-\boldsymbol{\beta}^1)\,]^2}{[\,J_{\text{vM}}(\boldsymbol{\sigma}-\boldsymbol{\beta}^1)\,]^2}\\
&= (\overset{\circ}{\varepsilon}_{\text{re}})^2\left[\frac{\langle J_{\text{vM}}(\boldsymbol{\sigma}-\boldsymbol{\beta}^1)-\beta^2-\sigma_{\text{tr}}\rangle}{\sigma_{\text{re}}}\right]^{2n}.
\end{aligned} \tag{9.105}$$

Manipulation (9.105) allows Rate (9.104) to be written in the form

$$\dot{\alpha}^2 = \rho_0\,\frac{\partial\varphi^{c}_{\text{mech}}}{\partial\beta^2} = -\frac{2}{3}\,J_{\text{vM}}(\dot{\boldsymbol{\varepsilon}}^{\text{v}}) + \frac{2}{3}\,C^{22}\left(\frac{\beta^2}{\sigma_{\text{re}}}\right)^m. \tag{9.106}$$

The expression for quantity $\dot{\alpha}^2$ obtained from Equation $(9.93)_2$ allows Equation (9.106) to be written as

$$\dot{\beta}^2 = n^1\,J_{\text{vM}}(\dot{\boldsymbol{\varepsilon}}^{\text{v}}) - n^2\left(\frac{\beta^2}{\sigma_{\text{re}}}\right)^m, \tag{9.107}$$

where the following is assumed:

$$C^{22} = \frac{n^2}{n^1}\,. \tag{9.108}$$

Evolution Equation (9.107) has the target form expressed in Equation (9.76).

Continuum thermodynamic formulation of the material model given by Expressions (9.74)–(9.78) is carried out. Next, the satisfaction of the Clausius–Duhem inequality is evaluated.

The mechanical part of Clausius–Duhem Inequality (6.250) yields

$$\boldsymbol{\sigma} : \dot{\boldsymbol{\varepsilon}}^{\mathrm{v}} + \boldsymbol{\beta}^1 : \dot{\boldsymbol{\alpha}}^1 + \beta^2 \, \dot{\alpha}^2 \geq 0 \,. \tag{9.109}$$

Substitution of Rates (9.100), (9.101), and (9.104) into Inequality (9.109) gives

$$\overset{\circ}{\varepsilon}_{\mathrm{re}} \left[ \frac{\left\langle J_{\mathrm{vM}}(\boldsymbol{\sigma} - \boldsymbol{\beta}^1) - \beta^2 - \sigma_{\mathrm{tr}} \right\rangle}{\sigma_{\mathrm{re}}} \right]^n \left[ \frac{(\mathbf{s} - \mathbf{b}^1) : (\boldsymbol{\sigma} - \boldsymbol{\beta}^1)}{J_{\mathrm{vM}}(\boldsymbol{\sigma} - \boldsymbol{\beta}^1)} - \tfrac{2}{3} \beta^2 \right]$$

$$+ \tilde{C}^{11} \left[ \frac{J_{\mathrm{vM}}(\boldsymbol{\beta}^1)}{\sigma_{\mathrm{re}}} \right]^k \frac{\boldsymbol{\beta}^1 : \mathbf{b}^1}{J_{\mathrm{vM}}(\boldsymbol{\beta}^1)} + \frac{2}{3} C^{22} \left( \frac{\beta^2}{\sigma_{\mathrm{re}}} \right)^m \beta^2 \geq 0 \,. \tag{9.110}$$

A sufficient condition for satisfaction of Inequality (9.110), is that all the terms are separately nonnegative. Therefore Inequality (9.110) is studied term by term.

The sign of the first term (line) of Clausius–Duhem Inequality (9.110), can be evaluated as discussed next.

Problem 8.2 and Definition (2.124) give

$$\tfrac{3}{2} \, (\boldsymbol{\sigma} - \boldsymbol{\beta}^1) : (\mathbf{s} - \mathbf{b}^1) = \tfrac{3}{2} \, (\mathbf{s} - \mathbf{b}^1) : (\mathbf{s} - \mathbf{b}^1) = \left[ J_{\mathrm{vM}}(\boldsymbol{\sigma} - \boldsymbol{\beta}^1) \right]^2 . \tag{9.111}$$

Double-dot product between two second-order tensors is commutative

$$\mathbf{c} : \mathbf{g} = \mathbf{g} : \mathbf{c} \,. \tag{9.112}$$

By multiplying the first term of Inequality (9.110) by a factor of 3/2 and noting Equations (9.111) and (9.112), the first term takes the following appearance:

$$\frac{3}{2} \overset{\circ}{\varepsilon}_{\mathrm{re}} \left[ \frac{\left\langle J_{\mathrm{vM}}(\boldsymbol{\sigma} - \boldsymbol{\beta}^1) - \beta^2 - \sigma_{\mathrm{tr}} \right\rangle}{\sigma_{\mathrm{re}}} \right]^n \left[ J_{\mathrm{vM}}(\boldsymbol{\sigma} - \boldsymbol{\beta}^1) - \beta^2 \right] . \tag{9.113}$$

The reference strain rate $\overset{\circ}{\varepsilon}_{\mathrm{re}}$, the reference stress $\sigma_{\mathrm{re}}$, and the threshold stress for creep $\sigma_{\mathrm{tr}}$ are nonnegative material parameters. Two different cases are evaluated here.

- If the term $J_{\mathrm{vM}}(\boldsymbol{\sigma} - \boldsymbol{\beta}^1) - \beta^2 - \sigma_{\mathrm{tr}} \leq 0$, due to the Macaulay brackets [see Definition (2.1)], the term $\left\langle J_{\mathrm{vM}}(\boldsymbol{\sigma} - \boldsymbol{\beta}^1) - \beta^2 - \sigma_{\mathrm{tr}} \right\rangle = 0$ and, therefore, Term (9.113) vanish.
- If the term $J_{\mathrm{vM}}(\boldsymbol{\sigma} - \boldsymbol{\beta}^1) - \beta^2 - \sigma_{\mathrm{tr}} > 0$ , the terms $\left\langle J_{\mathrm{vM}}(\boldsymbol{\sigma} - \boldsymbol{\beta}^1) - \beta^2 - \sigma_{\mathrm{tr}} \right\rangle$ and $J_{\mathrm{vM}}(\boldsymbol{\sigma} - \boldsymbol{\beta}^1) - \beta^2$ are nonnegative. Thus Term (9.113) is nonnegative.

The second term of Inequality (9.110) is evaluated next. Based on Problem 8.2, $\boldsymbol{\beta}^1 : \mathbf{b}^1 = \mathbf{b}^1 : \mathbf{b}^1$. The material parameter $\tilde{C}^{11}$ and the von Mises operator $J_{\mathrm{vM}}(\cdot)$ are nonnegative. The preceding facts imply that the second term of Inequality (9.110) is nonnegative.

The third term of Inequality (9.110) is considered next. The mathematical formulation of the force $\beta^2$, Expression (9.107), ensures that the force $\beta^2$

is nonnegative. The material parameter $C^{22}$ is nonnegative. Thus the third term of Inequality (9.110) is nonnegative. Since all three terms of Clausius–Duhem Inequality (9.110) are separately nonnegative, the sum is nonnegative, and therefore Clausius–Duhem Inequality (9.110) is satisfied.

## 9.7 Application of the Le Gac and Duval Model

The material model for creep by Le Gac and Duval [48] is applied for high-temperature material 10CrMo9-10 used in conventional power plant components. The creep data were obtained from a report by Mohrman and Riedel [64], and the values of the material parameters (see Table 9.1) were determined by the CURVEFIT code (see Santaoja [97]), which is based on the Levenberg–Marquardt method and contains a damping method that effectively prevents the values of the optimized material parameters from undergoing overly extensive changes during one iteration loop. With a damping method, the CURVEFIT provides much better values for the material parameters in cases in which there is a limited number of data for the curve-fitting process to determine these values. Based on Figure 9.8, the fit can be said to be good.

**Table 9.1** Fitted values of the material parameters of the creep model by Le Gac and Duval [48] for 10CrMo9-10 high-temperature power plant steel at 550 °C

| $\overset{\circ}{\varepsilon}_{\mathrm{re}}$ [1/s] | $\sigma_{\mathrm{re}}$ [MPa] | $\sigma_{\mathrm{tr}}$ [Pa] | $n$ | $m^1$ [GPa] |
|---|---|---|---|---|
| $1.060 \times 10^{-13}$ | 13.35 | 0.0173 | 5.284 | 0.3860 |
| $m^2$ [Pa/s] | $k$ | $n^1$ [GPa] | $n^2$ [Pa/s] | $m$ |
| $0.5730 \times 10^{-2}$ | 0.9198 | 0.2274 | 0.1801 | 4.042 |

## 9.8 Roles of Deviatoric Shear Stress and Internal Force

This section shows how the deviatoric stress tensor $\mathbf{s}$ and the deviatoric internal force $\mathbf{b}^1$ as the driving forces for thermoplasticity and creep ensure deformation without volume change. For deformation without volume change, the volume strain rate $\dot{e}$ is zero (see, e.g., Malvern [57], p. 134). For a small deformation, this is

$$\dot{e} = \mathbf{1} : \dot{\boldsymbol{\varepsilon}} = \dot{\varepsilon}_{11} + \dot{\varepsilon}_{22} + \dot{\varepsilon}_{33} = 0 \,. \tag{9.114}$$

In thermoplasticity and creep for metals, the driving force for deformation is the deviatoric shear stress $\mathbf{s}$, as Expressions (8.127), (8.138), (8.144), (8.152), (9.16), (9.43), (9.62), and (9.74) show. The expressions for inelastic strain rate tensor $\dot{\boldsymbol{\varepsilon}}^{\mathrm{i}}$ in the earlier-listed material models have the forms

$$\dot{\boldsymbol{\varepsilon}}^{\mathrm{i}} = f(\boldsymbol{\sigma}, \ldots)\,\mathbf{s} \qquad \text{or} \qquad \dot{\boldsymbol{\varepsilon}}^{\mathrm{i}} = g(\boldsymbol{\sigma}, \ldots)(\mathbf{s} - \mathbf{b}^1) \,, \tag{9.115}$$

where the inelastic strain rate tensor $\dot{\boldsymbol{\varepsilon}}^{\mathrm{i}}$ can be plastic strain rate tensor $\dot{\boldsymbol{\varepsilon}}^{\mathrm{p}}$ or the viscous strain rate tensor $\dot{\boldsymbol{\varepsilon}}^{\mathrm{v}}$ and where the functions $f(\boldsymbol{\sigma}, \ldots)$ and $g(\boldsymbol{\sigma}, \ldots)$

are scalar-valued functions of their arguments. The quantity $\mathbf{b}^1$ is the deviatoric part of the internal force $\boldsymbol{\beta}^1$.

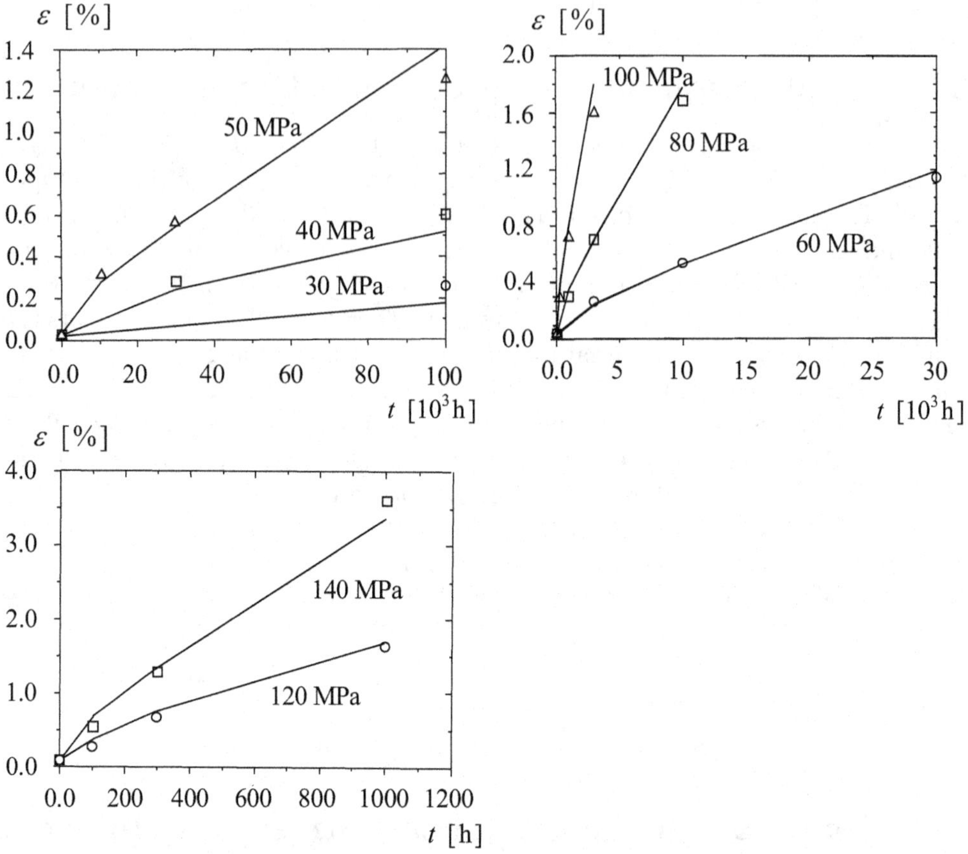


**Figure 9.8** Fitted creep curves for the 10 CrMo 9 10 steel of the model by Le Gac and Duval [48]. Redrawn from Mohrmann and Riedel [64]. First published in Santaoja ([97], Figure 14).

Equations (2.116) give

$$\mathbf{1}:\mathbf{s} = \mathbf{s}:\mathbf{1} \equiv 0 \qquad \text{and} \qquad \mathbf{1}:\mathbf{b}^1 = 0\,, \tag{9.116}$$

where Result $(9.116)_2$ utilizes the fact that Result $(9.116)_1$ is valid for all deviatoric second-order tensors. Equations (9.115) and (9.116) give

$$\mathbf{1}:\dot{\boldsymbol{\varepsilon}}^{\mathrm{i}} = f(\boldsymbol{\sigma},\ldots)\,\mathbf{1}:\mathbf{s} = 0 \quad \text{and} \quad \mathbf{1}:\dot{\boldsymbol{\varepsilon}}^{\mathrm{i}} = g(\boldsymbol{\sigma},\ldots)\,(\mathbf{1}:\mathbf{s} - \mathbf{1}:\mathbf{b}^1) = 0\,. \tag{9.117}$$

Comparison of Expressions (9.114) and (9.117) shows that Form (9.115) of the constitutive equation leads to a deformation without volume changes.

## 9.9 Material Model for Coupled Creep and Heat Conduction

This section derives a material model for coupled creep and heat conduction. The mechanical part of the material model is a simplified Le Gac and Duval [48] constitutive equation showing kinematic hardening only. The Le Gac and Duval creep model is evaluated in Section 9.6. The heat equation is that obtained from Fourier's law of heat conduction introduced in Section 8.2. The coupling means that the heat generated by material creep is taken into account as a heat source in the heat equation. This section does not include detailed derivation of the model but concentrates on the coupling between the mechanical model and the thermal model. Therefore it is suggested that the reader study Sections 8.2 and 9.6 before reading the present section.

The specific Helmholtz free energy $\psi$ is modelled as follows [cf. Expressions (9.80), (9.81), and $(9.90)_1$]:

$$\begin{aligned}\psi(\boldsymbol{\varepsilon}-\boldsymbol{\varepsilon}^{\mathrm{v}},\boldsymbol{\alpha},T) &= \psi_{\mathrm{r}}^{\mathrm{eT}} - s_{\mathrm{r}}\,(T-T_{\mathrm{r}}) + \frac{1}{2\,\rho_0}\,(\boldsymbol{\varepsilon}-\boldsymbol{\varepsilon}^{\mathrm{v}}):\mathbf{C}:(\boldsymbol{\varepsilon}-\boldsymbol{\varepsilon}^{\mathrm{v}}) \\ &\quad - \frac{E\,\hat{\alpha}}{\rho_0\,(1-2\,\nu)}\,\mathbf{1}:(\boldsymbol{\varepsilon}-\boldsymbol{\varepsilon}^{\mathrm{v}})\,(T-T_{\mathrm{r}}) - c_{\mathrm{r}}\,[T\,\ln(T/T_{\mathrm{r}}) - T] \\ &\quad + \frac{1}{2\,\rho_0}\,\boldsymbol{\alpha}:\mathbf{C}^1:\boldsymbol{\alpha}\,. \end{aligned} \tag{9.118}$$

As Material Model (9.118) shows, the controllable state variables are the strain tensor $\boldsymbol{\varepsilon}$ and the absolute temperature $T$. The internal variables are the creep strain tensor $\boldsymbol{\varepsilon}^{\mathrm{v}}$ and $\boldsymbol{\alpha}$, the latter being related to strain hardening. These state variables are the variables of the mechanical part of the problem. The thermal problem makes no contribution to the specific Helmholtz free energy $\psi$. In the present material model, it is assumed that the value of the stress tensor in the reference state $\boldsymbol{\sigma}_{\mathrm{r}}$ vanishes.

The specific dissipation function is assumed to be constructed on the mechanical part and on the thermal part. The mechanical part is expressed by the forces, that is, it is a specific complementary dissipation function $\varphi^{\mathrm{c}}_{\mathrm{mech}}$, whereas the thermal part is expressed by the flux, that is, it is a specific dissipation function $\varphi_{\mathrm{ther}}$. Therefore the following material model can be written [cf. Expressions (9.94) and (8.2)]:

$$\varphi^{\mathrm{c}}_{\mathrm{mech}}(\boldsymbol{\sigma},\boldsymbol{\beta}) = \varphi^{\mathrm{c1}}_{\mathrm{mech}}(\boldsymbol{\sigma},\boldsymbol{\beta}) + \varphi^{\mathrm{c2}}_{\mathrm{mech}}(\boldsymbol{\beta}) \quad \text{and} \quad \varphi_{\mathrm{ther}} = \phi_{\mathrm{ther}}(\vec{q};T)\,. \tag{9.119}$$

Material Model $(9.119)_2$ shows that the thermal part introduces one variable beyond the information given by the specific Helmholtz free energy $\psi$. This new variable is the heat conduction vector $\vec{q}$. The stress tensor $\boldsymbol{\sigma}$ and the internal force $\boldsymbol{\beta}$ are obtained from the specific Helmholtz free energy $\psi$ by applying the state equations.

Material Model (9.119) shows that, although a coupled creep–heat conduction model is evaluated, the dissipation potential is separated into two parts, that is, into a mechanical part and a thermal part. This is the consequence of a more fundamental separation: that of the Clausius–Duhem inequality into two separately nonnegative parts expressed by Inequalities (6.250) and (6.251), viz.

$$\Phi_{\text{mech}} = \boldsymbol{\sigma} : \dot{\boldsymbol{\varepsilon}}^{\text{v}} + \boldsymbol{\beta} : \dot{\boldsymbol{\alpha}} \;\; (\geq 0) \quad \text{and} \quad \Phi_{\text{ther}} = -\frac{\vec{\nabla} T}{T} \cdot \vec{q} \;\; (\geq 0) \,. \tag{9.120}$$

Separation (9.120) implies that in this coupled creep–heat conduction model, the dissipation potential is separated into a mechanical part and a thermal part. This means that the normality rule for the mechanical part of the coupled problem is separated from that of the thermal problem. Therefore the dissipation potentials for the mechanical part of the problem and for the thermal part of the problem can be homogeneous functions of different degree. This phenomenon provides more degrees of freedom in the modelling of the coupled mechanical-thermal problems.

The dissipation potentials on the right sides of Material Models (9.119) are [cf. Expressions (9.95), (9.96), and (8.2)]

$$\varphi^{\text{c1}}_{\text{mech}}(\boldsymbol{\sigma}, \boldsymbol{\beta}) = \frac{2\, \overset{\circ}{\varepsilon}_{\text{re}}\, \sigma_{\text{re}}}{3\, \rho_0\, (n+1)} \left[ \frac{\langle\, J_{\text{vM}}(\boldsymbol{\sigma} - \boldsymbol{\beta}) - \sigma_{\text{tr}} \,\rangle}{\sigma_{\text{re}}} \right]^{(n+1)} \tag{9.121}$$

and

$$\varphi^{\text{c2}}_{\text{mech}}(\boldsymbol{\beta}) = \frac{2\, \tilde{C}^{11}\, \sigma_{\text{re}}}{3\, \rho_0\, (k+1)} \left[ \frac{J_{\text{vM}}(\boldsymbol{\beta})}{\sigma_{\text{re}}} \right]^{(k+1)} , \tag{9.122}$$

and, finally,

$$\varphi_{\text{ther}}(\vec{q}; T) = \frac{\vec{q} \cdot \vec{q}}{2\, \mu_{\text{ther}}\, \rho_0\, \gamma(T)\, T} \,. \tag{9.123}$$

The mechanical part of the coupled material response equals that without coupling. Therefore, based on the investigation in Section 9.6, the following results can be written [cf. Expressions (9.77), $(9.88)_1$, $(9.87)_2$, (9.100), and (9.102)]. The strain tensor is assumed to be separable as follows:

$$\boldsymbol{\varepsilon} = \boldsymbol{\varepsilon}^{\text{e}} + \boldsymbol{\varepsilon}^{\text{Th}} + \boldsymbol{\varepsilon}^{\text{v}} \qquad \Rightarrow \qquad \boldsymbol{\varepsilon} - \boldsymbol{\varepsilon}^{\text{v}} = \boldsymbol{\varepsilon}^{\text{e}} + \boldsymbol{\varepsilon}^{\text{Th}} = \boldsymbol{\varepsilon}^{\text{eT}} \,, \tag{9.124}$$

where

$$\boldsymbol{\varepsilon}^{\text{e}} = \mathbf{S} : \boldsymbol{\sigma} \,, \qquad \boldsymbol{\varepsilon}^{\text{Th}} = \hat{\alpha}\, (T - T_{\text{r}})\, \mathbf{1} \qquad \text{and} \qquad \boldsymbol{\varepsilon}^{\text{eT}} = \mathbf{S} : \boldsymbol{\sigma} + \hat{\alpha}\, (T - T_{\text{r}})\, \mathbf{1} \,, \tag{9.125}$$

and, finally,

$$\dot{\boldsymbol{\varepsilon}}^{\text{v}} = \overset{\circ}{\varepsilon}_{\text{re}} \left[ \frac{\langle\, J_{\text{vM}}(\boldsymbol{\sigma} - \boldsymbol{\beta}) - \sigma_{\text{tr}} \,\rangle}{\sigma_{\text{re}}} \right]^{n} \frac{\mathbf{s} - \mathbf{b}}{J_{\text{vM}}(\boldsymbol{\sigma} - \boldsymbol{\beta})} \,. \tag{9.126}$$

The evolution equation for the internal force $\boldsymbol{\beta}$ is

$$\dot{\boldsymbol{\beta}} = m^1\, \dot{\boldsymbol{\varepsilon}}^{\mathrm{v}} - m^2 \left[ \frac{J_{\mathrm{vM}}(\boldsymbol{\beta})}{\sigma_{\mathrm{re}}} \right]^k \frac{\mathbf{b}}{J_{\mathrm{vM}}(\boldsymbol{\beta})}\,. \tag{9.127}$$

The thermal part of the problem is studied next.

Heat Equation (6.203) takes the following appearance:

$$\vec{\nabla}\cdot\vec{q} = \boldsymbol{\sigma}:\dot{\boldsymbol{\varepsilon}}^{\mathrm{v}} + \boldsymbol{\beta}:\dot{\boldsymbol{\alpha}} + \rho_0\, r + \rho_0\, T \left[ \frac{\partial^2 \psi}{\partial T \partial(\boldsymbol{\varepsilon} - \boldsymbol{\varepsilon}^{\mathrm{v}})} : (\dot{\boldsymbol{\varepsilon}} - \dot{\boldsymbol{\varepsilon}}^{\mathrm{v}}) + \frac{\partial^2 \psi}{\partial T \partial \boldsymbol{\alpha}} : \dot{\boldsymbol{\alpha}} + \frac{\partial^2 \psi}{\partial T^2} \dot{T} \right]. \tag{9.128}$$

Normality Rule (6.289) reads

$$-\frac{\vec{\nabla}T}{T} = \mu_{\mathrm{ther}}\, \rho_0\, \frac{\partial \varphi_{\mathrm{ther}}(\vec{q};\, \boldsymbol{\varepsilon} - \boldsymbol{\varepsilon}^{\mathrm{v}}, \boldsymbol{\alpha}, T, h)}{\partial \vec{q}}\,. \tag{9.129}$$

In Expressions (9.128) and (9.129), the inelastic strain tensor $\boldsymbol{\varepsilon}^{\mathrm{i}}$ is replaced with the viscous strain tensor $\boldsymbol{\varepsilon}^{\mathrm{v}}$ since in the present material model the inelastic strain tensor $\boldsymbol{\varepsilon}^{\mathrm{i}}$ is the viscous strain tensor $\boldsymbol{\varepsilon}^{\mathrm{v}}$.

Substitution of Material Model (9.123) into Normality Rule (9.129) gives

$$-\frac{\vec{\nabla}T}{T} = \frac{2\,\vec{q}}{2\,\gamma(T)\,T}\,, \qquad \text{which yields} \quad \vec{q} = -\,\gamma(T)\,\vec{\nabla}T\,. \tag{9.130}$$

Material Model (9.118) gives

$$\frac{\partial^2 \psi}{\partial T\, \partial(\boldsymbol{\varepsilon} - \boldsymbol{\varepsilon}^{\mathrm{v}})} = -\frac{E\,\hat{\alpha}}{\rho_0\,(1 - 2\,\nu)}\,\mathbf{1}\,, \quad \frac{\partial^2 \psi}{\partial T\, \partial \boldsymbol{\alpha}} = \mathbf{0}\,, \quad \text{and} \quad \frac{\partial^2 \psi}{\partial T^2} = -\frac{c_{\mathrm{r}}}{T}\,. \tag{9.131}$$

Substitution of Expressions $(9.130)_2$ and (9.131) into Equation (9.128) gives

$$\rho_0\, c_{\mathrm{r}}\, \dot{T} = \boldsymbol{\sigma}:\dot{\boldsymbol{\varepsilon}}^{\mathrm{v}} + \boldsymbol{\beta}:\dot{\boldsymbol{\alpha}} + \rho_0\, r + \frac{E\,\hat{\alpha}}{(1 - 2\,\nu)}\, T\, \mathbf{1}:(\dot{\boldsymbol{\varepsilon}} - \dot{\boldsymbol{\varepsilon}}^{\mathrm{v}}) + \vec{\nabla}\cdot[\gamma(T)\, \vec{\nabla}T]\,. \tag{9.132}$$

The heat source term $\rho_0\, r$ represents a non-mechanical heat source such as irradiation generated heat. In the present case it vanishes. Substitution of $r = 0$ into Expression (9.132) yields

$$\rho_0\, c_{\mathrm{r}}\, \dot{T} = \boldsymbol{\sigma}:\dot{\boldsymbol{\varepsilon}}^{\mathrm{v}} + \boldsymbol{\beta}:\dot{\boldsymbol{\alpha}} + \frac{E\,\hat{\alpha}}{(1 - 2\,\nu)}\, T\, \mathbf{1}:(\dot{\boldsymbol{\varepsilon}} - \dot{\boldsymbol{\varepsilon}}^{\mathrm{v}}) + \vec{\nabla}\cdot[\gamma(T)\, \vec{\nabla}T]\,. \tag{9.133}$$

Equation $(9.124)_2$ gives

$$\dot{\boldsymbol{\varepsilon}} - \dot{\boldsymbol{\varepsilon}}^{\mathrm{v}} = \dot{\boldsymbol{\varepsilon}}^{\mathrm{eT}} \tag{9.134}$$

Equation (9.134) allows Expression (9.133) to take the following appearance:

$$\rho_0\, c_{\mathrm{r}}\, \dot{T} = \boldsymbol{\sigma}:\dot{\boldsymbol{\varepsilon}}^{\mathrm{v}} + \boldsymbol{\beta}:\dot{\boldsymbol{\alpha}} + \frac{E\,\hat{\alpha}}{(1 - 2\,\nu)}\, T\, \mathbf{1}:\dot{\boldsymbol{\varepsilon}}^{\mathrm{v}} + \vec{\nabla}\cdot[\gamma(T)\, \vec{\nabla}T]\,. \tag{9.135}$$

The heat generated by creep (viscous) power is described by the first two terms on the right side of Equation (9.135). The third term gives the temperature change of the body related to the volume change due to thermoelastic deformation. The nonnegativeness of Clausius–Duhem inequalities need not be evaluated here, as this was done in Sections 8.2 and 9.6.

## 9.10 Summary

Some material models for creep were studied. The messages of Figures 9.1 and 9.2 should be known by everyone working on material modelling. Sections 9.2 and 9.3 introduced constitutive equations for the Maxwell solid and Kelvin–Voigt solid. The uniaxial versions of these models are based on spring–dashpot assembly. Since materials do not contain springs and dashpots, these elements cannot be used in the "verification" of these material models. However, as discussed in Sections 9.2 and 9.3, the three-dimensional counterparts of these material models may be useful for certain materials and, therefore, should be kept in mind when material modelling is done. Norton's law, studied in Section 9.4, describes the secondary creep only and has therefore too narrow a modelling capacity for engineering applications. Unfortunately, the time-hardening model introduced in Section 9.5 has been very popular among experimentalists, despite its severe problems, as demonstrated in Section 9.5. Therefore, in Section 9.5, the reader is guided to leave this model and replace it with the strain-hardening model, which is capable of describing primary creep. The background of the Le Gac and Duval [48] material model introduced in Sections 9.6 and 9.7 lies in the dislocation kinetics, as Section 10.5.10 shows. Furthermore, it is successfully applied to the high-temperature component material of power plants, as Section 9.7 shows, as well as to sea ice, as shown by Santaoja [98, 99] and Santaoja and Reddy [101]. Section 9.9 introduced a material model for coupled creep-heat conduction and derived Equation (9.133) for the description of heat conduction in a body showing Hookean deformation, thermal expansion, and creep.

# 10
# Damage

## 10.1 Introduction

This chapter studies damage. Section 10.2 introduces the concept of the representative volume element RVE, which creates a link between the noncontinuity of real materials and continuum mechanics, which is based on the assumption that the materials are continuous. Section 10.3 studies the response of a Hookean matrix deformation with penny-shaped microcracks, and Section 10.4 evaluates the same matrix deformation with spherical microvoids. The results of these two sections will be utilized in Section 10.5. Section 10.3 is quite cumbersome and, if the reader is not keen to study microcracks, can be left for later study, since Section 10.4 contains all the needed preliminary information for studying continuum damage mechanics. Sections 10.5.1–10.5.4 provide the core information on the continuum damage mechanics and should therefore be studied in depth. The uniaxial example on continuum damage mechanics given in Section 10.5.6 may give supporting information. Section 10.5.9 looks briefly at the Gurson–Tvergaard material model. At the end of this chapter a non-local material model for creep and damage is derived.

## 10.2 Representative Volume Element RVE

This section introduces the concept called the representative volume element RVE. In the theory of continuum thermodynamics and continuum mechanics, the field variables are assumed to be continuous functions. However, this assumption is violated when damage evolution creates microvoids or microcracks. Introduction of the representative volume element RVE solves this problem by replacing the discontinuous fields with continuous ones. Damage is not the only source for discontinuous fields in materials. Some materials have natural discontinuities.

Continuum mechanics and continuum thermodynamics describe the response of materials by continuous functions, the derivatives of which are continuous. This implies that the material has to be continuous, except of a finite number of discontinuity surfaces within the material.

Real materials do not fulfill the given continuity requirements. Material has a molecular structure. Furthermore, there are vacancies, dislocations, grain boundaries, for example, in crystalline materials. In polycrystalline ice, beyond the aforementioned flaws, there are also brine pockets and microcracks. Brine pockets are created during the freezing process of ice, when the dissolved salt is squeezed by the forming ice into pores. This results in pores containing water with a high salt content.

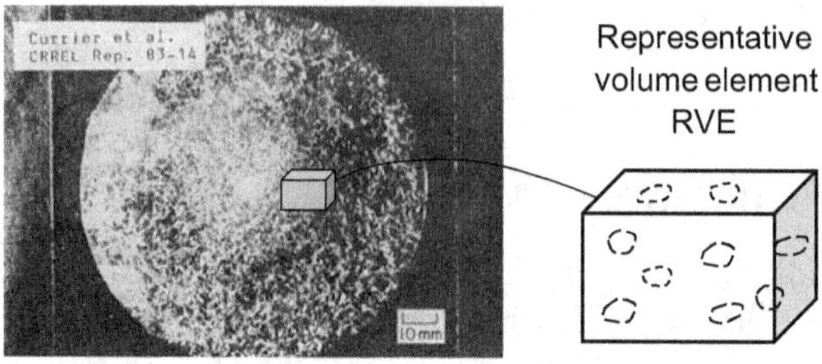


**Figure 10.1** Microcracks in laboratory ice (Currier et al. [19], Figure 32) and the representative volume element (RVE).

The preceding mismatch between real materials and the assumption of continuity is resolved by introducing representative volume element RVE (or representative elementary volume REV). The RVE is large enough to be statistically representative of the material properties to be modelled and small enough to be compared to the macroscopic structural dimensions. The latter means that the RVE has to be small enough to be treated as a material point in the analysis. By introducing the RVE, the material discontinuities are smoothed out and the governing variables and their derivatives of continuum mechanics and continuum thermodynamics can be assumed to be continuous functions. Representative volume element RVE is more or less a concept, but it does not have any strictly defined mathematical definition. However, Nemat-Nasser and Hori ([67], Section 1.3) give a more detailed discussion on RVE. Figure 10.1 shows a schematic view of a representative volume element RVE.

## 10.3 Response of a Hookean Matrix Deformation with Penny-Shaped Noninteracting Microcracks

This section evaluates the deformation of a material containing penny-shaped non-interacting microcracks in a matrix material the response of which is assumed to be linear elastic. The present material model can be used as part of a more extensive material model where also creep, for example, is modelled.

### 10.3.1 The Specific Gibbs Free Energy for a Material Containing Penny-Shaped Noninteracting Microcracks with a Hookean Matrix Deformation

This section gives an expression for the specific Gibbs free energy $g^{\mathrm{de}}$ for a material containing penny-shaped microcracks within a matrix that has a linear elastic deformation. According to Kachanov ([35], p. 1045), the complementary strain-energy density $w^{\mathrm{c}}$ for a material with penny-shaped non-interacting microcracks within a matrix that has a linear elastic response [see Figure 10.2(a)] is given by

$$w^{\mathrm{c}}(\boldsymbol{\sigma},\ldots)=\frac{1}{2}\left[\frac{1}{3\,(3\,\lambda+2\,\mu)}\,[\mathbf{1}:\boldsymbol{\sigma}]^2+\frac{1}{2\,\mu}\,\mathbf{s}:\mathbf{s}\right]+\frac{1}{V^{\mathrm{rve}}}\sum_{p=1}^{N}\frac{1}{2}\int_{A^p}(\vec{N}^p\cdot\boldsymbol{\sigma})\cdot\vec{b}^p\,\mathrm{d}A^p, \tag{10.1}$$

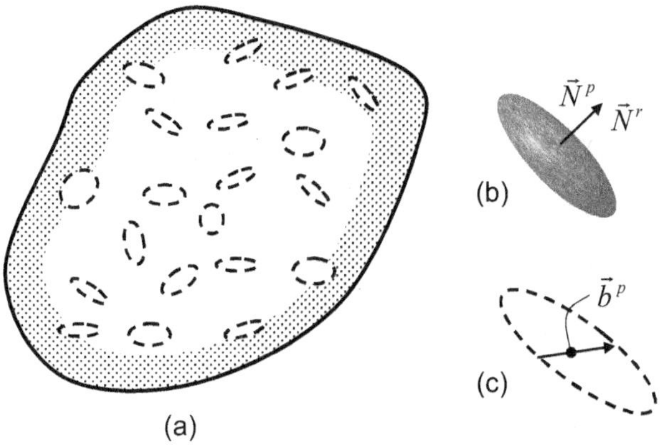


**Figure 10.2** (a) Penny-shaped microcracks in a solid. (b) The unit microcrack orientation vector is $\vec{N}^p$ ($\vec{N}^r$) and (c) the jump between the material points across the microcrack is $\vec{b}^p$.

where $V^{\mathrm{rve}}$ is the volume of the representative volume element RVE of the material in the initial microcrack-free configuration and $N$ is the number of microcracks within the volume $V^{\mathrm{rve}}$ (see Vakulenko and Kachanov [115], pp. 160 and 161). The area of the $p$th microcrack is denoted by $A^p$, and the unit normal vector to the surface $A^p$ is denoted by $\vec{N}^p$, as Figure 10.2(b) shows. The vector $\vec{b}^p$ describes the jump between the material points across the microcrack, as shown in Figure 10.2(c). It is worth noting that the value of the vector $\vec{b}^p$ varies on the surface of the microcrack $A^p$. The three dots in the expression $w^{\mathrm{c}}(\boldsymbol{\sigma},\ldots)$ indicate that the value of the complementary strain-energy density $w^{\mathrm{c}}$ also depends on the microcrack field, but the representative quantity of the microcrack field is not yet introduced. Here the appearance of the first part of Equation (10.1) is changed to be compatible with other equations in this book and the order of the terms within the summation Σ operator. The representative volume element $V^{\mathrm{rve}}$ of the material is introduced in Section 10.2.

Based on Kachanov ([35], p. 1045), the jump between the material points across the microcrack $\vec{b}^p$ reads

$$\vec{b}^p = \frac{16\,(1-\nu^2)}{(2-\nu)\,E\,\pi}\,\sqrt{(a^p)^2 - r^2}\,\Big[\boldsymbol{\sigma}\cdot\vec{N}^p - \frac{\nu}{2}\,\vec{N}^p\,(\vec{N}^p\cdot\boldsymbol{\sigma}\cdot\vec{N}^p)\Big]\,, \tag{10.2}$$

where $a^p$ is the radius of the $p$th microcrack and $r$ is the polar coordinate within a microcrack. In the original version for the vector $\vec{b}^p$, there were two misprints, which are corrected in Equation (10.2). In a further derivation, Kachanov [35] used Equation (10.2), not the one with the misprints. Santaoja [89, 90] added the Heaviside function $H(\vec{N}^p\cdot\boldsymbol{\sigma}\cdot\vec{N}^p)$ to Equation (10.2) to prevent penetration of the microcrack surfaces.

The complementary strain-energy density $w^{\mathrm{c}}(\boldsymbol{\sigma},\ldots)$ is proportional to the specific Gibbs free energy density $g(\boldsymbol{\sigma},\ldots,T)$ as follows (see, e.g., Fung [24], p. 349):

$$\rho_0\, g(\boldsymbol{\sigma},\ldots,T) \propto w^{\mathrm{c}}(\boldsymbol{\sigma},\ldots)\,. \tag{10.3}$$

In Expression (10.3), the dots ... represent a set of other (internal) variables.

When studying the effect of rectilinear noninteracting microcracks on the response of bodies, the quantities in the expression for the specific Gibbs free energy $g^{\mathrm{de}}$ were reorganized in such a way that allowed new internal variables called microcrack densities $Q^r$ to be introduced. Microcrack densities $Q^r$ are defined by (Santaoja [98, 99]; Santaoja and Reddy [101])

$$Q^r := \frac{m^r}{\rho_0\,V^{\mathrm{rve}}}, \qquad r = 1,\ldots,M\,. \tag{10.4}$$

Based on the foregoing discussion, the specific Gibbs free energy $g^{\mathrm{de}}$ for a Hookean matrix deformation with noninteracting penny-shaped microcracks is written as follows:

$$\begin{aligned} g^{\mathrm{de}}(\boldsymbol{\sigma},Q^r) = {} & \frac{1}{2\,\rho_0}\left[\frac{1}{3\,(3\,\lambda+2\,\mu)}\,[\mathbf{1}:\boldsymbol{\sigma}]^2 + \frac{1}{2\,\mu}\,\mathbf{s}:\mathbf{s}\right] \\ & + \frac{16\,(1-\nu^2)}{3\,(2-\nu)\,E\,\pi^{3/2}}\,\sum_{r=1}^{M} Q^r (A^r)^{3/2}\Big\{\vec{N}^r\cdot\boldsymbol{\sigma}\cdot\boldsymbol{\sigma}\cdot\vec{N}^r \\ & - \big[1 - H(\vec{N}^r\cdot\boldsymbol{\sigma}\cdot\vec{N}^r) + \tfrac{\nu}{2}\,H(\vec{N}^r\cdot\boldsymbol{\sigma}\cdot\vec{N}^r)\big]\,(\vec{N}^r\cdot\boldsymbol{\sigma}\cdot\vec{N}^r)^2\Big\}\,. \end{aligned} \tag{10.5}$$

In Definition (10.4), the notation $m^r$ is the number of microcracks within the $r$th microcrack group. Based on Definition (10.5), the microcrack densities $Q^r$ stand for the number of microcracks in a microcrack group $r$ per unit mass occupied by the representative volume element $V^{\mathrm{rve}}$. The meaning of the term *microcrack group* and the roles of the quantity $M$ and the Heaviside function $H(\vec{N}^r \cdot \boldsymbol{\sigma} \cdot \vec{N}^r)$ are studied later.

As Figure 10.3 shows, penny-shaped microcracks have two candidates for the direction of the unit microcrack orientation vector $\vec{N}^r$. Thus, the direction of the unit microcrack orientation vector $\vec{N}^r$ is not unique, but the opposite direction of $\vec{N}^r$ leads to the same value for the specific Gibbs free energy $g^{\mathrm{de}}$, since every term on the third line of Equation (10.5) has the unit microcrack orientation vector $\vec{N}^r$ twice.

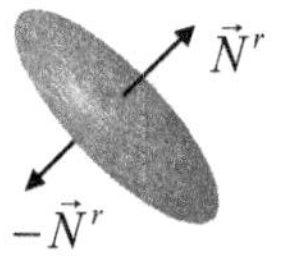


**Figure 10.3** Two candidates for the unit microcrack orientation vector $\vec{N}^r$.

In Expressions (10.1) and (10.5), the summation indices are different. The index $p$ in Expression (10.1) refers to the $p$th microcrack, whereas the index $r$ in Expression (10.5) refers to the $r$th microcrack group. It is important to distinguish these two meanings of the summation indices.

Equation (10.5) has four major enhancements to the original proposed by Kachanov [35]. First, microcracks are collected into microcrack groups by same size and orientation. The introduction of microcrack groups is of course only an approximation, given that the size and orientation of microcracks can be randomly distributed. However, microcrack groups make computation faster, and the approximation error can be neglected by increasing the number of microcrack groups. In Expressions (10.4) and (10.5), the number of microcrack groups is denoted by $M$.

The second enhancement involves the introduction of microcrack densities $Q^r$, which enter into the formulation of continuum thermodynamics as internal variables. The strong physical foundation of these internal variables makes them more attractive quantities for continuum damage mechanics than variable damage (scalar, vector, or tensor), the physical background of which is sometimes unclear.

The third enhancement is associated to the second one. As the expression for the microcrack densities rate $\dot{Q}^r$ will show, the microcrack densities $Q^r$ are assumed to be continuous functions, which simplifies mathematical analysis and computer coding. This choice is supported the fact, that from the continuum thermodynamics point of view, it is not enough to study the nonnegativeness of the Clausius–Duhem inequality just before and after the formation of a microcrack, but it is important to include the formation process in the study. Of course, this assumption leads to an odd number of microcracks, let's say, 40.27. The model is anyway an approximation of the physical reality, which makes the decimal number of microcracks acceptable.

Figure 10.4 shows a microcrack and two coordinate systems. The coordinate system $(X_1, X_2, X_3)$ is the initial material coordinate system, and the frame $(Z_1^r, Z_2^r, Z_3^r)$ is the microcrack coordinate system showing the orientation of a particular microcrack group, that is, the $r$th microcrack group, which implies that the orientation of the frame $(Z_1^r, Z_2^r, Z_3^r)$ may vary from microcrack group to microcrack group. The sizes of microcracks within two microcrack groups can differ, but they can have the same orientation. The basis of the microcrack coordinate system $(Z_1^r, Z_2^r, Z_3^r)$ is $(\vec{k}_1^r, \vec{k}_2^r, \vec{k}_3^r)$.

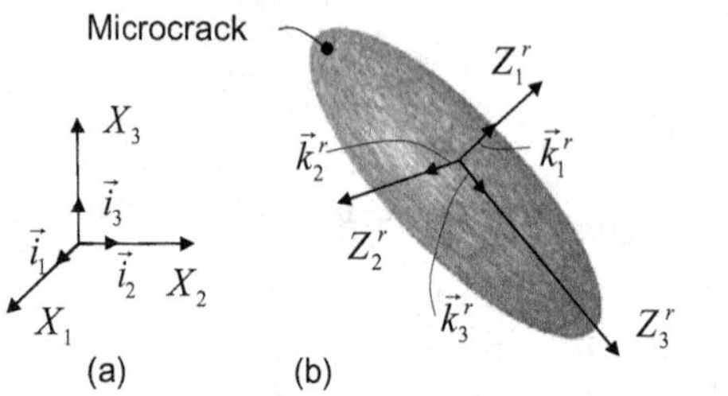


**Figure 10.4** (a) The initial material coordinate system and (b) the microcrack coordinate system.

The fourth enhancement involves the Heaviside function $H(\vec{N}^r \cdot \boldsymbol{\sigma} \cdot \vec{N}^r)$. Kachanov [35] wrote his expression for the complementary strain-energy density $w^c$ in a form that does not apply to compression. Santaoja [89, 90] introduced the Heaviside function $H(\vec{N}^r \cdot \boldsymbol{\sigma} \cdot \vec{N}^r)$ to extend the work by Kachanov [35] also to cover compression. The effect of the Heaviside function $H(\vec{N}^r \cdot \boldsymbol{\sigma} \cdot \vec{N}^r)$ is studied next. When expressed in the microcrack frame $(Z_1^r, Z_2^r, Z_3^r)$, the stress tensor $\boldsymbol{\sigma}$ and the unit microcrack orientation vector $\vec{N}^r$ take the following appearances:

$$\boldsymbol{\sigma} = \sigma_{ij}^r \, \vec{k}_i^r \vec{k}_j^r \,, \qquad \vec{N}^r = \vec{k}_1^r \,. \tag{10.6}$$

The superscript $r$ in Equation $(10.6)_1$ indicates that the scalar component is expressed in the microcrack coordinate system $(Z_1^r, Z_2^r, Z_3^r)$. Expressions (10.6) give

$$\vec{N}^r \cdot \boldsymbol{\sigma} \cdot \boldsymbol{\sigma} \cdot \vec{N}^r = \sigma_{11}^r \sigma_{11}^r + \sigma_{12}^r \sigma_{21}^r + \sigma_{13}^r \sigma_{31}^r \qquad \vec{N}^r \cdot \boldsymbol{\sigma} \cdot \vec{N}^r = \sigma_{11}^r \,. \tag{10.7}$$

Substitution of Equations (10.7) into Equation (10.5) gives

$$\begin{aligned} g^{\mathrm{de}}(\boldsymbol{\sigma}, Q^r) &\propto \left\{ \sigma_{11}^r \sigma_{11}^r + \sigma_{12}^r \sigma_{21}^r + \sigma_{13}^r \sigma_{31}^r - \left[1 - H(\sigma_{11}^r) + \tfrac{\nu}{2} H(\sigma_{11}^r)\right] \left[\sigma_{11}^r\right]^2 \right\} \\ &= \left(1 - \tfrac{\nu}{2}\right) H(\sigma_{11}^r)\, \sigma_{11}^r \sigma_{11}^r + \sigma_{12}^r \sigma_{21}^r + \sigma_{13}^r \sigma_{31}^r \,. \end{aligned} \tag{10.8}$$

As Expression (10.8) shows, the role of the Heaviside function $H(\vec{N}^r \cdot \boldsymbol{\sigma} \cdot \vec{N}^r)$ is to neglect the influence of the compressive normal stress $\sigma_{11}^r$ on the value of the specific Gibbs free energy $g^{\mathrm{de}}$. This implies that the Heaviside function $H(\vec{N}^r \cdot \boldsymbol{\sigma} \cdot \vec{N}^r)$ prevents the microcrack surfaces from penetrating each other under compression, that is, when $\sigma_{11}^r < 0$, as sketched in Figure 10.5. Section 10.3.7 gives a detailed derivation for the Heaviside function $H(\vec{N}^r \cdot \boldsymbol{\sigma} \cdot \vec{N}^r)$.

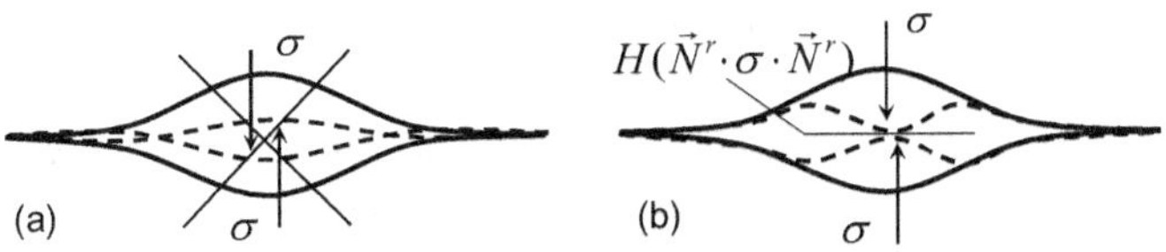


**Figure 10.5** (a) Surfaces of a microcrack penetrate each other under compression. (b) Penetration is prevented by the Heaviside function $H(\vec{N}^r \cdot \boldsymbol{\sigma} \cdot \vec{N}^r)$.

Equation (10.5) is not the only way to apply the Heaviside function denoted by $H(\vec{N}^r \cdot \boldsymbol{\sigma} \cdot \vec{N}^r)$ to enhance the expression for the complementary strain-energy density $w^{\mathrm{c}}$ by Kachanov [35]. The following enhancement can be introduced:

$$
\begin{aligned}
{}^1g^{\mathrm{de}}(\boldsymbol{\sigma}, Q^r) &= \frac{1}{2\,\rho_0}\left[\frac{1}{3\,(3\,\lambda + 2\,\mu)}\,[\mathbf{1}:\boldsymbol{\sigma}]^2 + \frac{1}{2\,\mu}\,\mathrm{s}:\mathrm{s}\right] + \frac{16\,(1-\nu^2)}{3\,(2-\nu)\,E\,\pi^{3/2}} \\
&\quad * \sum_{r=1}^{M} Q^r (A^r)^{3/2}\Big\{\,H(\vec{N}^r\cdot\boldsymbol{\sigma}\cdot\vec{N}^r)\,\big[\vec{N}^r\cdot\boldsymbol{\sigma}\cdot\boldsymbol{\sigma}\cdot\vec{N}^r - \frac{\nu}{2}\,(\vec{N}^r\cdot\boldsymbol{\sigma}\cdot\vec{N}^r)^2\big]\Big\}.
\end{aligned}
\tag{10.9}
$$

Substitution of Equations (10.7) into Equation (10.9) gives

$$
{}^1g^{\mathrm{de}}(\boldsymbol{\sigma}, Q^r) \propto \Big\{\,H(\sigma_{11}^r)\,\big[(1-\tfrac{\nu}{2})\,\sigma_{11}^r\,\sigma_{11}^r + \sigma_{12}^r\,\sigma_{21}^r + \sigma_{13}^r\,\sigma_{31}^r\,\big]\Big\}. \tag{10.10}
$$

Equation (10.10) shows that the first type of the specific Gibbs free energy ${}^1g^{\mathrm{de}}$ is for materials in which the forming surfaces of the microcracks are rough, so that even a small compressive stress prevents sliding between the microcrack surfaces. The second type of specific Gibbs free energy $g^{\mathrm{de}}$, Equation (10.5), is for materials with negligible friction between microcrack surfaces. Ice, for example, has low friction between microcrack surfaces. The specific Gibbs free energy $g^{\mathrm{de}}$ is applied in what follows.

### 10.3.2 Strain Tensor Difference $\boldsymbol{\varepsilon} - \boldsymbol{\varepsilon}^{\mathrm{i}}$ and Internal Force $Y^{\mathrm{r}}$

This section derives the explicit form for the strain tensor difference $\boldsymbol{\varepsilon} - \boldsymbol{\varepsilon}^{\mathrm{i}}$ and for the internal force $Y^r$ by assuming a general material model.

A material model is introduced by assuming the following general form for the specific Gibbs free energy $g$:

$$
g(\boldsymbol{\sigma}, Q^r, \boldsymbol{\alpha}, T) = g^{\mathrm{de}}(\boldsymbol{\sigma}, Q^r) + g^{\mathrm{rest}}(\boldsymbol{\alpha}, T), \tag{10.11}
$$

where $g^{\mathrm{rest}}(\boldsymbol{\alpha}, T)$ stands for the rest part of the specific Gibbs free energy $g(\boldsymbol{\sigma}, Q^r, \boldsymbol{\alpha}, T)$. In Material Model (10.11), the internal variable $\boldsymbol{\alpha}$ represents of the set of internal variables that are necessary for the presentation of the material response. Although here these variables are given as one second-order tensor $\boldsymbol{\alpha}$, they can be more, and they can be scalars, vectors, or tensors of any other order.

The following results are needed for the forthcoming derivation. Result (2.147) and that of Problem 2.26 yield

$$\frac{\partial}{\partial \boldsymbol{\sigma}}(\vec{a}\cdot\boldsymbol{\sigma}\cdot\boldsymbol{\sigma}\cdot\vec{b}) = \vec{a}\,\boldsymbol{\sigma}\cdot\vec{b} + \vec{a}\cdot\boldsymbol{\sigma}\,\vec{b} \qquad \frac{\partial}{\partial \boldsymbol{\sigma}}(\vec{a}\cdot\boldsymbol{\sigma}\cdot\vec{b}) = \vec{a}\,\vec{b}\,. \tag{10.12}$$

The variables $\vec{a}$ and $\vec{b}$ in Expression (10.12) are arbitrary vectors. Results (2.135) and (2.142) and Definition $(2.108)_1$ give

$$\frac{\partial[\mathbf{1}:\boldsymbol{\sigma}]^2}{\partial \boldsymbol{\sigma}} = 2\,[\mathbf{1}:\boldsymbol{\sigma}]\,\mathbf{1} \qquad \frac{\partial}{\partial \boldsymbol{\sigma}}(\mathbf{s}:\mathbf{s}) = 2\,\mathbf{s} = 2\,\mathbf{K}:\boldsymbol{\sigma}\,. \tag{10.13}$$

In the model investigated here, Expression (10.11), the specific Gibbs free energy $g^{\mathrm{de}}$ is the only part of the (total) specific Gibbs free energy $g$ that is dependent on the stress tensor $\boldsymbol{\sigma}$. Thus, based on State Equation $(6.181)_1$, Material Model (10.5), and Results (10.12) and (10.13), the following is obtained:

$$\begin{aligned} \boldsymbol{\varepsilon} - \boldsymbol{\varepsilon}^{\mathrm{i}} = \rho_0\,\frac{\partial g^{\mathrm{de}}(\boldsymbol{\sigma}, Q^r)}{\partial \boldsymbol{\sigma}} &= \frac{1}{3\,(3\,\lambda + 2\,\mu)}\,\mathbf{1}\,\mathbf{1}:\boldsymbol{\sigma} + \frac{1}{2\,\mu}\,\mathbf{K}:\boldsymbol{\sigma} \\ &+ \frac{16\,(1-\nu^2)\,\rho_0}{3\,(2-\nu)\,E\,\pi^{3/2}} \sum_{r=1}^{M} Q^r (A^r)^{3/2} \Big\{ \vec{N}^r\,\boldsymbol{\sigma}\cdot\vec{N}^r + \vec{N}^r\cdot\boldsymbol{\sigma}\,\vec{N}^r \\ &- \big[1 - H(\vec{N}^r\cdot\boldsymbol{\sigma}\cdot\vec{N}^r) + \frac{\nu}{2}\,H(\vec{N}^r\cdot\boldsymbol{\sigma}\cdot\vec{N}^r)\big]\,\big[2\,\vec{N}^r\,\vec{N}^r\,(\vec{N}^r\cdot\boldsymbol{\sigma}\cdot\vec{N}^r)\big]\Big\}\,. \end{aligned} \tag{10.14}$$

Based on the right side of Expression (10.14), the strain tensor difference $\boldsymbol{\varepsilon} - \boldsymbol{\varepsilon}^{\mathrm{i}}$ consists of two parts. The first line on the right side of Expression (10.14) gives the elastic strain tensor $\boldsymbol{\varepsilon}^{\mathrm{e}}$ (it will be proved later), and the two latter lines give the damage strain tensor $\boldsymbol{\varepsilon}^{\mathrm{d}}$. Thus, for the present material model, the following holds:

$$\boldsymbol{\varepsilon}^{\mathrm{de}} := \boldsymbol{\varepsilon}^{\mathrm{e}} + \boldsymbol{\varepsilon}^{\mathrm{d}} = \boldsymbol{\varepsilon} - \boldsymbol{\varepsilon}^{\mathrm{i}}\,, \tag{10.15}$$

where $\boldsymbol{\varepsilon}^{\mathrm{de}}$ is the damage-elastic strain tensor. The damage-elastic strain tensor $\boldsymbol{\varepsilon}^{\mathrm{de}}$ is needed for when it cannot be divided into two parts, as in the center, of Expression (10.15).

Based on Expressions (10.14) and (10.15) and the preceding discussion, the elastic strain tensor $\boldsymbol{\varepsilon}^{\mathrm{e}}$ reads

$$\boldsymbol{\varepsilon}^{\mathrm{e}} = \frac{1}{3\,(3\,\lambda + 2\,\mu)}\,\mathbf{1}\,\mathbf{1}:\boldsymbol{\sigma} + \frac{1}{2\,\mu}\,\mathbf{K}:\boldsymbol{\sigma}\,. \tag{10.16}$$

Definition $(2.108)_2$ is

$$\mathbf{K} := \mathbf{I} - \frac{1}{3}\,\mathbf{1}\,\mathbf{1}\,, \tag{10.17}$$

where $\mathbf{I}$ is the fourth-order identity tensor defined in Definition (2.69) and $\mathbf{1}$ is the second-order identity tensor defined in Definition (2.63). Definitions (2.69) and (2.106) give

$$\mathbf{I} : \boldsymbol{\sigma} = \boldsymbol{\sigma}, \qquad \mathbf{I}^{\mathrm{s}} : \boldsymbol{\sigma} = \frac{1}{2}(\boldsymbol{\sigma} + \boldsymbol{\sigma}^{\mathrm{T}}) = \boldsymbol{\sigma}, \tag{10.18}$$

where the symmetry of the stress tensor $\boldsymbol{\sigma}$ is exploited. Based on Equations (10.17) and (10.18), we arrive at the following result:

$$\mathbf{K} : \boldsymbol{\sigma} = \mathbf{I} : \boldsymbol{\sigma} - \frac{1}{3}\mathbf{1}\,\mathbf{1} : \boldsymbol{\sigma} = \mathbf{I}^{\mathrm{s}} : \boldsymbol{\sigma} - \frac{1}{3}\mathbf{1}\,\mathbf{1} : \boldsymbol{\sigma}. \tag{10.19}$$

Substitution of Result (10.19) into Expression (10.16) yields

$$\boldsymbol{\varepsilon}^{\mathrm{e}} = \frac{1}{3\,(3\lambda + 2\,\mu)}\,\mathbf{1}\,\mathbf{1}{:}\boldsymbol{\sigma} + \frac{1}{2\,\mu}\left[\mathbf{I}^{\mathrm{s}} : \boldsymbol{\sigma} - \frac{1}{3}\mathbf{1}\,\mathbf{1} : \boldsymbol{\sigma}\right]. \tag{10.20}$$

Expression (10.20) can be recast in the form

$$\begin{aligned}
\boldsymbol{\varepsilon}^{\mathrm{e}} &= \left[\frac{1}{3\,(3\,\lambda + 2\,\mu)} - \frac{1}{3 \times 2\,\mu}\right]\mathbf{1}\,\mathbf{1} : \boldsymbol{\sigma} + \frac{1}{2\,\mu}\,\mathbf{I}^{\mathrm{s}}{:}\boldsymbol{\sigma} \\
&= \frac{2\,\mu - (3\,\lambda + 2\,\mu)}{3 \times 2\,\mu\,(3\,\lambda + 2\,\mu)}\,\mathbf{1}\,\mathbf{1} : \boldsymbol{\sigma} + \frac{1}{2\,\mu}\,\mathbf{I}^{\mathrm{s}} : \boldsymbol{\sigma} \\
&= -\frac{\lambda}{2\,\mu\,(3\,\lambda + 2\,\mu)}\,\mathbf{1}\,\mathbf{1} : \boldsymbol{\sigma} + \frac{1}{2\,\mu}\,\mathbf{I}^{\mathrm{s}} : \boldsymbol{\sigma}.
\end{aligned} \tag{10.21}$$

The right side of Expressions (10.21) equals Equation (8.39). Thus Equation (8.41) can be applied in this case. It is

$$\boldsymbol{\varepsilon}^{\mathrm{e}} = \mathbf{S} : \boldsymbol{\sigma} \qquad \text{or} \qquad \boldsymbol{\varepsilon} - \boldsymbol{\varepsilon}^{\mathrm{i}} = \mathbf{S} : \boldsymbol{\sigma}, \tag{10.22}$$

where the compliance tensor for a Hookean deformation $\mathbf{S}$ is defined by [see Equation (8.42)]

$$\mathbf{S} := -\frac{\nu}{E}\,\mathbf{1}\,\mathbf{1} + \frac{1 + \nu}{E}\,\mathbf{I}^{\mathrm{s}}. \tag{10.23}$$

The elastic strain tensor $\boldsymbol{\varepsilon}^{\mathrm{e}}$ is the "average" deformation of the matrix material between the microcracks. Similarly to the preceding study, in the model studied here, the specific Gibbs free energy $g^{\mathrm{de}}$ is the only part of the (total) specific Gibbs free energy $g$ that is dependent on the microcrack densities $Q^r$. Thus, applying State Equation $(6.181)_2$ and using Material Model (10.5), the internal forces ($r = 1, \ldots, M$) are found to take the following appearance:

$$\begin{aligned}
Y^r = \rho_0\,\frac{\partial g^{\mathrm{de}}(\boldsymbol{\sigma}, Q^r)}{\partial Q^r} &= \frac{16\,(1 - \nu^2)\,\rho_0}{3\,(2 - \nu)\,E\,\pi^{3/2}}\,(A^r)^{3/2} \\
&* \left\{\vec{N}^r\!\cdot\boldsymbol{\sigma}\cdot\boldsymbol{\sigma}\cdot\vec{N}^r - \left[1 - H(\vec{N}^r\!\cdot\boldsymbol{\sigma}\cdot\vec{N}^r) + \frac{\nu}{2}\,H(\vec{N}^r\!\cdot\boldsymbol{\sigma}\cdot\vec{N}^r)\right](\vec{N}^r\!\cdot\boldsymbol{\sigma}\cdot\vec{N}^r)^2\right\}.
\end{aligned} \tag{10.24}$$

### 10.3.3 Scalar Components of the Compliance Tensor $\mathbf{S}^{dr}$

This section derives the expressions for scalar components of the fourth-order tensor $\mathbf{S}^{dr}$, which is the fourth-order compliance tensor related to a single microcrack group and expressed in the microcrack coordinate system $(Z_1^r, Z_2^r, Z_3^r)$.

One microcrack group, that is, a set of uniaxial equalized microcracks, is studied in the microcrack frame $(Z_1^r, Z_2^r, Z_3^r)$. Expressions (10.6) and $(10.7)_2$ are

$$\boldsymbol{\sigma} = \sigma_{ij}^r \vec{k}_i^r \vec{k}_j^r , \qquad \vec{N}^r = \vec{k}_1^r , \qquad \text{and} \qquad \vec{N}^r \cdot \boldsymbol{\sigma} \cdot \vec{N}^r = \sigma_{11}^r . \tag{10.25}$$

Based on Expressions $(10.25)_1$ and $(10.25)_2$, the following is achieved:

$$\vec{N}^r \boldsymbol{\sigma}^r \cdot \vec{N}^r = \sigma_{i1}^r \vec{k}_1^r \vec{k}_i^r = \sigma_{11}^r \vec{k}_1^r \vec{k}_1^r + \sigma_{21}^r \vec{k}_1^r \vec{k}_2^r + \sigma_{31}^r \vec{k}_1^r \vec{k}_3^r \tag{10.26}$$

and

$$\vec{N}^r \cdot \boldsymbol{\sigma}^r \vec{N}^r = \sigma_{1j}^r \vec{k}_j^r \vec{k}_1^r = \sigma_{11}^r \vec{k}_1^r \vec{k}_1^r + \sigma_{12}^r \vec{k}_2^r \vec{k}_1^r + \sigma_{13}^r \vec{k}_3^r \vec{k}_1^r . \tag{10.27}$$

Substitution of Expressions $(10.25)_3$, (10.26), and (10.27) into Equation (10.14) gives

$$\begin{aligned} \boldsymbol{\varepsilon}^{dr} = B_1 Q^r (A^r)^{3/2} \Big\{ & 2 \sigma_{11}^r \vec{k}_1^r \vec{k}_1^r + \sigma_{21}^r \vec{k}_1^r \vec{k}_2^r + \sigma_{12}^r \vec{k}_2^r \vec{k}_1^r + \sigma_{31}^r \vec{k}_1^r \vec{k}_3^r \\ & + \sigma_{13}^r \vec{k}_3^r \vec{k}_1^r - \left[1 - H(\sigma_{11}^r) + \tfrac{\nu}{2} H(\sigma_{11}^r)\right] \left[2 \sigma_{11}^r \vec{k}_1^r \vec{k}_1^r\right] \Big\} , \end{aligned} \tag{10.28}$$

where

$$B_1 := \frac{16 (1 - \nu^2) \rho_0}{3 (2 - \nu) E \pi^{3/2}} . \tag{10.29}$$

Expression (10.28) reduces to

$$\begin{aligned} \boldsymbol{\varepsilon}^{dr} = B_1 Q^r (A^r)^{3/2} \Big\{ & (2 - \nu) H(\sigma_{11}^r) \sigma_{11}^r \vec{k}_1^r \vec{k}_1^r + \sigma_{21}^r \vec{k}_1^r \vec{k}_2^r + \sigma_{12}^r \vec{k}_2^r \vec{k}_1^r \\ & + \sigma_{31}^r \vec{k}_1^r \vec{k}_3^r + \sigma_{13}^r \vec{k}_3^r \vec{k}_1^r \Big\} . \end{aligned} \tag{10.30}$$

In Equation (10.30), the superscript $r$ denotes that the damage strain tensor $\boldsymbol{\varepsilon}^{dr}$ is caused by the $r$th microcrack group (i.e., uniaxial microcrack field with equalized microcracks) and that it is expressed in the microcrack coordinate system $(Z_1^r, Z_2^r, Z_3^r)$ . The superscript $r$ in the stress tensor component $\sigma_{ij}^r$ indicates that it is given in the microcrack coordinate system $(Z_1^r, Z_2^r, Z_3^r)$. Since the stress tensor $\boldsymbol{\sigma}^r$ is symmetric, Equation (10.30) shows that the damage strain tensor $\boldsymbol{\varepsilon}^{dr}$ is symmetric as well. The superscript $r$ in the variables $Q^r$ and $A^r$ refers to the $r$th microcrack group. Since the variables $Q^r$ and $A^r$ are scalar-valued quantities, they are not referred to any coordinate system. The summation symbol $\Sigma$ is not present in Expressions (10.28) and (10.30), since it refers to the summation over the microcrack groups and now a single microcrack group is under consideration.

Equation (10.30) can be written in the following forms:

$$\boldsymbol{\varepsilon}^{\mathrm{d}r} = \mathbf{S}^{\mathrm{d}r} : \boldsymbol{\sigma}^{r} \qquad \text{or} \qquad \varepsilon^{\mathrm{d}r}_{ij} = S^{\mathrm{d}r}_{ijkl}\, \sigma^{r}_{kl} \,. \tag{10.31}$$

In Equation $(10.31)_1$, the notation $\mathbf{S}^{\mathrm{d}r}$ is the fourth-order compliance tensor describing the influence of the $r$th microcrack group expressed in the microcrack coordinate system $(Z^r_1, Z^r_2, Z^r_3)$.

Based on Equation (10.30), the compliance tensor $\mathbf{S}^{\mathrm{d}r}$ is not minor symmetric in the last pair of indices. However, the compliance tensor $\mathbf{S}^{\mathrm{d}r}$ can easily be made symmetric by utilizing the symmetry of the stress tensor $\boldsymbol{\sigma}$. This task is carried out in **Appendix J**, where, slightly misleadingly, also the new symmetric compliance tensor is denoted by $\mathbf{S}^{\mathrm{d}r}$. This was done to avoid introduction of new notations and may therefore be an acceptable act.

Based on **Appendix J**, Equation (10.30) gives the following expressions for the compliance tensor $\mathbf{S}^{\mathrm{d}r}$:

$$S^{\mathrm{d}r}_{1111} = (2-\nu)\, B_1\, Q^r\, (A^r)^{3/2}\, H(\sigma^r_{11}) \tag{10.32}$$

and

$$S^{\mathrm{d}r}_{1212} = S^{\mathrm{d}r}_{1221} = S^{\mathrm{d}r}_{2112} = S^{\mathrm{d}r}_{2121} = \frac{1}{2}\, B_1\, Q^r\, (A^r)^{3/2} \tag{10.33}$$

and finally

$$S^{\mathrm{d}r}_{1313} = S^{\mathrm{d}r}_{1331} = S^{\mathrm{d}r}_{3113} = S^{\mathrm{d}r}_{3131} = \frac{1}{2}\, B_1\, Q^r\, (A^r)^{3/2} \,. \tag{10.34}$$

Results (10.32)–(10.34) do not require any thermodynamics but can be applied with pure mechanical study.

### 10.3.4 Damage Strain Tensor due to a Multidirectional Microcrack Field

The previous section studied the influence of a single microcrack group on the deformation due to the damage and gave its results in the current microcrack coordinate system $(Z^r_1, Z^r_2, Z^r_3)$. This section extends the view by taking into account several differently oriented microcrack groups and by expressing the results in the initial material coordinate system $(X_1, X_2, X_3)$. This approach is possible, since the microcracks were assumed to be noninteracting ones, which makes it possible to study each microcrack group independently of the other ones and to obtain the response of the whole microcrack system as a sum of individual microcrack groups. If the material contains microcracks the sizes of which are different, every size of microcracks has to be described by a microcrack group.

Because of many single microcracks, there is in general a distance between the origin of the initial material coordinate system $(X_1, X_2, X_3)$ and that of the microcrack coordinate system $(Z_1^r, Z_2^r, Z_3^r)$, as shown in Figure 10.6. Since deformations are evaluated as derivatives of the displacements with respect to the coordinates, the influence of the translation of the coordinate system vanishes. The basis $(\vec{k}_1^r, \vec{k}_2^r, \vec{k}_3^r)$ of the microcrack coordinate system $(Z_1^r, Z_2^r, Z_3^r)$ is obtained by rotating the initial material coordinate system $(X_1, X_2, X_3)$ first about the $X_3$-axis by an angle $\alpha$ and second about the $X_2$-axis by an angle $\beta$, as shown in Figure 10.6.

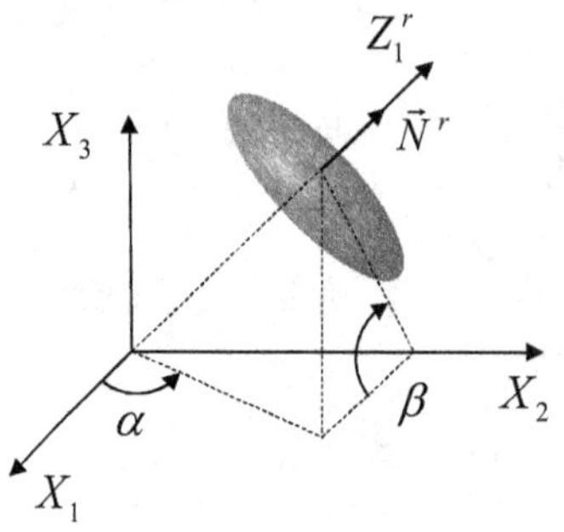


**Figure 10.6** Coordinate transformation between the initial material frame $(X_1, X_2, X_3)$ and the crack coordinate system $(Z_1^r, Z_2^r, Z_3^r)$. The direction of the unit normal $\vec{N}^r$ of the microcrack is defined by the angles $\alpha$ and $\beta$.

Let $\mathbf{A}^{\mathrm{X}}$ be an arbitrary second-order tensor and $A_{ij}^{\mathrm{X}}$ its component expressed in the initial material coordinate system $(X_1, X_2, X_3)$. In the microcrack coordinate system $(Z_1^r, Z_2^r, Z_3^r)$, these quantities are denoted by $\mathbf{A}^r$ and $A_{ij}^r$. The rule for changing second-order tensors under rotation of axes is given in Equation (K.10) of **Appendix K**,

$$\mathbf{A}^r = \mathbf{T}^{\mathrm{T}} \cdot \mathbf{A}^{\mathrm{X}} \cdot \mathbf{T} \,. \tag{10.35}$$

The scalar components of the coordinate "rotation tensor" $\mathbf{T}$ are given in Equation (K.4) of **Appendix K**.

The transformation of the stress tensor $\boldsymbol{\sigma}$ and the damage strain tensor due to microcrack group (parallel microcrack field with the same microcrack surface area) $\boldsymbol{\varepsilon}^{\mathrm{d}r}$ follow Expression (10.35). Thus the following can be written:

$$\boldsymbol{\varepsilon}^{\mathrm{d}r} = \mathbf{T}^{\mathrm{T}} \cdot \boldsymbol{\varepsilon}^{\mathrm{d}r\mathrm{X}} \cdot \mathbf{T} \qquad \text{and} \qquad \boldsymbol{\sigma}^r = \mathbf{T}^{\mathrm{T}} \cdot \boldsymbol{\sigma}^{\mathrm{X}} \cdot \mathbf{T} \,. \tag{10.36}$$

It is important to note the preceding two notations for the damage strain tensor $\boldsymbol{\varepsilon}^{\mathrm{d}}$. The notation $\boldsymbol{\varepsilon}^{\mathrm{d}r}$ stands for damage strain caused by the $r$th microcrack group and expressed in the microcrack coordinate system $(Z_1^r, Z_2^r, Z_3^r)$. The notation $\boldsymbol{\varepsilon}^{\mathrm{d}r\mathrm{X}}$, on the other hand, stands for damage strain caused by the $r$th microcrack group, but it is expressed in the initial material frame $(X_1, X_2, X_3)$. For clarity, the coordinate "rotation tensor" $\mathbf{T}$ is given without superscripts.

Substitution of Expression (10.36) into Equation $(10.31)_1$ leads to

$$\boldsymbol{\varepsilon}^{\mathrm{d}r} = \mathbf{S}^{\mathrm{d}r} : \boldsymbol{\sigma}^r \qquad \Rightarrow \qquad \mathbf{T}^{\mathrm{T}} \cdot \boldsymbol{\varepsilon}^{\mathrm{d}r\mathrm{X}} \cdot \mathbf{T} = \mathbf{S}^{\mathrm{d}r} : (\mathbf{T}^{\mathrm{T}} \cdot \boldsymbol{\sigma}^{\mathrm{X}} \cdot \mathbf{T}) \,. \tag{10.37}$$

Equation $(10.37)_2$ is multiplied by the $\mathbf{T}\cdot$ from the left and by the $\cdot\mathbf{T}^{\mathrm{T}}$ from the right. We arrive at the following:

$$\mathbf{T} \cdot (\mathbf{T}^{\mathrm{T}} \cdot \boldsymbol{\varepsilon}^{\mathrm{d}r\mathrm{X}} \cdot \mathbf{T}) \cdot \mathbf{T}^{\mathrm{T}} = \mathbf{T} \cdot \big[\, \mathbf{S}^{\mathrm{d}r} : (\mathbf{T}^{\mathrm{T}} \cdot \boldsymbol{\sigma}^{\mathrm{X}} \cdot \mathbf{T}) \,\big] \cdot \mathbf{T}^{\mathrm{T}} \,. \tag{10.38}$$

The order of performing the dot products $\cdot$ on the left side of Equation (10.38) can be changed, and therefore Expression (10.38) can be recast in the form

$$(\mathbf{T}\cdot\mathbf{T}^{\mathrm{T}})\cdot\boldsymbol{\varepsilon}^{\mathrm{d}r\mathrm{X}}\cdot(\mathbf{T}\cdot\mathbf{T}^{\mathrm{T}}) = \mathbf{T}\cdot\left[\,\mathbf{S}^{\mathrm{d}r}:(\mathbf{T}^{\mathrm{T}}\cdot\boldsymbol{\sigma}^{\mathrm{X}}\cdot\mathbf{T})\,\right]\cdot\mathbf{T}^{\mathrm{T}}. \tag{10.39}$$

According to Reddy ([84], Eq. $(2.5.20)_2$),

$$\mathbf{T}\cdot\mathbf{T}^{\mathrm{T}} = \mathbf{1} \qquad \Rightarrow \qquad \mathbf{T}^{\mathrm{T}}\cdot\mathbf{T} = \mathbf{1}. \tag{10.40}$$

Based on Expressions (10.40) and (2.66), the left side of Equation (10.39) simplifies and therefore Equation (10.39) reduces to

$$\boldsymbol{\varepsilon}^{\mathrm{d}r\mathrm{X}} = \mathbf{T}\cdot\left[\,\mathbf{S}^{\mathrm{d}r}:(\mathbf{T}^{\mathrm{T}}\cdot\boldsymbol{\sigma}^{\mathrm{X}}\cdot\mathbf{T})\,\right]\cdot\mathbf{T}^{\mathrm{T}}. \tag{10.41}$$

The goal is to derive a constitutive equation having the form

$$\boldsymbol{\varepsilon}^{\mathrm{d}r\mathrm{X}} = \mathbf{S}^{\mathrm{d}r\mathrm{X}}:\boldsymbol{\sigma}^{\mathrm{X}}. \tag{10.42}$$

In Expressions (10.41) and (10.42), the notation $\boldsymbol{\varepsilon}^{\mathrm{d}r\mathrm{X}}$ stands for the damage strain tensor associated to the $r$th microcrack group but expressed in the $(X_1, X_2, X_3)$ coordinate system. The same holds for the compliance tensor $\mathbf{S}^{\mathrm{d}r\mathrm{X}}$.

In order to express Equation (10.41) in the form of Equation (10.42), the tensors present on the right side of Equation (10.41) have to be written in the component form and the dot-products have to be calculated. This act is not carried out here, but the result is given. It is

$$S^{\mathrm{d}r\mathrm{X}}_{ijkl} = T_{io}\, S^{\mathrm{d}r}_{opqr}\, T^{\mathrm{T}}_{qk}\, T_{lr}\, T^{\mathrm{T}}_{pj}. \tag{10.43}$$

In order to continue derivation in the forthcoming derivation, Form (10.42) is used, although in practice the corresponding indicial form, that is, Expression (10.43), is used.

For a material containing multidirectional non-interacting microcracks the compliance tensor due to the damage $\mathbf{S}^{\mathrm{d}}$ is obtained as a sum of many microcrack groups, as follows:

$$\mathbf{S}^{\mathrm{dX}} = \sum_{r=1}^{M} \mathbf{S}^{\mathrm{d}r\mathrm{X}}, \tag{10.44}$$

where $M$ is the number of microcrack groups.

Based on Result (10.44), the damage strain tensor $\boldsymbol{\varepsilon}^{\mathrm{dX}}$ describing the influence of the multidirectional crack field can be obtained from the equation

$$\boldsymbol{\varepsilon}^{\mathrm{dX}} = \mathbf{S}^{\mathrm{dX}}:\boldsymbol{\sigma}^{\mathrm{X}} \qquad \Rightarrow \qquad \boldsymbol{\varepsilon}^{\mathrm{d}} = \mathbf{S}^{\mathrm{d}}:\boldsymbol{\sigma}. \tag{10.45}$$

Since Expression $(10.45)_1$ is a tensor expression, it is independent of the coordinate system, making Expression $(10.45)_2$ possible to write. The key point in selection of the coordinate system for Expression $(10.45)_2$ is the form of the coordinate rotation tensor $\mathbf{T}$. In the preceding study, the transformation is carried out from the coordinate system $(Z_1^r, Z_2^r, Z_3^r)$ to the frame $(X_1, X_2, X_3)$. However, the target coordinate system can be an arbitrary frame. Only a modified coordinate rotation tensor $\mathbf{T}$ is needed.

### 10.3.5 Damage-Elastic Strain Tensor $\varepsilon^{\mathrm{de}}$ due to a Multidirectional Microcrack Field

The present section gives the damage-elastic strain tensor $\boldsymbol{\varepsilon}^{\mathrm{de}}$ due to a multidirectional microcrack field that is constituted by many microcrack groups. As already said, the first line on the right side of Equation (10.14) gives the linear elastic strain tensor $\boldsymbol{\varepsilon}^{\mathrm{e}}$, whereas the rest of the right side is for the damage strain tensor $\boldsymbol{\varepsilon}^{\mathrm{d}}$.

According to Equation $(10.22)_1$, Hooke's law reads

$$\boldsymbol{\varepsilon}^{\mathrm{e}} = \mathbf{S} : \boldsymbol{\sigma} . \tag{10.46}$$

Equation (10.15) is recalled. It is

$$\boldsymbol{\varepsilon}^{\mathrm{de}} := \boldsymbol{\varepsilon}^{\mathrm{e}} + \boldsymbol{\varepsilon}^{\mathrm{d}} = \boldsymbol{\varepsilon} - \boldsymbol{\varepsilon}^{\mathrm{i}} . \tag{10.47}$$

Substitution of Expressions $(10.45)_2$ and (10.46) into Equation (10.47) gives

$$\boldsymbol{\varepsilon}^{\mathrm{de}} = \boldsymbol{\varepsilon}^{\mathrm{e}} + \boldsymbol{\varepsilon}^{\mathrm{d}} = (\mathbf{S} + \mathbf{S}^{\mathrm{d}}) : \boldsymbol{\sigma} . \tag{10.48}$$

Equation (10.48) can take the form

$$\boldsymbol{\varepsilon}^{\mathrm{de}} = \tilde{\mathbf{S}} : \boldsymbol{\sigma} , \qquad \text{where} \qquad \tilde{\mathbf{S}} := \mathbf{S} + \mathbf{S}^{\mathrm{d}} . \tag{10.49}$$

### 10.3.6 Damage-Elastic Strain Tensor Component $\varepsilon_{11}^{\mathrm{de}r}$ due to a Parallel Microcrack Field

This section derives an expression between the damage-elastic strain tensor component $\varepsilon_{11}^{\mathrm{de}r}$ and the stress tensor component $\sigma_{11}^{r}$ due to a parallel microcrack field. The normals of the microcracks are along the $X_1$-axis.

Equation $(10.31)_2$ is

$$\varepsilon_{ij}^{\mathrm{d}r} = S_{ijkl}^{\mathrm{d}r} \, \sigma_{kl}^{r} \qquad \Rightarrow \qquad \varepsilon_{11}^{\mathrm{d}r} = S_{1111}^{\mathrm{d}r} \, \sigma_{11}^{r} . \tag{10.50}$$

Equation $(10.22)_1$ is

$$\boldsymbol{\varepsilon}^{\mathrm{e}} = \mathbf{S} : \boldsymbol{\sigma} \qquad \Rightarrow \qquad \varepsilon_{ij}^{\mathrm{e}} = S_{ijkl} \, \sigma_{kl} \qquad \Rightarrow \qquad \varepsilon_{11}^{\mathrm{e}r} = S_{1111}^{r} \, \sigma_{11}^{r} . \tag{10.51}$$

Since the Hookean deformation is isotropic, that is, it does not show directional dependency, the following holds:

$$S_{1111}^{r} = S_{1111} . \tag{10.52}$$

Expression (2.64) is

$$\mathbf{1} = \delta_{ij} \, \vec{i}_i \, \vec{i}_j \qquad \Rightarrow \qquad 1_{ij} = \delta_{ij} . \tag{10.53}$$

Expression (2.107) is

$$\mathbf{I}^{\mathrm{s}} = \frac{1}{2}\,(\delta_{ik}\,\delta_{jl} + \delta_{il}\,\delta_{jk})\,\vec{i}_i\,\vec{i}_j\,\vec{i}_k\,\vec{i}_l \qquad \Rightarrow \qquad I^{\mathrm{s}}_{ijkl} = \frac{1}{2}\,(\delta_{ik}\,\delta_{jl} + \delta_{il}\,\delta_{jk})\,. \tag{10.54}$$

Equation (10.23) is recalled. It is

$$\mathbf{S} := -\frac{\nu}{E}\,\mathbf{1}\,\mathbf{1} + \frac{1+\nu}{E}\,\mathbf{I}^{\mathrm{s}}\,. \tag{10.55}$$

Based on Results $(10.53)_2$ and $(10.54)_2$, Expression (10.55) yields

$$S_{ijkl} = -\frac{\nu}{E}\,\delta_{ij}\,\delta_{kl} + \frac{1+\nu}{E}\,\frac{1}{2}\,(\delta_{ik}\,\delta_{jl} + \delta_{il}\,\delta_{jk})\,, \tag{10.56}$$

which leads to

$$S_{1111} = -\frac{\nu}{E}\,\delta_{11}\,\delta_{11} + \frac{1+\nu}{E}\,\frac{1}{2}\,(\delta_{11}\,\delta_{11} + \delta_{11}\,\delta_{11}) = -\frac{\nu}{E} + \frac{1+\nu}{E} = \frac{1}{E}\,, \tag{10.57}$$

where the fact that $\delta_{11} = 1$ [see Definition (2.4)] is exploited. Substitution of Result (10.57) into Equation $(10.51)_3$ leads to

$$\varepsilon^{\mathrm{e}r}_{11} = S_{1111}\,\sigma^r_{11} \qquad \Rightarrow \qquad \varepsilon^{\mathrm{e}r}_{11} = \frac{1}{E}\,\sigma^r_{11}\,. \tag{10.58}$$

Substitution of Equation (10.32) into Expression $(10.50)_2$ yields

$$\varepsilon^{\mathrm{d}=r}_{11} = S^{\mathrm{d}=r}_{1111}\,\sigma^r_{11} \qquad \Rightarrow \qquad \varepsilon^{\mathrm{d}=r}_{11} = (2-\nu)\,B_1\,Q^r\,(A^r)^{3/2}\,H(\sigma^r_{11})\,\sigma^r_{11}\,. \tag{10.59}$$

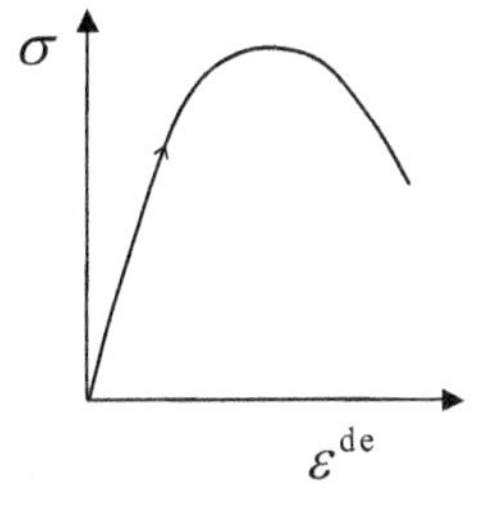


**Figure 10.7** Stress-strain (damage-elastic) curve due to microcracking .

Based on Definition (10.47), the following is achieved:

$$\boldsymbol{\varepsilon}^{\mathrm{de}} := \boldsymbol{\varepsilon}^{\mathrm{e}} + \boldsymbol{\varepsilon}^{\mathrm{d}} \qquad \Rightarrow \qquad \varepsilon^{\mathrm{de}r}_{11} = \varepsilon^{\mathrm{e}r}_{11} + \boldsymbol{\varepsilon}^{\mathrm{d}r}_{11}\,. \tag{10.60}$$

Substitution of Terms $(10.58)_2$ and $(10.59)_2$ into Expression $(10.60)_2$ gives

$$\varepsilon^{\mathrm{de}r}_{11} = \left[\frac{1}{E} + (2-\nu)\,B_1\,Q^r\,(A^r)^{3/2}\,H(\sigma^r_{11})\right]\sigma^r_{11}\,. \tag{10.61}$$

Equation (10.61) gives

$$\sigma^r_{11} = \frac{\varepsilon^{\mathrm{de}r}_{11}}{\frac{1}{E} + (2-\nu)\,B_1\,Q^r\,(A^r)^{3/2}\,H(\sigma^r_{11})}\,. \tag{10.62}$$

According to Definition (10.4), an increasing number of microcracks causes a higher value for microcrack densities $Q^r$. The increasing number of microcracks $m^r$ causes an increasing amount of damage strain $\varepsilon^{\mathrm{d}}$ and therefore an increasing amount of damage-elastic strain $\varepsilon^{\mathrm{de}}$. Thus, based on Equation (10.62), a response similar to that in Figure 10.7 is obtained. The superscripts and subscript are left out in Figure 10.7 for clarity.

### 10.3.7 Prevention of Microcrack Surfaces from Penetrating Each Other under Compression

In this section, we derive the expression that is used in preventing microcrack surfaces from penetrating each other under compression. The expression is already utilized in the previous pages of the present chapter.

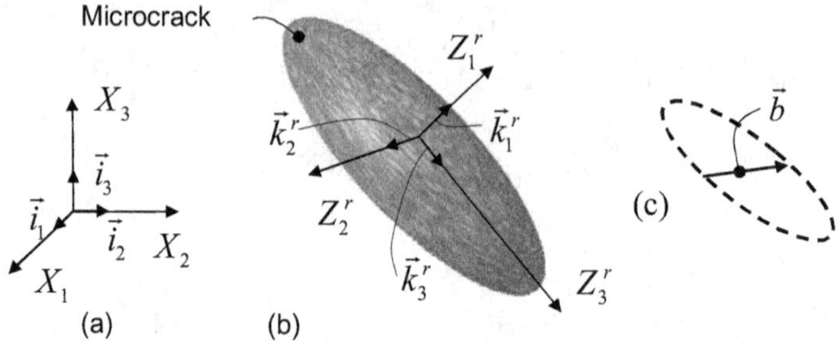


**Figure 10.8** (a) The initial material coordinate system $(X_1, X_2, X_3)$, (b) the microcrack coordinate system $(Z_1^r, Z_2^r, Z_3^r)$, and (c) jump between the material points across a microcrack.

Figure 10.8(c) is a copy of Figure 10.2(c), but without the superscript $p$, which in Equation (10.2), for example, refers to the $p$th microcrack. Equation (10.2) is recalled without the superscript $p$ and has the following appearance:

$$\vec{b} = \frac{16\,(1-\nu^2)}{(2-\nu)\,E\,\pi}\sqrt{(a)^2 - r^2}\,\left[\boldsymbol{\sigma}\cdot\vec{N} - \frac{\nu}{2}\,\vec{N}\,(\vec{N}\cdot\boldsymbol{\sigma}\cdot\vec{N})\right], \tag{10.63}$$

where $\vec{b}$ is the jump between the material points across the microcrack as shown in Figure 10.8(c). The form of the vector $\vec{b}$ is studied in the microcrack coordinate system $(Z_1^r, Z_2^r, Z_3^r)$ shown in Figure 10.8(b). Equations (10.6) are recalled:

$$\boldsymbol{\sigma} = \sigma_{ij}^r\,\vec{k}_i^r\,\vec{k}_j^r \quad\text{and}\quad \vec{N}^r = \vec{k}_1^r \quad\Rightarrow\quad \boldsymbol{\sigma}\cdot\vec{N} = \sigma_{ij}^r\,\vec{k}_i^r\,\vec{k}_j^r\cdot\vec{k}_1^r = \sigma_{i1}^r\,\vec{k}_i^r\,. \tag{10.64}$$

Equation $(10.7)_2$ reads

$$\vec{N}^r\cdot\boldsymbol{\sigma}\cdot\vec{N}^r = \sigma_{11}^r\,. \tag{10.65}$$

Since the vector $\vec{b}$ is written in the form

$$\vec{b} = b_i^r\,\vec{k}_i^r\,, \tag{10.66}$$

Expressions (10.64)–(10.66) allow Equation (10.63) to take the form

$$b_i^r\,\vec{k}_i^r = \frac{16\,(1-\nu^2)}{(2-\nu)\,E\,\pi}\sqrt{(a)^2 - r^2}\,\left[\sigma_{i1}^r\,\vec{k}_i^r - \frac{\nu}{2}\,\vec{k}_1^r\,(\sigma_{11}^r)\right]. \tag{10.67}$$

Based on Figures 10.8(b) and 10.8(c), the microcrack surfaces penetrate each other if the component $b_1^r$ of the vector $\vec{b}$ takes negative values. The other two components of the vector $\vec{b}$ represent the "shear displacements." Thus the component $b_1^r$ has to take nonnegative values. Based on Equation (10.67) this condition reads ($i = 1$)

$$b_1^r = \frac{16\,(1-\nu^2)}{(2-\nu)\,E\,\pi}\sqrt{(a)^2 - r^2}\left[\sigma_{11}^r - \frac{\nu}{2}\,(\sigma_{11}^r)\right] \geq 0\,, \tag{10.68}$$

which yields

$$b_1^r = \frac{16\,(1-\nu^2)}{(2-\nu)\,E\,\pi}\sqrt{(a)^2 - r^2}\left(1 - \frac{\nu}{2}\right)\sigma_{11}^r \geq 0\,. \tag{10.69}$$

Since the term in front of the square brackets and the term $1 - \nu/2$ are always nonnegative (see acceptable values for Poisson's ratio $\nu$ given in Section 8.3.2), the value of the stress tensor component $\sigma_{11}^r$ also has to be nonnegative; that is,

$$\sigma_{11}^r \geq 0 \qquad \Rightarrow \qquad \vec{N}^r \cdot \boldsymbol{\sigma} \cdot \vec{N}^r \geq 0\,, \tag{10.70}$$

where Equation (10.65) is exploited. Condition $(10.70)_2$ is satisfied in Expression (10.5) by adding the following term:

$$H(\vec{N}^r \cdot \boldsymbol{\sigma} \cdot \vec{N}^r)\,. \tag{10.71}$$

It is worth noting that Conditions (10.70) and (10.71) do not reject the study of tensile and shear stress only, but are introduced to prevent the microcrack surfaces from penetrating each other under compression. The influence of Conditions (10.70) and (10.71) is rejected for damage strain tensor, $\boldsymbol{\varepsilon}^{\mathrm{d}}$, and therefore they do not have any influence on the elastic strain tensor $\boldsymbol{\varepsilon}^{\mathrm{e}}$.

## 10.4 Deformation of a Hookean Matrix Response with Spherical Microvoids

This section evaluates the deformation of a material containing noninteracting spherical microvoids in a matrix material the response of which is assumed to be linear elastic. The present material model can be used as part of a more extensive material model where also creep, for example, is modelled.

### 10.4.1 Spherical Microvoids with the Hookean Matrix Response

This section evaluates the deformation of a material containing non-interacting spherical microvoids in a matrix material the response of which is assumed to be linear elastic. The density of the cavities is assumed to be so low that the interaction between them can be neglected.

Eshelby [21] studied the response of a material containing ellipsoidal inclusions. The deformation in both matrix and inclusions was assumed to follow

Hooke's law. As a special, case he determined the value for the complementary strain-energy density $w^{\mathrm{c}}$ of a material containing "a volume fraction $f$" of inhomogeneous spheres. For the purpose of this work, the inhomogeneous spheres are "replaced" with spherical cavities, as shown in Figure 10.9. This is done by assuming that the values for the elastic constants for the cavities vanish. The spherical cavity problem was approached earlier by Mackenzie [56], but the form of the result by Eshelby better fits the needs of this work. The complementary strain-energy density $w^{\mathrm{c}}(\boldsymbol{\sigma}, f)$ takes the form (Eshelby [21], p. 390)

$$w^{\mathrm{c}}(\boldsymbol{\sigma}, f) = \frac{1}{2}\left[\frac{1}{3\,(3\,\lambda + 2\,\mu)}\,(1 + \underline{A}\,f)\,[\mathbf{1} : \boldsymbol{\sigma}]^2 + \frac{1}{2\,\mu}\,(1 + \underline{B}\,f)\,\mathbf{s} : \mathbf{s}\right], \qquad (10.72)$$

where $\lambda$ and $\mu$ are the Lamé elastic constants of the matrix response given by Definitions (8.31) [see Eshelby [21], Eqs. (2.2) and (2.3)]. The second-order identity tensor $\mathbf{1}$ and the deviatoric stress tensor $\mathbf{s}$ are defined in Definitions (2.63) and (2.108).

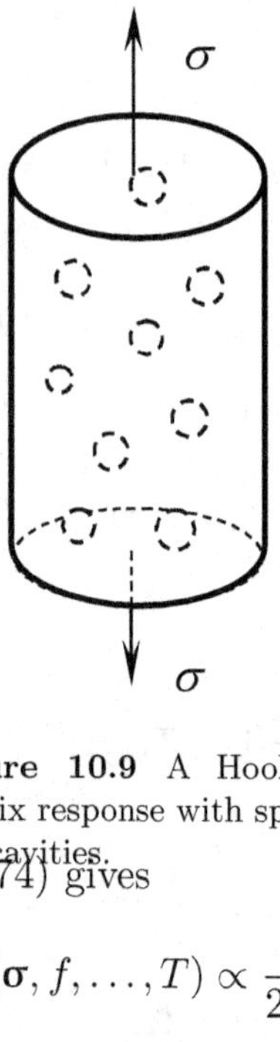


**Figure 10.9** A Hookean matrix response with spherical cavities.

For spherical voids, constants $\underline{A}$ and $\underline{B}$ take the forms (see Eshelby [21], pp. 389–390)].

$$\underline{A} = \frac{6\,\mu + 3\,\lambda}{4\,\mu} \qquad \text{and} \qquad \underline{B} = \frac{15\,(1-\nu)}{7 - 5\,\nu}. \qquad (10.73)$$

The complementary strain-energy density $w^{\mathrm{c}}(\boldsymbol{\sigma}, f)$ is proportional to the specific Gibbs free energy density $g(\boldsymbol{\sigma}, f, \ldots, T)$, as follows (see, e.g., Fung [24], p. 349):

$$\rho_0\, g(\boldsymbol{\sigma}, f, \ldots T) \propto w^{\mathrm{c}}(\boldsymbol{\sigma}, f)\,. \qquad (10.74)$$

In Equation (10.74), the dots ... represent a set of other (internal) variables.

Substitution of Equation (10.72) into Expression (10.74) gives

$$g(\boldsymbol{\sigma}, f, \ldots, T) \propto \frac{1}{2\,\rho_0}\left[\frac{1}{3\,(3\lambda + 2\,\mu)}\,(1 + \underline{A}\,f)\,[\mathbf{1} : \boldsymbol{\sigma}]^2 + \frac{1}{2\,\mu}(1 + \underline{B}\,f)\,\mathbf{s} : \mathbf{s}\right]. \qquad (10.75)$$

If also the swelling due to the microvoid nucleation and growth and other potential deformation mechanisms are taken into account, the material model for the present study takes the following format:

$$g(\boldsymbol{\sigma}, f, \ldots, T) = g^{\mathrm{de}}(\boldsymbol{\sigma}, f) + g^{\mathrm{sw}}(\boldsymbol{\sigma}, f) + g^{\mathrm{rest}}(\ldots, T)\,. \qquad (10.76)$$

The specific damage-elastic Gibbs free energy $g^{\mathrm{de}}(\boldsymbol{\sigma}, f)$ takes the form

$$g^{\mathrm{de}}(\boldsymbol{\sigma}, f) = \frac{1}{2\,\rho_0}\left[\frac{1}{3\,(3\,\lambda + 2\,\mu)}\,(1 + \underline{A}\,f)\,[\mathbf{1} : \boldsymbol{\sigma}]^2 + \frac{1}{2\,\mu}\,(1 + \underline{B}\,f)\,\mathbf{s} : \mathbf{s}\right]. \qquad (10.77)$$

Expression (10.77) is derived for a case in which the deformation mechanisms modelled in $g^{\rm rest}(\dots,T)$ are not present. Thus Separation (10.76) assumes that there is no interaction between the processes described by the models $g^{\rm de}(\boldsymbol{\sigma},f)$ and $g^{\rm rest}(\dots,T)$. The specific Gibbs free energy associated to porosity swelling $g^{\rm sw}(\boldsymbol{\sigma},f)$ is

$$g^{\rm sw}(\boldsymbol{\sigma},f) = \frac{f}{3\,\rho_0}\,\mathbf{1}:\boldsymbol{\sigma}\,. \tag{10.78}$$

The strain tensors $\boldsymbol{\varepsilon}$ and $\boldsymbol{\varepsilon}^{\rm i}$ and the void volume fraction $f$ are assumed to vanish at the reference state $\mathcal{E}_{\rm r}$. The terms $g^{\rm de}(\boldsymbol{\sigma},f)$ and $g^{\rm sw}(\boldsymbol{\sigma},f)$ are not added together, since they model different response mechanisms. The former describes the stiffness reduction due to the microvoids, and the latter models swelling due to the microvoid nucleation and growth.

The notation $g^{\rm rest}(\dots,T)$ stands for the rest part of the specific Gibbs free energy $g(\boldsymbol{\sigma},f,\dots,T)$. In the present study, it is assumed to be independent of the stress tensor $\boldsymbol{\sigma}$. It does not reject the generality of the following study. If the specific Gibbs free energy $g^{\rm rest}(\dots,T)$ has the stress tensor $\boldsymbol{\sigma}$ in the set of its arguments, the derivation follows the same steps taken here, but further terms will be obtained. The topic of this section is to study Expressions (10.77) and (10.78), and therefore other terms are not studied.

For the present set of state variables, State Equations (6.181) give

$$\boldsymbol{\varepsilon}-\boldsymbol{\varepsilon}^{\rm i} = \rho_0\,\frac{\partial g(\boldsymbol{\sigma},f,\dots,T)}{\partial\boldsymbol{\sigma}} \qquad \text{and} \qquad e = \rho_0\,\frac{\partial g(\boldsymbol{\sigma},f)}{\partial f}\,. \tag{10.79}$$

In State Equation $(10.79)_2$, the notation $e$ stands for the internal force associated with the void volume fraction $f$.

Definition (2.108) gives the deviatoric stress tensor $\mathbf{s}$,

$$\mathbf{s} := \mathbf{K}:\boldsymbol{\sigma}\,,\ \text{where the fourth-order tensor } \mathbf{K} := \mathbf{I}-\tfrac{1}{3}\,\mathbf{1}\,\mathbf{1}\,. \tag{10.80}$$

Results (2.135) and (2.142) and Definitions (10.80) give

$$\frac{\partial[\mathbf{1}:\boldsymbol{\sigma}]^2}{\partial\boldsymbol{\sigma}} = 2\,[\mathbf{1}:\boldsymbol{\sigma}]\,\mathbf{1} \ \text{ and } \ \frac{\partial}{\partial\boldsymbol{\sigma}}(\mathbf{s}:\mathbf{s}) = 2\,\mathbf{s} = 2\,\mathbf{K}:\boldsymbol{\sigma} = 2\,(\mathbf{I}-\tfrac{1}{3}\,\mathbf{1}\,\mathbf{1}):\boldsymbol{\sigma}\,. \tag{10.81}$$

For an arbitrary second-order tensor $\mathbf{c}$, Equation (2.149), that is, $\partial\mathbf{c}/\partial\mathbf{c} = \mathbf{I}$ and (2.69), that is, $\mathbf{c}:\mathbf{I} = \mathbf{c}$ , allow the following to be written:

$$\frac{\partial\mathbf{c}}{\partial\mathbf{c}} = \mathbf{I} \ \text{ and } \ \mathbf{c}:\mathbf{I} = \mathbf{c} \quad\Rightarrow\quad \frac{\partial[\mathbf{1}:\boldsymbol{\sigma}]}{\partial\boldsymbol{\sigma}} = \mathbf{1}:\frac{\partial\boldsymbol{\sigma}}{\partial\boldsymbol{\sigma}} = \mathbf{1}:\mathbf{I} = \mathbf{1}\,. \tag{10.82}$$

Substituting Models (10.77) and (10.78) into State Equation $(10.79)_1$ and taking Results (10.81) and $(10.82)_3$ into consideration gives

$$\begin{aligned}\boldsymbol{\varepsilon}-\boldsymbol{\varepsilon}^{\rm i} &= \rho_0\,\frac{\partial g(\boldsymbol{\sigma},f,\dots,T)}{\partial\boldsymbol{\sigma}} = \rho_0\,\frac{\partial g^{\rm de}(\boldsymbol{\sigma},f)}{\partial\boldsymbol{\sigma}} + \rho_0\,\frac{\partial g^{\rm sw}(\boldsymbol{\sigma},f)}{\partial\boldsymbol{\sigma}}\\ &= \frac{1+\underline{A}\,f}{3\,(3\,\lambda+2\,\mu)}\,\mathbf{1}:\boldsymbol{\sigma}\,\mathbf{1} + \frac{1+\underline{B}\,f}{2\,\mu}\,(\mathbf{I}-\frac{1}{3}\,\mathbf{1}\,\mathbf{1}):\boldsymbol{\sigma} + \frac{f}{3}\,\mathbf{1}\,.\end{aligned} \tag{10.83}$$

Definitions (2.69) and (2.106) give

$$\mathbf{I}:\boldsymbol{\sigma}=\boldsymbol{\sigma} \quad \text{and} \quad \mathbf{I}^{\mathrm{s}}:\boldsymbol{\sigma}=\frac{1}{2}\,(\boldsymbol{\sigma}+\boldsymbol{\sigma}^{\mathrm{T}})=\boldsymbol{\sigma} \quad \Rightarrow \quad \mathbf{I}:\boldsymbol{\sigma}=\mathbf{I}^{\mathrm{s}}:\boldsymbol{\sigma}\,, \tag{10.84}$$

where the symmetry of the stress tensor $\boldsymbol{\sigma}$ is exploited. Based on Result $(10.84)_3$, Constitutive Model (10.83) can be recast as

$$\boldsymbol{\varepsilon}-\boldsymbol{\varepsilon}^{\mathrm{i}}=\frac{1+Af}{3\,(3\,\lambda+2\,\mu)}\,\mathbf{1}\,\mathbf{1}:\boldsymbol{\sigma}+\frac{1+B\,f}{2\,\mu}\,(\,\mathbf{I}^{\mathrm{s}}-\frac{1}{3}\,\mathbf{1}\,\mathbf{1}\,):\boldsymbol{\sigma}+\frac{f}{3}\,\mathbf{1}\,. \tag{10.85}$$

Equation (10.85) can be written as

$$\boldsymbol{\varepsilon}-\boldsymbol{\varepsilon}^{\mathrm{i}}=\tilde{\mathbf{S}}(f):\boldsymbol{\sigma}+\frac{f}{3}\,\mathbf{1}\,, \tag{10.86}$$

where $\tilde{\mathbf{S}}(f)$ is the fourth-order compliance tensor for deformation of the Hookean matrix response with spherical microvoids. It is defined by

$$\tilde{\mathbf{S}}(f):=\frac{1+A\,f}{3\,(3\,\lambda+2\,\mu)}\,\mathbf{1}\,\mathbf{1}+\frac{1+B\,f}{2\,\mu}\,(\,\mathbf{I}^{\mathrm{s}}-\frac{1}{3}\,\mathbf{1}\,\mathbf{1}\,)\,. \tag{10.87}$$

The effective compliance tensor $\tilde{\mathbf{S}}(f)$ can be separated into two different parts,

$$\tilde{\mathbf{S}}(f):=\mathbf{S}+\mathbf{S}^{\mathrm{d}}(f)\,, \tag{10.88}$$

where $\mathbf{S}^{\mathrm{d}}(f)$ is the compliance tensor due to damage. Based on micromechanical evaluation, separation of the effective compliance tensor $\tilde{\mathbf{S}}$ into two parts given for microvoids in Result (10.87) is more general, as the work by Nemat-Nasser and Hori [[67], Eq. (4.3.6a)] shows. A similar separation is obtained for penny-shaped noninteracting microcracks, as Equation $(10.49)_2$ shows. Based on Equations (10.87) and (10.88), the compliance tensors $\mathbf{S}$ and $\mathbf{S}^{\mathrm{d}}(f)$ are given by

$$\mathbf{S}=\frac{1}{3\,(3\,\lambda+2\,\mu)}\,\mathbf{1}\,\mathbf{1}+\frac{1}{2\,\mu}\,(\,\mathbf{I}^{\mathrm{s}}-\frac{1}{3}\,\mathbf{1}\,\mathbf{1}\,) \tag{10.89}$$

and

$$\mathbf{S}^{\mathrm{d}}(f):=\frac{A\,f}{3\,(3\,\lambda+2\,\mu)}\,\mathbf{1}\,\mathbf{1}+\frac{B\,f}{2\,\mu}\,(\,\mathbf{I}^{\mathrm{s}}-\frac{1}{3}\,\mathbf{1}\,\mathbf{1}\,)\,. \tag{10.90}$$

The replacement of the fourth-order identity tensor $\mathbf{I}$ with the fourth-order symmetric identity tensor $\mathbf{I}^{\mathrm{s}}$ makes the effective compliance tensor $\tilde{\mathbf{S}}(f)$ minor and major symmetric. The replacement was possible because the stress tensor $\boldsymbol{\sigma}$ is symmetric. Furthermore, this leads to a form of the compliance tensor for a Hookean deformation $\mathbf{S}$ that is compatible with the one introduced in Definition (8.42). Finally, the replacement is necessary because the strain tensors $\boldsymbol{\varepsilon}$ and $\boldsymbol{\varepsilon}^{\mathrm{i}}$ and the stress tensor $\boldsymbol{\sigma}$ are symmetric.

Due to Separation (10.88), Constitutive Model (10.85) can be written as

$$\begin{aligned}\boldsymbol{\varepsilon} - \boldsymbol{\varepsilon}^{\mathrm{i}} &= \tilde{\mathbf{S}}(f) : \boldsymbol{\sigma} + \frac{f}{3}\mathbf{1} = \left[\mathbf{S} + \mathbf{S}^{\mathrm{d}}(f)\right] : \boldsymbol{\sigma} + \frac{f}{3}\mathbf{1} \\ &= \mathbf{S} : \boldsymbol{\sigma} + \mathbf{S}^{\mathrm{d}}(f) : \boldsymbol{\sigma} + \frac{f}{3}\mathbf{1}\,. \end{aligned} \tag{10.91}$$

Based on the second line of Expression (10.91), the following strain tensors can be introduced:

$$\boldsymbol{\varepsilon} - \boldsymbol{\varepsilon}^{\mathrm{i}} = \boldsymbol{\varepsilon}^{\mathrm{e}} + \boldsymbol{\varepsilon}^{\mathrm{d}} + \boldsymbol{\varepsilon}^{\mathrm{sw}} \quad \Rightarrow \quad \boldsymbol{\varepsilon} = \boldsymbol{\varepsilon}^{\mathrm{e}} + \boldsymbol{\varepsilon}^{\mathrm{d}} + \boldsymbol{\varepsilon}^{\mathrm{sw}} + \boldsymbol{\varepsilon}^{\mathrm{i}}\,, \tag{10.92}$$

where $\boldsymbol{\varepsilon}^{\mathrm{e}}$ is the elastic strain tensor, $\boldsymbol{\varepsilon}^{\mathrm{d}}$ is the damage strain tensor, $\boldsymbol{\varepsilon}^{\mathrm{sw}}$ is the porosity swelling strain tensor, and $\boldsymbol{\varepsilon}^{\mathrm{i}}$ is the inelastic strain tensor. Equations (10.91) and $(10.92)_1$ give

$$\boldsymbol{\varepsilon}^{\mathrm{e}} = \mathbf{S} : \boldsymbol{\sigma}\,, \qquad \boldsymbol{\varepsilon}^{\mathrm{d}} = \mathbf{S}^{\mathrm{d}}(f) : \boldsymbol{\sigma}\,, \qquad \text{and} \qquad \boldsymbol{\varepsilon}^{\mathrm{sw}} = \frac{f}{3}\mathbf{1}\,. \tag{10.93}$$

It is worth noting that Expression (10.92) does not include models for all potential deformation mechanisms. The thermal strain tensor $\boldsymbol{\varepsilon}^{\mathrm{Th}}$ could be added to the right side of Separation (10.92). Furthermore, the content of inelastic strain tensor $\boldsymbol{\varepsilon}^{\mathrm{i}}$ has been left open. Expressions (10.91), $(10.93)_1$, and $(10.93)_2$ allow the introduction of the damage-elastic strain tensor $\boldsymbol{\varepsilon}^{\mathrm{de}}$, defined by

$$\boldsymbol{\varepsilon}^{\mathrm{de}} := \boldsymbol{\varepsilon}^{\mathrm{e}} + \boldsymbol{\varepsilon}^{\mathrm{d}} = \left[\mathbf{S} + \mathbf{S}^{\mathrm{d}}(f)\right] : \boldsymbol{\sigma} = \tilde{\mathbf{S}}(f) : \boldsymbol{\sigma}\,, \text{ for damaged material.} \tag{10.94}$$

Expression $(10.93)_3$ is multiplied by $\mathbf{1}$ : from the left, and we arrive at the following:

$$\mathbf{1} : \boldsymbol{\varepsilon}^{\mathrm{sw}} = \frac{f}{3}\mathbf{1} : \mathbf{1} = f\,. \tag{10.95}$$

Equations (10.93)–(10.95) help to evaluate the physical meaning of the terms in Separation (10.92). The elastic strain tensor $\boldsymbol{\varepsilon}^{\mathrm{e}}$ gives the response of the spherical-free Hookean deformation. Thus the elastic strain tensor $\boldsymbol{\varepsilon}^{\mathrm{e}}$ is the "average" deformation of the matrix material. The damage strain tensor $\boldsymbol{\varepsilon}^{\mathrm{d}}$ gives the deformation due to the softening of the material caused by the microvoids. When a microcracked material is studied, the damage strain tensor $\boldsymbol{\varepsilon}^{\mathrm{d}}$ gives the deformation due to the softening of the material caused by the microcracks, as shown in Expressions (10.31) and (10.32)–(10.34). The role of the damage-elastic strain tensor $\boldsymbol{\varepsilon}^{\mathrm{de}}$ is discussed in Sections 10.4.2 and 10.5. The porosity swelling strain tensor $\boldsymbol{\varepsilon}^{\mathrm{sw}}$ gives the change of the relative volume of the body with changing porosity $f$, as shown in Equation (10.95).

The subsequent chapter shows the form of the stress–strain relation expressed by Equation (10.94). It is a special case, but the response of a material containing noninteracting penny-shaped microcracks follows Equation (10.94), as given in Equations (10.48) and (10.49).

Substitution of Material Models (10.77) and (10.78) into State Equation $(10.79)_2$ yields

$$e = \rho_0 \, \frac{\partial g(\boldsymbol{\sigma}, f)}{\partial f} = \frac{1}{2} \left[ \frac{A}{3\,(3\,\lambda + 2\,\mu)} \, [\mathbf{1} : \boldsymbol{\sigma}]^2 + \frac{B}{2\,\mu} \, \mathbf{s} : \mathbf{s} \right] + \frac{1}{3} \, \mathbf{1} : \boldsymbol{\sigma} \, . \tag{10.96}$$

Since the stress tensor $\boldsymbol{\sigma}$ is symmetric, Result $(10.84)_3$ holds, and we arrive at the following:

$$(\mathbf{I}^{\mathrm{s}} - \frac{1}{3} \mathbf{1}\,\mathbf{1}) : \boldsymbol{\sigma} = (\mathbf{I} - \frac{1}{3} \mathbf{1}\,\mathbf{1}) : \boldsymbol{\sigma} = \mathbf{K} : \boldsymbol{\sigma} = \mathbf{s} \, , \tag{10.97}$$

where Definitions (10.80) are utilized. Based on Manipulation (10.97), the multiplication of Definition (10.90) by $\boldsymbol{\sigma}:$ from the left and by the $:\boldsymbol{\sigma}$ from the right gives

$$\boldsymbol{\sigma} : \mathbf{S}^{\mathrm{d}}(f) : \boldsymbol{\sigma} = f \left[ \frac{A}{3\,(3\,\lambda + 2\,\mu)} \, \boldsymbol{\sigma} : \mathbf{1}\,\mathbf{1} : \boldsymbol{\sigma} + \frac{B}{2\,\mu} \, \boldsymbol{\sigma} : \mathbf{s} \right] . \tag{10.98}$$

Based on Problem 8.2, $\boldsymbol{\sigma} : \mathbf{s} = \mathbf{s} : \mathbf{s}$ . The dot products $\boldsymbol{\sigma} : \mathbf{1}$ and $\mathbf{1} : \boldsymbol{\sigma}$ are scalars. This yields

$$\boldsymbol{\sigma} : \mathbf{s} = \mathbf{s} : \mathbf{s} \qquad \text{and} \qquad \boldsymbol{\sigma} : \mathbf{1}\mathbf{1} : \boldsymbol{\sigma} = [\mathbf{1} : \boldsymbol{\sigma}]^2 \, . \tag{10.99}$$

Substitution of Results (10.99) into Expression (10.98) leads to

$$\boldsymbol{\sigma} : \mathbf{S}^{\mathrm{d}}(f) : \boldsymbol{\sigma} = f \left[ \frac{A}{3\,(3\,\lambda + 2\,\mu)} \, [\mathbf{1} : \boldsymbol{\sigma}]^2 + \frac{B}{2\,\mu} \, \mathbf{s} : \mathbf{s} \right] . \tag{10.100}$$

Comparison of Expressions (10.96) and (10.100) gives

$$e = \frac{1}{2\,f} \, \boldsymbol{\sigma} : \mathbf{S}^{\mathrm{d}}(f) : \boldsymbol{\sigma} + \frac{1}{3} \, \mathbf{1} : \boldsymbol{\sigma} \, . \tag{10.101}$$

The separated form of the Clausius–Duhem inequality is obtained from Expressions (6.264), which give

$$\boldsymbol{\sigma} : \dot{\boldsymbol{\varepsilon}}^{\mathrm{i}} + \boldsymbol{\beta} : \dot{\boldsymbol{\alpha}} \geq 0 \qquad \text{and} \qquad - \frac{\vec{\nabla} T}{T} \cdot \vec{q} \geq 0 \, . \tag{10.102}$$

For the present set of state variables, the mechanical part of Clausius–Duhem Inequality $(10.102)_1$ takes the following appearance:

$$\ldots + e \, \dot{f} \geq 0 \, . \tag{10.103}$$

Since no specific dissipation function $\varphi_{\mathrm{mech}}(\dot{f}; \boldsymbol{\varepsilon} - \boldsymbol{\varepsilon}^{\mathrm{i}}, f, \ldots, T)$ was given, the evolution equation for void volume fraction $f$ is not obtained. Therefore the satisfaction of Clausius–Duhem Inequality (10.103) cannot be checked.

According to Equation (10.87), the effective compliance tensor for deformation of a Hookean matrix response with spherical microvoids $\tilde{\mathbf{S}}(f)$ reads

$$\tilde{\mathbf{S}}(f) := \frac{1 + \underline{A}\, f}{3\,(3\,\lambda + 2\,\mu)}\,\mathbf{1}\,\mathbf{1} + \frac{1 + \underline{B}\, f}{2\,\mu}\,(\mathbf{I}^{\mathrm{s}} - \frac{1}{3}\,\mathbf{1}\,\mathbf{1})\,. \qquad (10.104)$$

Equation (10.104) is multiplied by $\boldsymbol{\sigma}:$ from the left and by $:\boldsymbol{\sigma}$ from the right, which with Manipulation (10.97) leads to

$$\boldsymbol{\sigma} : \tilde{\mathbf{S}}(f) : \boldsymbol{\sigma} = \frac{1 + \underline{A}\, f}{3\,(3\,\lambda + 2\,\mu)}\,\boldsymbol{\sigma} : \mathbf{1}\,\mathbf{1} : \boldsymbol{\sigma} + \frac{1 + \underline{B}\, f}{2\,\mu}\,\boldsymbol{\sigma} : \mathbf{s}\,. \qquad (10.105)$$

Based on Results (10.99), Equation (10.105) yields

$$\boldsymbol{\sigma} : \tilde{\mathbf{S}}(f) : \boldsymbol{\sigma} = \frac{1 + \underline{A}\, f}{3\,(3\,\lambda + 2\,\mu)}\,[\mathbf{1} : \boldsymbol{\sigma}]^2 + \frac{1 + \underline{B}\, f}{2\,\mu}\,\mathbf{s} : \mathbf{s}\,. \qquad (10.106)$$

Comparison of Expression (10.72) with Equation (10.106) gives

$$w^{\mathrm{c}}(\boldsymbol{\sigma}, f) = \frac{1}{2}\,\boldsymbol{\sigma} : \tilde{\mathbf{S}}(f) : \boldsymbol{\sigma}\,. \qquad (10.107)$$

### 10.4.2 Spherical Microvoids with the Hookean Matrix Response When $\nu = 0.2$

This section continues the topic of the previous section but studies the role of the damage-elastic strain tensor $\boldsymbol{\varepsilon}^{\mathrm{de}}$ as a special case by assuming that Poisson's ratio $\nu$ takes the value 0.2.

Equation (8.43) is recalled:

$$\mathbf{C} : \mathbf{S} = \mathbf{S} : \mathbf{C} = \mathbf{I}^{\mathrm{s}}\,, \qquad (10.108)$$

where $\mathbf{C}$ is the fourth-order constitutive tensor for a Hookean deformation. Based on Definition (2.106), the following holds true for every symmetric second-order tensor $\mathbf{c}$ (i.e., when $\mathbf{c}^{\mathrm{T}} = \mathbf{c}$ ):

$$\mathbf{I}^{\mathrm{s}} : \mathbf{c} = \frac{1}{2}\,(\mathbf{c} + \mathbf{c}^{\mathrm{T}}) = \mathbf{c}\,. \qquad (10.109)$$

Based on Equations (10.73), the following holds true:

$$\text{If} \quad \nu = 0.2\,, \quad \text{then} \quad \underline{A} = \underline{B} = 2.0\,. \qquad (10.110)$$

Substituting Value $(10.110)_2$ into Definition (10.90) gives with Expression (10.89)

$$\mathbf{S}^{\mathrm{d}}(f) = 2\,f\,\mathbf{S}\,. \qquad (10.111)$$

Substituting Result (10.111) into Equation (10.94) gives

$$\boldsymbol{\varepsilon}^{\mathrm{de}} = \Big[\mathbf{S} + \mathbf{S}^{\mathrm{d}}(f)\Big] : \boldsymbol{\sigma} \qquad \Rightarrow \qquad \boldsymbol{\varepsilon}^{\mathrm{de}} = (1 + 2\,f)\,\mathbf{S} : \boldsymbol{\sigma}\,. \qquad (10.112)$$

Equation (10.94) is as well

$$\boldsymbol{\varepsilon}^{\mathrm{de}} = \tilde{\mathbf{S}}(f) : \boldsymbol{\sigma} \,. \tag{10.113}$$

Comparison of Equation $(10.112)_2$ with Equation (10.113) yields

$$\tilde{\mathbf{S}}(f) = (1 + 2\,f\,)\,\mathbf{S} \,. \tag{10.114}$$

Equation $(10.112)_2$ is multiplied by $\mathbf{C}\,:$ from the left. By taking Equations (10.108) and (10.109) and the symmetry of the stress tensor $\boldsymbol{\sigma}$ into consideration, we arrive at the following:

$$\mathbf{C} : \boldsymbol{\varepsilon}^{\mathrm{de}} = (1 + 2\,f\,)\,\mathbf{C} : \mathbf{S} : \boldsymbol{\sigma} = (1 + 2\,f\,)\,\mathbf{I}^{\mathrm{s}} : \boldsymbol{\sigma} = (1 + 2\,f\,)\,\boldsymbol{\sigma} \,, \tag{10.115}$$

which gives

$$\boldsymbol{\sigma} = (1 + 2\,f\,)^{-1}\,\mathbf{C} : \boldsymbol{\varepsilon}^{\mathrm{de}} \,. \tag{10.116}$$

The Maclaurin series of the function $(1+x)^{\alpha}$ is defined to be (Abramowitz and Stegun [2], p. 15)

$$(1 + x)^{\alpha} = 1 + \alpha\,x + \frac{\alpha\,(\alpha - 1)}{2!}\,x^2 + \ldots \,, \tag{10.117}$$

where $-1 < x < 1$. Substitution of $\alpha = -1$ and $x = 2\,f$ into Series (10.117) yields

$$(1 + 2\,f\,)^{-1} = 1 - 2\,f + (2\,f)^2 + \ldots \,. \tag{10.118}$$

Since $2f$ is small compared to the unity, Series (10.118) can be truncated by taking the first two terms into account. This is

$$(1 + 2\,f\,)^{-1} \approx (1 - 2\,f\,) \,. \tag{10.119}$$

Equation (10.119) allows Expression (10.116) to be written in the form

$$\boldsymbol{\sigma} \approx (1 - 2\,f\,)\,\mathbf{C} : \boldsymbol{\varepsilon}^{\mathrm{de}} \,. \tag{10.120}$$

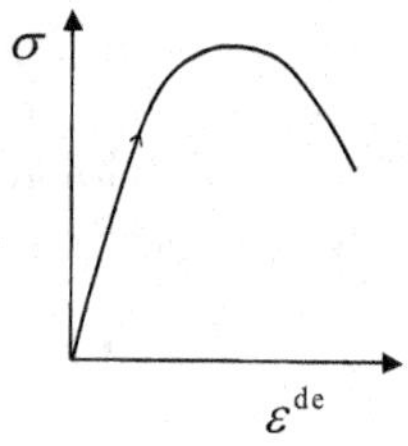


**Figure 10.10** Stress-strain (damage-elastic) curve due to evolution of void volume fraction $f$.

The damage-elastic strain $\boldsymbol{\varepsilon}^{\mathrm{de}}$ takes increasing value with an increasing value for the void volume fraction $f$. This means that damage evolution occurs. Thus, if the deformation of the matrix material obeys Hooke's law, according to Equation (10.120), the uniaxial stress–strain relation takes the form given by a nonlinear curve in Figure 10.10.

## 10.5 Continuum Damage Mechanics

This section gives an introduction to continuum damage mechanics. The present approach is a classical one describing the influence of microcracks and microvoids on the mechanical properties of materials. So, according to the present view, continuum damage mechanics describes the degradation of material due to microvoids and/or microcracks.

Although microcracks and microvoids are discrete objects forming jumps in the material response, their effect is averaged out by continuous functions over a finite volume. This is why the term *continuum damage* mechanics is adopted.

Microcracks and microvoids are not the only forms of damage that occur; other degradation mechanisms of materials can exist as well. This is discussed briefly at the end of this section.

In continuum damage mechanics, the variable damage denoted by $D$ is often introduced. It can be a scalar, vector, or tensor of any order. Instead of the variable $D$, it is preferable to introduce variables that are connected to the microstructure of the material, such as void volume fraction $f$, which was the topic of Section 10.4, and microcrack densities $Q^r$, which were introduced in Section 10.3. Both void volume fraction $f$ and microcrack densities $Q^r$ enter into the theory as internal variables, being therefore vital parts of the continuum thermodynamics. By introducing the variable $D$, the writer often expresses the fact that the source of the degradation of the material is unclear.

### 10.5.1 On Strain Tensors

This section discusses the definitions of the strain tensors. In order to simplify the evaluation, only damage-elastic deformation is studied. This means that the thermal expansion and inelastic deformation, for example, are omitted from the study. This does not mean that the evaluation is somehow restricted to covering damage-elastic deformation only, but the other deformation mechanisms can be included in the study by adding the corresponding strain terms.

When deformation of a Hookean matrix response with penny-shaped microcracks or spherical microvoids was studied, Expressions (10.62) and (10.120) and Figures 10.7 and 10.10 were introduced. Expressions (10.62) and (10.120) are

$$\sigma_{11}^r = \frac{\varepsilon_{11}^{\mathrm{der}}}{\frac{1}{E} + (2-\nu)\, B_1\, Q^r\, (A^r)^{3/2}\, H(\sigma_{11}^r)} \quad \text{and} \quad \boldsymbol{\sigma} \approx (1-2\,f)\,\mathbf{C} : \boldsymbol{\varepsilon}^{\mathrm{de}} . \quad (10.121)$$

Since Expression $(10.121)_2$ for microvoids allows a more educative approach, it is studied here. Based on Material Model $(10.121)_2$, Figure 10.10 was drawn.

In continuum damage mechanics, it is common practice to write the following constitutive equation (see, e.g., Lemaitre [50], Chaboche [14] p. 403, or Skrzypek and Ganczarski [103], Figure 3.5):

$$\sigma = (1-D)\, E\, \varepsilon^{\mathrm{e}} , \qquad (10.122)$$

where $D$ is the scalar-valued quantity called "damage." The relation between the void volume fraction $f$ and the damage $D$ is studied later in this book.

The message of Equation (10.122) in Skrzypek and Ganczarski ([103], Figure 3.5) is given by a drawing similar to that in Figure 10.11. The shape of their stress–strain curve is not the one shown in Figure 10.11, but it is not the key concept. The key concept is the difference between the horizontal axes in Figures 10.10 and 10.11.

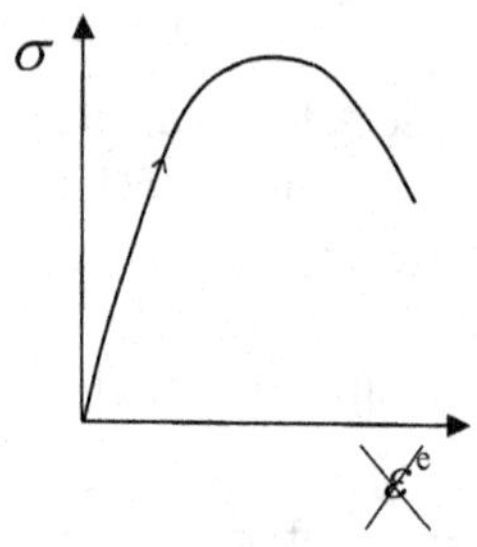


**Figure 10.11** Stress–strain curve due to damage evolution.

The shape of the stress–strain curve in Figure 10.11 does not fulfill the requirements set for the elastic deformation. The materials scientist would express the elastic deformation as follows: in elastic deformation, during unloading, the stress–strain curve takes the same path that it took during the loading. Furthermore, according to Malvern ([57], p. 278), a material is ideally elastic when a body formed of the material recovers its original form completely upon removal of the forces causing the deformation and there is a one-to-one relationship between the state of stress and the state of strain, for a given temperature. Based on the preceding discussion, the elastic response can follow the curves given in Figure 10.12.

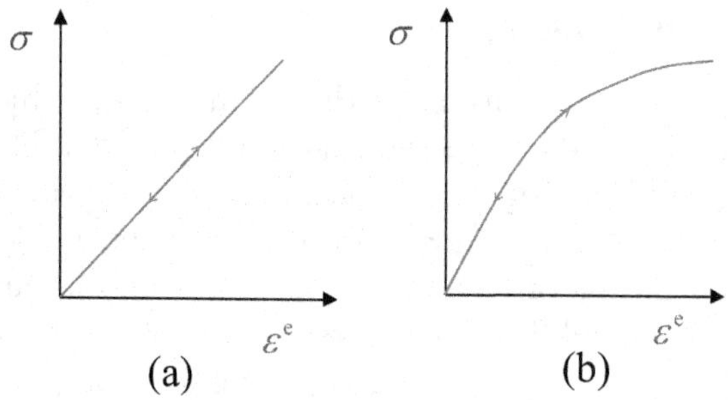


**Figure 10.12** (a) Linear and (b) nonlinear elastic stress–strain curve.

To avoid the earlier described problem, notations given in Expressions (10.94), $(10.93)_1$, and $(10.93)_2$ were introduced. They are

$$\boldsymbol{\varepsilon}^{\mathrm{de}} = \boldsymbol{\varepsilon}^{\mathrm{e}} + \boldsymbol{\varepsilon}^{\mathrm{d}}\,, \tag{10.123}$$

where

$$\boldsymbol{\varepsilon}^{\mathrm{e}} = \mathbf{S} : \boldsymbol{\sigma} \tag{10.124}$$

and

$$\boldsymbol{\varepsilon}^{\mathrm{d}} = \mathbf{S}^{\mathrm{d}}(f) : \boldsymbol{\sigma}\,. \tag{10.125}$$

When deformation of the matrix material obeys Hooke's law with the value $\nu = 0.2$ for Poisson's ratio, Expressions (10.123)–(10.125) lead to Expression $(10.121)_2$ for a Hookean matrix response with spherical microvoids.

The present representation satisfies the following necessary conditions. Expression $(10.121)_2$ has the same form as Expression (10.122), which is generally accepted in continuum damage mechanics, and the stress $\boldsymbol{\sigma}$ elastic strain $\boldsymbol{\varepsilon}^{\mathrm{e}}$ relation is unique, as shown in Equation (10.124). However, the elastic strain (tensor) $\boldsymbol{\varepsilon}^{\mathrm{e}}$ present in Expression (10.122) is replaced with the damage-elastic strain tensor $\boldsymbol{\varepsilon}^{\mathrm{de}}$ , which makes it possible to give a unique relation between the stress tensor $\boldsymbol{\sigma}$ and the elastic strain tensor $\boldsymbol{\varepsilon}^{\mathrm{e}}$. Thus, the introduction of the notion damage-elastic strain tensor $\boldsymbol{\varepsilon}^{\mathrm{de}}$ can be said to be a fruitful task. Since Expression (10.122) and the horizontal axis in Figure 10.11 do not fit the notions of the present book, they are crossed out.

Not only is the preceding discussion valid for a Hookean matrix response with spherical microvoids; it also covers the response of the same matrix material behavior with penny-shaped microcracks. The corresponding expressions to Equations (10.123)–(10.125) for the microcracks are given in Equations (10.47), (10.46), and $(10.45)_2$.

### 10.5.2 Damage Description by the Postulate of Strain Equivalence with the Effective Stress Concept

Tis section investigates the postulate of strain equivalence with the effective stress concept, which together form a solid foundation for description of damage of materials. In contrast to some other publications (see, e.g., Skrzypek and Ganczarski [103], Section 3.3)] the present work assumes that this postulate is for the damage-elastic response of the material. In order to obtain explicit results, a material model has to be introduced.

Sections 10.3, 10.4, and 10.5.1 formed a framework within which to introduce continuum damage mechanics for materials with microvoids and/or microcracks. This section gives an implicit formulation for continuum damage mechanics, whereas the following section provides explicit expressions, which are based on the results of Sections 10.3, 10.4, and 10.5.1.

The postulate of (damage-elastic) strain equivalence with the effective stress concept is not the only description for damage; rather, several other variants exist (see, e.g., Skrzypek and Ganczarski [103], Section 3.3). However, the foundation of the present description may be easiest to understand, and a clear physical interpretation for the effective stress is given here. Furthermore, it fits with the analytical results given in Sections 10.3 and 10.4. Therefore, the postulate of damage-elastic strain equivalence with the effective stress concept is adopted here.

The postulate of damage-elastic strain equivalence with the effective stress concept was introduced by Chaboche ([14], p. 19). Here the definition of the effective stress tensor $\tilde{\boldsymbol{\sigma}}$ by Chaboche is extended for a nonlinear material response as follows.

If the virgin (undamaged) material obeys the constitutive equation

$$\boldsymbol{\sigma} = f_1(\boldsymbol{\varepsilon}^{\mathrm{e}}, \mathrm{Virgin}), \qquad \text{For virgin material.} \tag{10.126}$$

then the effective stress tensor $\tilde{\boldsymbol{\sigma}}$ is defined by

$$\tilde{\boldsymbol{\sigma}} := f_1(\boldsymbol{\varepsilon}^{\mathrm{de}}, \mathrm{Virgin}), \qquad \text{For damaged material.} \tag{10.127}$$

It is important to note that Material Models (10.126) and (10.127) have an identical functional appearance. Determination of the function $f_1$ may be difficult in practice, since damage can occur immediately after loading and no loaded virgin state within which to determine the function $f_1$ exists. Such problems are not studied here.

The terms "Virgin" and "For damaged material" in Definition (10.127) seem to be contradictory, but this is not the case. "Virgin" indicates that the material properties are taken from an undamaged material, whereas "For damaged material" indicates that the definition is for the quantity, in this case, the effective stress tensor $\tilde{\boldsymbol{\sigma}}$ for the damaged material.

Based on the investigation carried out in Expressions $(10.49)_1$ and (10.94), the following assumption is made:

$$\boldsymbol{\varepsilon}^{\mathrm{de}} = f_2(\boldsymbol{\sigma}, \mathrm{Damaged}), \qquad \text{For damaged material.} \tag{10.128}$$

Definition (10.127) is the "effective stress concept." The "postulate of the strain equivalence" indicates that the damage-elastic strain tensors $\boldsymbol{\varepsilon}^{\mathrm{de}}$ in Expressions (10.127) and (10.128) are equal.

Note that the preceding damage description allows models to be introduced for other deformation mechanisms. In such a case, the damage description keeps the form presented earlier.

A clearer picture of the postulate of strain equivalence with the effective stress concept can be obtained when a particular material model is studied. This is the topic of the next section.

### 10.5.3 Analytical Relation between the Stress Tensors $\tilde{\boldsymbol{\sigma}}$ and $\boldsymbol{\sigma}$ for the Hookean Matrix Response

This section introduces explicit forms for the functions $f_1$ and $f_2$, and therefore derives the relationship between the effective stress tensor $\tilde{\boldsymbol{\sigma}}$ and the stress tensor $\boldsymbol{\sigma}$. It applies the postulate of strain equivalence with effective stress concept and the introduced functions $f_1$ and $f_2$. A circular hole in a plate is studied as an example to demonstrate why the uniaxial $\sigma_{22}$ stress induces a transverse effective stress $\tilde{\sigma}_{11}$. Finally, the obtained result is applied to porous material, where the deformation of its matrix material follows Hooke's law with a value of Poisson's ratio of $\nu = 0.2$.

In Section 10.5 starting from this point and ending with Section 10.5.10 the term *damage* refers to the damage due to microvoids and/or microcracks. In order to avoid long sentences the phrase "due to microvoids and/or microcracks" is not given.

Model 1 assumes that the elastic deformation of the matrix material obey's Hooke's law in its virgin state. Thus Equation (10.126) yields

$$\boldsymbol{\sigma} = f_1(\boldsymbol{\varepsilon}^{\mathrm{e}}, \text{Virgin}) \quad \Rightarrow \quad \boldsymbol{\sigma} = \mathbf{C} : \boldsymbol{\varepsilon}^{\mathrm{e}} . \qquad \text{For virgin material.} \tag{10.129}$$

Based on Model 1, the definition for the effective stress $\tilde{\boldsymbol{\sigma}}$ [Definition (10.127)] reads

$$\tilde{\boldsymbol{\sigma}} := f_1(\boldsymbol{\varepsilon}^{\mathrm{de}}, \text{Virgin}) \quad \Rightarrow \quad \tilde{\boldsymbol{\sigma}} = \mathbf{C} : \boldsymbol{\varepsilon}^{\mathrm{de}} . \qquad \text{For damaged material.} \tag{10.130}$$

In Expressions $(10.129)_2$ and $(10.130)_2$, the quantity $\mathbf{C}$ is the fourth-order constitutive tensor for a Hookean deformation. When penny-shaped microcracks or spherical microvoids with the Hookean matrix response were investigated, Expressions $(10.49)_1$ and (10.94) were derived. Now the results are extended to cover the damaged material in general by dropping the dependency of the effective compliance tensor $\tilde{\mathbf{S}}(f)$ on the void volume fraction $f$. When penny-shaped microcracks were studied, no independent variables for the effective compliance tensor $\tilde{\mathbf{S}}$ were shown, as can be seen in Equation (10.49). Thus the following model is written.

Model 2 is assumed to have the following appearance:

$$\boldsymbol{\varepsilon}^{\mathrm{de}} = f_2(\boldsymbol{\sigma}, \text{Damaged}) \quad \Rightarrow \quad \boldsymbol{\varepsilon}^{\mathrm{de}} = \tilde{\mathbf{S}} : \boldsymbol{\sigma} . \quad \text{For damaged material.} \tag{10.131}$$

Substitution of Equation $(10.131)_2$ into Definition $(10.130)_2$ leads to

$$\tilde{\boldsymbol{\sigma}} = \mathbf{C} : \tilde{\mathbf{S}} : \boldsymbol{\sigma} \quad \Rightarrow \quad \tilde{\boldsymbol{\sigma}} = \mathbf{M} : \boldsymbol{\sigma} , \quad \text{where} \quad \mathbf{M} := \mathbf{C} : \tilde{\mathbf{S}} , \tag{10.132}$$

where $\mathbf{M}$ is the damage effect tensor. Expressions (10.132) give an important result for continuum damage mechanics. They show that in the case of non-interacting microvoids and non-interacting microcracks, it is possible to derive an exact analytical relation between the effective stress tensor $\tilde{\boldsymbol{\sigma}}$ and the stress tensor $\boldsymbol{\sigma}$. Furthermore, as the following section shows the effective stress tensor $\tilde{\boldsymbol{\sigma}}$ is symmetric. Thus no ad hoc model is needed.

Model $(10.131)_2$ is valid for a Hookean matrix response with penny-shaped microcracks [cf. Equation (10.49)] and spherical microvoids [cf. Equation (10.94)]. Therefore Equations (10.132) are valid for these types of material behaviors. This was the reason for setting Model $(10.131)_2$ as a definition for the damaged material in general.

Equation (8.43) is recalled. It reads

$$\mathbf{C} : \mathbf{S} = \mathbf{S} : \mathbf{C} = \mathbf{I}^{\mathrm{s}} , \tag{10.133}$$

where $\mathbf{S}$ is the fourth-order compliance tensor for a Hookean deformation. Based on Definition (2.106), the following holds true for every symmetric second-order tensor $\mathbf{c}$ (i.e., when $\mathbf{c}^{\mathrm{T}} = \mathbf{c}$):

$$\mathbf{I}^{\mathrm{s}} : \mathbf{c} = \frac{1}{2}\left(\mathbf{c} + \mathbf{c}^{\mathrm{T}}\right) = \mathbf{c} \,. \tag{10.134}$$

By multiplying Equation $(10.130)_2$ by $\mathbf{S}:$ from the left and by taking Equations (10.133) and (10.134) into consideration, we arrive at the following:

$$\mathbf{S} : \tilde{\boldsymbol{\sigma}} := \mathbf{S} : \mathbf{C} : \boldsymbol{\varepsilon}^{\mathrm{de}} \quad \Rightarrow \quad \mathbf{S} : \tilde{\boldsymbol{\sigma}} = \mathbf{I}^{\mathrm{s}} : \boldsymbol{\varepsilon}^{\mathrm{de}} \quad \Rightarrow \quad \boldsymbol{\varepsilon}^{\mathrm{de}} = \mathbf{S} : \tilde{\boldsymbol{\sigma}} \,, \tag{10.135}$$

where the symmetry of the damage-elastic strain tensor $\boldsymbol{\varepsilon}^{\mathrm{de}}$ is exploited.

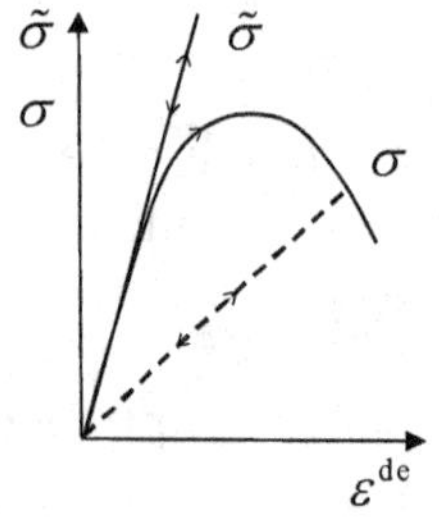


**Figure 10.13** Stress–strain curves.

According to Material Model 1, that is, Expression $(10.130)_2$ [see Expression $(10.135)_3$ as well], the relationship between the effective stress $\tilde{\sigma}$ and the damage-elastic strain $\varepsilon^{\mathrm{de}}$ is linear. This is shown in Figure 10.13 by the solid linear line. It is important to note that the relationship between the effective stress $\tilde{\sigma}$ and the damage-elastic strain $\varepsilon^{\mathrm{de}}$ is also linear in the case of damage evolution.

The response of microcracked or microvoided matrix material the response of which obeys Hooke's law takes the form of the solid nonlinear curve shown in Figures 10.7 and 10.13. It is a modification of Figure 10.10. Now, the damage is assumed to be due to the microcracks. When progressive damage occurs, the strain is not pure elastic strain $\varepsilon^{\mathrm{e}}$. The microcrack formation is a dissipative, that is, irreversible process. The elastic strain $\varepsilon^{\mathrm{e}}$, on the other hand, describes a reversible process in which the unloading path in the stress–strain space takes the same curve, but opposite to, the one during the loading.

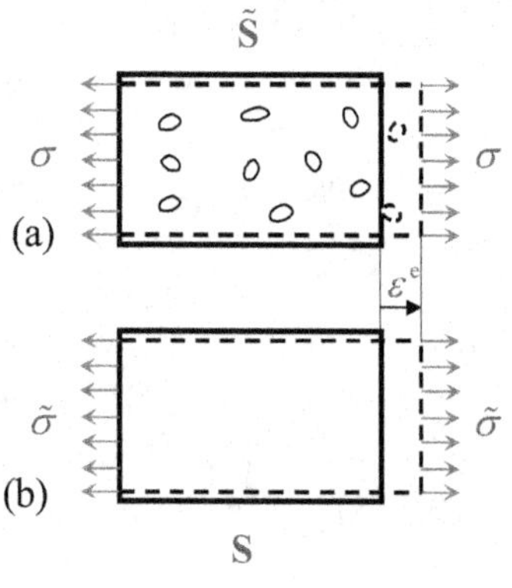


**Figure 10.14** Postulate of strain equivalence for Hookean matrix response.

If the microcracking starts after a certain threshold value of deformation, during loading, the response of the material is first linear elastic, then, when the microcracking occurs, the value for the damage strain $\varepsilon^{\mathrm{d}}$ tends to take increasing values and the curve will deviate from a straight line and finally go downhill, as Figure 10.13 illustrates. So, in the beginning of the unloading, there are the elastic strain $\varepsilon^{\mathrm{e}}$ and the damage strain $\varepsilon^{\mathrm{d}}$ within the material. During the unloading procedure the deformation of the matrix material, that is, the elastic strain $\varepsilon^{\mathrm{e}}$ , tends toward zero and the microcracks close, that is, the damage strain $\varepsilon^{\mathrm{d}}$ tends toward zero. Both processes are linear, and therefore the unloading path is linear, but it is not pure elastic.

Figure 10.14 sketches the postulate of strain equivalence in a case in which the deformation of the virgin material obeys Hooke's law. In Figure 10.14(a), there is a block of damaged material under tensile stress $\sigma$ . Due to the microvoids (or microcracks) the stiffness of the material is reduced and therefore the effective compliance tensor takes a raised value of $\tilde{\mathbf{S}}$.

However, there is no damage evolution during the loading cases studied in Figure 10.14. The stress–strain relation for the damage-elastic response of the material shown in Figure 10.14(a) follows Equation $(10.131)_2$, and therefore the value of the damage-elastic strain is $\varepsilon^{\mathrm{de}}$. Figure 10.14(b) clarifies the definition of the effective stress tensor $\tilde{\boldsymbol{\sigma}}$ given by Definition $(10.130)_2$ [see Equation $(10.135)_3$ as well]. If the effective stress $\tilde{\sigma}$ acted on the same block of material in its virgin state (i.e., when no microvoids or microcracks would existed), the damage-elastic strain $\varepsilon^{\mathrm{de}}$ (the damage strain $\varepsilon^{\mathrm{d}}$ is zero) would take the same value as in the case shown in Figure 10.14(a).

In the case of uniaxial stress $\sigma_{11}$, Equation $(10.132)_2$ gives

$$\tilde{\sigma}_{11} = M_{1111}\,\sigma_{11}\,, \tag{10.136}$$

where

$$\begin{aligned} M_{1111} &= C_{1111}\,\tilde{S}_{1111} + C_{1112}\,\tilde{S}_{1211} + C_{1113}\,\tilde{S}_{1311} \\ &\quad + C_{1121}\,\tilde{S}_{2111} + C_{1122}\,\tilde{S}_{2211} + C_{1123}\,\tilde{S}_{2311} \\ &\quad + C_{1131}\,\tilde{S}_{3111} + C_{1132}\,\tilde{S}_{3211} + C_{1133}\,\tilde{S}_{3311}\,. \end{aligned} \tag{10.137}$$

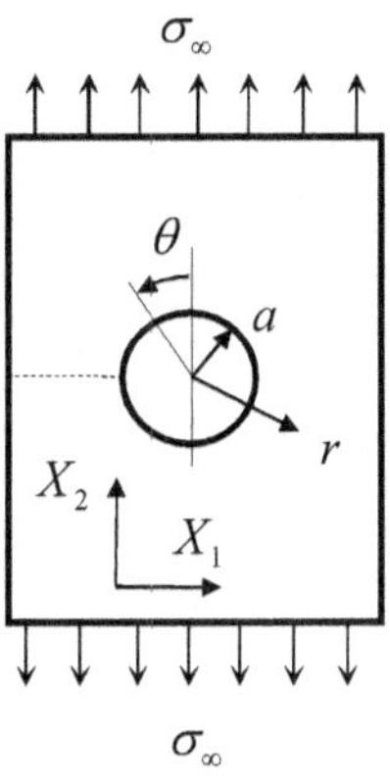


**Figure 10.15** Plate with a circular hole subjected to uniform tensile stress.

It is important to note that applied uniaxial stress $\sigma_{22}$, for example, induces a multiaxial effective stress $\tilde{\boldsymbol{\sigma}}$ field. This phenomenon can be illustrated by studying a two-dimensional infinite plate with a circular hole of radius $a$. According to Figure 10.15, the plate is assumed to be under uniaxial stress $\sigma_{22} = \sigma_\infty$ acting at infinity. According to Reddy ([84], p. 308), the in-plane stresses are

$$\begin{aligned} \sigma_{\mathrm{rr}} &= \frac{\sigma_\infty}{2}\left(1 - \frac{a^2}{r^2}\right) + \frac{\sigma_\infty}{2}\left(1 + \frac{3\,a^4}{r^4} - \frac{4\,a^2}{r^2}\right)\cos 2\theta\,, \\ \sigma_{\theta\theta} &= \frac{\sigma_\infty}{2}\left(1 + \frac{a^2}{r^2}\right) - \frac{\sigma_\infty}{2}\left(1 + \frac{3\,a^4}{r^4}\right)\cos 2\theta\,, \\ \sigma_{\mathrm{r}\theta} &= -\frac{\sigma_\infty}{2}\left(1 - \frac{3\,a^4}{r^4} + \frac{2\,a^2}{r^2}\right)\sin 2\theta\,. \end{aligned} \tag{10.138}$$

The stress component $\sigma_{11}$ equals $\sigma_{\mathrm{rr}}$ when in the foregoing equations the angle $\theta$ takes the value $\pi/2$. As Equation $(10.138)_1$ shows, the stress component $\sigma_{11}$

takes a nonzero value along the dashed horizontal line, except for the surface of the hole. According to Equations (L.29) and (L.30) of Appendix L, the maximum value of the stress component $\sigma_{11}$ on the dashed line is at the point $r = \sqrt{2}\,a$, where $\sigma_{11} = 3\,\sigma_{22}\,/\,8 = 0.375\,\sigma_{22}$. Based on expression (L.25) of Appendix L, if Poisson's ratio is $\nu = 0.3$ and the void volume fraction is $f = 0.1$, the present continuum damage mechanics theory gives $\tilde{\sigma}_{11} = 0.143\,\sigma_{22}$.

As an example, deformation of a Hookean matrix response with spherical microvoids is studied. Equation (10.88), that is, $\tilde{\mathbf{S}} = \mathbf{S} + \mathbf{S}^{\mathrm{d}}$, and Expression (10.133), that is, $\mathbf{S}:\mathbf{C} = \mathbf{I}^{\mathrm{s}}$, allow Definition $(10.132)_3$ to take the form

$$\mathbf{M} := \mathbf{C}:\tilde{\mathbf{S}} = \mathbf{C}:(\mathbf{S} + \mathbf{S}^{\mathrm{d}}) = \mathbf{I}^{\mathrm{s}} + \mathbf{C}:\mathbf{S}^{\mathrm{d}}\,. \tag{10.139}$$

Expression (10.139) gives

$$\begin{aligned} M_{1111} = 1 &+ C_{1111}\,S^{\mathrm{d}}_{1111} + C_{1112}\,S^{\mathrm{d}}_{1211} + C_{1113}\,S^{\mathrm{d}}_{1311} \\ &+ C_{1121}\,S^{\mathrm{d}}_{2111} + C_{1122}\,S^{\mathrm{d}}_{2211} + C_{1123}\,S^{\mathrm{d}}_{2311} \\ &+ C_{1131}\,S^{\mathrm{d}}_{3111} + C_{1132}\,S^{\mathrm{d}}_{3211} + C_{1133}\,S^{\mathrm{d}}_{3311}\,. \end{aligned} \tag{10.140}$$

Next, Poisson's ratio is assumed to be $\nu = 0.2$. Equations (10.116) and $(10.135)_3$ are

$$\mathbf{C}:\boldsymbol{\varepsilon}^{\mathrm{de}} = (1 + 2\,f)\,\boldsymbol{\sigma} \qquad \text{and} \qquad \boldsymbol{\varepsilon}^{\mathrm{de}} = \mathbf{S}:\tilde{\boldsymbol{\sigma}}\,. \tag{10.141}$$

Since the effective stress tensor $\tilde{\boldsymbol{\sigma}}$ is symmetric, Equations (10.141) give

$$\mathbf{C}:\mathbf{S}:\tilde{\boldsymbol{\sigma}} = (1 + 2\,f)\,\boldsymbol{\sigma} \quad \Rightarrow \quad \tilde{\boldsymbol{\sigma}} = (1 + 2\,f)\,\boldsymbol{\sigma}\,, \tag{10.142}$$

where Equation (10.133), $\mathbf{C}:\mathbf{S} = \mathbf{I}^{\mathrm{s}}$, and Equation (10.134), $\mathbf{I}^{\mathrm{s}}:\tilde{\boldsymbol{\sigma}} = \tilde{\boldsymbol{\sigma}}$ (this requires the symmetry of the effective stress tensor $\tilde{\boldsymbol{\sigma}}$, which is shown in the next section), are exploited. Result (10.119) is

$$(1 + 2\,f)^{-1} \approx (1 - 2\,f) \quad \Rightarrow \quad (1 + 2\,f) \approx (1 - 2\,f)^{-1}\,. \tag{10.143}$$

Substitution of Approximation $(10.143)_2$ into Equation $(10.142)_2$ yields

$$\tilde{\boldsymbol{\sigma}} \approx \frac{\boldsymbol{\sigma}}{1 - 2\,f}\,. \tag{10.144}$$

### 10.5.4 Effective Stress Tensor $\tilde{\boldsymbol{\sigma}}$ Is Symmetric

This section shows that the effective stress tensor $\tilde{\boldsymbol{\sigma}}$ is symmetric. Equation (10.132) reads

$$\tilde{\boldsymbol{\sigma}} = \mathbf{C}:\tilde{\mathbf{S}}:\boldsymbol{\sigma} \quad \Rightarrow \quad \tilde{\boldsymbol{\sigma}} = \mathbf{M}:\boldsymbol{\sigma}\,, \quad \text{where} \quad \mathbf{M} := \mathbf{C}:\tilde{\mathbf{S}}\,. \tag{10.145}$$

**Case A**: An expression for the component $\tilde{\sigma}_{ij}$ of the effective stress tensor $\tilde{\boldsymbol{\sigma}}$ is derived.

The following component forms are introduced:

$$\tilde{\boldsymbol{\sigma}} = \tilde{\sigma}_{ij}\,\vec{i}_i\,\vec{i}_j \qquad \text{and} \qquad \mathbf{C} = C_{ijuv}\,\vec{i}_i\,\vec{i}_j\,\vec{i}_u\,\vec{i}_v \tag{10.146}$$

and

$$\tilde{\mathbf{S}} = \tilde{S}_{mnqr}\,\vec{i}_m\,\vec{i}_n\,\vec{i}_q\,\vec{i}_r \quad \text{and} \quad \boldsymbol{\sigma} = \sigma_{st}\,\vec{i}_s\,\vec{i}_t\,. \tag{10.147}$$

Component Forms (10.146) and (10.147) are substituted into Equation $(10.145)_1$. This gives

$$\tilde{\boldsymbol{\sigma}} = \mathbf{C}{:}\tilde{\mathbf{S}}{:}\boldsymbol{\sigma} \quad \Rightarrow \quad \tilde{\sigma}_{ij}\,\vec{i}_i\,\vec{i}_j = C_{ijuv}\,\vec{i}_i\,\vec{i}_j\,\vec{i}_u\,\vec{i}_v : \tilde{S}_{mnqr}\,\vec{i}_m\,\vec{i}_n\,\vec{i}_q\,\vec{i}_r{:}\sigma_{st}\,\vec{i}_s\,\vec{i}_t\,. \tag{10.148}$$

When performing the double-dot products, Equation $(10.148)_2$ takes the following appearance (detailed instructions for performing double-dot products are given in **Example 2.6**, for example):

$$\begin{aligned} \tilde{\sigma}_{ij}\,\vec{i}_i\,\vec{i}_j = C_{ijuv}\,\vec{i}_i\,\vec{i}_j\,\vec{i}_u\,\vec{i}_v : \tilde{S}_{mnqr}\,\vec{i}_m\,\vec{i}_n\,\vec{i}_q\,\vec{i}_r : \sigma_{st}\,\vec{i}_s\,\vec{i}_t \\ \Rightarrow \quad \tilde{\sigma}_{ij}\,\vec{i}_i\,\vec{i}_j = C_{ijmn}\,\tilde{S}_{mnst}\,\sigma_{st}\,\vec{i}_i\,\vec{i}_j\,. \end{aligned} \tag{10.149}$$

Equation $(10.149)_2$ gives

$$\tilde{\sigma}_{ij}\,\vec{i}_i\,\vec{i}_j = C_{ijmn}\,\tilde{S}_{mnst}\,\sigma_{st}\,\vec{i}_i\,\vec{i}_j \quad \Rightarrow \quad \tilde{\sigma}_{ij} = C_{ijmn}\,\tilde{S}_{mnst}\,\sigma_{st}\,. \tag{10.150}$$

**Case B**: An expression for the component $\tilde{\sigma}_{ji}$ of the effective stress tensor $\tilde{\boldsymbol{\sigma}}$ is derived.

When the indices $i$ and $j$ are interchanged in Expression $(10.150)_2$, we arrive at the following:

$$\tilde{\sigma}_{ji} = C_{jimn}\,\tilde{S}_{mnst}\,\sigma_{st}\,. \tag{10.151}$$

According to Theorem 4 of **Appendix I**, the constitutive tensor for a Hookean deformation $\mathbf{C}$ is minor symmetric, that is, $C_{jimn} = C_{ijmn}$. Thus Equation (10.151) can be written in the following form:

$$\tilde{\sigma}_{ji} = C_{ijmn}\,\tilde{S}_{mnst}\,\sigma_{st}\,. \tag{10.152}$$

**Comparison of Cases A and B**: Since the right sides of Expressions $(10.150)_2$ and (10.152) are equal, their left sides are equal; that is,

$$\tilde{\sigma}_{ij} = \tilde{\sigma}_{ji} \qquad \text{or} \qquad (\tilde{\boldsymbol{\sigma}})^{\mathrm{T}} = \tilde{\boldsymbol{\sigma}}\,. \tag{10.153}$$

Expression (10.153) shows that the effective stress tensor $\tilde{\boldsymbol{\sigma}}$ is symmetric.

### 10.5.5 Extended Rabotnov Effective Stress Concept

This section studies the effective stress concept introduced by Rabotnov and its three-dimensional extension.

When studying creep damage, Rabotnov ([82], p. 344; see also Rabotnov [81]) introduced the concept of the effective stress $\tilde{\sigma}$ using the following uniaxial definition:

$$\tilde{\sigma} := \frac{\sigma}{1-D} \qquad \Rightarrow \qquad \tilde{\sigma}_{11} = \frac{\sigma_{11}}{1-D}, \tag{10.154}$$

where the scalar-valued quantity $D$ is today called damage. For a virgin material $D = 0$, and for a fully damaged material $D = 1$.

Comparison of Definition $(10.154)_1$ with Analytical Expression $(10.132)_1$ provides a tool for deriving the analytical expression for scalar-valued damage $D$ for a Hookean matrix response with spherical microvoids. This is done next.

Equations (10.136) and (10.137) are

$$\begin{aligned}\tilde{\sigma}_{11} = M_{1111}\,\sigma_{11}, \quad \text{where} \quad M_{1111} &= C_{1111}\,\tilde{S}_{1111} + C_{1112}\,\tilde{S}_{1211} + C_{1113}\,\tilde{S}_{1311}\\ &+ C_{1121}\,\tilde{S}_{2111} + C_{1122}\,\tilde{S}_{2211} + C_{1123}\,\tilde{S}_{2311}\\ &+ C_{1131}\,\tilde{S}_{3111} + C_{1132}\,\tilde{S}_{3211} + C_{1133}\,\tilde{S}_{3311}\,.\end{aligned} \tag{10.155}$$

Comparison of Equation $(10.154)_2$ with Expression $(10.155)_1$ yields

$$\frac{1}{1-D} = M_{1111}, \qquad \text{which gives} \qquad D = \frac{M_{1111}-1}{M_{1111}}. \tag{10.156}$$

Expressions $(10.156)_2$ and $(10.155)_2$ allow calculation of the value for the damage $D$ when it describes spherical microvoids in a matrix the response of which obeys Hooke's law. Result $(10.156)_2$ is written for the relation between the uniaxial (normal) stresses $\tilde{\sigma}_{11}$ and $\sigma_{11}$.

The concept by Rabotnov, Definition $(10.154)_1$, is often extended for a three-dimensional state of stress and for isotropic damage as follows:

$$\tilde{\boldsymbol{\sigma}} = \frac{\boldsymbol{\sigma}}{1-D} \qquad \Rightarrow \qquad \tilde{\boldsymbol{\sigma}} := \frac{\mathbf{I}:\boldsymbol{\sigma}}{1-D}, \tag{10.157}$$

where Definition (2.69) is exploited.

Equation $(10.132)_1$ is recalled:

$$\tilde{\boldsymbol{\sigma}} = \mathbf{C}:\tilde{\mathbf{S}}:\boldsymbol{\sigma}\,. \tag{10.158}$$

In order for Equations $(10.157)_2$ and (10.158) to be equal, the following should hold:

$$\mathbf{C}:\tilde{\mathbf{S}} = \mathbf{I}\,(1-D)^{-1}. \tag{10.159}$$

Equation (10.159) is multiplied by $\mathbf{S}:$ from the left, Property (10.133), that is, $\mathbf{S}:\mathbf{C}=\mathbf{I}^{\mathrm{s}}$, Property (I.36) of **Appendix I**, that is, $\mathbf{I}^{\mathrm{s}}:\tilde{\mathbf{S}}=\tilde{\mathbf{S}}$, and Property $(2.71)_2$, that is, $\mathbf{S}:\mathbf{I}=\mathbf{S}$, are exploited. This gives

$$\mathbf{S}:\mathbf{C}:\tilde{\mathbf{S}}=\mathbf{S}:\mathbf{I}\,(1-D)^{-1} \qquad \Rightarrow \qquad \tilde{\mathbf{S}}=\mathbf{S}\,(1-D)^{-1}\,. \tag{10.160}$$

According to Equations (10.110), (10.111), and (10.114), only when $\nu=0.2$ does the following exact result hold true for a Hookean matrix response with noninteracting spherical microvoids:

$$\mathbf{S}^{\mathrm{d}}(f)=2\,f\,\mathbf{S} \qquad \text{and} \qquad \tilde{\mathbf{S}}(f)=(1+2f)\,\mathbf{S}\,. \tag{10.161}$$

It is important to note that the linear relationship between the compliance tensors $\tilde{\mathbf{S}}$ and $\mathbf{S}$ given in Equation $(10.161)_2$ holds true only when Poisson's ratio takes the value of $\nu=0.2$. In such a case, Equations $(10.160)_2$ and $(10.161)_2$ coincide, and the extended Rabotnov effective stress concept, Equation $(10.157)_1$, is valid.

The result derived earlier means that the extended Rabotnov concept of effective stress, Expressions $(10.157)_1$, is not generally valid for a porous material. This means that, at least for a porous material, Equations $(10.157)_1$ is an approximation and therefore should be interpreted as a model, as discussed by Santaoja and Kuistiala [100]. This is an important point, since Expression $(10.157)_1$ is given by many publications, such as those by Lemaitre and Chaboche ([51], p. 331), Lemaitre ([50], p. 42), and Skrzypek and Ganczarski [[103], Eq. (3.22)]. Actually, the isotropic damage description requires two damage variables, as pointed out, for example, in Cauvin and Testa ([12], p. 758), Chow and Wei ([15], p. 359), and Yang and Leng ([118], p. 179).

The relationship between the void volume fraction $f$ and damage $D$ is evaluated. A porous material with a linear elastic matrix response is studied as an example.

Expressions $(10.160)_2$ and $(10.161)_2$ give, for a material with the value $\nu=0.2$ for Poisson's ratio, the result

$$(1-D)^{-1}=(1+2\,f) \qquad \Rightarrow \qquad D=\frac{2\,f}{1+2\,f}\,. \tag{10.162}$$

Result $(10.162)_2$ implies that for a low void volume fraction $f$ [Eshelby's theory is based on this assumption], and when Poisson's ratio $\nu=0.2$, the following holds:

$$D\approx 2\,f\,. \tag{10.163}$$

The definition for the effective stress $\tilde{\boldsymbol{\sigma}}$, Definition $(10.130)_2$, and the extended Rabotnov effective stress concept, Expression $(10.157)_1$, are

$$\tilde{\boldsymbol{\sigma}}:=\mathbf{C}:\boldsymbol{\varepsilon}^{\mathrm{de}} \qquad \text{and} \qquad \tilde{\boldsymbol{\sigma}}=\frac{\boldsymbol{\sigma}}{1-D}\,. \tag{10.164}$$

Equations (10.164) give

$$\mathbf{C} : \boldsymbol{\varepsilon}^{\rm de} = \frac{\boldsymbol{\sigma}}{1-D} \qquad \Rightarrow \qquad \boldsymbol{\sigma} = (1-D)\,\mathbf{C} : \boldsymbol{\varepsilon}^{\rm de}\,. \tag{10.165}$$

Result $(10.165)_2$ is widely used in continuum damage mechanics, although the damage-elastic strain tensor $\boldsymbol{\varepsilon}^{\rm de}$ is usually (if not always) replaced by the elastic strain tensor $\boldsymbol{\varepsilon}^{\rm e}$. However, Form $(10.165)_2$ is not acceptable, since the variable damage $D$ does not have any specific physical meaning. Instead, Form (10.120), that is, $\boldsymbol{\sigma} \approx (1-2\,f)\,\mathbf{C} : \boldsymbol{\varepsilon}^{\rm de}$, should be used, since the void volume fraction $f$ is a well-defined quantity.

### 10.5.6 Uniaxial Bar Model for Interpretation of the Variables Related to Damage

In the present section, special attention is given to clarifying the roles of the stress tensor $\boldsymbol{\sigma}$ and the effective stress tensor $\tilde{\boldsymbol{\sigma}}$. This is done by studying their uniaxial scalar-valued counterparts, which makes interpretation of the roles of these stress measures easier. Some other continuum damage mechanics variables are investigated with the same evaluation. It is important to note that the stress tensors $\boldsymbol{\sigma}$ and $\tilde{\boldsymbol{\sigma}}$ are quantities for small deformation theory.

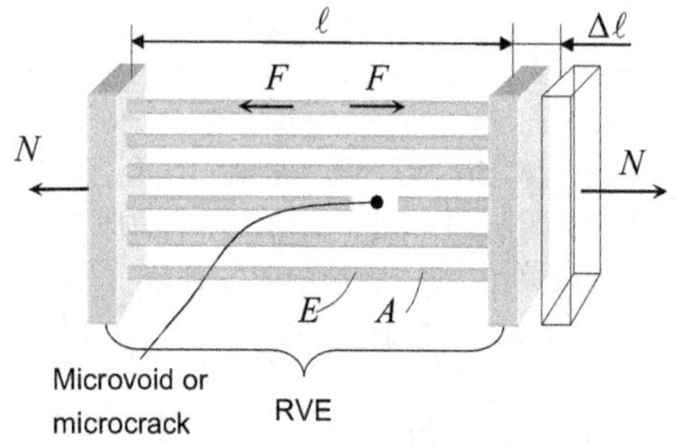


**Figure 10.16** Uniaxial bar model and RVE for evaluation of the stresses $\sigma$ and $\tilde{\sigma}$.

In order to demonstrate the roles of the stress $\sigma$ and the effective stress $\tilde{\sigma}$ , the uniaxial model for a damaged material, shown in Figure 10.16, is evaluated. The model is assumed to be based on $n$ equal tensile bars. The cross-sectional area of a single bar is denoted by $A$. The length of the bars is given by $\ell$. Young's modulus of a single bar is $E$. Variable $m$ gives the number of broken bars. As shown in Figure 10.16, the tensile force along an uncracked bar is denoted by $F$, and the force over the entire system is denoted by $N$. The elongation of the bars due to the loading is denoted by $\Delta\ell$. The vertical walls are assumed to form the vertical "boundaries" for the representative volume element RVE. The horizontal boundaries are on the top and bottom of the vertical walls. In order to keep the figure simple, they are not shown in Figure 10.16. The concept for the representative volume element RVE was introduced in Section 10.2.

The comments on the right sides of the following expressions require a brief clarification. If the equation holds for a single unbroken bar, the right-side comment will be "for unbroken bar" or simply "for bar." If the equation is for the representative volume element RVE, the comment reads "for RVE."

Sometimes it is good practice to show whether the quantity refers to a bar or to the RVE. In that case, the comment "RVE = RVE × bar" shows that the quantity on the left side of the equation refers to RVE. The same holds for the first quantity on the right side, whereas the second quantity refers to the bar. We hope that these comments will help the reader to follow the derivation.

Based on Figure 10.16, we arrive at the following:

$$\sigma^* = E\,\varepsilon^{\mathrm{e}}\,, \quad \text{where } \sigma^* = \frac{F}{A}\,, \quad \varepsilon^{\mathrm{e}} = \frac{\Delta\ell}{\ell}\,. \quad \text{(For unbroken bar).} \tag{10.166}$$

In Equations (10.166), the notation $\sigma^*$ stands for the stress in unbroken bars. Since it is evident that all the expressions written for bars are for unbroken bars, the phrase "unbroken" will be left out. Equations (10.166) give

$$\sigma^* = E\,\frac{\Delta\ell}{\ell}\,, \quad \frac{F}{A} = E\,\frac{\Delta\ell}{\ell}\,, \quad \text{and} \quad \frac{F}{A} = E\,\varepsilon^{\mathrm{e}}\,. \quad \text{(For bar).} \tag{10.167}$$

The (total) force over the RVE is denoted by $N$ and it is the sum of the forces along the unbroken bars. This reads

$$\underset{\uparrow}{N} = (n - \underset{\uparrow}{m})\,\underset{\Uparrow}{F}\,. \quad \text{(RVE = RVE} \times \text{bar).} \tag{10.168}$$

The relation between the RVE quantities and a microscopic quantity is obtained in Equation (10.168). The force $N$ on the left side of Equation (10.168) and the term $(n - m)$ on the right side are for the representative volume element RVE, whereas the force $F$ on the right side of Equation (10.168) is for a bar. The RVE quantities are marked by $\uparrow$ and the bar-level quantities by $\Uparrow$. Thus Equation (10.168) is a connection between the RVE (macroscopic) quantities and the microscopic quantities.

Equation (10.168) is divided by $n\,A$, which gives

$$\frac{N}{n\,A} = \frac{(n-m)}{n}\,\frac{F}{A}\,. \quad \text{(RVE = RVE} \times \text{bar).} \tag{10.169}$$

Substitution of Equation $(10.167)_2$ and $(10.167)_3$ into Equation (10.169) yields

$$\frac{N}{n\,A} = \frac{(n-m)}{n}\,E\,\frac{\Delta\ell}{\ell}\,. \quad \text{(RVE = RVE} \times \text{bar} \times \text{bar).} \tag{10.170}$$

and

$$\frac{N}{n\,A} = \frac{(n-m)}{n}\,E\,\varepsilon^{\mathrm{e}}\,. \quad \text{(RVE = RVE} \times \text{bar} \times \text{bar).} \tag{10.171}$$

The RVE (macroscopic) stress $\sigma$ is defined to be

$$\sigma := \frac{N}{n\,A}\,. \quad \text{(For RVE).} \tag{10.172}$$

It is worth noting that the denominator in Definition (10.172) is a sum of the areas of all the bars, although some of them are broken. Thus the stress $\sigma$ can be interpreted to be a macroscopic averaged stress in the material of the RVE.

Substitution of Definition (10.172) into Equations (10.170) and (10.171) gives

$$\sigma = \frac{(n-m)}{n} E \frac{\Delta \ell}{\ell} \quad \text{and} \quad \sigma = \frac{(n-m)}{n} E\, \varepsilon^{\mathrm{e}} . \qquad (\text{RVE} = \text{RVE} \times \text{bar} \times \text{bar}). \tag{10.173}$$

Figure 10.16 gives

$$\varepsilon^{\mathrm{de}} = \frac{\Delta \ell}{\ell} . \qquad (\text{For RVE}). \tag{10.174}$$

Substitution of Term (10.174) into Expression $(10.173)_1$ gives

$$\sigma = \frac{(n-m)}{n} E\, \varepsilon^{\mathrm{de}} . \qquad (\text{RVE} = \text{RVE} \times \text{bar} \times \text{RVE}). \tag{10.175}$$

The effective Young's modulus $\tilde{E}$ is defined to be

$$\tilde{E} := \frac{(n-m)}{n} E . \qquad (\text{RVE} = \text{RVE} \times \text{bar}). \tag{10.176}$$

Definition (10.176) fulfills the following two necessary conditions: for a virgin material $\tilde{E} = E$ and for a fully damaged material $\tilde{E} = 0$. The effective compliance $\tilde{S}$ is

$$\tilde{S} := 1/\tilde{E} . \qquad (\text{For RVE}). \tag{10.177}$$

Definitions (10.176) and (10.177) allow Expression (10.175) to take the following form:

$$\varepsilon^{\mathrm{de}} = 1/\tilde{E}\, \sigma \quad \Rightarrow \quad \varepsilon^{\mathrm{de}} = \tilde{S}\, \sigma . \qquad (\text{For RVE}). \tag{10.178}$$

Result $(10.178)_2$ is the uniaxial counterpart of the expression derived for the noninteracting microcracks given by Expression $(10.49)_1$ and of the expression for the noninteracting microvoids given by Expression (10.94).

For virgin representative volume element RVE, the number of broken bars is zero, that is, $m = 0$, and therefore Expression $(10.173)_2$ reduces to

$$\sigma = E\, \varepsilon^{\mathrm{e}} . \qquad (\text{For virgin RVE}). \tag{10.179}$$

According to the postulate of strain equivalence with the effective stress concept, the effective stress $\tilde{\sigma}$ is defined by a material model having an appearance similar to the right side of Expression (10.179). Thus the effective stress $\tilde{\sigma}$ is obtained from [cf. Definition $(10.130)_2$, i.e., $\tilde{\boldsymbol{\sigma}} = \mathbf{C} : \boldsymbol{\varepsilon}^{\mathrm{de}}$ ]

$$\tilde{\sigma} := E\, \varepsilon^{\mathrm{de}} . \qquad (? = \text{bar} \times \text{RVE}). \tag{10.180}$$

The question mark in Definition (10.180) shows that it is not known whether the effective stress $\tilde{\sigma}$ is defined for the bar or for the RVE. In Expression (10.179), Young's modulus $E$ was for the RVE (and for the bar, of course), since the material was virgin. In Definition (10.180), on the other hand, the effective stress $\tilde{\sigma}$ is for damaged material, but the material parameters are for a virgin material, as stated in Equation (10.127).

Substitution of Expression (10.174) into Definition (10.180) yields

$$\tilde{\sigma} = E\,\frac{\Delta\ell}{\ell}\,. \qquad (? = \text{bar} \times \text{RVE}). \quad (10.181)$$

Since besides the RVE the quantity $\Delta\ell/\ell$ is valid for a bar as well, Equation (10.181) can be written in the form

$$\tilde{\sigma} = E\,\frac{\Delta\ell}{\ell}\,. \qquad (? = \text{bar} \times \text{bar}). \quad (10.182)$$

Equations (10.167) are written for a (unbroken) bar. This means that Equations (10.167) have the form "bar = bar × bar." Equation (10.182), on the other hand, has the form "? = bar × bar."

The right sides of Equations $(10.167)_1$ and (10.182) are equal. The same holds for the right-most comments in these two equations. They are "= bar × bar." Thus, the comments and the physical meanings on the left sides have to be equal, which gives

$$\tilde{\sigma} = \sigma^* \qquad \text{and} \qquad \tilde{\sigma} = E\,\frac{\Delta\ell}{\ell}\,. \qquad (\text{For bar}). \quad (10.183)$$

Based on Equation $(10.183)_1$, the effective stress $\tilde{\boldsymbol{\sigma}}$ is a microscopic stress in the matrix material between microvoids and/or microcracks. Since the effective stress $\tilde{\boldsymbol{\sigma}}$ takes the same value over all the unbroken bars, the effective stress $\tilde{\boldsymbol{\sigma}}$ is a homogeneous stress field. This means that the effective stress $\tilde{\boldsymbol{\sigma}}$ is an averaged microscopic stress. The "real" microscopic stress showing stress peaks at microcrack tips, for example, is called microstress and is denoted by $\sigma^{\mathrm{m}}$.

Substitution of Equation $(10.183)_2$ into Expression $(10.173)_1$ yields

$$\sigma = \frac{(n-m)}{n}\,\tilde{\sigma} \qquad \Rightarrow \qquad \tilde{\sigma} = \frac{n}{(n-m)}\,\sigma\,. \qquad (10.184)$$

Based on Result $(10.184)_1$, the stress $\sigma$ is obtained as averaging the effective stress $\tilde{\sigma}$ over the entire volume of the representative volume element RVE. Since the effective stress $\tilde{\sigma}$ is defined as the average value of the microstress $\sigma^{\mathrm{m}}$ between the microvoids and microcracks, the stress $\sigma$ is a macroscopic stress but is defined as the average of the microstress $\sigma^{\mathrm{m}}$ over the whole volume of the RVE. This interpretation agrees with the result by Dormieux and Kondo [[20], Eq. (8.17)]. The role of the microstress $\sigma^{\mathrm{m}}$ is discussed in more detail in Section 10.5.8.

Void volume fraction $f$ can be defined to be

$$f := \frac{m}{n} \quad \Rightarrow \quad 1 - f = \frac{n}{n} - \frac{m}{n} = \frac{n-m}{n}. \qquad \text{(For RVE)}. \quad (10.185)$$

For this particular example, it is reasonable to define the quantity damage $D$ as follows: Damage $D$ is the number of broken bars divided by the number of all bars. This is

$$D := \frac{m}{n} \quad \Rightarrow \quad 1 - D = \frac{n}{n} - \frac{m}{n} = \frac{n-m}{n}. \qquad \text{(For RVE)}. \quad (10.186)$$

Substitution of Expressions $(10.185)_2$ and $(10.186)_2$ into Equation $(10.184)_2$ leads to

$$\tilde{\sigma} = \frac{1}{1-f}\,\sigma \quad \text{and} \quad \tilde{\sigma} = \frac{1}{1-D}\,\sigma\,. \qquad (\text{bar} = \text{RVE}). \quad (10.187)$$

Result $(10.187)_1$ is similar to Expression (10.144), although in the latter expression, the latter quantity in the denominator is $2f$ instead of $f$. The difference is due to the modelling inaccuracy in the bar model. In the elastic response of the material, there are stress peaks on surfaces of the microvoids that are not present in the bar model. Result $(10.187)_2$ equals Expression $(10.154)_1$ . Based on Equation (10.187), $D = f$, which contradicts the result in Result (10.163), which gave $D \approx 2f$. As earlier, the difference is due to the modelling inaccuracy.

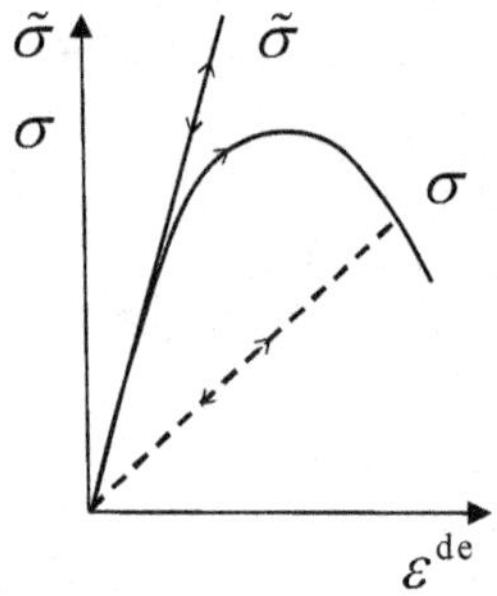


**Figure 10.17** Stress-strain curves.

Equation (10.175) is recalled:

$$\sigma = \frac{(n-m)}{n}\,E\,\varepsilon^{\mathrm{de}} \;\Rightarrow\; \sigma = \left(1 - \frac{m}{n}\right) E\,\varepsilon^{\mathrm{de}}. \qquad (\text{RVE} = \text{RVE} \times \text{bar} \times \text{RVE}). \quad (10.188)$$

From a physics point of view, the damage-elastic strain $\varepsilon^{\mathrm{de}}$ takes increasing values with a growing number of broken bars, that is, with the growing number of $m$. This implies with Equation $(10.188)_2$ that the $\sigma\,\varepsilon^{\mathrm{de}}$ relation is nonlinear, as shown in Figure 10.17. Expression (10.188) and the explanation after it do not imply the descendant behavior of the curve. This requires a progressive model for breakage of bars. Damage evolution equations are not a topic for this example and are therefore not given here. The unloading curve shown by the dashed line is for a material where the microcracks close during unloading.

Equation $(10.178)_2$ and Definition (10.180) are recalled:

$$\varepsilon^{\mathrm{de}} = \tilde{S}\,\sigma \qquad \text{and} \qquad \tilde{\sigma} := E\,\varepsilon^{\mathrm{de}}, \qquad (10.189)$$

which give

$$\tilde{\sigma} := E\,\tilde{S}\,\sigma\,. \qquad (\text{bar} = \text{bar} \times \text{RVE} \times \text{RVE}). \quad (10.190)$$

Equation (10.190) gives the link between the averaged microscopic stress $\tilde{\sigma}$ and the averaged macroscopic (RVE level) stress $\sigma$. It is a uniaxial counterpart for Equation $(10.132)_1$.

Next, the strain-energy density of the representative volume element RVE $w(\varepsilon^{\mathrm{e}})$ is studied. The strain-energy density of an unbroken bar $w^{\mathrm{bar}}(\varepsilon^{\mathrm{e}})$ is

$$w^{\mathrm{bar}}(\varepsilon^{\mathrm{e}}) = \tfrac{1}{2}\,\varepsilon^{\mathrm{e}}\,E\,\varepsilon^{\mathrm{e}}\,. \qquad \text{(For bar)}. \quad (10.191)$$

Since the value for the elastic strain $\varepsilon^{\mathrm{e}}$ is the same through an unbroken bar, the strain energy of an unbroken bar $W^{\mathrm{bar}}(\varepsilon^{\mathrm{e}})$ is the strain-energy density of the bar strain energy of an unbroken bar $W^{\mathrm{bar}}(\varepsilon^{\mathrm{e}})$ is the strain-energy density of the bar $w^{\mathrm{bar}}(\varepsilon^{\mathrm{e}})$ times the volume of the bar $\ell A$. This is

$$W^{\mathrm{bar}}(\varepsilon^{\mathrm{e}}) = w^{\mathrm{bar}}(\varepsilon^{\mathrm{e}})\,\ell\,A = \frac{1}{2}\,\varepsilon^{\mathrm{e}}\,E\varepsilon^{\mathrm{e}}\,\ell\,A\,. \qquad \text{(For bar)}. \quad (10.192)$$

Since the strain energies of the broken bars are zero, the strain energy of the RVE $W(\varepsilon^{\mathrm{e}})$ is the sum of the strain energies of the unbroken bars $W^{\mathrm{bar}}(\varepsilon^{\mathrm{e}})$, i.e.,

$$W(\varepsilon^{\mathrm{e}}) = W^{\mathrm{bar}}(\varepsilon^{\mathrm{e}})\,(n-m) = \tfrac{1}{2}\,\varepsilon^{\mathrm{e}}\,E\,\varepsilon^{\mathrm{e}}\,(n-m)\,\ell\,A\,. \qquad \text{(For RVE)}. \quad (10.193)$$

The strain-energy density of the RVE $w(\varepsilon^{\mathrm{e}})$ is the strain energy of the RVE $W(\varepsilon^{\mathrm{e}})$ divided by the volume of the RVE $\ell\,n\,A$. Therefore Equation (10.193) gives

$$w(\varepsilon^{\mathrm{e}}) = \frac{W(\varepsilon^{\mathrm{e}})}{\ell\,n\,A} = \frac{1}{2}\,\varepsilon^{\mathrm{e}}\,E\,\varepsilon^{\mathrm{e}}\,\frac{(n-m)\,\ell\,A}{\ell\,n\,A} = \frac{1}{2}\,\varepsilon^{\mathrm{e}}\,E\,\varepsilon^{\mathrm{e}}\,\frac{(n-m)}{n}\,. \qquad \text{(For RVE)}. \quad (10.194)$$

Substitution of the definition for the effective Young's modulus $\tilde{E} = E(n-m)/n$, Definition (10.176), into the right side of Expression (10.194) gives

$$w(\varepsilon^{\mathrm{e}}) = \frac{1}{2}\,\varepsilon^{\mathrm{e}}\,E\,\varepsilon^{\mathrm{e}}\,\frac{n-m}{n} = \frac{1}{2}\,\varepsilon^{\mathrm{e}}\,\tilde{E}\,\varepsilon^{\mathrm{e}}\,. \qquad \text{(For RVE)}. \quad (10.195)$$

Next, the complementary strain-energy density of the RVE $w^{\mathrm{c}}(\sigma)$ is studied. The complementary strain-energy density of an unbroken bar $w^{\mathrm{c,bar}}(\tilde{\sigma})$ reads

$$w^{\mathrm{c,bar}}(\tilde{\sigma}) = \tfrac{1}{2}\,\tilde{\sigma}\,S\,\tilde{\sigma}\,. \qquad \text{(For bar)}. \quad (10.196)$$

where $S$ is the compliance of the material of the bars. Since the effective stress $\tilde{\sigma}$ takes equal value everywhere within an unbroken bar, the complementary strain energy of a bar $W^{\mathrm{c}}(\tilde{\sigma})$ is the complementary strain-energy density of a bar $w^{\mathrm{c,bar}}(\tilde{\sigma})$ multiplied by the volume of the bar $\ell\,A$. This is

$$W^{\mathrm{c,bar}}(\tilde{\sigma}) = w^{\mathrm{c,bar}}(\tilde{\sigma})\,\ell\,A = \frac{1}{2}\,\tilde{\sigma}\,S\,\tilde{\sigma}\,\ell\,A\,. \qquad \text{(For bar)}. \quad (10.197)$$

Since the complementary strain energies of the broken bars is zero, the complementary strain energy of the representative volume element RVE $W^{\mathrm{c}}(\tilde{\sigma})$ is the sum of the complementary strain energies of the unbroken bars $W^{\mathrm{c,bar}}(\tilde{\sigma})$. Based on Expression (10.197), this is

$$W^{\mathrm{c}}(\tilde{\sigma}) = (n-m)\, W^{\mathrm{c,bar}}(\tilde{\sigma}) = \frac{1}{2}\,(n-m)\,\ell\, A\, \tilde{\sigma}\, S\, \tilde{\sigma}\,. \qquad \text{(For RVE)}. \quad (10.198)$$

The complementary strain-energy density of the RVE $w^{\mathrm{c}}(\tilde{\sigma})$ is obtained by dividing the complementary strain energy of the RVE $W^{\mathrm{c}}(\tilde{\sigma})$ by the volume of the RVE $\ell\, n\, A$. Together with Equation (10.198), this leads to

$$w^{\mathrm{c}}(\tilde{\sigma}) = \frac{W^{\mathrm{c}}(\tilde{\sigma})}{\ell\, n\, A} = \frac{1}{2}\,\frac{(n-m)\,\ell\, A}{\ell\, n\, A}\,\tilde{\sigma}\, S\, \tilde{\sigma} = \frac{1}{2}\,\frac{(n-m)}{n}\,\tilde{\sigma}\, S\, \tilde{\sigma}\,. \qquad \text{(For RVE)}. \quad (10.199)$$

The effective stress $\tilde{\sigma}$ is not a quantity for the representative volume element RVE, but it is a microscopic quantity. Thus it is not for description of the complementary strain-energy density of the RVE $w^{\mathrm{c}}$. Therefore the following manipulation is needed: when the right side of Equation (10.199) is divided and multiplied by $(n-m)/n$, we arrive at the following:

$$w^{\mathrm{c}}(\tilde{\sigma}) = \frac{1}{2}\,\frac{(n-m)}{n}\,\tilde{\sigma}\, S\, \tilde{\sigma} = \frac{1}{2}\,\frac{(n-m)}{n}\,\tilde{\sigma}\, S\, \tilde{\sigma}\,\frac{(n-m)/n}{(n-m)/n}\,. \qquad \text{(For RVE)}. \quad (10.200)$$

Expression $(10.184)_2$, that is, $\sigma = \tilde{\sigma}(n-m)/n$, allows Equation (10.200) to take the following appearance:

$$w^{\mathrm{c}}(\sigma) = \frac{1}{2}\,\frac{(n-m)}{n}\,\tilde{\sigma}\, S\, \tilde{\sigma}\,\frac{(n-m)/n}{(n-m)/n} = \frac{1}{2}\,\sigma\,\frac{S}{(n-m)/n}\,\sigma\,. \qquad \text{(For RVE)}. \quad (10.201)$$

Since $S = 1/E$, Equation (10.201) can be written in the form

$$w^{\mathrm{c}}(\sigma) = \frac{1}{2}\,\sigma\,\frac{S}{(n-m)/n}\,\sigma = \frac{1}{2}\,\sigma\,\frac{1}{E\,(n-m)/n}\,\sigma\,. \qquad \text{(For RVE)}. \quad (10.202)$$

Definition (10.176) of the effective Young's modulus $\tilde{E}$, that is, $\tilde{E} := E(n-m)/n$, is substituted into the right side of Equation (10.202), and the following expression is obtained:

$$w^{\mathrm{c}}(\sigma) = \frac{1}{2}\,\sigma\,\frac{1}{E\,(n-m)/n}\,\sigma = \frac{1}{2}\,\sigma\,\frac{1}{\tilde{E}}\,\sigma = \frac{1}{2}\,\sigma\,\tilde{S}\,\sigma\,. \qquad \text{(For RVE)}, \quad (10.203)$$

where Definition (10.177), that is, $\tilde{S} = 1/\tilde{E}$ , is exploited. Expression (10.203) is the uniaxial counterpart for Expression (10.107).

The bar model was used in the derivation of several uniaxial results, the multiaxial counterparts of which are introduced earlier in this chapter. The results were (a) the fact that the local stress $\sigma^*$ and the effective stress $\tilde{\sigma}$ are equal, Equation $(10.183)_1$; (b) the relation between the effective stress $\tilde{\sigma}$ and the stress $\sigma$ in terms of the variables void volume fraction $f$ and damage $D$, Equation (10.187); (c) the relation between the stress $\sigma$ and the damage-elastic strain $\varepsilon^{\mathrm{de}}$, Equation (10.178); (d) the relation between the effective stress $\tilde{\sigma}$ and the stress $\sigma$ in terms of the variables Young's modulus $E$ and the effective compliance $\tilde{S}$, Equation (10.190); and finally, (e) the form for the complementary strain-energy density of the RVE $w^{\mathrm{c}}(\sigma)$, Expression (10.203). Thus, the mechanical model given by Figure 10.16 and by the equations introduced in this section can be interpreted to provide a mechanical acceptance for the extended postulate of the strain equivalence with the effective stress concept.

### 10.5.7 Tube Model for Interpretation of the Variables Related to Damage

In this section, special attention is given to clarifying the roles of the stress tensor $\boldsymbol{\sigma}$ and the effective stress tensor $\tilde{\boldsymbol{\sigma}}$. This is done by studying their uniaxial scalar-valued counterparts, which makes interpreting of the roles of these stress measures easier. Several other continuum damage mechanics variables are investigated using the same evaluation. It is important to note that the stress tensors $\boldsymbol{\sigma}$ and $\tilde{\boldsymbol{\sigma}}$ are quantities for small deformation theory. The previous section utilized a bar model, whereas a tube model under uniaxial loading is used here.

The present section derives many results similar to those obtained in the previous section with the uniaxial bar model. However, this approach is slightly different and gives a different perspective of the results and description of continuum damage mechanics. Also, the derivation with the current rod example is shorter, which makes it easier to understand the results. In the present evaluation, the representative volume element of the material RVE is compared with a model of it. The concept for the representative volume element RVE was introduced in Section 10.2.

A rod under tensile load $N$, shown in Figure 10.18(a), is assumed to have microvoids. The rod is assumed to form a representative volume element RVE of the material. The response of the material between the microvoids is linear elastic, the Young's modulus of which is denoted by $E$. The area of the cross section is $A$, and the length of the RVE is $\ell$. The void volume fraction of the material is denoted by $f$. The elongation of the RVE due to the tensile force $N$ is $\Delta\ell$.

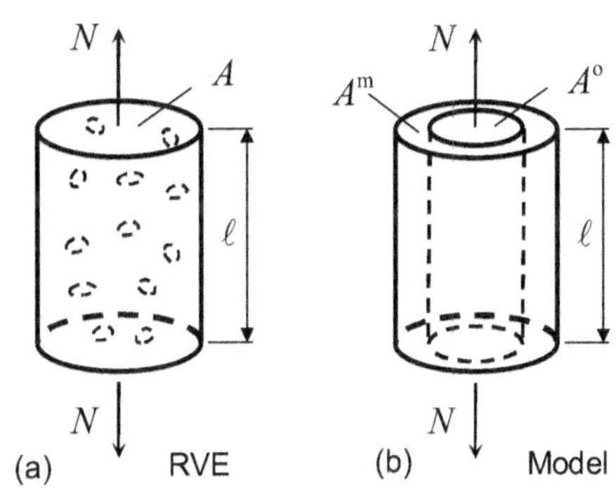


**Figure 10.18** (a) RVE with microvoids is modelled as a (b) tube with continuous matter.

The material of the representative volume element RVE (of the rod) is simplified by concentrating all the microvoids around the centerline of the RVE. This means that the RVE is modelled as a tube, as sketched in Figure 10.18(b). Thus, in the model, the wall is continuous matter where no microvoids exist. The cross-sectional area of the wall material of the model is denoted by $A^{\mathrm{m}}$, where the superscript m refers to the matrix. The cross-sectional area of the model equals that of the RVE, and it is obtained from

$$A = A^{\mathrm{m}} + A^{\mathrm{o}}\,, \qquad \text{(For model)}. \quad (10.204)$$

where the area $A^{\mathrm{o}}$ refers to the area of the hole in the model.

When the tube is considered, the loading $N$ cannot affect the central area of the tube, since there is no matter, but it is distributed over the wall, as shown in Figure 10.18(b). The distributed load is the stress $\sigma^{\blacktriangle}$ the resultant force of which is $N$; that is,

$$\sigma^{\blacktriangle} := \frac{N}{A^{\mathrm{m}}}\,. \qquad \text{(For wall)}. \quad (10.205)$$

Since the response of the material of the wall follows Hooke's law, we arrive at the following:

$$\sigma^{\blacktriangle} = E\,\varepsilon^{\mathrm{e}}, \quad \text{where} \quad \varepsilon^{\mathrm{e}} = \frac{\Delta\ell}{\ell} \quad \Rightarrow \quad \sigma^{\blacktriangle} = E\,\frac{\Delta\ell}{\ell}\,. \qquad \text{(For wall)}. \quad (10.206)$$

Based on Figure 10.18(b), the following holds:

$$\varepsilon^{\mathrm{de}} = \frac{\Delta\ell}{\ell}\,. \qquad \text{(For wall and for RVE)}. \quad (10.207)$$

Effective stress $\tilde{\sigma}$ is defined to be [cf. Definition $(10.130)_2$, i.e., $\tilde{\boldsymbol{\sigma}} = \mathbf{C} : \boldsymbol{\varepsilon}^{\mathrm{de}}$]

$$\tilde{\sigma} := E\,\varepsilon^{\mathrm{de}}\,. \qquad (10.208)$$

Substitution of Equation (10.207) into Definition (10.208) yields

$$\tilde{\sigma} = E\,\frac{\Delta\ell}{\ell}\,. \qquad (10.209)$$

Comparison of Equation $(10.206)_3$ with Equation (10.209) gives

$$\sigma^{\blacktriangle} = \tilde{\sigma}\,. \qquad \text{(For wall)}. \quad (10.210)$$

Expressions (10.210) and (10.205) give

$$\tilde{\sigma} = \frac{N}{A^{\mathrm{m}}}\,. \qquad \text{(For wall)}. \quad (10.211)$$

Based on Results (10.210) and (10.211), the effective stress $\tilde{\boldsymbol{\sigma}}$ is a microscopic stress in the matrix material between microvoids. Since the effective stress $\tilde{\boldsymbol{\sigma}}$

takes the same value everywhere over the wall of the tube, the effective stress $\tilde{\sigma}$ is a homogeneous stress field. This means that the effective stress $\tilde{\sigma}$ is an averaged microscopic stress.

The void volume fraction $f$ is defined as follows:

$$f := \frac{V^{\mathrm{o}}}{V} \quad \Rightarrow \quad f = \frac{V^{\mathrm{o}}}{V} = \frac{\ell\, A^{\mathrm{o}}}{\ell\, A} = \frac{A^{\mathrm{o}}}{A}\,. \qquad \text{(For RVE)}. \quad (10.212)$$

In Expression (10.212), $V^{\mathrm{o}}$ is the volume of the hole and $V$ is the volume of the tube including the volume of the wall and that of the hole.

Expressions (10.204) and $(10.212)_2$ give

$$A^{\mathrm{m}} = A - A^{\mathrm{o}} = A\left(1 - \frac{A^{\mathrm{o}}}{A}\right) = A\,(1 - f)\,. \qquad \text{(wall = RVE)}. \quad (10.213)$$

Substitution of the area $A^{\mathrm{m}}$ from Equation (10.213) into Result (10.211) gives

$$\tilde{\sigma} = \frac{N}{A^{\mathrm{m}}} = \frac{N}{A\,(1-f)}\,. \qquad \text{(wall = wall = RVE)}. \quad (10.214)$$

The stress $\sigma$ is defined to be

$$\sigma := \frac{N}{A}\,. \qquad \text{(For RVE)}. \quad (10.215)$$

It is worth noting that the denominator in Definition (10.215) is the whole cross-sectional area of the rod $A$ (tube as well). Thus the stress $\sigma$ is a macroscopic stress, that is, the stress measure for the representative volume element RVE, as written in Definition (10.215). The volumes of some of the microvoids are crossed by the horizontal surfaces of the rod. This allows the interpretation that the stress $\sigma$ is a macroscopic "averaged" stress measure. Furthermore, the model in Figure 10.18(b) supports this view, since the denominator in Definition (10.215) is the area $A$, which covers the area of the hole $A^{\mathrm{o}}$ as well.

Substitution of the stress $\sigma$ from Definition (10.215) into Expression (10.214) leads to

$$\tilde{\sigma} = \frac{N}{A\,(1-f)} = \frac{\sigma}{(1-f)}\,. \qquad \text{(wall = RVE = RVE)}. \quad (10.216)$$

The Maclaurin series of the function $(1+x)^{\alpha}$ is defined to be (Abramowitz and Setgun [2], p. 15)

$$(1+x)^{\alpha} = 1 + \alpha\, x + \frac{\alpha\,(\alpha-1)}{2!}\, x^2 + \cdots\,, \quad \text{where} \quad -1 < x < 1. \quad (10.217)$$

Substitution of $\alpha = -1$ and $x = -f$ into Series (10.217) yields

$$(1-f)^{-1} = 1 + f + (f)^2 + \cdots\,. \quad (10.218)$$

For a low void volume fraction $f$, Series (10.218) can be approximated by taking into account the first two terms only. This is

$$(1-f)^{-1} \approx 1+f\,. \qquad \text{For a low value of } f. \quad (10.219)$$

Substitution of Approximation (10.219) into Equation (10.216) gives

$$\tilde{\sigma} = \frac{\sigma}{(1-f)} = (1-f)^{-1}\,\sigma \approx (1+f)\,\sigma\,. \qquad (\text{wall} = \text{RVE}). \quad (10.220)$$

Expression (10.220) is the uniaxial counterpart to the multiaxial relation between the effective stress tensor $\tilde{\boldsymbol{\sigma}}$ and the stress tensor $\boldsymbol{\sigma}$ given by Equation $(10.142)_2$, although the multiplier in the multiaxial expression is $(1+2\,f)$. The difference is due to the modelling inaccuracy in the tube model. In the elastic response of the material there are stress peaks on the surfaces of the microvoids that are not present in the tube model.

Substitution of Definition (10.208), that is, $\tilde{\sigma} := E\,\varepsilon^{\text{de}}$, into the first equation of Equation (10.220) leads to

$$E\,\varepsilon^{\text{de}} = \frac{\sigma}{(1-f)} \qquad \Rightarrow \qquad \varepsilon^{\text{de}} = \frac{1}{E\,(1-f)}\,\sigma\,. \qquad (10.221)$$

Equation $(10.221)_2$ is a constitutive equation for the RVE material, although the Young's modulus $E$ is that for the matrix response. Thus the following is achieved:

$$\varepsilon^{\text{de}} = \tilde{S}\,\sigma\,, \qquad \text{where} \qquad \tilde{S} := \frac{1}{E\,(1-f)}\,. \qquad (\text{For RVE}). \quad (10.222)$$

Equation $(10.221)_1$ is the uniaxial counterpart for the multiaxial expression given by Equation $(10.131)_2$.

Since $S := 1/E$, Approximation (10.219) can be used for the low void volume fraction $f$, and the effective compliance of the RVE material $\tilde{S}$, Definition $(10.222)_2$, can be approximated as follows:

$$\tilde{S} := \frac{1}{E\,(1-f)} \quad \text{for low } f \quad \tilde{S} = (1+f)\,S\,. \quad (\text{RVE} = \text{RVE} \times \text{wall}). \quad (10.223)$$

Equation $(10.223)_2$ is the uniaxial counterpart to the multiaxial expression, Equation $(10.161)_2$, although the multiplier in the multiaxial expression is $(1+2f)$.

The right side of Equation (10.220) is multiplied and divided by the Young's modulus $E$. Thus the following is achieved:

$$\tilde{\sigma} = \frac{\sigma}{(1-f)} = \frac{E\,\sigma}{E\,(1-f)} = E\,\tilde{S}\,\sigma\,. \qquad (\text{wall} = \text{wall} \times \text{RVE}). \quad (10.224)$$

Manipulation (10.224) utilizes the definition for the effective compliance $\tilde{S}$, Definition $(10.222)_2$. Expression (10.224) is the uniaxial counterpart to Expression $(10.132)_1$.

The derivation of the expressions for the strain-energy density of the RVE $w(\varepsilon^{\mathrm{e}})$ and for the complementary strain-energy density of the RVE $w^{\mathrm{c}}(\sigma)$ follows similar steps to those taken in the previous section and is therefore not covered here.

The tube model was used in the derivation of several uniaxial results the multiaxial counterparts to which were introduced earlier in this chapter. The results were (a) the relation between the effective stress $\tilde{\sigma}$ and the stress $\sigma$ in terms of the void volume fraction $f$, Equation (10.220); (b) the relation between the damage-elastic strain $\varepsilon^{\mathrm{de}}$ and the stress $\sigma$, Equation (10.222); (c) the relation between the effective compliance $\tilde{S}$ and the compliance $S$, Equation $(10.223)_2$; and (d) the relation between the effective stress $\tilde{\sigma}$ and the stress $\sigma$ in terms of the variables Young's modulus $E$ and the effective compliance $\tilde{S}$, Equation (10.224). Therefore the mechanical model given by Figure 10.18 and by the equations introduced in this section can be interpreted to provide a mechanical acceptance for the extended postulate of the strain equivalence with the effective stress concept.

### 10.5.8 Interpretation of the Roles of the Stress Tensor $\boldsymbol{\sigma}$ and the Effective Stress Tensor $\tilde{\boldsymbol{\sigma}}$ by Integration

In this section, we study the roles of the stress tensor $\boldsymbol{\sigma}$ and the effective stress tensor $\tilde{\boldsymbol{\sigma}}$, based mainly on the models introduced in the previous two sections, although some new results will be derived.

The two (micro) mechanical models introduced in Sections 10.5.6 and 10.5.7 provide a mechanical interpretation for the stress tensor $\boldsymbol{\sigma}$ and for the effective stress tensor $\tilde{\boldsymbol{\sigma}}$. Based on Definitions (10.172) and (10.215) and Equation $(10.184)_2$, the stress tensor $\boldsymbol{\sigma}$ is a stress measure for the macroscopic level, that is, for the representative volume element RVE, and it is obtained as an averaged value of the microstress tensor $\boldsymbol{\sigma}^{\mathrm{m}}$ calculated over the entire RVE, that is, the averaging volume covers the volumes of the microvoids as well. Microstress tensor $\boldsymbol{\sigma}^{\mathrm{m}}$ is assumed to be the "real" stress between the microvoids. Due to the stress concentrations on the surfaces of the microvoids (and at the tips of the microcracks), the microstress stress tensor $\boldsymbol{\sigma}^{\mathrm{m}}$ takes different values within the matrix in a representative volume element RVE. As Figure 10.19 shows, the components of the stress tensor $\boldsymbol{\sigma}$ are acting on the surfaces of the representative volume element RVE.

The effective stress tensor $\tilde{\boldsymbol{\sigma}}$, on the other hand, is the local stress between microvoids and microcracks, and it therefore has a microscopic character. Thus the effective stress tensor $\tilde{\boldsymbol{\sigma}}$ is the driving force for the microscopic processes in damaged materials. Such processes are, for example, plastic yield, dislocation creep, grain boundary sliding and microcracking. The limitation of the effective stress tensor $\tilde{\boldsymbol{\sigma}}$ is that it is an averaged quantity and therefore incapable of modelling nonhomogeneous stress states that may play important role for certain material models.

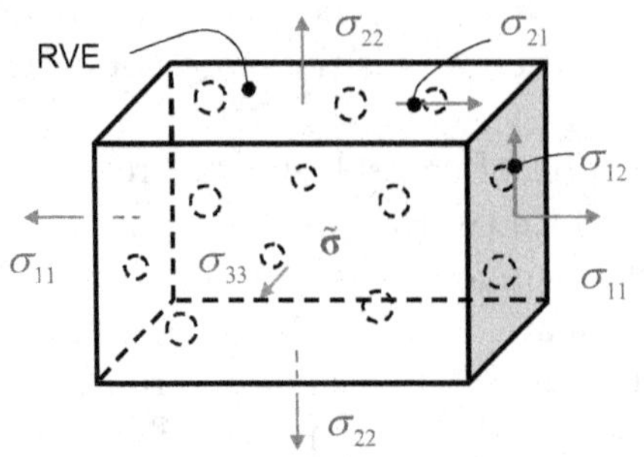


**Figure 10.19** Macroscopic stress tensor $\boldsymbol{\sigma}$ and microscopic effective stress tensor $\tilde{\boldsymbol{\sigma}}$.

Besides the bar and tube models, a relation between the stress tensor $\boldsymbol{\sigma}$ and the effective stress tensor $\tilde{\boldsymbol{\sigma}}$ can be studied in the following pure mathematical way.

A porous material with a linear elastic matrix response is investigated. According to the preceding two models, the stress tensor $\boldsymbol{\sigma}$ is the "averaged" macroscopic stress defined for the representative volume element RVE. As written, the stress tensor $\boldsymbol{\sigma}$ has an averaged character and is therefore defined as follows:

$$\boldsymbol{\sigma} := \frac{1}{V^{\mathrm{rve}}} \int_{V^{\mathrm{rve}}} \boldsymbol{\sigma}^{\mathrm{m}} \, \mathrm{d}V \,, \tag{10.225}$$

where $V^{\mathrm{rve}}$ is the volume of the representative volume element RVE.

Since the microstress tensor $\boldsymbol{\sigma}^{\mathrm{m}}$ takes zero values within the microvoids, the integration in Definition (10.225) can be replaced by the integration over the volume of the matrix material within the representative volume element $V^{\mathrm{rve,m}}$. Thus we arrive at the following:

$$\boldsymbol{\sigma} = \frac{1}{V^{\mathrm{rve}}} \int_{V^{\mathrm{rve,m}}} \boldsymbol{\sigma}^{\mathrm{m}} \, \mathrm{d}V \,. \tag{10.226}$$

It is reasonable to define the effective stress tensor $\tilde{\boldsymbol{\sigma}}$ as the averaged microstress $\boldsymbol{\sigma}^{\mathrm{m}}$ within the matrix material within the representative volume element $V^{\mathrm{rve,m}}$; that is,

$$\tilde{\boldsymbol{\sigma}} := \frac{1}{V^{\mathrm{rve,m}}} \int_{V^{\mathrm{rve,m}}} \boldsymbol{\sigma}^{\mathrm{m}} \, \mathrm{d}V \,. \tag{10.227}$$

Based on Equation (10.226) and Definition (10.227), the following result is obtained:

$$\tilde{\boldsymbol{\sigma}} \, V^{\mathrm{rve,m}} = V^{\mathrm{rve}} \, \boldsymbol{\sigma} \qquad \Rightarrow \qquad \tilde{\boldsymbol{\sigma}} = \frac{V^{\mathrm{rve}}}{V^{\mathrm{rve,m}}} \, \boldsymbol{\sigma} \,. \tag{10.228}$$

The void volume fraction $f$ within the RVE is defined to be

$$f := \frac{V^{\mathrm{rve}} - V^{\mathrm{rve,m}}}{V^{\mathrm{rve}}} \qquad \Rightarrow \qquad \frac{V^{\mathrm{rve}}}{V^{\mathrm{rve,m}}} = (1-f)^{-1} \,. \tag{10.229}$$

Substituting Result $(10.229)_2$ into Equation $(10.228)_2$ yields

$$\tilde{\boldsymbol{\sigma}} = \frac{\boldsymbol{\sigma}}{1-f} \,. \tag{10.230}$$

Result (10.230) coincides with the uniaxial results, Expressions $(10.187)_1$ and (10.216). Thus the bar model, the tube example, and the integral method lead to the same relationship between the stress $\sigma$ and the effective stress $\tilde{\sigma}$.

The roles of the stress tensors $\boldsymbol{\sigma}$ and $\tilde{\boldsymbol{\sigma}}$ are difficult to determine exactly, as Chapter 2 of the book by Nemat-Nasser and Hori [67] shows.

### 10.5.9 Gurson–Tvergaard Material Model

This section briefly discusses the Gurson–Tvergaard material model. The model describes the influence of the microvoids on the plastic yield. A more detailed evaluation is described elsewhere by Santaoja [94].

The present chapter began to evaluate the effect of microvoids on the deformation when the response of the matrix material obeys Hooke's law. Now the view is extended to cover the influence of microvoids on plastic yield.

The following evaluation focuses mainly on the enhancement of the Gurson [27] model done by Tvergaard [112]. In his mathematical derivation, Tvergaard [112] used finite deformation theory. The following comments relate to the corresponding expressions for small deformations and rotations. Several other small changes will be made for the sake of compatibility with the notations of this book.

Tvergaard [[112], Eq. (5.1)] enhanced the material model by Gurson [[27], Eqs. (3.12) and (4.9)] by introducing the following yield function:

$$F(\boldsymbol{\sigma}, \sigma^{\mathrm{M}}, f) = \left(\frac{J_{\mathrm{vM}}(\boldsymbol{\sigma})}{\sigma^{\mathrm{M}}}\right)^2 + 2\, f\, q^1 \cosh\left(\frac{q^2\, \mathbf{1} : \boldsymbol{\sigma}}{2\, \sigma^{\mathrm{M}}}\right) - (1 + q^3 f^2)\,, \quad (10.231)$$

where $q^1$, $q^2$, and $q^3$ are material constants. According to Tvergaard [112], the variable $\sigma^{\mathrm{M}}$ represents the equivalent tensile flow stress in the matrix material, "disregarding the local stress variations." Thus, based on the notations of the present chapter, the variable $\sigma^{\mathrm{M}}$ could be replaced with the variable $J_{\mathrm{vM}}(\tilde{\boldsymbol{\sigma}})$. However, the effective stress tensor $\tilde{\boldsymbol{\sigma}}$ is a microscopic quantity and is not a viable candidate for continuum thermodynamics. Based on Equation $(10.132)_2$, $\tilde{\boldsymbol{\sigma}}$ can be replaced by $\mathbf{M} : \boldsymbol{\sigma}$.

Tvergaard [112] further added to the model elastic deformation, which he assumed to obey Hooke's law, as given by Equation $(10.129)_2$, viz.

$$\boldsymbol{\sigma} = \mathbf{C} : \boldsymbol{\varepsilon}^{\mathrm{e}}\,. \quad (10.232)$$

Tvergaard [112] introduced his modification to obtain a better fit between the numerical results of his microscopic finite element analysis and the macroscopic continuum model. By assuming that the constants $q^1 = q^2 = q^3 = 1$, that the variable $\sigma^{\mathrm{M}}$ takes constant value $\sigma^{\mathrm{Y}}$ (microscopic equivalent yield stress), and that the elastic deformation can be neglected as a small quantity, the Gurson–Tvergaard material model reduces to the Gurson model.

Tvergaard ([112], p. 403) stated that as the matrix material is assumed to be plastically incompressible, the evolution equation for the void volume fraction $f$ is given by

$$\dot{f} = (1 - f)\,\mathbf{1} : \dot{\boldsymbol{\varepsilon}}^{\mathrm{psw}}\,. \tag{10.233}$$

In Equation (10.233), $\dot{\boldsymbol{\varepsilon}}^{\mathrm{psw}}$ is the void-plastic strain rate tensor. It models the plastic strain of the matrix material and the strain due to the porosity change.

Based on Santaoja ([94], Section 4.2), the void-plastic strain rate tensor $\dot{\boldsymbol{\varepsilon}}^{\mathrm{psw}}$ can be separated as follows:

$$\dot{\boldsymbol{\varepsilon}}^{\mathrm{psw}} = \dot{\boldsymbol{\varepsilon}}^{\mathrm{p}} + \dot{\boldsymbol{\varepsilon}}^{\mathrm{sw}}\,, \tag{10.234}$$

where $\dot{\boldsymbol{\varepsilon}}^{\mathrm{sw}}$ is the swelling rate tensor introduced in Section 10.4.1.

Chu and Needleman [16] were not satisfied with Equation (10.233) and replaced it with the following:

$$\dot{f} = \dot{f}_{\mathrm{growth}} + \dot{f}_{\mathrm{nucleation}}\,, \tag{10.235}$$

where the first term is related to the growth of the existing voids and the second one is related to the increase of void volume fraction due to the nucleation of new voids. The first term equals the right side of Equation (10.233),

$$\dot{f}_{\mathrm{growth}} = (1 - f)\,\mathbf{1} : \dot{\boldsymbol{\varepsilon}}^{\mathrm{psw}}\,. \tag{10.236}$$

Addition of the second term violates the axiom of conservation of mass, as shown by Santaoja ([94], Section 4.2).

Tvergaard and Needleman ([113], p. 159) adopted Equations (10.235) and (10.236). Furthermore, they introduced a new modification for $f$ by replacing it in the yield function $F$ with the quantity $f^*$, which is defined as

$$f^* := \begin{cases} f, & \text{when } f \leq f_{\mathrm{c}}\,, \\ f_{\mathrm{c}} + K\,(f - f_{\mathrm{c}}), & \text{when } f > f_{\mathrm{c}}. \end{cases} \tag{10.237}$$

In Equation (10.237), $f_{\mathrm{c}}$ is the critical value for the void volume fraction, and $K$ is a constant related to the ultimate value of $f^*$ at which the macroscopic stress-carrying capacity vanishes.

The foregoing model is available in the Abaqus finite element code (Abaqus Analysis User's Guide [1], Section 23.2.9), although Expression (10.237) is replaced by a more complicated model. One of the aims of introducing this model was to show that sometimes a simplified, theoretically well-expressed model has to be drastically modified to be suitable for engineering purposes. In the preceding model, the original constitutive equation is "enhanced" by adding several ad hoc terms the exact roles of which are more or less unclear, at least to the authors. Sometimes the modified model does not satisfy strict rules by nature, as the axiom of conservation of mass in this case, but engineers apply it nonetheless.

## 10.5.10 Candidates for Damage

This section discusses candidates for damage that do not follow the concepts already presented in this chapter.

The investigation already covered here focused on studying damage caused by microvoids or microcracks. The obtained expressions focus on cases in which increasing deterioration of the elastic stiffness of the material leads to an increasing value for effective stress tensor $\tilde{\boldsymbol{\sigma}}$. Since the effective stress tensor $\tilde{\boldsymbol{\sigma}}$ is the driving force behind microscopic processes, due to the growing value for the effective stress tensor $\tilde{\boldsymbol{\sigma}}$, some micromechanical processes, such as plastic yield and dislocation creep, take increasing values.

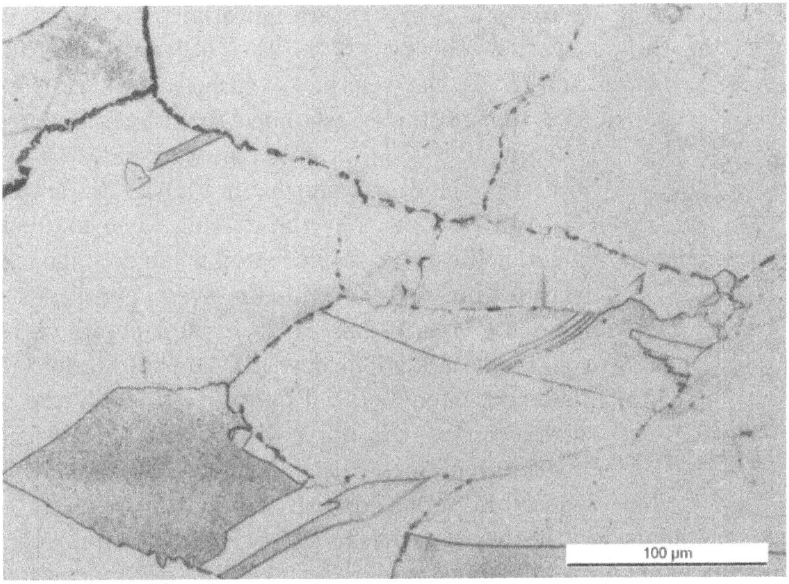


**Figure 10.20** Creep cavities at grain boundaries and their coalescence to form microcracks in front of a macrocrack in a Cu-OFHC CT-specimen. Vertical stress direction. Etched. Photograph by J. Rantala, VTT Technical Research Centre of Finland.

There are, however, some other potential sources for damage. The steam pipes of traditional power plants operate at 550 °C and higher. In such conditions, in power plant materials, creep is due mainly to dislocation creep, and creep resistance is often due to carbides, which form obstacles to dislocation glide. The uniaxial material model for dislocation creep presented in Section 7.6.1 can now be extended for demonstration. Thus, the model under consideration reads

$$\varepsilon = \varepsilon^{\mathrm{de}} + \varepsilon^{\mathrm{Th}} + \varepsilon^{\mathrm{v}} \,. \tag{10.238}$$

According to Constitutive Equation (10.238), the (total) strain $\varepsilon$ consists of the damage-elastic strain $\varepsilon^{\mathrm{de}}$, thermal strain $\varepsilon^{\mathrm{Th}}$, and creep strain $\varepsilon^{\mathrm{v}}$. In contrast to the model presented in Section 7.6.1, the present model is written for tertiary creep as well. The damage of the high-temperature component material has several mechanisms. First, there will be cavities (microvoids) on the grain boundaries, shown in Figure 10.20. When damage proceeds, the grain boundary cavities coalesce and form microcracks on grain boundaries. These damage mechanisms can be modelled by the expressions given in earlier sections of this chapter and are included in the damage-elastic strain $\varepsilon^{\mathrm{de}}$. It is not the topic of the current section, in which the focus is on the creep strain $\varepsilon^{\mathrm{v}}$.

The creep model introduced in Section 7.6.1 is now extended by writing

$$\dot{\varepsilon}^{\mathrm{v}} = \overset{\circ}{\varepsilon}_{\mathrm{re}} \left( \frac{\tilde{\tau} - \beta}{\sigma_{\mathrm{re}}} \right)^{n} \quad \text{and} \quad \dot{\beta} = e\,\dot{\varepsilon}^{\mathrm{v}} - b \left( \frac{\omega}{\omega_{\mathrm{re}}} \right) \left( \frac{\beta}{\sigma_{\mathrm{re}}} \right)^{m} e^{\frac{E_{\mathrm{a}}}{R}\left(\frac{1}{T_{\mathrm{r}}} - \frac{1}{T}\right)} . \quad (10.239)$$

The notations $\overset{\circ}{\varepsilon}_{\mathrm{re}}, \sigma_{\mathrm{re}}, n, e, b, \omega_{\mathrm{re}}$, and $m$ are material parameters. The parameter $E_{\mathrm{a}}$ is the activation energy, $R$ is the universal gas constant, and $T_{\mathrm{r}}$ is the reference temperature. The quantity $\beta$ is the obstacle resistance to dislocation creep, and the quantity $\omega$ is the spacing of carbides (obstacles in the slip plane), as shown in Figure 10.21. Equation $(10.239)_1$ shows that the driving force for dislocation creep is the shear stress $\tau$, and when damage is present, the driving force is the effective shear stress $\tilde{\tau}$. The quantity $b^*$ in Figure 10.21 is the Burgers vector.

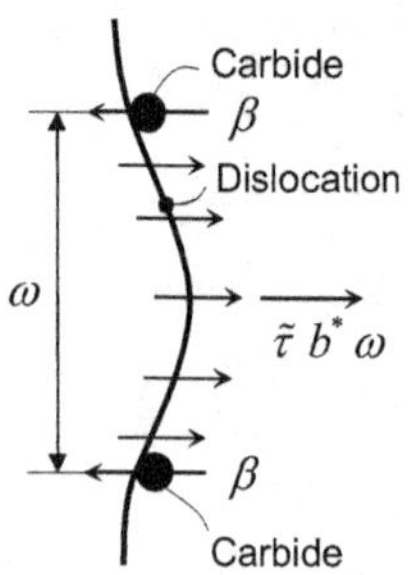


**Figure 10.21** Dislocation hits carbides.

There is a small flaw in Material Model $(10.239)_1$. As mentioned below Equation (10.231), the effective stress tensor $\tilde{\boldsymbol{\sigma}}$ is not a viable variable for continuum thermodynamics with internal variables and therefore should be replaced with the quantity $\mathbf{M}:\boldsymbol{\sigma}$. The same argumentation holds for the effective shear stress $\tilde{\tau}$. Unfortunately, for the scalar-valued quantity $\tilde{\tau}$, no corresponding expression to $\mathbf{M}:\boldsymbol{\sigma}$ exists. For pedagogical reasons, the notation $\tilde{\tau}$ is acceptable. The short materials science description for Material Model (10.239) is as follows.

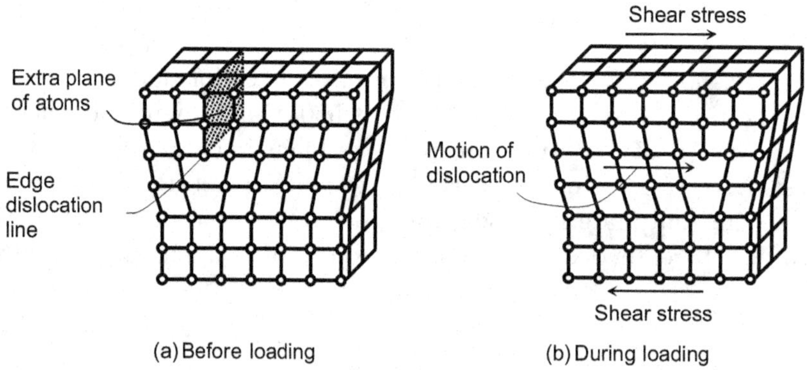


**Figure 10.22** Shear stress-driven dislocation glide. Atomic structure of the matter (a) before and (b) during loading.

Figure 10.22 sketches the concept of dislocation glide, which is the process behind plastic yield and dislocation creep. In Figure 10.22(a), there is an extra plane of atoms, the lower end of which [in Figure 10.22(a)] is called the edge dislocation. The effective shear stress $\tilde{\tau}$ induces the motion of dislocation, as shown in Figure 10.22(b). The block of matter in Figure 10.22(b) is not in equilibrium, since for clarity, only the loading that causes the dislocation movement is shown. Since the driving force for the dislocation movement is the shear stress between the microvoids, the quantity in Equation $(10.239)_1$ is the effective shear stress $\tilde{\tau}$.

Since in the present model the deformation due to dislocation glide is described by the viscous strain $\varepsilon^{\mathrm{v}}$, the first term on the right side of Equation $(10.239)_2$, that is, $e\,\dot{\varepsilon}^{\mathrm{v}}$, displays the fact that the more dislocations glide, the more they hit the obstacles, producing stronger forces on the carbides. Based on Newton's law of action and reaction, the carbides push back against the dislocations by the same but opposite force $\beta$. Therefore the value of the quantity $\beta$ is growing with growing viscous strain $\varepsilon^{\mathrm{v}}$. The role of the denominators $\sigma_{\mathrm{re}}$ and $\omega_{\mathrm{re}}$ is to make the quantities in parentheses dimensionless, which is very important when a quantity within parentheses is raised to a power. This was discussed in Section 9.4.

The middle part of the second term on the right side of Equation $(10.239)_2$, that is, $(\beta/\sigma_{\mathrm{re}})^m$, indicates that the more the dislocations are pushed against carbides, the more they climb over them, with the result that the force on the carbides ($\beta$) diminishes with the decreasing number of dislocations pushed on carbides. Therefore the second term on the right side of Equation $(10.239)_2$ has a minus sign.

The spacing of carbides $\omega$ is not a constant, but carbide coarsening is a vital deformation mechanism at elevated temperatures. Since no new material for carbides is available, the spacing of carbides $\omega$ grows with increasing carbide coarsening. Since the resultant force has the form $\tilde{\tau}\,b^*\,\omega$, as shown in Figure 10.21, the resultant force on the carbides ($\beta$) grows with carbide coarsening. As the resultant force $\tilde{\tau}\,b^*\,\omega$ grows, the more dislocations climb over the carbides, which finally leads to a lower value for the force $\beta$. Therefore the quantity $\omega$ is placed after the minus sign in Equation $(10.239)_2$.

The last exponential term in Equations (10.239) is the Arrhenius equation having the form

$$e^{\frac{E_{\mathrm{a}}}{R}\left(\frac{1}{T_{\mathrm{r}}}-\frac{1}{T}\right)}\,. \tag{10.240}$$

It describes the temperature dependence of the rates of processes. For the present case, it describes the vibrations of atoms, otherwise known as temperature $T$. The more the atoms vibrate, the higher the temperature $T$. The increasing vibration of atoms allows dislocations to climb more easily over carbides, which leads to a higher creep strain rate $\dot{\varepsilon}^{\mathrm{v}}$. At the same time, the carbide coarsening rate $\dot{\omega}$ takes higher values, since it is dependent on the amount of vibrations of atoms. Thus, the model for the carbide coarsening rate $\dot{\omega}$ should have the form

$$\dot{\omega} \propto e^{\frac{E_{\mathrm{a}}}{R}\left(\frac{1}{T_{\mathrm{r}}}-\frac{1}{T}\right)}. \tag{10.241}$$

Increasing operational temperature $T$ can be interpreted to enhance damage in the components of traditional power plants by these two processes as well.

The preceding examples demonstrate the variety of potential damage mechanisms. They cannot be modelled just by introducing a $(1-D)$ term; a detailed investigation of the microscopic processes behind every single mechanism is necessary, and the macroscopic model has to be a description of these microscopic processes.

The nominator $\tilde{\tau}-\beta$ in Material Model $(10.239)_1$ expresses the fact that the effective stress $\tilde{\tau}$ is the driving force behind dislocation glide, that is, dislocation creep, whereas the force $\beta$ creates an obstacle to it. The effective shear stress $\tilde{\tau}$ is replaced in the three-dimensional model with the deviatoric effective stress tensor $\tilde{\mathbf{s}}$ and the von Mises value of the effective stress tensor $J_{\mathrm{vM}}(\tilde{\boldsymbol{\sigma}})$, which should be vital parts of the three-dimensional counterpart to Constitutive Model (10.239).

Based on the preceding discussion and Material Model (10.239), the carbide coarsening increases the creep rate $\dot{\varepsilon}^{\mathrm{v}}$, and it can therefore be interpreted to describe damage to the material. However, the physical processes behind carbide coarsening and their mathematical representations deviate substantially from those that are usually assumed to cause damage to materials. Usually, when creep damage in high-temperature power plant materials is modelled, the damage and therefore the tertiary creep is associated to the grain boundary cavitation and microcracking. The foregoing discussion does not take a stand as to which are the deformation mechanisms behind creep damage, but it demonstrates how to prepare material models and, in this case, how to obtain the response shown in Figure 10.23.

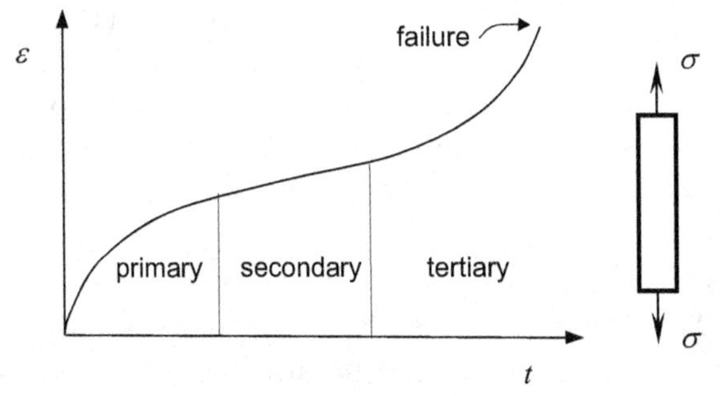


**Figure 10.23** Creep curve.

## 10.5.11 Conclusions

This section covered some of the topics that do not fall solely under any particular earlier section. In these studies, the response of the matrix material was assumed to obey Hooke's law. These models do not exclude other deformation mechanisms in the matrix material, since, with acceptable accuracy, the deformation of the matrix material can show plastic yield, creep, and so on, as well. This topic was discussed briefly in Section 10.4.1. The second comment relates to the notations at the representative volume element level and at microscopic level. The bar model for the representative volume element RVE is used for studying these two scales of quantities. Therefore, Figure 10.16 is copied to Figure 10.24.

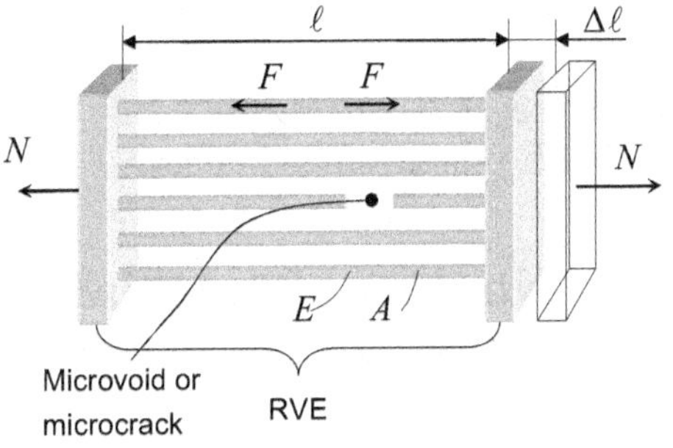


**Figure 10.24** Uniaxial bar model and RVE for evaluating the stresses $\sigma$ and $\tilde{\sigma}$.

Definition (10.172) and Expressions $(10.178)_2$ and (10.195) are recalled:

$$\sigma := \frac{N}{n\,A}, \qquad \varepsilon^{\mathrm{de}} = \tilde{S}\,\sigma \qquad \text{and} \qquad w(\varepsilon^{\mathrm{e}}) = \frac{1}{2}\,\varepsilon^{\mathrm{e}}\,\tilde{E}\,\varepsilon^{\mathrm{e}}, \qquad \text{(For RVE).} \tag{10.242}$$

where $\tilde{S}$ is the effective compliance and $\tilde{E}$ is the effective Young's modulus. Both values express the reduced rigidity of the material due to the effect of microvoids or microcracks. The reduced values are expressed by the notation $\tilde{}$ above the quantities. Thus the notation $\tilde{}$ expresses the macroscopic variables, that is, variables for the representative volume element. Due to the approach of the "extended postulate of strain equivalence with the effective stress concept," the present theory does not distinguish between microscopic and macroscopic strain measures. The stress measures, however, are different. Equations $(10.166)_2$ and $(10.183)_1$ are recalled:

$$\sigma^* = \frac{F}{A} \qquad \text{and} \qquad \tilde{\sigma} = \sigma^* \qquad \Rightarrow \qquad \tilde{\sigma} = \frac{F}{A}. \qquad \text{(For bar).} \tag{10.243}$$

Comparison of Expressions $(10.242)_1$ and $(10.243)_3$ shows that the microscopic stress $\tilde{\sigma}$ is expressed with $\tilde{}$, whereas the stress measure for RVE, that is, $\sigma$, is not. Even their names are interchanged. The stress measure $\tilde{\sigma}$ is called "effective stress" (cf. effective stress tensor $\tilde{\boldsymbol{\sigma}}$), and the stress measure $\sigma$ is called "stress" (cf. stress tensor $\boldsymbol{\sigma}$ ).

The third comment relates to the previous discussion on the names and notations of the stress measures. The axioms and basic laws should be written using microscopic quantities. Furthermore, these quantities should be exact; that is, they should be defined without any averaging procedure. This is not the case in continuum damage mechanics, but the basic laws and axioms are expressed in the variables for the representative volume element RVE.

In Equation $(10.239)_1$ the evolution law is written with the effective stress $\tilde{\sigma}$ (with the effective stress tensor $\tilde{\boldsymbol{\sigma}}$). Although the effective stress $\tilde{\sigma}$ is a microscopic stress between the microvoids and/or microcracks, its use is not entirely correct, since it is an averaged stress neglecting the stress peaks. Due to the exponent $n$ in Equation $(10.239)_1$, the stress peaks cause high strain rates $\dot{\varepsilon}^{\mathrm{v}}$ which are not compensated for by the lower strain rates $\dot{\varepsilon}^{\mathrm{v}}$ in those parts of the body where the stress takes lower values. This effect cannot be described by the averaged effective stress $\tilde{\sigma}$.

## 10.6 Nonlocal Material Model for Creep and Damage

As an example of the gradient theory, a material model showing Hooke's law and the Le Gac and Duval (see Section 9.6) type of creep behavior in a virgin state is demonstrated. Material damage is described traditionally, that is, by the "effective stress concept" and by a scalar-valued damage quantity denoted by $D$. The damage evolution equation contains a gradient term (actually a Laplacian term) to avoid localization. This chapter follows the work by Santaoja [96].

The thermodynamical formulation for the mechanical part of the material model is

$$\psi = \frac{1}{2\,\rho_0}\,(1-D)\,(\boldsymbol{\varepsilon}-\boldsymbol{\varepsilon}^{\mathrm{i}}):\mathbf{C}:(\boldsymbol{\varepsilon}-\boldsymbol{\varepsilon}^{\mathrm{i}})+\frac{k}{4\,\rho_0\,e}\,(\vec{\nabla}D)\cdot(\vec{\nabla}D)+\frac{1}{2\,\rho_0}\,\boldsymbol{\alpha}^1:\mathbf{C}^1:\boldsymbol{\alpha}^1 \qquad (10.244)$$

and

$$\varphi^{\mathrm{c}}_{\mathrm{mech}} = \frac{e}{\rho_0}\eta^2+\frac{2\,\overset{\circ}{\varepsilon}_{\mathrm{re}}\,\sigma_{\mathrm{re}}}{3\,\rho_0\,(n+1)}\left[\frac{\langle\,J_{\mathrm{vM}}(\boldsymbol{\sigma}-\boldsymbol{\beta}^1)\,\rangle}{\sigma_{\mathrm{re}}}\right]^{(n+1)}+\frac{2\,\tilde{C}^{11}\,\sigma_{\mathrm{re}}}{3\,\rho_0\,(k+1)}\left[\frac{J_{\mathrm{vM}}(\boldsymbol{\beta}^1)}{\sigma_{\mathrm{re}}}\right]^{(k+1)}. \qquad (10.245)$$

In Model (10.244), $\mathbf{C}$ stands for the constitutive tensor for a Hookean deformation, $e$ is a material parameter, and $\mathbf{C}^1$ is a fourth-order constitutive tensor. In Model (10.245), $\overset{\circ}{\varepsilon}_{\mathrm{re}}$ is the reference strain rate, $\sigma_{\mathrm{re}}$ is the reference stress, and $n$, $\tilde{C}^{11}$, and $k$ are material constants. The von Mises operator $J_{\mathrm{vM}}(\cdot)$ and the Macaulay brackets $\langle\cdot\rangle$ are given by Definitions (2.124) and (2.1), respectively.

The fourth-order constitutive tensor for a Hookean deformation $\mathbf{C}$ (i.e., for Hooke's law) is defined by [see Definition (8.30)]

$$\mathbf{C} := \lambda\,\mathbf{1}\,\mathbf{1} + 2\,\mu\,\mathbf{I}^{\mathrm{s}}\,, \qquad (10.246)$$

where $\lambda$ and $\mu$ are Lamé elastic constants given by Definition (8.31). The second-order identity tensor $\mathbf{1}$ and the fourth-order symmetric identity tensor $\mathbf{I}^{\mathrm{s}}$ are given by Definitions (2.63) and (2.106).

Substitution of Material Model (10.244) into State Equations (7.74) and (7.75) yields

$$\boldsymbol{\sigma} := \rho_0\,\frac{\partial\,\psi(\boldsymbol{\varepsilon}-\boldsymbol{\varepsilon}^{\mathrm{i}}, D, \vec{\nabla}D, \boldsymbol{\alpha}^1)}{\partial(\boldsymbol{\varepsilon}-\boldsymbol{\varepsilon}^{\mathrm{i}})} = (1-D)\,\mathbf{C}:(\boldsymbol{\varepsilon}-\boldsymbol{\varepsilon}^{\mathrm{i}}) \qquad (10.247)$$

and

$$\upsilon := -\rho_0 \frac{\partial \psi(\boldsymbol{\varepsilon} - \boldsymbol{\varepsilon}^{\mathrm{i}}, D, \vec{\nabla} D, \boldsymbol{\alpha}^1)}{\partial D} = \frac{1}{2}(\boldsymbol{\varepsilon} - \boldsymbol{\varepsilon}^{\mathrm{i}}) : \mathbf{C} : (\boldsymbol{\varepsilon} - \boldsymbol{\varepsilon}^{\mathrm{i}}), \tag{10.248}$$

and furthermore,

$$\vec{\zeta} := -\rho_0 \frac{\partial \psi(\boldsymbol{\varepsilon} - \boldsymbol{\varepsilon}^{\mathrm{i}}, D, \vec{\nabla} D, \boldsymbol{\alpha}^1)}{\partial \vec{\nabla} D} = -\frac{k}{2\,e}(\vec{\nabla} D), \tag{10.249}$$

and, finally,

$$\boldsymbol{\beta}^1 := -\rho_0 \frac{\partial \psi(\boldsymbol{\varepsilon} - \boldsymbol{\varepsilon}^{\mathrm{i}}, D, \vec{\nabla} D, \boldsymbol{\alpha}^1)}{\partial \boldsymbol{\alpha}^1} = -\mathbf{C}^1 : \boldsymbol{\alpha}^1, \tag{10.250}$$

which yields

$$\dot{\boldsymbol{\beta}}^1 = -\mathbf{C}^1 : \dot{\boldsymbol{\alpha}}^1. \tag{10.251}$$

Substitution of Forces (10.248) and (10.249) into Definition $(7.95)_1$ gives

$$\eta := \langle \upsilon - \vec{\nabla} \cdot \vec{\zeta} \rangle = \left\langle \frac{1}{2}(\boldsymbol{\varepsilon} - \boldsymbol{\varepsilon}^{\mathrm{i}}) : \mathbf{C} : (\boldsymbol{\varepsilon} - \boldsymbol{\varepsilon}^{\mathrm{i}}) + \frac{k}{2\,e}(\nabla^2 D) \right\rangle. \tag{10.252}$$

Substitution of Model (10.245) into Normality Rule $(7.97)_2$ provides

$$\dot{D} = \rho_0 \frac{\partial \varphi^{\mathrm{c}}_{\mathrm{mech}}(\boldsymbol{\sigma}, \eta, \boldsymbol{\beta}^1; \boldsymbol{\varepsilon} - \boldsymbol{\varepsilon}^{\mathrm{i}}, D, \vec{\nabla} D, \boldsymbol{\alpha}^1)}{\partial \eta} = 2\,e\,\eta. \tag{10.253}$$

Substituting Expression (10.252) into Rate (10.253) gives

$$\dot{D} = \left\langle e(\boldsymbol{\varepsilon} - \boldsymbol{\varepsilon}^{\mathrm{i}}) : \mathbf{C} : (\boldsymbol{\varepsilon} - \boldsymbol{\varepsilon}^{\mathrm{i}}) + k(\nabla^2 D) \right\rangle. \tag{10.254}$$

Substitution of Model (10.245) into Normality Rules $(7.97)_1$ and (7.98a) gives

$$\begin{aligned} \dot{\boldsymbol{\varepsilon}}^{\mathrm{i}} &= \rho_0 \frac{\partial \varphi^{\mathrm{c}}_{\mathrm{mech}}(\boldsymbol{\sigma}, \eta, \boldsymbol{\beta}^1; \boldsymbol{\varepsilon} - \boldsymbol{\varepsilon}^{\mathrm{i}}, D, \vec{\nabla} D, \boldsymbol{\alpha}^1)}{\partial \boldsymbol{\sigma}} \\ &= \overset{\circ}{\varepsilon}_{\mathrm{re}} \left[ \frac{\langle J_{\mathrm{vM}}(\boldsymbol{\sigma} - \boldsymbol{\beta}^1) \rangle}{\sigma_{\mathrm{re}}} \right]^n \frac{\mathbf{s} - \mathbf{b}^1}{J_{\mathrm{vM}}(\boldsymbol{\sigma} - \boldsymbol{\beta}^1)} \end{aligned} \tag{10.255}$$

and

$$\begin{aligned} \dot{\boldsymbol{\alpha}}^1 &= \rho_0 \frac{\partial \varphi^{\mathrm{c}}_{\mathrm{mech}}(\boldsymbol{\sigma}, \eta, \boldsymbol{\beta}^1; \boldsymbol{\varepsilon} - \boldsymbol{\varepsilon}^{\mathrm{i}}, D, \vec{\nabla} D, \boldsymbol{\alpha}^1)}{\partial \boldsymbol{\beta}^1} \\ &= \dot{\boldsymbol{\varepsilon}}^{\mathrm{i}} + \tilde{C}^{11} \left[ \frac{J_{\mathrm{vM}}(\boldsymbol{\beta}^1)}{\sigma_{\mathrm{re}}} \right]^k \frac{\mathbf{b}^1}{J_{\mathrm{vM}}(\boldsymbol{\beta}^1)}. \end{aligned} \tag{10.256}$$

In Rate Equations (10.255) and (10.256), $\mathbf{s}$ and $\mathbf{b}^1$ are the deviatoric parts of the stress tensor $\boldsymbol{\sigma}$ and the internal force $\boldsymbol{\beta}^1$, respectively. Derivation of Rate Equations (10.255) and (10.256) exploits the following results [see Expressions (A.2) and (A.3) of **Appendix A**]:

$$\frac{\partial J_{\mathrm{vM}}(\boldsymbol{\sigma}-\boldsymbol{\beta}^1)}{\partial \boldsymbol{\sigma}} = \frac{3}{2}\frac{\mathbf{s}-\mathbf{b}^1}{J_{\mathrm{vM}}(\boldsymbol{\sigma}-\boldsymbol{\beta}^1)} \qquad \text{and} \qquad \frac{\partial J_{\mathrm{vM}}(\boldsymbol{\beta}^1)}{\partial \boldsymbol{\beta}^1} = \frac{3}{2}\frac{\mathbf{b}^1}{J_{\mathrm{vM}}(\boldsymbol{\beta}^1)}\,. \tag{10.257}$$

Substitution of Rate (10.256) into Expression (10.251) leads to

$$\dot{\boldsymbol{\beta}}^1 = m^1\,\dot{\boldsymbol{\varepsilon}}^{\mathrm{i}} - m^2\left[\frac{J_{\mathrm{vM}}(\boldsymbol{\beta}^1)}{\sigma_{\mathrm{re}}}\right]^k \frac{\mathbf{b}^1}{J_{\mathrm{vM}}(\boldsymbol{\beta}^1)}\,, \tag{10.258}$$

where the following was assumed:

$$\mathbf{C}^1 = m^1\,\mathbf{I} \qquad \text{and} \qquad \tilde{C}^{11} = \frac{m^2}{m^1}\,. \tag{10.259}$$

In Equation (10.258), $m^1$ and $m^2$ are material parameters.

A vital part of the thermodynamical study of a material model is evaluation of the satisfaction of the Clausius–Duhem inequality. Based on Clausius–Duhem Inequality (7.96), the following is obtained for the present material model:

$$\boldsymbol{\sigma}\colon\dot{\boldsymbol{\varepsilon}}^{\mathrm{i}} + \eta\,\dot{D} + \boldsymbol{\beta}^1:\dot{\boldsymbol{\alpha}}^1 \geq 0\,, \tag{10.260}$$

where the thermal part is dropped out.

Based on Rates (10.255) and (10.256), the following is achieved:

$$\boldsymbol{\sigma}:\dot{\boldsymbol{\varepsilon}}^{\mathrm{i}} + \boldsymbol{\beta}^1:\dot{\boldsymbol{\alpha}}^1$$
$$= \overset{\circ}{\varepsilon}_{\mathrm{re}}\left[\frac{\langle J_{\mathrm{vM}}(\boldsymbol{\sigma}-\boldsymbol{\beta}^1)\rangle}{\sigma_{\mathrm{re}}}\right]^n \frac{(\mathbf{s}-\mathbf{b}^1):(\boldsymbol{\sigma}-\boldsymbol{\beta}^1)}{J_{\mathrm{vM}}(\boldsymbol{\sigma}-\boldsymbol{\beta}^1)} + \tilde{C}^{11}\left[\frac{J_{\mathrm{vM}}(\boldsymbol{\beta}^1)}{\sigma_{\mathrm{re}}}\right]^k \frac{\mathbf{b}^1:\boldsymbol{\beta}^1}{J_{\mathrm{vM}}(\boldsymbol{\beta}^1)}\,. \tag{10.261}$$

Problem 8.2 gives

$$(\mathbf{s}-\mathbf{b}^1):(\boldsymbol{\sigma}-\boldsymbol{\beta}^1) = (\mathbf{s}-\mathbf{b}^1):(\mathbf{s}-\mathbf{b}^1) \quad \text{and} \quad \mathbf{b}^1:\boldsymbol{\beta}^1 = \mathbf{b}^1:\mathbf{b}^1\,. \tag{10.262}$$

Definition (2.124) for the von Mises operator $J_{\mathrm{vM}}(\cdot)$ reads

$$J_{\mathrm{vM}}(\boldsymbol{\gamma}) := \sqrt{\tfrac{3}{2}\,\mathbf{g}:\mathbf{g}}\,, \tag{10.263}$$

where $\mathbf{g}$ denotes the deviatoric tensor of any second-order tensor $\boldsymbol{\gamma}$.

Based on Equalities (10.262) and Definition (10.263), we arrive at the following:

$$(\mathbf{s}-\mathbf{b}^1):(\boldsymbol{\sigma}-\boldsymbol{\beta}^1) = \tfrac{2}{3}\left[J_{\mathrm{vM}}(\boldsymbol{\sigma}-\boldsymbol{\beta}^1)\right]^2 \quad \text{and} \quad \mathbf{b}^1:\boldsymbol{\beta}^1 = \tfrac{2}{3}\left[J_{\mathrm{vM}}(\boldsymbol{\beta}^1)\right]^2. \tag{10.264}$$

Substitution of Forms (10.264) into Expression (10.261) yields

$$\boldsymbol{\sigma} : \dot{\boldsymbol{\varepsilon}}^{\mathrm{i}} + \boldsymbol{\beta}^1 : \dot{\boldsymbol{\alpha}}^1$$

$$= \tfrac{2}{3}\,\overset{\circ}{\varepsilon}_{\mathrm{re}} \left[ \frac{\langle J_{\mathrm{vM}}(\boldsymbol{\sigma} - \boldsymbol{\beta}^1) \rangle}{\sigma_{\mathrm{re}}} \right]^n J_{\mathrm{vM}}(\boldsymbol{\sigma} - \boldsymbol{\beta}^1) + \tfrac{2}{3}\,\tilde{C}^{11} \left[ \frac{J_{\mathrm{vM}}(\boldsymbol{\beta}^1)}{\sigma_{\mathrm{re}}} \right]^k J_{\mathrm{vM}}(\boldsymbol{\beta}^1) (\geq 0). \tag{10.265}$$

Since the material parameters $\overset{\circ}{\varepsilon}_{\mathrm{re}}$ and $\tilde{C}^{11}$ are nonnegative, the left side of Expression (10.265) takes nonnegative values. This phenomenon is set into Expression (10.265).

Based on Rate (10.253), the following is obtained:

$$\eta\,\dot{D} = 2\,e\,\eta^2 \qquad (\geq 0)\,. \tag{10.266}$$

Since the material parameter $e$ is nonnegative, the left side of Expression (10.266) takes nonnegative values. This feature is substituted into Expression (10.266).

Inequalities (10.265) and (10.266) prove that the present material model [Expressions (10.244) and (10.245)] satisfy Clausius–Duhem Inequality (10.260). This means that this constitutive equation does not violate the basic laws and axioms of continuum thermodynamics and is therefore an acceptable model.

In the preceding study, the creep model was a simplified one. Investigation of the full creep model by Le Gac and Duval with a nonlocal damage evolution equation is of course still possible.

In order to make the material model clearer for the reader, the derived constitutive model is rewritten. Equations (10.247), (10.253), (10.255), and (10.258) give

$$\boldsymbol{\sigma} = (1 - D)\,\mathbf{C} : (\boldsymbol{\varepsilon} - \boldsymbol{\varepsilon}^{\mathrm{i}}) \tag{10.267}$$

and

$$\dot{D} = \Big\langle\, e\,(\boldsymbol{\varepsilon} - \boldsymbol{\varepsilon}^{\mathrm{i}}) : \mathbf{C} : (\boldsymbol{\varepsilon} - \boldsymbol{\varepsilon}^{\mathrm{i}}) + k\,(\nabla^2 D) \,\Big\rangle, \tag{10.268}$$

and furthermore,

$$\dot{\boldsymbol{\varepsilon}}^{\mathrm{i}} = \overset{\circ}{\varepsilon}_{\mathrm{re}} \left[ \frac{\langle\, J_{\mathrm{vM}}(\boldsymbol{\sigma} - \boldsymbol{\beta}^1) \,\rangle}{\sigma_{\mathrm{re}}} \right]^n \frac{\mathbf{s} - \mathbf{b}^1}{J_{\mathrm{vM}}(\boldsymbol{\sigma} - \boldsymbol{\beta}^1)}, \tag{10.269}$$

and finally,

$$\dot{\boldsymbol{\beta}}^1 = m^1\,\dot{\boldsymbol{\varepsilon}}^{\mathrm{i}} - m^2 \left[ \frac{J_{\mathrm{vM}}(\boldsymbol{\beta}^1)}{\sigma_{\mathrm{re}}} \right]^k \frac{\mathbf{b}^1}{J_{\mathrm{vM}}(\boldsymbol{\beta}^1)}\,. \tag{10.270}$$

## 10.7 Summary

This chapter studied damage. Some preliminary results were derived for materials containing noninteracting penny-shaped microcracks or noninteracting microvoids within a matrix, the response of which was assumed to obey Hooke's law. Using these preliminary results, continuum damage mechanics was introduced using the extended postulate of (damage-elastic) strain equivalence with the effective stress concept. There are other postulates for damage, but the authors' view is that this one is the best; furthermore, it fits very well the results obtained from studies on microcracks and microvoids.

Readers who prefer to evaluate mechanical models rather than almost pure mathematical derivation should study Sections 10.5.6 and 10.5.7, in which a uniaxial bar model and a tube model are investigated and many of the results derived in this chapter are obtained. Based on the derivation carried out in Section 10.5.4, the effective stress tensor $\tilde{\boldsymbol{\sigma}}$ is symmetric, which is an important phenomenon. Section 10.5.8 showed that the effective stress tensor $\tilde{\boldsymbol{\sigma}}$ is an averaged microstress $\boldsymbol{\sigma}^{\mathrm{m}}$ within the matrix material within the representative volume element $V^{\mathrm{rve,m}}$. Thus the effective stress tensor $\tilde{\boldsymbol{\sigma}}$ is the stress measure for crystal-scale deformation mechanisms, such as creep, and plastic yield. Often, damage of materials is related to microcracks and microvoids. If damage is connected to the deterioration of materials in general, other mechanisms should be connected to damage. Therefore Section 10.5.10 discussed the possibility of interpreting carbide coarsening to describe damage. Finally, a nonlocal material model for creep and damage was introduced.

## Problem

**10.1** A material model for pure tension is evaluated. It is given by the following two functions. The specific Helmholtz free energy $\psi$ is

$$\begin{aligned}\psi = \psi^{\mathrm{eT}} - s_{\mathrm{r}}\,(T - T_{\mathrm{r}}) + \frac{1}{2\,\rho_0}\,(1 - D)\,E\,(\varepsilon - \varepsilon^{\mathrm{v}})^2 + \frac{e}{2\,\rho_0}\,\alpha^2 \\ - \frac{\hat{\alpha}\,E}{\rho_0}\,(1 - D)\,(\varepsilon - \varepsilon^{\mathrm{v}})\,(T - T_{\mathrm{r}}) - \frac{1}{2\,\rho_0}\,d\,D^2 - c_{\mathrm{r}}\,[T \ln (T/T_{\mathrm{r}}], \quad (1)\end{aligned}$$

where $\rho_0$, $E$, $e$, $\hat{\alpha}$, $d$ and $c_{\mathrm{r}}$ are material constants. The specific complementary dissipation function $\varphi^{\mathrm{c}}_{\mathrm{mech}}$ is

$$\begin{aligned}\varphi^{\mathrm{c}}_{\mathrm{mech}} = \frac{\overset{\circ}{\varepsilon}_{\mathrm{re}}\,\sigma_{\mathrm{re}}\,(1 - D)}{\rho_0\,(n + 1)} \left[\frac{\sigma - \beta}{\sigma_{\mathrm{re}}\,(1 - D)}\right]^{n+1} \\ + \frac{b\,\sigma_{\mathrm{re}}}{\rho_0\,e\,(m + 1)} \left(\frac{\beta}{\sigma_{\mathrm{re}}}\right)^{m+1} e^{\frac{E_{\mathrm{a}}}{R}\left(\frac{1}{T_{\mathrm{r}}} - \frac{1}{T}\right)} \\ + \frac{r}{\rho_0}\,Y \left[\varepsilon - \varepsilon^{\mathrm{v}} - \hat{\alpha}\,(T - T_{\mathrm{r}})\right]^p, \quad (2)\end{aligned}$$

where $\overset{\circ}{\varepsilon}_{\mathrm{re}}$, $\sigma_{\mathrm{e}}$, $n$, $b$, $e$, $m$, $E_a$, $R$, $r$, and $p$ are material constants. The parameter $E_{\mathrm{a}}$ is the activation energy, $R$ is the universal gas constant and $T_{\mathrm{r}}$ is the reference temperature.

In Material Models (1) and (2), the variable $D$ refers to damage and the quantity $Y$ is the corresponding internal force. The other variables in Models (1) and (2) are explained in Section 7.6.1. Take the following steps presented in Sections 7.6 and 7.6.1:

**1** Select the set of state variables for this particular material model. Although modelling does not require study of the specific entropy $s$, include it in the set of state functions. Write the following expressions:

$$\psi = \psi(\textit{state variables})$$
$$\varphi^{\mathrm{c}}_{\mathrm{mech}} = \varphi^{\mathrm{c}}_{\mathrm{mech}}(\textit{state functions};\ \textit{parameters}). \quad (3)$$

**2** Write the explicit form for a potential describing the relationships between the sets *state variables* and *state functions* for the material model.

**3** Apply the state equations.

**4** Based on Step 3, write the (total) strain tensor $\boldsymbol{\varepsilon}$ in terms of the other strain tensors in the form $\boldsymbol{\varepsilon} = \boldsymbol{\varepsilon}^{\mathrm{e}} + \cdots$ .

**5** Write the Clausius–Duhem inequalities for this particular material model.

**6a** Write the explicit form of the second potential for the material model.

**7a** Apply the normality rule.

**8a, 8b** Validate the material model by satisfying the Clausius–Duhem inequality.

**8a** The dissipation potential is a nonnegative homogeneous function.

**8b** The dissipation potential is not a nonnegative homogeneous function or no dissipation potential exists.

Based on the equations derived in Steps 4 and 7a, show the equations that form the material model and should therefore be implemented in the Abaqus UMAT subroutine. Take some time to think through this question, but it is mainly for the professor.

# 11

# Introduction to Material Models in Fluid Mechanics

## 11.1 Introduction

This chapter is dedicated to the foundations of fluid mechanics with some preliminary results. First, some general expressions for fluid mechanics are given. Section 11.2 gives a continuum thermodynamic derivation for the Navier–Poisson law of a Newtonian fluid, and Section 11.5 studies the Navier–Stokes equations. Finally, the ideal gas law is constructed separately for continuum thermodynamics and for thermostatics.

The following discussion is modified from Malvern ([57], pp. 295–296). Experience indicates that a fluid at rest or in uniform flow cannot sustain a shear stress. Hence, in a fluid at rest or in uniform flow, the maximum shear stress magnitude is zero, and the stress is a purely hydrostatic state of stress,

$$\sigma_{ij} = -p^0_{\text{mech}}\,\delta_{ij}, \qquad \text{or} \qquad \boldsymbol{\sigma} = -p^0_{\text{mech}}\,\mathbf{1}\,, \tag{11.1}$$

where $p^0_{\text{mech}}$ is the static mechanical pressure. The mechanical pressure $p_{\text{mech}}$ is defined by Definition (2.117),

$$p_{\text{mech}} := -\frac{1}{3}\,\mathbf{1} : \boldsymbol{\sigma}\,. \tag{11.2}$$

The subscript "mech" is added to distinguish the mechanical pressure $p_{\text{mech}}$ from the thermodynamics pressure $p$, and the superscript zero is added to indicate that it pertains to the condition of rest or uniform flow. In thermodynamics, the static pressure $p^0_{\text{mech}}$ in a fluid pure substance is assumed to be related to the density $\rho$ and the absolute temperature $T$ by an equation of state

$$F(p^0_{\text{mech}}, \rho, T) = 0 \tag{11.3}$$

in a thermodynamic equilibrium. In fluid mechanics, the thermodynamic pressure $p$ is defined as a quantity given by the same functional relation to $\rho$ and $T$ that gives the static mechanical pressure $p^0_{\text{mech}}$ in an equilibrium state, that is,

**Def**
$$F(p, \rho, T) = 0\,. \tag{11.4}$$

In fluid mechanics, Equation (11.4) is referred to as an equation of state, sometimes called the "kinetic equation of state" for the fluid to distinguish it from the caloric equation of state, that is $u = u(\rho, s)$. It is not convenient to write material models in terms of the specific entropy $s$. Therefore, in the material models, the specific internal energy $u$ is replaced with the specific Helmholtz free energy $\psi$,

$$\psi = \psi(\rho, T) \,. \tag{11.5}$$

Equation (11.5) was already given by Expression $(6.191)_1$ .

With Definition (11.4) of the thermodynamic pressure $p$, there is no guarantee that $p$ will be equal to $-\mathbf{1} : \boldsymbol{\sigma}^{\mathrm{d}}/3$, and it will be shown that it often is not. An example of an equation of state is the ideal gas law (authors' remark: it is a material model)

$$p = \rho \, R_{\mathrm{spec}} \, T \,, \tag{11.6}$$

where $R_{\mathrm{spec}}$ is the gas constant for the particular gas (specific gas constant). Equations (11.4)–(11.6) are material models.

## 11.2 Navier–Poisson Law of a Newtonian Fluid

This section derives the Navier–Poisson law of a Newtonian fluid. It must be kept in mind that the Navier–Poisson law is not a law but a material model. Furthermore, its form differs from those prepared for solids, as will be seen in this section.

Since the Navier–Poisson law of a Newtonian fluid does not take on a stand the form of the relation between the thermodynamic pressure $p$ and the density $\rho$ (or the absolute temperature $T$), no form for the specific Helmholtz free energy $\psi$ is given in this section.

The mechanical part of the specific dissipation function $\varphi_{\mathrm{mech}}$ is modelled as follows:

$$\varphi_{\mathrm{mech}}(\overset{\circ}{\boldsymbol{\varepsilon}}; \rho, T) = \frac{\overline{\mu}(\rho, T)}{\mu \, \rho} \, \overset{\circ}{\boldsymbol{\varepsilon}} : \overset{\circ}{\boldsymbol{\varepsilon}} + \frac{\overline{\lambda}(\rho, T)}{2 \, \mu \, \rho} \, (\mathbf{1} : \overset{\circ}{\boldsymbol{\varepsilon}}) \, (\mathbf{1} : \overset{\circ}{\boldsymbol{\varepsilon}}) \,, \tag{11.7}$$

where $\overline{\lambda}$ and $\overline{\mu}$ are two independent variables characterizing the viscosity of the fluid. The coefficient $\mu$ is related to the normality rule (see Section 6.7.1) and Lamé elastic constant (see Section 8.3.1).

Normality Rule (6.331) is recalled:

$$\boldsymbol{\sigma}^{\mathrm{d}} = \mu \, \rho \, \frac{\partial \varphi_{\mathrm{mech}}(\overset{\circ}{\boldsymbol{\varepsilon}}; \rho, T)}{\partial \overset{\circ}{\boldsymbol{\varepsilon}}} \,, \tag{11.8}$$

where $\boldsymbol{\sigma}^{\mathrm{d}}$ is the dissipative stress tensor. Substitution of Material Model (11.7) into Normality Rule (11.8) gives [see Equation (6.149) and Definition (2.69)]

$$\boldsymbol{\sigma}^{\mathrm{d}} = \mu \, \rho \, \frac{\partial \varphi_{\mathrm{mech}}(\overset{\circ}{\boldsymbol{\varepsilon}}; \rho, T)}{\partial \overset{\circ}{\boldsymbol{\varepsilon}}} = 2 \, \overline{\mu}(\rho, T) \, \overset{\circ}{\boldsymbol{\varepsilon}} + \overline{\lambda}(\rho, T) \, (\mathbf{1} : \overset{\circ}{\boldsymbol{\varepsilon}}) \, \mathbf{1} \,. \tag{11.9}$$

The fluid strain rate tensor $\overset{\circ}{\boldsymbol{\varepsilon}}$ is separated into the spherical part $\overset{\circ}{\boldsymbol{\varepsilon}}^{\mathrm{sp}}$ and the deviatoric part $\overset{\circ}{\boldsymbol{\varepsilon}}^{\mathrm{dev}}$. Expression (2.121) gives

$$\overset{\circ}{\boldsymbol{\varepsilon}}^{\mathrm{sp}} = \frac{1}{3}\,\mathbf{1}\,\mathbf{1} : \overset{\circ}{\boldsymbol{\varepsilon}}\,. \tag{11.10}$$

Definition (2.119) is

$$\mathbf{a} := \mathbf{a}^{\mathrm{sp}} + \mathbf{a}^{\mathrm{dev}}\,, \qquad \text{which gives} \qquad \overset{\circ}{\boldsymbol{\varepsilon}} = \overset{\circ}{\boldsymbol{\varepsilon}}^{\mathrm{sp}} + \overset{\circ}{\boldsymbol{\varepsilon}}^{\mathrm{dev}}\,. \tag{11.11}$$

Substitution of Equation (11.10) into Equation $(11.11)_2$ leads to

$$\overset{\circ}{\boldsymbol{\varepsilon}} = \frac{1}{3}\,\mathbf{1}\,\mathbf{1} : \overset{\circ}{\boldsymbol{\varepsilon}} + \overset{\circ}{\boldsymbol{\varepsilon}}^{\mathrm{dev}}\,. \tag{11.12}$$

Substitution of Equation (11.12) into the expression of the dissipative stress tensor $\boldsymbol{\sigma}^{\mathrm{d}}$, Equation (11.9), gives

$$\boldsymbol{\sigma}^{\mathrm{d}} = 2\,\overline{\mu}(\rho,T)\,\overset{\circ}{\boldsymbol{\varepsilon}}^{\mathrm{dev}} + \left[\overline{\lambda}(\rho,T) + \tfrac{2}{3}\,\overline{\mu}(\rho,T)\right](\mathbf{1} : \overset{\circ}{\boldsymbol{\varepsilon}})\,\mathbf{1}\,. \tag{11.13}$$

In Equation (6.75), the fluid stress tensor $\boldsymbol{\sigma}^{\mathrm{f}}$ was assumed to be separable into quasi-conservative and dissipative parts, as follows:

$$\boldsymbol{\sigma}^{\mathrm{f}} = -p\,\mathbf{1} + \boldsymbol{\sigma}^{\mathrm{d}}\,. \tag{11.14}$$

Substitution of the dissipative stress tensor $\boldsymbol{\sigma}^{\mathrm{d}}$ from Equation (11.9) into Separation (11.14) yields

$$\boldsymbol{\sigma}^{\mathrm{f}} = -p\,\mathbf{1} + 2\,\overline{\mu}(\rho,T)\,\overset{\circ}{\boldsymbol{\varepsilon}} + \overline{\lambda}(\rho,T)\,(\mathbf{1} : \overset{\circ}{\boldsymbol{\varepsilon}})\,\mathbf{1}\,. \tag{11.15}$$

According to Malvern [[57], Eq. (6.3.10)], Equation (11.15) is the Navier–Poisson law of a Newtonian fluid.

## 11.3 Relation between the Pressures $p$ and $p_{\mathrm{mech}}$

In this section, relations between the thermodynamic pressure $p$ and mechanical pressure $p_{\mathrm{mech}}$ will be derived. Expression $(3.123)_2$ and Problem 3.1 are to show that

$$\overset{\circ}{\boldsymbol{\varepsilon}} = \frac{1}{2}\,(\vec{v}\overleftarrow{\nabla} + \vec{\nabla}\vec{v}) \qquad \text{and} \qquad \mathbf{1} : \overset{\circ}{\boldsymbol{\varepsilon}} = \vec{\nabla}\cdot\vec{v}\,. \tag{11.16}$$

Substitution of Equations (11.16) into Material Model (11.15) gives

$$\boldsymbol{\sigma}^{\mathrm{f}} = -p\,\mathbf{1} + \overline{\mu}(\rho,T)\,(\vec{\nabla}\vec{v} + \vec{v}\overleftarrow{\nabla}) + \overline{\lambda}(\rho,T)\,(\vec{\nabla}\cdot\vec{v})\,\mathbf{1}\,. \tag{11.17}$$

Form (11.17) is given by White [([116], Eq. (2-27)]. White ([116], p. 66) wrote that this deformation law [Expression (11.17)] was first given by Stokes in 1845.

The dissipative stress tensor $\boldsymbol{\sigma}^{\mathrm{d}}$ is substituted from Equation (11.9) into Separation (11.14). We arrive at the following:

$$\boldsymbol{\sigma}^{\mathrm{f}} = -p\,\mathbf{1} + 2\,\overline{\mu}(\rho,T)\,\overset{\circ}{\boldsymbol{\varepsilon}}^{\mathrm{dev}} + \Big[\,\overline{\lambda}(\rho,T) + \tfrac{2}{3}\,\overline{\mu}(\rho,T)\Big]\,(\mathbf{1}:\overset{\circ}{\boldsymbol{\varepsilon}})\,\mathbf{1}\,. \tag{11.18}$$

Multiplication of Material Model (11.18) by $\frac{1}{3}\,\mathbf{1}$: from the left gives

$$\tfrac{1}{3}\,\mathbf{1}{:}\boldsymbol{\sigma}^{\mathrm{f}} = -p\,\tfrac{1}{3}\,\mathbf{1}{:}\mathbf{1} + \tfrac{2}{3}\,\tilde{\mu}(\rho,T)\,\mathbf{1}{:}\overset{\circ}{\boldsymbol{\varepsilon}}^{\mathrm{dev}} + \tfrac{1}{3}\Big[\,\tilde{\lambda}(\rho,T) + \tfrac{2}{3}\,\tilde{\mu}(\rho,T)\Big]\,(\mathbf{1}{:}\overset{\circ}{\boldsymbol{\varepsilon}})\,\mathbf{1}{:}\mathbf{1}\,. \tag{11.19}$$

In order to recast Expression (11.19) in a new form, two results are needed, as follows: Equation $(2.116)_2$: For every deviatoric second-order tensor $\mathbf{a}^{\mathrm{dev}}$, expression $\mathbf{1}:\mathbf{a}^{\mathrm{dev}} \equiv 0$ holds and Expression $(2.65)_2$: $\mathbf{1}:\mathbf{1} = 3$. Thus Expression (11.19) reduces to

$$\frac{1}{3}\,\mathbf{1}:\boldsymbol{\sigma}^{\mathrm{f}} = -p + \Big[\,\overline{\lambda}(\rho,T) + \tfrac{2}{3}\,\overline{\mu}(\rho,T)\Big]\,(\mathbf{1}:\overset{\circ}{\boldsymbol{\varepsilon}})\,. \tag{11.20}$$

Combining Definition (11.2), that is, $p_{\mathrm{mech}} = -\,\mathbf{1}:\boldsymbol{\sigma}^{\mathrm{f}}/3$, with Equation (11.20) yields

$$p_{\mathrm{mech}} = p - \Big[\,\overline{\lambda}(\rho,T) + \tfrac{2}{3}\,\overline{\mu}(\rho,T)\Big]\,(\mathbf{1}:\overset{\circ}{\boldsymbol{\varepsilon}})\,. \tag{11.21}$$

As Result (11.21) shows, the thermodynamic pressure $p$ does not equal the mechanical pressure $p_{\mathrm{mech}}$.

Stokes [104] assumed that (see White [116], p. 67, or Malvern [57], p. 299)

$$\overline{\lambda}(\rho,T) + \tfrac{2}{3}\,\overline{\mu}(\rho,T) = 0, \qquad \text{which yields} \qquad p_{\mathrm{mech}} = p\,. \tag{11.22}$$

Thus Stokes's Assumption $(11.22)_1$ leads to the results that the mechanical pressure $p_{\mathrm{mech}}$ equals the thermodynamic pressure $p$. The limitations of Assumption $(11.22)_1$ have been addressed by White ([116], p. 67) and Malvern ([57], p. 300).

If the fluid is incompressible, the following holds [see Continuity Equations (4.69) and (4.70) and Equation $(11.16)_2$ ]:

$$\vec{\nabla}\cdot\vec{v} = 0 \qquad \text{and} \qquad \mathbf{1}:\overset{\circ}{\boldsymbol{\varepsilon}} = 0\,. \tag{11.23}$$

Substitution of Condition $(11.23)_2$ into Equation (11.21) gives

$$\text{Incompressible fluid} \qquad \Rightarrow \qquad p_{\mathrm{mech}} = p\,. \tag{11.24}$$

Thus also the assumption of incompressibility leads to the result that the mechanical pressure $p_{\mathrm{mech}}$ equals the thermodynamic pressure $p$. White ([116], p. 67) and Malvern ([57], p. 300) briefly discuss the incompressibility of fluid flow.

Since during fluid flow matter is neither created nor destroyed, the material derivative of mass $m$ disappears (**AX 1**, *axiom of conservation of mass*). The mass is given by expression $m = \rho V$ and therefore the following can be written:

$$\dot{m} = \dot{\rho}\, V + \rho\, \dot{V} = 0\,, \qquad \text{which gives} \qquad \frac{\dot{\rho}}{\rho} = -\frac{\dot{V}}{V}\,. \tag{11.25}$$

Equation (6.80) and Problems 2.21 and 3.1 are to show that

$$\mathbf{1}:\vec{\nabla}\vec{v} = -\frac{\dot{\rho}}{\rho}\,, \qquad \mathbf{1}:\vec{\nabla}\vec{v} = \vec{\nabla}\cdot\vec{v}\,, \qquad \text{and} \qquad \mathbf{1}:\overset{\circ}{\boldsymbol{\varepsilon}} = \vec{\nabla}\cdot\vec{v}\,. \tag{11.26}$$

Equations $(11.25)_2$ and (11.26) give

$$\mathbf{1}:\overset{\circ}{\boldsymbol{\varepsilon}} = \vec{\nabla}\cdot\vec{v} = \mathbf{1}:\vec{\nabla}\vec{v} = -\frac{\dot{\rho}}{\rho} = \frac{\dot{V}}{V}\,. \tag{11.27}$$

Bulk viscosity, denoted by $\kappa(\rho, T)$, is defined by

$$\kappa(\rho, T) := \overline{\lambda}(\rho, T) + \tfrac{2}{3}\,\overline{\mu}(\rho, T)\,. \tag{11.28}$$

Substitution of Equation (11.27) and Definition (11.28) into Equation (11.21) gives

$$p_{\text{mech}} = p - \Big[\overline{\lambda}(\rho, T) + \tfrac{2}{3}\,\overline{\mu}(\rho, T)\Big]\,(\mathbf{1}:\overset{\circ}{\boldsymbol{\varepsilon}}\,) = p - \kappa(\rho, T)\,\frac{\dot{V}}{V}\,. \tag{11.29}$$

## 11.4 Fluid Stress Tensor and Restrictions on Viscosity Parameters

In this section, the physical meanings of the terms of the fluid stress tensor $\boldsymbol{\sigma}^r$ are discussed. Furthermore, the restrictions on the values of the viscosity parameters $\bar{\lambda}$ and $\bar{\mu}$ are derived.

Definitions (2.120) and $(2.108)_2$ are

$$\mathbf{a}^{\text{dev}} := \mathbf{K}:\mathbf{a}, \quad \text{where the fourth-order tensor} \quad \mathbf{K} := \mathbf{I} - \tfrac{1}{3}\,\mathbf{1}\,\mathbf{1}\,. \tag{11.30}$$

Definitions (11.30) give

$$\mathbf{s}^{\text{f}} = \boldsymbol{\sigma}^{\text{f;dev}} = \mathbf{K}:\boldsymbol{\sigma}^{\text{f}} = \mathbf{I}:\boldsymbol{\sigma}^{\text{f}} - \tfrac{1}{3}\,\mathbf{1}\,\mathbf{1}:\boldsymbol{\sigma}^{\text{f}} = \boldsymbol{\sigma}^{\text{f}} - \tfrac{1}{3}\,\mathbf{1}\,\mathbf{1}:\boldsymbol{\sigma}^{\text{f}}\,, \tag{11.31}$$

where the notation $\mathbf{s}^{\text{f}}$ stands for the fluid deviatoric stress tensor and where Definition (2.69), that is, $\mathbf{I}:\mathbf{c} = \mathbf{c}$, was exploited.

Substitution of Equations (11.18) and (11.20) into Equation (11.31) yields

$$\begin{aligned} \mathbf{s}^{\text{f}} &= -p\,\mathbf{1} + 2\,\overline{\mu}(\rho, T)\,\overset{\circ}{\boldsymbol{\varepsilon}}{}^{\text{dev}} + \Big[\overline{\lambda}(\rho, T) + \tfrac{2}{3}\,\overline{\mu}(\rho, T)\Big]\,(\mathbf{1}:\overset{\circ}{\boldsymbol{\varepsilon}}\,)\,\mathbf{1} \\ &\quad + p\,\mathbf{1} - \Big[\overline{\lambda}(\rho, T) + \tfrac{2}{3}\,\overline{\mu}(\rho, T)\Big]\,(\mathbf{1}:\overset{\circ}{\boldsymbol{\varepsilon}}\,)\,\mathbf{1} = 2\,\overline{\mu}(\rho, T)\,\overset{\circ}{\boldsymbol{\varepsilon}}{}^{\text{dev}}\,. \end{aligned} \tag{11.32}$$

Expressions (11.22), (11.32), and (11.29) allow the following interpretation to be made of the role of the terms in Material Model (11.18). Material Model (11.18) is

$$\boldsymbol{\sigma}^{\mathrm{f}} = -p\,\mathbf{1} + 2\,\overline{\mu}(\rho,T)\,\overset{\circ}{\boldsymbol{\varepsilon}}^{\mathrm{dev}} + \left[\,\overline{\lambda}(\rho,T) + \tfrac{2}{3}\,\overline{\mu}(\rho,T)\,\right](\mathbf{1}:\overset{\circ}{\boldsymbol{\varepsilon}})\,\mathbf{1}\,. \tag{11.33}$$

The first term, that is, the thermodynamic pressure $p$, gives the instantaneous reversible volumetric response of the fluid. The second term represents the effect of the shear flow of the fluid on the shear stress of the fluid. The last term of Material Model (11.33) [i.e., Material Model (11.18)] gives the influence of the volumetric viscous flow of the fluid on the stress state of the fluid.

Equation (6.261) gives the mechanical part of the Clausius–Duhem inequality for fluids. It is

$$\boldsymbol{\sigma}^{\mathrm{d}} : \overset{\circ}{\boldsymbol{\varepsilon}} \geq 0\,. \tag{11.34}$$

Substitution of the dissipative stress tensor $\boldsymbol{\sigma}^{\mathrm{d}}$ from Equation (11.9) into Inequality (11.34) yields

$$\boldsymbol{\sigma}^{\mathrm{d}} : \overset{\circ}{\boldsymbol{\varepsilon}} = 2\,\tilde{\mu}(\rho,T)\,\overset{\circ}{\boldsymbol{\varepsilon}}^{\mathrm{dev}} : \overset{\circ}{\boldsymbol{\varepsilon}}^{\mathrm{dev}} + \left[\,\overline{\lambda}(\rho,T) + \tfrac{2}{3}\,\overline{\mu}(\rho,T)\,\right](\mathbf{1}:\overset{\circ}{\boldsymbol{\varepsilon}})\,(\mathbf{1}:\overset{\circ}{\boldsymbol{\varepsilon}}) \;\geq\; 0\,. \tag{11.35}$$

Inequality (11.35) exploited the result $\overset{\circ}{\boldsymbol{\varepsilon}}^{\mathrm{dev}} : \overset{\circ}{\boldsymbol{\varepsilon}} = \overset{\circ}{\boldsymbol{\varepsilon}}^{\mathrm{dev}} : \overset{\circ}{\boldsymbol{\varepsilon}}^{\mathrm{dev}}$, which can be obtained from Problem 8.2.

At least in principle, it is possible to create processes whereby either $\overset{\circ}{\boldsymbol{\varepsilon}}^{\mathrm{dev}} : \overset{\circ}{\boldsymbol{\varepsilon}}^{\mathrm{dev}} = 0$ or $(\mathbf{1}:\overset{\circ}{\boldsymbol{\varepsilon}})\,(\mathbf{1}:\overset{\circ}{\boldsymbol{\varepsilon}}) = 0$. Since these terms are nonnegative, according to Inequality (11.35) the sufficient condition for the satisfaction of the Clausius–Duhem inequality reads

$$\overline{\mu}(\rho,T) \;\geq\; 0 \quad \text{and} \quad \overline{\lambda}(\rho,T) + \tfrac{2}{3}\,\overline{\mu}(\rho,T) \;\geq\; 0\,. \tag{11.36}$$

## 11.5 Navier–Stokes Equations

This section derives the Navier–Stokes equations. Equation $(4.135)_1$ is recalled:

$$\vec{\nabla}\cdot\boldsymbol{\sigma}^{\mathrm{f}} + \rho\,\vec{b} = \rho\,\frac{\mathrm{D}\vec{v}}{\mathrm{D}t}\,. \tag{11.37}$$

Equation (11.17) is

$$\boldsymbol{\sigma}^{\mathrm{f}} = -p\,\mathbf{1} + \overline{\mu}\,(T)\,(\vec{\nabla}\vec{v} + \vec{v}\,\overleftarrow{\nabla}) + \overline{\lambda}(T)\,(\vec{\nabla}\cdot\vec{v})\,\mathbf{1}\,. \tag{11.38}$$

The fluid stress tensor $\boldsymbol{\sigma}^{\mathrm{f}}$ from Equation (11.38) is substituted into Equation (11.37), and the terms in the obtained expression are reordered. This procedure leads to

$$\rho\,\frac{\mathrm{D}\vec{v}}{\mathrm{D}t} = \rho\,\vec{b} - p\,\overleftarrow{\nabla}\cdot\mathbf{1} + \vec{\nabla}\cdot\left[\,\overline{\mu}(\rho,T)\,(\vec{\nabla}\vec{v} + \vec{v}\overleftarrow{\nabla}) + \overline{\lambda}(\rho,T)\,(\vec{\nabla}\cdot\vec{v})\,\mathbf{1}\,\right]. \tag{11.39}$$

Based on Definition (2.63), that is, $\mathbf{1}\cdot\vec{u} = \vec{u}$, the following holds:

$$\overleftarrow{\nabla}\cdot\mathbf{1} \equiv \overleftarrow{\nabla}\,, \qquad \text{which gives} \qquad p\overleftarrow{\nabla}\cdot\mathbf{1} \equiv p\overleftarrow{\nabla} = \vec{\nabla}p\,. \tag{11.40}$$

Equation $(11.40)_2$ allows Equation (11.39) to be written in the form

$$\rho\,\frac{\mathrm{D}\vec{v}}{\mathrm{D}t} = \rho\,\vec{b} - \vec{\nabla}p + \vec{\nabla}\cdot\Big[\overline{\mu}(\rho,T)\,(\vec{\nabla}\vec{v} + \vec{v}\overleftarrow{\nabla}) + \overline{\lambda}(\rho,T)\,(\vec{\nabla}\cdot\vec{v})\,\mathbf{1}\Big]\,. \tag{11.41}$$

Equations (11.41) equal that of White [[116], Eq. (2-29b)]. White ([116], p. 68) writes that these [Equations (11.41)] are the Navier–Stokes equations, fundamental to the subject of viscous-fluid flow.

If the values for the parameters $\overline{\mu}$ and $\overline{\lambda}$ are assumed to be independent of the density $\rho$, and on the absolute temperature $T$, the following holds:

$$\overline{\mu}(\rho,T) = \overline{\mu} \qquad \text{and} \qquad \overline{\lambda}(\rho,T) = \overline{\lambda}\,. \tag{11.42}$$

Expressions (11.42) show that the values of the parameters $\overline{\mu}$ and $\overline{\lambda}$ do not depend on the space coordinates $(x_1, x_2, x_3)$. Thus $\vec{\nabla}\overline{\mu} \equiv 0$ and $\vec{\nabla}\overline{\lambda} \equiv 0$. Substitution of Assumptions (11.42) into Navier–Stokes Equations (11.41) yields

$$\rho\,\frac{\mathrm{D}\vec{v}}{\mathrm{D}t} = \rho\,\vec{b} - \vec{\nabla}p + \overline{\mu}\Big[\vec{\nabla}\cdot(\vec{\nabla}\vec{v}) + \vec{\nabla}\cdot(\vec{v}\overleftarrow{\nabla})\Big] + \overline{\lambda}\,\vec{\nabla}(\vec{\nabla}\cdot\vec{v})\,. \tag{11.43}$$

Problem 2.22 is to show that

$$\vec{\nabla}(\vec{\nabla}\cdot\vec{v}) = \vec{\nabla}\cdot(\vec{v}\overleftarrow{\nabla})\,. \tag{11.44}$$

Substitution of Equality (11.44) into Expression (11.43) yields

$$\rho\,\frac{\mathrm{D}\vec{v}}{\mathrm{D}t} = \rho\,\vec{b} - \vec{\nabla}p + \overline{\mu}\,\vec{\nabla}\cdot(\vec{\nabla}\vec{v}) + (\overline{\mu} + \overline{\lambda})\,\vec{\nabla}(\vec{\nabla}\cdot\vec{v})\,. \tag{11.45}$$

In the case of incompressible fluid flow, Continuity Equation (4.69) reduces to Form (4.70), $\vec{\nabla}\cdot\vec{v} = 0$, and therefore Equation (11.45) reduces to the following expression:

$$\rho\,\frac{\mathrm{D}\vec{v}}{\mathrm{D}t} = \rho\,\vec{b} - \vec{\nabla}p + \overline{\mu}\,\vec{\nabla}\cdot(\vec{\nabla}\vec{v}) \quad \text{or} \quad \rho\,\frac{\mathrm{D}\vec{v}}{\mathrm{D}t} = \rho\,\vec{b} - \vec{\nabla}p + \overline{\mu}\,\nabla^2\vec{v}\,. \tag{11.46}$$

Equations (11.46) are the vector forms of the Navier–Stokes equations for constant viscosity and density (White [116], p. 68).

## 11.6 Ideal Gas Law

The present section introduces the specific Helmholtz free energy $\psi$ for ideal gas law. The specific Helmholtz free energy $\psi$ is assumed to have the following form:

$$\psi(\rho, T) = \frac{1}{\rho_{\rm re}} R_{\rm spec} \, T \ln \frac{\rho}{\rho_{\rm re}} \,, \tag{11.47}$$

where $R_{\rm spec}$ is the specific gas constant. In Material Model (11.47), the role of the quantity $\rho_{\rm re}$ is to make $\rho/\rho_{\rm re}$ dimensionless. The state equations for fluids are given by Expressions (6.193) and $(6.192)_1$ :

$$p = \rho^2 \frac{\partial \psi(\rho, T)}{\partial \rho} \qquad \text{and} \qquad s = -\frac{\partial \psi(\rho, T)}{\partial T} \,. \tag{11.48}$$

Substitution of Material Model (11.47) into State Equation $(11.48)_1$ yields

$$p = \rho^2 \frac{\partial \psi(\rho, T)}{\partial \rho} = \rho \, R_{\rm spec} \, T \qquad \Rightarrow \qquad p = \rho \, R_{\rm spec} \, T \,. \tag{11.49}$$

Material Model $(11.49)_2$ equals that at which we were aiming in Expression (11.6). Since Equation $(11.49)_2$ is a field equation, it does not have the appearance often seen in this context. Usually, instead of Equation $(11.49)_2$, expression for a homogeneous system is derived, which is the task of the following derivation.

The state $\mathcal{E}$ of the system $\mathrm{m^b}$ is assumed to be homogeneous, that is, the values of the pressure $p$, density $\rho$, and temperature $T$ take equal values everywhere within the system $\mathrm{m^b}$. Both sides of Equation $(11.49)_2$ are integrated over the volume of the system $\mathrm{m^b}$, that is, over the volume $V$. The notation $V$ there is not the same as in this chapter on fluid dynamics, since in fluid dynamics, the control volume $\mathrm{V^{cv}}$ is generally used. However, here thermostatics is studied; thus the volume of the system $\mathrm{m^b}$ is denoted by $V$, as in Section 5.3. Integration of Equation $(11.49)_2$ gives

$$\int_V p \, \mathrm{d}V = \int_V \rho \, R_{\rm spec} \, T \, \mathrm{d}V \qquad \Rightarrow \qquad p \int_V \mathrm{d}V = \rho \, R_{\rm spec} \, T \int_V \mathrm{d}V \,, \tag{11.50}$$

which yields

$$p \, V = \rho \, R_{\rm spec} \, T \, V \,. \tag{11.51}$$

Result (11.51) is obvious, since the state of the system $\mathcal{E}$ was assumed to be homogeneous. Equation (11.51) can be obtained just by multiplying both sides of Equation $(11.49)_2$ by $V$.

The amount of gas when expressed by the unit mole $n$ and the density of the material $\rho$ are given by

$$n = \frac{\mathrm{m^b}}{M} \quad \text{and} \quad \rho = \frac{\mathrm{m^b}}{V} \qquad \Rightarrow \qquad \rho = \frac{nM}{V} \,, \tag{11.52}$$

where $M$ is the molar mass of the gas. Substitution of Expression $(11.52)_3$ into Equation (11.51) yields

$$p V = n M R_{\text{spec}} T . \tag{11.53}$$

The relation between the universal gas constant $R$ and the specific gas constant $R_{\text{spec}}$ reads

$$R = M R_{\text{spec}} . \tag{11.54}$$

Substitution of Expression (11.54) into Equation (11.53) gives

$$p V = n R T . \tag{11.55}$$

Form (11.55) is the form for the ideal gas at which we were aiming.

Material Models $(11.49)_2$ and (11.55) have different characters. Material Model $(11.49)_2$ is a field equation, whereas Material Model (11.55) is for the whole system $\mathrm{m}^{\mathrm{b}}$. Thus, the former is for continuum thermodynamics whereas the latter is for thermostatics.

## 11.7 Summary

In this chapter, fluid mechanics was studied. Some traditional equations of fluid mechanics were investigated: the Navier–Poisson law of a Newtonian fluid, well known as the Navier–Stokes equations, and the ideal gas law.

# *Appendix A*
# On Partial Derivatives of the von Mises Operator $J_{\mathrm{vM}}$ Acting on a Tensorial Variable

This appendix investigates partial derivatives of a quantity obtained when the von Mises operator $J_{\mathrm{vM}}$ acts on a tensorial quantity $\boldsymbol{\alpha} - \boldsymbol{\tau}$. Some special cases are evaluated at the end of the appendix.

The lemma introduced in Expression (2.127) reads as follows: let $f(x, \cdot) \in C^1$ be a function of the independent variable $x$ and some other independent variables denoted by $\cdot$. By extending Marsden and Hughes ([58], p. 185), the following can be expressed: The partial directional derivative of $f(x, \cdot)$ in the direction of $u$ is denoted by $\partial_{\mathrm{x}} f(x, \cdot) * u$ and obeys the expression

$$\frac{\partial f(x, \cdot)}{\partial x} * u = \partial_{\mathrm{x}} f(x, \cdot) * u = \frac{\partial}{\partial h} f(x + h\, u, \cdot)\Big|_{h=0}, \tag{A.1}$$

where the variable $u$ is a tensor of the same degree as the variable $x$ and $h$ is a scalar-valued quantity. The notation $*$ stands for multiplication, a dot product, a double-dot product, and so on, depending on the form of the quantities $x$ and $u$. For scalars $x$ and $u$, the notation $*$ stands for multiplication; for vectors $x$ and $u$, the notation $*$ stands for a dot product; for second-order tensors $x$ and $u$, the notation $*$ stands for a double-dot product, and so on.

Here $C^m(R)$ denotes the set of functions $f(x)$ whose derivatives up to the $m$th order inclusive are continuous in the domain $R$. When the argument $R$ is not included, the domain is understood in the context of the discussion (e.g., $R$ can be a line or a two-dimensional domain).

In the following study, the quantity $\mathbf{u}$ is replaced with the quantity $\boldsymbol{\gamma}$, since its deviatoric part $\mathbf{g}$ plays an important role.

**Theorem 1:** The following equations hold:

$$\frac{\partial J_{\mathrm{vM}}(\boldsymbol{\sigma} - \boldsymbol{\beta}^1)}{\partial \boldsymbol{\sigma}} = \frac{3}{2} \frac{\mathbf{s} - \mathbf{b}^1}{J_{\mathrm{vM}}(\boldsymbol{\sigma} - \boldsymbol{\beta}^1)} \tag{A.2}$$

and

$$\frac{\partial J_{\mathrm{vM}}(\boldsymbol{\sigma} - \boldsymbol{\beta}^1)}{\partial \boldsymbol{\beta}^1} = -\frac{3}{2} \frac{\mathbf{s} - \mathbf{b}^1}{J_{\mathrm{vM}}(\boldsymbol{\sigma} - \boldsymbol{\beta}^1)}, \tag{A.3}$$

where $\mathbf{b}^1$ is the deviatoric part of the second-order tensor $\boldsymbol{\beta}^1$.

**Proof:** The von Mises operator $J_{\mathrm{vM}}$ acts on the tensorial variable $\boldsymbol{\alpha}-\boldsymbol{\tau}$. Based on Definition (2.124) for the von Mises operator $J_{\mathrm{vM}}$, the following is obtained:

$$J_{\mathrm{vM}}(\boldsymbol{\alpha}-\boldsymbol{\tau}) := \sqrt{\tfrac{3}{2}\,[\mathbf{a}-\mathbf{t}]:[\mathbf{a}-\mathbf{t}]}\,. \tag{A.4}$$

Based on Definition (2.120), the deviatoric tensors $\mathbf{a}$ and $\mathbf{t}$ are defined by

$$\mathbf{a} := \mathbf{K}:\boldsymbol{\alpha} \qquad \text{and} \qquad \mathbf{t} := \mathbf{K}:\boldsymbol{\tau}\,, \tag{A.5}$$

where the fourth-order tensor $\mathbf{K}$ is defined to be [see Definition $(2.108)_2$]

$$\mathbf{K} := \mathbf{I} - \frac{1}{3}\,\mathbf{1}\,\mathbf{1}\,. \tag{A.6}$$

Based on Expressions (A.4) and (A.5), the following is obtained:

$$\frac{\partial J_{\mathrm{vM}}(\boldsymbol{\alpha}-\boldsymbol{\tau})}{\partial \boldsymbol{\alpha}} = \frac{\partial}{\partial \boldsymbol{\alpha}}\sqrt{\tfrac{3}{2}\,[\mathbf{a}-\mathbf{t}]:[\mathbf{a}-\mathbf{t}]} = \frac{\partial}{\partial \boldsymbol{\alpha}}\sqrt{\tfrac{3}{2}\,[\mathbf{K}:\boldsymbol{\alpha}-\mathbf{t}]:[\mathbf{K}:\boldsymbol{\alpha}-\mathbf{t}]}\,. \tag{A.7}$$

Expressions (A.1) and (A.7) yield

$$\begin{aligned}
\frac{\partial J_{\mathrm{vM}}(\boldsymbol{\alpha}-\boldsymbol{\tau})}{\partial \boldsymbol{\alpha}}:\boldsymbol{\gamma} &= \frac{\partial}{\partial h}\sqrt{\tfrac{3}{2}\,[\mathbf{K}:(\boldsymbol{\alpha}+h\,\boldsymbol{\gamma})-\mathbf{t}]:[\mathbf{K}:(\boldsymbol{\alpha}+h\,\boldsymbol{\gamma})-\mathbf{t}]}\,\Bigg|_{h=0} \\
&= \frac{\tfrac{1}{2}\tfrac{3}{2}\{(\mathbf{K}:\boldsymbol{\gamma}):[\mathbf{K}:(\boldsymbol{\alpha}+h\,\boldsymbol{\gamma})-\mathbf{t}]+[\mathbf{K}:(\boldsymbol{\alpha}+h\,\boldsymbol{\gamma})-\mathbf{t}]:(\mathbf{K}:\boldsymbol{\gamma})\}}{\sqrt{\tfrac{3}{2}\,[\mathbf{K}:(\boldsymbol{\alpha}+h\,\boldsymbol{\gamma})-\mathbf{t}]:[\mathbf{K}:(\boldsymbol{\alpha}+h\,\boldsymbol{\gamma})-\mathbf{t}]}}\,\Bigg|_{h=0} \\
&= \frac{\tfrac{1}{2}\tfrac{3}{2}\{(\mathbf{K}:\boldsymbol{\gamma}):[\mathbf{K}:\boldsymbol{\alpha}-\mathbf{t}]+[\mathbf{K}:\boldsymbol{\alpha}-\mathbf{t}]:(\mathbf{K}:\boldsymbol{\gamma})\}}{\sqrt{\tfrac{3}{2}\,[\mathbf{K}:\boldsymbol{\alpha}-\mathbf{t}]:[\mathbf{K}:\boldsymbol{\alpha}-\mathbf{t}]}} \\
&= \frac{\tfrac{1}{2}\tfrac{3}{2}\{\mathbf{g}:[\mathbf{a}-\mathbf{t}]+[\mathbf{a}-\mathbf{t}]:\mathbf{g}\}}{\sqrt{\tfrac{3}{2}\,[\mathbf{a}-\mathbf{t}]:[\mathbf{a}-\mathbf{t}]}}\,.
\end{aligned} \tag{A.8}$$

In Manipulation (A.8), the notation $\mathbf{g}$ stands for the deviatoric part of the second-order tensor $\boldsymbol{\gamma}$. Since $\mathbf{g}:\mathbf{a} = \mathbf{a}:\mathbf{g}$ and $\mathbf{g}:\mathbf{t} = \mathbf{t}:\mathbf{g}$, the terms of the numerator on the last line of Manipulation (A.8) are equal. Thus the following is obtained:

$$\begin{aligned}
\frac{\partial J_{\mathrm{vM}}(\boldsymbol{\alpha}-\boldsymbol{\tau})}{\partial \boldsymbol{\alpha}}:\boldsymbol{\gamma} &= \frac{\tfrac{1}{2}\tfrac{3}{2}\{\mathbf{g}:[\mathbf{a}-\mathbf{t}]+[\mathbf{a}-\mathbf{t}]:\mathbf{g}\}}{\sqrt{\tfrac{3}{2}\,[\mathbf{a}-\mathbf{t}]:[\mathbf{a}-\mathbf{t}]}} \\
&= \frac{3}{2}\,\frac{[\mathbf{a}-\mathbf{t}]:\mathbf{g}}{\sqrt{\tfrac{3}{2}\,[\mathbf{a}-\mathbf{t}]:[\mathbf{a}-\mathbf{t}]}} = \frac{3}{2}\,\frac{[\mathbf{a}-\mathbf{t}]:\mathbf{K}}{\sqrt{\tfrac{3}{2}\,[\mathbf{a}-\mathbf{t}]:[\mathbf{a}-\mathbf{t}]}}:\boldsymbol{\gamma}\,.
\end{aligned} \tag{A.9}$$

Expression (A.9) gives

$$\frac{\partial J_{\text{vM}}(\boldsymbol{\alpha}-\boldsymbol{\tau})}{\partial \boldsymbol{\alpha}} = \frac{3}{2}\frac{[\mathbf{a}-\mathbf{t}]:\mathbf{K}}{\sqrt{\frac{3}{2}[\mathbf{a}-\mathbf{t}]:[\mathbf{a}-\mathbf{t}]}}\,. \tag{A.10}$$

Definitions $(A.5)_1$ and (A.6) yield

$$\mathbf{a}:\mathbf{K} = \mathbf{a}:[\mathbf{I}-\tfrac{1}{3}\mathbf{1}\,\mathbf{1}] = \mathbf{a}:\mathbf{I}-\tfrac{1}{3}\mathbf{a}:\mathbf{1}\,\mathbf{1} = \mathbf{a}\,. \tag{A.11}$$

In Manipulation (A.11), identities $\mathbf{a}:\mathbf{I}=\mathbf{a}$ and $\mathbf{a}:\mathbf{1}=0$ [see Definition (2.69) and Result $(2.116)_2$] are exploited. A similar result holds for $\mathbf{t}:\mathbf{K}$. Based on Result (A.11), Derivative (A.10) reduces to

$$\frac{\partial J_{\text{vM}}(\boldsymbol{\alpha}-\boldsymbol{\tau})}{\partial \boldsymbol{\alpha}} = \frac{3}{2}\frac{[\mathbf{a}-\mathbf{t}]}{\sqrt{\frac{3}{2}[\mathbf{a}-\mathbf{t}]:[\mathbf{a}-\mathbf{t}]}} = \frac{3}{2}\frac{\mathbf{a}-\mathbf{t}}{J_{\text{vM}}(\boldsymbol{\alpha}-\boldsymbol{\tau})}\,. \tag{A.12}$$

Analogously to the preceding derivation, the following is thus obtained:

$$\frac{\partial J_{\text{vM}}(\boldsymbol{\alpha}-\boldsymbol{\tau})}{\partial \boldsymbol{\tau}} = -\frac{3}{2}\frac{\mathbf{a}-\mathbf{t}}{J_{\text{vM}}(\boldsymbol{\alpha}-\boldsymbol{\tau})}\,. \tag{A.13}$$

Replacing the second-order tensors $\boldsymbol{\alpha}$ and $\boldsymbol{\tau}$ with the stress tensor $\boldsymbol{\sigma}$ and the internal force tensor $\boldsymbol{\beta}^1$, Derivatives (A.12) and (A.13) give

$$\frac{\partial J_{\text{vM}}(\boldsymbol{\sigma}-\boldsymbol{\beta}^1)}{\partial \boldsymbol{\sigma}} = \frac{3}{2}\frac{\mathbf{s}-\mathbf{b}^1}{J_{\text{vM}}(\boldsymbol{\sigma}-\boldsymbol{\beta}^1)} \tag{A.14}$$

and

$$\frac{\partial J_{\text{vM}}(\boldsymbol{\sigma}-\boldsymbol{\beta}^1)}{\partial \boldsymbol{\beta}^1} = -\frac{3}{2}\frac{\mathbf{s}-\mathbf{b}^1}{J_{\text{vM}}(\boldsymbol{\sigma}-\boldsymbol{\beta}^1)}\,. \tag{A.15}$$

Results (A.14) and (A.15) show that **Theorem 1** holds.

# *Appendix B*
# Chain Rules

Theorems for chain rules are established in this appendix. The proofs are too short for separate appendices. The vectors and tensors are represented in the orthogonal rectangular Cartesian coordinate system, the basis of which is $(\vec{i}_1, \vec{i}_2, \vec{i}_3)$. Since it is convenient to express vectors and tensors in terms of their components for calculating dot products, in the following derivations the vectors and tensors are represented in the component form.

**Theorem 1:** The derivative of a second-order tensor $\mathbf{a}(\mathbf{e})$ with respect a second-order tensor $\mathbf{c}$, where $\mathbf{e} = f(\mathbf{c})$, is obtained by the following chain rule:

$$\frac{\partial \mathbf{a}(\mathbf{e})}{\partial \mathbf{c}} = \frac{\partial \mathbf{e}}{\partial \mathbf{c}} : \frac{\partial \mathbf{a}(\mathbf{e})}{\partial \mathbf{e}} . \tag{B.1}$$

**Proof:** Definition $(2.173)_1$ is

**Def**
$$\frac{\partial}{\partial \mathbf{c}} := \vec{i}_i \, \vec{i}_n \, \frac{\partial}{\partial c_{in}} , \tag{B.2}$$

where $\mathbf{c}$ is an arbitrary second-order tensor.

Based on Definition (B.2), the left side of Derivative (B.1) takes the form

$$\frac{\partial \mathbf{a}(\mathbf{e})}{\partial \mathbf{c}} = \vec{i}_i \, \vec{i}_n \, \frac{\partial (a_{uw} \, \vec{i}_u \, \vec{i}_w)}{\partial c_{in}} = \vec{i}_i \, \vec{i}_n \, \frac{\partial a_{uw}}{\partial c_{in}} \, \vec{i}_u \, \vec{i}_w = \frac{\partial a_{uw}}{\partial c_{in}} \, \vec{i}_i \, \vec{i}_n \, \vec{i}_u \, \vec{i}_w . \tag{B.3}$$

For the scalar component $\partial a_{uw} / \partial c_{in}$, the "standard" chain rule can be applied, which leads to the following expression:

$$\frac{\partial \mathbf{a}(\mathbf{e})}{\partial \mathbf{c}} = \frac{\partial a_{uw}}{\partial c_{in}} \, \vec{i}_i \, \vec{i}_n \, \vec{i}_u \, \vec{i}_w = \frac{\partial e_{st}}{\partial c_{in}} \, \frac{\partial a_{uw}}{\partial e_{st}} \, \vec{i}_i \, \vec{i}_n \, \vec{i}_u \, \vec{i}_w . \tag{B.4}$$

The order of the partial derivatives in Expression (B.4) stems from the fact that scalar components have to be in the same order as the base vectors.

Based on Definitions (B.2) and (2.52) and Expression (2.6), the right side of Derivative (B.1) can be written as follows:

$$\frac{\partial \mathbf{e}}{\partial \mathbf{c}} : \frac{\partial \mathbf{a}(\mathbf{e})}{\partial \mathbf{e}} = \vec{i}_i\,\vec{i}_n\,\frac{\partial(e_{st}\,\vec{i}_s\,\vec{i}_t)}{\partial c_{in}} : \vec{i}_j\,\vec{i}_m\,\frac{\partial(a_{uw}\,\vec{i}_u\,\vec{i}_w)}{\partial e_{jm}}$$

$$= \vec{i}_i\,\vec{i}_n\,\frac{\partial e_{st}}{\partial c_{in}}\,\vec{i}_s\,\vec{i}_t : \vec{i}_j\,\vec{i}_m\,\frac{\partial a_{uw}}{\partial e_{jm}}\,\vec{i}_u\,\vec{i}_w = \vec{i}_i\,\vec{i}_n\,\frac{\partial e_{st}}{\partial c_{in}}\,\delta_{sj}\,\delta_{tm}\,\frac{\partial a_{uw}}{\partial e_{jm}}\,\vec{i}_u\,\vec{i}_w$$

$$= \vec{i}_i\,\vec{i}_n\,\frac{\partial e_{st}}{\partial c_{in}}\,\frac{\partial a_{uw}}{\partial e_{st}}\,\vec{i}_u\,\vec{i}_w = \frac{\partial e_{st}}{\partial c_{in}}\,\frac{\partial a_{uw}}{\partial e_{st}}\,\vec{i}_i\,\vec{i}_n\,\vec{i}_u\,\vec{i}_w\,. \tag{B.5}$$

Comparison of Results (B.4) and (B.5) shows that Theorem 1 is correct.

**Theorems 2.1 and 2.2:** The following expressions hold:

$$\frac{\partial \vec{a}(\mathbf{e})}{\partial \mathbf{c}} = \frac{\partial \mathbf{e}}{\partial \mathbf{c}} : \frac{\partial \vec{a}(\mathbf{e})}{\partial \mathbf{e}} \qquad \text{and} \qquad \frac{\partial a(\mathbf{e})}{\partial \mathbf{c}} = \frac{\partial \mathbf{e}}{\partial \mathbf{c}} : \frac{\partial a(\mathbf{e})}{\partial \mathbf{e}}\,, \tag{B.6}$$

where $\vec{a}$ is a vector, $a$ is a scalar, and $\mathbf{e}$ and $\mathbf{c}$ are second-order tensors.

**Proof:** According to Proof (B.5), the base vectors of the second-order tensor $\mathbf{a}(\mathbf{e})$ does not operate with other base vectors. Thus Manipulation (B.5) proves **Theorems 2.1** and **2.2**.

**Theorems 3.1 and 3.2:** The following expressions hold:

$$\frac{\partial \vec{a}(\vec{e})}{\partial \vec{b}} = \frac{\partial \vec{e}}{\partial \vec{b}} \cdot \frac{\partial \vec{a}(\vec{e})}{\partial \vec{e}} \qquad \text{and} \qquad \frac{\partial a(\vec{e})}{\partial \vec{b}} = \frac{\partial \vec{e}}{\partial \vec{b}} \cdot \frac{\partial a(\vec{e})}{\partial \vec{e}}\,, \tag{B.7}$$

where $\vec{a}$ is a vector, $a$ is a scalar, and $\vec{e}$ and $\vec{b}$ are vectors.

**Proof:** The proof is similar to that given in Proof (B.5).

**Theorem 4:** The material derivative of the scalar-valued function $\underline{\theta} = \underline{\theta}(\underline{\vec{x}}(t), t)$ has the form

$$\frac{\mathrm{D}\underline{\theta}(\underline{\vec{x}}(t), t)}{\mathrm{D}t} = \frac{\partial \underline{\theta}(\underline{\vec{x}}(t), t)}{\partial t} + \frac{\mathrm{D}\underline{\vec{x}}(t)}{\mathrm{D}t} \cdot \frac{\partial \underline{\theta}(\underline{\vec{x}}(t), t)}{\partial \underline{\vec{x}}(t)}\,. \tag{B.8}$$

**Proof:** Theorem 4 can be proven by applying the chain rule and Expression $(B.7)_2$ (which also is a chain rule).

# *Appendix C*
# Material Derivative of the Jacobian Determinant

This appendix derives the expression for the material derivative of the Jacobian determinant $J(\vec{X}, t)$ between the current material coordinates $\underline{x}_i(t)$ and the initial material coordinates $X_i$. The text is based on the lecture notes of Tiihonen ([109], pp. 6 and 7). The original version of the derivation is given by Allen et al. ([4], pp. 7 and 8).

According to Equation (3.128), the Jacobian determinant $J(\vec{X}, t)$ between the current material coordinates $\underline{x}_i(t)$ and initial material coordinates $X_i$ is defined to be

$$J(\vec{X}, t) := \begin{vmatrix} \frac{\partial \underline{x}_1(t)}{\partial X_1} & \frac{\partial \underline{x}_1(t)}{\partial X_2} & \frac{\partial \underline{x}_1(t)}{\partial X_3} \\ \frac{\partial \underline{x}_2(t)}{\partial X_1} & \frac{\partial \underline{x}_2(t)}{\partial X_2} & \frac{\partial \underline{x}_2(t)}{\partial X_3} \\ \frac{\partial \underline{x}_3(t)}{\partial X_1} & \frac{\partial \underline{x}_3(t)}{\partial X_2} & \frac{\partial \underline{x}_3(t)}{\partial X_3} \end{vmatrix} . \tag{C.1}$$

For derivation purposes, the Jacobian determinant $J(\vec{X}, t)$ is written in the following simplified form:

$$J(\vec{X}, t) = \begin{vmatrix} J_{11} & J_{12} & J_{13} \\ J_{21} & J_{22} & J_{23} \\ J_{31} & J_{32} & J_{33} \end{vmatrix} . \tag{C.2}$$

Based on Expression (C.2), the material derivative of the Jacobian determinant $J(\vec{X}, t)$ takes the following form:

$$\begin{aligned} \frac{\mathrm{D}}{\mathrm{D}t} J(\vec{X}, t) = & \begin{vmatrix} \frac{\mathrm{D}J_{11}}{\mathrm{D}t} & \frac{\mathrm{D}J_{12}}{\mathrm{D}t} & \frac{\mathrm{D}J_{13}}{\mathrm{D}t} \\ J_{21} & J_{22} & J_{23} \\ J_{31} & J_{32} & J_{33} \end{vmatrix} + \begin{vmatrix} J_{11} & J_{12} & J_{13} \\ \frac{\mathrm{D}J_{21}}{\mathrm{D}t} & \frac{\mathrm{D}J_{22}}{\mathrm{D}t} & \frac{\mathrm{D}J_{23}}{\mathrm{D}t} \\ J_{31} & J_{32} & J_{33} \end{vmatrix} \\ & + \begin{vmatrix} J_{11} & J_{12} & J_{13} \\ J_{21} & J_{22} & J_{23} \\ \frac{\mathrm{D}J_{31}}{\mathrm{D}t} & \frac{\mathrm{D}J_{32}}{\mathrm{D}t} & \frac{\mathrm{D}J_{33}}{\mathrm{D}t} \end{vmatrix} . \end{aligned} \tag{C.3}$$

The material time derivative of a term $J_{ij}$ present in Expression (C.3) is studied next. Based on Definition (C.1), the following is written:

$$\frac{\mathrm{D}}{\mathrm{D}t}J_{ij}(\vec{X},t) = \frac{\mathrm{D}}{\mathrm{D}t}\frac{\partial \underline{x}_i(t)}{\partial X_j} = \frac{\partial}{\partial X_j}\frac{\mathrm{D}\underline{x}_i(t)}{\mathrm{D}t} = \frac{\partial \underline{v}_i(\vec{\underline{x}}(t),t)}{\partial X_j} = \frac{\partial \underline{x}_k(t)}{\partial X_j}\frac{\partial \underline{v}_i(\vec{\underline{x}}(t),t)}{\partial \underline{x}_k(t)}$$
$$= J_{kj}(\vec{X},t)\frac{\partial \underline{v}_i(\vec{\underline{x}}(t),t)}{\partial \underline{x}_k(t)}. \tag{C.4}$$

The order of the terms in the last term of the first line of Equation (C.4) is discussed in Theorem 1 of Appendix B.

In order to simplify the notations in the following derivation, the independent variables $\vec{\underline{x}}(t)$ and $t$ are dropped. Based on Result (C.4), the first term on the right side of Expression (C.3) takes the following appearance:

$$\begin{vmatrix} \frac{\mathrm{D}J_{11}}{\mathrm{D}t} & \frac{\mathrm{D}J_{12}}{\mathrm{D}t} & \frac{\mathrm{D}J_{13}}{\mathrm{D}t} \\ J_{21} & J_{22} & J_{23} \\ J_{31} & J_{32} & J_{33} \end{vmatrix} = \sum_{k=1}^{3} \begin{vmatrix} J_{k1}\frac{\partial \underline{v}_1}{\partial \underline{x}_k} & J_{k2}\frac{\partial \underline{v}_1}{\partial \underline{x}_k} & J_{k3}\frac{\partial \underline{v}_1}{\partial \underline{x}_k} \\ J_{21} & J_{22} & J_{23} \\ J_{31} & J_{32} & J_{33} \end{vmatrix}$$
$$= \begin{vmatrix} J_{11}\frac{\partial \underline{v}_1}{\partial \underline{x}_1} & J_{12}\frac{\partial \underline{v}_1}{\partial \underline{x}_1} & J_{13}\frac{\partial \underline{v}_1}{\partial \underline{x}_1} \\ J_{21} & J_{22} & J_{23} \\ J_{31} & J_{32} & J_{33} \end{vmatrix} + \begin{vmatrix} J_{21}\frac{\partial \underline{v}_1}{\partial \underline{x}_2} & J_{22}\frac{\partial \underline{v}_1}{\partial \underline{x}_2} & J_{23}\frac{\partial \underline{v}_1}{\partial \underline{x}_2} \\ J_{21} & J_{22} & J_{23} \\ J_{31} & J_{32} & J_{33} \end{vmatrix}$$
$$+ \begin{vmatrix} J_{31}\frac{\partial \underline{v}_1}{\partial \underline{x}_3} & J_{32}\frac{\partial \underline{v}_1}{\partial \underline{x}_3} & J_{33}\frac{\partial \underline{v}_1}{\partial \underline{x}_3} \\ J_{21} & J_{22} & J_{23} \\ J_{31} & J_{32} & J_{33} \end{vmatrix}. \tag{C.5}$$

In arriving at Manipulation (C.5), the summation rule of determinants is exploited. Based on the multiplication rule of determinants by a scalar, the derivative terms of the last two lines of Manipulation (C.5) can be written in the new format. Thus the following result is obtained:

$$\begin{vmatrix} \frac{\mathrm{D}J_{11}}{\mathrm{D}t} & \frac{\mathrm{D}J_{12}}{\mathrm{D}t} & \frac{\mathrm{D}J_{13}}{\mathrm{D}t} \\ J_{21} & J_{22} & J_{23} \\ J_{31} & J_{32} & J_{33} \end{vmatrix} = \begin{vmatrix} J_{11} & J_{12} & J_{13} \\ J_{21} & J_{22} & J_{23} \\ J_{31} & J_{32} & J_{33} \end{vmatrix}\frac{\partial \underline{v}_1}{\partial \underline{x}_1} + \begin{vmatrix} J_{21} & J_{22} & J_{23} \\ J_{21} & J_{22} & J_{23} \\ J_{31} & J_{32} & J_{33} \end{vmatrix}\frac{\partial \underline{v}_1}{\partial \underline{x}_2}$$
$$+ \begin{vmatrix} J_{31} & J_{32} & J_{33} \\ J_{21} & J_{22} & J_{23} \\ J_{31} & J_{32} & J_{33} \end{vmatrix}\frac{\partial \underline{v}_1}{\partial \underline{x}_3}. \tag{C.6}$$

On the right side of Equation (C.6), there are two identical rows in the last two determinants. Therefore, they vanish, and Equation (C.6) reduces to

$$\begin{vmatrix} \frac{\mathrm{D}J_{11}}{\mathrm{D}t} & \frac{\mathrm{D}J_{12}}{\mathrm{D}t} & \frac{\mathrm{D}J_{13}}{\mathrm{D}t} \\ J_{21} & J_{22} & J_{23} \\ J_{31} & J_{32} & J_{33} \end{vmatrix} = \begin{vmatrix} J_{11} & J_{12} & J_{13} \\ J_{21} & J_{22} & J_{23} \\ J_{31} & J_{32} & J_{33} \end{vmatrix}\frac{\partial \underline{v}_1}{\partial \underline{x}_1}. \tag{C.7}$$

Based on Result (C.4), the second term on the right side of Expression (C.3) takes the following appearance:

$$
\begin{aligned}
\begin{vmatrix} J_{11} & J_{12} & J_{13} \\ \frac{\mathrm{D}J_{21}}{\mathrm{D}t} & \frac{\mathrm{D}J_{22}}{\mathrm{D}t} & \frac{\mathrm{D}J_{23}}{\mathrm{D}t} \\ J_{31} & J_{32} & J_{33} \end{vmatrix}
&= \begin{vmatrix} J_{11} & J_{12} & J_{13} \\ J_{k1}\frac{\partial \underline{v}_2}{\partial \underline{x}_k} & J_{k2}\frac{\partial \underline{v}_2}{\partial \underline{x}_k} & J_{k3}\frac{\partial \underline{x}_2}{\partial \underline{x}_k} \\ J_{31} & J_{32} & J_{33} \end{vmatrix} \\
&= \begin{vmatrix} J_{11} & J_{12} & J_{13} \\ J_{11}\frac{\partial \underline{v}_2}{\partial \underline{x}_1} & J_{12}\frac{\partial \underline{v}_2}{\partial \underline{x}_1} & J_{13}\frac{\partial \underline{v}_2}{\partial \underline{x}_1} \\ J_{31} & J_{32} & J_{33} \end{vmatrix}
+ \begin{vmatrix} J_{11} & J_{12} & J_{13} \\ J_{21}\frac{\partial \underline{v}_2}{\partial \underline{x}_2} & J_{22}\frac{\partial \underline{v}_2}{\partial \underline{x}_2} & J_{23}\frac{\partial \underline{x}_2}{\partial \underline{x}_2} \\ J_{31} & J_{32} & J_{33} \end{vmatrix} \\
&+ \begin{vmatrix} J_{11} & J_{12} & J_{13} \\ J_{31}\frac{\partial \underline{v}_2}{\partial \underline{x}_3} & J_{32}\frac{\partial \underline{v}_2}{\partial \underline{x}_3} & J_{33}\frac{\partial \underline{v}_2}{\partial \underline{x}_3} \\ J_{31} & J_{32} & J_{33} \end{vmatrix} .
\end{aligned}
\tag{C.8}
$$

As earlier, the derivatives are moved in front of the determinants. Then the first and third determinans will vanish, since they have two identical rows. Thus the following result is obtained:

$$
\begin{vmatrix} J_{11} & J_{12} & J_{13} \\ \frac{\mathrm{D}J_{21}}{\mathrm{D}t} & \frac{\mathrm{D}J_{22}}{\mathrm{D}t} & \frac{\mathrm{D}J_{23}}{\mathrm{D}t} \\ J_{31} & J_{32} & J_{33} \end{vmatrix}
= \begin{vmatrix} J_{11} & J_{12} & J_{13} \\ J_{21} & J_{22} & J_{23} \\ J_{31} & J_{32} & J_{33} \end{vmatrix} \frac{\partial \underline{v}_2}{\partial \underline{x}_2} . \tag{C.9}
$$

Based on Results (C.7) and (C.9), the last term of Equation (C.3) takes the following appearance:

$$
\begin{vmatrix} J_{11} & J_{12} & J_{13} \\ J_{21} & J_{22} & J_{23} \\ \frac{\mathrm{D}J_{31}}{\mathrm{D}t} & \frac{\mathrm{D}J_{32}}{\mathrm{D}t} & \frac{\mathrm{D}J_{33}}{\mathrm{D}t} \end{vmatrix}
= \begin{vmatrix} J_{11} & J_{12} & J_{13} \\ J_{21} & J_{22} & J_{23} \\ J_{31} & J_{32} & J_{33} \end{vmatrix} \frac{\partial \underline{v}_3}{\partial \underline{x}_3} . \tag{C.10}
$$

Substitution of Terms (C.7), (C.9), and (C.10) into Equation (C.3) yields

$$
\begin{aligned}
\frac{\mathrm{D}}{\mathrm{D}t} J(\vec{X}, t) &= \begin{vmatrix} J_{11} & J_{12} & J_{13} \\ J_{21} & J_{22} & J_{23} \\ J_{31} & J_{32} & J_{33} \end{vmatrix} \frac{\partial \underline{v}_1}{\partial \underline{x}_1} + \begin{vmatrix} J_{11} & J_{12} & J_{13} \\ J_{21} & J_{22} & J_{23} \\ J_{31} & J_{32} & J_{33} \end{vmatrix} \frac{\partial \underline{v}_2}{\partial \underline{x}_2} \\
&+ \begin{vmatrix} J_{11} & J_{12} & J_{13} \\ J_{21} & J_{22} & J_{23} \\ J_{31} & J_{32} & J_{33} \end{vmatrix} \frac{\partial \underline{v}_3}{\partial \underline{x}_3} = \begin{vmatrix} J_{11} & J_{12} & J_{13} \\ J_{21} & J_{22} & J_{23} \\ J_{31} & J_{32} & J_{33} \end{vmatrix} \left( \frac{\partial \underline{v}_1}{\partial \underline{x}_1} + \frac{\partial \underline{v}_2}{\partial \underline{x}_2} + \frac{\partial \underline{v}_3}{\partial \underline{x}_3} \right) \\
&= J(\vec{X}, t)\, \vec{\nabla}(\vec{x}) \cdot \vec{v}(\underline{\vec{x}}(t), t) .
\end{aligned}
\tag{C.11}
$$

Expression (C.11) can be written in the form

$$
\frac{\dot{J}(\vec{X}, t)}{J(\vec{X}, t)} = \vec{\nabla}(\underline{\vec{x}}) \cdot \underline{\vec{v}}(\underline{\vec{x}}(t), t) . \tag{C.12}
$$

# *Appendix D*
# Double-Dot Product between Skew-Symmetric Third-Order Tensor and Symmetric Second-Order Tensor

**Theorem 1:** The following relation holds:

$$\mathbf{c} : \mathbf{h} \equiv \vec{0}\,, \tag{D.1}$$

where $\mathbf{c}$ is a third-order tensor that is skew-symmetric in the last two indices and $\mathbf{h}$ is a symmetric second-order tensor. The notation $\vec{0}$ refers to a zero vector, which means that the values of all the components of the vector $\vec{0}$ vanish.

**Proof:** The vectors and tensors are represented in the orthogonal rectangular Cartesian coordinate system the basis of which is $(\vec{i}_1, \vec{i}_2, \vec{i}_3)$. It is convenient to write vectors and tensors in terms of their components for calculation of dot products. Thus

$$\mathbf{c} = c_{ijk}\, \vec{i}_i\, \vec{i}_j\, \vec{i}_k \qquad \text{and} \qquad \mathbf{h} = h_{st}\, \vec{i}_s\, \vec{i}_t\,. \tag{D.2}$$

The definitions for the skew symmetry in the last two indices of the tensor $\mathbf{c}$ and for the symmetry of the tensor $\mathbf{h}$ read

$$c_{ijk} = -\, c_{ikj} \tag{D.3}$$

and

$$\mathbf{h} = \mathbf{h}^{\mathrm{T}} \qquad \text{or} \qquad h_{ij} = h_{ji}\,. \tag{D.4}$$

Definition (D.3) leads to

$$c_{i11} = c_{i22} = c_{i33} \equiv 0\,. \tag{D.5}$$

Substitution of Forms (D.2) into the left side of Theorem (D.1) yields

$$\begin{aligned} \mathbf{c} : \mathbf{h} &= c_{ijk}\, \vec{i}_i\, \vec{i}_j\, \vec{i}_k : h_{st}\, \vec{i}_s\, \vec{i}_t = (c_{ist}\, h_{st})\, \vec{i}_i \\ &= (\, c_{i11}\, h_{11} + c_{i12}\, h_{12} + c_{i13}\, h_{13} + c_{i21}\, h_{21} + c_{i22}\, h_{22} + c_{i23}\, h_{23}\,)\, \vec{i}_i \\ &+ (\, c_{i31}\, h_{31} + c_{i32}\, h_{32} + c_{i33}\, h_{33}\,)\, \vec{i}_i\,. \end{aligned} \tag{D.6}$$

Based on Properties (D.3)–(D.5) of the tensors **c** and **h**, the following reduced form for Quantity (D.6) is obtained:

$$\mathbf{c}:\mathbf{h} = (\,c_{i12}\,h_{12} + c_{i13}\,h_{13} - c_{i12}\,h_{12} + c_{i23}\,h_{23} - c_{i13}\,h_{13} - c_{i23}\,h_{23}\,)\,\vec{\imath}_i = \vec{0}\,. \tag{D.7}$$

Result (D.7) establishes the result of Theorem 1.

# *Appendix E*
# Legendre Transformations

## E.1 Legendre Transformation

The derivation of the Legendre transformation carried out here follows the presentation given by Lanczos ([42], pp. 161–163), except for the introduction of coefficients $a$ and $b$ in Equations (E.2) and (E.3) and the assumption that the variables are second-order tensors.

The investigation is started with a given scalar-valued function $F$ of $m$ second-order tensorial variables $\mathbf{u}^1, \ldots, \mathbf{u}^m$:

$$F = F(\mathbf{u}^1, \ldots, \mathbf{u}^m)\,. \tag{E.1}$$

A new set of second-order tensors $\boldsymbol{\gamma}^1, \ldots, \boldsymbol{\gamma}^{\mathrm{m}}$ is introduced by means of the following transformation:

$$\boldsymbol{\gamma}^i := a\,\frac{\partial F(\mathbf{u}^1, \ldots, \mathbf{u}^m)}{\partial \mathbf{u}^i}\,, \qquad i = 1, \ldots, m\,, \tag{E.2}$$

where the coefficient $a$ is independent of the variables $\mathbf{u}^{\mathrm{i}}$ and $\boldsymbol{\gamma}^{\mathrm{i}}$ $(i = 1, \ldots, m)$. In the present derivation, variables $\mathbf{u}^{\mathrm{i}}$ and $\boldsymbol{\gamma}^{\mathrm{i}}$ are assumed to be second-order tensors but certainly can be tensors of any order.

The so-called Hessian – the determinant formed by the second partial derivatives of $F$ – is assumed to be different from zero, guaranteeing the independence of the $m$ variables $\boldsymbol{\gamma}^{\mathrm{i}}$. In that case, Equations (E.2) are solvable for $\mathbf{u}^{\mathrm{i}}$ as a function of $\boldsymbol{\gamma}^{\mathrm{i}}$.

The Legendre transformation $\Omega$ of the function $F$ is defined as

$$b\,\Omega \overset{\text{Legendre}}{=} \sum_{i=1}^{m} \mathbf{u}^i : \boldsymbol{\gamma}^i - a\,F\,, \tag{E.3}$$

where $b$ is a coefficient independent of the variables $\mathbf{u}^{\mathrm{i}}$ and $\boldsymbol{\gamma}^{\mathrm{i}}$ $(i = 1, \ldots, m)$. Besides Expression (E.3), Derivatives (E.2) are a vital part of the Legendre transformation. Without Derivatives (E.2), no Legendre transformation can be performed. This means that the Legendre transformation consists of two parts: Derivatives (E.2) and Expression (E.3). The reason for adding "Legendre" above the equals sign in Expression (E.3) is to show that Expression (E.3) is not a normal equation.

The variables $\mathbf{u}^{\mathrm{i}}$ as expressed in terms of tensors $\boldsymbol{\gamma}^{\mathrm{i}}$ [Equation (E.2)] are substituted into Equation (E.3). The function $\Omega$ can then be expressed in terms of the new variables $\boldsymbol{\gamma}^{\mathrm{i}}$ alone as follows:

$$\Omega = \Omega(\boldsymbol{\gamma}^{i}, \ldots, \boldsymbol{\gamma}^{m}) . \tag{E.4}$$

The infinitesimal variation of $b\,\Omega$, produced by arbitrary infinitesimal variations of $\boldsymbol{\gamma}^{\mathrm{i}}$, is considered next. Because the coefficients $a$ and $b$ are independent of the variables $\mathbf{u}^{\mathrm{i}}$ and $\boldsymbol{\gamma}^{\mathrm{i}}$, they are constants with respect to the variation. The combination of Equations (E.3) and (E.4) gives

$$\delta(b\,\Omega) = \sum_{i=1}^{m} b\,\frac{\partial \Omega}{\partial \boldsymbol{\gamma}^{i}} : \delta\boldsymbol{\gamma}^{i} = \sum_{i=1}^{m} (\mathbf{u}^{i} : \delta\boldsymbol{\gamma}^{i} + \delta\mathbf{u}^{i} : \boldsymbol{\gamma}^{i}) - a\,\frac{\partial F}{\partial \mathbf{u}^{i}} : \delta\mathbf{u}^{i} . \tag{E.5}$$

According to Equation $(2.62)_2$, the scalar product of two second-order tensors is commutative:

$$\delta\mathbf{u}^{i} : \boldsymbol{\gamma}^{i} = \boldsymbol{\gamma}^{i} : \delta\mathbf{u}^{i} \qquad i = 1, \ldots m, \text{with no summation on } i\,. \tag{E.6}$$

Applying Equation (E.6), Equation (E.5) can be rewritten in the form

$$\sum_{i=1}^{m} b\,\frac{\partial \Omega}{\partial \boldsymbol{\gamma}^{i}} : \delta\boldsymbol{\gamma}^{i} = \sum_{i=1}^{m} \left[ \mathbf{u}^{i} : \delta\boldsymbol{\gamma}^{i} + \left( \boldsymbol{\gamma}^{i} - a\,\frac{\partial F}{\partial \mathbf{u}^{i}} \right) : \delta\mathbf{u}^{i} \right] . \tag{E.7}$$

Since $\Omega$ is a function of the tensors $\boldsymbol{\gamma}^{\mathrm{i}}$ alone, the variables $\mathbf{u}^{\mathrm{i}}$ should be expressed as functions of the variables $\boldsymbol{\gamma}^{\mathrm{i}}$. This expresses the variations of the tensors $\mathbf{u}^{\mathrm{i}}$ in terms of the variations of the variables $\boldsymbol{\gamma}^{\mathrm{i}}$. However, examination of Equation (E.7) shows that this elimination is rendered unnecessary by the fact that the coefficient of the variation $\delta\mathbf{u}^{\mathrm{i}}$ is automatically zero, since the variables $\boldsymbol{\gamma}^{\mathrm{i}}$ are defined according to Equation (E.2). However, Equation (E.7) at once gives

$$\mathbf{u}^{i} = b\,\frac{\partial \Omega(\boldsymbol{\gamma}^{1}, \ldots, \boldsymbol{\gamma}^{m})}{\partial \boldsymbol{\gamma}^{i}} , \qquad i = 1, \ldots, m\,. \tag{E.8}$$

## E.2 Legendre Partial Transformation

In this work, the Legendre transformation containing both active and passive variables is referred to as the Legendre partial transformation. Here only details that differ from those of Section E.1 are presented. The study in this section follows the concept of Lanczos ([42], pp. 163 and 164).

In this case, the scalar-valued function $F$ is assumed to be a function of two independent sets of tensorial variables, namely, $\mathbf{u}^1, \ldots, \mathbf{u}^m$ and $\mathbf{w}^1, \ldots, \mathbf{w}^n$:

$$F = F(\mathbf{u}^{1}, \ldots, \mathbf{u}^{m}, \mathbf{w}^{1}, \ldots, \mathbf{w}^{n}) . \tag{E.9}$$

The new independent set of second-order tensorial variables $\boldsymbol{\gamma}^1, \ldots, \boldsymbol{\gamma}^m$ is assumed to be defined by

$$\boldsymbol{\gamma}^i := a \, \frac{\partial F(\mathbf{u}^1, \ldots, \mathbf{u}^m, \, \mathbf{w}^1, \ldots, \mathbf{w}^n)}{\partial \mathbf{u}^i} \,, \qquad i = 1, \ldots, m \,, \tag{E.10}$$

where $a$ is a coefficient independent of $\mathbf{u}^i$, $\mathbf{w}^j$, and $\boldsymbol{\gamma}^i$ ($i = 1, \ldots, m$ and $j = 1, \ldots, n$). The variables $\mathbf{u}^i$ are called the active variables and the variables $\mathbf{w}^j$ are called the passive variables of the transformation. A new function $\Omega$, called the Legendre partial transformation, is introduced. It is defined by

$$b \, \Omega(\boldsymbol{\gamma}^1, \ldots, \boldsymbol{\gamma}^m, \, \mathbf{w}^1, \ldots, \mathbf{w}^n) \overset{\text{Legendre}}{=} \sum_{i=1}^{m} \boldsymbol{\gamma}^i : \mathbf{u}^i - a \, F(\mathbf{u}^1, \ldots, \mathbf{u}^m, \mathbf{w}^1, \ldots, \mathbf{w}^n) \,. \tag{E.11}$$

The variables $\mathbf{w}^j$ and $\boldsymbol{\gamma}^i$ are given arbitrary variations $\delta \mathbf{w}^j$ and $\delta \boldsymbol{\gamma}^i$. Besides Expression (E.11), Derivatives (E.10) are a vital part of the Legendre transformation. Without Derivatives (E.10), no Legendre transformation can be performed. This means that the Legendre transformation consists of two parts: Derivatives (E.10) and Expression (E.11). The reason for adding "Legendre" above the equals sign in Expression (E.11) is to show that Expression (E.11) is not a normal equation.

Thus Expression (E.11) gives

$$\begin{aligned} \delta b \, \Omega(\boldsymbol{\gamma}^1, \ldots, \boldsymbol{\gamma}^m, \mathbf{w}^1, \, \ldots, \mathbf{w}^n) &= \sum_{i=1}^{m} b \, \frac{\partial \Omega}{\partial \boldsymbol{\gamma}^i} : \delta \boldsymbol{\gamma}^i + \sum_{j=1}^{n} b \, \frac{\partial \Omega}{\partial \mathbf{w}^j} : \delta \mathbf{w}^j \\ &= \sum_{i=1}^{m} (\boldsymbol{\gamma}^i : \delta \mathbf{u}^i + \delta \boldsymbol{\gamma}^i : \mathbf{u}^i) - \sum_{i=1}^{m} a \, \frac{\partial F}{\partial \mathbf{u}^i} : \delta \mathbf{u}^i - \sum_{j=1}^{n} a \, \frac{\partial F}{\partial \mathbf{w}^j} : \delta \mathrm{w}^j \,. \end{aligned} \tag{E.12}$$

Substitution of Derivatives (E.10) into Expression (E.11) yields

$$\begin{aligned} \sum_{i=1}^{m} b \, \frac{\partial \Omega}{\partial \boldsymbol{\gamma}^i} : \delta \boldsymbol{\gamma}^i + \sum_{i=1}^{n} b \, \frac{\partial \Omega}{\partial \mathbf{w}^j} : \delta \mathbf{w}^j &= \sum_{i=1}^{m} \left( \boldsymbol{\gamma}^i - a \, \frac{\partial F}{\partial \mathbf{u}^i} \right) : \delta \mathbf{u}^i \\ &\quad + \sum_{i=1}^{m} \mathbf{u}^i : \delta \boldsymbol{\gamma}^i - \sum_{j=1}^{n} a \, \frac{\partial F}{\partial \mathbf{w}^j} : \delta \mathbf{w}^j \,. \end{aligned} \tag{E.13}$$

According to Equation (E.10), the first term on the right side of Equation (E.13) vanishes, giving the following equations:

$$\mathbf{u}^i = b \, \frac{\partial \Omega(\boldsymbol{\gamma}^1, \ldots, \boldsymbol{\gamma}^m, \mathbf{w}^1, \ldots, \mathbf{w}^n)}{\partial \boldsymbol{\gamma}^i} \,, \qquad i = 1, \ldots, m \tag{E.14}$$

and

$$a \, \frac{\partial F(\mathbf{u}^1, \ldots, \mathbf{u}^m, \, \mathrm{w}^1, \ldots, \mathrm{w}^n)}{\partial \mathrm{w}^j} = -b \, \frac{\partial \Omega(\boldsymbol{\gamma}^1, \ldots, \boldsymbol{\gamma}^m, \, \mathrm{w}^1, \ldots, \mathrm{w}^n)}{\partial \mathrm{w}^j} \tag{E.15}$$

for $j = 1, \ldots, n$.

## E.3 Legendre Transformation of a Homogeneous Function

The Legendre transformation of a homogeneous function is investigated here. It is shown that the Legendre transformation of a homogeneous function is a homogeneous function. This does not hold if the original function is a homogeneous function of degree 1.

A scalar-valued function $F$ of $m$ different tensorial variables $\mathbf{u}^1, \cdots, \mathbf{u}^m$ is studied. Function $F$, expressed as

$$F = F(\mathbf{u}^1, \ldots, \mathbf{u}^m)\,, \tag{E.16}$$

is assumed to be a homogeneous function of degree $\omega$, and therefore it satisfies the following definition and equation:

$$F(k\,\mathbf{u}^1, \ldots, k\,\mathbf{u}^m) := k^{\omega}\, F(\mathbf{u}^1, \ldots, \mathbf{u}^m) \tag{E.17}$$

and

$$\omega\, F(\mathbf{u}^1, \ldots, \mathbf{u}^m) = \left( \frac{\partial F}{\partial \mathbf{u}^1} : \mathbf{u}^1 + \cdots + \frac{\partial F}{\partial \mathbf{u}^m} : \mathbf{u}^m \right)\,, \tag{E.18}$$

where $k$ is an arbitrary positive real number (see, e.g., Widder [117], pp. 19 and 20).

Next, $m$ second-order tensors $\boldsymbol{\gamma}^1, \ldots, \boldsymbol{\gamma}^m$ are introduced by defining

$$\boldsymbol{\gamma}^i := a\, \frac{\partial F}{\partial \mathbf{u}^i}\,, \qquad i = 1, \ldots, m\,, \tag{E.19}$$

where $a$ is an arbitrary coefficient independent of both $\mathbf{u}^i$ and $\boldsymbol{\gamma}^i$ $(i = 1, \ldots, m)$.

The Legendre transformation $\Omega$ of the function $F$ has two parts, which are the derivative(s) [cf. Derivatives (E.2)] copied to Derivatives (E.19) and Expression (E.3)], that is,

$$b\, \Omega(\boldsymbol{\gamma}^1, \ldots, \boldsymbol{\gamma}^m) \overset{\text{Legendre}}{=} \sum_{i=1}^{m} \mathbf{u}^i : \boldsymbol{\gamma}^i - a\, F(\mathbf{u}^1, \ldots, \mathbf{u}^m)\,, \tag{E.20}$$

where the coefficient $b$ does not depend on the tensorial variables $\mathbf{u}^i$ and $\boldsymbol{\gamma}^i$ $(i = 1, \ldots, m)$.

Substitution of Definition (E.19) into Expression (E.18) gives

$$\omega\, F(\mathbf{u}^1, \ldots, \mathbf{u}^m) = \frac{1}{a} \sum_{i=1}^{m} \mathbf{u}^i : \boldsymbol{\gamma}^i\,, \tag{E.21}$$

which yields

$$\sum_{i=1}^{m} \mathbf{u}^i : \boldsymbol{\gamma}^i = a\, \omega\, F(\mathbf{u}^1, \ldots, \mathbf{u}^m)\,. \tag{E.22}$$

Substituting Equation (E.22) into Legendre Transformation (E.20) gives

$$F(\mathbf{u}^1,\ldots,\mathbf{u}^m) = c\,\Omega(\boldsymbol{\gamma}^1,\ldots,\boldsymbol{\gamma}^m)\,, \tag{E.23}$$

where the coefficient $c$ is

$$c = \frac{b}{a\,\omega - a}\,. \tag{E.24}$$

Once again, the definition of variables $\boldsymbol{\gamma}^i$ is used. Thus the variables $\boldsymbol{\gamma}^i$ in Equation (E.19) are substituted into the arguments of $\Omega$ on the right side of Equation (E.23), and the following equation is obtained:

$$F(\mathbf{u}^1,\ldots,\mathbf{u}^m) = c\,\Omega\left(a\,\frac{\partial F(\mathbf{u}^1,\ldots,\mathbf{u}^m)}{\partial \mathbf{u}^1},\ldots,a\,\frac{\partial F(\mathbf{u}^1,\ldots,\mathbf{u}^m)}{\partial \mathbf{u}^m}\right). \tag{E.25}$$

If the variables in Equation (E.25) were changed by replacing $\mathbf{u}^i$ by $k\,\mathbf{u}^i$, Equation (E.25) would take the following form:

$$F(k\,\mathbf{u}^1,\ldots,k\,\mathbf{u}^m) = c\,\Omega\left(a\,\frac{\partial F(k\,\mathbf{u}^1,\ldots,k\,\mathbf{u}^m)}{\partial (k\,\mathbf{u}^1)},\ldots,a\,\frac{\partial F(k\,\mathbf{u}^1,\ldots,k\,\mathbf{u}^m)}{\partial (k\,\mathbf{u}^m)}\right). \tag{E.26}$$

The definition of a homogeneous function given by Definition (E.17) allows Equation (E.26) to be written in the form

$$k^\omega\,F(\mathbf{u}^1,\ldots,\mathbf{u}^m) = c\,\Omega\left(a\,\frac{\partial\left[k^\omega\,F(\mathbf{u}^1,\ldots,\mathbf{u}^m)\right]}{\partial (k\,\mathbf{u}^1)},\ldots,a\,\frac{\partial\left[k^\omega\,F(\mathbf{u}^1,\ldots,\mathbf{u}^m)\right]}{\partial (k\,\mathbf{u}^m)}\right). \tag{E.27}$$

For the partial derivatives of the arguments of the Legendre transformation $\Omega$ on the right side of Equation (E.27), the following equations hold:

$$k^{\omega-1}\,\frac{\partial F(\mathbf{u}^1,\ldots,\mathbf{u}^m)}{\partial \mathbf{u}^i} = \frac{\partial\left[k^\omega\,F(\mathbf{u}^1,\ldots,\mathbf{u}^m)\right]}{\partial (k\,\mathbf{u}^i)}\,, \qquad i = 1,\ldots,m\,. \tag{E.28}$$

Substitution of Definition (E.19) into the left side of Equation (E.28) yields

$$\frac{1}{a}\,k^{\omega-1}\,\boldsymbol{\gamma}^i = \frac{\partial\left[k^\omega\,F(\mathbf{u}^1,\ldots,\mathbf{u}^m)\right]}{\partial (k\,\mathbf{u}^i)}\,, \qquad i = 1,\ldots,m\,. \tag{E.29}$$

Substitution of Equation (E.23) into the left side of Equation (E.27) and Equation (E.29) into the right side of Equation (E.27) gives the following equality:

$$k^\omega\,\Omega(\boldsymbol{\gamma}^1,\ldots,\boldsymbol{\gamma}^m) = \Omega(k^{\omega-1}\boldsymbol{\gamma}^1,\ldots,k^{\omega-1}\boldsymbol{\gamma}^m)\,. \tag{E.30}$$

Changing the variables by replacing $k^{\omega-1}$ with $t$ allows Equation (E.30) to be written in the form

$$t^{\frac{\omega}{\omega-1}}\,\Omega(\boldsymbol{\gamma}^1,\ldots,\boldsymbol{\gamma}^m) = \Omega(t\boldsymbol{\gamma}^1,\ldots,t\boldsymbol{\gamma}^m)\,. \tag{E.31}$$

Equation (E.31) therefore shows the Legendre transformation $\Omega(\boldsymbol{\gamma}^1, \ldots, \boldsymbol{\gamma}^m)$ to be a homogeneous function of degree $\omega/(\omega-1)$, where $\omega$ is the degree of the original function $F$. This does not hold for the case $\omega = 1$, as can be seen in Equations (E.24) and (E.31).

If the original function $F$ were a homogeneous function of degree $\mu = 1/\kappa$, the function $\Omega$ would be a homogeneous function of degree $1/(1-\mu)$. As earlier, this does not hold for the case $\mu = 1$.

# *Appendix F*
# Heat Equation for Solids in Terms of the Specific Gibbs Free Energy

This appendix derives the heat equation for a case in which the material model is expressed by the specific Gibbs free energy $g$. Definition (6.179) reads

$$\rho_0\, g\big(\boldsymbol{\sigma}, \boldsymbol{\alpha}, T, h(\vec{X})\big) \overset{\text{Legendre}}{=} \boldsymbol{\sigma} : (\boldsymbol{\varepsilon} - \boldsymbol{\varepsilon}^{\mathrm{i}}) - \rho_0\, \psi\big(\boldsymbol{\varepsilon} - \boldsymbol{\varepsilon}^{\mathrm{i}}, \boldsymbol{\alpha}, T, h(\vec{X})\big)\,,$$
$$\text{with} \qquad \boldsymbol{\sigma} = \rho_0\, \frac{\partial \psi(\boldsymbol{\varepsilon} - \boldsymbol{\varepsilon}^{\mathrm{i}}, \boldsymbol{\alpha}, T, h(\vec{X}))}{\partial(\boldsymbol{\varepsilon} - \boldsymbol{\varepsilon}^{\mathrm{i}})}\,. \tag{F.1}$$

Definition $(\mathrm{F}.1)_1$ gives

$$\rho_0\, \psi = \boldsymbol{\sigma} : (\boldsymbol{\varepsilon} - \boldsymbol{\varepsilon}^{\mathrm{i}}) - \rho_0\, g\,. \tag{F.2}$$

Based on Equation (F.2), the rate $\rho_0\, \dot{\psi}$ is

$$\rho_0\, \dot{\psi} = \dot{\boldsymbol{\sigma}} : (\boldsymbol{\varepsilon} - \boldsymbol{\varepsilon}^{\mathrm{i}}) + \boldsymbol{\sigma} : (\dot{\boldsymbol{\varepsilon}} - \dot{\boldsymbol{\varepsilon}}^{\mathrm{i}}) - \rho_0\, \dot{g}\,. \tag{F.3}$$

Equation (6.197) is recalled:

$$\rho_0\, \dot{\psi} + \rho_0\, \dot{T}\, s + \rho_0\, T\, \dot{s} - \boldsymbol{\sigma} : \dot{\boldsymbol{\varepsilon}} - \rho_0\, r + \vec{\nabla} \cdot \vec{q} = 0\,. \tag{F.4}$$

Substitution of the rate $\rho_0\, \dot{\psi}$ obtained from Equation (F.3) into Equation (F.4) gives

$$\dot{\boldsymbol{\sigma}} : (\boldsymbol{\varepsilon} - \boldsymbol{\varepsilon}^{\mathrm{i}}) + \boldsymbol{\sigma} : (\dot{\boldsymbol{\varepsilon}} - \dot{\boldsymbol{\varepsilon}}^{\mathrm{i}}) - \rho_0\, \dot{g} + \rho_0\, \dot{T}\, s + \rho_0\, T\, \dot{s} - \boldsymbol{\sigma} : \dot{\boldsymbol{\varepsilon}} - \rho_0\, r + \vec{\nabla} \cdot \vec{q} = 0\,. \tag{F.5}$$

Based on the chain rule, the material time derivative of the specific Gibbs free energy $g(\boldsymbol{\sigma}, \boldsymbol{\alpha}, T, h(\vec{X}))$ takes the following appearance:

$$\dot{g}(\boldsymbol{\sigma}, \boldsymbol{\alpha}, T, h(\vec{X})) = \frac{\partial g}{\partial \boldsymbol{\sigma}} : \dot{\boldsymbol{\sigma}} + \frac{\partial g}{\partial \boldsymbol{\alpha}} : \dot{\boldsymbol{\alpha}} + \frac{\partial g}{\partial T}\, \dot{T}\,. \tag{F.6}$$

Equation (F.6) exploits the fact that $\dot{h}(\vec{X})$ vanishes. State Equations (6.180) and (6.181) are

$$s = \frac{\partial g}{\partial T}\,, \qquad \boldsymbol{\varepsilon} - \boldsymbol{\varepsilon}^{\mathrm{i}} = \rho_0\, \frac{\partial g}{\partial \boldsymbol{\sigma}}\,, \qquad \text{and} \qquad \boldsymbol{\beta} = \rho_0\, \frac{\partial g}{\partial \boldsymbol{\alpha}}\,. \tag{F.7}$$

Derivative (F.6) is multiplied by the density $\rho_0$, which yields

$$\rho_0\,\dot{g}(\boldsymbol{\sigma},\boldsymbol{\alpha},T,h(\vec{X})) = \rho_0\,\frac{\partial g}{\partial \boldsymbol{\sigma}}:\dot{\boldsymbol{\sigma}} + \rho_0\,\frac{\partial g}{\partial \boldsymbol{\alpha}}:\dot{\boldsymbol{\alpha}} + \rho_0\,\frac{\partial g}{\partial T}\,\dot{T}\,. \tag{F.8}$$

Substitution of State Equations (F.7) into Derivative (F.8) gives

$$\rho_0\,\dot{g} = (\boldsymbol{\varepsilon} - \boldsymbol{\varepsilon}^{\mathrm{i}}):\dot{\boldsymbol{\sigma}} + \boldsymbol{\beta}:\dot{\boldsymbol{\alpha}} + \rho_0\,s\,\dot{T}\,. \tag{F.9}$$

State Equation $(\text{F.7})_1$ and Rate (F.6) give

$$\dot{s} = \frac{\partial \dot{g}}{\partial T} = \frac{\partial^2 g}{\partial T\,\partial \boldsymbol{\sigma}}:\dot{\boldsymbol{\sigma}} + \frac{\partial^2 g}{\partial T\,\partial \boldsymbol{\alpha}}:\dot{\boldsymbol{\alpha}} + \frac{\partial^2 g}{\partial T^2}\,\dot{T}\,. \tag{F.10}$$

Substitution of Derivatives (F.9) and (F.10) into Equation (F.5) yields

$$\begin{aligned}&\dot{\boldsymbol{\sigma}}:(\boldsymbol{\varepsilon} - \boldsymbol{\varepsilon}^{\mathrm{i}}) + \boldsymbol{\sigma}:(\dot{\boldsymbol{\varepsilon}} - \dot{\boldsymbol{\varepsilon}}^{\mathrm{i}}) - (\boldsymbol{\varepsilon} - \boldsymbol{\varepsilon}^{\mathrm{i}}):\dot{\boldsymbol{\sigma}} - \boldsymbol{\beta}:\dot{\boldsymbol{\alpha}} - \rho_0\,s\,\dot{T} + \rho_0\,\dot{T}\,s \\ &\quad + \rho_0\,T\left[\frac{\partial^2 g}{\partial T\,\partial \boldsymbol{\sigma}}:\dot{\boldsymbol{\sigma}} + \frac{\partial^2 g}{\partial T\,\partial \boldsymbol{\alpha}}:\dot{\boldsymbol{\alpha}} + \frac{\partial^2 g}{\partial T^2}\,\dot{T}\right] - \boldsymbol{\sigma}:\dot{\boldsymbol{\varepsilon}} - \rho_0\,r + \vec{\nabla}\cdot\vec{q} = 0\,.\end{aligned} \tag{F.11}$$

Equation (F.11) reduces to

$$\vec{\nabla}\cdot\vec{q} = \boldsymbol{\sigma}:\dot{\boldsymbol{\varepsilon}}^{\mathrm{i}} + \boldsymbol{\beta}:\dot{\boldsymbol{\alpha}} + \rho_0\,r - \rho_0\,T\left[\frac{\partial^2 g}{\partial T\,\partial \boldsymbol{\sigma}}:\dot{\boldsymbol{\sigma}} + \frac{\partial^2 g}{\partial T\,\partial \boldsymbol{\alpha}}:\dot{\boldsymbol{\alpha}} + \frac{\partial^2 g}{\partial T^2}\,\dot{T}\right]. \tag{F.12}$$

Equation (F.12) is the heat equation when the material model is expressed by the specific Gibbs free energy $g$.

# *Appendix G* Clausius–Duhem Inequality When the Material Model Is Expressed by the Specific Helmholtz Free Energy

In this appendix, we derive the Clausius–Duhem inequality when the material model is expressed by the specific Helmholtz free energy. Inequality (6.263) reads

$$\boldsymbol{\sigma} : \dot{\boldsymbol{\varepsilon}} - \rho_0\, \dot{u} + \rho_0\, T\, \dot{s} - \frac{\vec{\nabla} T}{T} \cdot \vec{q} \geq 0\,. \tag{G.1}$$

Definition (6.171) gives the specific Helmholtz free energy $\psi(\boldsymbol{\varepsilon} - \boldsymbol{\varepsilon}^{\mathrm{i}}, \boldsymbol{\alpha}, T, h(\vec{X}))$. It has the following appearance:

$$\begin{aligned} \psi(\boldsymbol{\varepsilon} - \boldsymbol{\varepsilon}^{\mathrm{i}}, \boldsymbol{\alpha}, T, h(\vec{X})) &\overset{\text{Legendre}}{=} u(\boldsymbol{\varepsilon} - \boldsymbol{\varepsilon}^{\mathrm{i}}, \boldsymbol{\alpha}, s, h(\vec{X})) - T\, s\,, \\ \text{with} \quad T &= \frac{\partial u(\boldsymbol{\varepsilon} - \boldsymbol{\varepsilon}^{\mathrm{i}}, \boldsymbol{\alpha}, s, h(\vec{X}))}{\partial s}\,, \end{aligned} \tag{G.2}$$

which yields

$$\dot{u} = \dot{\psi} + \dot{T}\, s + T\, \dot{s}\,. \tag{G.3}$$

Substitution of Equation (G.3) into Equation (G.1) yields

$$\boldsymbol{\sigma} : \dot{\boldsymbol{\varepsilon}} - \rho_0\, \dot{\psi} - \rho_0\, \dot{T}\, s - \frac{\vec{\nabla} T}{T} \cdot \vec{q} \geq 0\,. \tag{G.4}$$

Based on the chain rule, the material time derivative of the specific Helmholtz free energy $\psi(\varepsilon - \boldsymbol{\varepsilon}^{\mathrm{i}}, \boldsymbol{\alpha}, T, h(\vec{X}))$ takes the following appearance:

$$\dot{\psi}(\boldsymbol{\varepsilon} - \boldsymbol{\varepsilon}^{\mathrm{i}}, \boldsymbol{\alpha}, T, h(\vec{X})) = \frac{\partial \psi}{\partial(\boldsymbol{\varepsilon} - \boldsymbol{\varepsilon}^{\mathrm{i}})} : (\dot{\boldsymbol{\varepsilon}} - \dot{\boldsymbol{\varepsilon}}^{\mathrm{i}}) + \frac{\partial \psi}{\partial \boldsymbol{\alpha}} : \dot{\boldsymbol{\alpha}} + \frac{\partial \psi}{\partial T}\, \dot{T}\,. \tag{G.5}$$

Derivative (G.5) exploits the fact that $\dot{h}(\vec{X})$ vanishes.

State Equations (6.172) and (6.174) are recalled:

$$s = -\frac{\partial \psi}{\partial T}, \qquad \boldsymbol{\sigma} = \rho_0 \frac{\partial \psi}{\partial (\boldsymbol{\varepsilon} - \boldsymbol{\varepsilon}^{\mathrm{i}})}, \qquad \text{and} \qquad \boldsymbol{\beta} = -\rho_0 \frac{\partial \psi}{\partial \boldsymbol{\alpha}}. \tag{G.6}$$

Derivative (G.5) is multiplied by the density $\rho_0$, and State Equations (G.6) are substituted into the obtained form. This gives

$$\rho_0 \dot{\psi} = \boldsymbol{\sigma} : (\dot{\boldsymbol{\varepsilon}} - \dot{\boldsymbol{\varepsilon}}^{\mathrm{i}}) - \boldsymbol{\beta} : \dot{\boldsymbol{\alpha}} - \rho_0 \, s \, \dot{T}. \tag{G.7}$$

Substituting Derivative (G.7) into Expression (G.4) gives

$$\boldsymbol{\sigma} : \dot{\boldsymbol{\varepsilon}}^{i} + \boldsymbol{\beta} : \dot{\boldsymbol{\alpha}} - \frac{\vec{\nabla} T}{T} \cdot \vec{q} \geq 0. \tag{G.8}$$

Inequality (G.8) is the Clausius–Duhem inequality.

# *Appendix H* Normality Rules for the Dissipation $\Phi$

## H.1 Normality Rule for the Nonseparated Dissipation $\Phi$

In this section of the appendix, we evaluate the principle of maximum dissipation in a case in which the dissipation $\Phi$ is not separated into mechanical and thermal parts. At the end of this appendix, the results are compared with those obtained when the dissipation $\Phi$ is separated into mechanical part $\Phi_{\text{mech}}$ and thermal part $\Phi_{\text{ther}}$. The derivation carried out here is very short and requires information from Chapter 6.

The nonseparated form of the Clausius–Duhem inequality is given by Equation (6.263):

$$\Phi = \boldsymbol{\sigma} : \dot{\boldsymbol{\varepsilon}}^{\mathrm{i}} + \boldsymbol{\beta} : \dot{\boldsymbol{\alpha}} - \frac{\vec{\nabla} T}{T} \cdot \vec{q} \quad (\geq 0) \,. \tag{H.1}$$

By following the concept given by Section 6.7.1, the principle of maximum rate of dissipation for nonseparated dissipation $\Phi$ is written in the following mathematical form:

maximize with respect to the fluxes $(\dot{\boldsymbol{\varepsilon}}^{\mathrm{i}}, \dot{\boldsymbol{\alpha}}, \vec{q})$,

$$\Phi = \boldsymbol{\sigma} : \dot{\boldsymbol{\varepsilon}}^{\mathrm{i}} + \boldsymbol{\beta} : \dot{\boldsymbol{\alpha}} - \frac{\vec{\nabla} T}{T} \cdot \vec{q} \,, \tag{H.2}$$

subject to

$$\tau = \rho_0 \, \varphi(\dot{\boldsymbol{\varepsilon}}^{\mathrm{i}}, \dot{\boldsymbol{\alpha}}, \vec{q}; \boldsymbol{\varepsilon}, -\boldsymbol{\varepsilon}^{\mathrm{i}}, \boldsymbol{\alpha}, T, h) - \Phi(\dot{\boldsymbol{\varepsilon}}^{\mathrm{i}}, \dot{\boldsymbol{\alpha}}, \vec{q}, \boldsymbol{\sigma}, \boldsymbol{\beta}, \vec{\nabla} T / T) = 0 \,, \tag{H.3}$$

where $\tau = 0$ is a constraint and $\varphi$ is the specific dissipation function.

Both $\Phi$ and $\tau$ are assumed to have at least continuous second partial derivatives with respect to the arguments $(\dot{\boldsymbol{\varepsilon}}^{\mathrm{i}}, \dot{\boldsymbol{\alpha}}, \vec{q})$. It should be pointed out that also Inequality (H.1) must be satisfied. Applying the first-order sufficient condition (see Luenberger [55], p. 225) for the point $(\dot{\boldsymbol{\varepsilon}}^{\mathrm{i}}, \dot{\boldsymbol{\alpha}}, \vec{q})$ to be a local maximum is

$$\frac{\partial}{\partial \dot{\boldsymbol{\varepsilon}}^{\mathrm{i}}} (\Phi + \lambda \tau) = \mathbf{0} \qquad \text{and} \qquad \frac{\partial}{\partial \dot{\boldsymbol{\alpha}}} (\Phi + \lambda \tau) = \mathbf{0} \,, \tag{H.4}$$

and furthermore,

$$\frac{\partial}{\partial \vec{q}}(\Phi + \lambda\,\tau) = \vec{0} \qquad \text{and} \qquad \tau = 0\,, \tag{H.5}$$

where $\lambda$ is the Lagrange multiplier. As mentioned by Arfken ([7], p. 946), the method based on Lagrange multipliers will fail if in Expressions (H.4) and $(\text{H.5})_1$ the coefficients of $\lambda$ vanish at the extremum. Therefore also special points where

$$\frac{\partial \tau}{\partial \dot{\boldsymbol{\varepsilon}}^{\mathrm{i}}} = \mathbf{0}\,, \qquad \frac{\partial \tau}{\partial \dot{\boldsymbol{\alpha}}} = \mathbf{0}\,, \qquad \text{and} \qquad \frac{\partial \tau}{\partial \vec{q}} = \vec{0} \tag{H.6}$$

must be studied.

The preceding indicates that there are two different cases for evaluation of the local maximum: utilization of Expressions (H.4) and (H.5), referred to as case A, and the special case described by Expression (H.6), referred to as case B.

The derivation continues by following the same steps that were taken in Section 6.7.1. Therefore only the final result is sketched here. As a result, the following normality rules were achieved:

$$\boldsymbol{\sigma} = \mu\,\rho_0\,\frac{\partial \varphi(\dot{\boldsymbol{\varepsilon}}^{\mathrm{i}}, \dot{\boldsymbol{\alpha}}, \vec{q};\, \boldsymbol{\varepsilon} - \boldsymbol{\varepsilon}^{\mathrm{i}}, \boldsymbol{\alpha}, T, h)}{\partial \dot{\boldsymbol{\varepsilon}}^{\mathrm{i}}} \tag{H.7}$$

and

$$\boldsymbol{\beta} = \mu\,\rho_0\,\frac{\partial \varphi\,(\dot{\boldsymbol{\varepsilon}}^{\mathrm{i}}, \dot{\boldsymbol{\alpha}}, \vec{q};\, \boldsymbol{\varepsilon} - \boldsymbol{\varepsilon}^{\mathrm{i}}, \boldsymbol{\alpha}, T, h)}{\partial \dot{\boldsymbol{\alpha}}}\,, \tag{H.8}$$

and, finally,

$$-\frac{\vec{\nabla} T}{T} = \mu\,\rho_0\,\frac{\partial \varphi(\dot{\boldsymbol{\varepsilon}}^{\mathrm{i}}, \dot{\boldsymbol{\alpha}}, \vec{q};\, \boldsymbol{\epsilon} - \boldsymbol{\varepsilon}^{\mathrm{i}}, \boldsymbol{\alpha}, T, h)}{\partial \vec{q}}\,. \tag{H.9}$$

The specific dissipation function $\varphi$ has to satisfy the following condition:

$$\varphi(\dot{\boldsymbol{\varepsilon}}^{\mathrm{i}}, \dot{\boldsymbol{\alpha}}, \vec{q};\, \boldsymbol{\varepsilon} - \boldsymbol{\varepsilon}^{\mathrm{i}}, \boldsymbol{\alpha}, T, h) = \mu\left(\frac{\partial \varphi}{\partial \dot{\boldsymbol{\varepsilon}}^{\mathrm{i}}} : \dot{\boldsymbol{\varepsilon}}^{\mathrm{i}} + \frac{\partial \varphi}{\partial \dot{\boldsymbol{\alpha}}} : \dot{\boldsymbol{\alpha}} + \frac{\partial \varphi}{\partial \vec{q}} \cdot \vec{q}\right)\,. \tag{H.10}$$

The first-order sufficient condition for the point $(\dot{\boldsymbol{\varepsilon}}^{\mathrm{i}}, \dot{\boldsymbol{\alpha}}, \vec{q})$ to be a local maximum is that Equations (H.7)–(H.9) hold and that the specific dissipation function $\varphi$ is a homogeneous function of degree $1/\mu$. The latter property is obtained if the coefficient $\mu$ in Expression (H.10) is a constant. If the multiplier $\mu$ is not a constant but $\mu = \mu(\boldsymbol{\varepsilon} - \boldsymbol{\varepsilon}^{\mathrm{i}}, \boldsymbol{\alpha}, T, h)$, the specific dissipation function $\varphi$ is not a homogeneous function and the value for $\mu$ is obtained from Equation (H.10).

## H.2 Normality Rule for the Separated Dissipation $\Phi$

This part of the appendix is dedicated to two interpretations for the principle of maximum dissipation in a case in which the dissipation $\Phi$ is separated into mechanical part $\Phi_{\text{mech}}$ and thermal part $\Phi_{\text{ther}}$. It will be shown that these interpretations lead to the same result. Since this appendix utilizes several results derived in Sections 6.7.1 and 6.7.3, the reader is referred to these sections before reading the present appendix.

The Clausius–Duhem inequalities for the separated dissipation $\Phi$ are given by Expression (6.264):

$$\Phi_{\text{mech}} = \boldsymbol{\sigma} : \dot{\boldsymbol{\varepsilon}}^{\text{i}} + \boldsymbol{\beta} : \dot{\boldsymbol{\alpha}} \quad (\geq 0) \qquad \text{and} \qquad \Phi_{\text{ther}} = -\frac{\vec{\nabla} T}{T} \cdot \vec{q} \ (\geq 0) \,. \tag{H.11}$$

By following Separation (H.11), the specific dissipation function $\varphi$ is also separated, as follows:

$$\varphi(\dot{\boldsymbol{\varepsilon}}^{\text{i}}, \dot{\boldsymbol{\alpha}}, \vec{q}; \boldsymbol{\varepsilon} - \boldsymbol{\varepsilon}^{\text{i}}, \boldsymbol{\alpha}, T, h) = \varphi_{\text{mech}}(\dot{\boldsymbol{\varepsilon}}^{\text{i}}, \dot{\boldsymbol{\alpha}}; \boldsymbol{\varepsilon} - \boldsymbol{\varepsilon}^{\text{i}}, \boldsymbol{\alpha}, T, h) + \varphi_{\text{ther}}(\vec{q}; \boldsymbol{\varepsilon} - \boldsymbol{\varepsilon}^{\text{i}}, \boldsymbol{\alpha}, T, h). \tag{H.12}$$

The principle of maximum dissipation is written in the following two mathematical forms:

Case A: Consider two different problems, which are as follows:

maximize with respect to the fluxes, $(\dot{\boldsymbol{\varepsilon}}^{\text{i}}, \dot{\boldsymbol{\alpha}})$,

$$\Phi_{\text{mech}}(\dot{\boldsymbol{\varepsilon}}^{\text{i}}, \dot{\boldsymbol{\alpha}}, \boldsymbol{\sigma}, \boldsymbol{\beta}) = \boldsymbol{\sigma} : \dot{\boldsymbol{\varepsilon}}^{\text{i}} + \boldsymbol{\beta} : \dot{\boldsymbol{\alpha}} \,, \tag{H.13}$$

subject to

$$\tau_{\text{mech}} = \rho_0 \, \varphi_{\text{mech}}(\dot{\boldsymbol{\varepsilon}}^{\text{i}}, \dot{\boldsymbol{\alpha}}; \boldsymbol{\varepsilon} - \boldsymbol{\varepsilon}^{\text{i}}, \boldsymbol{\alpha}, T, h) - \Phi_{\text{mech}}(\dot{\boldsymbol{\varepsilon}}^{\text{i}}, \dot{\boldsymbol{\alpha}}, \boldsymbol{\sigma}, \boldsymbol{\beta}) = 0 \,, \tag{H.14}$$

where $\tau_{\text{mech}} = 0$ is a constraint,

and

maximize with respect to the flux $(\vec{q})$,

$$\Phi_{\text{ther}}(\vec{q}, \vec{\nabla} T / T) = \frac{\vec{\nabla} T}{T} \cdot \vec{q} \,, \tag{H.15}$$

subject to

$$\tau_{\text{ther}} = \rho_0 \, \varphi_{\text{ther}}(\vec{q}; \boldsymbol{\varepsilon} - \boldsymbol{\varepsilon}^{\text{i}}, \boldsymbol{\alpha}, T, h) - \Phi_{\text{ther}}(\vec{q}, \vec{\nabla} T / T) = 0 \,, \tag{H.16}$$

where $\tau_{\text{ther}} = 0$ is a constraint.

Case B: Consider the following problem:

maximize with respect to the fluxes $(\dot{\boldsymbol{\varepsilon}}^{\mathrm{i}}, \dot{\boldsymbol{\alpha}}, \vec{q})$,

$$\Phi = \Phi_{\text{mech}}(\dot{\boldsymbol{\varepsilon}}^{\mathrm{i}}, \dot{\boldsymbol{\alpha}}, \boldsymbol{\sigma}, \boldsymbol{\beta}) + \Phi_{\text{ther}}(\vec{q}, \vec{\nabla}T/T) = \boldsymbol{\sigma} : \dot{\boldsymbol{\varepsilon}}^{\mathrm{i}} + \boldsymbol{\beta} : \dot{\boldsymbol{\alpha}} - \frac{\vec{\nabla}T}{T} \cdot \vec{q}\,, \tag{H.17}$$

subject to the constraints

$$\tau_{\text{mech}} = \rho_0\, \varphi_{\text{mech}}(\dot{\boldsymbol{\varepsilon}}^{\mathrm{i}}, \dot{\boldsymbol{\alpha}}; \boldsymbol{\varepsilon} - \boldsymbol{\varepsilon}^{\mathrm{i}}, \boldsymbol{\alpha}, T, h) - \Phi_{\text{mech}}(\dot{\boldsymbol{\varepsilon}}^{\mathrm{i}}, \dot{\boldsymbol{\alpha}}, \boldsymbol{\sigma}, \boldsymbol{\beta}) = 0 \tag{H.18}$$

and

$$\tau_{\text{ther}} = \rho_0\, \varphi_{\text{ther}}(\vec{q}; \boldsymbol{\varepsilon} - \boldsymbol{\varepsilon}^{\mathrm{i}}, \boldsymbol{\alpha}, T, h) - \Phi_{\text{ther}}(\vec{q}, \vec{\nabla}T/T) = 0\,. \tag{H.19}$$

Case A was studied in Sections 6.7.1 and 6.7.3. Therefore case B is studied now.

Applying Luenberger ([55], p. 225), the first-order sufficient condition for the point $(\dot{\boldsymbol{\varepsilon}}^{i}, \dot{\boldsymbol{\alpha}}, \vec{q})$ to be a local maximum is

$$\frac{\partial}{\partial \dot{\boldsymbol{\varepsilon}}^{\mathrm{i}}}\left( \Phi_{\text{mech}} + \Phi_{\text{ther}} + \lambda_{\text{mech}}\, \tau_{\text{mech}} + \lambda_{\text{ther}}\, \tau_{\text{ther}} \right) = \mathbf{0} \tag{H.20}$$

and

$$\frac{\partial}{\partial \dot{\boldsymbol{\alpha}}}\left( \Phi_{\text{mech}} + \Phi_{\text{ther}} + \lambda_{\text{mech}}\, \tau_{\text{mech}} + \lambda_{\text{ther}}\, \tau_{\text{ther}} \right) = \mathbf{0}\,, \tag{H.21}$$

and furthermore,

$$\frac{\partial}{\partial \vec{q}}\left( \Phi_{\text{mech}} + \Phi_{\text{ther}} + \lambda_{\text{mech}}\, \tau_{\text{mech}} + \lambda_{\text{ther}}\, \tau_{\text{ther}} \right) = \vec{0}\,, \tag{H.22}$$

finally,

$$\tau_{\text{mech}} = 0 \qquad \text{and} \qquad \tau_{\text{ther}} = 0\,. \tag{H.23}$$

In Expressions (H.20)–(H.22), the notations $\lambda_{\text{mech}}$ and $\lambda_{\text{ther}}$ are Lagrange multipliers. As mentioned by Arfken ([7], p. 946), the method based on Lagrange multipliers will fail if in Expressions (H.20)–(H.22) the coefficients of $\lambda_{\text{mech}}$ and $\lambda_{\text{ther}}$ vanish at the extremum. Therefore also special points where

$$\frac{\partial \tau_{\text{mech}}}{\partial \dot{\boldsymbol{\varepsilon}}^{\mathrm{i}}} = \mathbf{0}\,, \qquad \frac{\partial \tau_{\text{mech}}}{\partial \dot{\boldsymbol{\alpha}}} = \mathbf{0}\,, \qquad \text{and} \qquad \frac{\partial \tau_{\text{ther}}}{\partial \vec{q}} = \vec{0} \tag{H.24}$$

must be studied.

Substitution of Equations (H.17)–(H.19) into Equation (H.20) gives

$$\boldsymbol{\sigma} + \lambda_{\text{mech}} \left( \rho_0\, \frac{\partial \varphi_{\text{mech}}}{\partial \dot{\boldsymbol{\varepsilon}}^{\mathrm{i}}} - \boldsymbol{\sigma} \right) = \mathbf{0}\,, \tag{H.25}$$

which yields the following result:

$$\boldsymbol{\sigma} = \frac{\lambda_{\text{mech}}}{\lambda_{\text{mech}} - 1}\,\rho_0\,\frac{\partial \varphi_{\text{mech}}}{\partial \dot{\boldsymbol{\varepsilon}}^{\text{i}}}\,. \tag{H.26}$$

Similarly, Equations (H.21) and (H.22) give

$$\boldsymbol{\beta} = \frac{\lambda_{\text{mech}}}{\lambda_{\text{mech}} - 1}\,\rho_0\,\frac{\partial \varphi_{\text{mech}}}{\partial \dot{\boldsymbol{\alpha}}} \tag{H.27}$$

and

$$-\frac{\vec{\nabla} T}{T} = \frac{\lambda_{\text{ther}}}{\lambda_{\text{ther}} - 1}\,\rho_0\,\frac{\partial \varphi_{\text{ther}}(\vec{q};\, \boldsymbol{\varepsilon} - \boldsymbol{\varepsilon}^{\text{i}}, \boldsymbol{\alpha}, T, h)}{\partial \vec{q}}\,. \tag{H.28}$$

By substituting $\Phi_{\text{mech}}$ obtained from Equation (H.17) into Constraint (H.18) [that is, Expression $(\text{H.23})_1$], Forces (H.26) and (H.27) into the obtained expression and rearranging the terms, the following is obtained:

$$\varphi_{\text{mech}}(\dot{\boldsymbol{\varepsilon}}^{\text{i}}, \dot{\boldsymbol{\alpha}};\, \boldsymbol{\varepsilon} - \boldsymbol{\varepsilon}^{\text{i}}, \boldsymbol{\alpha}, T, h) = \frac{\lambda_{\text{mech}}}{\lambda_{\text{mech}} - 1}\left(\frac{\partial \varphi_{\text{mech}}}{\partial \dot{\boldsymbol{\varepsilon}}^{\text{i}}} : \dot{\boldsymbol{\varepsilon}}^{\text{i}} + \frac{\partial \varphi_{\text{mech}}}{\partial \dot{\boldsymbol{\alpha}}} : \dot{\boldsymbol{\alpha}}\right). \tag{H.29}$$

Correspondingly to the preceding, by substituting $\Phi_{\text{ther}}$ obtained from Equation (H.17) into Constraint (H.19) [that is, Expression $(\text{H.23})_2$], Force (H.28) into the obtained expression and rearranging the terms, the following result is obtained:

$$\varphi_{\text{ther}}(\vec{q};\, \boldsymbol{\varepsilon} - \boldsymbol{\varepsilon}^{\text{i}}, \boldsymbol{\alpha}, T, h) = \frac{\lambda_{\text{ther}}}{\lambda_{\text{ther}} - 1}\,\frac{\partial \varphi_{\text{ther}}}{\partial \vec{q}} \cdot \vec{q}\,. \tag{H.30}$$

Comparison of Results (H.26), (H.27), and (H.29) with Expressions $(6.273)_2$, (6.274), and (6.275) shows that they are equal. A similar answer is obtained for the heat problem expressed by Equations (H.28), (H.30), (6.289), and (6.290), since [cf. Definition $(6.278)_1$]

$$\mu_{\text{ther}} := \frac{\lambda_{\text{ther}}}{\lambda_{\text{ther}} - 1}\,. \tag{H.31}$$

Thus cases A and B lead to the same result.

## H.3 Normality Rule for the Separated Specific Dissipation Function $\varphi$ Having the Same Internal Variables in Several Parts

In this part of the appendix, we evaluate cases in which the specific dissipation function $\varphi$ is separated into several parts but the same internal variable exists in more than one specific dissipation function. In order to make the evaluation easy to follow, the simplest possible case is evaluated here, namely, the case in which the specific dissipation function $\varphi_{\text{mech}}$ is assumed to be separated into two parts.

The Clausius–Duhem inequality for the mechanical part was given by Expression $(6.264)_1$. It has the following appearance:

$$\Phi_{\text{mech}} = \boldsymbol{\sigma} : \dot{\boldsymbol{\varepsilon}}^{\text{i}} + \boldsymbol{\beta} : \dot{\boldsymbol{\alpha}} \qquad (\geq 0)\,. \tag{H.32}$$

The mechanical part of the specific dissipation function $\varphi_{\text{mech}}$ is assumed to be separated into two parts, as follows:

$$\varphi_{\text{mech}}(\dot{\boldsymbol{\varepsilon}}^{\text{i}}, \dot{\boldsymbol{\alpha}}; \boldsymbol{\varepsilon}-\boldsymbol{\varepsilon}^{\text{i}}, \boldsymbol{\alpha}, T, h) = \varphi^1_{\text{mech}}(\dot{\boldsymbol{\varepsilon}}^{\text{i}}, \dot{\boldsymbol{\alpha}}; \boldsymbol{\varepsilon}-\boldsymbol{\varepsilon}^{\text{i}}, \boldsymbol{\alpha}, T, h) + \varphi^2_{\text{mech}}(\dot{\boldsymbol{\alpha}}; \boldsymbol{\varepsilon}-\boldsymbol{\varepsilon}^{\text{i}}, \boldsymbol{\alpha}, T, h). \tag{H.33}$$

As Separation (H.33) shows, the internal variable $\dot{\boldsymbol{\alpha}}$ is on both parts of the specific dissipation function $\varphi$.

According to Expressions (6.267) and (6.268) of Section 6.7.1, the principle of maximum dissipation reads as follows:

maximize with respect to the fluxes $(\dot{\boldsymbol{\varepsilon}}^{\text{i}}, \boldsymbol{\alpha})$,

$$\Phi_{\text{mech}}(\dot{\boldsymbol{\varepsilon}}^{\text{i}}, \dot{\boldsymbol{\alpha}}, \boldsymbol{\sigma}, \boldsymbol{\beta}) = \boldsymbol{\sigma} : \dot{\boldsymbol{\varepsilon}}^{\text{i}} + \boldsymbol{\beta} : \dot{\boldsymbol{\alpha}}\,, \tag{H.34}$$

subject to

$$\tau_{\text{mech}} = \rho_0\, \varphi_{\text{mech}}(\dot{\boldsymbol{\varepsilon}}^{\text{i}}, \dot{\boldsymbol{\alpha}}; \boldsymbol{\varepsilon} - \boldsymbol{\varepsilon}^{\text{i}}, \boldsymbol{\alpha}, T, h) - \Phi_{\text{mech}}(\dot{\boldsymbol{\varepsilon}}^{\text{i}}, \dot{\boldsymbol{\alpha}}, \boldsymbol{\sigma}, \boldsymbol{\beta}) = 0\,, \tag{H.35}$$

where $\tau_{\text{mech}} = 0$ is a constraint and $\varphi_{\text{mech}}$ is the specific dissipation function for mechanical behavior.

The investigation can be continued by following the same procedure as in Section 6.7.1. This means that the normality rule follows Normality Rules (6.282) and (6.283):

$$\boldsymbol{\sigma} = \mu\, \rho_0 \frac{\partial \varphi_{\text{mech}}(\dot{\boldsymbol{\varepsilon}}^{\text{i}}, \dot{\boldsymbol{\alpha}}; \text{state})}{\partial \dot{\boldsymbol{\varepsilon}}^{\text{i}}} = \mu\, \rho_0 \frac{\partial \varphi^1_{\text{mech}}(\dot{\boldsymbol{\varepsilon}}^{\text{i}}, \dot{\boldsymbol{\alpha}}; \text{state})}{\partial \dot{\boldsymbol{\varepsilon}}^{\text{i}}} \tag{H.36}$$

and

$$\begin{aligned} \boldsymbol{\beta} &= \mu\, \rho_0 \frac{\partial \varphi_{\text{mech}}(\dot{\boldsymbol{\varepsilon}}^{\text{i}}, \dot{\boldsymbol{\alpha}}; \text{state})}{\partial \dot{\boldsymbol{\alpha}}} \\ &= \mu\, \rho_0 \frac{\partial \varphi^1_{\text{mech}}(\dot{\boldsymbol{\varepsilon}}^{\text{i}}, \dot{\boldsymbol{\alpha}}; \text{state})}{\partial \dot{\boldsymbol{\alpha}}} + \mu\, \rho_0 \frac{\partial \varphi^2_{\text{mech}}(\dot{\boldsymbol{\alpha}}; \text{state})}{\partial \dot{\boldsymbol{\alpha}}}\,. \end{aligned} \tag{H.37}$$

In Normality Rules (H.36) and (H.37), the set of state variables $(\boldsymbol{\varepsilon} - \boldsymbol{\varepsilon}^{\text{i}}, \boldsymbol{\alpha}, T)$ and function $h$ are expressed by the notation *state*. It is worth noting that the approach given by Expressions (H.34) and (H.35) leads to the result that the parameters $\mu$ related to the specific dissipation functions $\varphi^1_{\text{mech}}$ and $\varphi^2_{\text{mech}}$ are the same [as shown in Expressions (H.36) and (H.37)].

# *Appendix I*
# Constitutive Tensor C and Compliance Tensor S

In the present appendix, we derive some results for the constitutive tensor for a Hookean deformation $\mathbf{C}$ and the compliance tensor for a Hookean deformation $\mathbf{S}$.

**Theorem 1:** The following holds:

$$\mathbf{I}^{\mathrm{s}} : \mathbf{I}^{\mathrm{s}} = \mathbf{I}^{\mathrm{s}} . \tag{I.1}$$

**Proof:** Expression (2.107) is

$$\mathbf{I}^{\mathrm{s}} = \frac{1}{2}(\delta_{ik}\,\delta_{jl} + \delta_{il}\,\delta_{jk})\,\vec{i}_i\,\vec{i}_j\,\vec{i}_k\,\vec{i}_l\,. \tag{I.2}$$

Based on Expression (I.2), we arrive at the following:

$$\begin{aligned}
\mathbf{I}^{\mathrm{s}} : \mathbf{I}^{\mathrm{s}} &= \tfrac{1}{2}\,(\,\delta_{ik}\,\delta_{jl} + \delta_{il}\,\delta_{jk}\,)\,\vec{i}_i\,\vec{i}_j\,\vec{i}_k\,\vec{i}_l : \tfrac{1}{2}\,(\,\delta_{su}\,\delta_{tv} + \delta_{sv}\,\delta_{tu}\,)\,\vec{i}_s\,\vec{i}_t\,\vec{i}_u\,\vec{i}_v \\
&= \tfrac{1}{4}\,(\,\delta_{ik}\,\delta_{jl} + \delta_{il}\,\delta_{jk}\,)\,\vec{i}_i\,\vec{i}_j\,(\,\delta_{su}\,\delta_{tv} + \delta_{sv}\,\delta_{tu}\,)\,\delta_{ks}\,\delta_{lt}\,\vec{i}_u\,\vec{i}_v \\
&= \tfrac{1}{4}\,(\,\delta_{ik}\,\delta_{jl} + \delta_{il}\,\delta_{jk}\,)\,\vec{i}_i\,\vec{i}_j\,(\,\delta_{ku}\,\delta_{lv} + \delta_{kv}\,\delta_{lu}\,)\,\vec{i}_u\,\vec{i}_v \\
&= \tfrac{1}{4}\,(\,\delta_{iu}\,\delta_{jv} + \delta_{iv}\,\delta_{ju} + \delta_{iv}\,\delta_{ju} + \delta_{iu}\,\delta_{jv}\,)\,\vec{i}_i\,\vec{i}_j\,\vec{i}_u\,\vec{i}_v \\
&= \tfrac{1}{2}\,(\,\delta_{iu}\,\delta_{jv} + \delta_{iv}\,\delta_{ju}\,)\,\vec{i}_i\,\vec{i}_j\,\vec{i}_u\,\vec{i}_v = \mathbf{I}^{\mathrm{s}}\,.
\end{aligned} \tag{I.3}$$

**Theorem 2:** The following expressions for the compliance tensor $\mathbf{S}$ coincide:

$$\mathbf{S} := -\frac{\nu}{E}\,\mathbf{1}\,\mathbf{1} + \frac{1+\nu}{E}\,\mathbf{I}^{\mathrm{s}} \qquad \text{and} \qquad \mathbf{C} : \mathbf{S} = \mathbf{S} : \mathbf{C} = \mathbf{I}^{\mathrm{s}}\,. \tag{I.4}$$

**Proof:** Equation (8.30) is recalled:

$$\mathbf{C} := \lambda\,\mathbf{1}\,\mathbf{1} + 2\,\mu\,\mathbf{I}^{\mathrm{s}}\,. \tag{I.5}$$

Form $(\text{I.4})_1$ is obtained from Definition (I.5) by a pure mathematical derivation, as seen in Section 8.3.1 [see Expression (8.42)]. Inspired by Definition $(\text{I.4})_1$, the following guess is made for the compliance tensor $\mathbf{S}$ in Property $(\text{I.4})_2$:

$$\mathbf{S} = -a\,\mathbf{1}\,\mathbf{1} + c\,\mathbf{I}^{\mathrm{s}}\,. \tag{I.6}$$

Substitution of Expressions (I.5) and (I.6) into Property $(\text{I.4})_2$ gives

$$\begin{aligned}\mathbf{C}:\mathbf{S} &= (\lambda\,\mathbf{1}\,\mathbf{1} + 2\,\mu\,\mathbf{I}^{\mathrm{s}}):(-a\,\mathbf{1}\,\mathbf{1} + c\,\mathbf{I}^{\mathrm{s}}) \\ &= -a\,\lambda\,\mathbf{1}\,\mathbf{1}:\mathbf{1}\,\mathbf{1} + c\,\lambda\,\mathbf{1}\,\mathbf{1}:\mathbf{I}^{\mathrm{s}} - 2\,a\,\mu\,\mathbf{I}^{\mathrm{s}}:\mathbf{1}\,\mathbf{1} + 2\,c\,\mu\,\mathbf{I}^{\mathrm{s}}:\mathbf{I}^{\mathrm{s}}\,.\end{aligned} \tag{I.7}$$

Equation $(2.65)_2$, Definition (2.106), and the symmetry of the second-order identity tensor $\mathbf{1}$ [see Equations (2.64) and (2.4)] give

$$\mathbf{1}:\mathbf{1} = \delta_{ii} = 3 \qquad \text{and} \qquad \mathbf{I}^{\mathrm{s}}:\mathbf{1} = \frac{1}{2}\,(\mathbf{1} + \mathbf{1}^{\mathrm{T}}) = \mathbf{1}\,. \tag{I.8}$$

Similar to Equation $(\text{I.8})_2$, the following relation can be derived:

$$\mathbf{1}:\mathbf{I}^{\mathrm{s}} = \frac{1}{2}\,(\mathbf{1} + \mathbf{1}^{\mathrm{T}}) = \mathbf{1}\,. \tag{I.9}$$

Expressions (I.1), (I.8), and (I.9) allow Equation (I.7) to take the following appearance:

$$\begin{aligned}\mathbf{C}:\mathbf{S} &= -a\,\lambda\,\mathbf{1}\,\mathbf{1}:\mathbf{1}\,\mathbf{1} + c\,\lambda\,\mathbf{1}\,\mathbf{1}:\mathbf{I}^{\mathrm{s}} - 2\,a\,\mu\,\mathbf{I}^{\mathrm{s}}:\mathbf{1}\,\mathbf{1} + 2\,c\,\mu\,\mathbf{I}^{\mathrm{s}}:\mathbf{I}^{\mathrm{s}} \\ &= -3\,a\,\lambda\,\mathbf{1}\,\mathbf{1} + c\,\lambda\,\mathbf{1}\,\mathbf{1} - 2\,a\,\mu\,\mathbf{1}\,\mathbf{1} + 2\,c\,\mu\,\mathbf{I}^{\mathrm{s}} \\ &= (-3\,a\,\lambda + c\,\lambda - 2\,a\,\mu)\,\mathbf{1}\,\mathbf{1} + 2\,c\,\mu\,\mathbf{I}^{\mathrm{s}}\,.\end{aligned} \tag{I.10}$$

Substitution of Expression (I.10) into Equation $(\text{I.4})_2$ yields

$$\mathbf{C}:\mathbf{S} = \mathbf{I}^{\mathrm{s}} \quad \Rightarrow \quad (-3\,a\,\lambda + c\,\lambda - 2\,a\,\mu)\,\mathbf{1}\,\mathbf{1} + 2\,c\,\mu\,\mathbf{I}^{\mathrm{s}} = \mathbf{I}^{\mathrm{s}}\,. \tag{I.11}$$

Equation $(\text{I.11})_2$ gives the following two equations:

$$2\,c\,\mu = 1 \qquad \Rightarrow \qquad c = \frac{1}{2\,\mu} \tag{I.12}$$

and

$$-3\,a\,\lambda + c\,\lambda - 2\,a\,\mu = 0 \qquad \Rightarrow \qquad a\,(3\,\lambda + 2\,\mu) = c\,\lambda\,, \tag{I.13}$$

which with the help of Equation $(\text{I.12})_2$ leads to

$$a = \frac{\lambda}{2\,\mu\,(3\,\lambda + 2\,\mu)}\,. \tag{I.14}$$

Lamé elastic constants $\lambda$ and $\mu$ are given by Definitions (8.31):

$$\lambda := \frac{\nu\,E}{(1+\nu)\,(1-2\,\nu)} \qquad \text{and} \qquad \mu := \frac{E}{2\,(1+\nu)}\,. \tag{I.15}$$

Substitution of Lamé elastic constants $\lambda$ and $\mu$ from Equations (I.15) into Equations $(\text{I.12})_2$ and (I.14) leads to

$$c = \frac{1}{2\,\mu} = \frac{1+\nu}{E} \tag{I.16}$$

and

$$a = \frac{\lambda}{2\,\mu\,(3\,\lambda + 2\,\mu)} = \frac{\frac{\nu\,E}{(1+\nu)\,(1-2\,\nu)}}{\frac{E}{(1+\nu)}\left[\frac{3\,\nu\,E}{(1+\nu)\,(1-2\,\nu)} + \frac{E}{(1+\nu)}\right]}$$

$$= \frac{\frac{\nu\,E}{(1+\nu)\,(1-2\,\nu)}}{\frac{E}{(1+\nu)}\left[\frac{3\,\nu\,E}{(1+\nu)\,(1-2\,\nu)} + \frac{(1-2\,\nu)\,E}{(1+\nu)\,(1-2\,\nu)}\right]} \tag{I.17}$$

$$= \frac{\nu\,E}{\frac{E}{(1+\nu)}[(1+\nu)\,E]} = \frac{\nu}{E}\,.$$

Substitution of constants $c$ and $a$ from Equations (I.16) and (I.17) into Equation (I.6) gives

$$\mathbf{S} = -\,a\,\mathbf{1}\,\mathbf{1} + c\,\mathbf{I}^{\mathrm{s}} = -\frac{\nu}{E}\,\mathbf{1}\,\mathbf{1} + \frac{1+\nu}{E}\,\mathbf{I}^{\mathrm{s}}\,. \tag{I.18}$$

Comparison of Expressions $(\text{I}.4)_1$ and (I.18) shows that Expression $(\text{I}.4)_1$ is correct. Since Expressions $(\text{I}.12)_1$, and $(\text{I}.13)_1$ are shown to be correct, Equation (I.11) proves $(\text{I}.4)_2$. A similar evaluation can be carried out to show that Expression $(\text{I}.4)_1$ and the second part of Equation $(\text{I}.4)_2$, $\mathbf{S}:\mathbf{C} = \mathbf{I}^{\mathrm{s}}$, coincide. Thus Theorem 2 is proved to be correct.

**Theorem 3:** Constitutive tensor $\mathbf{C}$ has a major symmetry. This is

$$C_{ijkl} = C_{klij}\,. \tag{I.19}$$

**Proof:** Material Model (8.29) is

$$\psi^{\mathrm{e}}(\boldsymbol{\varepsilon} - \boldsymbol{\varepsilon}^{\mathrm{i}}) = \psi^{\mathrm{e}}_{\mathrm{r}} + \frac{1}{2\,\rho_0}\,(\boldsymbol{\varepsilon} - \boldsymbol{\varepsilon}^{\mathrm{i}}):\mathbf{C}:(\boldsymbol{\varepsilon} - \boldsymbol{\varepsilon}^{\mathrm{i}})\,. \tag{I.20}$$

Equation (I.20) can be written in the following index form:

$$\psi^{\mathrm{e}}(\boldsymbol{\varepsilon} - \boldsymbol{\varepsilon}^{\mathrm{i}}) = \psi^{\mathrm{e}}_{\mathrm{r}} + \frac{1}{2\,\rho_0}\,(\varepsilon_{ij} - \varepsilon^{\mathrm{i}}_{ij})\,C_{ijkl}\,(\varepsilon_{kl} - \varepsilon^{\mathrm{i}}_{kl})\,. \tag{I.21}$$

The stress tensor $\boldsymbol{\sigma}$ is obtained from State Equation $(8.32)_1$:

$$\boldsymbol{\sigma} = \rho_0\,\frac{\partial\psi^{\mathrm{e}}(\boldsymbol{\varepsilon} - \boldsymbol{\varepsilon}^{\mathrm{i}})}{\partial(\boldsymbol{\varepsilon} - \boldsymbol{\varepsilon}^{\mathrm{i}})} = \mathbf{C}:(\boldsymbol{\varepsilon} - \boldsymbol{\varepsilon}^{\mathrm{i}}) \quad \Rightarrow \quad \sigma_{ij} = C_{ijst}\,(\varepsilon_{st} - \varepsilon^{\mathrm{i}}_{st})\,. \tag{I.22}$$

Equation $(\text{I}.22)_1$ gives

$$\sigma_{ij} = \frac{\partial\psi^{\mathrm{e}}(\boldsymbol{\varepsilon} - \boldsymbol{\varepsilon}^{\mathrm{i}})}{\partial(\varepsilon_{ij} - \varepsilon^{\mathrm{i}}_{ij})} \quad \Rightarrow \quad \frac{\partial\sigma_{ij}}{\partial(\varepsilon_{kl} - \varepsilon^{\mathrm{i}}_{kl})} = \frac{\partial^2\psi^{\mathrm{e}}(\boldsymbol{\varepsilon} - \boldsymbol{\varepsilon}^{\mathrm{i}})}{\partial(\varepsilon_{kl} - \varepsilon^{\mathrm{i}}_{kl})\,\partial(\varepsilon_{ij} - \varepsilon^{\mathrm{i}}_{ij})} \tag{I.23}$$

and

$$\sigma_{kl} = \frac{\partial\psi^{\mathrm{e}}(\boldsymbol{\varepsilon} - \boldsymbol{\varepsilon}^{\mathrm{i}})}{\partial(\varepsilon_{kl} - \varepsilon^{\mathrm{i}}_{kl})} \quad \Rightarrow \quad \frac{\partial\sigma_{kl}}{\partial(\varepsilon_{ij} - \varepsilon^{\mathrm{i}}_{ij})} = \frac{\partial^2\psi^{\mathrm{e}}(\boldsymbol{\varepsilon} - \boldsymbol{\varepsilon}^{\mathrm{i}})}{\partial(\varepsilon_{ij} - \varepsilon^{\mathrm{i}}_{ij})\,\partial(\varepsilon_{kl} - \varepsilon^{\mathrm{i}}_{kl})}\,. \tag{I.24}$$

Since the order of the derivation can be interchanged, the right sides of Expressions $(I.23)_2$ and $(I.24)_2$ are equal. Therefore also the left sides of Expressions $(I.23)_2$ and $(I.24)_2$ are equal. This is

$$\frac{\partial \sigma_{ij}}{\partial(\varepsilon_{kl} - \varepsilon^{\mathrm{i}}_{kl})} = \frac{\partial \sigma_{kl}}{\partial(\varepsilon_{ij} - \varepsilon^{\mathrm{i}}_{ij})} \,. \tag{I.25}$$

Based on Constitutive Equation $(I.22)_2$, we arrive at the following:

$$\frac{\partial \sigma_{ij}}{\partial(\varepsilon_{kl} - \varepsilon^{\mathrm{i}}_{kl})} = \frac{\partial}{\partial(\varepsilon_{kl} - \varepsilon^{\mathrm{i}}_{kl})} \left[ C_{ijst} \left( \varepsilon_{st} - \varepsilon^{\mathrm{i}}_{st} \right) \right] = C_{ijst}\, \delta_{ks}\, \delta_{lt} = C_{ijkl} \tag{I.26}$$

and

$$\frac{\partial \sigma_{kl}}{\partial(\varepsilon_{ij} - \varepsilon^{\mathrm{i}}_{ij})} = \frac{\partial}{\partial(\varepsilon_{ij} - \varepsilon^{\mathrm{i}}_{ij})} \left[ C_{klst} \left( \varepsilon_{st} - \varepsilon^{\mathrm{i}}_{st} \right) \right] = C_{klst}\, \delta_{is}\, \delta_{jt} = C_{klij} \,. \tag{I.27}$$

Comparison of Equations (I.26) and (I.27) with Expression (I.25) gives

$$C_{ijkl} = C_{klij} \,. \tag{I.28}$$

Result (I.28) proves the major symmetry of the constitutive tensor $\mathbf{C}$.

**Theorem 4:** Constitutive tensor $\mathbf{C}$ has a minor symmetry in the first pair of indices and in the second pair of indices. This is

$$C_{ijkl} = C_{jikl} \qquad \text{and} \qquad C_{ijkl} = C_{ijlk} \,. \tag{I.29}$$

**Proof:** Hooke's Law $(I.22)_2$ can be written in the following two forms:

$$\sigma_{ij} = C_{ijkl} \left( \varepsilon_{kl} - \varepsilon^{\mathrm{i}}_{kl} \right) \qquad \text{and} \qquad \sigma_{ji} = C_{jikl} \left( \varepsilon_{kl} - \varepsilon^{\mathrm{i}}_{kl} \right) . \tag{I.30}$$

Since the stress tensor $\boldsymbol{\sigma}$ is symmetric, Hooke's Law $(I.30)_2$ can be written in the following form:

$$\sigma_{ij} = C_{jikl} \left( \varepsilon_{kl} - \varepsilon^{\mathrm{i}}_{kl} \right) . \tag{I.31}$$

Form (I.31) is subtracted from Form $(I.30)_1$, and the following result is obtained:

$$0 = \left[ C_{ijkl} - C_{jikl} \right] \left( \varepsilon_{kl} - \varepsilon^{\mathrm{i}}_{kl} \right) \qquad \Rightarrow \qquad C_{ijkl} = C_{jikl} \,. \tag{I.32}$$

Equation $(I.32)_2$ proves Expression $(I.29)_1$ in Theorem 4. By exploiting the symmetry of the strain tensors $\boldsymbol{\varepsilon}$ and $\boldsymbol{\varepsilon}^{\mathrm{i}}$, one can prove Expression $(I.29)_2$ in Theorem 4. Thus constitutive tensor $\mathbf{C}$ is minor symmetric in pairs $ij$ and $kl$.

**Theorem 5:** The following holds:

$$\mathbf{I}^{\mathrm{s}} : \mathbf{C} = \mathbf{C} : \mathbf{I}^{\mathrm{s}} = \mathbf{C} \qquad \text{and} \qquad \mathbf{I}^{\mathrm{s}} : \mathbf{S} = \mathbf{S} : \mathbf{I}^{\mathrm{s}} = \mathbf{S} \,. \tag{I.33}$$

**Proof:** Based on Equation (2.107), the following can be written:

$$\begin{aligned}\mathbf{I}^{\mathrm{s}} : \mathbf{A} &= \frac{1}{2}\,(\delta_{ik}\,\delta_{jl} + \delta_{il}\,\delta_{jk})\,\vec{i}_i\,\vec{i}_j\,\vec{i}_k\,\vec{i}_l : A_{stuw}\,\vec{i}_s\,\vec{i}_t\,\vec{i}_u\,\vec{i}_w \\ &= \frac{1}{2}\,(\delta_{ik}\,\delta_{jl} + \delta_{il}\,\delta_{jk})\,\vec{i}_i\,\vec{i}_j\,A_{stuw}\,\delta_{ks}\,\delta_{lt}\,\vec{i}_u\,\vec{i}_w \\ &= \frac{1}{2}\,(\delta_{ik}\,\delta_{jl} + \delta_{il}\,\delta_{jk})\,\vec{i}_i\,\vec{i}_j\,A_{kluw}\,\vec{i}_u\,\vec{i}_w = \frac{1}{2}\,(A_{ijuw} + A_{jiuw})\,\vec{i}_i\,\vec{i}_j\,\vec{i}_u\,\vec{i}_w\,.\end{aligned} \tag{I.34}$$

According to **Theorem 4**, the constitutive tensor $\mathbf{C}$ has a minor symmetry. Therefore Expression (I.34) gives

$$\mathbf{I}^{\mathrm{s}} : \mathbf{C} = \mathbf{C}\,. \tag{I.35}$$

The other equalities present in Expression (I.33) can be shown by taking similar steps to those taken in Manipulation (I.34).

**Theorem 6:** The following holds:

$$\mathbf{I}^{\mathrm{s}} : \tilde{\mathbf{S}} = \tilde{\mathbf{S}}\,, \tag{I.36}$$

where $\tilde{\mathbf{S}}$ is the effective compliance tensor for deformation of a material containing noninteracting spherical microvoids within a matrix material having a linear elastic response.

**Proof:** Theorem 5 shows that Expression (I.36) holds if the effective compliance tensor $\tilde{\mathbf{S}}$ is minor symmetric in the first pair of indices.

Equation (10.88) is

$$\tilde{\mathbf{S}}(f) := \mathbf{S} + \mathbf{S}^{\mathrm{d}}(f)\,, \tag{I.37}$$

where the superscript d refers to damage. Expression (I.37) gives

$$\mathbf{I}^{\mathrm{s}} : \tilde{\mathbf{S}} = \mathbf{I}^{\mathrm{s}} : [\,\mathbf{S} + \mathbf{S}^{\mathrm{d}}(f)\,] = \mathbf{I}^{\mathrm{s}} : \mathbf{S} + \mathbf{I}^{\mathrm{s}} : \mathbf{S}^{\mathrm{d}}(f) = \mathbf{S} + \mathbf{I}^{\mathrm{s}} : \mathbf{S}^{\mathrm{d}}(f)\,, \tag{I.38}$$

where Expression $(\mathrm{I}.33)_2$ is exploited. Definition (10.90) reads

$$\mathbf{S}^{\mathrm{d}}(f) := \frac{A\,f}{3\,(3\,\lambda + 2\,\mu)}\,\mathbf{1}\,\mathbf{1} + \frac{B\,f}{2\,\mu}\,(\mathbf{I}^{\mathrm{s}} - \tfrac{1}{3}\,\mathbf{1}\,\mathbf{1})\,. \tag{I.39}$$

Equation (2.64) is recalled. It has the following appearance:

$$\mathbf{1} = \delta_{ij}\,\vec{i}_i\,\vec{i}_j\,. \tag{I.40}$$

Based on Property (2.6), $\delta_{ij} = \delta_{ji}$, which implies that the second-order identity tensor $\mathbf{1}$ is symmetric and therefore the tensor $\mathbf{1}\,\mathbf{1}$ is minor symmetric in the first pair of indices.

Expression (2.107) is

$$\mathbf{I}^{\mathrm{s}} = \frac{1}{2}\left(\delta_{ik}\,\delta_{jl} + \delta_{il}\,\delta_{jk}\right)\vec{\imath}_i\,\vec{\imath}_j\,\vec{\imath}_k\,\vec{\imath}_l\,. \tag{I.41}$$

Interchanging the first two indices leads to the following form for the right side of Expression (I.41):

$$\ldots = \frac{1}{2}\left(\delta_{jk}\,\delta_{il} + \delta_{jl}\,\delta_{ik}\right)\vec{\imath}_i\,\vec{\imath}_j\,\vec{\imath}_k\,\vec{\imath}_l\,. \tag{I.42}$$

Since the order of the Kronecker deltas $\delta_{ij}$ can be interchanged, the right sides of Expressions (I.41) and (I.42) are equal. This means that the fourth-order symmetric identity tensor $\mathbf{I}^{\mathrm{s}}$ is minor symmetric in the first pair of indices.

The preceding derivation leads to the conclusion that the compliance tensor $\mathbf{S}^{\mathrm{d}}(f)$ is minor symmetric in the first pair of indices, and therefore Equation (I.38) yields

$$\mathbf{I}^{\mathrm{s}} : \tilde{\mathbf{S}} = \mathbf{S} + \mathbf{I}^{\mathrm{s}} : \mathbf{S}^{\mathrm{d}}(f) = \mathbf{S} + \mathbf{S}^{\mathrm{d}}(f) = \tilde{\mathbf{S}}\,. \tag{I.43}$$

Result (I.43) proves Theorem (I.36).

# *Appendix J* Scalar Components of the Compliance Tensor $\mathbf{S}^{\mathrm{d}r}$

This appendix studies penny-shaped microcracks, shown in Figure J.1, and derives the expressions for scalar components of the fourth-order tensor $\mathbf{S}^{\mathrm{d}r}$, which is the fourth-order compliance tensor related to a single microcrack group and expressed in the microcrack coordinate system $(Z_1^r, Z_2^r, Z_3^r)$.

According to Expressions (10.30), (10.4), and (10.29), the damage strain tensor $\boldsymbol{\varepsilon}^{\mathrm{d}r}$ is obtained from the following equation:

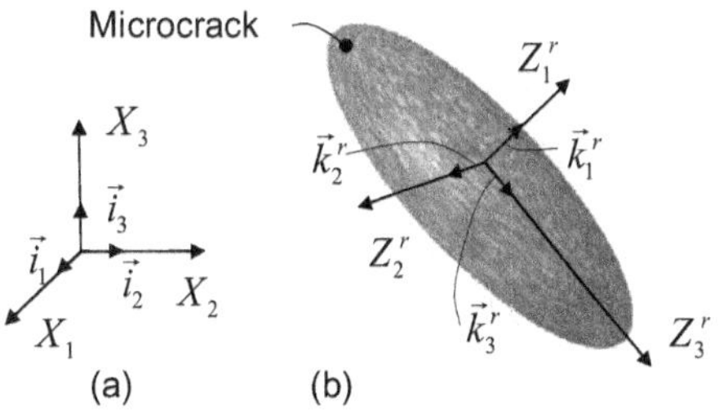


**Figure J.1** Penny-shaped microcrack: (a) the reference frame and (b) the microcrack frame.

$$\boldsymbol{\varepsilon}^{\mathrm{d}r} = B_1\, Q^r\, (A^r)^{3/2} \Big[ (2-\nu)\, H(\sigma_{11}^r)\, \sigma_{11}^r\, \vec{k}_1^r\, \vec{k}_1^r + \sigma_{21}^r\, \vec{k}_1^r\, \vec{k}_2^r + \sigma_{12}^r\, \vec{k}_2^r\, \vec{k}_1^r + \sigma_{31}^r\, \vec{k}_1^r\, \vec{k}_3^r + \sigma_{13}^r\, \vec{k}_3^r\, \vec{k}_1^r \Big] \tag{J.1}$$

where

$$Q^r := \frac{m^r}{\rho_0\, V^{\mathrm{rve}}} \quad (r = 1, \cdots, M), \qquad B_1 := \frac{16}{3}\, \frac{(1-\nu^2)\, \rho_0}{(2-\nu)\, E\, \pi^{3/2}}. \tag{J.2}$$

Equation (J.1) gives

$$\varepsilon_{11}^{\mathrm{d}r} = B_1\, Q^r\, (A^r)^{3/2}\, (2-\nu)\, H(\sigma_{11}^r)\, \sigma_{11}^r \tag{J.3}$$

and

$$\varepsilon_{12}^{\mathrm{d}r} = B_1\, Q^r\, (A^r)^{3/2}\, \sigma_{21}^r = B_1\, Q^r\, (A^r)^{3/2}\, \tfrac{1}{2}\, (\sigma_{12}^r + \sigma_{21}^r)\,, \tag{J.4}$$

and furthermore,

$$\varepsilon_{21}^{\mathrm{d}r} = B_1\, Q^r\, (A^r)^{3/2}\, \sigma_{12}^r = B_1\, Q^r\, (A^r)^{3/2}\, \tfrac{1}{2}\, (\sigma_{12}^r + \sigma_{21}^r)\,, \tag{J.5}$$

and furthermore,

$$\varepsilon_{13}^{\mathrm{dr}} = B_1\, Q^r\, (A^r)^{3/2}\, \sigma_{31}^r = B_1\, Q^r (A^r)^{3/2}\, \tfrac{1}{2}\, (\sigma_{13}^r + \sigma_{31}^r)\,, \tag{J.6}$$

finally,

$$\varepsilon_{31}^{\mathrm{dr}} = B_1\, Q^r (A^r)^{3/2}\, \sigma_{13}^r = B_1\, Q^r (A^r)^{3/2}\, \tfrac{1}{2}\, (\sigma_{13}^r + \sigma_{31}^r)\,. \tag{J.7}$$

The symmetry of the stress tensor $\boldsymbol{\sigma}$ was exploited in Expressions (J.4)–(J.7).

Expressions (J.3)–(J.7) give

$$S_{1111}^{\mathrm{dr}} = (2-\nu)\, B_1\, Q^r\, (A^r)^{3/2}\, H(\sigma_{11}^r) \tag{J.8}$$

and

$$S_{1212}^{\mathrm{dr}} = S_{1221}^{\mathrm{dr}} = S_{2112}^{\mathrm{dr}} = S_{2121}^{\mathrm{dr}} = \frac{1}{2}\, B_1\, Q^r\, (A^r)^{3/2} \tag{J.9}$$

and finally,

$$S_{1313}^{\mathrm{dr}} = S_{1331}^{\mathrm{dr}} = S_{3113}^{\mathrm{dr}} = S_{3131}^{\mathrm{dr}} = \frac{1}{2}\, B_1\, Q^r\, (A^r)^{3/2}\,. \tag{J.10}$$

When the Voigt notation is used (see Section 2.9 or Reddy [84], Section 6.2.2), the following replacements of the stress components are used:

$$\sigma_1^r = \sigma_{11}^r\,, \qquad \text{and} \qquad \sigma_4^r = \sigma_{12}^r \qquad \text{and finally,} \qquad \sigma_5^r = \sigma_{13}^r\,. \tag{J.11}$$

Similarly, the corresponding strain components are

$$\varepsilon_1^{\mathrm{dr}} = \varepsilon_{11}^{\mathrm{dr}}\,, \qquad \text{and} \qquad \varepsilon_5^{\mathrm{dr}} = 2\,\varepsilon_{13}^{\mathrm{dr}} \qquad \text{and finally,} \qquad \varepsilon_6^{\mathrm{dr}} = 2\,\varepsilon_{12}^{\mathrm{dr}}\,. \tag{J.12}$$

Based on Expressions (J.3), (J.4), and (J.6) and Notations (J.11) and (J.12), the following is obtained:

$$\varepsilon_s^{\mathrm{dr}} = S_{st}^{\mathrm{dr}}\, \sigma_t^r\,, \qquad \text{where} \qquad s,t = 1, 5, \text{ and } 6\,. \tag{J.13}$$

The nonzero components of the compliance matrix $[S^{\mathrm{dr}}]$ are

$$S_{11}^{\mathrm{dr}} = (2-\nu)\, B_1\, Q^r\, (A^r)^{3/2}\, H(\sigma_{11}^r) \tag{J.14}$$

and

$$S_{44}^{\mathrm{dr}} = S_{55}^{\mathrm{dr}} = 2\, B_1\, Q^r\, (A^r)^{3/2}\,. \tag{J.15}$$

# *Appendix K* Change of the Coordinate Systems

The transformation between two rectangular Cartesian coordinate systems shown in Figure K.1 is examined in this appendix. The change of the coordinate system is formed by a rotation of the axes without any translation, so the origins of the coordinate systems coincide. The axes of the bases $(\vec{i}_1, \vec{i}_2, \vec{i}_3)$ and $(\vec{k}_1^r, \vec{k}_2^r, \vec{k}_3^r)$ are denoted by $(X_1, X_2, X_3)$ and $(Z_1^r, Z_2^r, Z_3^r)$, respectively. The basis $(\vec{k}_1^r, \vec{k}_2^r, \vec{k}_3^r)$ is obtained by rotating the basis $(\vec{i}_1, \vec{i}_2, \vec{i}_3)$ first about the $X_3$-axis by an angle $\alpha$ and second about the $X_2$-axis by an angle $\beta$, as shown in Figure K.1.

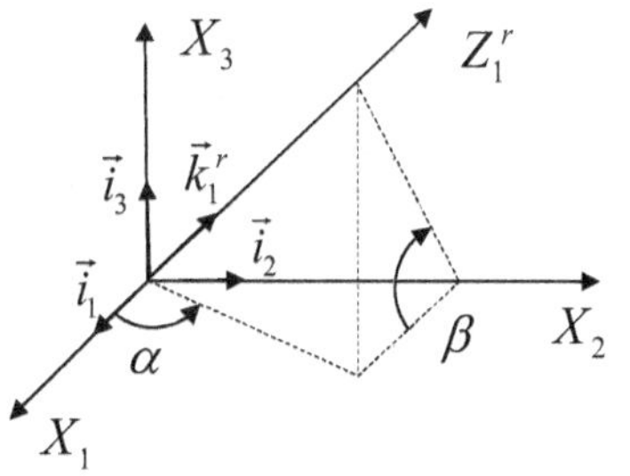


**Figure K.1** Two rectangular Cartesian coordinate systems studied in this appendix.

The mutual dependencies of the scalar components of the vector $\vec{u}$ with respect to the coordinate systems $(X_1, X_2, X_3)$ and $(Z_1^r, Z_2^r, Z_3^r)$ are investigated next. Figure K.2 shows the rotation projections of the scalar components $u_{Z1}$, $u_{Z2}$, and $u_{Z3}$ of the vector $\vec{u}$.

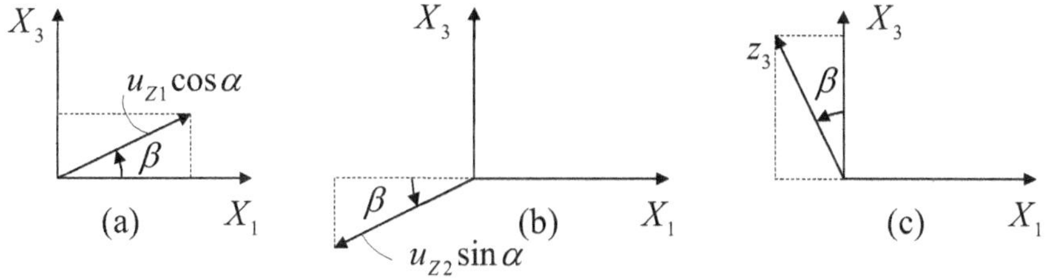


**Figure K.2** Projections of the scalar components $u_{Z1}$ , $u_{Z2}$, and $u_{Z3}$ after two rotations.

Based on Figure K.2, the following expressions are obtained:

$$u_{X1} = u_{Z1} \cos\alpha \cos\beta - u_{Z2} \sin\alpha \cos\beta - u_{Z3} \sin\beta \tag{K.1}$$

and

$$u_{X2} = u_{Z1}\,\sin\alpha - u_{Z2}\,\cos\alpha \tag{K.2}$$

and finally,

$$u_{X3} = u_{Z1}\,\cos\alpha\,\sin\beta - u_{Z2}\,\sin\alpha\,\sin\beta + u_{Z3}\,\cos\beta \tag{K.3}$$

Equations (K.1)–(K.3) can be written in the following matrix form:

$$\begin{Bmatrix} u_{X1} \\ u_{X2} \\ u_{X3} \end{Bmatrix} = \begin{bmatrix} \cos\alpha\,\cos\beta & -\sin\alpha\,\cos\beta & -\sin\beta \\ \sin\alpha & \cos\alpha & 0 \\ \cos\alpha\,\sin\beta & -\sin\alpha\,\sin\beta & \cos\beta \end{bmatrix} \begin{Bmatrix} u_{Z1} \\ u_{Z2} \\ u_{Z3} \end{Bmatrix}$$
$$\rightarrow \quad \{u_X\} = [T]\,\{u_Z\}. \tag{K.4}$$

When tensor notation is used, the message of Expression $(K.4)_1$ is

$$u_k = T_{km}\,u^r_m\,, \tag{K.5}$$

where the superscript $r$ refers to the coordinate system $(Z^r_1, Z^r_2, Z^r_3)$. Equation (K.5) holds for the base vectors as well. Thus the following can be written:

$$\vec{i}_k = T_{km}\,\vec{k}^r_m\,. \tag{K.6}$$

An arbitrary second-order tensor **A** takes the same value independently of the coordinate system. Thus

$$A_{ij}\,\vec{i}_i\,\vec{i}_j = A^r_{st}\,\vec{k}^r_s\,\vec{k}^r_t\,. \tag{K.7}$$

By changing the indices and by substituting Expression (K.6) into Equation (K.7), the following result is obtained:

$$A_{ij}\,T_{is}\,\vec{k}^r_s\,T_{jt}\,\vec{k}^r_t = A^r_{st}\,\vec{k}^r_s\,\vec{k}^r_t \quad \Rightarrow \quad (\,A_{ij}\,T_{is}\,T_{jt} - A^r_{st}\,)\,\vec{k}^r_s\,\vec{k}^r_t = \mathbf{0}\,. \tag{K.8}$$

Since the base vectors $\vec{k}^r_s$ and $\vec{k}^r_t$ are not zero vectors, the term between the parentheses has to vanish, giving

$$A_{ij}\,T_{is}\,T_{jt} - A^r_{st} = 0 \quad \Rightarrow \quad A^r_{st} = A_{ij}\,T_{is}\,T_{jt}\,. \tag{K.9}$$

When expressed in tensor form, Expression $(K.9)_2$ becomes

$$\mathbf{A}^r = \mathbf{T}^{\mathrm{T}}\cdot\mathbf{A}\cdot\mathbf{T}\,. \tag{K.10}$$

# *Appendix L* Effective Stress Tensor Component $\tilde{\sigma}_{11}$ Caused by the Uniaxially Applied Stress $\sigma_{22}$

According to Equation (10.132), the relation between the effective stress tensor $\tilde{\boldsymbol{\sigma}}$ and the stress tensor $\boldsymbol{\sigma}$ reads

$$\tilde{\boldsymbol{\sigma}} = \mathbf{C} : \tilde{\mathbf{S}} : \boldsymbol{\sigma} \quad \Rightarrow \quad \tilde{\boldsymbol{\sigma}} = \mathbf{M} : \boldsymbol{\sigma}, \quad \text{where} \quad \mathbf{M} := \mathbf{C} : \tilde{\mathbf{S}} \tag{L.1}$$

and where $\mathbf{M}$ is the damage effect tensor.

Based on Equation $(\text{L.1})_1$, the relation between the effective stress tensor component $\tilde{\sigma}_{11}$ and the stress tensor component $\sigma_{22}$ is

$$\tilde{\sigma}_{st} = C_{stkm}\, \tilde{S}_{kmuv}\, \sigma_{uv} \quad \Rightarrow \quad \tilde{\sigma}_{11} = C_{11km}\, \tilde{S}_{km22}\, \sigma_{22}\,. \tag{L.2}$$

Expression $(\text{L.2})_2$ gives

$$\begin{aligned} \tilde{\sigma}_{11} = \Big[ & C_{1111}\, \tilde{S}_{1122}(f) + C_{1112}\, \tilde{S}_{1222}(f) + C_{1113}\, \tilde{S}_{1322}(f) \\ & + C_{1121}\, \tilde{S}_{2122}(f) + C_{1122}\, \tilde{S}_{2222}(f) + C_{1123}\, \tilde{S}_{2322}(f) \\ & + C_{1131}\, \tilde{S}_{3122}(f) + C_{1132}\, \tilde{S}_{3222}(f) + C_{1133}\, \tilde{S}_{3322}(f) \Big]\, \sigma_{22}\,. \end{aligned} \tag{L.3}$$

Equation (10.87) gives the effective compliance tensor $\tilde{\mathbf{S}}(f)$ for the deformation of a material containing noninteracting spherical microvoids in a matrix material the response of which is assumed to be linear elastic. The effective compliance tensor $\tilde{\mathbf{S}}(f)$ is

$$\tilde{\mathbf{S}}(f) := \frac{1 + \underline{A}\, f}{3\,(3\,\lambda + 2\,\mu)}\, \mathbf{1}\,\mathbf{1} + \frac{1 + \underline{B}\, f}{2\,\mu} (\mathbf{I}^{\mathrm{s}} - \tfrac{1}{3}\, \mathbf{1}\,\mathbf{1})\,. \tag{L.4}$$

Expressions (2.64) and (2.107) are

$$\mathbf{1} = \delta_{ij}\, \vec{i}_i\, \vec{i}_j \qquad \text{and} \qquad \mathrm{I}^{\mathrm{s}} = \frac{1}{2}\, (\,\delta_{ik}\, \delta_{jl} + \delta_{il}\, \delta_{jk}\,)\, \vec{i}_i\, \vec{i}_j\, \vec{i}_k\, \vec{i}_l\,. \tag{L.5}$$

Lamé elastic constants $\lambda$ and $\mu$ are defined by Definitions (8.31), viz.

$$\lambda := \frac{\nu\, E}{(1+\nu)\,(1-2\,\nu)} \qquad \text{and} \qquad \mu := \frac{E}{2\,(1+\nu)}\,. \tag{L.6}$$

Definitions (L.6) give

$$3\,\lambda + 2\,\mu = \frac{3\,\nu\, E + (1-2\,\nu)\, E}{(1+\nu)\,(1-2\,\nu)} = \frac{(1+\nu)\, E}{(1+\nu)\,(1-2\,\nu)} = \frac{E}{(1-2\,\nu)}\,. \tag{L.7}$$

Expressions (L.4)–(L.7) yield

$$\begin{aligned}\tilde{S}_{ijkl}(f) &= \frac{1-2\,\nu}{3\,E}\,(1+\underline{A}\,f)\,\delta_{ij}\,\delta_{kl} \\ &+ \frac{1+\nu}{E}\,(1+\underline{B}\,f)\Big(\frac{1}{2}\,(\delta_{ik}\,\delta_{jl}+\delta_{il}\,\delta_{jk}) - \tfrac{1}{3}\,\delta_{ij}\,\delta_{kl}\Big)\,.\end{aligned} \tag{L.8}$$

Expression (L.8) gives

$$\tilde{S}_{1122}(f) = \frac{1-2\,\nu}{3\,E}\,(1+\underline{A}\,f) \;-\; \frac{1+\nu}{E}\,(1+\underline{B}\,f)\,\frac{1}{3} \tag{L.9}$$

and

$$\tilde{S}_{2222}(f) = \frac{1-2\,\nu}{3\,E}\,(1+\underline{A}\,f) + \frac{1+\nu}{E}\,(1+\underline{B}\,f)\,\frac{2}{3}\,, \tag{L.10}$$

and furthermore,

$$\tilde{S}_{3322}(f) = \frac{1-2\,\nu}{3\,E}\,(1+\underline{A}\,f) \;-\; \frac{1+\nu}{E}\,(1+\underline{B}\,f)\,\frac{1}{3}\,, \tag{L.11}$$

finally,

$$\tilde{S}_{1222}(f) = \tilde{S}_{2322}(f) = \tilde{S}_{2122}(f) = \tilde{S}_{2322}(f) = \tilde{S}_{3122}(f) = \tilde{S}_{3222}(f) = 0. \tag{L.12}$$

Based on Terms (L.9)–(L.12), Relation (L.3) reduces to

$$\tilde{\sigma}_{11} = \Big[\,C_{1111}\,\tilde{S}_{1122}(f) + C_{1122}\,\tilde{S}_{2222}(f) + C_{1133}\,\tilde{S}_{3322}(f)\,\Big]\,\sigma_{22}\,. \tag{L.13}$$

Based on Equation (8.30), the fourth-order constitutive tensor $\mathbf{C}$ for a Hookean deformation (i.e., for Hooke's law) is defined by

$$\mathrm{C} := \lambda\,\mathbf{1}\,\mathbf{1} + 2\,\mu\,\mathrm{I}^{\mathrm{s}}\,. \tag{L.14}$$

Based on Definitions (L.6) for Lamé elastic constants $\lambda$ and $\mu$, Definition (L.14) takes the following appearance:

$$\mathbf{C} = \frac{\nu\, E}{(1+\nu)\,(1-2\,\nu)}\,\mathbf{1}\,\mathbf{1} + \frac{E}{(1+\nu)}\,\mathbf{I}^{\mathrm{s}}. \tag{L.15}$$

Expressions (L.5) and (L.15) let the term $C_{ijkl}$ be written in the following form:

$$C_{ijkl} = \frac{\nu\, E}{(1+\nu)\,(1-2\nu)}\,\delta_{ij}\,\delta_{kl} + \frac{E}{2\,(1+\nu)}\,(\,\delta_{ik}\,\delta_{jl} + \delta_{il}\,\delta_{jk}\,)\,. \tag{L.16}$$

Equation (L.16) gives

$$\begin{aligned} C_{1111} &= \frac{\nu\, E}{(1+\nu)\,(1-2\,\nu)} + \frac{E}{2\,(1+\nu)}\,(1+1) \\ &= \frac{\nu\, E}{(1+\nu)\,(1-2\,\nu)} + \frac{E}{(1+\nu)} \\ &= \frac{\nu\, E + (1-2\,\nu)\,E}{(1+\nu)\,(1-2\,\nu)} = \frac{(1-\nu)\,E}{(1+\nu)\,(1-2\,\nu)} \end{aligned} \tag{L.17}$$

and

$$C_{1122} = C_{1133} = \frac{\nu\, E}{(1+\nu)\,(1-2\,\nu)}\,. \tag{L.18}$$

Substitution of Terms (L.9)–(L.12) and (L.16)–(L.17) into Expression (L.13) yields

$$\begin{aligned} \tilde{\sigma}_{11} &= \Big[\, C_{1111}\,\tilde{S}_{1122}(f) + C_{1122}\,\tilde{S}_{2222}(f) + C_{1133}\,\tilde{S}_{3322}(f) \,\Big]\,\sigma_{22} \\ &= \left\{ \frac{(1-\nu)\,E}{(1+\nu)\,(1-2\,\nu)} \left[ \frac{1-2\,\nu}{3\,E}\,(1+\underline{A}\,f) - \frac{1+\nu}{E}(1+\underline{B}\,f)\,\frac{1}{3} \right] \right. \\ &\quad + \frac{\nu\, E}{(1+\nu)\,(1-2\,\nu)} \left[ \frac{1-2\,\nu}{3\,E}\,(1+\underline{A}\,f) + \frac{1+\nu}{E}\,(1+\underline{B}\,f)\,\frac{2}{3} \right] \\ &\quad \left. + \frac{\nu\, E}{(1+\nu)\,(1-2\,\nu)} \left[ \frac{1-2\,\nu}{3\,E}\,(1+\underline{A}\,f) - \frac{1+\nu}{E}\,(1+\underline{B}\,f)\frac{1}{3} \right] \right\} \sigma_{22}\,. \end{aligned} \tag{L.19}$$

Equation (L.19) leads to

$$\begin{aligned} \tilde{\sigma}_{11} &= \Big[\, C_{1111}\,\tilde{S}_{1122}(f) + C_{1122}\,\tilde{S}_{2222}(f) + C_{1133}\,\tilde{S}_{3322}(f) \,\Big]\,\sigma_{22} \\ &= \left[ \frac{(1-\nu)}{3\,(1+\nu)}\,(1+\underline{A}\,f) - \frac{1-\nu}{3\,(1-2\,\nu)}\,(1+\underline{B}\,f) \right. \\ &\quad + \frac{\nu}{3\,(1+\nu)}\,(1+\underline{A}\,f) + \frac{2\,\nu}{3\,(1-2\,\nu)}\,(1+\underline{B}\,f) \\ &\quad \left. + \frac{\nu}{3\,(1+\nu)}\,(1+\underline{A}\,f) - \frac{\nu}{3\,(1-2\,\nu)}\,(1+\underline{B}\,f) \right] \sigma_{22} \end{aligned} \tag{L.20}$$

and finally,

$$\begin{aligned}\tilde{\sigma}_{11} &= \Big[ C_{1111}\, \tilde{S}_{1122}(f) + C_{1122}\, \tilde{S}_{2222}(f) + C_{1133}\, \tilde{S}_{3322}(f) \Big]\, \sigma_{22} \\ &= \left[ \frac{1}{3}\,(1 + \underline{A}\, f) - \frac{1}{3}\,(1 + \underline{B}\, f) \right] \sigma_{22} = \frac{(\underline{A} - \underline{B})f}{3}\, \sigma_{22}\,. \end{aligned} \tag{L.21}$$

Equations (10.73) give the multipliers $\underline{A}$ and $\underline{B}$ . They are

$$\underline{A} = \frac{6\,\mu + 3\,\lambda}{4\,\mu} \qquad \text{and} \qquad \underline{B} = \frac{15\,(1-\nu)}{7 - 5\,\nu}\,, \tag{L.22}$$

Substitution of the Lamé elastic constant from Equation (L.6) into Equation $(\text{L.22})_1$ yields

$$\underline{A} = \frac{3\,(1-\nu)}{2\,(1-2\,\nu)} \qquad \text{and} \qquad \underline{B} = \frac{15\,(1-\nu)}{7 - 5\,\nu}\,. \tag{L.23}$$

Based on Expressions (L.23), Relation (L.21) takes the form

$$\tilde{\sigma}_{11} = \frac{(\underline{A} - \underline{B})\, f}{3}\, \sigma_{22} = -\frac{9\,(1 - 6\,\nu + 5\,\nu^2)}{2\,(1-2\,\nu)\,(7-5\,\nu)}\, \sigma_{22}\,. \tag{L.24}$$

If the Poisson's ratio $\nu = 0.3$ and the void volume fraction $f = 0.1$, Relation (L.24) gives

$$\nu = 0.3 \quad \Rightarrow \quad \tilde{\sigma}_{11} = 1.43\, f\, \sigma_{22} \quad \Rightarrow \quad f = 0.1 \quad \Rightarrow \quad \tilde{\sigma}_{11} = 0.143\, \sigma_{22}\,. \tag{L.25}$$

Section 10.5.3 studies the analytical relation between the stress tensor $\tilde{\boldsymbol{\sigma}}$ and $\boldsymbol{\sigma}$ with a hole in an infinite plate. The text in Section 10.5.3 reads that it is important to note that applied uniaxial stress $\sigma_{22}$, for example, induces a multiaxial effective stress $\tilde{\boldsymbol{\sigma}}$ field. This phenomenon can be illustrated by studying a two-dimensional infinite plate with a circular hole of radius $a$. According to Figure 10.15, which is reproduced as Figure L.1 for convenience, the plate is assumed to be under uniaxial stress $\sigma_{22} = \sigma_\infty$ acting at infinity. According to Reddy ([84], p. 308), the in-plane stresses are

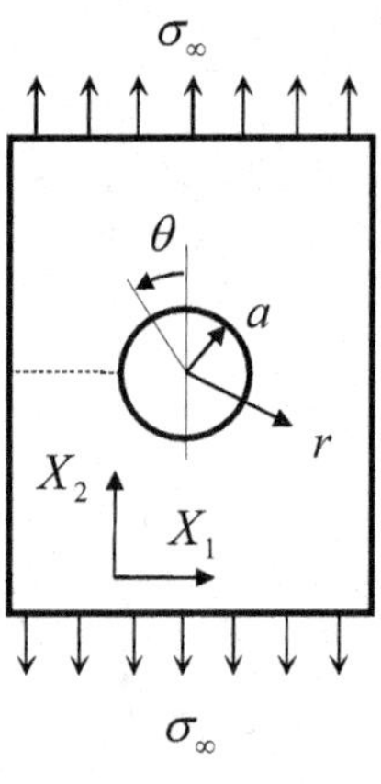


**Figure L.1** Plate with a circular hole subjected to uniform tensile stress.

$$\sigma_{\text{rr}} = \frac{\sigma_\infty}{2}\left(1 - \frac{a^2}{r^2}\right) + \frac{\sigma_\infty}{2}\left(1 + \frac{3\,a^4}{r^4} - \frac{4\,a^2}{r^2}\right)\cos 2\theta$$

$$\sigma_{\theta\theta} = \frac{\sigma_\infty}{2}\left(1 + \frac{a^2}{r^2}\right) - \frac{\sigma_\infty}{2}\left(1 + \frac{3\,a^4}{r^4}\right)\cos 2\theta\,,$$

$$\sigma_{\text{r}\theta} = -\frac{\sigma_\infty}{2}\left(1 - \frac{3\,a^4}{r^4} + \frac{2\,a^2}{r^2}\right)\sin 2\theta\,. \tag{L.26}$$

The stress component $\sigma_{11}$ equals $\sigma_{\mathrm{rr}}$ when in the preceding equations, the angle $\theta$ takes the value $\pi/2$. As Equation $(\mathrm{L}.26)_1$ [the original one is Equation $(10.138)_1$] shows, the stress component $\sigma_{11}$ takes a nonzero value along the dashed horizontal line, except for the surface of the hole. Based on Equation $(\mathrm{L}.26)_1$, the value of the stress component $\sigma_{11}$ along the dashed line reads

$$\sigma_{11} = \sigma_{\mathrm{rr}}(\theta = \pi/2) = \frac{\sigma_\infty}{2}\left(1 - \frac{a^2}{r^2}\right) - \frac{\sigma_\infty}{2}\left(1 + \frac{3\,a^4}{r^4} - \frac{4\,a^2}{r^2}\right). \tag{L.27}$$

The maximum value of $\sigma_{11}$ is obtained at a point where its derivative with respect to the $r$ coordinate vanishes. We arrive at the following:

$$\begin{aligned}\frac{\partial \sigma_{11}}{\partial r} &= \sigma_\infty\, a^2\, r^{-3} + 6\,\sigma_\infty\, a^4\, r^{-5} - 4\,\sigma_\infty\, a^2\, r^{-3} \\ &= 6\,\sigma_\infty\, a^4\, r^{-5} - 3\,\sigma_\infty\, a^2\, r^{-3} = 2\,a^2\, r^{-2} - 1 = 0\,.\end{aligned} \tag{L.28}$$

Expression (L.28) yields

$$r^2 = 2\,a^2 \qquad \Rightarrow \qquad r = \sqrt{2}\,a\,. \tag{L.29}$$

Result $(\mathrm{L}.29)_2$ gives

$$\sigma_{11}(r = \sqrt{2}a) = \frac{\sigma_\infty}{2}\left(1 - \frac{1}{2}\right) - \frac{\sigma_\infty}{2}\left(1 + \frac{3}{4} - \frac{4}{2}\right) = \frac{3}{8}\,\sigma_\infty \approx 0.375\,\sigma_\infty\,. \tag{L.30}$$

# References


1. Abaqus. 2016. Analysis User's Guide. Dassault Systems.
2. Abramowitz, M., and Stegun, J. A. 1972. *Handbook of Mathematical Functions* (Dover edition). New York, NY: Dover. 1046 p.
3. Aifantis, E. C. 1992. On the role of gradients in the localization of deformation and fracture. *International Journal of Engineering Sciences,* **30**(10), 1279–1299.
4. Allen, M. B., Herrera, I., and Pinder, G. F. 1988. *Numerical Modeling in Science and Engineering.* New York, NY: John Wiley, 418 p.
5. Andrade, C. 1910. On the viscous flow in metals, and allied phenomena. *Proceedings of the Royal Society, A*, **84**, 1–12.
6. Apostol, T. M. 1957. *Mathematical Analysis: A Modern Approach to Advanced Calculus.* Reading, MA: Addison-Wesley. 559 p.
7. Arfken, G. 1985. *Mathematical Methods for Physicists.* 3rd ed. New York, NY: Academic Press. 985 p.
8. Astarita, G. 1989. *Thermodynamics; An Advanced Textbook for Chemical Engineers.* New York, NY: Plenum Press. 444 p.
9. Bailey, R. W. 1929. *Transactions of the World Power Conference.* Tokyo.
10. Bažant, Z. P. 1994. Nonlocal damage theory based on micromechanics of crack interactions. *Journal of Engineering Mechanics*, **120**(3), 593–617.
11. Boas, M. 2006. *Mathematical Methods in the Physical Sciences.* 3rd ed. New York: John Wiley. 837 p.
12. Cauvin, A., and Testa, R. B. 1999. Damage mechanics: Basic variables in continuum theories. *International Journal of Solids and Structures*, **36**, 747–761.
13. Cattaneo, C. 1948. *Sulla Conduzione Del Calore.* Atti Semin. Mat. Fis. della, Università di Modena, Reggio Emilia, **3**(3), 83–101.
14. Chaboche, J.-L. 1978. *Description thermodynamique et phénoménologique de la viscoplasticité cyclique avec endommagement.* Chatillon, France: Office National d'Etudes et de Recherches Aérospatiales (in French). Publication no. 1978-3. 156 p.
15. Chow, C. L., and Wei, Y. 1999. Constitutive modelling of material damage for fatigue failure prediction. *International Journal for Damage Mechanics*, **8**, 355–375.
16. Chu, C. C., and Needleman, A. 1980. Void nucleation effects in biaxially stretched sheets. *International Journal of Materials and Technology*, **102**, 249–256.

17. Coburn, N. 1955. *Vector and Tensor Analysis*. New York, NY: Macmillan. 341 p.
18. Coleman, D. C. 1964. Thermodynamics of materials with memory. In: Truesdel, C. (ed.), *Archive for Rational Mechanics and Analysis*. Berlin: Springer. pp. 1–46.
19. Currier, J. H., Schulson, E. M., and St. Lawrence, W. F. 1983. A study on the tensile strength of ice as a function of grain size. Hanover, NH: Gold Regions Research & Engineering Laboratory. Report 83-14. 38 p.
20. Dormieux, L., and Kondo, D. 2016. *Micromechanics of Fracture and Damage*. New York, NY: John Wiley.
21. Eshelby, J. D. 1957. The determination of the elastic field of an ellipsoidal inclusion, and related problems. *Proceedings of the Royal Society of London, A* 241, 376–396.
22. Flügge, W. 1975. *Viscoelasticity*. 2nd ed. Berlin: Springer. 194 p.
23. Fulks, W. 1961. *Advanced Calculus: An Introduction to Analysis*. New York, NY: John Wiley. 592 p.
24. Fung, Y. C. 1965. *Foundations of Solid Mechanics*. Englewood Cliffs, NJ: Prentice Hall. 525 p.
25. Germain, P., Nguyen Q. S., and Suquet, P. 1983. Continuum thermodynamics. *Journal of Applied Mechanics*, **50**(4b), 1010–1020.
26. Gill, P. 2011. When should we change the definition of the second? *Philosophical Transactions of the Royal Society, A*, **369**(1953), 4109–4130 (also see https://en.wikipedia.org/wiki/Second).
27. Gurson, A. L. 1977. Continuum theory of ductile rupture by void nucleation and growth. Part I: Yield criteria and flow rules for porous ductile media. *Journal of Engineering Materials and Technology,* **99**(1), 2–15.
28. Hertzberg, R. W., Vinci, R. P., and Hertzberg, J. L. 2012. *Deformation and Fracture Mechanics of Engineering Materials*. 5th ed. New York, NY: John Wiley. 800 p.
29. Holzapfel, G. A. 2010. *Nonlinear Solid Mechanics*. Chichester, UK: John Wiley. 455 p.
30. Huang, T. C. 1969. *Engineering Mechanics: Volume I. Statics*. Reading, MA: Addison-Wesley. 419 p.
31. Hult, J. A. H. 1966. *Creep in Engineering Structures*. Waktham, MA: Blaisdel. 115 p.
32. Incropera, F. P., and DeWitt, D. P. 1996. *Fundamentals of Heat and Mass Transfer*. 4th ed., New York, NY: John Wiley. 886 p.
33. Isachenko, V. P., Osipova, V. A., and Sukomel, A. S. 1977. *Heat Transfer*, 3rd ed. Moscow: Mir. 493 p.

34. Jou, D., Casas-Vázquez, J., and Lebon, G. 1993. *Extended Irreversible Thermodynamics.* Berlin: Springer. 319 p.
35. Kachanov, M. 1980. Continuum model of medium with cracks. *Journal of the Engineering Mechanics Division*, **106**, 1039–1051.
36. Kanervo, K. 2000. Private discussions on October 20. Helsinki University of Technology.
37. Kestin, J. 1978. *A Course in Thermodynamics.* Vols. 1–2. New York, NY: Hemisphere. 725 p. + 617 p.
38. Krajcinovic, D. 2003. *Damage Mechanics.* Amsterdam: Elsevier. 761 p.
39. Kukudžanov, V. N., Bourago, N. G., Kovshov, A. N., Ivanov, V. L., and Schneiderman, D. N. 1995. *On the Problem of Damage and Localization of Strains.* Göteborg, Sweden: Chalmers University of Technology, Department of Structural Mechanics. Publication 95:11. 35 p.
40. Kukudžanov, V. N. 1996. Private discussions on August 8. The Institute of Problems in Mechanics, Russian Academy of Sciences.
41. Kundu, P. K., and Cohen, I. M. 2008. *Fluid Mechanics.* 4th ed. Amsterdam: Elsevier. 840 p. + 4 p.
42. Lanczos, C. 1949. *The Variational Principles of Mechanics.* Toronto, Canada: University of Toronto Press. 307 p.
43. Landau, L. D., and Lifshitz, E. M. 1998. *Course of Theoretical Physics: Vol. 1, Mechanics.* 3rd ed. Oxford: Butterworth-Heinemann. 170 p.
44. Lavenda, B. H. 1993. *Thermodynamics of Irreversible Processes.* New York, NY: Dover. 182 p. [Originally published: London: Macmillan, 1978.]
45. Lebon, G. 1992. *Extended Thermodynamics.* Lecture notes on the course entitled "Non-equilibrium thermodynamics with application to solids," held at the International Centre for Mechanical Sciences. Udine, Italy, September 28–October 2. 62 p.
46. Lebon, G., Jou, D., and Casas-Vázquez, J. 1992. Questions and answers about a thermodynamic theory of the third type. *Contemporary Physics*, **33**(1), 41–51.
47. Lebon, G., Jou, D., and Casas-Vázquez, J. 2008. *Understanding Non-equilibrium Thermodynamics.* Berlin: Springer. 325 p.
48. Le Gac, H., and Duval, P. 1980. Constitutive relations for the non-elastic deformation of polycrystalline ice. In: Tryde, P., *Physics and Mechanics of Ice.* Berlin: Springer. pp. 51–59.
49. Leigh, D. C. 1968. *Nonlinear Continuum Mechanics.* New York, NY: McGraw-Hill. 240 p.
50. Lemaitre, J. 1992. *A Course on Damage Mechanics.* Berlin: Springer. 210 p.

51. Lemaitre, J., and Chaboche, J.-L. 1990. *Mechanics of Solid Materials.* New York: Ny Cambridge University Press. 556 p. [Originally published in French: *Mécanique des matériaux solides.* Paris: Dunod (and Bordas) 1985].
52. Lin, T. H. 1968. *Theory of Inelastic Structures.* New York, NY: John Wiley. 454 p.
53. Lubliner, J. 1972. On the thermodynamic foundations of non-linear solid mechanics. *International Journal of Non-linear Mechanics*, **7**, 237–254.
54. Lubliner, J. 1990. *Plasticity Theory.* Singapore: Macmillan. 495 p.
55. Luenberger, D. G. 1973. *Introduction to Linear and Nonlinear Programming.* Reading, MA: Addison-Wesley. 356 p.
56. Mackenzie, J. K. 1950. The elastic constants of a solid containing spherical holes. *Proceedings of the Royal Society of London, B*, **63**(2), 2–11.
57. Malvern, L. E. 1969. *Introduction to the Mechanics of a Continuous Medium.* Englewood Cliffs, NJ: Prentice Hall. 713 p.
58. Marsden, J. E., and Hughes, T. J. R. 1994. *Mathematical Foundation of Elasticity.* New York, NY: Dover.
59. Maugin, G. A. 1990. Internal variables and dissipative structures. *Journal of Non-Equilibrium Thermodynamics*, **15**, 173–192.
60. Maugin, G. A. 1992. *The Thermomechanics of Plasticity and Fracture.* Cambridge: Cambridge University Press. 350 p.
61. Maugin, G. A., and Muschik, W. 1994. Thermodynamics with internal variables. Part I. General concepts. *Journal of Non-Equilibrium Thermodynamics*, **19**, 217–249.
62. McGill, D. J., and King, W. W. 1984. *Engineering Mechanics: An Introduction to Dynamics.* Monterey, CA: Wadsworth. 608 p.
63. Mikkola, T. 2023. Private discussion on February 1. Aalto University.
64. Mohrmann, R., and Riedel, H. 1992. *Implementation of the Constitutive Model Proposed by Le Gac and Duval and Parameter Adjustment to 10 CrMo 9 10 Creep Data at* 550° *C.* Freiburg, Germany: IWM-Fraunhofer Institut für Werkstoffmechanik. Report V 63/92. 4 p.
65. Muschik, W. 1993. Fundamentals of Nonequilibrium Thermodynamics. In: Muschik, E. (ed.), *Non-Equilibrium Thermodynamics with Applications to Solids.* CISM Courses and Lectures no 336, Udine, Italy. International Centre for Mechanical Sciences. Vienna: Springer. pp. 1–64.
66. Narasimhan, M. N. L. 1993. *Principles of Continuum Mechanics.* New York, NY: John Wiley. 567 p.
67. Nemat-Nasser, S., and Hori, M. 1993. *Micromechanics: Overall Properties of Heterogeneous Materials.* Amsterdam: Elsevier. 687 p.
68. Norton, F. H. 1929. *The Creep of Steel at High Temperatures.* New York, NY: McGraw-Hill. 90 p.

69. Nguyen, Q. S. 2000. Private discussions on September 6. Laboratoire de Mécanique des Solides, Ecole Polytechnique, Paris.
70. Nguyen, Q. S., and Bui, H. D. 1974. Sur les matériaux élastoplastiques à écrouissage positif ou negatif. *Journal de Mécanique*, **13**(2), 321–342.
71. Odqvist, F. K. G. 1933. Die Verfestigung von flußeisenähnlichen Körpern. *Zeitschrift für Angewandte Mathematik und Mechanik*, **13**(5), 360–363.
72. Odqvist, F. K. G. 1933. *Plasticitetsteori med tillämpningar* (in Swedish) *Plasticity Theory with Applications.* Royal Swedish Academy of Engineering Sciences. Stockholm: Generalstabens Litografiska Anstalts Förlag. 80 p.
73. Onsager, L. 1931. Reciprocal relations in irreversible processes I. *The Physical Review*, **37**, 405–426.
74. Onsager, L. 1931. Reciprocal relations in irreversible processes II. *The Physical Review*, **38**, 2265–2279.
75. Parkus, H. 1976. *Thermoelasticity.* 2nd ed. Vienna: Springer. 119 p.
76. Piskunov, N. 1974. *Differential and Integral Calculus.* Vols. I and II. Moscow: Mir. 471 p. + 576 p.
77. Prager, W. 1955. The theory of plasticity: A survey of recent achievements. *Proceedings of the Institute for Mechanical Engineers*, **169**, 41–57.
78. Prager, W. 1956. A new method of analyzing stresses and strains in work-hardening plasticsolids. *Journal of Applied Mechanics*, **23**, 493–496.
79. Prigogine, I. 1947. *Etude thermodynamique des phénomènes irréversibles.* Doctoral dissertation, Université libre de Bruxelles. 143 p.
80. Prigogine, I. 1961. *Introduction to Thermodynamics of Irreversible Processes.* 2nd rev. ed. New York, NY: Interscience. 119 p.
81. Rabotnov, Y. N. 1959. On the mechanism of gradual failure. In: *Questions of Strength of Materials and Construction* (in Russian). Moscow. Academy of Sciences. pp. 5–7
82. Rabotnov, Y. N. 1968. Creep rupture. In: Hetényi., M., and Vincent, W. G. (eds.), *Proceedings of the XII International Congress of Applied Mechanics* Stanford. Berlin: Springer, 1969. pp. 342–349.
83. Ramaswamy, N., and Aravas, N. 1998. Finite element implementation of gradient plasticity models; Part I: Gradient-dependent yield functions. *Computer Methods in Applied Mechanics and Engineering*, **163**(1–4), 11–32.
84. Reddy, J. N. 2013. *An Introduction to Continuum Mechanics.* 2nd ed. New York, NY: Cambridge University Press. 450 p.
85. Reddy, J. N. 2017. *Energy Principles and Variational Methods in Applied Mechanics.* 3rd ed. New York, NY: John Wiley. 730 p.

86. Salonen, E.-M. 1999. *Statiikka* (in Finnish). Espoo, Finland: Otatieto Oy. 282 p.
87. Santaoja, K. 1987. An effective approximate algorithm to predict the delayed elastic strain. In: Sackinger, W. M., and Jeffries, M. O. (eds.), *Proceedings of the Ninth International Conference on Port and Ocean Engineering under Arctic Conditions*, Vol. III. pp. 31–43.
88. Santaoja, K. 1988. Continuum damage mechanics approach to describe the uniaxial microcracking of ice. In: Saeki, H. S., and Hirayama, K. (eds.), *The 9th International Symposium for Ice,* Sapporo, Japan, August 23–27. pp. 138–151.
89. Santaoja, K. 1989. Continuum damage mechanics approach to describe multidirectional microcracking of ice. In: Sinha, N. K., Sodhi, D. S., and Chung, J. S. (eds.), *Proceedings of the Eight International Conference on Offshore Mechanics and Arctic Engineering*, The Hague, March 19-23, 1989. The American Society of Mechanical Engineers. Vol. **IV**. pp. 55–65.
90. Santaoja, K. 1990. *Mathematical Modelling of Deformation Mechanisms in Ice.* Doctoral dissertation, Helsinki University of Technology. VTT Research Reports 676. Espoo, Finland: VTT. 215 p. + app. 13 p.
91. Santaoja, K. 1994. *Thermomechamics of creep equation proposed by Le Gac and Duval.* VTT Publications 193. Espoo, Finland: Technical Research Centre of Finland. 130 p. + app. 17 p.
92. Santaoja, K. 2000. *Thermomechanics of a gradient theory.* Research Reports - TKK-LO-28. Espoo, Finland: Helsinki University of Technology Laboratory for Mechanics of Materials. 57 p. + app. 9 p.
93. Santaoja, K. 2000. Thermomechanical formulation for the gradient theory. In: Koski, J., and Virtanen, S. (eds.), VII *Suomen Mekaniikkapäivät* (VII *Finnish Mechanics Days*), Tampere University of Technology. pp. 113–123.
94. Santaoja, K. 2002. Evaluation of the Gurson–Tvergaard material model by using damage mechanics and thermomechanics. *Proceedings of the Estonian Academy of Sciences,* **8**(4), 248–269.
95. Santaoja, K. 2003. Derivation of the normality rule for time-dependent deformation using the principle of maximal rate of entropy production. *Archives of Mechanics*, **55**(5–6), 501–518.
96. Santaoja, K. 2004. Gradient theory from the thermomechanics point of view. *Engineering Fracture Mechanics*, **71**(4–6), 557–566.
97. Santaoja, K. 2012. *Extended Levenberg–Marquardt Method for Determination of Values for Material Parameters.* Espoo, Finland: Sasata. 42 p.
98. Santaoja, K. 2014. Thermodynamic formulation of a material model for microcracking applied to creep damage. *Procedia Materials Science*, **3**, 1179–1184.

99. Santaoja, K. 2015. Thermodynamics of a material model showing creep and damage. In: Kleiber, M., et al. (eds.), *3rd Polish Congress of Mechanics* and *21st International Conference Computer Methods in Mechanics.* Gdansk, Poland: Polish Academy of Sciences.
100. Santaoja, K., and Kuistiala, A. 2004. Material models for Hookean materials with voids or cracks. In: Gutkowski, W., and Kowalewski, T. A. (eds.), *Proceedings of the 21st International Congress of Theoretical and Applied Mechanics*, Warsaw, August 15–21. p. 237.
101. Santaoja, K., and Reddy, J. N. 2016. Material model for creep-assisted microcracking applied to S2 sea ice. *Journal of Applied Mechanics*, **83**(11), article 111002.
102. Siikonen, T. 2010. Private communication on September 8. Aalto University.
103. Skrzypek, J., and Ganczarski, A. 1999. *Modeling of Material Damage and Failure of Structures.* Berlin: Springer. 326 p.
104. Stokes, G. G. 1845. On the theories of internal friction of fluids in motion. *Transactions of the Cambridge Philosophical Society*, **8**, 287–305.
105. Surana, K. S. 2015. *Advanced Mechanics of Continua.* Boca Raton, FL: CRC Press. 759 p.
106. Synge, J. L., and Grifith, B. A. 1959. *Principles of Mechanics.* New York, NY: McGraw-Hill. 552 p.
107. Symon, K. R. 1971. *Mechanics.* Amsterdam: Addison-Wesley. 639 p.
108. Szabó, J. 1970. Foreword by the editor. In: *Non-equilibrium Thermodynamics: Field Theory and Variational Principles.* Berlin: Springer. 184 p.
109. Tiihonen, T. 1991. *Matemaattinen mallittaminen* (in Finnish). Lecture Note 16. Jyväskylä, Finland: Department of Mathematics, University of Jyväskylä 93 p.
110. Truesdell, C. 1988. *Rational Thermodynamics.* 2nd enlarged ed. New York, NY: Springer. 578 p. [First edition published New York, NY: McGraw-Hill 1969]
111. Truesdell, C., and Noll, W. 1965. The non-linear field theories of mechanics. In: Flügge, S. (ed.), *Encyclopedia of Physics.* Vol. III/3. Berlin: Springer. 1–579.
112. Tvergaard, V. 1981. Influence of voids on shear band instabilities under plane strain conditions. *International Journal of Fracture*, **17**(4), 389–407.
113. Tvergaard, V., and Needleman, A. 1984. Analysis of the cup-cone fracture in a round tensile bar. *Acta Metallurgiga*, **32**, 157–169.

114. von Mises, R. 1913. Mechanik der festen Körper im plastisch-deformablen Zustand. *Nachrichten von der Gesellschaft der Wissenschaften zu Göttingen (Mathematisch-physikalische Klasse)* **1**. Göttingen: University of Göttingen, 582–592.
115. Vakulenko, A. A., and Kachanov, M. L. 1971. Kontinual'naya teoria sredy s treshchinami . [Continuum model of medium with cracks] (in Russian). *Mehanika Tverdogo Tela*, **4**, 159–166.
116. White, F. M. 2006. *Viscous Fluid Flow.* 3rd ed. Singapore: McGraw-Hill.
117. Widder, D. V. 1989. *Advanced Calculus.* 2nd ed. New York, NY: Dover. 544 p.
118. Yang, Q., and Leng, K. 2014. A microplane-based anisotropic damage effective stress. *International Journal of Damage Mechanics*, **23**(2), 178–191.
119. Young, H. D., and Freedman, R. A. 2016. *University Physics with Modern Physics.* 14th ed. London: Pearson Education. 1593 p.
120. Ziegler, H. 1963. Some extremum principles in irreversible thermodynamics with application to continuum mechanics. In: Sneddon, I. W., and Hill, R. (eds.), *Progress in Solid Mechanics.* Vol. 4, 93–193, Amsterdam: North-Holland.
121. Ziegler, H. 1983. *An Introduction to Thermomechanics.* 2nd ed. Amsterdam: North-Holland. 355 p.
122. Ziegler, H., and Wehrli, C. 1987. The derivation of constitutive relations from the free energy and the dissipation function. *Advances in Applied Mechanics*, **25**, 183–238.

# Index

Printed in the United States
by Baker & Taylor Publisher Services